TIMBER CONSTRUCTION MANUAL

TIMBER CONSTRUCTION MANUAL

THIRD EDITION
1985

AMERICAN INSTITUTE OF TIMBER CONSTRUCTION
Englewood, Colorado

A WILEY-INTERSCIENCE PUBLICATION

JOHN WILEY & SONS

New York · Chichester · Brisbane · Toronto · Singapore

Published by John Wiley & Sons, Inc.
1966, 1974, 1985 by American Institute of Timber Construction.

Library of Congress Cataloging in Publication Data:

Main entry under title:

Timber construction manual.

Includes index.
1. Building, Wooden—Handbooks, manuals, etc.
I. American Institute of Timber Construction.
TA666.T47 1985 694 85-7165
ISBN 0-471-82758-4

Printed in the United States of America
10 9 8 7 6 5 4 3 2

This edition of the Timber Construction Manual is dedicated to Thomas E. Brassell who for over 25 years has provided technical guidance to AITC and the laminating industry.

PREFACE

This Third Edition has been prepared to update the AITC Timber Construction Manual to reflect current timber design methods. Part I of the manual contains general design data and construction information. Part II contains information on loads and the design of structural elements and their fastenings. Part III contains reference information and AITC recommended standards and specifications for engineered timber construction.

The work of the preparation of the *Timber Construction Manual* was guided by the AITC Technical Advisory Committee and was carried out by AITC staff engineers and by engineers and technical representatives of AITC member firms.

Suggestions for the improvement of this manual will be welcomed and will receive consideration in the preparation of future editions.

The *Timber Construction Manual* has been adopted by the American Institute of Timber Construction as its official recommendation.

The American Institute of Timber Construction has developed this *Timber Construction Manual* for convenient reference by architects, engineers, contractors, teachers, and the laminating and fabricating industry, and all others having need for up-to-date technical data and recommendations on engineered timber construction.

While these data have been prepared in accordance with recognized engineering principles and are based on the most accurate and reliable technical data available, they should not be used or relied upon for any general or specific application without competent professional examination and verification of their accuracy, suitability, and applicability by a licensed professional engineer, designer, or architect. By the publication of this manual, AITC intends no representation or warranty, expressed or implied, that the information contained herein is suitable for any general or specific use or is free from infringement of any patent or copyright. Any user of this information assumes all risk and liability arising from such use.

PREFACE TO SECOND EDITION

The first edition of the AITC *Timber Construction Manual* was published in 1966. Changes in the wood products industry and technological advances and improvements in the structural timber fabricating industry have necessitated this revised edition of the *Manual*.

New lumber sizes and revisions in grading requirements for lumber and glued laminated timber are reflected in this second edition. Improved and refined design procedures are also incorporated.

The *Timber Construction Manual* was prepared by the AITC engineering staff with the guidance of the Institute's Technical Advisory Committee. The valuable assistance provided from many sources in developing technical data for the *Manual* is gratefully acknowledged.

PREFACE TO FIRST EDITION

In recent years, technical developments and the establishment of an engineered timber fabricating and laminating industry have had a profound effect on construction. Long clear spans of timber trusses, girders, arches, and decking are now commonplace. Engineered timber is widely used in such diversified construction as schools, churches, commercial buildings, industrial buildings, residences and farm buildings, highway and railway bridges, towers, theater screens, ships, and military and marine installations.

Modern practices combine engineering, quality control, and careful grading with the use of proper working stresses, dependable adhesives, and efficient mechanical fastenings to produce reliable construction. Laminating with strong, durable adhesives permits the manufacture of curved and variable shaped members and thus increases the versatility of timber construction.

The American Institute of Timber Construction is a nonprofit, technical, industrial association of manufacturers and fabricators who may design, plant-laminate, fabricate, assemble, and erect load-carrying sawn and glued timber framing and decking for roofs and other structural parts of schools, churches, commercial, industrial, and other buildings, and for other structures such as bridges, towers, and marine installations.

The American Institute of Timber Construction has developed this *Timber Construction Manual* for convenient reference by architects, engineers, contractors, teachers, the laminating and fabricating industry, and all others having a need for reliable, up-to-date technical data and recommendations on engineered timber construction. The information and the recommendations herein are based on the most reliable technical data available and reflect the commercial practices found to be most practical. Their application results in structurally sound construction.

The *Manual* has been arranged primarily for convenient use by designers, detailers, and fabricators of engineered timber construction. To avoid repetition, material which pertains to more than one area will be found in only one section. Suitable cross references are made in the other pertinent sections.

Information of an engineering textbook nature, such as derivations of formulae, is not included, since the purpose of the *Manual* is to present data for design and construction application by those familiar with engineering procedures.

Part I of the *Manual* contains design data and construction information. Part II contains AITC recommended standards and specifications which will aid the designer in preparing plans and specifications for engineered timber construction.

Material has been compiled from many sources. Where it has been possible to identify the author of the material reproduced, it is used with the author's permission.

Every precaution has been taken to assure that all the data and information included are as accurate as possible. However, the Institute cannot assume responsibility for errors or omissions resulting from the use of this *Manual* in the preparation of plans or specifications. The Institute does not prepare engineering plans.

The work of the preparation of the *Timber Construction Manual* was guided by the AITC Technical Advisory Committee and was carried out by AITC staff engineers and by engineers and technical representatives of AITC member firms.

Suggestions for the improvement of this *Manual* will be welcomed and will receive consideration in the preparation of future editions.

The *Timber Construction Manual* has been adopted by the American Institute of Timber Construction as its official recommendation.

CONTENTS

Part II DESIGN

GENERAL NOMENCLATURE

The following abbreviations and symbols are in general use throughout this manual. Deviations from these notations are indicated where they occur.

ABBREVIATIONS

AASHTO	American Association of State Highway and Transportation Officials
AREA	American Railway Engineers Association
AITC	American Institute of Timber Construction
ANSI	American National Standards Institute
APA	American Plywood Association
ASCE	American Society of Civil Engineers
ASTM	American Society for Testing and Materials
AWPA	American Wood-Preservers' Association
Btu	British thermal unit
DL	Dead load (psf)
EL	Earthquake load (psf)
EMC	Equilibrium moisture content (%)
FPL	Forest Products Laboratory, U.S. Forest Service
ft, ft^2, ft^3	feet, square feet, cubic feet
G	Specific gravity
hr	Hour
in., in.^{2}, in.^{3}, in.^{4}...	inches, square inches, cubic inches, inches to the fourth power
in.-lb	Inch-pounds
k	Kip (one thousand pounds)
KD	Kiln dried
lb	Pound
LL	Live Load (psf)
MC	Moisture content (%)
min	Minimum
MSR	Machine stress rated

NA	Neutral axis
NDS	*National Design Specification for Wood Construction*
o.c.	On centers
°F	Degrees Fahrenheit
pcf	Pounds per cubic foot
plf	Pounds per lineal foot
psf	Pounds per square foot
psi	Pounds per square inch
SL	Snow load (psf)
TL	Total load (psf)
USDA	United States Department of Agriculture
WL	Wind Load (psf)

SYMBOLS

A	Area of cross section (in.^{2})
A_1	In fastener group analysis, cross-sectional area of main wood member(s) before boring and grooving (in.^{2})
A_2	In fastener group analysis, sum of cross-sectional areas of wood or metal side member(s) before boring or drilling (in.^{2})
a	Dimension of member (in.)
a	Distance to load for bracket columns (in.)
A_c	Area of concrete footing in pole design (ft^2)
A_s	Area of steel member (in.^{2})
b	Breadth (width) of rectangular member (in.)
b	Smaller side of beam or column before exposure to fire (in.)
b	Width of column flange (in.)
C	Compressive force (lb)
C	Pole circumference at point of maximum moment (in.)
C	Thermal conductance (Btu/hr ft^2 °F)
c	Distance from neutral axis to outer surface of beam (in.)
C_C	Curvature factor
C_{co}	Seasoning conditioning modification factor for poles
C_{cs}	Critical section modification factor for poles
C_D	Duration-of-load factor
C_d	Depth-of-embedment factor
C_{dt}	Constant for tapered beam deflection
C_e	Fastener edge distance factor

C_f	Form factor
C_F	Size factor
C_g	Group action factor
C_I	Interaction stress factor
C_k	For bending members, largest value of C_s at which intermediate-beam formula applies
C_L	Lateral stability-of-beams factor
C_{lb}	Lag bolt modifying factor
C_M	Moisture content factor
C_M	Steel stress coefficient for bridge dowel design (psi)
C_n	Fastener end distance factor
C_P	Lateral stability-of-columns factor
C_p	Ponding magnification factor
C_R	Fire-retardant treatment factor
C_R	Steel stress coefficient for bridge dowel design (psi)
C_r	Reduction factor for double-tapered curved beams
C_s	Fastener spacing factor
C_s	Slenderness factor for beam stability
C_{SF}	Modifier for safety factor for poles
C_{st}	Fastener steel side plate factor
C_t	Temperature factor
C_x	Spaced column fixity factor
C_y	Factor for tapered beam deflection
D	Diameter (in.)
d	Bridge dowel diameter (in.)
d	Depth of rectangular member (in.)
d	Least dimension of compression member (in.)
d	Larger side of beam or column before exposure to fire (in.)
d_b	Arch depth at base (in.)
d_c	Depth of cross section at centerline (in.)
d_{cb}	Approximate centerline depth for double-tapered curved beams (in.)
d_{crt}	Minimum centerline depth due to radial tension for double-tapered curved beams (in.)
d_{eb}	Factor for calculating depth of double-tapered curved beams (in.)
d_{eff}	Approximate effective centerline deflection for double-tapered curved beams (in.)
D_H	Diameter of hole for pole design (ft)

d_t	Depth of tangent point (in.)
E	Modulus of elasticity (psi)
e	Eccentricity of load (in.)
f	Dimensionless factor from Figure 1.2
F_b	Design value in bending (psi)
f_b	Bending stress (psi)
F_c	Design value in compression parallel to grain (psi)
f_c	Compression parallel to grain stress (psi)
$F_{c\perp}$	Design value in compression perpendicular to grain (psi)
$f_{c\perp}$	Compression perpendicular to grain stress (psi)
f_{cr}	Ultimate column buckling strength (psi)
F_g	Design value for end grain in bearing (psi)
f_0	Reference stress for double-tapered curved beams (psi)
f_r	Radial stress (psi)
F_{rt}	Design value in radial tension (psi)
f_{rt}	Radial tension stress (psi)
f_s	Torsional stress (psi)
F_t	Design value in tension parallel to grain (psi)
f_t	Tension parallel to grain stress (psi)
F_v	Design value in horizontal shear (psi)
f_v	Horizontal shear stress (psi)
G	Shear modulus (modulus of rigidity) (psi)
h	Height of crown of arch (ft)
h_a	Height of apex for double-tapered curved beams (in.)
h_s	Height of soffit at midspan for double-tapered curved beams (in.)
I	Initial moisture content (below 30%) (%)
I	Moment of inertia (in.^{4})
I_K/I_G	Ratio of moment of inertia of knots to moment of inertia of gross cross section
J_x, J_y	Factor for column stability check
K	Constant for bridge deck design
K	Factor for intermediate columns
K	Bending stress factor for double-tapered curved beams
k	Change in member thickness for arch deflection (%)
k	Thermal conductivity (Btu in./hr ft^2 °F)
K_e	Effective buckling length factor
K_R	Factor for round columns
K_r	Radial stress factor

K_1, K_2	Coefficients for truss deflection
L	Span (ft)
l	Span length of beam or unsupported length of column (in.)
l/d	Span-to-depth ratio
L_c	Length between tangent points for double-tapered curved beams (ft)
L_e	Effective length for shear (ft)
l_e	Unsupported column length (in.)
l_e	Effective length of beam (in.)
l_t	Length of tapered leg for double-tapered curved beams (in.)
l_u	Unsupported beam length (in.)
M	Moment capacity (in. lb)
m	Final moisture content (below 30) (%)
M_D	Moment capacities for dowel bridge design (in. lb)
M_s	Bending moment due to unit load (in. lb)
M_y	Total secondary moment for dowel bridge design (in. lb)
N	Fastener value for angle with direction of grain (lb)
n	Number of dowels for bridge deck design
P	Axial load (lb)
P	Design wheel load for bridge design (lb)
p	Allowable passive soil pressure for poles (psf)
p	Fastener value for load acting parallel to grain (lb)
Q	Fastener value for load acting perpendicular to grain (lb)
Q	Statical moment of area (in.^{3})
R	Radius of curvature of inside face of lamination (in.)
r	Radius of gyration (in.)
R_D	Shear capacities for dowel bridge design (lb)
R_H	Horizontal reaction (lb)
R_m	Radius of curvature of centerline of curved member (in.)
R_T, R_1, R_2, ...	Thermal resistance (hr ft^2 °F/Btu)
R_v	Vertical reaction (lb)
R_y	Total secondary shear for dowel bridge design (lb)
S	Section modulus (in.^{3})
s	Effective bridge deck span (in.)
s	Length of arch segment (in.)

S_B	Allowable soil-bearing capacity for poles (psf)
S_m	Shrinkage from initial moisture condition to final moisture content m (%)
S_n	Section modulus times size factor (in.^{3})
S_0	Total shrinkage from Table 2.3 (%)
S_0, S_1, S_3, S_4	Allowable lateral soil-bearing pressure for poles (psf)
T	Applied torque (in. lb)
T	Tensile force (lb)
t	Bridge deck thickness (in.)
t	Fire resistance rating (min)
t	Thickness of column flange (in.)
t	Thickness of lamination (in.)
U	Overall heat transfer coefficient (Btu/hr ft^2 °F)
u	Force in truss member caused by unit load (lb)
V	Vertical shear force (lb)
W	Total uniform load (lb)
w	Uniform load (pounds per unit length)
W'	Total load of 1 in. of water (lb/in.)
X	Distance (ft)
x	Distance (in.)
x	Horizontal location (ft)
y	Vertical location (ft)
y	Wall height of arch (ft)
α	Angle measure (degrees)
α_r	Radial coefficient of thermal expansion
α_t	Tangental coefficient of thermal expansion
Δ_H	Horizontal movement (in.)
Δ	Deflection (in.)
Δ_c	Centerline deflection (in.)
θ	Angle measure (degrees)
π	Pi
σ_{PL}	Proportional limit stress for bridge dowel design (psi)
ϕ	Angle measure (degrees)
Ω	Coefficient of variation

TIMBER CONSTRUCTION MANUAL

Part I

GENERAL

SECTION 1

DESIGN CONSIDERATIONS IN THE USE OF STRUCTURAL TIMBER

INTRODUCTION

The American Institute of Timber Construction (AITC) has developed this *Timber Construction Manual* to provide state-of-the-art technical data and recommendations on engineered timber construction.

Section 1 includes basic information related to the use of structural timber framing. Topics include economy, durability, fire safety, erection, and detailing. With an understanding of these areas, the designer can more effectively utilize the advantages of wood construction. The unique characteristics of wood, design information, and standard practices are covered in subsequent sections.

This manual applies to two types of engineered timber construction—sawn lumber and structural glued laminated timber (glulam). Sawn lumber is the product of lumber mills and is produced from many species. Glued laminated members are produced in laminating plants by gluing together dry lumber, normally of 2- or 1-in. nominal thickness, under controlled conditions of temperature and pressure. Members with a wide variety of sizes, profiles, and lengths can be produced having superior characteristics of strength, serviceability, and appearance. Glued laminated members are manufactured from several species, primarily Douglas Fir-Larch and Southern Pine, but also lesser amounts of Hem-Fir, Western Woods, and California Redwood are used. For cost-efficient design of structural glued laminated timber members, stress values as determined by the design should be specified rather than a particular species or stress combination.

DESIGNING FOR ECONOMY

The economic success of the construction of a project may be greatly influenced by design. Important elements of design include, but are not limited to, the layout of the framing, proper selection of materials and design of all components, ease of construction, serviceability for the intended use, and durability.

The best economy in timber construction is generally realized when standard-size members can be utilized in a repetitious arrangement. However, timber framing, especially glulam, can be custom fabricated to provide a nearly infinite variety of unique but cost-effective architectural forms and arrangements.

Standard Sizes and Grades

The use of standard sizes and grades will result in maximum economy. For glued laminated members, standard sizes as given in Table 7.2 are generally most economical. Any length, up to the maximum length limited by transportation and handling restrictions, is available. For sawn lumber, the sizes given in Table 7.1 are more economical than special sizes and special lengths. Lengths are generally available in even 2-ft increments, and there is a limit on the maximum length normally available from local suppliers.

Standard Connection Details

Specially designed connecting hardware should be avoided. A great variety of steel connecting devices have proved their effectiveness in permanent construction. Typical connection details are given in *Typical Construction Details,* AITC 104 (1), included in Part III of this manual. Also, see fastener manufacturers' catalogs.

Framing Systems, Sawn Timbers, and Glued Laminated Timbers

A great variety of structural timber framing systems are available. The relative economy of any one system over another will depend on the particular requirements of a specific job. Consideration of the overall structure, intended use, geographical location, required configuration, and other factors play an important part in determining the framing system to be used on a job. Table 1.1 may be used for preliminary design purposes to determine the economical span ranges for various timber framing systems. It must be emphasized that the table is to be used for preliminary planning purposes only. All systems require a very close analysis for final design.

The following additional considerations, when applied to timber framing system design, tend to reduce costs. Joints as simple and as few in number as practicable should be used. Splices should be so placed as to minimize design, fabrication, and erection problems. Unnecessary variations in members should be avoided; that is, the identical member design should be used repetitively where practical and the number of variations kept to a minimum. Certain roof profiles will affect the amount and type of load on a structure and may, therefore, affect economy. Better economy usually results from specifying the required design values rather than the lumber grades to be used. Judicious use of multiple continuous spans or cantilever systems with suspended spans tends to balance positive and negative moments and may lower costs.

Appearance Grades for Glued Laminated Timber

An additional consideration related to economy of design is appearance. AITC has developed specifications for three standard appearance grades of glued laminated members. These are given in *Standard Appearance Grades for Structural Glued Laminated Timber,* AITC 110 (2), included in Part III of this manual. These appearance grades are not related to strength. It is more economical to specify the finish or appearance grade best suited for each job than to require the "premium" appearance grade for all jobs.

TABLE 1.1

Economical Span Ranges for Various Timber Framing Systems

Type of System	Economical Span Range (ft)	Considerations
A. Primary Framing Systems		
	Roof Framing Systems	
Beams		Beam systems are frequently used where a low-pitched roof shape is desired
Simple spans		
Straight beams		
Glued laminated	10–100	
Sawn	6–32	
I Joists	12–40	
Tapered beams	25–100	
Double tapered-pitched beams	25–100	
Curved beams	25–100	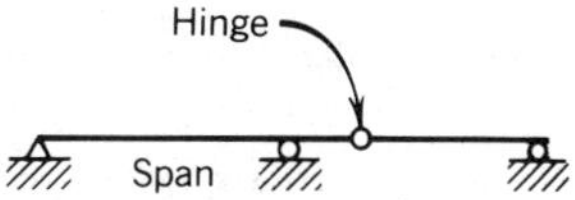
Cantilevered systems		
Glued laminated	Up to 90	
Sawn	Up to 24	Usually more economical than simple spans when span is over 40 ft
Continuous spans		
Glued laminated	10–32	
Sawn	Up to 16	
Girders	40–80	
Arches		
Three-hinged arches		For relatively high rise applications
Gothic	40–90	
Tudor	20–120	Provides required vertical wall frame
A-Frame	20–100	
Three centered	40–250	
Parabolic	40–250	
Radial	40–250	
Two-hinged arches		For relatively low rise applications
Radial	50–200	
Parabolic	50–200	

(continued)

TABLE 1.1 (*Continued*)

Type of System	Economical Span Range (ft)	Considerations
Trusses (Heavy)		Provide openings for passage of wiring, piping, etc.
Flat or parallel chord	50–150	Low roof profile
Triangular or pitched	50–90	For pitched roofs requiring flat surfaces
Bowstring (continuous chord)	50–200	Provide greatest clearance with least wall height
Carrying	40–60	
Trusses (light)		Most light trusses commonly used within these ranges are based on proprietary connections and fabrication methods
Flat or parallel chord	20–50	
Triangular or pitched	20–75	
Tied arches	50–200	Good where no ceiling is wanted; give clear open appearance for low-rise curve; normally more expensive than bowstring; buttress not required
Dome structures	50–500+	
	Floor Framing Systems	
Beams		
Simple span		
Glued laminated	6–40	
Sawn	6–20	
I Joists	12–30	
Continuous	25–40	
B. Secondary Framing Systems		
	Roof Framing Systems	
Sheathing and decking		
1-in. sheathing applied directly to primary system	1–4	Check deflection on spans greater than 32 in.
2-in. roof deck applied directly to primary system	6–10	Check deflection on spans greater than 8 ft
3-in. roof deck applied directly to primary system	8–15	2-, 3-, and 4-in. decking provide good insulation, fire resistance, appearance; easy to erect

TABLE 1.1 (*Continued*)

Type of System	Economical Span Range (ft)	Considerations
Sheathing and decking (Continued)		
4-in. roof deck applied directly to primary system	12–20	
Plywood or structural panel sheathing applied directly to primary system	1–4	
Stressed skin panels	8–40	
Joists with sheathing	16–24	
Purlins with sheathing	16–36	
Beams	20–40	
	Floor Framing Systems	
Plank decking		Floor and ceiling in one
Edge to edge	4–16	
Wide face to wide face	4–16	
Joists with sheathing	10–24	

DESIGNING FOR PERMANENCE

Permanent timber structures should be built not only to be structurally adequate but also to be durable with a minimum of maintenance. With proper design details, construction procedures, and usage, wood is a permanent construction material. Certain conditions affect durability and maintenance costs. If proper consideration is given to these in the design phases of a project, there will be greater assurance that the structure will be durable and that maintenance costs will be minimal. Untreated wood has a proven performance of indefinitely long service if it is kept below 20% moisture content. When wood is exposed to the weather and not properly protected by a roof, eave overhang, or similar covering, or subjected to other conditions of free water or high relative humidity where decay is possible, a preservative treatment is required unless the heartwood of a naturally decay resistant species such as Redwood or Cedar is used. See *Wood Handbook* (3) for a listing of domestic woods that are resistant or very resistant to heartwood decay. The need for preservatively treated wood is a design consideration based on the conditions intended for the wood in service. See *Evaluation, Maintenance and Upgrading of Wood Structures* (16) for information on existing structures.

Wood–Moisture Relationships

Dimensional changes in wood result primarily from a gain or loss of moisture. Once the moisture content (MC) has been lowered to the fiber saturation point (approximately 25–30%), further loss of moisture results in wood shrinkage. This shrinkage continues almost linearly down to 0% moisture content. Eventually, wood assumes a condition near equilibrium with its environment. Wood shrinks and swells most significantly in a direction perpendicular to the grain. Consideration must be given for changes in dimension if moisture changes can occur.

Wood absorbs water in free liquid form and as a water vapor and gives off water again when the humidity of the surrounding air is lowered. If wood becomes wet after being installed dry, its swelling results in increased dimensions and sometimes in distortion and twisting. If installed wet, wood may dry and shrink in service with resulting checking, movement, or distortion as a function of the final moisture content.

Significance of Checking

Checking is the result of rapid lowering of surface moisture content combined with differential moisture contents of the inner and outer portions of the piece. As wood loses moisture content to the surrounding atmosphere, the outer cells of the members lose moisture at a more rapid rate than do the inner cells. As the outer cells try to shrink, they are restrained by the inner portion of the member, which has a higher moisture content. The more rapid the rate of drying, the greater will be the differential in shrinkage between the outer and inner fibers and the higher will be the shrinkage stresses. Some species exhibit more checking than others because they may gain or lose moisture faster or have higher shrinkage rates.

In sawn lumber and timber, controlling the rate of drying will avoid or minimize checking. Many monumental buildings with great historical value have been constructed in such a way as to exploit the natural beauty of large sawn timbers. In these structures, seasoning checks are accepted as an inherent characteristic of the material.

One of the principal advantages of glued laminated timber members is their freedom from major checking; however, seasoning checks may occur in glued laminated members for the same reasons as in sawn timbers, but generally the range of moisture content permitted by industry standards approximates the moisture content in normal-use conditions, thereby minimizing checking that might occur. Moisture content of lumber at the time of gluing is, therefore, of great importance in the control of checking in service. However, serious rapid changes in moisture content after gluing will result in shrinkage or swelling of the wood and may develop stresses in both glued joints and wood that will cause checking. If glued laminated timbers are not carefully protected during shipping, storage, and erection, they may pick up moisture during this period. Subsequent drying may result in more checking than would have occurred had the members remained at the moisture content existing at the time of manufacture. Differentials in the shrinkage rate of individual laminations tend to concentrate shrinkage stresses at or near glue lines. The presence of wood fiber separation indicates adequate glue bond and not delamination.

Structural Considerations.

Glued Laminated Timber. In general, checks have little effect on the strength of laminated members. Glued laminated members are made from laminations that are thin enough to season readily without developing checks on the edges. Checks or splits that occur in drying generally are on the wide faces and do not materially affect the shear strength of bending members loaded perpendicular to the wide faces of the laminations (horizontally laminated beams), which is the most common use of laminated members. If bending members are designed with the load applied parallel to the wide faces of the laminations (vertically laminated members), the occurrence of checks on the wide faces of laminations may affect the shear strength of a beam. For this reason, the design value in shear, F_{vy}, for vertically laminated members has been reduced as shown in the tables in AITC 117—Design included in Part III of this manual. Seasoning checks in bending members affect the horizontal shear value only. They are usually not of structural importance unless the checks are significant in depth, occur in the midheight of the member and near the support, and shear governs the design of the members. All these factors must be considered in appraising checks from a structural viewpoint. In general, the reduction in shear strength is directly proportional to the ratio of depth of check to the width of the bending member. Minor checking may be disregarded, as there is an ample factor of safety in design values.

Sawn Lumber and Timber. Checks affect the horizontal shear strength of lumber, and in establishing design values, checks are anticipated by a large reduction factor applied to test values in recognition of stress concentrations at the ends of the checks. The published design values for horizontal shear for the lumber grades are again adjusted for the amount of checking permissible in the various grades at the time of the grading. Because the strength properties of wood increase with dryness, checks may enlarge somewhat with increased dryness from the time of shipment without decreasing shear strength materially. Actually, a fully seasoned timber may be checked in excess of the grade limitations without affecting its adequacy in the structure because the grading rules are set up with anticipation that some checking beyond the grade limitations at the moisture content graded may occur with seasoning. This is particularly true if loads producing shear stresses are low. Furthermore, only checks that occur near the supports and in the midhalf of the depth of the member are important to shear strength.

Even though the grading rules would exclude such a piece, a column may be checked nearly through without having its strength seriously reduced because the main consideration is that the member act as a unit without splitting entirely into two parts because the l/d ratio would then be reduced.

Cross-grain checks and splits that tend to run out the side of a piece or excessive checks and splits that tend to enter connection areas may be serious and may require repair or replacement. Details for minimizing and controlling the effects of checking in connection areas should be incorporated into the design details. To avoid excessive splitting between rows of bolts due to shrinkage during the seasoning period of timbers, the rows of bolts should not be spaced more than 5 in.

apart or a saw kerf, terminating in a bored hole, should be provided between the lines of bolts. Alternatively, the use of two splice plates, one for each row of bolts, rather than a single splice plate, will lessen the probability of splitting. Whenever possible, maximum end distances for connections should be specified to minimize the effect of checks running into the joint area. Some designs require stitch bolts or fully threaded lag bolts in members with multiple connections loaded at an angle to the grain. Stitch bolts, which should be kept tight, will reinforce pieces where checking is excessive.

The final decision about whether or not shrinkage checks are detrimental to the strength requirements of any particular structure should be made by a competent engineer experienced in timber construction.

Protection

Glued Laminated Timber. The application of selected sealers, paints, and similar protective measures to the surface of the members can retard, but not prevent, checking under extreme conditions. Laminated members for use under conditions that may produce checking more serious than slight surface checks should receive protective coatings or other protection should be required in accordance with *Recommended Practice for Protection of Structural Glued Laminated Timber during Transit, Storage and Erection,* AITC 111, (4) included in Part III of this manual.

Sawn Lumber and Timber. Because lumber loses moisture approximately 10 times faster through the end grain than through flat grain or vertical grain faces, the application of sealer to the end grain will help to minimize the shrinkage checks or splits. The extent of checking is related to the steepness of the moisture gradient between the surfaces and the interior of the piece. Unseasoned lumber should be protected from hot sun and dry winds. Because attic spaces are frequently higher in temperature and lower in humidity than the surrounding atmosphere, they should be properly vented to prevent lumber from rapid drying and resultant checking.

Wood-Destroying Organisms and Their Control

Mold, Stain, and Decay

Decay of wood is caused by low forms of plant life (fungi) that develop and grow from microscopic spores that are present wherever wood is used [see *Wood Handbook* (3)]. The fungi convert wood substance into food. If deprived of any one of the four essentials of life (food, air, moisture, or favorable temperature), decay growth is prevented or stopped and the wood remains sound, retaining its existing strength with no further deterioration. Wood will not be attacked by fungi if it is submerged in water, thereby excluding air, kept continuously below 20% moisture or maintained at temperatures below freezing or much above 100°F. Growth can begin or resume whenever climatic conditions are favorable. The *Standard for Preservative Treatment of Structural Glued Laminated Timber,* AITC 109 (5), included in Part III of this manual, provides information about preservative treatment of glued laminated timber. For treatment of sawn lumber, see the American Wood-Preservers' Association standards (6).

Molds and stains are confined largely to sapwood and are of various colors.

Molds generally do not stain the wood but produce surface blemishes varying from white or light colors to black that can often be brushed off. Fungus stains penetrate the sapwood and normally cannot be removed by scraping or sanding.

Molds and stains should not be considered as stages of decay because these fungi do not attack the wood substance appreciably. For most uses in which appearance is not a factor, wood strength is practically unimpaired by stains and molds. Ordinarily, their only effect is confined to those properties that determine shock resistance or toughness. The only danger is that the early stages of decay may also be hidden in the discolored areas of molds or stain.

Decay-producing fungi, under conditions that favor their growth, may attack both heartwood and sapwood. Heartwood is more resistant to attack than sapwood. Fresh surface growths are usually fluffy or cottony, seldom powdery like the surface growths of molds. The early stages of decay are often accompanied by a discoloration of the wood, which is more evident on freshly exposed surfaces of unseasoned wood than on dry wood. However, many fungi produce early stages of decay that are similar in color to that of normal wood or they give the wood a water-soaked appearance. Later stages of decay are easy to recognize because the wood has undergone definite changes in color and texture.

Decay from fungi reduces the strength of wood. In later stages, all decay fungi seriously reduce the strength of wood and also its fire resistance.

Brown, crumbly rot, in the dry condition, is sometimes called dry rot, but the term is incorrect because decay fungi must have some source of moisture for development, even though the wood may have subsequently become dry.

In some cases, wood submerged in water may be attacked by bacteria resulting in a slow decay process [see *Wood Handbook* (3)].

Prevention and Control of Decay. Timber to be used where conditions favorable to the growth of decay-producing organisms are unavoidable should be pressure treated with a wood preservative, unless the heartwood of naturally decay-resistant species is available and is considered adequate in view of the decay hazard encountered. See *Wood Handbook* (3) for a list of woods resistant or very resistant to decay. Where serious decay problems are present in buildings, they are almost always signs of faulty design or construction or of lack of reasonable maintenance. Design and construction principles that will assure long service and avoid decay hazards in buildings include

1. positive site and building drainage,
2. adequate separation of wood from known moisture sources, and
3. ventilation and condensation control in enclosed spaces.

Building sites should always be graded to provide positive drainage away from foundation walls. All stumps, wood debris, stakes, or wood concrete forms should be removed from the immediate vicinity of a building before backfilling and before placing floor or slabs on grade.

All exposed wood surfaces and adjoining areas should be pitched to assure rapid runoff of water. Construction details that tend to trap moisture in or near wood members should be avoided. The prevention of decay in timber structures rests largely in designs that prevent the entrance and retention of rainwater.

A fairly wide roof overhang with gutters and downspouts that are properly designed is desirable.

Adequate separation of wood from known sources of moisture is necessary to prevent excessive absorption of moisture and to facilitate periodic inspection. When it is impossible or impractical to provide adequate separation, it is recommended that preservatively treated wood or the heartwood of naturally durable species be used.

Wood in contact with concrete near the ground should be protected by a moistureproof membrane such as heavy asphalt paper. In many cases, preservative treatment of the wood in actual contact with the concrete is advisable and required by building codes. Girder and joist openings in masonry walls should be big enough to assure that there will be an air space around the sides and ends of these wood members, and if the members are below the outside soil level, moistureproofing of the outer face of the walls is essential.

Unventilated and inaccessible spaces under buildings should be avoided. Wetting of the wood by condensation may result in serious decay damage. A crawl space with at least an 18-in. clearance should be left under wood joists and girders. Condensation can be minimized by providing openings on opposite foundation walls for cross ventilation. Laying heavy roll roofing or polyethylene membrane with lapped joints on the soil provides an effective moisture barrier.

Porches, breezeways, patios, and other appurtenances may present a decay hazard that cannot be fully avoided by construction practices. Therefore, it is recommended that preservatively treated wood be used for members exposed to decay hazards in these applications.

Sheathing papers on the cold surface of the walls should be of a "breathing" or vapor-permeable type. Vapor barriers, if installed, should be near the warm face of insulated walls and ceilings [see ASHRAE recommendations (7)]. Attic ventilation should be provided.

Where highly humid conditions are present in buildings, as in textile mills, pulp and paper mills, controlled atmosphere plants or storage buildings, and enclosed pools and shower rooms, preservatively treated wood should be used unless the heartwood of naturally decay resistant species is available and considered adequate in view of the decay hazard encountered. When the humid condition involves chemical fumes as well as water vapor, the compatibility of the environment with the treatment and the wood itself should be investigated.

To supplement good design and construction practices, periodic inspections of a stucture will provide assurance that decay-preventive measures are being maintained and that additional decay hazards are not present. These inspections should reveal indications of moisture penetration or condensation, and if detected, corrective measures should be taken to avoid significant damage.

Insects

In terms of economic loss, the most destructive insect to attack wood buildings is the subterranean termite. In certain localities, nonsubterranean termites are also very destructive. Other insects attack timber buildings, but, ordinarily, these occurrences are rather rare and their damage is slight. In many cases, these insects can be controlled by the methods used for termites.

Subterranean Termites. The occurrence of subterranean termites and the damage caused by them is much greater in southern states than in northern states, where lower temperatures do not favor their development (see Fig. 1.1). However, damage to individual buildings may be just as great in northern states as in southern states.

Subterranean termites develop and maintain colonies in the ground from which they build their tunnels through the earth and around obstructions to get at the wood they need for food. Each colony shuts itself off and lives in the dark; but unless they have a constant source of moisture, the termites will die. The worker members of the colony cause the destruction of wood. At certain seasons of the year, male and female winged forms swarm from the colony, fly a short time, lose their wings, mate and, if they succeed in locating suitable places, start new colonies. The appearance of "flying ants" or of their shed wings is an indication that a termite colony may be near and causing serious damage. Not all flying ants are termites; therefore, suspected insects should be identified before money is spent for their eradication.

Subterranean termites do not establish themselves in buildings by being carried there in lumber, but by entering from the ground nests after the building has been

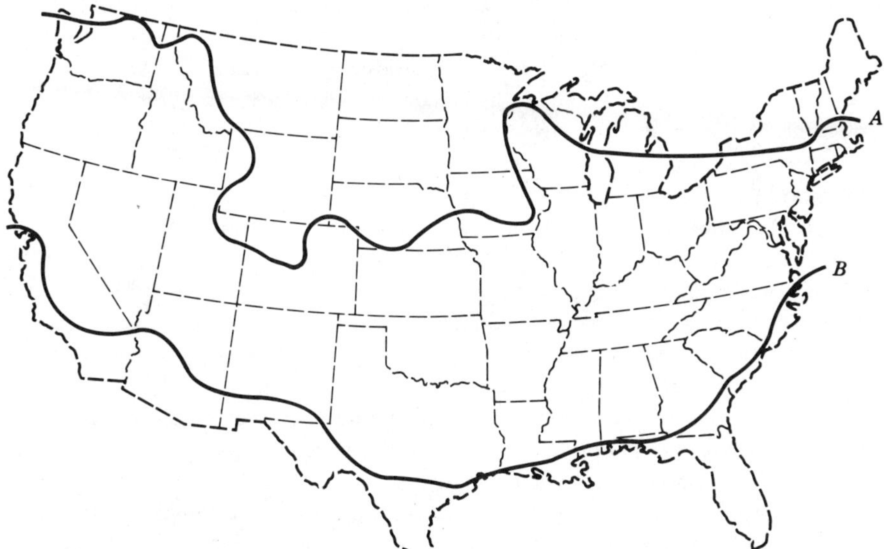

FIGURE 1.1 Limits of Termite Damage. *A,* the northern limit of recorded damage done by subterranean termites in the United States; *B,* the northern limit of damage done by drywood or nonsubterranean termites. Source: *Wood Handbook,* U.S. Department of Agriculture Handbook No. 72, 1974.

constructed. They must maintain contact with a source of moisture such as the soil. Telltale signs of the presence of termites are the earthen tubes, or runways, built by these insects over the surfaces of foundation walls to reach the wood above. Another sign is the swarming of winged adults early in the spring or fall. In wood itself, the termites make galleries that follow the grain, leaving a shell of sound wood to conceal their activities. Because the galleries seldom show on the wood surface, probing with an ice pick or knife is advisable if the presence of termites is suspected.

Protection and Control. Where subterranean termites are prevalent, the best protection is to build so as to prevent their gaining access to the building. The foundations should be of preservatively treated wood, concrete, or other solid material through which the termites cannot penetrate. With brick, stone, or concrete blocks, cement mortar should be used, for termites can work through some other kinds of mortar.

Wood that is not impregnated with an effective preservative must be kept away from the ground. If there is a basement, it should preferably be floored with concrete. Posts supporting the first-floor beams must be thoroughly treated if they bear directly on the ground or on wood blocking. Untreated posts should rest on concrete piers extending at least 1 in. above the basement floor if that floor is of concrete; if the basement floor is of earth, the concrete piers should extend at least 18 in. above it. If the earth is not excavated beneath the building and if the floor is of wood construction, the floor joists and other woodwork, unless adequately treated with preservative, should be kept at lease 18 in. above the earth and good ventilation should be provided beneath the floor.

Moisture condensation on the floor joists and subflooring, which may cause conditions favorable to decay and thus make the wood more attractive to termites, can be avoided by covering the soil with a waterproof membrane. For buildings with concrete slab floors laid over the soil or directly over a sand or gravel fill, the soil under the floor should be treated before the concrete is placed. Furthermore, when insulation containing cellulose is to be used as a filler in expansion joints, it should be impregnated with a chemical toxic to termites. Sealing the top $\frac{1}{2}$ in. of expansion joints and other openings will provide additional protection from ground-nesting termites.

All concrete forms, stakes, stumps, and waste wood should be removed from the building site at the time of construction because they are possible sources of infestation. Generally, precautions that are effective against decay are also helpful against subterranean termites.

Where protection is needed in addition to that obtained by structural methods, the soil adjacent to the foundation walls and piers beneath the building should be thoroughly treated with a recognized insecticide.

To control termites already in a building, break contact between the termite colony in the soil and the wood. This can be done by mechanically blocking the runways from soil to wood, by treating the soil, or by both of these methods. Possible reinfestations should be guarded against by frequent inspections for the telltale signs previously given.

The Formosa termite has become established in the United States. It is more active and voracious than the subterranean termites native to the United States. However, the conventional methods of protection appear to be effective.

Nonsubterranean Termites. Nonsubterranean termites have been found only in a narrow strip of territory extending from central California around the southern edge of the continental United States to Virginia and also in the West Indies and Hawaii (see Fig. 1.1). Their damage is confined to an area in southern California, to parts of southern Florida, notably Key West, and to Hawaii. The nonsubterranean dry-wood termites are fewer in number, and their depredations are not rapid but if they are allowed to work unmolested for a few years, they can occasionally ruin timbers with their tunnelings.

Protection and Control. In the principal damage areas, careful examination of wood is needed to avoid the occurrence of infestations during the construction of a building. All exterior wood can be protected by placing fine-mesh screen over all holes in the walls or roof of the building. If a building is found to be infested by dry-wood termites, badly damaged wood must be replaced. Further termite activity can be arrested by approved chemical treatments applied under proper supervision which will provide for safety of people, domestic animals, and wildlife. Where practical, fumigation is another method of destroying insects.

Other Wood-Inhabiting Insects. Large wood-boring beetles and wood wasps infect green wood but may complete their development in seasoned wood. They do not reinfest dry wood. They occur in all forest areas. The borers in timber or lumber can be killed by heating the wood to a center temperature of 130°F for 1 hr or by fumigation. If infested wood is used in constructing a building, the emerging adult borers will chew $\frac{1}{8}$ to $\frac{1}{2}$-in. holes to the surface, penetrating insulation, vapor barriers, siding, or interior surface materials. The surface holes can be plugged and the damaged spots finished or refinished to make them inconspicuous. Because the borers do not reinfest dry wood, extermination treatments are not required in buildings.

Powder-post beetles can infest and reinfest dry wood, reducing it to floury sawdust. The *Lyctus* powder-post beetles, which are encountered most frequently, attack large-pored hardwoods. Their attacks may be recognized by tunnels packed with floury sawdust and numerous emergence holes $\frac{1}{32}$ to $\frac{1}{8}$ in. in diameter. Heat or fumigation treatments will kill the beetles but will not prevent reinfestation. It can be prevented by a surface application of an approved insecticide in a light-oil solution. Any finishing material that plugs the surface holes of wood will also immunize the wood from *Lyctus* attack. Usually, infestations in buildings result from the use of infested wood, and insecticidal treatment or fumigation may be needed to eliminate them.

Carpenter ants chew nesting galleries in wood. The principal species are large dark-colored ants, and many individuals in the colony are $\frac{1}{2}$ in. long. They exist throughout the United States. Because the ants require a nearly saturated atmosphere in their nest, an ant infestation may indicate a moisture problem in the wood that could also result in decay damage. Ant infestations may be controlled by insecticides.

Marine Borers. Fixed or floating wood structures in salt or brackish water are subject to attack by marine borers. Marine borers include shipworms such as *Teredo* and *Bankia,* the pholads *Martesia* and *Zylophaga,* and *Limnoria* and *Sphaeroma.* Almost all attack wood as free-swimming organisms in the early part of their lives. Shipworms and pholads bore an entrance hole generally at the waterline, attach themselves, and grow in size as they bore tunnels into the wood. *Limnoria* and *Sphaeroma* generally burrow just below the surface of the wood.

Protection and Control. For areas where shipworm and pholad attack are known or expected and where *Limnoria* attack is not expected, the wood should be pressure treated with a creosote and/or creosote–coal tar solution. For areas where *Limnoria* and pholad attack are known or expected, a dual treatment of waterborne salts and creosote is recommended. Where *Limnoria* attack is known or expected and where pholads are absent, either a dual treatment or waterborne salts preservatives may be used.

Effects of Temperature on Wood

The effect of temperature on the dimensional stability of wood is discussed in Section 2. Temperature also has an important effect on the strength of wood.

Design values given in tables in Part III apply to glued laminated timber or sawn lumber used under ordinary temperature conditions. Some reduction of these values may be necessary for members subjected to elevated temperatures for repeated or prolonged periods of time, especially where the high temperature is associated with a high moisture content in the wood.

The temperature effect on strength is immediate, and its magnitude depends on the moisture content of the wood and, when the temperature is elevated, on the time of exposure. When wood is exposed to temperatures above normal for a limited period and the temperature is not excessive, the wood can be expected to recover essentially all its original strength when the temperature is reduced to normal. However, when wood is exposed to temperatures above normal for a limited period and is expected to carry design loads during this period, reduction in design values should be considered. Experiments indicate that dry wood (12% MC) can probably be exposed to temperatures up to nearly 150°F for a year or more without an important permanent loss in most of its strength properties, but while heated, its strength will be temporarily reduced compared to its strength at normal temperature. However, if wood is heated to temperatures of up to 150°F for extended periods of time, adjustment of design values may be necessary as indicated in Table 3.3.

Tests of wood conducted at about −300°F show that the important strength properties of dry wood in bending and in compression, including stiffness and shock resistance, are much higher at the extremely low temperature than at normal temperatures.

The approximate immediate effect of temperature on most of the static strength properties of dry wood (12% MC) within the range 0–150°F can be estimated as an increase or a decrease in the strength at 70°F of about $\frac{1}{3}$ to $\frac{1}{2}$% for each 1°F decrease or increase in temperature. The change in properties will be greater if the moisture content is high and less if the moisture content is low. In some geographical locations, fairly high temperatures are commonly experienced, but the accompanying relative humidity is ordinarily quite low. Wood exposed to such conditions generally has a low moisture content, and the immediate effect of the high temperature is not significant.

When wood is exposed to temperatures of 150°F or higher for extended periods of time, it is permanently weakened, even though the temperature is subsequently reduced and the wood is used at normal temperatures. The permanent or nonrecoverable strength loss depends on a number of factors, including the moisture content and temperature of the wood, the heating medium and time of exposure, and to some extent on the species and the size of the piece. In special cases, timbers that are exposed to elevated temperatures for extended lengths of time may need adjustments to the design values. Refer to Appendix C, *National Design Specification* (NDS) (8).

Glued laminated members are normally cured at temperatures of less than 150°F. Therefore, no reduction in their design values due to temperature effects during manufacturing is necessary.

Adhesives used under standard specifications for structural glued laminated members (i.e., casein, resorcinol resin, phenol resin, and melamine resin adhesives) are not affected substantially by the high temperatures that char wood. The use of other adhesives that might deteriorate at lower temperatures is not permitted by standard specifications for structural glued laminated timber. Low temperatures appear to have no significant effect on the strength of glue joints.

Effects of Chemical Processes or Stored Chemicals on Wood

Wood is often superior to many other common construction materials in its resistance to chemical attack. For this reason, wood is used for storage buildings and containers for many chemicals and in processing plants in which structural members are subjected to spillage, leakage, or condensation of chemicals.

Wooden tanks are commonly employed for the storage of water or chemicals that deteriorate other materials. Experience has shown that the heartwood of Cypress, Douglas Fir-Larch, Southern Pine, and California Redwood is the most suitable for water tanks and that the heartwood of the first three of these species is most suitable for tanks when resistance to chemicals in appreciable concentrations is an important factor. The heartwood of each of the four species combines moderate to high resistance to water penetration with moderate to high natural resistance to decay and hydrolysis. Structural members, floors, stairways, catwalks, and docks are frequently exposed to spillage or leakage of chemicals. Volatile chemicals may attack roof supports, ventilating ducts, and stacks in chemical processing plants. Wood should be considered for these applications.

Chemical actions of three general types may affect the strength of wood. The first causes swelling and the resultant weakening of the wood. Liquids such as water, alcohols, and some other organic liquids swell wood. This action is almost completely reversible; hence, if the swelling liquid or solution is removed by evaporation or by extraction followed by evaporation of the solvent, the original dimensions and strength are practically restored. Liquids such as petroleum oils and creosote do not swell wood. The second type of action brings about permanent changes in the wood, such as hydrolysis of the cellulose by acids or acid salts. The third type of action, which is also permanent, involves delignification of the wood and dissolving of hemicelluloses by alkalies.

Experience and available data indicate species and conditions where wood is equal or superior to other materials in resisting the degradative action of chemicals. In general, heartwood of such species as Cypress, Douglas Fir-Larch, Southern Pine, California Redwood, Maple, and White Oak is quite resistant to attack by dilute mineral and organic acids. Oxidizing acids, such as nitric acid, have a greater degradative action than nonoxidizing acids. Alkaline solutions are more destructive than acidic solutions, and hardwoods are more susceptible to attack by both acids and alkalies than softwoods.

Highly acidic salts tend to hydrolyze wood when present in high concentrations. Even relatively low concentrations of such salts have shown signs that the salt may migrate to the surface of railroad ties, which are occasionally wet and dried in a hot, arid region. This migration, combined with the high concentrations of salt relative to the small amount of water present, causes an acidic condition sufficient to make wood brittle.

Iron salts, which develop at points of contact with plates, bolts, and the like, have a degradative action on wood, especially in the presence of moisture. In addition, iron salts probably precipitate toxic extractives and thus lower the natural decay resistance of wood. The softening and discoloration of wood around corroded iron fastenings is a commonly observed phenomenon; it is especially pronounced in acidic woods, such as Oak, and in woods such as California Redwood, which contain considerable tannin and related compounds. The oxide layer formed on iron is transformed through reaction with wood acids into soluble iron salts, which not only degrade the surrounding wood but probably catalyze the further corrosion of the metal. The action is accelerated by moisture; oxygen may also play an important role in the process. This effect is not encountered with well-dried wood used in dry locations. Under damp-use conditions, it can be avoided or minimized by using corrosion-resistant fastenings.

Many substances have been employed as impregnants to enhance the natural resistance of wood to chemical degradation. One of the more economical treatments involves pressure impregnation with a viscous coke-oven coal tar to retard liquid penetration. Acid resistance of wood is increased by impregnation with phenolic resin solutions followed by appropriate drying and curing. Treatment with furfuryl alcohol has been used to increase resistance to alkaline solutions. Another procedure involves massive impregnation with a monomeric resin, such as methyl methacrylate, followed by polymerization.

DESIGNING FOR FIRE SAFETY

Neither building materials alone, nor building features alone, nor detection and fire extinguishing equipment alone can provide adequate safety from fire in buildings. A proper combination of these should provide the necessary degree of protection for the occupants and for the property. Some of the more important design considerations that should be investigated are listed.

General considerations:

1. Use, occupancy, or activity taking place in the building.
2. Fire stopping, draft stopping, and the elimination or proper protection of concealed spaces.
3. Separation of areas in which hazardous processes or operations take place, such as boiler rooms and workshops.
4. Location of structure within the property lines.
5. Height of structure, allowable floor areas, automatic alarms, and sprinkler systems.

Safety to life considerations:

1. The number, size, type, and accessibility of exit ways (particularly stairways) and their distance from each other.
2. The installation of automatic alarm systems.
3. Enclosure of stairwells and use of self-closing fire doors.
4. Interior finishes that will assure that the surfaces will not spread flame at hazardous rates.
5. Ventilation systems or self-contained breathing apparatus.

Safety to property considerations:

1. The installation of automatic sprinkler systems.
2. Proper placement of firewalls and proper protection of openings in them.
3. Interior finishes to assure that surfaces will not spread flame at hazardous rates.
4. Roof venting equipment or provision for draft curtains.

Protection of the occupants of a building and of the property itself can be achieved by taking advantage of the fire-endurance properties of wood and by giving careful attention to the above-mentioned details that make a building fire-safe, that is, fire-safe rather than so-called fireproof.

Wood, when exposed to heat and/or flame, forms a self-insulating surface layer of char. Although the surface chars, the undamaged inner wood below the char retains its strength and will support loads equivalent to the capacity of the uncharred section. Very often, heavy timber members will retain their structural integrity through long periods of fire exposure and still remain serviceable after the surface has been cleaned and refinished. The fire endurance and excellent performance of heavy timber is attributable to the size of the wood members and to the slow rate at which the charring penetrates.

Timber, unlike most other construction materials, is not appreciably distorted by high temperatures; therefore, it is not likely that walls will be displaced or pushed over by expanding members, as often happens with some other materials.

Size of timber members is of particular importance with respect to fire endurance. Building codes specify minimum nominal dimensions for exposed heavy timber structural members. For fire-resistance classification purposes, buildings of wood construction are generally classified as heavy timber construction, ordinary construction, or wood frame construction. *Heavy timber construction* is that type in which fire resistance is attained by placing limitations on the minimum size, thickness, or composition of all load-carrying wood members; by avoiding concealed spaces under floors or roofs; by using approved fastenings, construction details, and adhesives; and by providing the required degree of fire resistance in the exterior and interior walls. *Ordinary construction* has exterior masonry walls and wood framing members of sizes smaller than heavy timber sizes. *Wood frame construction* has wood-framed walls and structural framing of sizes smaller than heavy timber sizes. Depending on the occupancy of a building or the hazard of the operations within it, a building of wood frame or ordinary construction may have its members covered with fire-resistive coverings.

All of the nationally recognized building codes and most other codes recognize heavy timber construction by allowing larger building areas and uses for which ordinary construction and wood frame is not permitted. The requirements for heavy timber construction are contained in *Standard for Heavy Timber Construction*, AITC 108 (9), in Part III of this manual.

Fire-Rated Timber Construction

In some instances, a specific fire rating is required by code for structural members in a building. Common stud walls when covered with a fire-rated gypsum board can provide a rating of 1 hr or more. Larger members may also be covered with gypsum board to increase the resistance to fire. The fire rating of wood members may be calculated by taking into account the depth, breadth, and number of sides exposed to fire and the ratio of the actual load on a member to the allowable design load. For instance, a member loaded to 50% of its design load has a greater fire rating than a member fully loaded. The following formulas may be used to calculate the fire-resistance rating of wood members, both sawn and glued laminated timber. This method is based on a procedure contained in "A Method for Assessing the Fire Resistance of Laminated Beams and Columns" (10) by T. T. Lie. The fire-resistance rating t, in minutes, of beams and columns can be calculated by use of the following formulas.

For beams exposed to fire on four sides,

$$t = 2.54Zb\left(4 - \frac{2b}{d}\right) \tag{1-1}$$

For beams exposed to fire on three sides,

$$t = 2.54\ Zb\left(4 - \frac{b}{d}\right) \tag{1-2}$$

For columns exposed to fire on four sides,

$$t = 2.54\ Zd\left(3 - \frac{d}{b}\right) \tag{1-3}$$

For columns exposed to fire on three sides,

$$t = 2.54\ Zd\left(3 - \frac{d}{2b}\right) \tag{1-4}$$

where Z = dimensionless load factor from Figure 1.2,
b = breadth (width) of a beam or larger side of a column before fire exposure (in.), and
d = larger side of a beam or smaller side of a column before exposure to fire (in.).

Equation (1-4) is valid only if the unexposed face is the smaller side of a column. There is no experimental data to justify this equation when one of the larger faces is not exposed to the fire. If the column is recessed into a wall, Eq. (1-4) may be used, but the full dimensions of the column apply.

The value of the effective length factor K_e is obtained from Table 5.4. In Figure 1.2, l_e is the unsupported length of the column or compression member, l, multiplied by K_e.

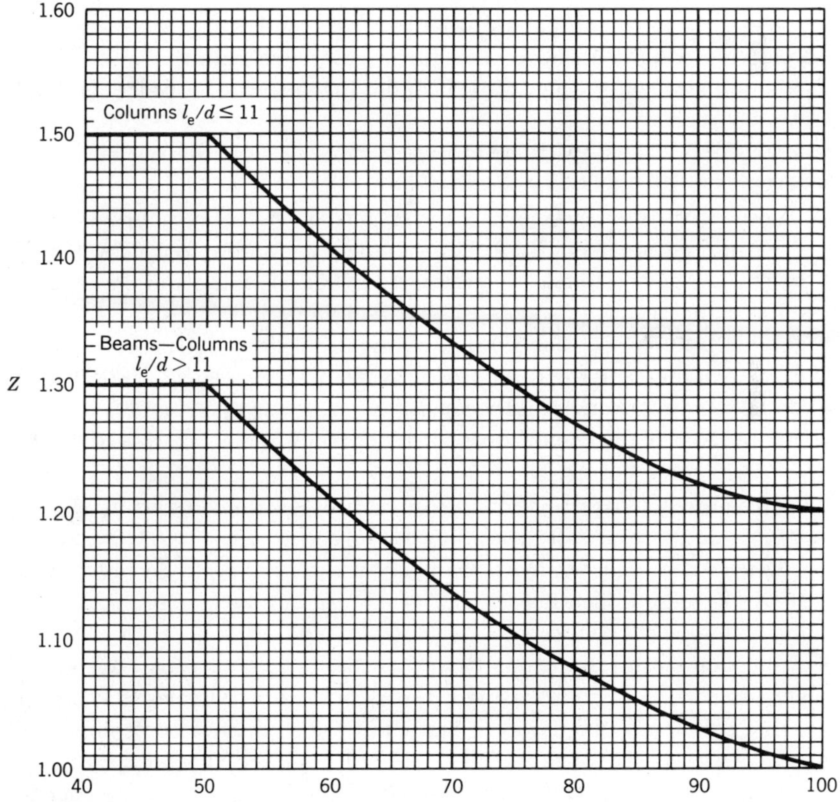

FIGURE 1.2 Load on member as a percentage of the allowable load.

The allowable design loads on beams and columns are determined by procedures in Section 5. Also, an example of calculation of the fire-resistance rating is included in Section 5. Fastenings for fire-resistance-rated members are discussed in Section 6. For additional information on calculation of fire resistance, see (11) and (12). For more information on this procedure and requirements for fastenings, see (13).

Fire Safety in Buildings

Regulations governing the construction of buildings have evolved over the years mainly to protect property against the ravages of fire. The life safety and fire protection provisions found in modern building codes provide what are considered minimum requirements for fire safety based on experience accumulated from fires and on research in the fire protection field. Such provisions become law when adopted by local, state, or regional jurisdictions. Typical provisions are found in all nationally recognized model building codes.

The extent of structural fire endurance desirable beyond building code requirments is a matter of economic balance between the risk of loss and the available fire protection facilities. Experience indicates that too much emphasis on the fire resistance of the structural frame results in a false sense of security. The emphasis is wrongly placed on so-called fireproof buildings rather than on fire-safe buildings. Good building design, adequate structural endurance, and automatic fire detection equipment and extinguishing facilities may provide fire safety at reasonable cost.

Safety from fire begins with the fire prevention measures that are taken. They include a properly designed building, including the electrical and mechanical systems. However, the fire record over the years has shown that most fires have their beginning in the contents of the building rather than in the structure itself. For this reason, fire prevention is related largely to housekeeping and maintenance, which are in turn directly associated with building administration. Thus, preventing a fire from starting has little or nothing to do with the structural materials used in the building, and the word *fireproof* as related to structural materials often deludes the building planner and the occupant by giving them an illusory feeling of safety. Because of the combustible nature of the contents of a building, no occupied building is fireproof.

Even with the best of prevention measures, fires will and do start in buildings, perhaps through unforseeable conditions and events or uncontrollable acts of nature. The first consideration then is the safety of the occupants of the building and, after their evacuation, the safety of the firefighters and then the protection of the building and adjacent property.

The most important protection factors are the prompt detection of a fire, immediate alarm, and the rapid extinguishment of the fire. The fact that there are people in a building is no guarantee that a fire will be discovered promptly. The promptness with which a fire is discovered and the speed with which it is extinguished have an important bearing on life saftey and on the extent of damage both to the contents and to the structure. Automatic detectors and extinguishing systems can provide a high degree of protection, but they should not be considered a substitute for other protection measures such as the proper number, location, and type of exit ways. The Life Safety Code, promulgated by the National Fire Protection Association, is widely used as a guide for safety to life from fire in buildings.

TIMBER DETAILING

Typical Construction Details, AITC 104 (1) is included in Part III of this manual as a suggested guide for designers to use in their work and to give greater assurance of quality with a minimum of maintenance for engineered timber structures. Also, see Section 6 of this manual for recommendations on connection design and detailing.

The architect or engineer must select, modify, and design the details best suited to fit the requirements of a particular job. In doing so, the designer should keep in mind ease and economy of assembly and erection, as well as minimizing maintenance requirements and increasing durability through design and construction practices. Allowances should be made for differential movements resulting from dimensional changes, especially shrinkage perpendicular to grain. Whenever possible, connections should be designed to avoid putting wood members in tension perpendicular to grain. Moisture barriers, flashing, and other features should be used where necessary to avoid moisture or water traps. Durable structures result when such details are incorporated with other good design and construction, and protective treatment and coatings for wood and construction hardware.

When not otherwise required, the overall side clearance between nonbearing surfaces of steel assemblies and timber members should be not less than $\frac{1}{8}$ in. ($\frac{1}{16}$ in. each side). Unless shown and required on the shop details, welds should not be located where they will interfere with the assembly of the connection. Hardware and steel accessories should be painted or coated to prevent staining of timbers during assembly and erection in work where appearance is a factor.

HANDLING, STORAGE, ERECTION, AND SEASONING OF STRUCTURAL TIMBER

The erection of structural timber framing requires experienced erection crews and adequate lifting equipment to protect lives and property and to assure that the framing is not improperly assembled or damaged during handling. The unloading and storage of structural timber framing before erection also demands care and good judgment. It is suggested that a shipment of structural timber framing, on receipt at the job site, be checked for tally and damage. The following general precautions apply.

Precautions during Unloading

Structural timber framing is subject to surface marring and damage when not properly handled and protected. At the erection site, the following precautions are suggested.

1. Lift members or roll them on dollies or rollers out of railroad cars; do not drag or drop them. Unload trucks by lifting from the truck; do not dump or drop members.
2. If unloading with lifting equipment, use wide fabric or plastic belts or other slings that will not mar wood. If chains or cables are used, provide protective blocking or padding to sharp edges or sharp corners.
3. Guard against soiling, dirt, footprints, or abrasions. If members are wrapped, avoid tearing or damaging the protective material.

Precautions during Storage

If structural timber framing is to be stored before erection, it should be placed on blocks well off the ground, and individual members should be separated by strips so that air may circulate around all four sides. The top and all sides of storage piles should be covered with moisture-resistant material. Clear polyethylene films should not be used because wood members are subject to bleaching from sunlight. Individual wrappings should be slit or punctured on the lower side to permit drainage of water that may have accumulated.

Water-resistant wrapping used for the in-transit protection of glued laminated members should be left intact until the members are enclosed within the building. If wrapping has to be removed at certain connection points during the erection, it should be replaced after the connection is made. If it is impractical to replace the wrapping, all of it should be removed to avoid the nonuniform appearance caused by sun and weather exposure.

Precautions during Erection

Assembly. Trusses are usually shipped partially or completely disassembled and are assembled on the ground at the site before erection. Arches, which are generally shipped in halves, may be assembled on the ground or connections may be made after the half arches are in position. When trusses and arches are assembled on the ground at the site, they should be assembled on level blocking to permit connections to be fitted properly and tightened securely without damage. The end compression joints should be brought into full bearing and compression plates installed where specified.

Before erection, the assembly should be checked for prescribed overall dimensions, prescribed camber, and accuracy of anchorage connections. Erection should be planned and executed in such a way that the close fit and neat appearance of joints and the structure as a whole will not be impaired.

Anchor bolts should be checked prior to start of erection. Before erection begins, all supports and anchors should be complete, accessible, and free of obstructions. The weights and balance points of the structural timber framing should be determined before lifting begins so that proper equipment and lifting methods may be employed. When long members or timber trusses of long span are raised from a flat to a vertical position preparatory to lifting, stresses entirely different from the normal design stress may be introduced. The magnitude and distribution of these stresses will vary, depending on such factors as the weight, dimensions, and type of member. A competent rigger should consider these factors in determining how much suspension and stiffening, if any, is required and where it should be located.

Bracing

All framing must be true and plumbed. Permanent bracing is bracing so designed and installed as to form an integral part of the final structure. Erection bracing is bracing installed to hold the framing in a safe position until sufficient permanent bracing is in place to provide full stability. Proper and adequate temporary erection bracing is introduced whenever necessary to take care of all loads to which the structure may be subjected during erection, including equipment and its operation. This bracing is left in place as long as may be required for safety. Part or all of the permanent bracing may also act as erection bracing. Erection bracing serves to plumb the framing during erection and gives it adequate stability to receive purlins, joists, and roofing materials. It may include sway bracing, guy ropes, tieing off framing nearest to end walls, steel tie rods with turnbuckle take-ups, struts, shoes, and similar items. As erection progresses, bracing is securely fastened in place to take care of all dead load, erection stresses, and normal weather conditions. Excessive concentrated construction loads, such as bundles of sheathing, piles of purlins, roofing, or other materials, should be avoided.

Final Alignment

Final tightening of alignment bolts should not be completed until the structure has been properly aligned.

Removal of Temporary Bracing

Temporary erection bracing should be removed only after diaphragms and permanent bracing are installed, the structure has been properly aligned, and connections and fastenings have been finally tightened. Retightening of connections prior to final completion or closing in of inaccessible connections is recommended.

Field Connections

The joining, holding, and welding of steel connections in the field are performed according to the requirements for shop work of such operations, except where such requirements apply to shop conditions only. Steel connections should comply with the specifications of the American Institute of Steel Construction (14) and the American Welding Society (15).

Protection of Field Cuts

All field cuts of timbers should be coated with an approved moisture seal if the member was initially coated unless otherwise specified. All field framing is done in accordance with the requirements of shop practice except where such requirements apply to shop conditions only. If timber framing has been pressure treated, field framing after treatment must be avoided or at least, insofar as possible, held to a minimum. When field cuts in pressure-treated material are unavoidable, additional treatment should be provided in accordance with AWPA Standard M4 (6).

Protection against Moisture

During erection operations, all timber framing that requires moisture content control, whether sawn or glued laminated timbers, should be protected against moisture pickup. Any fabricated structural materials to be stored for an extended period of time before erection should, insofar as is practicable, be assembled into subassemblies for storage purposes.

Seasoning Period

Heat should not be fully turned on as soon as the structure is enclosed; otherwise, excessive checking may occur due to rapid lowering of the relative humidity in the building. A gradual seasoning period at moderate temperature should be provided.

REFERENCES

1. American Institute of Timber Construction, *Typical Construction Details*, AITC 104, Englewood, CO, 1984.
2. American Institute of Timber Construction, *Standard Appearance Grades for Structural Glued Laminated Timber*, AITC 110, Englewood, CO, 1984.
3. United States Department of Agriculture, Forest Service, Forest Products Laboratory, *Wood Handbook: Wood as an Engineering Material*, Agriculture Handbook No. 72, Madison, WI, 1974.
4. American Institute of Timber Construction, *Recommended Practice for Protection of Structural Glued Laminated Timber During Transit, Storage and Erection*, AITC 111, Englewood, CO, 1979.
5. American Institute of Timber Construction, *Standard for Preservative Treatment of Structural Glued Laminated Timber*, AITC 109, Englewood, CO, 1984.
6. American Wood-Preservers' Association, *Book of Standards*, Stevensville, MD, 1984.
7. American Society of Heating, Refrigerating and Air-Conditioning Engineers, Inc., *ASHRAE Fundamentals Handbook*, Atlanta, GA, 1981.
8. National Forest Products Association, *National Design Specification for Wood Construction*, Washington, DC, 1982.
9. American Institute of Timber Construction, *Standard for Heavy Timber Construction*, AITC 108, Englewood, CO, 1980.
10. T. T. Lie, "A Method for Assessing the Fire Resistance of Laminated Beams and Columns." National Research Council of Canada, Division of Building Research, DBR 718, Ottawa, Ontario, 1977.
11. United States Department of Agriculture, Forest Service, Forest Products Laboratory, *Strength Validation and Fire Endurance of Glued Laminated Beams*, Madison, WI, 1985.
12. Warnock-Hersey, *Report of Standard Fire Endurance Test, Nominal $8\frac{3}{4}$ in. by $16\frac{1}{2}$ in. Glulam Wood Beam*, Report No. WH1-694-0069, Severna Park, MD, 1982.
13. American Institute of Timber Construction, *Calculation of Fire Resistance of Glued Laminated Timbers*, Technical Note No. 7, Englewood, CO, 1984.
14. American Institute of Steel Construction, *Manual of Steel Construction*, 8th ed., Chicago, IL, 1980.
15. American Welding Society, *Structural Welding Code*, D1.1 Miami, Fl, 1983.
16. American Society of Civil Engineers, *Evaluation, Maintenance and Upgrading of Wood Structures*, New York, 1982.

SECTION 2

PHYSICAL STRUCTURE OF WOOD

INTRODUCTION

Wood is a cellular organic material made up principally of cellulose, which comprises the structural units (cells), and lignin, which cements the structural units together. It also contains hemicelluloses, extractives, and ash-forming minerals. Wood cells are hollow, and they vary from about 0.04 to 0.33 in. in length and from 0.0004 to 0.0033 in. in diameter. Most cells are elongated and are oriented vertically in the growing tree, but some, called rays, are oriented horizontally and extend from the bark toward the center or pith of the tree.

Hardwoods and Softwoods

Trees are divided into two broad classes: hardwoods, which have broad leaves, and softwoods or conifers, which have needlelike or scalelike leaves. Most hardwoods shed their leaves at the end of each growing season, and most softwoods are evergreens. The terms *hardwood* and *softwood* are often misleading because they do not directly indicate the hardness or softness of wood. Some hardwoods are softer than certain softwoods and some softwoods are harder than some hardwoods.

Heartwood and Sapwood

The cross section of a tree shows several distinct zones: the bark; a light-colored zone called sapwood; an inner zone, generally of darker color, called heartwood; and, at the center, the pith.

A tree increases in diameter by adding new layers of cells from the pith outward. For a time, this new layer functions as living cells that conduct sap and store food, but eventually, as the tree increases in diameter, cells toward the center become inactive and serve only as support for the tree. The inactive inner layer is the heartwood; the outer layer containing living cells is the sapwood. There is no consistent difference between the weight and strength properties of heartwood and sapwood. In some species, heartwood is significantly more resistant to decay fungi than is sapwood, although there is a great range in the durability of heartwood among the various species.

Annual Rings

In climates where temperature limits the growing season of a tree, each annual increment of growth usually is readily distinguishable. Such an increment is known as an annual growth ring, or annual ring, and consists of an earlywood and a latewood band.

Earlywood (Springwood) and Latewood (Summerwood)

In many wood species, large, thin-walled cells are formed in the spring when growth is greatest, whereas smaller, thicker-walled cells are formed later in the year. The areas of fast growth are called earlywood, and the areas of slower growth, latewood. In annual rings, the inner, lighter-colored area is the earlywood and the outer, darker layer is the latewood.

Latewood contains more solid wood substance than does earlywood and, therefore, is denser and stronger. The proportion of width of latewood to width of annual ring as well as the number of rings per inch is used as one of the visual measures of the quality and strength of wood for some species.

Grain and Texture

The terms *grain* and *texture* are used in many ways to describe the characteristics of wood and, in fact, do not have a definite meaning. Grain often refers to the width of the annual rings, as in "close-grained" or "coarse-grained." Sometimes it indicates whether the fibers are parallel to or at an angle with the sides of the pieces, as in "straight-grained" or "cross-grained." Texture usually refers to the fineness of wood structure rather than to the annual rings. When these terms are used in connection with wood, the meaning intended should be defined.

Moisture Content

A tree develops in the presence of moisture, and throughout its life the tree remains moist or "green." The amount of moisture in a living tree varies among species, in individual trees within the same species, in different parts of the same tree, and between heartwood and sapwood.

Moisture content (MC) is the weight of the water contained in wood, expressed as a percentage of the weight of the oven-dry wood. (An oven-dry condition is reached when no further loss of weight is experienced on subsequent oven drying.) As wood loses moisture, the water in the cell cavity is evaporated first. The condition at which the water in the cell cavity has been evaporated but the cell wall is still saturated is known as the fiber saturation point. The fiber saturation point is quite variable between species and among individual pieces of the same species but averages about 30% MC. Further drying of wood below its fiber saturation point results in loss of moisture from the cell walls, which in turn causes shrinkage of the wood.

Wood in use gives off or takes on moisture from the surrounding atmosphere with changes in temperature and relative humidity until it attains a balance relative to the atmospheric conditions. The moisture content at this point of balance is known as the equilibrium moisture content (EMC). At constant temperature, the equilibrium moisture content depends entirely on the relative humidity of the atmosphere surrounding the wood and the hygroscopicity of the wood or contained materials. (See Table 2.1, which, for practical purposes, applies to the wood of any species.)

Growth Characteristics

Wood contains certain natural growth characteristics such as knots, slope of grain, compression wood, and shakes, which may, depending on their size, number, and location in a structural member, adversely affect the strength properties of that member. These characteristics are discussed in detail in *Wood Handbook* (1) by the U.S. Forest Products Laboratory. Structural grading rules take into account the effects of these growth characteristics on the strength of wood in establishing design values for lumber and glued laminated timber.

Directional Properties

Wood is nonisotropic because of the orientation of its cells and the manner in which it increases in diameter. It has different mechanical properties with respect to its three principal axes of symmetry: longitudinal, radial, and tangential (see Fig. 2.1). Strength and elastic properties corresponding to these three axes may be used in design.

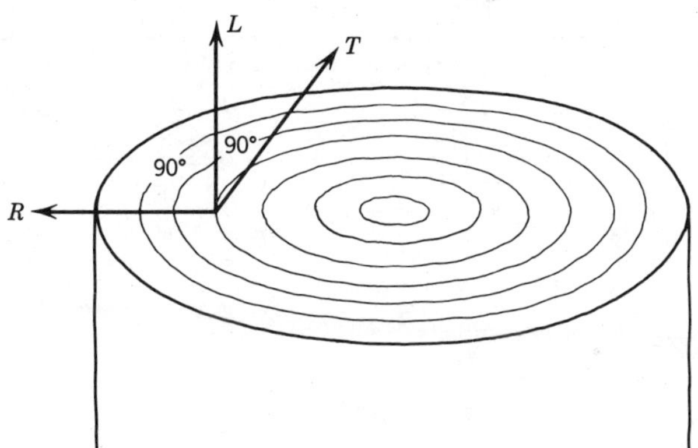

FIGURE 2.1 The Three Principal Axes of Wood. *L*, longitudinal (parallel to grain); *R*, radial (perpendicular to grain, radial to annual rings); *T*, tangential (perpendicular to grain, tangential to annual rings).

The difference between properties in the radial and tangential directions is seldom of practical importance in most structural designs; for structural purposes, it is usually sufficient to differentiate only between properties parallel and perpendicular to the grain.

WEIGHT AND SPECIFIC GRAVITY OF COMMERCIAL LUMBER SPECIES

Table 2.2 gives average weights, in pounds per cubic foot, of various commercial lumber species based on their weight and volume at 12% and 20% MC. It

TABLE 2.1

Moisture Content of Wood in Equilibrium With Stated Dry-Bulb Temperature and Relative Humidity[a]

Temperature Dry-bulb, °F	Relative Humidity, Percent																		
	5	10	15	20	25	30	35	40	45	50	55	60	65	70	75	80	85	90	95	98
30	1.4	2.6	3.7	4.6	5.5	6.3	7.1	7.9	8.7	9.5	10.4	11.3	12.4	13.5	14.9	16.5	18.5	21.0	24.3	26.9
40	1.4	2.6	3.7	4.6	5.5	6.3	7.1	7.9	8.7	9.5	10.4	11.3	12.3	13.5	14.9	16.5	18.5	21.0	24.3	26.9
50	1.4	2.6	3.6	4.6	5.5	6.3	7.1	7.9	8.7	9.5	10.3	11.2	12.3	13.4	14.8	16.4	18.4	20.9	24.3	26.9
60	1.3	2.5	3.6	4.6	5.4	6.2	7.0	7.8	8.6	9.4	10.2	11.1	12.1	13.3	14.6	16.2	18.2	20.7	24.1	26.8
70	1.3	2.5	3.5	4.5	5.4	6.2	6.9	7.7	8.5	9.2	10.1	11.0	12.0	13.1	14.4	16.0	17.9	20.5	23.9	26.6
80	1.3	2.4	3.5	4.4	5.3	6.1	6.8	7.6	8.3	9.1	9.9	10.8	11.7	12.9	14.2	15.7	17.7	20.2	23.6	26.3
90	1.2	2.3	3.4	4.3	5.1	5.9	6.7	7.4	8.1	8.9	9.7	10.5	11.5	12.6	13.9	15.4	17.3	19.8	23.3	26.0
100	1.2	2.3	3.3	4.2	5.0	5.8	6.5	7.2	7.9	8.7	9.5	10.3	11.2	12.3	13.6	15.1	17.0	19.5	22.9	25.6
110	1.1	2.2	3.2	4.0	4.9	5.6	6.3	7.0	7.7	8.4	9.2	10.0	11.0	12.0	13.2	14.7	16.6	19.1	22.4	25.2
120	1.1	2.1	3.0	3.9	4.7	5.4	6.1	6.8	7.5	8.2	8.9	9.7	10.6	11.7	12.9	14.4	16.2	18.6	22.0	24.7
130	1.0	2.0	2.9	3.7	4.5	5.2	5.9	6.6	7.2	7.9	8.7	9.4	10.3	11.3	12.5	14.0	15.8	18.2	21.5	24.2
140	0.9	1.9	2.8	3.6	4.3	5.0	5.7	6.3	7.0	7.7	8.4	9.1	10.0	11.0	12.1	13.6	15.3	17.7	21.0	23.7
150	0.9	1.8	2.6	3.4	4.1	4.8	5.5	6.1	6.7	7.4	8.1	8.8	9.7	10.6	11.8	13.1	14.9	17.2	20.4	23.1
160	0.8	1.6	2.4	3.2	3.9	4.6	5.2	5.8	6.4	7.1	7.8	8.5	9.3	10.3	11.4	12.7	14.4	16.7	19.9	22.5
170	0.7	1.5	2.3	3.0	3.7	4.3	4.9	5.6	6.2	6.8	7.4	8.2	9.0	9.9	11.0	12.3	14.0	16.2	19.3	21.9
180	0.7	1.4	2.1	2.8	3.5	4.1	4.7	5.3	5.9	6.5	7.1	7.8	8.6	9.5	10.5	11.8	13.5	15.7	18.7	21.3
190	0.6	1.3	1.9	2.6	3.2	3.8	4.4	5.0	5.5	6.1	6.8	7.5	8.2	9.1	10.1	11.4	13.0	15.1	18.1	20.7
200	0.5	1.1	1.7	2.4	3.0	3.5	4.1	4.6	5.2	5.8	6.4	7.1	7.8	8.7	9.7	10.9	12.5	14.6	17.5	20.0
210	0.5	1.0	1.6	2.1	2.7	3.2	3.8	4.3	4.9	5.4	6.0	6.7	7.4	8.3	9.2	10.4	12.0	14.0	16.9	19.3

16% 19%

[a]Numbers to the right of the heavy line represent wet-use service conditions (16% or greater for glued laminated timber and greater than 19% for sawn timbers.)

also gives specific gravities based on weight when oven dry and volume at 12% MC and when green.

For design purposes, it is satisfactory to use weights at 12% MC for glued laminated timber and sawn timber that will remain dry in service. The weight for unseasoned sawn material when it is installed and before it dries in service should be taken at 20% MC for relatively small sizes of less than 3 in. in thickness. The weight based on the equilibrium moisture content with allowance for preservative treatments, if applied, should be employed for unprotected members in exterior or other wet conditions of use. Adhesives used in laminating do not have an appreciable effect on the weight.

Effect of Moisture Content

A change in moisture content will result in a corresponding change in the weight of the wood. Values for weights at 12 and 20% MC are given in Table 2.2. For other moisture contents, the weight may be determined by the designer if this value is critical by using the formulas given in the Appendix to ASTM D2395 or Figure 1 of ASTM D 2395 (3).

Effect of Treatments

Some preservative treatments may add significantly to the weight of wood, depending on the retentions obtained and, in the case of waterborne salts, on the extent to which the wood is seasoned after treatment. Weight increases due to preservative salts are small because retentions of the dry salt are in the range 0.30–1.25 pcf. However, some salt treatments are hygroscopic, and wood treated with these salts may tend to increase in moisture content and thus increase in weight during service conditions. In pressure treatments with oil-borne preservatives, retentions are higher and the weight increase may be from 5 to 20 pcf or more. For specific values of recommended retentions for various preservatives in different species, see *Standard for Preservative Treatment of Structural Glued Laminated Timber*, AITC 109 (5), included in Part III of this manual.

DIMENSIONAL STABILITY

Effect of Temperature

Wood, like most other solids, expands on heating and contracts on cooling. In most structural designs, the increase in size of wood for a rise in temperature is negligible, and, as a result, the secondary stresses due to temperature changes may, in most cases, be neglected. This increase in size is important only in certain structures that are subjected to considerable temperature changes or in members with very long spans.

The coefficient of linear thermal expansion differs in wood's three structural directions. In the longitudinal direction (parallel to grain), this coefficient appears

TABLE 2.2

Weights and Specific Gravities of Commercial Lumber Species

Species[a]	Specific Gravity, *G*, Based on Oven-Dry Weight and Volume		Weight (pcf) Based on Weight and Volume	
	At 12% MC[b]	When Green[c]	At 12% MC	At 20% MC
Softwoods				
Balsam Fir	0.34	0.32	23.8	24.7
California Redwood	0.36	0.34	25.2	26.2
Coast Sitka Spruce	0.41	0.38	28.7	29.2
Coast Species	0.44	0.41 (0.36–0.55)	30.8	32.2
Douglas Fir-Larch	0.49	0.45 (0.45–0.48)	34.3	35.2
Douglas Fir-Larch (North)	0.50	0.46 (0.45–0.55)	35.0	36.0
Douglas Fir South	0.46	0.43	32.2	33.7
Eastern Hemlock	0.42	0.39	29.4	30.7
Eastern Hemlock–Tamarack	0.43	0.40 (0.39–0.44)	30.1	31.5
Eastern Hemlock–Tamarack (North)	0.45	0.42 (0.40–0.48)	31.5	33.0
Eastern Softwoods	0.38	0.36 (0.32–0.49)	26.6	27.7
Eastern Spruce	0.38	0.36 (0.33–0.38)	26.6	27.7
Eastern White Pine	0.37	0.35	25.9	27.0
Eastern White Pine (North)	0.38	0.36	26.6	27.7
Eastern Woods	0.40	0.37 (0.32–0.49)	28.0	28.5
Engelmann Spruce–Alpine Fir	0.34	0.32 (0.31–0.33)	23.8	24.7
Engelmann Spruce–Lodgepole Pine	0.38	0.36 (0.33–0.39)	26.6	27.7
Hem-fir	0.42	0.39 (0.35–0.42)	29.4	30.7
Hem-Fir (North)	0.43	0.40 (0.36–0.41)	30.1	31.5
Idaho White Pine	0.37	0.35	25.9	27.0
Lodgepole Pine	0.42	0.39	29.4	30.7
Mountain Hemlock	0.45	0.42	31.5	33.0

Mountain Hemlock–Hem-Fir	0.42	0.39 (0.35–0.42)	29.4	30.7
Northern Pine	0.46	0.43 (0.40–0.47)	32.2	33.7
Northern Species	0.41	0.38 (0.31–0.55)	28.7	29.2
Northern White Cedar	0.30	0.29	21.0	22.5
Ponderosa Pine	0.42	0.39	29.4	30.7
Ponderosa Pine–Sugar Pine	0.40	0.37 (0.34–0.39)	28.0	28.5
Poonderosa Pine–Lodgepole Pine	0.42	0.39	29.4	30.7
Red Pine	0.45	0.42	31.5	33.0
Sitka Spruce	0.41	0.38	28.7	29.2
Southern Pine	0.52	0.48 (0.47–0.54)	36.4	37.5
Spruce–Pine–Fir	0.41	0.38 (0.33–0.42)	28.7	29.2
Virginia Pine–Pond Pine	0.52	0.48 (0.46–0.51)	36.4	37.5
Western Cedars	0.35	0.33 (0.31–0.42)	24.5	25.5
Western Cedars (North)	0.34	0.32 (0.31–0.42)	23.8	24.7
Western Hemlock	0.45	0.42	31.5	33.0
Western Hemlock (North)	0.44	0.41	30.8	32.2
Western White Pine	0.38	0.36	26.6	27.7
White Woods (Western Woods)	0.41	0.38 (0.31–0.42)	28.7	29.2
Hardwoods				
Aspen	0.37	0.35 (0.35–0.36)	25.9	27.0
Black Cottonwood	0.32	0.30	22.4	23.2
Cottonwood	0.40	0.37	28.0	28.2
Northern Aspen	0.40	0.37 (0.37–0.39)	28.0	28.5
Yellow Poplar	0.43	0.40	30.1	31.5

[a]Species listed are those commonly used for structural purposes as listed in Supplement to the *National Design Specification* (2).

[b]Obtained by conversion from specific gravity when green by using Figure 1 ASTM D 2395, *Standard Test Methods for Specific Gravity of Wood and Wood-Base Materials* (3).

[c]Unseasoned condition, obtained from Tables 1 and 2, ASTM D 2555, *Standard Methods for Establishing Clear Wood Strength Values* (4). Species groups show weighted average specific gravities for the groups based on standing timber volumes shown in Tables 4 and 5, ASTM D 2555. The range of average specific gravities of species included in the species grouping is shown to indicate the variation to be expected.

to be independent of specific gravity and species and ranges from about 1.7×10^{-6} to 2.5×10^{-6} per 1°F for oven-dry wood for both hardwoods and softwoods. This value is about one-tenth to one-third of those for other common structural materials. For that reason, consideration must be given to the differential thermal expansion of materials used in conjunction with wood.

Coefficients of linear thermal expansion across the grain (radial and tangential) are proportional to wood's specific gravity G. These coefficients range from about 5 to more than 10 times greater than the parallel-to-grain coefficients. The radial coefficient of thermal expansion α_r, can be approximated by the formula

$$\alpha_r = (32G + 9.9) \times 10^{-6} \tag{2-1}$$

per 1°F for oven-dry specific gravities in the range of about 0.1 to 0.8. The tangential coefficient of thermal expansion, α_t, over the same oven-dry specific gravity range is approximated by the formula

$$\alpha_t = (33G + 18.4) \times 10^{-6} \tag{2-2}$$

per 1°F. Coefficients of thermal expansion vary slightly within the temperature range of −60 to +130°F but for most practical purposes can be considered constant.

Wood containing moisture reacts to varying temperature differently than does dry wood. As moist wood is heated, it tends to expand due to temperature but shrinks due to loss of moisture. Unless the wood is very dry initially (3 or 4% MC or less), shrinkage due to moisture loss on heating will exceed thermal expansion, resulting in a negative net dimensional change due to heating. At the intermediate moisture levels (about 8–20%), wood will expand when first heated but will shrink gradually to a volume smaller than the initial volume as the wood gradually loses moisture while in the heated condition. In the longitudinal direction, where dimensional change due to moisture change is very small, shrinkage still predominates over thermal expansion unless the wood is initially very dry.

Effect of Moisture Content

Between the fiber saturation point and zero moisture content, wood shrinks as it loses moisture and swells as it absorbs moisture. Above the fiber saturation point, there is no dimensional change with variation in moisture content. The amount of shrinkage and swelling differs in the tangential, radial and longitudinal directions of the piece. Engineering design should consider shrinkage and swelling in the detailing and use of timbers.

Shrinkage occurs when the moisture content is reduced to a value below the fiber saturation point (for purposes of dimensional change, commonly assumed to be 30% MC) and is proportional to the amount of moisture lost below this point. Swelling occurs when the moisture content is increased until the fiber saturation point is reached. The total swelling is equal numerically to the total shrinkage. Shrinking and swelling are expressed as percentages based on the green dimensions of the wood. Wood shrinks most in a direction tangent to the annual growth rings and somewhat less in the radial direction or across these rings. In general, shrinkage is greater in heavier pieces than in lighter pieces of the same species and greater in hardwoods than in softwoods.

As a piece of green or wet wood dries, the surface is reduced to a moisture content below the fiber saturation point much sooner than is the interior. Thus, the piece may show some shrinkage before the average moisture content reaches the fiber saturation point.

Table 2.3 gives the average tangential, radial, and volumetric shrinkage values for various species during drying from the green condition to 0% MC. Because the faces of the pieces of lumber are seldom so oriented that the annual growth rings are exactly tangent and radial to the faces of the piece, it is customary in determining cross-sectional dimensional changes to use an intermediate or average value between the tangential and radial values. The values in Table 2.3 can be used to estimate shrinkage by using the formula

$$S_m = \frac{S_0(I - m)}{30} \tag{2-3}$$

where S_m = shrinkage from initial moisture condition to final moisture content m(%),
S_0 = total shrinkage from Table 2.3(%),
m = final moisture content (below 30%) (%), and
I = initial moisture content (below 30%) (%).

Values for longitudinal shrinkage with a change in moisture content are not tabulated in Table 2.3 because they are ordinarily negligible. The total longitudinal shrinkage of commonly used species from fiber saturation to oven-dry condition usually ranges from 0.1 to 0.2% of the green dimension. Abnormal longitudinal shrinkage may occur in compression wood, wood with steep slope of grain, and exceptionally lightweight wood of any species.

Because there is considerable variation in shrinkage for any species, it is difficult to predict the shrinkage of an individual piece of wood. The values given in Table 2.3 may be used to estimate the average shrinkage of a quantity of pieces.

There is a normal tendency for more flat grain lumber to be used in glued laminated timber than edge grain lumber. Therefore, the shrinkage in the direction perpendicular to the glue line is more comparable to radial shrinkage than tangential shrinkage. Shrinkage can be estimated by using formula (2-3) and Table 2-3. A rule of thumb commonly used is to assume that 1% shrinkage occurs for every 5% change in moisture content.

The effects of dimensional changes due to a change in moisture content have been considered in the development of the details shown in *Typical Construction Details*, AITC 104 (6), in Part III of this manual.

THERMAL INSULATING PROPERTIES

Thermal insulation, commonly referred to simply as insulation, is the inherent characteristic of a substance to prevent or retard the transfer of heat through its body. Heat transfer can occur in several ways: solid conduction, gas conduction,

TABLE 2.3

Average Shrinkage Values of Wood Based on Dimensions when Green

Species[a]	Percentage of Shrinkage from Green to Oven-Dry Moisture Content[b]		
	Radial	Tangential	Volumetric
Softwoods			
Balsam Fir	2.9	6.9	11.2
California Redwood	2.2	4.9	7.0
Douglas Fir-Larch[c]	4.6 (3.8–4.8)	7.6 (6.9–9.1)	12.0 (10.7–14.0)
Eastern Hemlock	3.0	6.8	9.7
Eastern Hemlock–Tamarack[c]	3.1 (3.0–3.7)	6.9 (6.8–7.4)	10.2 (9.7–13.6)
Eastern Softwoods[c]	3.2 (2.1–4.1)	7.1 (6.1–7.8)	10.5 (8.2–13.6)
Eastern White Pine	2.1	6.1	8.2
Engelmann Spruce-Alpine Fir[c]	3.4 (2.6–3.8)	7.2 (7.1–7.4)	10.5 (9.4–11.0)
Engelmann Spruce-Lodgepole Pine[c]	4.1 (3.8–4.3)	6.9 (6.7–7.1)	11.1 (11.0–11.1)
Hem-Fir[c]	3.9 (3.5–4.5)	7.9 (7.0–9.2)	11.7 (9.8–13.0)
Idaho White Pine	4.1	7.4	11.8
Lodgepole Pine	4.3	6.7	11.1
Mountain Hemlock	4.4	7.1	11.1
Mountain Hemlock-Hem-Fir[c]	3.9 (3.3–4.5)	7.7 (7.0–9.2)	11.8 (9.8–13.0)
Northern White Cedar	2.2	4.9	7.2
Ponderosa Pine-Sugar Pine[c]	3.8 (2.9–3.9)	6.1 (5.6–6.2)	9.5 (7.9–9.7)
Ponderosa Pine-Lodgepole Pine[c]	4.0 (3.9–4.3)	6.3 (6.2–6.7)	10.2 (9.7–11.1)
Sitka Spruce	4.3	7.5	11.5
Southern Pine[c]	4.8 (4.6–5.4)	7.6 (7.4–7.7)	12.3 (12.1–12.3)
Virginia Pine-Pond Pine[c]	4.5 (4.2–5.1)	7.1 (7.1–7.2)	11.6 (11.2–11.9)
Western Cedars[c]	2.8 (2.4–4.6)	5.2 (5.0–6.9)	7.2 (6.8–10.1)
Western Hemlock	4.2	7.8	12.4
White Woods (Western Woods)[c]	4.0 (2.9–4.5)	7.2 (5.6–9.2)	11.1 (7.9–13.0)
Hardwoods			
Aspen[c]	3.5 (3.3–3.5)	7.0 (6.7–7.4)	11.6 (11.5–11.8)
Eastern Cottonwood	3.9	9.2	13.9
Yellow Poplar	4.6	8.2	12.7

[a]Species listed are those commonly used for structural purposes as listed in Supplement to the *NDS* (2).

[b]Values are from Table 3.5, *Wood Handbook* (1).

[c]Species values for combinations are the weighted average based on the standing timber volumes shown in Table 4, ASTM D2555, *Standard Methods for Establishing Clear Wood Strength Values* (4). The range of values within the combination is shown in parentheses.

convection, and radiation. These mechanisms may occur separately or in combination. Wood is considered to be a very good insulator due to its physical structure. Each wood fiber or cell contains a void, which in living wood is used to transfer or store sap. The sawn and dried wood thus contains microscopic voids or air spaces that retard heat transfer.

Thermal conductivity k is defined as the rate of heat transmission by conduction only through a unit area of an infinite slab, in a direction perpendicular to the slab surface, when the temperature gradient between the two surfaces of the slab is unity. This definition gives rise to several possible values of thermal conductivity for the same material. The first and most common, based on a temperature gradient of 1°F/in., has units of

$$\frac{\text{Btu}}{\text{hr ft}^2\ {}^\circ\text{F/in.}} \text{ or } \frac{\text{Btu in.}}{\text{hr ft}^2\ {}^\circ\text{F}}$$

The second, based on a temperature gradient of 1°F/ft,has units of

$$\frac{\text{Btu}}{\text{hr ft}^2\ {}^\circ\text{F/ft}} \text{ or } \frac{\text{Btu}}{\text{hr ft }{}^\circ\text{F}}$$

The relationship between the two values for a given material is

$$k(\text{Btu in./hr ft}^2\ {}^\circ\text{F}) = 12\ k(\text{Btu/hr ft }{}^\circ\text{F})$$

The thermal conductivity of wood varies with (1) the direction of grain, (2) specific gravity, (3) moisture content, (4) extractives present in the wood, and (5) growth characteristics such as knots, slope of grain, seasoning checks, and growth rings.

Thermal conductivity is approximately the same in radial and tangential directions but is about $2\frac{1}{2}$ times greater along the grain. To determine thermal conductivity values across the grain for wood at various moisture contents and specific gravities, use the formula

$$k = G\,(1.39 + 0.028\ \text{MC}) + 0.165$$

where k = thermal conductivity (Btu in./hr ft^2 °F),
G = specific gravity based on volume at the existing moisture content and oven-dry weight, and
MC = the existing moisture content (%).

A similar term, thermal conductance C, is defined as the rate of heat transmission through a unit area of a particular body or assembly having defined surfaces when unit average temperature difference is established between the surfaces. Thermal conductance values are measured across the total thickness of the material or assembly of materials, whatever that thickness may be.

A direct measure of the insulating value of a material or assembly of materials is the reciprocal of thermal conductance and is termed thermal resistance, R. Thermal resistance values are most commonly used to indicate the insulating qualities of a material or a type of construction. Tables 2.4–2.6 show approximate ther-

TABLE 2.4

Thermal Design Values: Surface Conductances and Resistances[a]

Position of Surface	Direction of Heat Flow	Nonreflective $\epsilon = 0.90$[b] h_i	Nonreflective $\epsilon = 0.90$[b] R	Reflective $\epsilon = 0.20$[c] h_i	Reflective $\epsilon = 0.20$[c] R	Reflective $\epsilon = 0.05$[d] h_i	Reflective $\epsilon = 0.05$[d] R
Still air							
Horizontal	Upward	1.63	0.61	0.91	1.10	0.76	1.32
Sloping—45°	Upward	1.60	0.62	0.88	1.14	0.73	1.37
Vertical	Horizontal	1.46	0.68	0.74	1.35	0.59	1.70
Sloping—45°	Downward	1.32	0.76	0.60	1.67	0.45	2.22
Horizontal	Downward	1.08	0.92	0.37	2.70	0.22	4.55
Moving Air (Any position)							
15 mph wind (for winter)	Any	6.00[e]	0.17				
7.5 mph wind (for summer)	Any	4.00[e]	0.25				

[a]Reprinted with permission of the American Society of Heating, Refrigerating and Air-Conditioning Engineers, Inc., Atlanta, GA (7).

[b]Emittance typical of building materials such as wood, paper, masonry, nonmetallic paints, and regular glass.

[c]Emittance typical of aluminum-coated paper or bright galvanized steel.

[d]Emittance typical of bright aluminum foil.

[e]These values are h_o.

TABLE 2.5

Thermal Design Values: Thermal Resistances (R) of Air Spaces[a, b]

Position of Air Space	Direction of Heat Flow	$\epsilon = 0.03$[c]	$\epsilon = 0.82$[d]
Horizontal	Upward	2.84	0.80
Sloping—45°	Upward	3.18	0.82
Vertical	Horizontal	3.69	0.85
Sloping—45°	Downward	4.81	0.90
Horizontal	Downward	10.07	1.00

[a]Reprinted with permission of the American Society of Heating, Refrigerating and Air-Conditioning Engineers, Inc., Atlanta, GA (7).

[b]Values are for 3.5 in. dead air space, 90°F mean temperature, 10°F temperature difference.

[c]Effective emittance ϵ typical of air space surfaces of bright aluminum foil.

[d]Effective emittance ϵ typical of air space surfaces of wood, paper, masonry, or nonmetallic paints.

TABLE 2.6

Thermal Design Values: Typical Building Materials[a]

Material	Conductivity k per in. Thickness	Conductance C for Thickness Listed	Resistance R	
			Per in. Thickness ($1/k$)	For Thickness Listed ($1/C$)
Wood				
Hardwoods: maple, oak	1.10		0.91	
Softwoods: fir, pine	0.80		1.25	
0.75 in.		1.06		0.94
1.5 in.		0.53		1.88
3.5 in.		0.23		4.38
Insulation				
Blanket, mineral fiber 3 to $3\frac{1}{2}$ in.		0.091		11
Boards				
Polystyrene, extruded	0.20		5.00	
Polystyrene, molded beads	0.26			
Glass fiber	0.25		4.00	
Polyurethane	0.16		6.25	
Mineral fiber	0.29		3.45	
Loose fill				
Cellulosic insulation	0.30		3.33	
Perlite, expanded	0.37		2.70	
Mineral fiber, $7\frac{1}{2}$ to 10 in.				22
Vermiculite, exfoliated	0.47		2.13	
Preformed roof insulation above deck, range for available thicknesses		0.12–0.72		8.33–1.39

(continued)

TABLE 2.6 (*Continued*)

Material	Conductivity k per in. Thickness	Conductance C for Thickness Listed	Resistance R	
			Per in. Thickness ($1/k$)	For Thickness Listed ($1/C$)
Concrete				
Sand and gravel aggregate 140 lb/ft^3	12.0		0.08	
Lightweight aggregate 100 lb/ft^3	3.6		0.28	
Masonry units				
Brick, common, 120 lb/ft^3	5.0		0.20	
Brick, face, 130 lb/ft^3	9.0		0.11	
Concrete blocks				
Sand and gravel aggregate, 4 in.		1.40		0.71
Sand and gravel aggregate, 8 in.		0.90		1.11
Lightweight aggregate, 4 in.		0.67		1.50
Lightweight aggregate, 8 in.		0.50		2.00
Metals				
Aluminum	128			
Steel	26.2			
Earth, dry and packed	0.037			
Water at 68°F	0.348			
Building boards				
Plywood	0.80		1.25	
Gypsum board, $\frac{1}{2}$ in.		2.22		0.45
Particle board, medium density	0.94		1.06	

Building membranes				
Vapor-permeable felt		16.70		0.06
Vapor-seal plastic film				negl.
Plastering materials				
Cement plaster, sand aggregate	5.0		0.20	
Gypsum plaster, sand aggregate	5.6		0.18	
Roofing				
Asbestos cement shingles		4.76		0.21
Asbestos roll roofing		6.50		0.15
Asphalt shingles		2.27		0.44
Built-up roofing, $\frac{3}{8}$ in.		3.00		0.33
Wood shingles		1.06		0.94
Siding Materials				
Asbestos cement shingles		4.76		0.21
Wood shingles, 16 in., $7\frac{1}{2}$ in. exposure		1.15		0.87
Asphalt insulating siding		0.69		1.46
Hardboard siding, $\frac{7}{16}$ in.	1.49		0.67	
Wood, drop, 1 × 8 in.		1.27		0.79
Wood, bevel, $\frac{1}{2}$ × 8 in. lapped		1.23		0.81
Wood, bevel, $\frac{3}{4}$ × 10 in. lapped		0.95		1.05
Wood,plywood, $\frac{3}{8}$ in. lapped		1.59		0.59
Aluminum or steel over sheathing		1.61		0.61

[a]Reprinted with permission of the American Society of Heating, Refrigerating and Air-Conditioning Engineers, Inc., Atlanta, GA (7).

mal design values for some common building materials. Thermal resistance values for plane constructions, such as flat walls or roofs that have essentially one dimensional heat flow, can be approximated as

$$R_T = R_1 + R_2 + R_3 + \cdots + R_n \tag{2-5}$$

where R_T = total thermal resistance of the assembly (hr ft^2 °F/Btu) and

$R_1, R_2, \cdots$ = individual resistances of the components in the assembly (hr ft^2 °F/Btu).

The overall heat transfer coefficient or overall transmittance of the assembly is defined by

$$U = \frac{1}{R_T} \tag{2-6}$$

with U having the same units as conductance.

Normally, the first and last resistances in formula (2-5) are due to the thin film of air that tends to adhere to the solid surfaces of the structure. These surface resistances are the reciprocals of the surface conductance (k). Surface conductance (or film coefficient) has the same units as C, but the symbol k is used to distinguish the fluid-film conductance from solid-body conductance. The value of k, and therefore film resistance, depends on orientation of the surface, air movement, direction of heat flow, and reflectivity of the surface.

The thermal resistance of the enclosed air spaces depends on orientation of the space, direction of heat flow, thickness of space, mean air temperature in the space, and temperature difference across the space. Values of thermal resistance for $3\frac{1}{2}$-in.-thick air spaces are given in Table 2.5. A typical determination of approximate total thermal resistance and overall heat transfer coefficient for a frame wall is shown below.

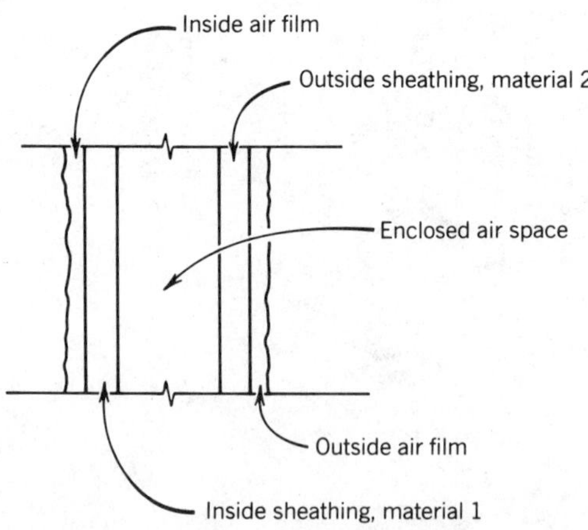

$$R_T = \frac{1}{h_i} + \frac{L_1}{k_1} + R_a + \frac{L_2}{k_2} + \frac{1}{h_o} \qquad (2\text{-}7)$$

$$U = \frac{1}{R_T}$$

where h_i = inside surface conductance (Btu/hr ft^2 °F),
h_o = outside surface conductance (Btu/hr ft^2 °F),
L_1, L_2 = individual thicknesses of materials (in.),
k_1, k_2 = individual thermal conductivities of the materials (Btu in./hr ft^2°F),
R_a = thermal resistance of air space (hr ft^2 °F/Btu), and
U = overall heat transfer coefficient (Btu/hr ft^2 °F).

Tables 2.7 and 2.8 show typical examples of numerical calculations of the above type, but in tabular form. Note that in these examples the overall heat transfer coefficient is calculated for both the area between framing (U_i) and for the area backed by framing members (U_s). A weighted average is then taken to determine an average value of U for the actual construction. For more detailed information, refer to *ASHRAE Fundamentals Handbook* (7).

ACOUSTICAL PROPERTIES

The acoustical properties of a material or composite construction are determined by its sound insulation and sound absorption abilities. Sound insulation abilities are measured in terms of the reduction in intensity of sound when it passes through a barrier. Sound absorption refers to the amount of incident sound on a surface that is not reflected by that surface.

Sound Insulation

Sound insulating values for materials of construction are related to the sound transmission loss for the construction measured in decibels at various frequencies. The insulating values depend on the intensity of sound on the opposite face of a barrier, the mass and stiffness of the material making up a barrier, and the nature of its design and fastenings. Sound insulation may be designed into a barrier by using materials with a high mass per unit area, using materials with low rigidity, or using barriers that employ air spaces and that avoid rigid ties from one face to the other. Like most common construction materials, wood alone does not provide good sound insulation but, when combined with other materials in typical constructions, will provide a structural unit of satisfactory sound-insulating ability.

Sound Absorption

The effectiveness of any material with respect to absorbing sound is given by its absorption coefficient at the frequency range specified. The sound absorption coefficient for a material is used to determine the total magnitude of the absorption

TABLE 2.7

Examples of Calculations for Thermal Resistance and Overall Heat Transfer Coefficient[a]

COEFFICIENTS OF TRANSMISSION (U) OF FRAME WALLS[b]

Construction 1	1 Between Framing	1 At Framing	2 Between Framing	2 At Framing
	Resistance R			
1. Outside surface (15-mph wind)	0.17	0.17	0.17	0.17
2. Siding, wood, 0.5 × 8 in. lapped (average)	0.81	0.81	0.81	0.81
3. Sheathing, 0.5 in. asphalt impregnated	1.32	1.32	1.32	1.32
4. Nonreflective air space, 3.5 in. (50°F mean; 10°F temperature difference)	1.01	—	11.00	—
5. Nominal 2 × 4-in. wood stud	—	4.35	—	4.35
6. Gypsum wallboard, 0.5 in.	0.45	0.45	0.45	0.45
7. Inside surface (still air)	0.68	0.68	0.68	0.68
Total Thermal Resistance (R)	$R_i = 4.44$	$R_s = 7.78$	$R_i = 14.43$	$R_s = 7.78$

Construction 1: $U_i = 1/4.44 = 0.225$; $U_s = 1/7.78 = 0.128$. With 20% framing (typical of 2 × 4-in. studs at 16-in. o.c.), $U_{av} = (0.8)(0.225) + (0.2)(0.128) = 0.206$.

Construction 2: $U_i = 1/14.43 = 0.069$; $U_s = 0.128$. With framing unchanged, $U_{av} = (0.8)(0.069) + (0.2)(0.128) = 0.081$.

Coefficients of Transmission (U) of Pitched Roofs[c,d]

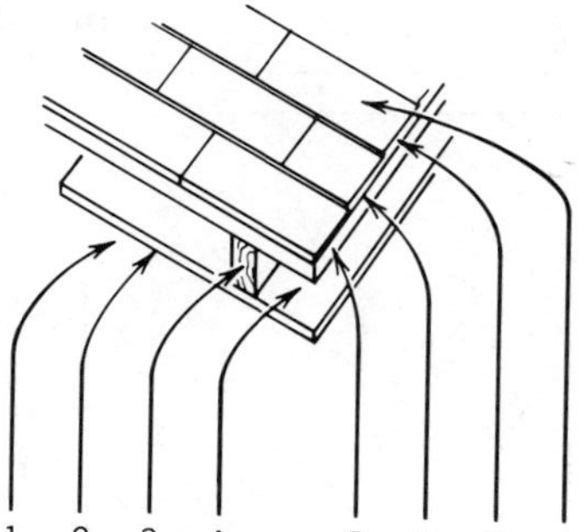

Construction 1 (Reflective Air Space)	1 Heat Flow Up (Winter) Between Rafters	1 Heat Flow Up (Winter) At Rafters	2 Heat Flow Down (Summer) Between Rafters	2 Heat Flow Down (Summer) At Rafters
1. Inside surface (still air)	0.62	0.62	0.76	0.76
2. Gypsum wallboard 0.5 in., foil backed	0.45	0.45	0.45	0.45
3. Nominal 2 × 4-in. ceiling rafter	—	4.35	—	4.35
4. 45° slope reflective air space, 3.5 in. (50°F mean, 30°F temperature difference)	2.17	—	4.33	—
5. Plywood sheathing, $\frac{5}{8}$ in.	0.77	0.77	0.77	0.77
6. Felt building membrane	0.06	0.06	0.06	0.06
7. Asphalt shingle roofing	0.44	0.44	0.44	0.44
8. Outside surface (15 mph wind)	0.17	0.17	0.25[e]	0.25[e]
Total Thermal Resistance (R)	$R_i = 4.68$	$R_s = 6.86$	$R_i = 7.06$	$R_s = 7.08$

Construction 1: $U_i = 1/4.68 = 0.214$; $U_s = 1/6.86 = 0.146$. With 10% framing (typical of 2-in. rafters at 16-in. o.c.), $U_{av} = (0.9)(0.214) + (0.1)(0.146) = 0.207$.

Construction 2: $U_i = 1/7.06 = 0.142$; $U_s = 1/7.08 = 0.141$. With framing unchanged, $U_{av} = (0.9)(0.142) + (0.1)(0.141) = 0.142$.

TABLE 2.7 (*Continued*)

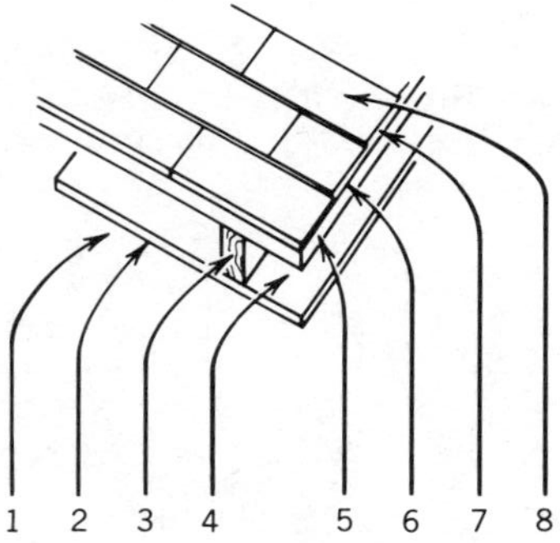

Construction 3 (Nonreflective Air Space)	3 Heat Flow Up (Winter)		4 Heat Flow Down (Summer)	
	Between Rafters	At Rafters	Between Rafters	At Rafters
1. Inside surface (still air)	0.62	0.62	0.76	0.76
2. Gypsum wallboard, 0.5 in.	0.45	0.45	0.45	0.45
3. Nominal 2 × 4-in. ceiling rafter	—	4.35	—	4.35
4. 45° slope, nonreflective air space, 3.5 in. (50°F mean; 10°F temperature difference)	0.96	—	0.90[f]	—
5. Plywood sheathing, $\frac{5}{8}$ in.	0.77	0.77	0.77	0.77
6. Felt building membrane	0.06	0.06	0.06	0.06
7. Asphalt shingle roofing	0.44	0.44	0.44	0.44
8. Outside surface (15-mph wind)	0.17	0.17	0.25[e]	0.25[e]
Total Thermal Resistance (R)	$R_i = 3.47$	$R_s = 6.86$	$R_i = 3.63$	$R_s = 7.08$

Construction 3: $U_i = 1/3.47 = 0.288$; $U_s = 1/6.86 = 0.146$. With 10% framing (typical of 2-in. rafters at 16-in. o.c.), $U_{av} = (0.9)(0.288) + (0.1)(0.146) = 0.274$.

Construction 4: $U_i = 1/3.63 = 0.275$; $U_s = 1/7.08 = 0.141$. With framing unchanged, $U_{av} = (0.9)(0.275) + (0.1)(0.141) = 0.262$.

[a]Reprinted with permission of the American Society of Heating, Refrigerating and Air-Conditioning Engineers, Inc., Atlanta, GA (7).

[b]Coefficients expressed in Btu per (hour) (square foot) (degree Fahrenheit difference in temperature between air on two sides) and are based on outside wind velocity of 15 mph. Replace air space with 3.5-in. R-11 blanket insulation for construction 2.

[c]Coefficients are expressed in Btu per (hour) (square foot) (degree Fahrenheit difference in temperature between the air on the two sides) and are based on an outside wind velocity of 15 mph for heat flow upward and 7.5 mph for heat flow downward.

[d]Pitch of roof—45°.

[e]7.5 mph wind.

[f]Air space value at 90°F mean, 10°F temperature difference.

TABLE 2.8

Approximate Thermal Insulation Design Values for Typical Plywood Roof Deck Systems

3/8" B.U. roof
Outside air
1/2" plywood
Inside air
2 × 4 Stiffeners
24" o.c. (typical)

NO INSULATION

Construction	Heat Flow Up (Winter)	Heat Flow Down (Summer)
1. Outside surface (15-mph wind)	0.17	0.25
2. $\frac{3}{8}$ in. B.U. roofing	0.33	0.33
3. $\frac{1}{2}$ in. plywood	0.62	0.62
4. Inside surface (still air)	0.61	0.92
Total Thermal Resistance (R)	1.73	2.12
$U = 1/R$	0.578	0.472

3/8" B.U. roofing
Outside air
3½" Air space
Reflective side
1/2" Plywood
Inside air
Paper[a]
2 × 4 Stiffeners
24" o.c. 8% Framing[d]
***Framing occupies 8% of the roof surface

REFLECTIVE INSULATION[a]

Construction	Heat Flow Up (Winter)		Heat Flow Down (Summer)	
	Between Framing	At Framing	Between Framing	At Framing
1. Outside surface (15-mph wind)	0.17	0.17	0.25	0.25
2. $\frac{3}{8}$ in. B.U. roofing	0.33	0.33	0.33	0.33
3. $\frac{1}{2}$ in. plywood	0.62	0.62	0.62	0.62
4. $3\frac{1}{2}$ in. air space[b]	2.01	—	8.17	—
5. 2 × 4 in. stiffener	—	4.35	—	4.35
6. Aluminum paper	—	—	—	—
7. Inside surface (still air)	0.61	0.61	0.92	0.92
Total Thermal Resistance (R)	3.74	6.08	10.29	6.47

Winter: $U_i = 1/3.74 = 0.267$, $U_s = 1/6.08 = 0.164$; $U_{avg} = (0.92)(0.267) + (0.08)(0.164) = 0.259$.
Summer: $U_i = 1/10.29 = 0.097$, $U_s = 1/6.47 = 0.155$; $U_{avg} = (0.92)(0.097) + (0.08)(0.155) = 0.102$.

TABLE 2.8 (*Continued*)

FLEXIBLE INSULATION

	Construction	Heat Flow Up (Winter)		Heat Flow Down (Summer)	
		Between Framing	At Framing	Between Framing	At Framing
1.	Outside surface (15-mph wind)	0.17	0.17	0.25	0.25
2.	$\frac{3}{8}$ in. B.U. roofing	0.33	0.33	0.33	0.33
3.	$\frac{1}{2}$ in. plywood	0.62	0.62	0.62	0.62
4.	R11 insulation	11.0	—	11.0	—
5.	2 × 4 in. stiffener	—	4.35	—	4.35
6.	Aluminum paper	—	—	—	—
7.	Inside surface (still air)	0.61	0.61	0.92	0.92
	Total Thermal Resistance (R)	12.73	6.08	13.12	6.47

Winter: $U_i = 1/12.73 = 0.078$, $U_s = 1/6.08 = 0.164$; $U_{av} = (0.92)(0.078) + (0.08)(0.164) = 0.085$.
Summer: $U_i = 1/13.12 = 0.076$, $U_s = 1/6.47 = 0.155$; $U_{av} = (0.92)(0.076) + (0.08)(0.155) = 0.082$.

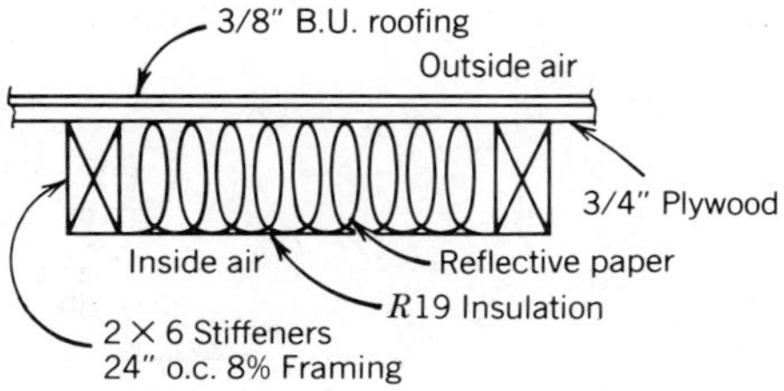

Flexible Insulation

	Construction	Heat Flow Up (Winter)		Heat Flow Down (Summer)	
		Between Framing	At Framing	Between Framing	At Framing
1.	Outside surface (15-mph wind)	0.17	0.17	0.25	0.25
2.	$\frac{3}{8}$ in. B.U. roofing	0.33	0.33	0.33	0.33
3.	$\frac{3}{4}$ in. plywood	0.93	0.93	0.93	0.93
4.	R19 insulation	19.0	—	19.0	—
5.	2 × 6 in. stiffener	—	6.88	—	6.88
6.	Aluminum paper	—	—	—	—
7.	Inside surface (still air)	1.32	1.32	4.55	4.55
	Total Thermal Resistance (R)	21.75	9.63	25.06	12.94

Winter: $U_i = 1/21.75 = 0.046$, $U_s = 1/9.63 = 0.104$; $U_{av} = (0.92)(0.046) + (0.08)(0.104) = 0.051$.
Summer: $U_i = 1/25.06 = 0.040$, $U_s = 1/12.94 = 0.077$; $U_{av} = (0.92)(0.040) + (0.08)(0.077) = 0.043$.

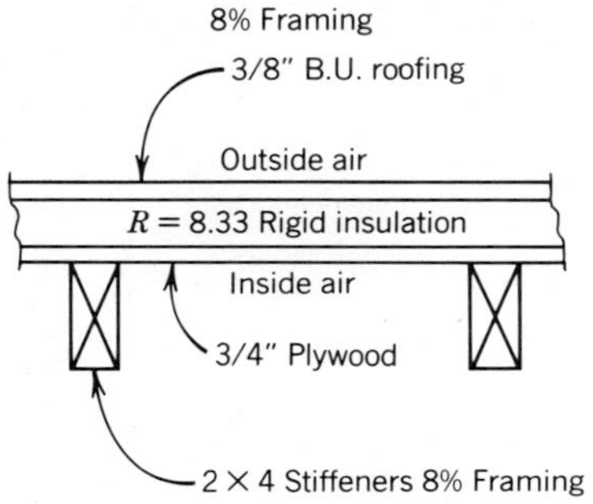

Rigid Insulation

	Construction	Heat Flow Up (Winter)	Heat Flow Down (Summer)
1.	Outside surface (15-mph wind)	0.17	0.25
2.	$\frac{3}{8}$ in. B.U. roofing	0.33	0.33
3.	Insulation, $R = 8.33$	8.33	8.33
4.	$\frac{3}{4}$ in. plywood	0.93	0.93
5.	Inside surface (still air)	0.61	0.92
	Total Thermal Resistance (R)	10.37	10.76
	$U = 1/R$	0.096	0.093

TABLE 2.8 (*Continued*)

HEAVY TIMBER ROOF DECK SYSTEMS

Type of Deck		Type of Rigid Insulation											
		None		Polystyrene[c] or Fiberglass Thickness (in.) ($k = 0.25$)				Polystyrene[c] Thickness (in.) ($k = 0.20$)				Polyurethane Thickness (in.) ($k = 0.14$)	
				1		$1\frac{1}{2}$		1		$1\frac{1}{2}$		1	
		Winter	Summer	Winter	Summer	Winter	Summer	Winter	Summer	Winter	Summer	Winter	Summer
$1\frac{1}{8}$ in. thick plywood													
	U	0.397	0.344	0.154	0.145	0.117	0.112	0.133	0.126	0.100	0.096	0.104	0.100
	R	2.52	2.91	6.49	6.90	8.55	8.93	7.52	7.94	10.00	10.42	9.62	10.00
2 in. nominal wood deck ($1\frac{1}{2}$ in. actual)													
	U	0.335	0.296	0.143	0.136	0.111	0.107	0.125	0.119	0.095	0.092	0.099	0.095
	R	2.99	3.38	6.99	7.35	9.01	9.35	8.00	8.40	10.53	10.87	10.10	10.53

3 in. nominal wood deck ($2\frac{1}{2}$ in. actual)													
	U	0.236	0.216	0.121	0.116	0.098	0.094	0.108	0.104	0.085	0.082	0.088	0.085
	R	4.24	4.63	8.26	8.62	10.20	10.64	9.26	9.62	11.76	12.20	11.36	11.76
4 in. nominal wood deck ($3\frac{1}{2}$ in. actual)													
	U	0.182	0.170	0.105	0.101	0.087	0.084	0.095	0.092	0.077	0.075	0.079	0.077
	R	5.49	5.88	9.52	9.90	11.49	11.90	10.53	10.87	12.99	13.33	12.66	12.99

[a]Aluminum-coated paper, reflective inside only.

[b]50°F mean, 30°F temperature differential, $\epsilon = 0.05$. Venting may be required for high-humidity conditions.

[c]Verify k value of polystyrene with manufacturer.

[d]Framing occupies 8% of the roof surface.

property of the material expressed in units of sabins. Thus, the total sound absorption capacity of a wall, ceiling, floor, and so forth in sabins is determined by multiplying the sound absorption coefficient of the material involved by its total area in square feet.

Sound absorption values for any material may be compared to that of an open window, which is assumed to be the most complete absorber of sound. An open window, 1 ft square, is given a sound absorption coefficient of unity. A piece of hairfelt of equal size may absorb half as much sound as the open window and, therefore, has a coefficient of 0.50. Sound absorption coefficients are measured for various frequencies of sound, and the absorption ability of a material may vary widely with differences in frequency. Sound-absorption values for wood vary with moisture content, direction of grain, and density.

An important acoustical property of wood is that it absorbs low-frequency sounds more readily than high-frequency sounds.

ELECTRICAL PROPERTIES

The most important electrical properties of wood are its resistance to the passage of an electric current and its dielectric properties. The electrical resistance of wood is utilized in electric moisture meters used to determine moisture content. The dielectric properties of wood are utilized in the high-frequency curing of adhesives in glued laminated members and in the seasoning of wood.

The electrical resistance of wood varies with moisture content, density, direction of travel of the current with respect to the direction of the grain, and temperature. It varies greatly with moisture content, especially below the fiber saturation point, decreasing with an increase in moisture content. Electrical resistance varies inversely with the density of wood, although this effect is slight compared to the variance due to moisture content and is greater across the grain than along it. The electrical resistance of wood approximately doubles for each drop in temperature of 22.5°F, but this relationship varies considerably with the level of moisture content. There is also a variation in electrical resistance between species, which is possibly caused by minerals or electrolytes in the wood itself or dissolved in the water present in the wood.

Metallic salts, such as used in preservative and fire-retardant treatments, may lower the electrical resistance of the wood considerably. Use of wood containing such salts should be avoided for applications where electrical resistance is critical and in processes involving dielectric heating. Electric moisture meters may give erroneous readings for such wood.

Wood at a low moisture content is normally classified as an electric insulator, or dielectric, rather than as a conductor. A dielectric can be heated by using it as the medium between electrodes carrying charges of oscillating high-frequency electricity or by placing it in an electric field of like nature. The dielectric constant of wood (the ratio of the capacitance of a wood condenser to the capacitance of a similar condenser employing a vacuum) is proportional to density at a given moisture content. The dielectric constant parallel to grain is significantly greater than the corresponding constant perpendicular to grain. The constant also decreases with an increase in frequency of the oscillating current.

The power factor of wood is the ratio of the power absorbed in the wood per cycle of oscillation of an electric current to the total apparent power stored in the wood during that cycle. The power factor generally increases with an increase in moisture content. The parallel-to-grain power factors are usually greater than the corresponding perpendicular-to-grain factors.

COEFFICIENT OF FRICTION

The coefficient of friction depends on the moisture content of the wood and surface roughness. It varies little with species except for those species that contain abundant oily or waxy extractives.

Coefficients of static friction for wood on unpolished steel have been reported to be approximately 0.70 for dry wood and 0.40 for green wood. Coefficients of static friction for smooth wood on smooth wood are 0.60 for dry wood and 0.83 for green wood.

It is not usual practice to use friction to resist forces in the design of timber connections.

PROPERTIES OF SECTIONS

Sawn Lumber and Timber

Table 7.1 in Part III of this manual lists sizes and properties of sections for sawn lumber and timber and includes values for net cross-sectional area, section modulus, moment of inertia, weight per lineal foot, and board measure per lineal foot.

The use of standard sizes is more economical than the use of special sizes or lengths. Lengths are generally available in even 2-ft increments. There is a limit on the maximum length normally available.

Glued Laminated Timber

Table 7.2 in Part III of this manual gives section properties for structural glued laminated timbers based on industry-recommended standard thicknesses and widths.

The most efficient and economical production of glued laminated structural members results when standard lumber sizes are used for the laminates. Industry-recommended practice uses nominal 2-in.-thick lumber of standard nominal width to produce straight members and curved members where the radius of curvature is within the bending radius limits for that thickness of the species. Nominal 1-in.-thick boards are normally used when the bending radius is too sharp to permit use of nominal 2-in.-thick laminations. These standard practices are subject to modification for specific job requirements and plant procedures. The use of nominal 1- and 2-in.-thick laminations will generally be the most economical, and, therefore, conformance with these usual thicknesses is recommended for all normal uses. Exceptions should be made only when the shape of the structure requires nonstandard laminations.

For laminating, dimension lumber of 2 in. nominal thickness is normally surfaced to a net thickness ranging from $1\frac{1}{2}$ to $1\frac{3}{8}$ in. The section properties included in Part III of this manual are based on $1\frac{1}{2}$ in. thickness. However, other thicknesses of laminations may be used to achieve the desired member depth. Nominal boards of 1 in. usually are surfaced to $\frac{3}{4}$ in. net thickness. Finished depths of members are normally increments of these net thicknesses, as illustrated in the Table 2.9.

TABLE 2.9

	Net Depth of Member (in.)		
		Nominal 2-in. Laminations	
No. of Laminations	Nominal 1-in. Laminations	$1\frac{1}{2}$ in. Actual	$1\frac{3}{8}$ in. Actual
4	3	6	$5\frac{1}{2}$
5	$3\frac{3}{4}$	$7\frac{1}{2}$	$6\frac{7}{8}$
6	$4\frac{1}{2}$	9	$8\frac{1}{4}$
7	$5\frac{1}{4}$	$10\frac{1}{2}$	$9\frac{5}{8}$
8	6	12	11
etc.	etc.	etc.	etc.

The use of laminations of special thicknesses because of the bending radius of the member or the mixing of thicknesses for special purposes may result in net finished depths that may differ from those indicated in Table 2.9.

It is necessary to surface the sides of laminated members after gluing to remove the glue squeeze-out and to provide a uniformly smooth surface. Therefore, the net finished width of the glued laminated member is less than the net finished width of industry standard boards and dimension. Normal standard net finished widths for glued laminated structural members are as follows. Other finished widths may be used to meet the size requirements of a design or to meet other special requirements.

Nominal width, in.	3	4	6	8	10	12	14	16
Net finished width, in. (western softwoods)	$2\frac{1}{8}$	$3\frac{1}{8}$	$5\frac{1}{8}$	$6\frac{3}{4}$	$8\frac{3}{4}$	$10\frac{3}{4}$	$12\frac{1}{4}$	$14\frac{1}{4}$
Net finished width, in. (Southern Pine)	$2\frac{1}{8}$	3	5	$6\frac{3}{4}$	$8\frac{3}{4}$	$10\frac{3}{4}$	$12\frac{1}{4}$	$14\frac{1}{4}$

Industry standards permit the following tolerances at the time of manufacture:

1. Width of $\pm\frac{1}{16}$ in. of the specified width.
2. Depth of $+\frac{1}{8}$ in. per ft of specified depth, $-\frac{1}{8}$ in., or $-\frac{1}{16}$ in. per ft of specified depth, whichever is the larger.

3. Length of up to 20 ft, $\pm\frac{1}{16}$ in.; over 20 ft, $\pm\frac{1}{16}$ in. per 20 ft except where length dimensions are not specified or critical.
4. Squareness: the cross section of all glued laminated structural members shall be square within $\pm\frac{1}{8}$ in. per ft of specified depth of member unless a specially shaped section is specified.
5. Camber or straightness of $\pm\frac{1}{4}$ in. for beams up to 20 ft in length and $\pm\frac{1}{8}$ in. per 20 ft or fraction thereof for beams exceeding 20 ft in length, provided this amount does not exceed a total of $\pm\frac{3}{4}$ in. These camber tolerances apply at the time of manufacture without allowance for dead-load deflection. They are intended for straight or slightly cambered beams only and are not applicable to curved laminated members such as arches.

REFERENCES

1. United States Department of Agriculture, Forest Service, Forest Products Laboratory, *Wood Handbook: Wood as an Engineering Material,* Agriculture Handbook No. 72, Madison, WI, 1974.
2. National Forest Products Association, *National Design Specification for Wood Construction,* Washington, DC, 1982.
3. American Society for Testing and Materials, *Standard Test Methods for Specific Gravity of Wood and Wood-Base Materials,* ASTM D 2395, Philadelphia, PA, 1983.
4. American Society for Testing and Materials, *Standard Methods for Establishing Clear Wood Strength Values,* ASTM D 2555, Philadelphia, PA, 1984.
5. American Institute of Timber Construction, *Standard for Preservative Treatment of Structural Glued Laminated Timber,* AITC 109, Englewood, CO, 1984.
6. American Institute of Timber Construction, *Typical Construction Details,* AITC 104, Englewood, CO, 1984.
7. American Society of Heating, Refrigerating and Air-Conditioning Engineers, Inc., *ASHRAE Fundamentals Handbook,* Atlanta, GA, 1981.

SECTION 3

MECHANICAL PROPERTIES OF WOOD

LUMBER GRADING AND DESIGN VALUES

Lumber, as it is sawn from a log, is quite variable in its mechanical properties. Individual pieces may differ in strength by as much as several hundred percent. For simplicity and economy in use, pieces of lumber of similar mechanical properties can be placed in a single class known as a grade. The properties of a particular grade depend on the sorting criteria used and on additional factors independent of the sorting criteria.

Grades and their associated design properties are found in grading rules published by rules-writing grading agencies and these design values are also included in *National Design Specification for Wood Construction* (1). Not all grades listed are available in commercial quantities, and availability varies with market conditions and by geographical area. The designer should check on availability before specifying.

Visual Grading

Visual grading is the oldest lumber-grading method. It is based on the premise that mechanical properties of lumber differ from mechanical properties of clear wood because of naturally occurring characteristics that can be seen and judged by eye. These visual characteristics are used to sort the lumber into grades and include, but are not limited to, density, slope of grain, size and location of knots, and checks and splits.

The derivation of mechanical properties of visually graded lumber is based on clear-wood properties and on the presence of lumber characteristics allowed by the visual sorting criteria as well as on the effects of size, duration of load, and moisture content. Clear-wood properties are determined in accordance with the American Society for Testing and Materials, ASTM D 2555, *Standard Methods for Establishing Clear Wood Strength Values* (2).

Each strength property of a piece of lumber is derived from the product of the clear-wood strength for the species and the limiting strength ratio. The strength ratio is a hypothetical ratio of the strength of a piece of lumber with visible strength-reducing characteristics to its strength if these characteristics are absent. Adjustments of clear-wood strength values for such factors as density, slope of grain, knots, shakes, checks and splits, size, variability, duration of load, and moisture content are given in *Standard Methods for Establishing Structural Grades and Related Allowable Properties for Visually Graded Lumber,* ASTM D 245 (3).

In visual grading, the modulus of elasticity E assigned is an estimate of the mean modulus of the lumber grade. The average E value for clear wood of the species, as given in ASTM D 2555, is used as a base. The clear-wood value is multiplied by "quality factors" representing the reduction in E that occurs by lumber grade. This procedure is outlined in ASTM D 245.

Machine Stress-Rated (MSR) Grading or E-Rated Grading

MSR grading is based on an observed relationship between modulus of elasticity E and bending strength, tensile strength, or other strength properties. The E value of lumber is thus a major sorting criterion for this method of grading.

The E used as a sorting criterion for mechanical properties can be measured in a variety of ways. Usually, the apparent E or a stiffness-related deflection is measured. Most MSR grading machines in the United States are designed to measure the flatwise bending stiffness that occurs in any span of approximately 4 ft and to use this measurement in the sorting criterion.

Each grade derived from MSR lumber relates allowable strength in bending, compression, and tension parallel to grain to the E levels by which the grade is identified. As in visual grading, the E value assigned to a grade for design represents the average for the grade. However, the stress-rating machines can be adjusted so that the coefficient of variation of E for a MSR grade is less than for a visual grade.

In machine stress rating, strengths for horizontal shear and compression perpendicular to grain are handled in relationship to clear-wood properties and visual grading characteristics.

Commercial machine stress-rating practices in the United States combine edge characteristic size restriction (such as knots, knot holes, burls, distorted grain, or decay) with E as a predictor of grade properties. It has been shown that these concepts used together provide a better strength prediction than either alone.

Design Values for Sawn Lumber

Design values for visually graded and MSR lumber are given in Tables 7.3 and 7.4 in Part III of this manual. The values in those tables are reprinted from *Design Values for Wood Construction*, a supplement to the *National Design Specification for Wood Construction* (1). The availability of particular grades and species should be determined prior to the use of these design values.

Design Values for Bearing on End Grain

Design values for end grain in bearing parallel to grain on a rigid surface are given in Table 3.1. The values in Table 3.1 apply to the net area in bearing.

When the design value for end grain in bearing exceeds 75% of the values given in Table 3.1, the bearing should be on a metal plate or strap or on other rigid material of adequate strength.

The values in Table 3.1 apply to end-to-end bearing of compression members, for example as in a column, provided there is adequate lateral support and the end cuts are accurately squared and parallel. When a rigid insert is required, it should be of not less than 20-gage metal plate or equivalent and should be snugly placed between abutting ends.

Modification of Design Values

The design values tabulated in Tables 3.1, 7.3, and 7.4 are for normal duration of load under the moisture or service conditions noted. Modifications to these design values for frequently encountered service conditions commonly accounted for in design follow. For other conditions, the designer has the responsibility to relate design assumptions and design values and to make modifications of design values appropriate to the end use.

Moisture Service Condition

Design values tabulated in Tables 3.1, 7.3, and 7.4 should be modified for moisture service conditions in accordance with the applicable footnotes to the tables.

Duration of Load

The design values tabulated in Tables 3.1, 7.3, and 7.4 are for normal duration of loading. Normal load duration anticipates fully stressing a member to the full design value by the application of the full design load for a duration of approximately 10 years (applied either continuously or cumulatively). If the member is designed to be fully stressed by maximum design loads for long-term loading conditions (greater than 10 years either continuously or cumulatively), the design values are taken as 90% of the values tabulated.

For other durations of load, either continuously or intermittently applied, that result in a cumulative effect, the appropriate factor C_D taken from Table 3.2 or determined from Figure 3.1 should be applied to adjust the tabulated design values. However, these duration-of-load modifications, including the 0.90 reduction for permanent loading, are not applicable to modulus of elasticity or compression perpendicular to grain. These provisions are applicable to the modification of the design values for mechanical fastenings when the wood (not the strength of the metal fastening) controls the load capacity.

If loads of different durations are applied simultaneously, the size of member required is determined for the total of all loads applied at the design value modified by the factor for the load of shortest duration in the combination. In like manner, but neglecting the load of shortest duration, the size of member required to support the remaining loads at the design value modified by the factor for the load of next shortest duration is determined. By repeating this procedure for all the remaining loads, the size of member required for the controlling duration-of-load condition is obtained. When the permanently applied load is less than or equals 90% of the total normal load (including the permanently applied load), the normal loading condition will control the size of member required.

Size Factor

When the depth of a rectangular sawn bending member exceeds 12 in., the design value for extreme fiber in bending, F_b, should be decreased by multiplying

TABLE 3.1

Design Values for End Grain in Bearing Parallel to Grain, F_g, on a Rigid Surface for Sawn Lumber (psi)[a]

Species	Wet-Service Conditions[b]	Dry-Service Conditions[c,d] More than 4 in. Thick	Dry-Service Conditions[c,d] Not more than 4 in. Thick
Ash, Commercial White	1370	1510	2060
Aspen	740	820	1110
Balsam Fir	890	980	1330
Beech	1190	1310	1780
Birch, Sweet and Yellow	1150	1260	1720
Black Cottonwood	620	690	930
California Redwood	1560	1720	2270
California Redwood, open grain	1150	1270	1670
Coast Sitka Spruce	950	1040	1420
Coast Species	950	1040	1420
Cottonwood, Eastern	765	840	1150
Douglas Fir-Larch (dense)[e]	1570	1730	2360
Douglas Fir-Larch[d]	1340	1480	2020
Douglas Fir (South)	1220	1340	1820
Eastern Hemlock-Tamarack[e]	1150	1270	1730
Eastern Hemlock	1140	1260	1710
Eastern Softwoods	900	1000	1360
Eastern Spruce	970	1070	1460
Eastern White Pine[e]	900	1000	1360
Eastern Woods	820	900	1230
Engelmann Spruce–Alpine Fir	810	890	1220
Hem-Fir[e]	1110	1220	1670
Hickory and Pecan	1370	1510	2050
Idaho White Pine	930	1020	1390
Lodgepole Pine	970	1060	1450
Maple, Black and Sugar	1140	1260	1710
Mountain Hemlock	1070	1170	1600
Mountain Hemlock, Hem-Fir	1070	1170	1600
Northern Aspen	740	810	1110
Northern Pine	1040	1150	1570
Northern Species	800	970	1320
Northern White Cedar	740	810	1110
Oak, Red and White	1060	1160	1590
Ponderosa Pine-Sugar Pine	910	1000	1370
Red Pine	880	970	1320
Sitka Spruce	990	1090	1480
Southern Cypress	1330	1460	1990
Southern Pine (dense)	1540	1690	2310
Southern Pine	1320	1450	1970
Spruce–Pine–Fir	940	1040	1410

TABLE 3.1 (*Continued*)

Species	Wet-Service Conditions[b]	Dry-Service Conditions[c,d] More than 4 in. Thick	Dry-Service Conditions[c,d] Not more than 4 in. Thick
Sweetgum and Tupelo	1020	1120	1530
Virginia Pine–Pond Pine	1270	1390	1900
Western Cedars[e]	1040	1140	1520
Western Hemlock	1240	1360	1860
Western White Pine	930	1030	1400
White Woods (Western Woods)	810	890	1220
West Coast Woods (Mixed Species)	810	890	1220
Yellow Poplar	890	980	1340

[a]Source: *National Design Specification for Wood Construction* (1).

[b]Wet-service conditions are defined as exceeding 19% MC for sawn lumber.

[c]Applies to sawn lumber members that are at a 19% MC or less when full design load is applied, regardless of moisture content at time of manufacture.

[d]See applicable values in Annex A, AITC 117—Design (4) for glued laminated timber.

[e]Values also apply when species name includes the designation "North."

by the size factor C_F as determined from the formula

$$C_F = \left(\frac{12}{d}\right)^{1/9} \tag{3-1}$$

where C_F = size factor and
d = depth of member (in.).

The size factor does not apply to MSR lumber. It also does not apply to visually graded lumber 2–4 in. thick except for flatwise use. Footnote 6 to Table 7.3 provides size factors for increasing F_b when visually graded lumber 2–4 in. thick is loaded

TABLE 3.2

Usual Modification Factors for Duration of Loading

Duration of Load	Modification Factor C_D
2 months (as for snow)	1.15
7 days	1.25
Wind or earthquake	1.33
Impact	2.00

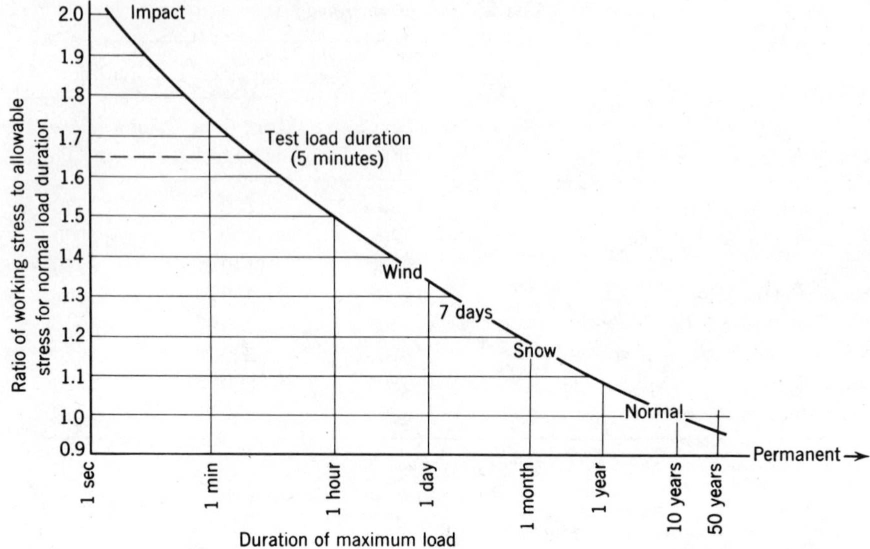

FIGURE 3.1 Duration of load factors, derived from Forest Products Laboratory Report No. R 1916.

flatwise. These size factors are based on Eq. (3-1). For beams of circular cross section that have a diameter greater than 13.5 in. or for square beams 12 in. or larger loaded in the plane of the diagonal, C_F may be determined on the basis of an equivalent conventionally loaded square beam of thc same cross-sectional area.

Modification of design values for extreme fiber in bending, F_b, for size factor is cumulative with the modification for form factor but is not cumulative with the modification for slenderness as defined in Section 5.

Form Factor

For bending members with circular cross sections or square cross sections loaded in the plane of the diagonal, the bending design value, F_b, should be modified by a form factor, C_f, as given in the formula

$$M = \frac{F_b C_f C_F I}{c} \tag{3-2}$$

where M = moment capacity (in. lb),
C_f = form factor,
C_F = size factor,
F_b = design value for extreme fiber in bending (psi),
I = moment of inertia (in.^{4}), and
c = distance from neutral axis to outer surface of beam (in.).

The form factor C_f of a bending member with a circular cross section is 1.18. If a circular member is tapered, it should be considered a beam of variable cross section.

The form factor C_f of a square bending member that is loaded in the plane of a diagonal is 1.414.

Temperature

The *National Design Specification for Wood Construction* (1) indicates that the design values for sawn lumber tabulated in Tables 3.1, 7.3, and 7.4 are applicable to members used under ordinary ranges of temperature and occasionally heated in use to temperatures up to 150°F. Wood increases in strength when cooled below normal temperatures and decreases in strength when heated. Members heated in use to temperatures up to 150°F will return essentially to original strength when cooled. Prolonged temperatures above 150°F may result in permanent loss of strength. The reduction in design values indicated in Table 3.3 may be necessary in specific applications to account for the temporary decrease in strength occurring when sawn lumber is heated to elevated temperatures up to 150°F when these temperatures occur simultaneously with maximum design load. More information on the effect of temperature on design values is given in *Wood Handbook* (5) by the U.S. Forest Products Laboratory and Appendix C of *National Design Specification for Wood Construction* (1).

Preservative Treatments

The design values tabulated in Tables 3.1, 7.3, and 7.4 apply to sawn lumber pressure impregnated with preservatives by an approved process as recommended by the *American Wood-Preservers' Association* (AWPA) (6). For sawn lumber pressure impregnated with preservative salts to the heavy retentions required for "marine" exposure as defined by the AWPA, the duration-of-load factor for impact does not apply.

Fire-Retardant Treatments

The National Forest Products Association (1) recommends that the effects of fire-retardant chemical treatments on strength be considered and that information on this effect be obtained from the company providing the treating and redrying service.

TABLE 3.3

Percent Increase or Decrease in Design Values for Each 1°F Decrease or Increase in Temperature (%)[a]

Property	Moisture Content	Cooling below 68°F (Minimum −300°F)	Heating above 68°F (Maximum 150°F)
Modulus of elasticity	0	+0.04	−0.04
	12	+0.15	−0.21
Other properties	0	+0.17	−0.17
	12	+0.32	−0.49

[a]Source: *National Design Specification for Wood Construction* (1).

STRUCTURAL GLUED LAMINATED TIMBER DESIGN VALUES

Establishment of Design Values

The term *structural glued laminated timber* refers to an engineered, stress-rated product of a timber laminating plant comprising assemblies of suitably selected and prepared wood laminations bonded together with adhesives. The grain of all laminations is approximately parallel longitudinally. The separate laminations do not exceed 2 in. in net thickness. They may be comprised of pieces end joined to form any length, of pieces placed or glued edge to edge to make wider ones, or of pieces bent to curved form during gluing. The design values for glued laminated timber are established by following the procedures given in *Standard Method for Establishing Stresses for Structural Glued Laminated Timber* (*Glulam*), ASTM D 3737 (7). The following discussion covers the method used to establish design values in bending. (See ASTM D 3737 for procedures for setting other values.)

The bending strength of horizontally laminated timbers (bending members that are loaded perpendicular to the wide faces of the laminations) depends on the position of various grades of laminations within the member. High-grade laminations are normally placed in the outer portions of the member where higher strength is effectively used; lower-grade laminations are placed in the inner portion where lower strength will not greatly affect the overall strength of the member. Also, due to the random dispersal of knots and other strength-reducing characteristics (the laminating effect), improved strength can be obtained.

The principal determinants of bending strength in glued laminated structural members are the clear-wood strength and grade characteristics such as knots and slope of grain. The effects of these grade characteristics, called grade factors, are not cumulative; that is, the lowest of the grade factors controls the strength. Where other factors such as curvature or size of member are applicable, they are applied in addition to those for knots and slope of grain.

Knots

The effect of knots on the bending strength and stiffness of laminated timbers depends on their frequency within a cross section and their size and position with respect to the neutral axis of the member. The sum of the moments of inertia of areas occupied by knots within 6 in. of a critical cross section is represented by the symbol I_K. The moment of inertia for the full cross section of the member is represented by I_G. Tests of laminated timbers containing knots in various concentrations established a relationship between the I_K/I_G ratio and bending strength of the members. The procedures for calculating the I_K/I_G ratio are given in USDA Forest Service Research Paper FPL 292, *Improved Utilization of Lumber in Glued Laminated Beams* (8), and in ASTM D 3737 (7).

The I_K/I_G ratio and its relationship to bending strength yields reliable results if specially selected tension laminations are used. AITC has established these tension lamination grades for various design value levels and they are included in AITC 117—Manufacturing (9).

Slope of Grain

As the slope of grain becomes steeper in a wood member, the strength decreases. For laminated timbers, it is possible to vary the slope-of-grain limitations at different points in the depth of a beam in accordance with the stress requirements. That is, steeper slopes of grain may be permitted in the interior laminations than in the outer laminations.

End Joints

For most glued laminated timbers manufactured, pieces of lumber must be joined end to end to provide laminations of sufficient length because of the size of the member. Both plane scarf joints and finger joints can be manufactured with adequate strength for structural glued laminated timber. The required end joint strength is monitored by physical testing procedures contained in industry standards. Butt joints generally can transmit no tensile stress and can transmit compressive stress only after considerable deformation. They also cause concentrations of both shear stress and longitudinal stress. For these reasons, they are not permitted for use in structural glued laminated timbers.

Evaluations of timbers vertically laminated from dimension lumber (members loaded parallel to the wide faces of the lamination) indicate that the effect of the lamination grade and number of laminations may be accounted for by procedures recommended in USDA Forest Service Research Paper FPL 333, *Bending Strength of Vertically Glued Laminated Beams with One to Five Plies* (10), and in ASTM D 3737 (7).

Design Values

Design values for structural glued laminated timber are given in Tables 1 and 2 of *Standard Specifications for Structural Glued Laminated Timber of Softwood Species*, AITC 117—Design (4), included in Part III of this manual.

When curved members are subject to bending moment, radial stresses are induced in a direction parallel to the radius of curvature of the centerline of the member (perpendicular to grain). If the moment decreases the curvature (tends to straighten the member), the stress is tension; if it increases the curvature (makes the member more sharply curved), the stress is compression. For these members, when the moment induces a stress in tension across the grain, this tensile stress is limited to one-third the design value in horizontal shear for Douglas Fir-Larch, Douglas Fir South, Hem-Fir, and Western Woods and Canadian softwood species for wind or earthquake loadings and for Southern Pine for all conditions of loading. The limit is 15 psi for Douglas Fir-Larch, Douglas Fir South, Hem-Fir, and Western Woods and Canadian softwood species for other types of loads. For Douglas Fir-Larch, Douglas Fir South, Hem-Fir, or Western Woods and Canadian softwood species where the calculated stress exceeds this 15-psi value, mechanical reinforcing designed in accordance with the procedure on page 5-229 of this manual (or an equivalent design method) should be used and should be sufficient to resist all radial tension stresses. In no case shall the calculated radial stress exceed one-third of the design value in horizontal shear. When mechanical reinforcing is used, the maximum moisture content of the laminations at the time of manufacture should not exceed 12%.

When the moment is in a direction causing a stress in compression across the grain, this stress is limited to the design value in compression perpendicular to the grain for the species involved.

Design Values for Bearing on End Grain

Design values for bearing on the end grain of structural glued laminated timbers are based on the clear-wood strength of the species and are given in Tables A-1 and A-2 of AITC 117—Design (4) included in Part III of this manual.

Modification of Design Values

The design values tabulated in Tables 1 and 2 of AITC 117—Design (4) are for dry-service conditions and normal duration of load.

Wet-Service Conditions

Design values tabulated in Tables 1 and 2 of AITC 117-Design (4) should be modified for wet-service conditions (equilibrium moisture content equal to or exceeding 16% in service) by the factors given in Table 3.4.

Duration of Load

Design values tabulated in Tables 1 and 2 of AITC 117-Design (4) are for normal duration of loading. The discussion of duration of load in the section "Modification of Design Values for Sawn Lumber" applies also to structural glued laminated timber.

Size Factor

When the depth of a rectangular glued laminated timber bending member exceeds 12 in., the design value for extreme fiber in bending, F_b, should be decreased by multiplying by the size factor C_F as determined from Eq. (3-1).

$$C_F = \left(\frac{12}{d}\right)^{1/9}$$

TABLE 3.4

Modification Factors for Wet-Service Conditions, C_M

Type of Design Value							
		Compression					
Extreme Fiber in Bending F_b	Tension Parallel to Grain F_t	Parallel to Grain F_c	End Bearing F_g	Perpendicular to Grain $F_{c\perp}$	Radial Tension F_{rt}	Horizontal Shear F_v	Modulus of Elasticity E
0.80	0.80	0.73	0.57	0.53	0.875	0.875	0.833

where C_F = size factor and
d = depth of member (in.).

For glued laminated members less than 12 in. deep measured parallel to the wide faces of the laminations, the footnotes to Tables 1 and 2 of AITC 117—Design (4) provide size factors for increasing F_{by}. These factors are based on Eq. (3-1). The size factor relationship is based on a bending member that is uniformly loaded and simply supported with a span-to-depth (l/d) ratio of 21.

Table 3.5 presents a tabulation of size factors in 1-in. increments for members satisfying the basic design conditions. Equation (3-1) may be used for depths other than those given in Table 3.5 or C_F may be obtained with sufficient accuracy by applying straight-line interpolation.

TABLE 3.5

Size Factors[a]

Depth (in.)	C_F	Depth (in.)	C_F	Depth (in.)	C_F
12 or less	1.00	35	0.89	58	0.84
13	0.99	36	0.88	59	0.84
14	0.98	37	0.88	60	0.84
15	0.98	38	0.88	61	0.83
16	0.97	39	0.88	62	0.83
17	0.96	40	0.87	63	0.83
18	0.96	41	0.87	64	0.83
19	0.95	42	0.87	65	0.83
20	0.94	43	0.87	66	0.83
21	0.94	44	0.87	67	0.83
22	0.93	45	0.86	68	0.82
23	0.93	46	0.86	69	0.82
24	0.93	47	0.86	70	0.82
25	0.92	48	0.86	71	0.82
26	0.92	49	0.86	72	0.82
27	0.91	50	0.85	73	0.82
28	0.91	51	0.85	74	0.82
29	0.90	52	0.85	75	0.82
30	0.90	53	0.85	76	0.81
31	0.90	54	0.85	77	0.81
32	0.90	55	0.84	78	0.81
33	0.89	56	0.84	79	0.81
34	0.89	57	0.84	80	0.81

[a]Applicable to uniformly loaded simple-span beams with $l/d = 21$. For conditions of loading other than a uniformly distributed load, the tabulated values for C_F may be adjusted by applying the percentages given in AITC 117—Design (4). For span-to-depth ratios other than 21, the tabulated values for C_F may be adjusted by applying the percentages given in AITC 117—Design (4). See Table 7.2 for additional values of C_F.

The size factor obtained from Eq. (3-1) can be applied with reasonable accuracy to most commonly encountered design situations. This is the procedure used in the examples given in Section 5 of this manual. Where greater accuracy is desired, modifications for span-to-depth ratios other than 21 and loading conditions other than uniform may be obtained from Section 4.4.2.2, AITC 117—Design contained in Section 7 of this manual.

For rectangular bending members with variable cross section along their length, such as tapered beams, d for determination of the size factor should be taken as that depth at which the stresses are being analyzed.

For continuous and cantilevered members, determine the size factor by assuming the members to be equivalent simply supported members with a uniformly distributed load. This will result in a slightly conservative size factor being applied to the design of span types other than simple spans.

The design value in bending, F_b, is modified by applying the appropriate size factor C_F to the formula for determining moment as follows:

$$M = F_b S C_F = F_b S_n \tag{3-3}$$

where M = bending moment (in. lb),
S = section modulus (in.^{3}),
S_n = section modulus times C_F (in.^{3}),
F_b = design value for extreme fiber in bending (psi), and
C_F = size factor.

For convenience in design, the size factor C_F may be multiplied by the section modulus S and the resulting product SC_F is identified as S_n and is used to select the member size. See Table 7.2 for values of S_n.

For members subjected to combined bending and axial compression, see page 5-199.

Curvature Factor

Stress is induced when laminations are bent to curved forms. Although much of this stress is quickly relieved, some remains and tends to reduce the strength of a curved member. Also, in a curved member, the extreme outer fiber bending stress is greater than that of a straight prismatic member subjected to the same moment. Therefore, to account for these factors, the bending design value F_b must be modified by multiplying by the curvature factor C_c:

$$C_c = 1 - 2000\left(\frac{t}{R}\right)^2 \tag{3-4}$$

where t = thickness of lamination (in.) and
R = radius of curvature of inside face of lamination (in.).

The ratio t/R may not exceed 1/100 for hardwoods and Southern Pine nor 1/125 for softwoods other than Southern Pine. The curvature factor is not applied to design values in the straight portion of a member regardless of curvature in other portions.

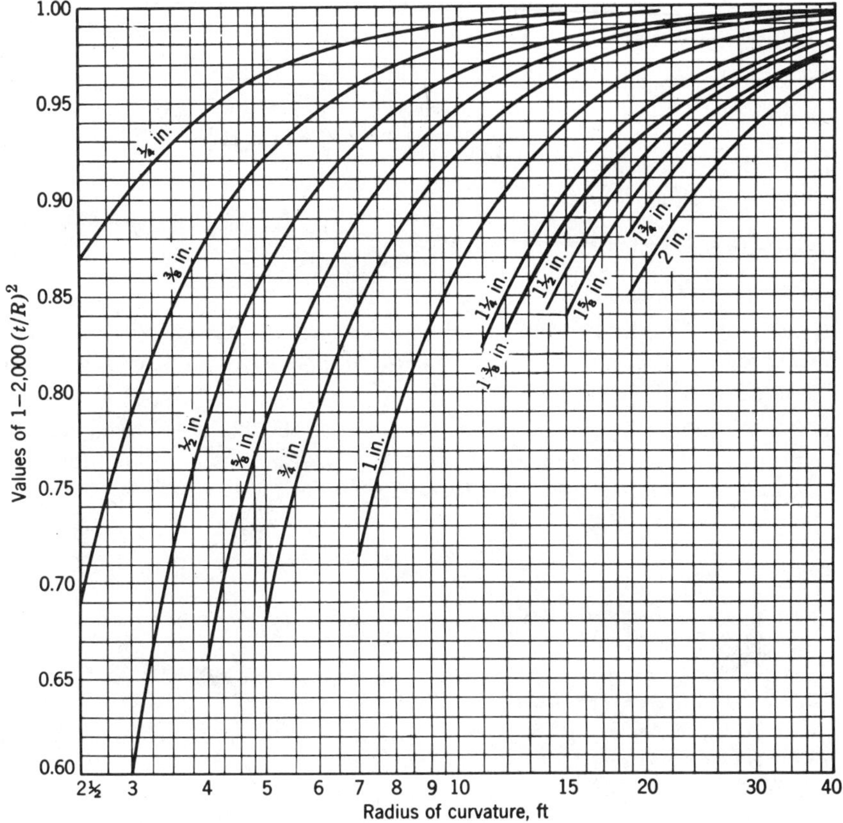

FIGURE 3.2 Curvature factors.

Figure 3.2 may be used to determine curvature factors for several different lamination thicknesses.

Lateral Stability

The design values for bending tabulated in AITC 117—Design (4) are applicable to members that are adequately braced. When deep, slender members not adequately braced are used, a reduction to the tabulated design values in bending due to lateral stability considerations must be applied. The check of lateral stability should be made in design as shown in Section 5 of this manual.

A reduction in the design value in bending for lateral stability is not cumulative with a reduction in design value due to the application of size factor. See Section 5 of this manual for further discussion.

Temperature

Design values tabulated in Tables 1 and 2 of AITC 117—Design (4) are applicable to structural glued laminated timbers used under ordinary ranges of temperature and occasionally heated in use to temperatures up to 150°F. The discusion of temperature effects on strength on page 000 applies also to structural glued laminated timber.

Preservative Treatments

Design values tabulated in Tables 1 and 2 of AITC 117—Design (4) are applicable to glued laminated timbers pressure impregnated by processes and preservatives as recommended by the American Wood-Preservers' Association (6). The provisions in *Standard for Preservative Treatment of Structural Glued Laminated Timber*, AITC 109 (11), included in Part III of this manual, also apply.

Fire-Retardant Treatments

The design values for fire-retardant treated structural glued laminated timber treated before or after gluing are dependent on the species and treatment combinations involved. The effect on strength must be determined for each treatment; however, indications are that a 10–25% reduction in the design values in bending is applicable. The manufacturer of the treatment should be contacted for more specific information on stress adjustments for all design values.

REFERENCES

1. National Forest Products Association, *National Design Specification for Wood Construction*, Washington, DC, 1982.
2. American Society for Testing and Materials, *Standard Methods for Establishing Clear Wood Strength Values*, ASTM D 2555, Philadelphia, PA, 1981.
3. American Society for Testing and Materials, *Standard Methods for Establishing Structural Grades and Related Allowable Properties for Visually Graded Lumber*, ASTM D 245, Philadelphia, PA, 1981.
4. American Institute of Timber Construction, *Standard Specifications for Structural Glued Laminated Timber of Softwood Species*, AITC 117—Design, Englewood, CO, 1984.
5. United States Department of Agriculture, Forest Service, Forest Products Laboratory, *Wood Handbook: Wood as an Engineering Material*, Agriculture Handbook No. 72, Madison, WI, 1974.
6. American Wood-Preservers' Association, *Book of Standards*, Stevensville, MD, 1984.
7. American Society for Testing and Materials, *Standard Method for Establishing Stresses for Structural Glued Laminated Timber (Glulam)*, ASTM D 3737, Philadelphia, PA, 1983.
8. United States Department of Agriculture, Forest Service, Forest Products Laboratory, *Improved Utilization of Lumber in Glued Laminated Beams*, Research Paper FPL 292, Madison, WI, 1977.
9. American Institute of Timber Construction, *Standard Specifications for Structural Glued Laminated Timber of Softwood Species*, AITC 117—Manufacturing, Englewood, CO, 1984.
10. United States Department of Agriculture, Forest Service, Forest Products Laboratory, *Bending Strength of Vertically Glued Laminated Beams with One to Five Plies*, Research Paper FPL 333, Madison, WI, 1979.
11. American Institute of Timber Construction, *Standard for Preservative Treatment of Structural Glued Laminated Timber*, AITC 109, Englewood, CO, 1984.

Part II

DESIGN

SECTION 4

LOADS

LOADS

Buildings or other structures and all parts thereof should be designed to safely support all loads, including dead loads, that may reasonably be expected to affect the structure during its service life. These loads should be as stipulated by the governing building code or, in the absence of such a code, the loads, forces, and combination of loads should be in accordance with accepted engineering practice for the geographical area under consideration.

DEAD LOADS

Dead load is defined as the vertical load due to all permanent structural and nonstructural components of a structure, such as walls, floors, roofs, partitions, stairways, and fixed service equipment.

During the life of a building or other structure, additional loads may be applied that actually become dead loads but in all probability are not treated as such in the original design. If it is anticipated that such loads will be added at a later date, modifications should be made in the original design. For example, consideration should be given to the possible reroofing of the structure, and allowance appropriate to the type of roofing involved should be made.

The actual weights of the various materials and constructions should be used in design if this information is available. In the absence of such information, values satisfactory to the building official may be assumed. Table 7.10 (Section 7) gives the approximate weight of various construction and other materials as a guide to the designer.

LIVE LOADS

Live load is defined as the load superimposed by the use and occupancy of the building or other structure, not including the wind load, snow load, earthquake load, or dead load.

Floor Live Loads

In the absence of a governing building code, the minimum uniformly distributed live loads or concentrated loads in *Minimum Design Loads for Buildings and Other Structures*, ANSI A58.1 (1) are recommended.

Roof Live Loads

The minimum roof live loads should be as stipulated by the governing building code. Roof live loads used in design should represent the designer's determination of the particular service requirements for the structure, but in the absence of a governing code, in no case should they be less than the recommended minimum.

Minimum roof live loads for flat, pitched, or curved roofs are recommended in ANSI A58.1 (1). Roofs should be designed to resist either the tabulated minimum live loads, applied as balanced or full unbalanced, or the snow load, whichever produces the greater stress. Roofs to be used for special purposes should be designed for appropriate anticipated loads.

SNOW LOADS

Ground Snow Loads

Snow loads vary widely throughout the United States. Factors affecting snow load accumulation on roofs include climatic variables, geographic location, roof exposure, roof slope, roof thermal condition, snow drifting, and sliding snow. Snowfall varies from year to year, and either a mean recurrence interval must be established for design purposes or design should be based on the maximum recorded snow load for which data is available. Snow loads should be as stipulated by the governing building code, but in the absence of such a code, snow loading used for design should be based on local experience or the use of accepted snow load maps.

Although the analysis of roofs for snow loading is complex because of the many variables involved, recent technical data have provided the designer with sufficient information to make a realistic analysis.

Roof design loads are obtained from ground snow loads. Several researchers and agencies have measured ground snow load distribution and plotted appropriate isogram maps depicting these loads. Data from the National Weather Service are used in preparing those maps. Snow loads for mountainous areas, especially in the Western states, should be established based on local experience. Actual snow pack of over 700 psf has been recorded in isolated regions and some of the inhabited regions have snow loads ranging up to 300 psf.

Roof Snow Loads

Roof snow loads may be based on a determination of the ground snow load as specified by the governing building code or the roof snow loads may be specified in the code. Maximum snow load maps may also be used to determine the actual snow loads to be expected on the roof surface. As previously indicated, the roof snow load is a function of various factors.

These factors can be accounted for in design by applying appropriate snow load coefficients to the basic ground snow loads. Specific snow load coefficients have been developed to relate roof snow load to ground snow load based on comprehensive surveys of actual conditions.

For the design of both ordinary and multiple series roofs, either flat, pitched, or curved, minimum roof snow loads may be determined by multiplying the ground snow load by the appropriate coefficients based on the roof geometry and climatic conditions. Specific coefficients for these roof configurations are given in ANSI A58.1 (1).

Because unbalanced loading can occur as the result of drifting, sliding, melting, and refreezing or physical removal of snow, structural roof members should be designed to resist increased loading in certain areas or other unbalanced loading conditions. For example, the roof should resist the full snow load as defined above distributed over the entire roof area, the full snow load distributed on any one portion of the area, and dead load only on the remainder of the area, depending on which load produces the greatest stress on the member considered.

Duration of load is the cumulative time during which the maximum design load is on the structure over its entire life. A 2-month duration is generally recognized as the proper design level for snow loads. Although some snow remains on roofs for periods exceeding 2 months in a single year, such snow loads seldom approach the full design load. In geographical areas where near-maximum snow load remains on the roof each winter for long periods of time or in buildings that are unheated, have heavily insulated roofs, or are used for cold storage, the use of the 2-month duration-of-load increase may not be advisable.

WIND LOADS

Much research has been conducted to evaluate wind effects on various structures and has resulted in the establishment of design coefficients that account for building shape and wind direction. In addition, extensive studies of basic wind velocities related to geographical location have resulted in the development of detailed wind speed maps for the United States. Other studies of surface resistance relative to the degree of land development and gust characteristics at a given location have provided a method for a further refinement of the basic wind velocity and its effect on structures. Additional work relates the dynamic behavior of structures to wind forces that are attributable to gusting and turbulence.

Sources of wind load analysis information include *Minimum Design Loads for Buildings and Other Structures*, ANSI A58.1 (1), which contains a number of other references on this subject.

Basic Wind Speed

Basic wind speeds for observed air flows in open, level country at a height of 33 ft above the ground have been developed. For the design of most permanent structures, a basic wind speed with a 50-yr mean recurrence interval should be applied. However, if in the judgment of the engineer or authority having jurisdiction the structure presents an unusually high degree of hazard to life and property in case of failure, the basic wind speed may be modified appropriately. Similarly, for temporary structures or structures having negligible risk of human life in case of failure, the design wind speed may be reduced.

Basic wind speeds are adjusted for height and site exposure conditions. In addition, gust response factors are applied to account for the response of the structure or building to the fluctuating nature of wind. Because of the complex nature of this subject, the designer is referred to the appropriate sections of ANSI A58.1 (1).

Wind Pressure

To determine the design wind pressure distribution acting on a building or structural element, the calculated velocity pressure is multiplied by an appropriate pressure coefficient. These pressure coefficients thus define the wind pressure acting normally on the surface of a building or element thereof and are dependent on the external shape of the structure and its orientation with the wind. Pressure coefficients are considered to be either positive, representing an inward pressure, or negative, indicating an outward force on the structural element being analyzed. Therefore, depending on orientation to the wind and existence of openings, a building or element may be subjected to a pressure difference between opposite sides or faces, and it is thus the total resultant wind pressure that must be accounted for in design.

BASIC MINIMUM ROOF LOAD COMBINATIONS

In designing, the most severe realistic distribution, concentration, and combination of roof loads and forces should be taken into consideration. Table 4.1 gives examples of roof load design combinations that should be checked and examples of the application of the duration-of-load factor for different load combinations.

Each type of load should be determined individually, and the effect of all possible combinations should be investigated. All possibilities must be investigated for unsymmetrical buildings. Special consideration should be given to structures of great span and/or height and to trusses bearing on very long columns. In addition to analyzing the structure for the combinations of loading shown in Table 4.1, the designer should check to see that all loading combinations as required by the governing building code have been met.

Structural members or systems should be designed to resist the stresses caused by partially or fully unbalanced live or snow loads and wind loads, including uplift, in combination with dead loads on the member or system. Such loading may result in reversal of stresses or stresses greater in some portion than the stresses produced when the entire roof is fully loaded. The occupancy or use of the structure, its configuration, heating and insulating considerations, local climatic conditions,

TABLE 4.1

Examples of Load Combinations for Design[a]

Roof Load Combinations

wind→

Windward Side	Leeward Side	Duration of Load	Duration of Load Factor C_D
DL	DL	Permanent	0.90
DL + LL	DL + LL	7 days	1.25
DL[b]	DL + LL[b]	7 days	1.25
DL + SL	DL + SL	2 months	1.15
DL + $\frac{1}{2}$SL[c]	DL + SL[c]	2 months	1.15
DL + WL	DL + WL	Wind	1.33
DL + EL	DL + EL	Earthquake	1.33

[a]Symbols: DL = dead load, LL = live load, SL = snow load, WL = wind load, EL = earthquake load. Special configurations, locations, and occupancies of structures may require investigation of (a) full unbalanced SL or (b) combinations of WL with LL and DL, or WL with partial SL and DL. It may be assumed that wind and earthquake loads will not occur simultaneously.

[b]Full unbalanced loading.

[c]Half unbalanced loading.

or other considerations may also cause partial or full unbalanced loading on the structure.

In order to determine the critical load combination and associated duration-of-load factor, each load combination may be divided by the applicable duration-of-load factors. The highest value determines the critical combination of loads. The following example illustrates this method.

Example. A beam is to be designed for the following loads. Determine the critical combination of loads and applicable duration-of-load factor.

$$
\begin{aligned}
\text{DL} &= 10 \text{ psf} \qquad & C_D &= 0.9 \\
\text{LL} &= 12 \text{ psf} & &= 1.25 \\
\text{SL} &= 20 \text{ psf} & &= 1.15 \\
\text{WL} &= 5 \text{ psf} & &= 1.33
\end{aligned}
$$

$$
\begin{aligned}
&\text{DL}/0.9 = 10/0.9 = 11.1 \\
&(\text{DL} + \text{LL})/1.25 = (10 + 12)/1.25 = 17.6 \\
&(\text{DL} + \text{SL})/1.15 = (10 + 20)/1.15 = 26.1 \\
&(\text{DL} + \text{WL})/1.33 = (10 + 5)/1.33 = 11.3 \\
&(\text{DL} + \text{WL} + \text{SL}/2)/1.33 = (10 + 5 + 20/2)/1.33 = 18.8 \\
&(\text{DL} + \text{SL} + \text{WL}/2)/1.33 = (10 + 20 + 5/2)/1.33 = 24.4
\end{aligned}
$$

For these loads and the loading combinations checked, the combination of dead load plus snow load with a duration-of-load factor of 1.15 is critical.

EARTHQUAKE LOADS

In areas subject to earthquake shocks, every building or other structure and every portion thereof should be so designed and constructed as to resist stresses produced by lateral forces as provided by the governing building code regulations. Sources of earthquake load design information include the current *Uniform Building Code* (2) by the International Conference of Building Officials, *Recommended Lateral Force Requirements* (3) by the Seismology Committee of the Structural Engineers Association of California, and *Tentative Provisions for the Development of Seismic Regulations for Buildings* (4) by the Applied Technology Council.

HIGHWAY LOADING

As indicated by the excellent performance of timber highway bridges over the years, wood bridges meet all the general bridge design requirements for modern highway traffic loads. In addition, wood offers the important advantage of being able to absorb impact stresses to the degree that impact loads may be neglected in the design of wood highway structures; however, they must be considered in the metal-to-metal connections. Highway design loads and their application should be in accordance with *Standard Specifications for Highway Bridges*, adopted by the American Association of State Highway and Transportation Officials (AASHTO) (5). Except for bridge railing loading, which is a short-time load, normal duration of loading is currently being used for design.

RAILWAY LOADING

Timber railway bridges and trestles have been long and extensively used in this country. With the advent of pressure preservative treatments, a service life of 50 years is commonplace. In the design of timber bridges and trestles, the recommendations in *Manual of Recommended Practice* of the American Railway Engineering Association (AREA) (6) are ordinarily applied. Recognizing the increased strength of wood under quickly applied loading, the AREA specifications permit omission of impact loads in determining stresses in the wood. Impact must be considered, however, in metal-to-metal connections. It should be pointed out that AREA specifications assume that all railway bridges and trestles are under "long-time" loading.

CRANE BEAM LOADING

Timber is often used for crane beams and girders because of its ability to absorb impact forces. In designing crane beams, wheel loads recommended by crane manufacturers should be used if available.

The ANSI A58.1 (1) recommends that for the design of all crane runways, the design loads should be increased for impact as

1. a vertical force equal to 25% of the maximum wheel load;

2. a lateral force equal to 20% of the weight of trolley and lifted load only, applied one-half at the top of each rail; and
3. a longitudinal force of 10% of the maximum wheel loads of the crane applied at the top of the rail.

Some crane manufacturers specify wheel loads that include a percentage for impact. This factor should be taken into account in designing because it is unnecessary to add impact to wheel loads for timber designs unless there is a steel connection through which the load must pass or unless impact exceeds 100% of other normal loads. Impact must be considered for metal-to-metal connections. Normal duration of loading should be used for design.

Crane runway rails should be designed to prevent undue vertical and lateral deflection in accordance with crane manufacturers' recommendations.

CYCLIC LOADING

Tests to date indicate that wood is less sensitive to repeated loads than are crystalline structural materials such as metals. In general, the fatigue strength of wood is higher in proportion to the ultimate strength values than are the endurance limits for most structural metals. In present design practice, no factor is applied for fatigue in deriving working stress values for wood, nor is it considered necessary to do so. For these reasons, normal duration of load is used for cyclic loading design.

Where estimation of repetitions of design or near-design loads are indicated to approximate one million or more cycles, the designer should investigate the possibility of fatigue failures in shear or tension perpendicular to grain. When shear governs design and fatigue is a distinct possibility, shear stresses should be reduced 10% or changes should be made on the basis of a detailed analysis as indicated by examination of USDA Forest Products Laboratory Report No. 2236, *Fatigue Resistance of Quarter-Scale Bridge Stringers in Flexure and Shear* (7) and/or ASCE Paper No. 2470, *Design Considerations for Fatigue in Timber Structures* (8). Analyses made on the research available indicate that bending failure due to fatigue is not usual and need not be considered in ordinary design situations. Tension perpendicular to grain loading of wood should be avoided, especially when cyclic loading is expected.

VIBRATION

Vibration is closely related to impact and cyclic loading. The effects of vibration can usually be neglected in timber structures, except in those portions made of materials for which impact or vibration forces may be critical or where the vibration may be objectionable to the human occupancy. Vibration may cause loosening of threaded connections used in timber structures. If there is a possibility of fatigue failure due to vibration, it should be considered and care should be taken to avoid notches, eccentric connections, and similar design conditions.

BLAST LOADS

Data on the design of structures resistant to blast loading such as may be caused by nuclear weapon explosions is beyond the scope of this manual. Such data may be found in the American Society of Civil Engineers Manual No. 42, *Design of Structures to Resist Nuclear Weapons Effects* (9).

PONDING

When there is the possibility of water ponding, which may cause excessive loads and additional progressive deflection, each component of the roof system, including decking, purlins, beams, girders, or other principal structural supports, should be designed accordingly. Continuous or cantilevered components should be designed for balanced or unbalanced load, whichever produces the more critical condition. Adequate drainage capacity and proper construction details should be provided.

Ponding is usually a greater problem on roofs with long-span structural members than for roofs with short-span beams because the "spring constant" of a long-span system is likely to be larger. The spring constant of a roof can be expressed as the deflection in inches per arbitrary unit of load. It is convenient to express spring constant as inches of deflection per 5 lb of load per square foot because the weight of a 1-in. depth of water is approximately equal to a load of 5 psf. (The actual weight of a 1-in. depth of water is 5.2 psf.) Roofs with large tributary drainage areas are, in general, more susceptible to ponding than roofs with smaller areas.

Ponding problems may occur in all parts of the country. They appear to be greater in areas where small design live loads are used. Ponding failures have occurred in semiarid regions because rainstorms of high intensity occur, although the annual rainfall in these regions is small. Problems have also been found in colder climates. Roofs in these areas are generally designed for large snow loads, which reduces the spring constant of the system, but unusual weather conditions can occur that may result in the blockage of drain paths by packed snow or ice. In such cases, the ponding load must be added to that of the snow or ice on the roof.

The most effective method of preventing problems due to ponding is to provide adequate roof slope and drainage. It is recommended that roof surfaces have a minimum residual slope or camber of not less than $\frac{1}{4}$ in./ft of horizontal distance between the level of the drain and any point of the roof to minimize the ponding of water. The minimum slope should remain after long-term deflection of the members has taken place. In most cases, this should provide for adequate drainage. In selecting the roof slope, the designer should take into account the unevenness of the type of roofing used, the combined effect of all members within the framing system, and the possibility of strong winds blowing the water up-slope and the effect of member shrinkage on drainage. The designer should also make sure that *all elements of the roof maintain the minimum slope*. This is sometimes overlooked in side or end bays where uncambered members are parallel to the slope.

Many ponding problems are caused by inadequate drainage. Consideration should be given to providing adequate drainage to handle the maximum intensity rainfall expected. (Maximum storm intensity for 50- or 100-yr storms are commonly used.) The location of drains is critical. They should be located away from points that remain at high levels when the roof deflects, such as at columns. Quite often, gravel stops will cause a dam that can prevent drainage. Improper sizing and placement of scuppers and roof drainage can allow water to accumulate to depths not anticipated in the design. Wood members may shrink somewhat in reaching moisture equilibrium in a structure, causing movement of the roof. Therefore, it is preferable to use flexible connections for roof drains so that they may move with the roof. After the roof is completed and the building is in use, an ongoing maintenance program for checking the roof drains should be established. Roof drains should be kept clean of debris or opened if they are frozen over to assure their proper function.

Camber is normally recommended for glued laminated timbers so they do not appear to be sagging after the application of dead load. Experience has shown that, in order to offset inelastic deflection or "permanent set" as well as the calculated deflection, a camber of $1\frac{1}{2}$ times the dead load deflection is appropriate. This camber is intended to provide a relatively straight beam. Where it is used to obtain the required slope to prevent ponding, additional camber is required. Also, when camber is used to provide slope for drainage, the middle third of the member may be relatively flat. This condition should be considered in design. Extreme care should also be taken when camber is used to compensate for dead-load deflection in a cantilever beam system. Too little as well as too much camber is to be avoided.

Beam deflection should not exceed the recommended limitations given in Table 5.8. Building codes vary in their requirements for deflection limitations. Deflection limitations should *not* be considered as being adequate for preventing a potential ponding condition or objectionable vibration. They merely set a minimum stiffness that should be used under certain conditions. The recommended limitations given in Table 5.8 are primarily for appearance purposes. In addition, they provide a reasonable minimum stiffness where ponding or vibration are not problems.

When roof slope is not adequate to prevent ponding, the designer should use one of the several design methods available to make sure that the roof is stiff enough to prevent the condition. The following are some suggested considerations in the use of these methods.

a. The moduli of elasticity of wood contained in the tables of design values are average values. All strength properties of wood, including modulus of elasticity, are variable. This variability is expressed by the coefficient of variation Ω, which is a measure of the extent of the variation. The *National Design Specification for Wood Construction* (10) published by the National Forest Products Association lists the coefficients of variation for modulus of elasticity, E, values of wood products as follows:

Visually graded sawn lumber	0.25
Machine stress-rated sawn lumber	0.11
Glued laminated timber (six or more laminations)	0.10

This variation should be considered when designing for ponding because members with lower E values deflect more, creating the potential for larger ponding loads than would occur by using average E values in calculations. The variations in the modulus of elasticity of wood approach a "normal distribution" from a statistical standpoint. If a designer wishes to know the estimated value of modulus of elasticity at some percentile of the total population, such as the fifth percentile (19 pieces out of 20 will have a higher E than the estimated value), it can be calculated as follows:

$$E_{0.05} = E - 1.645E\,\Omega \tag{4-1}$$

where $E_{0.05}$ = estimated modulus of elasticity at the lower fifth percentile (psi),
E = average modulus of elasticity (psi),
Ω = coefficient of variation, and
1.645 = a statistical constant for determination of values at the fifth percentile (constants for other percentiles may be obtained from standard statistical tables).

Example. The average E of a grade of visually graded sawn lumber is 1,800,000 psi, the E value at the fifth percentile is $E_{0.05}$ = 1,800,000 − (1,800,000)(0.25) (1.645) = 1,060,000 psi

b. The inelastic deflection, creep or permanent set of glued laminated timber averages approximately one-half of the calculated dead-load deflection. This deflection can be offset by camber, as previously stated. The permanent set of unseasoned sawn lumber often used for secondary framing is approximately equal to the calculated dead-load deflection. Therefore, the total long-time deflection is approximately 2 times the calculated initial elastic deformation. The additional ponding load resulting from the permanent set of all members within a particular framing system should be considered in the design analysis.

c. The deflection of all elements in a roof system should be considered. In a typical panelized wood roof system, this includes the glued laminated timber beams, the purlins, the subpurlins (or stiffeners), and the plywood roof sheathing.

When low-pitched roofs have insufficient slope for drainage (less than $\frac{1}{4}$ in./ft), the stiffness of supporting members should be such that a 5-psf load will cause no more than $\frac{1}{2}$ in. deflection (11). A simplified approach is based on the assumption that the roof structure between the supporting members is relatively stiff and only a small amount of deflection will occur between supporting members. This method permits some additional load due to deflection of purlins, stiffeners, or sheathing as well as some variation in the modulus of elasticity, and where drainage is adequate and no unusual conditions occur, this rule of thumb works reasonably well.

Where the deflection of the secondary roof framing system between the supporting members is large enough to produce significant ponding loads, a complete analysis should be made. It must be recognized that some water has to accumulate to start deflection. This accumulation can be from a number of causes—roughness of the surface, blowing winds, high gravel stops, drains placed slightly above the level of the roof, and so on.

Example. A low-pitched roof is supported by simple-span glued laminated timber beams that have spring constants of $\frac{1}{2}$ in. deflection for a uniform load of 5 psf. A $\frac{1}{2}$-in.-high gravel stop is around the perimeter of the roof. No drains are provided. The deflection of a secondary framing system between the beams is small. Assume that $\frac{1}{2}$ in. of water, held by the gravel stops, is spread uniformly over the roof. The load from this water causes $\frac{1}{4}$ in. of deflection, which in turn causes an additional load of water to accumulate, resulting in $\frac{1}{8}$ in. of deflection and so forth until equilibrium is reached at about $\frac{1}{2}$ in. of deflection. The total depth of water at midspan is $\frac{1}{2} + \frac{1}{2} = 1$ in., or approximately a 5-psf load. Thus, a roof system limited to $\frac{1}{2}$ in. of deflection for a 5-psf load should reach equilibrium before the ponding load becomes excessive.

The ponding load caused by a deflection of 1 in. in a member with equal height supports is somewhat less than 5 psf because the load caused by the deflection is not uniform but in the approximate shape of a curve, defined by the equation

$$\Delta = \Delta_{max} \sin \frac{\pi x}{L}$$

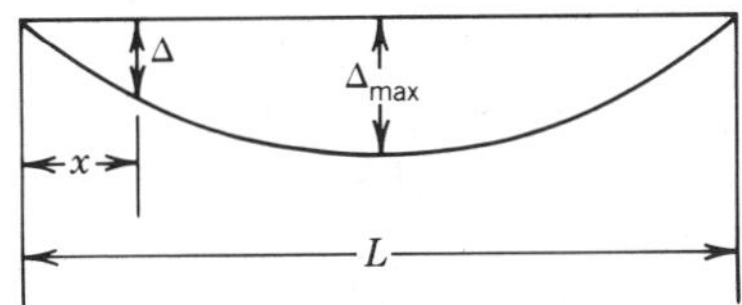

Actual measurements of pools of ponded water on roofs indicate that large variations from the theoretical deflection curve can be expected and that a close approximation can be obtained by assuming the shape of the deflection curve to be parabolic. A parabolic load is roughly equivalent to a uniform load with a depth equal to 0.83 times the depth of the parabola. This method can be used to estimate deflections of individual portions of a roof system.

Example. For the roof system used in the previous example, the deflection of the supporting member is calculated as follows: The $\frac{1}{2}$-in.-depth uniform load of water causes a deflection of $(\frac{1}{2})$ (0.5) = 0.25 in. where the beam deflects $\frac{1}{2}$ in. per inch depth of water. This 0.25-in. deflection in turn results in a parabolic load equivalent to a uniform load with a depth of (0.83)(0.25) = 0.208 in., which causes an additional deflection of (0.208)(0.5) = 0.104 in. The complete calculations are shown as follows:

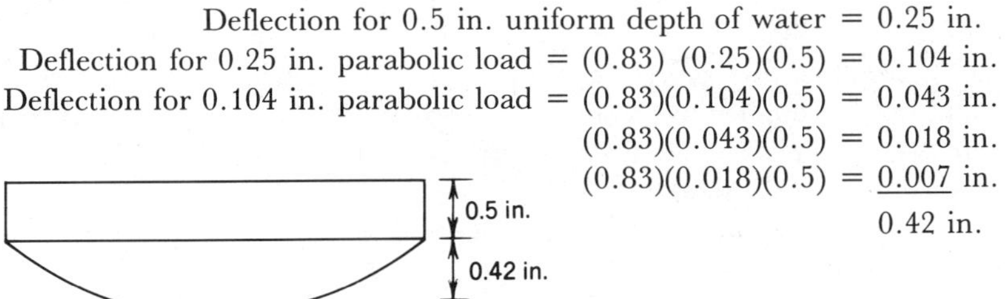

Deflection for 0.5 in. uniform depth of water = 0.25 in.
Deflection for 0.25 in. parabolic load = (0.83) (0.25)(0.5) = 0.104 in.
Deflection for 0.104 in. parabolic load = (0.83)(0.104)(0.5) = 0.043 in.
(0.83)(0.043)(0.5) = 0.018 in.
(0.83)(0.018)(0.5) = 0.007 in.
0.42 in.

At equilibrium, this results in a total depth of water of 0.92 in. weighing approximately 4.8 psf.

The added refinement of assuming parabolic loading is not suggested for use where the rule of thumb limiting deflection to $\frac{1}{2}$ in. for a 5-psf load is used. The method can be used to calculate the deflection of all portions of a roof system by using a series of successive calculations.

Magnification Factor

The effect of ponding of water on a low-pitched roof system supported by sawn beams or glued laminated beams without camber is to increase deflections and stresses. This increase can be expressed by the formula

$$C_p = \frac{1}{1 - W'\, l^3/\pi^4\, EI} \tag{4-2}$$

where C_p = factor for multiplying stresses and deflections under existing loads to determine stresses and deflections under existing loads plus ponding and is known as the magnification factor,
W' = total load of 1-in. depth of water on the roof area supported by the beam or deck (lb/in.),
l = span length of beam or deck (in.),
E = modulus of elasticity of beam or deck (psi), and
I = moment of inertia of beam or deck (in.^{4}).

The value of E frequently used in design calculations is the tabular design value, which is an average E of a grade of lumber or laminating combination. If the E at the lower fifth percentile is desired for use in Eq. (4-2), Eq. (4-1) may be used.

It is noted that as the term $W'l^3/\pi^4EI$ approaches unity, the magnification factor approaches infinity and the deflection and bending stresses are increased indefinitely. The derivation of this magnification factor was based on an approximate analysis verified closely by experiment (12). The analysis assumed elastic behavior and did not account for stresses or deflections caused by creep. The factor applies to each element of a roof system, including decking, purlins, beams, and girders. Design values as modified for duration of loading, size factor, and other applicable factors, and deflection limits may not be exceeded after application of the magnification factor to existing stresses and deflections.

As a further illustration of the potential effect of ponding as it relates to increasing bending stress and deflection, Fig. 4.1 represents a graphic presentation of the magnification factor as a function of span length, deflection criteria, and ratio of ponding load to design load. By examining this plot, it can readily be seen that for a given span and deflection criteria, roofs designed for a relatively low value of dead load plus applied live load (i.e., a high W'/W ratio, where W' is as previously defined and W is the total design dead load plus live load) will be most susceptible to possible ponding problems.

The use of the magnification factor is a simplified approach that should be used with judgment. It is valid only for roofs and single members that are flat and do

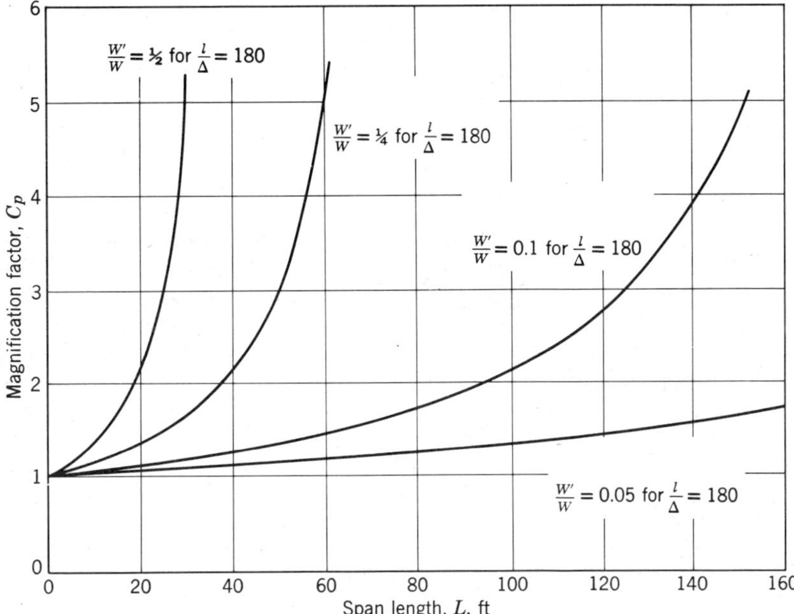

FIGURE 4.2 Magnification factor, C_p, relating deflection or bending values under ponding to values without ponding, for simply supported beams under initial uniformly distributed load.

not contain camber (disregarding camber gives a conservative answer). It disregards the deflection of secondary members. The magnification factor will increase stresses caused by dead and live loads assumed to be acting at the time ponding can occur. There are no uniformly recognized criteria for determination of loads to be magnified, and the designer should make a reasonable judgment as to what snow, water, or other load is to be included. If any conservatism is obtained by using full design load, it could be counterbalanced to some extent by this omission. Theoretically, all subsystems in the roof should be investigated. This is a very laborious calculation, and methods making use of computers are suggested where greater accuracy is desired.

REFERENCES

1. American National Standards Institute, *Minimum Design Loads for Buildings and Other Structures*, ANSI A58.1, New York, 1982.
2. International Conference of Building Officials, *Uniform Building Code*, Whittier, CA, 1982.
3. Seismology Committee of the Structural Engineers Association of California, *Recommended Lateral Force Requirements*, San Francisco, CA, 1984.
4. Applied Technology Council, *Tentative Provisions for the Development of Seismic Regulations for Buildings*, ATC 3-06, amended, Berkeley, CA, 1978.
5. American Association of State Highway and Transportation Officials, *Standard Specifications for Highway Bridges*, Washington, DC, 1983.

6. American Railway Engineering Association, *Manual of Recommended Practice*, Washington, DC, 1969.

7. United States Department of Agriculture, Forest Service, Forest Products Laboratory, *Fatigue Resistance of Quarter-Scale Bridge Stringers in Flexure and Shear*, Report No. 2236, Madison, WI, 1962.

8. American Society of Civil Engineers, "Design Considerations for Fatigue in Timber Structures," *Journal of Structural Division*, ASCE Paper No. 2470, Vol. 86, pp. 15–23, New York, 1960.

9. American Society of Civil Engineers, *Design of Structures to Resist Nuclear Weapons Effects*, Manual No. 42, 1985.

10. National Forest Products Associations, *National Design Specification for Wood Construction*, Washington, DC, 1982.

11. Haussler, R. W., *Roof Deflection Caused by Rainwater Pools*, Civil Engineering, ASCE, Vol. 32, No. 10, October 1962, pp. 58–59.

12. E. W. Kuenzi and B. Bohannan, "Increases in Deflection and Stresses Caused by Ponding of Water on Roofs," *Forest Products Journal*, **14**(9), 421–424 (September 1964).

SECTION 5

DESIGN OF STRUCTURAL ELEMENTS

INTRODUCTION

This section covers the basic types of structural members as well as several of the more common wood structural systems. Under the headings herein are contained general information on the design features of the member or system, tabular data, a typical design procedure, and, in most cases, a design example.

The procedures presented have been found to be most suited to the particular member or system in question; however, other procedures may be used if substantiated by tests or by sound engineering principles. The examples are illustrative only. In actual designs, all the conditions that might be expected to have a bearing on the ability of the member or structure to support all anticipated loads safely should be considered.

In general, in the example calculations no rounding was done during intermediate steps to obtain the final answer. For purposes of illustration, intermediate steps were recorded using rounded figures, which may result in slightly different answers when the rounded intermediate figures are used.

Abbreviations and Symbols

The abbreviations and symbols used in this section are contained in the List of Symbols immediately following the Contents. Deviations from these notations are identified where they occur.

TABULAR DESIGN VALUES

The tabular design values in Table 7.3 for sawn lumber, AITC 117—Design (1) for glued laminated timber of softwood species and AITC 119(2) for glued laminated timber of hardwood species, are defined as the maximum allowable stresses based on standard conditions as set forth in the tables. These tabular design values are designated as F_b for bending, F_c for compression parallel to grain, F_v for shear, and so on. The symbol F denotes a tabular design value for a given strength property and the lowercase subscript denotes the strength property. For some conditions, these values must be modified before they can be used in design. The modifiers are denoted by C and a subscript letter indicating the reason for modification, such as C_F for size factor and C_D for duration of load. Some of these modifiers are applicable to all strength properties while others are applicable to specific strength properties; for example, the curvature factor C_C is applicable to bending in curved members only.

When all applicable modifiers have been applied to the tabular design value, the allowable value to use in design is denoted by F', such as F'_b for bending, F'_c for compression parallel to grain, and F'_v for horizontal shear.

Calculations involving columns and beams in the intermediate unsupported length range frequently require an intermediate design value which is the tabular design value adjusted by all applicable modifiers except the modification for slenderness or size. These values are designated as F'', such as F''_c and F''_b. They are used only as interim values in calculations.

The actual unit stresses of a member under load are denoted by the same letters as the tabular design value, except that a lowercase f if used, such as f_b, f_c, f_v.

Glued Laminated Timber

The tabular design values for glued laminated timber in AITC 117—Design (1) and AITC 119 (2) are based on dry conditions of use, a depth of 12 in., and loads of normal duration. These values are based on an average moisture content of 12%. The modifying factors for moisture content for glued laminated timber are different than for sawn lumber, which has several bases for moisture content modifications. For glued laminated timber, moisture content factors are listed at the bottoms of the columns of design values in Tables 1 and 2 of AITC 117—Design (1) and in Table 5.1.

Glued laminated timbers may have different tabular design values when loaded in bending about the *x–x* axis and the *y–y* axis depending on the grades and arrangement of grades of lumber within the member. Therefore, Tables 1 and 2 in AITC 117—Design (1) have strength properties listed for loading in bending about both the *x–x* and *y–y* axes. The tabular design values when bending is about the *x–x* axis are denoted by the subscript *x* such as F_{bx}, F_{vx}, and E_x. For bending about the *y–y* axis, the subscript *y* is used. When no subscripts are used, the tabular design values about the *x–x* axis are assumed. For clarity in specifying, it is recommended that the subscripts be used, particularly when the member is loaded in bending about the *y–y* axis. Note that the subscripts *xx* and *yy* are sometimes used to indicate stresses about the *x–x* and *y–y* axes. The single-letter subscript is the preferred form.

Sawn Lumber

Sawn lumber is available as visually graded lumber in all sizes and as machine stress-rated (MSR) lumber with 2 in. nominal thickness. The design values for visually graded lumber are contained in Table 7.3. These design values were taken from *National Design Specification (NDS)* (3) and were current at the time of publication of this manual. For current information on grades and design values, the designer is referred to the latest NDS or the governing building code, which may include the applicable tabular design values.

Modifiers to tabular design values are not always the same for visually graded lumber and MSR lumber and are listed separately in Tables 5.1 and 5.2.

TABLE 5.1

Moisture Content Factors C_M[a]

Strength Property	F_b	F_t	F_c	$F_{c\perp}$	F_v	F_g	E	F_{rt}
GLUED LAMINATED TIMBER—ALL SPECIES								
dry condition of use (MC 16% or less)	1.00	1.00	1.00	1.00	1.00	1.00	1.00	1.00
wet conditions of use (MC greater than 16%)	0.80	0.80	0.73	0.53	0.875	0.57	0.833	0.875
VISUALLY GRADED SAWN LUMBER—ALL SPECIES EXCEPT SOUTHERN PINE OR VIRGINIA PINE-POND PINE								
Sawn lumber all thicknesses surfaced dry or surfaced green and used at 19% maximum MC	1.00	1.00	1.00	1.00	1.00	[e]	1.00	
Sawn lumber—nominal 4 in. or less in thickness surfaced dry or green and used at MC greater than 19%	0.86	0.84	0.70	0.67	0.97	[e]	0.97	
Sawn lumber—nominal 4 in. or less in thickness surfaced at 15% maximum MC and used at MC of 15% or less[d]	1.08	1.08	1.17[b]	1.00	1.05	[e]	1.05[c]	
Sawn lumber—nominal 5 in. and thicker used where MC exceeds 19%	1.00	1.00	0.91	0.67	1.00	[e]	1.00	
VISUALLY GRADED SAWN LUMBER—SOUTHERN PINE OR VIRGINIA PINE–POND PINE								
Sawn lumber—nominal 4 in. or less in thickness surfaced at 15% maximum MC and used at 15% maximum MC	1.00	1.00	1.00	1.00	1.00	1.00	1.00	
Sawn lumber—nominal 4 in. or less in thickness surfaced at 15% maximum MC and used where MC exceeds 15% but less than 19%	Use tabular values for corresponding grades surfaced dry, used at 19% maximum MC							
Sawn lumber—nominal 4 in. or less in thickness surfaced at 15% maximum MC and used where MC exceeds 19%	Use tabular values for corresponding grades of nominal $2\frac{1}{2}$ to 4-in.-thick Southern Pine surfaced green, used at any condition							

(*continued*)

TABLE 5.1 (*Continued*)

Strength Property	F_b	F_t	F_c	$F_{c\perp}$	F_v	F_g	E	F_{rt}
Sawn lumber—nominal 4 in. or less in thickness surfaced dry and used at 19% maximum MC	1.00	1.00	1.00	1.00	1.00	1.00	1.00	
Sawn lumber—nominal 4 in. or less thickness surfaced dry and used where MC exceeds 19%	Use tabular values for corresponding grades of nominal $2\frac{1}{2}$ to 4-in.-thick Southern Pine surfaced green, used at any condition							
Sawn lumber—all thicknesses, surfaced green and used at any moisture content condition	1.00	1.00	1.00	1.00	1.00	1.00	1.00	
MACHINE STRESS-RATED SAWN LUMBER								
All species used where MC in use is 19% or less, 2 in. thick or less	1.00	1.00	1.00	1.00	1.00	1.00	1.00	
All species used where MC in use exceeds 19%, 2 in. thick or less	0.86	0.84	0.70	0.67	0.97	[e]	0.97	

[a]See page 2-32 for discussion of moisture content (MC).

[b]For California Redwood only, use 1.15.

[c]For California Redwood only, use 1.04.

[d]When decking graded to WWPA rules is surfaced at 15% maximum MC and used where the MC will exceed 15% for an extended period of time, the tabulated design values for decking surfaced at 15% maximum MC shall be multiplied by the following factors: Extreme fiber in bending F_b, 0.79; modulus of elasticity E, 0.92.

[e]Use tabulated values for wet-service conditions in Table 3.1.

The tabular design values for visually graded lumber are based on various moisture contents and conditions of service depending on the size of the lumber and in some cases on the species. Therefore, the moisture content factor C_M varies as is shown in Table 5.1. The effect of size is discussed on page 5-105. The fire-retardant treatment factor C_R should be obtained from the manufacturer of the treatment.

Visually graded lumber is manufactured with various moisture contents. The design values in Table 7.3 are listed under headings that indicate the moisture content at time of surfacing and the moisture content in use. Lumber with nominal sizes of 4 in. and less in thickness can be dried prior to surfacing or it can be surfaced in the green state. Usually dried lumber is available in thicknesses of 2 in. nominal and less. Lumber with nominal sizes over 4 in. in thickness usually

TABLE 5.2

Size Factor for Lumber Used Flatwise, C_F

VISUALLY GRADED DIMENSION LUMBER USED FLATWISE

Nominal Width	Nominal Thickness (in.) 2	3	4
2 to 4 in.	1.10	1.04	1.00
5 in. and wider	1.22	1.16	1.11

MSR LUMBER USED FLATWISE[a]

Nominal width, in.	3	4	5	6	8	10	12	14
Size factor, C_F	1.06	1.10	1.12	1.15	1.19	1.22	1.25	1.28

[a]Based on nominal 2-in.-thick lumber.

is not kiln dried (KD) but occasionally is air dried prior to surfacing. When lumber 4 in. and less in nominal thickness is surfaced green, the surfaced size is slightly larger than lumber surfaced dry to compensate for the additional shrinkage that is to be expected in normal use.

The tabular design values in Table 7.3 are based on the following moisture conditions:

1. All species except Southern Pine, Virginia Pine-Pond Pine.

(a) Surfaced dry or surfaced green used at 19% maximum moisture content (identified by footnote *a* in Table 7.3). The lumber that has been surfaced dry (nominal 4 in. or less in thickness) has been conditioned by kiln drying (or in some cases by air drying to a specified moisture content). The moisture content of the lumber that has been surfaced green is approximately the moisture content of the log at the time of sawing and may vary considerably. Usually it is close to the fiber saturation point.

In both cases, the design values set are applicable for lumber used where the equilibrium moisture content does not exceed 19%.

When lumber of 4 in. nominal thickness or less is used where the moisture content will exceed 19% for an extended length of time, the tabular design values are reduced by the modifying factors C_M shown in the footnotes to Table 7.3 or in Table 5.1.

When lumber of nominal 5 in. thickness and greater (designated as "Beams and stringers" or "Posts and timbers" in Table 7.3) is used where the moisture content exceeds 19% for an extended period of time, the tabular design values for compression perpendicular and parallel to grain are adjusted by use of the modifying factors C_M as shown in the footnotes to Table 7.3 or in Table 5.1.

(b) Surfaced at 15% maximum moisture content (MC 15). This condition is used for decking where the equilibrium moisture content in use does not exceed 15%. The values apply to lumber of 4 in. nominal thickness and less and is probably not available in thicknesses greater than 2 in. nominal. The tabular design values shown in Tables 7.3 and 7.4 may be adjusted by application of the modifying factor C_M shown in the footnotes to Table 7.3.

2. Southern Pine, Virginia Pine-Pond Pine. The tabular design values for these species in Table 7.3 are based on different moisture content considerations. Theses are:

(a) Surfaced at 15% maximum moisture content (KD 15). This condition is used where the equilibrium moisture content does not exceed 15% (applies to lumber of 4 in. nominal thickness or less) (identified by footnote *u* in Table 7.3). When this lumber is used where the equilibrium moisture content exceeds 15% for an extended period of time, lower design values must be used. When this lumber is used where the equilibrium moisture content exceeds 15% but does not exceed 19%, use the appropriate design values, identified by footnote *v*. When this lumber is used where the equilibrium moisture content exceeds 19% for an extended period of time, the tabular design values for corresponding grades of nominal $2\frac{1}{2}$ to 4-in.-thick surfaced green Southern Pine should be used (identified by footnote *w*).

(b) Surfaced dry, or KD 19, used where the equilibrium moisture content does not exceed 19% (identified by footnote *v* in Table 7.3). This requirement applies to lumber of 4 in. nominal thickness or less. When this lumber is used where the equilibrium moisture content exceeds 19% for an extended period of time, the tabular design values for corresponding grades of nominal $2\frac{1}{2}$ to 4-in.-thick surfaced green Southern Pine should be used.

(c) Surfaced green, used under any condition (identified by footnote *w* in Table 7.3). This requirement applies to all thicknesses and the tabular design values apply for any equilibrium moisture content reached in service.

Table 5.1 summarizes the moisture content modification factors from the footnotes to Table 7.3 for all species.

Machine stress-rated lumber is commercially available in nominal 2-in.-thick lumber in all widths. The designer, however, should check on availability of the wider widths prior to specifying. Machine stress-rated lumber is available in a number of grade designations. Some of these grades are available in most species, but others are restricted to certain species as indicated in Table 7.4. Design values for compression perpendicular to grain, $F_{c\perp}$, and horizontal shear F_v are species dependent and are the same as those listed in Table 7.3 for No. 2 grade of the species for visually graded lumber, except for mixed species for which design values are listed in the footnotes to Table 7.4 for MSR lumber. When MSR lumber is used under conditions where the equilibrium moisture content does not exceed 19%, the modifier for moisture content C_M is 1 for all properties. When it is used under conditions where the equilibrium moisture content is greater than 19%, the values of C_M in the footnotes to Table 7.4 or in Table 5.1 should be used.

Modifications to Tabular Design Values

Some of the modifiers such as duration of load are applicable to all strength properties (except modulus of elasticity and compression perpendicular to grain), but other modifiers may have different values for different strength properties or are applicable only to a specific property. In general, they are cumulative, but in some cases the more restrictive of two modifiers apply such as the size factor C_F and the lateral stability of beams factor C_L for bending members.

Duration-of-Load Factor C_D

Duration-of-load factor C_D is applied to tabular design values of all strength properties and fastening values except the modulus of elasticity and compression perpendicular to grain. See page 3-65 for additional information.

Size Factor C_F

Sawn Lumber. The tabular design values in bending for visually graded dimension lumber 2–4 in. thick in Table 7.3 are based on edgewise loading. For this lumber and loading condition, the effect of size has been included in the tabular design value and no adjustment for size is required. When sawn members over 4 in. thick are used as bending members over 12 in. deep, the size factor is applied as a reduction to the tabular design value in bending. When sawn members (except for decking) are loaded flatwise, the design values in bending may be increased to account for the size factor as indicated by footnote *g* of Table 7.3. (See Table 5.2).

The size factor does not apply to MSR lumber except as provided for flatwise bending in footnote *g* to Table 7.4.

Glued Laminated Timber. The size factor is applied as a reduction to tabular design values for bending about the *x–x* axis, F_{bx}, for members over 12 in. deep. For bending about the *y–y* axis, the design values F_{by} may be increased to account for the size factor as indicated by footnote *r* to Table 1 and footnote *f* to Table 2 of AITC 117—Design (1). This increase in F_{by} is cumulative with the lateral stability of beams factor C_L.

See pages 3-65 and 3-72 for additional information on size factor. When the size factor is less than 1, it is not cumulative with the lateral stability factor, C_L, because the size factor relates to a reduction in tensile strength on the tension side of a bending member and the lateral stability factor relates to a reduction in strength on the compression side of a bending member due to buckling. It is also not cumulative with the interaction stress factor when the tapered cut is on the compression side of the member. However, it is cumulative when the tapered cut is on the tension side of the member. (Taper cuts on the tension side of glued laminated timber bending members are not recommended.)

Lateral Stability of Beams Factor C_L

Lateral stability of beams factor C_L is applied to the tabular design value in bending of intermediate beams. It is not cumulative with the size factor (except when C_F is unity or greater) in all beams or the interaction stress factor in tapered beams when the tapered cut is on the tension side of a bending member. (Taper cuts on the tension side of glued laminated timber bending members are not recommended.) For short beams, the lateral stability of beams factor is unity and is not customarily used in calculations. The design value of long beams is determined directly by calculation and C_L is not ordinarily used for long beams. See page 5-164.

Lateral Stability of Columns Factor C_P

This factor is applied to the tabular design values for compression parallel to grain of all members loaded in compression that are classified as intermediate columns. For short columns, the lateral stability of columns factor is unity and is not ordinarily used in calculations. The design value for long columns is determined directly by calculation, and C_P is not ordinarily used for long columns. See page 5-113 for additional details.

Curvature Factor C_C

The curvature factor is a modifier of the design value in bending for curved members only. See page 3-74 for additional information for its use. It takes into account the difference in extreme outer fiber stress between a curved member and a straight prismatic member as well as any residual stresses that may remain in a lamination that has been bent to the stated curvature. It is not used in straight beams or cambered beams because it has very little effect at the radii of curvature used for cambering. It is not applied in the design of double-tapered pitched and curved beams because this effect is accounted for in the K_θ factor. The effect is less than 1% for members with $1\frac{1}{2}$-in.-thick laminations and a radius of 56 ft or more.

Moisture Content Factor C_M

The moisture content factor is applied to all tabular design values for strength and modulus of elasticity. It varies for different strength properties and conditions of use and is tabulated at the bottom of each column of Tables 1 and 2, AITC 117—Design (1).

Fire-Retardant Treatment Factor C_R

The fire-retardant treatment factor applies to all tabular design values including modulus of elasticity. It also applies to tabular design values for fastenings. The effect of fire-retardant treatment varies for both sawn lumber and glued laminated timber. The manufacturer of the treatment should be contacted for specific information on the value of C_R for all design values.

Interaction Stress Factor C_I

The interaction stress factor applies to the bending stress of taper cut glued laminated timber bending members. It is not applied to arches or double-tapered pitched and curved beams. When the tapered cut is on the compression side of a bending member, it is not cumulative with the size factor C_F. However, it is cumulative when the taper cut is on the tension side. (Taper cuts on the tension side of beams are not recommended.) It is cumulative with the lateral stability of beams factor C_L when the taper cut is on the compression side. See page 5-210 for additional information.

Form Factor C_f

The form factor applies only to the tabular design values in bending for round wood members and rectangular or square members loaded on a diagonal. It is cumulative with all other applicable modifiers, including size factor. See page 3-68 for additional details.

Temperature Factor C_t

Most wood is used under conditions where temperature is not a significant factor. When an elevated temperature of a member is anticipated for an extended period of time or when elevated temperatures are expected to occur simultaneously with maximum design loads, a reduction in design values may be necessary. See page 3-69 for additional information.

Other Modifiers

The preceding discussion covers modifiers usually applicable to timber design. For special cases such as poles and piles, other modifiers may be necessary.

Application of Modifiers

The following examples illustrate the application of modifiers to various design values:

Bending $F'_b = F_b C_D C_F C_L C_C C_M C_R C_I C_f C_t$
Tension parallel to grain $F'_t = F_t C_D C_M C_R C_t$*
Compression parallel to grain $F'_c = F_c C_D C_P C_M C_R C_t$
Compression perpendicular to grain $F'_{c\perp} = F_{c\perp} C_M C_R C_t$
Horizontal Shear $F'_v = F_v C_D C_M C_R C_t$
End Grain in Bearing $F'_g = F_g C_D C_M C_R C_t$
Modulus of Elasticity $E' = E C_M C_R C_t$
Radial Tension $F'_{rt} = F_{rt} C_D C_M C_R C_t$

The symbol for the modified design value is shown with a prime (such as F'_b) and the tabular value is shown without the prime (such as F_b). In many cases, the modifying factors are unity and are usually not shown in design calculations. It is customary to show only those applicable modifiers that change the tabular design values (see Table 5.3).

Example. Bending. A straight simply supported glued laminated beam 24 in. deep is laterally supported along the top and at the ends. It is used in a dry location at normal temperature to support a snow load. The dead load is less than 0.90/1.15 times the total load, therefore the duration of load for snow controls. The beam will not be fire-retardant treated.

$$F'_b = F_b C_D C_F C_L C_C C_M C_R C_I C_f C_t$$

$$C_L,\ C_C,\ C_M,\ C_R,\ C_I,\ C_f, \text{ and } C_t = 1$$

$$C_D = 1.15 \text{ (Table 3.2);}$$

$$C_F = 0.93 \text{ (Table 7.2); and}$$

$$F'_b = F_b(1.15)(0.93) = 1.07 F_b.$$

Example. A simply supported curved beam 12 in. deep is laterally supported along the top and at the ends. The radius of curvature, R, is 35 ft and the thickness of laminations, t, is 1.5 in. It is used in a wet location at normal temperature. The

*For sawn lumber, footnote *d* to Table 7.3 also applies.

TABLE 5.3

Applicability of Modification Factors

Modifying Factor	Design Value							
	F_b	F_t	F_c	$F_{c\perp}$	F_v	F_g	E	F_{rt}
Duration of load, C_D	Yes	Yes	Yes	No	Yes	Yes	No	Yes
Size, C_F[a]	Yes	No	No	No	No	No	No	No
Lateral stability of beams, C_L[a]	Yes	No	No	No	No	No	No	No
Lateral stability of columns, C_p	No	No	Yes	No	No	No	No	No
Curvature, C_C	Yes	No	No	No	No	No	No	No
Moisture content, C_M	Yes	Yes	Yes	Yes	Yes	Yes	Yes	Yes
Fire-retardant treatment, C_R	Yes	Yes	Yes	Yes	Yes	Yes	Yes	Yes
Interaction stress, C_I[a]	Yes	No	No	No	No	No	No	No
Form, C_f	Yes	No	No	No	No	No	No	No
Temperature, C_t	Yes	Yes	Yes	Yes	Yes	Yes	Yes	Yes

[a]Factors not always cumulative. See page 5-106.

maximum bending stress is caused by wind. The beam has been made from preservatively treated lumber to prevent decay.

$$F'_b = F_b C_D C_F C_L C_C C_M C_R C_I C_f C_t$$

$$C_F,\ C_L,\ C_R,\ C_I,\ C_f, \text{ and } C_t = 1$$

$$C_D = 1.33$$

$$C_C = 1 - 2000(t/R)^2 = 1 - 2000\,[1.5/(35)(12)]^2 = 0.97$$

$$C_M = 0.80 \text{ (from Table 1, AITC 117—Design)}$$

$$F'_b = F_b(1.33)(0.97)(0.80) = 1.03\,F_b$$

COLUMNS AND COMPRESSION MEMBERS

The term *column* is generally applied to all compression members including truss members, posts, or other structural components stressed in compression. Columns are divided into three general types consisting of (1) simple columns, (2) spaced columns, and (3) built-up columns. Simple wood columns consist of a single piece of sawn lumber, post, timber, pole, or glued laminated timber. Spaced columns consist of two or more individual members with their longitudinal axes parallel, separated at their ends and midpoints by blocking, and joined at the ends by connectors capable of developing the required shear resistance. Built-up columns consist of two or more pieces of lumber placed side by side and joined with mechanical fasteners.

Effective Column Length

Wood columns are further divided into three categories, depending on the ratio of the unbraced length to depth (l_e/d). They consist of short, intermediate, and long columns. The unbraced length of a column or compression member is the distance between two points along its length between which the member is assumed to buckle. For a laterally unsupported simple column with assumed pinned ends, the effective length l_e is equal to the total length of the column. For columns with different degrees of fixity at the ends or with intermediate lateral support, as illustrated in Table 5.4 and Figure 5.1, the effective length varies. The K_e value should be determined by good engineering judgment following the guidelines in Table 5.4. The effective buckling length factors K_e are shown as theoretical K_e values and recommended values of K_e to use for design. The recommended values of K_e for design take into account the lack of perfect fixity in use. However, when compression members depend on the rigidity of other in-plane members entering the joint to provide fixity and the combined stiffness of these members is relatively small compared to the unbraced column segments, K_e can exceed the values shown in Table 5.4. Determine the effective length by multiplying K_e by the length of the member between points of lateral support. Columns with intermediate bracing points may have different theoretical K_e factors.

TABLE 5.4

Effective Column Length for Various End Conditions[a]

Buckling Modes						
Theoretical K_e value	0.5	0.7	1.0	1.0	2.0	2.0
Recommended design K_e when ideal conditions approximated	0.65	0.8	1.2	1.0	2.1	2.4

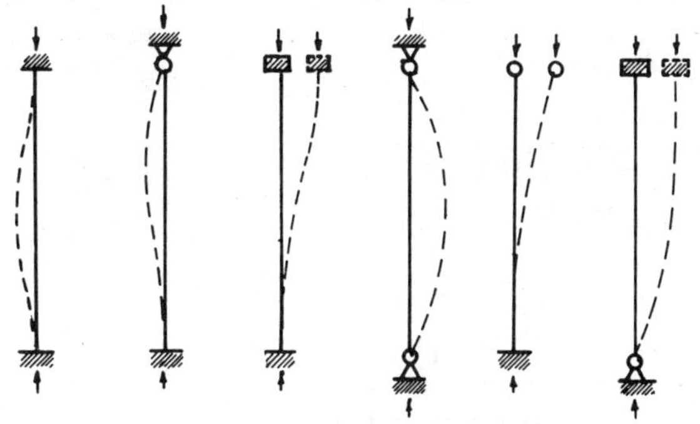

End condition code 
- Rotation fixed, translation fixed
- Rotation free, translation fixed
- Rotation fixed, translation free
- Rotation free, translation free

[a]Source: *National Design Specification* (3).

Column End Conditions

The column formulas are based on pin end conditions with the ends fixed against translation. They assume that the column buckling curve approximates the shape of a sine wave. These formulas are commonly used for columns with square cut ends, which is a less critical condition than the pin end condition. Conditions are sometimes encountered in design where the restraint is either more or less than the standard conditions.

Simple Solid Rectangular Columns

Short Columns

Short columns are those whose strength is controlled by the compression parallel to grain strength. When tested to failure, they fail in crushing. The lateral stability of columns factor C_P for short columns is 1. For short rectangular columns, the length–depth ratio l_e/d is 11 or less. The design value for a short column is equal to the tabular design value in compression parallel to grain modified for duration of load, moisture, and so on, as explained on pages 5-104 to 5-107.

$$F'_c = F_c C_D C_M C_R C_t$$

Long Columns

Long columns are columns controlled by stiffness and the strength is thus a function of the modulus of elasticity. When tested to failure, they fail in buckling. The l_e/d ratio of a long solid column is equal to K or greater but is limited to a maximum of 50 for a simple solid column. See page 5-112 for the definition of K.

The Euler formula is used for the design of long columns. The basic Euler formula is used to determine the ultimate buckling strength, f_{cr}, of a solid pin-ended column:

$$f_{cr} = \frac{\pi^2 E}{(l_e/r)^2} \tag{5-1}$$

where E = true bending modulus of elasticity (psi),
l_e = unsupported column length (in.), multiplied by the effective buckling length factor K_e, and
r = radius of gyration (in.).

For design of solid wood columns made from visually graded lumber, E is reduced by dividing by 2.74. According to *NDS* (3), for visually graded sawn lumber, the 2.74 reduction factor for modulus of elasticity represents an approximate 5% lower exclusion value (5th percentile) on pure bending modulus of elasticity, which results in a 1.66 factor of safety. For glued laminated timber and MSR lumber, the 2.74 combined reduction factor represents less than a 0.01% lower exclusion value with a 1.66 factor of safety. The design values for sawn lumber obtained from the column formulas for short, intermediate, and long columns are those usually used for most construction. Where unusual hazards exist, a larger reduction factor may be appropriate. The basic Euler equation reduced by the 2.74 factor results in the equation

$$F'_c = \frac{3.6E'}{(l_e/r)^2} \tag{5-2}$$

where f'_c = design value in compression parallel to grain (psi),
E' = tabular modulus of elasticity (psi) adjusted by applicable modifiers,
and l_e and r are as previously defined.

For rectangular shaped columns, it is more convenient to use the least dimension of the column (d) rather than the radius of gyration. When there are different unbraced lengths for the different axes of buckling (as shown in Figure 5.1), the larger l_e/d ratio should be used in calculation of F'_c. In the long-column formula used for design, $d/\sqrt{12}$ is substituted for the radius of gyration resulting in the equation

$$F'_c = \frac{0.30E'}{(l_e/d)^2} \tag{5-3}$$

where E' = tabular modulus of elasticity (psi) adjusted by applicable modifiers,
l_e = unsupported length of column multiplied by effective buckling length factor K_e (see Table 5.4) (in.), and
d = dimension of column in plane of assumed buckling (in.).

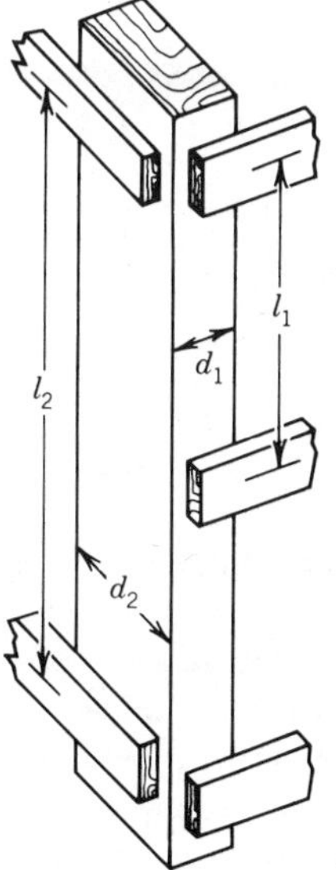

FIGURE 5.1 Simple solid column [from *National Design Specification* (3)]. l_1 and l_2 are the distances between points of lateral support of column in planes 1 and 2 (in.); d_1 and d_2 are the dimensions of column in planes of lateral support, (in.).

The minimum l_e/d ratio at which a rectangular column can be expected to perform as a long column is where F'_c, determined by the long-column formula, is approximately two-thirds of the tabular design value in compression parallel to grain adjusted by applicable modifiers. This ratio is denoted as K and is determined by the equation

$$K = 0.671\sqrt{\frac{E'}{F''_c}} \tag{5-4}$$

where E' = tabular modulus of elasticity (psi) adjusted by applicable modifiers and

F''_c = tabular design value in compression parallel to grain (psi) adjusted by applicable modifiers. (No modification is included for l_e/d ratio. This value is equal to the design value in compression parallel to grain for a short column.)

For columns of other than rectangular shape where the radius of gyration r is used rather than the least dimension d, K determined by the Eq. (5-4) should be increased by multiplying by $\sqrt{12}$.

The tabular design value of modulus of elasticity E for wood is the average value for the grade and species or laminating combination. The coefficient of variation of E is approximately 0.25 for visually graded sawn lumber, 0.11 for MSR lumber, and 0.10 for glued laminated timber with six or more laminations.

For the usual design of long columns, the designer is encouraged to use Eq. (5-3). For columns with less variability than visually graded lumber, such as glued laminated timber and MSR lumber, the designer may wish to use Eqs. (5-5) and (5-6) from Appendix G of *NDS* (3). These formulas provide the same factor of safety at the 5% exclusion value as do the above formulas, which are applicable to visually graded sawn lumber.

$$F'_c = \frac{0.418E'}{(l_e/d)^2} \tag{5-5}$$

$$K = 0.792\sqrt{E'/F''_c} \tag{5-6}$$

Intermediate Columns

For many years, designers have customarily designed columns using only the short- and long-column formulas. The design values calculated by use of the long-column formula in the intermediate range can be as much as 10% higher than those calculated by the intermediate-column formula. In 1977, the National Forest Products Association (3) recommended that the intermediate-column formula be used to replace the simplified method.

If the compressive strength of a column were infinite, the Euler formula could be used for all lengths of columns. However, in wood and other materials, as the ratio l_e/d decreases, a point is reached where the compressive strength is less than that indicated by the Euler formula. When l_e/d decreases to K, the compressive strength begins to influence the ultimate strength and from less than K to an l_e/d of approximately 11, both compressive strength and modulus of elasticity affect the strength. Columns in this range are called intermediate columns. Their design values are obtained by the formula

$$F'_c = F''_c\left[1 - \frac{1}{3}\left(\frac{l_e/d}{K}\right)^4\right] \tag{5-7}$$

where F'_c = design value for compression parallel to grain of an intermediate column (psi),

F''_c = tabular design value in compression parallel to grain (psi) adjusted by applicable modifiers. (No modification is included for l_e/d ratio. This value is equal to the design value in compression parallel to grain for a short column.)

l_e = unsupported length of column multiplied by effective buckling length factor K_e (in.) (see Table 5.4),

d = dimension of column in plane of assumed buckling (in.), and

K = minimum value of l_e/d at which column can be expected to perform as Euler column [see Eq. (5-4) for K].

Figure 5.2 displays graphically compression parallel to grain, F'_c, for an axially loaded member for one grade of lumber under one condition of use. The curves for other grades and species are similar. Note that the transition of the design curves between long columns and intermediate columns is continuous. However, there is a discontinuity between the transition from the intermediate columns to the short columns at $l_e/d = 11$, which by definition is the upper limit of a short column. This amounts to a difference of approximately 5% in the design values between intermediate and short columns at l_e/d of 11. The definition of an intermediate column includes those columns with an l_e/d greater than 11 but less than K.

The intermediate-column formula [Eq. 5-7] is presented in this form in *NDS* (3) and standard textbooks. In order to follow a consistent rational procedure for application of modifying factors, the portion of the intermediate formula

$$\left[1 - \frac{1}{3}\left(\frac{l_e/d}{K}\right)^4\right]$$

can be designated as the lateral stability of columns factor C_P. The design value of an intermediate column can then be determined by applying the applicable modifiers directly to the tabular design value

$$F'_c = F_c C_D C_P C_M C_R C_t$$

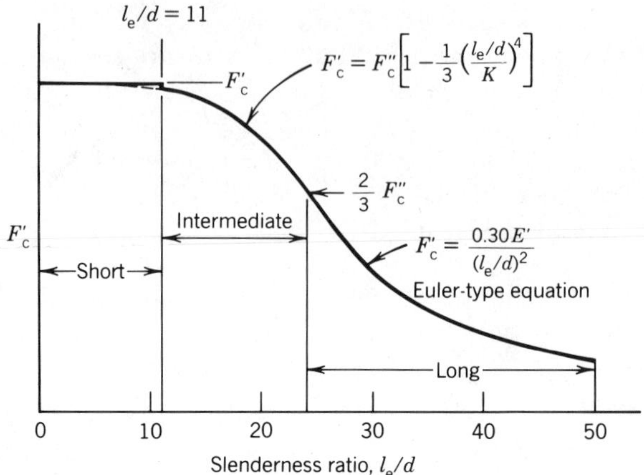

FIGURE 5.2*a* Basic provisions for concentrically loaded columns.

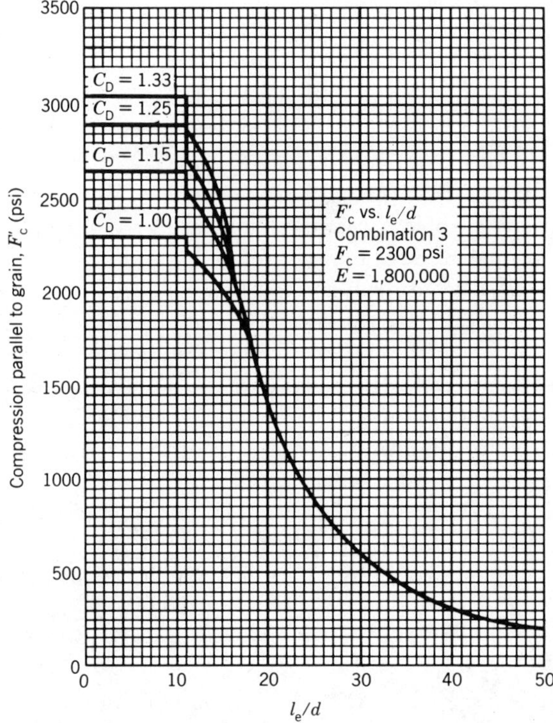

FIGURE 5.2*b* Compression parallel to grain versus l_e/d.

where

$$C_P = \left[1 - \frac{1}{3}\left(\frac{l_e/d}{K}\right)^4\right] \tag{5-8}$$

and other terms are as defined previously. Alternatively, F'_c may be expressed as shown in Eq. (5-7).

Example. Determine the axial snow load capacity P of a 12-ft-long 6 × 6 in. No. 1 Hem-Fir Posts and Timbers grade column. The column is located in a dry location. The top and bottom of the column are held to prevent translation and no lateral support is provided along the length.

$$F_c = 850 \text{ psi}, \; F''_c = F_c C_D = (850)(1.15) = 978 \text{ psi}$$

$$E = E' = 1{,}300{,}000 \text{ psi}$$

$$A = 30.25 \text{ in.}^2$$

$$l_e/d = \frac{(12)(12)}{5.5} = 26.2$$

$$K = 0.671\sqrt{\frac{E'}{F''_c}} = 0.671\sqrt{\frac{1{,}300{,}000}{978}} = 24.5$$

Since $50 > l_e/d > K$, use the long-column formula:

$$F'_c = \frac{0.30E'}{(l_e/d)^2} = \frac{(0.30)(1{,}300{,}000)}{(26.2)^2} = 569 \text{ psi}$$

$$P = AF'_c = (30.25)(569) = 17{,}200 \text{ lb}$$

Check the same-size column with a length of 10 ft:

$$l_e/d = \frac{(10)(12)}{5.5} = 21.8$$

Since $11 < l_e/d < K$, use the intermediate-column forumula:

$$F'_c = F''_c\left[1 - \frac{1}{3}\left(\frac{l_e/d}{K}\right)^4\right] = 978\left[1 - \frac{1}{3}\left(\frac{21.8}{24.5}\right)^4\right] = 772 \text{ psi}$$

$$P = AF'_c = (30.25)(772) = 23{,}300 \text{ lb}$$

Check the same-size column with a length of 4 ft 6 in.:

$$l_e/d = \frac{(4.5)(12)}{5.5} = 9.8$$

Since $l_e/d < 11$, use the short-column formula:

$$F'_c = F''_c = 978 \text{ psi}$$

$$P = AF'_c = (30.25)(978) = 29{,}600 \text{ lb}$$

Example. Determine the axial snow load capacity P of a 20-ft-long $6\frac{3}{4} \times 9$-in. column made from combination 16F-V1 SP/SP. (Note that Table 2, AITC 117—Design combinations are usually used for columns, but this combination was selected to illustrate the application of E_x and E_y.) The top and bottom of the column are held to prevent translation, and lateral support is provided at midheight about the y axis. The column is used in a dry location. From Table 1, AITC 117—Design, the following design values are obtained:

$$F_c = 1450 \text{ psi}, \; F_c'' = F_c C_D = (1450)(1.15) = 1668 \text{ psi}$$

$$E_x = E_x' = 1{,}400{,}000 \text{ psi}$$

$$E_y = E_y' = 1{,}300{,}000 \text{ psi}$$

$$A = (6.75)(9) = 60.75 \text{ in.}^2$$

From Table 5.4, $K_e = 1$.

$$(l_e/d)_x = \frac{(20)(12)}{9} = 26.67$$

$$(l_e/d)_y = \frac{(10)(12)}{6.75} = 17.78$$

$$K_x = 0.671\sqrt{\frac{E_x'}{F_c''}} = 0.671\sqrt{\frac{1{,}400{,}000}{1668}} = 19.44$$

$$K_y = 0.671\sqrt{\frac{E_y'}{F_c''}} = 0.671\sqrt{\frac{1{,}300{,}000}{1668}} = 18.74$$

Where $50 > (l_e/d)_x > K_x$, use long-column formula; where $11 < (l_e/d)_y < K_y$, use intermediate-column formula. For buckling about the x axis, check strength using the long-column formula

$$F_c' = \frac{0.30E_x'}{(l_e/d)_x^2} = \frac{(0.30)(1{,}400{,}000)}{(26.67)^2} = 590 \text{ psi}$$

For buckling about the y axis, check strength using the intermediate-column formula

$$F_c' = F_c''\left[1 - \frac{1}{3}\left(\frac{(l_e/d)_y}{K_y}\right)^4\right]$$

$$= 1668\left[1 - \frac{1}{3}\left(\frac{17.78}{18.73}\right)^4\right] = 1220 \text{ psi}$$

Buckling about the x axis controls, and the capacity is

$$P = AF_c' = (60.75)(590) = 35{,}880 \text{ lb}$$

Round Columns

A round column is a simple solid column with a circular cross section. Its load-carrying capacity equals that of a square column with the same cross-sectional area. For design purposes, it is usually convenient to determine the required size of a square column and use a circular column of the same cross-sectional area. The diameter of the equivalent round column, D, is $1.128d$, where d is the dimension of a side of a square column with the same cross-sectional area. When this method is used, K is the same as that determined for a rectangular column. This results in a slightly conservative value for long and intermediate columns.

It is frequently more convenient to work directly with the diameter D of round columns rather than converting to equivalent square columns. Pole sizes are generally designated by the circumference, from which the diameter is readily obtained. A short column may also be defined as one in which the l_e/D ratio does not exceed 9.75. The basic Euler long-column formula converts into

$$F'_c = \frac{0.225E'}{(l_e/D)^2} \tag{5-9}$$

where E' = tabular value for modulus of elasticity (psi) adjusted by applicable modifiers,
l_e = unsupported length of the column multiplied by K_e (in.), and
D = diameter of a round column (in.).

The l_e/D ratio for a long column should not exceed 44. The value of K for round columns is designated as K_R and is determined by

$$K_R = 0.58\sqrt{\frac{E'}{F''_c}} \tag{5-10}$$

where E' and F''_c are as previously defined. The formula for an intermediate column with a circular cross section is

$$F'_c = F''_c\left[1 - \frac{1}{3}\left(\frac{l_e/0.866D}{K}\right)^4\right] = F''_c\left[1 - \frac{1}{3}\left(\frac{l_e/D}{K_R}\right)^4\right] \tag{5-11}$$

where F''_c, D, l_e, and K_R are as defined previously.

Tapered Columns

Tapered columns may be tapered from the larger cross section toward one end or both ends. They are designed as simple columns with the least dimension, d, of columns of rectangular cross section or the minimum diameter, D, of columns with round cross section. These dimensions are modified to determine the slenderness ratio, l_e/d or l_e/D, to use in design as follows: For rectangular columns tapered from one or both ends, the dimension, d, in each plane (face of column) is calculated by taking the sum of the minimum d plus one-third of the difference between the minimum and maximum d in that plane. The design value, F'_c, determined for the tapered column applies to the cross-sectional area corresponding to the dimension d used in the calculations of F'_c.

For tapered columns of circular cross section, the minimum diameter, D, to be used in design is taken as the smallest diameter of the column plus one-third the difference between the smallest diameter and the largest diameter but in no case should D exceed 1.5 times the smallest diameter. The design value for compression parallel to grain, F'_c, should not be exceeded at any cross section along the column. The small end of the column should also be investigated to make sure that the design value for bearing in end grain, F'_g, is not exceeded.

Example. A wood pole used in a dry location is stayed against translation at the top and bottom. It is laterally unsupported along its length. Determine the load P, which is primarily a snow load, that can be supported by the pole:

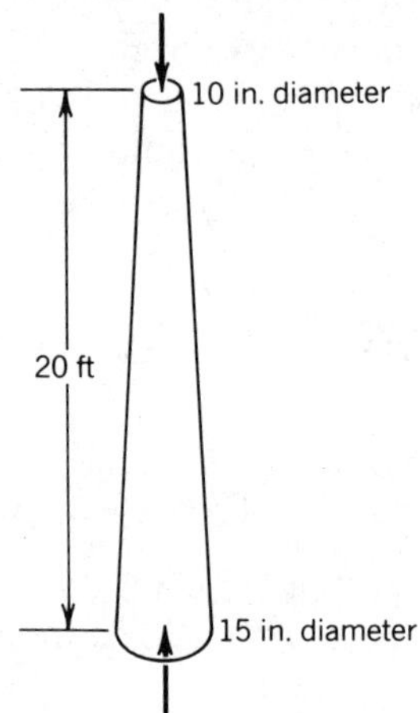

$$E = E' = 1{,}500{,}000 \text{ psi}$$

$$F_c = 1250 \text{ psi}, \; C_D = 1.15$$

$$F''_c = (1250)(1.15) = 1440 \text{ psi}$$

$$\text{Design diameter } D = 10 + \frac{15 - 10}{3} = 11.7 \text{ in.}$$

$$K_R = 0.58 \sqrt{\frac{E'}{F''_c}} = 0.58 \sqrt{\frac{1{,}500{,}000}{1440}} = 18.7$$

$$l_e/D = \frac{(20)(12)}{11.7} = 20.6$$

Since $18.7 < 20.6 < 44$, use the long-column formula:

$$F'_c = \frac{0.225E'}{(l_e/D)^2} = \frac{(0.225)(1{,}500{,}000)}{(20.6)^2} = 798 \text{ psi}$$

$$P = F'_c\, A = \frac{(798)\pi(11.7)^2}{4} = 85{,}300 \text{ lb}$$

Check small end of column:

$$P = F_c''A = \frac{(1440)\pi(10)^2}{4} = 113{,}000 \text{ lb} \quad \text{O.K.}$$

Use $P = 85{,}300$ lb. The design value, F_c', calculated by the intermediate- or long-column formula is applied to the cross-sectional area at which D was calculated.

In addition, the design value must not exceed the tabular design value adjusted by applicable modifiers, F_c'', based on the net column cross section at any point in the column.

Spaced Columns

Spaced columns have increased load-carrying capacity in a direction perpendicular to the wide faces of the individual members (parallel to dimension d_1, Fig. 5.3) due to the fixity at the ends, which provides shear resistance that tends to make

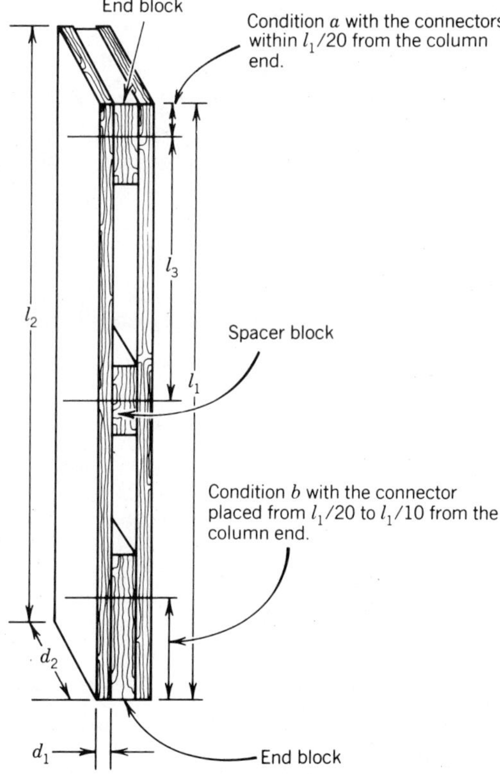

FIGURE 5.3 Spaced column, connector joined [from the NDS (3)] l_1 and l_2 are the distances from center to center of lateral supports of continuous spaced columns, and from end to end of simple spaced columns (in.); l_3 is the distance from center of connectors, in end blocks, to center of spacer block, (in.); d_1 is the dimension of narrowest side of individual member, (in.); and d_2 is the dimension of wide face of individual member (in.).

the column function as a unit in this direction. The load capacity in the direction parallel to the wide faces of the pieces (parallel to dimension d_2 as shown in Fig. 5.3) is not increased due to the fixity and the load capacity in this direction is the sum of the load capacities of the individual members designed as simple solid columns using l_2/d_2.

The individual members in a spaced column are considered to act together to carry the total column load. Each member is designed separately on the basis of its l_e/d ratio. Because of end fixity developed in spaced columns, a greater l_e/d ratio than allowed for simple solid columns is permitted. These limitations for individual members of a spaced column are

1. l_1/d_1 is limited to 80 where l_1 is the distance between lateral supports that provide restraint perpendicular to the wide faces of the individual members,
2. l_2/d_2 is limited to 50 where l_2 is the distance between lateral supports that provide restraint parallel to the wide faces of the individual members, and
3. l_3/d_1 is limited to 40 where l_3 is the distance between the centroid of connectors in an end block and the center of the spacer block.

The dimensions l_1, l_2, l_3, d_1, and d_2 are as shown in Figure 5.3. The design values for spaced columns are a function of the fixity condition *a* and *b* as shown in Figure 5.3 and the slenderness ratio of the individual members of the spaced column.

Because the following design procedure is based on tests of spaced columns using split rings for the connections at the ends, it is recommended that split rings be used for these connections. For short columns with l_1/d_1 ratios of individual members of 11 or less, the design value is

$$F_c' = F_c C_D C_M C_R C_t$$

For intermediate columns with l_1/d_1 ratios of individual members of greater than 11 but less than K, the design value is

$$F_c' = F_c C_D C_P C_M C_R C_t$$

where

$$C_P = \left[1 - \frac{1}{3}\left(\frac{l_1/d_1}{K}\right)^4\right]$$

and l_1, d_1, and K are as previously defined.

For long columns with l_1/d_1 of K or greater, the design value is:

$$F_c' = \frac{0.3 C_x E'}{(l_1/d_1)^2} \qquad (5\text{-}12)$$

where l_1, d_1 = dimensions shown in Figure 5.3 (in.),
C_x = 2.5 for fixity condition a,
C_x = 3.0 for fixity condition b, and

E' = tabular modulus of elasticity (psi) adjusted by applicable modifiers.

The value of K for spaced columns is determined as

$$K = 0.671\sqrt{C_x E'/F_c''} \tag{5-13}$$

where C_x, E', and F_c'' are as previously defined. *Note*: Formulas (5-12) and (5-13) are usually used for both sawn lumber and glued laminated timber. However, the designer may wish to change the values 0.30 and 0.671 in formulas (5-12) and (5-13), respectively, to 0.418 and 0.792 for both MSR lumber and glued laminated timber as shown in Appendix G of *NDS* (3).

When different grades, species, or thicknesses of members are used, the lesser value of F_c' determined for either member is applied to both members. Connectors are not required for a single spacer block located in the middle tenth of the column length l_2. Connectors are required for multiple spacer blocks, and the distance between two adjacent blocks may not exceed one-half the distance between centers of connectors in the end blocks. When spaced columns are used as truss compression members, panel points that are restrained laterally are considered as the ends of the spaced column. The portion of the web members between the individual pieces of which the spaced column is comprised may be considered as the end blocks in the end of the column. (See Fig. 5.3.) The total load capacity determined by using the spaced column formulas should be checked against the sum of the load capacities of the individual members taken as simple solid columns without regard to fixity; their greater d and the l between lateral supports that provide restraint in a direction parallel to the greater d should be used.

The design values for F_c' determined by the formulas should not exceed the tabular design values for compression parallel to grain, F_c, for the species, as adjusted by applicable modifiers.

Spacer and end block thickness should not be less than that of the individual members of the spaced column, nor shall thickness, width, and length of spacer and end block be less than required for connectors of a size and number capable of carrying the computed load.

To obtain spaced column action, the connectors in mutually contacting surfaces of end blocks and individual members at each end of a spaced column must be of a size and number to provide a load capacity equal to the required cross-sectional area of one of the individual members times the appropriate constant from Table 5.5. For spaced columns that are part of a truss system, the connectors required by joint design should be checked against the values obtained by using the constants from Table 5.5.

Example. Select a two-member spaced column of No. 1 Southern Pine (group B species) from Table 7.3 to meet the following requirements:

Two-month (snow load) duration of load factor, $C_D = 1.15$
Dry conditions of use, $C_M = 1.00$
No fire-retardant treatment, $C_R = 1.00$

TABLE 5.5

End Block Constants for Connector Joined Spaced Columns

l/d Ratio of Individual Members of Spaced Column[d]	End Spacer Block Constant for Connector Load Groups[b,c]			
	Group A[e]	Group B[e]	Group C[e]	Group D[e]
0 to 11	0	0	0	0
15	38	33	27	21
20	86	73	61	48
25	134	114	94	75
30	181	155	128	101
35	229	195	162	128
40	277	236	195	154
45	325	277	229	181
50	372	318	263	208
55	420	358	296	234
60 to 80	468	399	330	261

[a] Source: *National Design Specification* (3).

[b] Constants are based on area in square inches and load in pounds.

[c] When Eqs. (5-5) and (5-6) are used for the design of MSR lumber and glued laminated members, multiply values by 1.4.

[d] Constants for intermediate l/d ratios may be obtained by straight-line interpolation.

[e] See Table 6.1 for species in each group.

Normal temperature, $C_t = 1.00$

$$F'_c = F_c C_D C_P C_M C_R C_t = F_c C_D C_P$$

$$P = 40{,}000 \text{ lb} \qquad E = 1{,}700{,}000 \text{ psi}$$

$$L = 12 \text{ ft} \qquad F_c = 1250 \text{ psi} \qquad E' = E$$

$$F''_c = F_c C_D = (1250)(1.15) = 1440 \text{ psi}$$

Check end fixity: Distance from end of column to center of 4-in. split rings in end blocks = x = 12 in.

$$\frac{l}{20} = \frac{(12)(12)}{20} = 7.2 \text{ in.} \qquad \frac{l}{10} = \frac{(12)(12)}{10} = 14.4 \text{ in.}$$

Column has fixity condition b since $\frac{l}{20} < x < \frac{l}{10}$. $C_x = 3.0$. Try two 3 × 6 members, $A = (2)(2.5)(5.5) = 27.5 \text{ in.}^2$.

$$l_1/d_1 = \frac{(12)(12)}{2.5} = 57.6$$

$$K = 0.671\sqrt{\frac{C_x E'}{F''_c}} = 0.671\sqrt{\frac{(3.0)(1{,}700{,}000)}{1440}} = 40$$

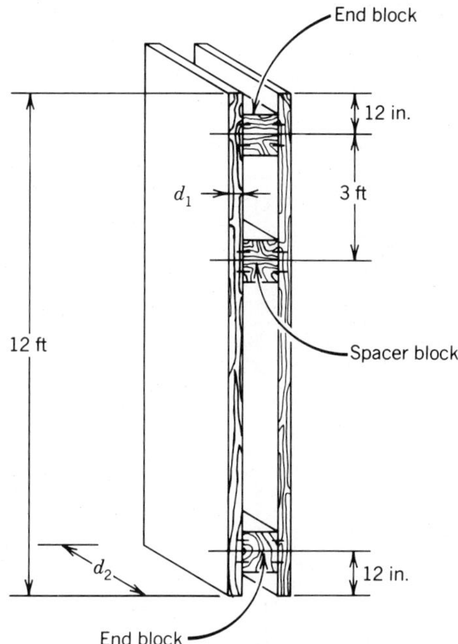

Since 57.6 > 40, use the long-column formula.

$$F_c' = \frac{0.30C_xE'}{(l_1/d_1)^2} = \frac{(0.30)(3.0)(1{,}700{,}000)}{(57.6)^2} = 461 \text{ psi}$$

$$\text{Required } A = \frac{P}{F_c'} = \frac{40{,}000}{461} = 87 \text{ in.}^2 > 27.5 \text{ in.}^2 \quad \text{N.G.}$$

Try two 4 × 8 members:

$$A = (2)(3.5)(7.25) = 50.75 \text{ in.}^2$$

$$l_1/d_1 = \frac{(12)(12)}{3.5} = 41.1 > 40$$

Again use the long-column formula.

$$F_c' = \frac{(0.30)(3.0)(1{,}700{,}000)}{(41.1)^2} = 904 \text{ psi}$$

$$\text{Required } A = \frac{P}{F_c'} = \frac{40{,}000}{904} = 44.2 \text{ in.}^2 < 50.75 \quad \text{O.K.}$$

Check as a solid column using the larger $d = d_2 = 7.25$ in.

$$K = 0.671\sqrt{\frac{E'}{F_c''}} = 0.671\sqrt{\frac{1{,}700{,}000}{1440}} = 23.1$$

$$l_2/d_2 = \frac{(12)(12)}{7.25} = 19.9$$

Since 11 < 19.9 < 23.1, use the intermediate-column formula.

$$F'_c = F''_c \; C_P$$

$$C_P = \left[1 - \frac{1}{3}\left(\frac{l/d}{K}\right)^4\right] = \left[1 - \frac{1}{3}\left(\frac{19.9}{23.1}\right)^4\right] = 0.82$$

$$F'_c = F''_c C_P = (1440)(0.82) = 1170 \text{ psi}$$

$$P = AF'_c = (50.75)(1170) = 59{,}600 > 40{,}000 \text{ lb} \quad \text{O.K.}$$

Check the connectors for spaced column action. From Table 5.5, the end block constant for $l_1/d_1 = 41.4$ in group B species is 245. Total connector load = end block constant × area = (245)(3.5)(7.25) = 6220 lb. Allowable load for a 4-in. split ring from Table 6.11 is (5260)(1.15) = 6050 lb < 6220 lb. One connector is slightly overstressed (less than 3%). Use one connector.

Built-Up Columns

Built-up columns of two or more pieces of wood fastened together with mechanical fasteners such as nails, bolts, or other fasteners will not carry as much load as a solid column of the same size because of shear distortion that can occur in the fastening joints. The nailing of cover plates to the edge of the built-up layers improves the load-carrying capacity, according to *Wood Handbook* (4). The percentage of strength shown in Figure 5.4 can be obtained for built-up columns with edges fastened together or a boxed solid cover.

Strength data is limited on other configurations of mechanically fastened built-up columns, but they should be somewhat stronger than the sum of the strengths of the individual pieces.

Built-up columns may also have the members separated by spacer blocks. Un-

(*a*) Boxed solid core

(*b*) Edges fastened together with cover plates

l/d ratio	Strength of single piece column (%)
6	82
10	77
14	71
18	65
22	74
26	82

FIGURE 5.4 Built-Up Columns; Mechanically Fastened; data from *Wood Handbook* (4).

less the procedure for spaced columns as shown on pages 5-119 to 5-124 is followed, the load capacity should be taken as the sum of the loads of the individual pieces acting as simple solid columns if test data for the specific arrangement of members is not available.

Light Truss Compression Chords

When a light truss compression chord has plywood sheathing attached, the sheathing, in addition to preventing buckling of the chords due to axial compression loads about the weak axis, helps to prevent buckling of the chord about the strong axis. This increased chord-buckling stiffness can be accounted for by the buckling stiffness factor, C_T. Equation (5-14) applies only to light truss compression chords 2 × 4 in. or smaller that have effective buckling lengths of 96 in. or less, $\frac{3}{8}$ in. or thicker plywood sheathing nailed to the narrow faces of the chord, and chords seasoned to 19% maximum moisture content at the time the plywood is applied.

$$C_T = 1 + \frac{2300 l_e}{E'_{0.05}} \tag{5-14}$$

where C_T = buckling stiffness factor,
l_e = effective buckling length of chord about its strong axis used in design of chord for compression loading (in.),
$E'_{0.05}$ = $0.589E'$ for visually graded lumber (psi) (the lower 5% exclusion limit value of E),
= $0.819E'$ for MSR lumber (psi), and
E' = tabular modulus of elasticity (psi) from Table 7.3 or 7.4 adjusted by applicable modifiers.

For wood that is unseasoned or partially seasoned at the time of plywood attachment, the buckling stiffness factor is determined from the formula

$$C_T = 1 + \frac{1200 l_e}{E'_{0.05}} \tag{5-15}$$

The buckling stiffness factor should be applied only to chords that are subject to combined flexure and compression. It should not be applied for wet conditions of use.

The buckling factor C_T is not applicable to the short-column formula. It is applied to the intermediate and long-column formulas as follows:

Intermediate columns $$F'_c = F''_c\left[1 - \frac{1}{3}\left(\frac{l_e/d}{K}\right)^4\right]$$

where $$K = 0.671\sqrt{\frac{C_T E'}{F''_c}}$$

Long columns $$F_c' = \frac{0.30C_T E'}{(l_e/d)^2} \tag{5-16}$$

where terms are as defined previously.

Columns With Flanges

Glued laminated and built-up columns are usually square or rectangular but can also be made with flanges. Because of fabrication and handling difficulties, these shapes are not common and it is recommended that the designer check with the fabricator prior to designing a column with flanges. The strength of flanged columns is somewhat less than would be determined by use of the standard column formula because the thin outstanding flanges are subject to buckling and there is incomplete shear transfer when mechanical fastenings are used. A formula that can be converted to a design formula by dividing by 2.74 (the same factor used for reducing the ultimate buckling strength of a long column) can be found in *Wood Handbook* (5). A design value, F_c', can be obtained for the flanges of H and I shapes (as shown in Fig. 5.6) as follows:

$$F_c' = 0.016E'\left(\frac{t}{b}\right)^2 \tag{5-17}$$

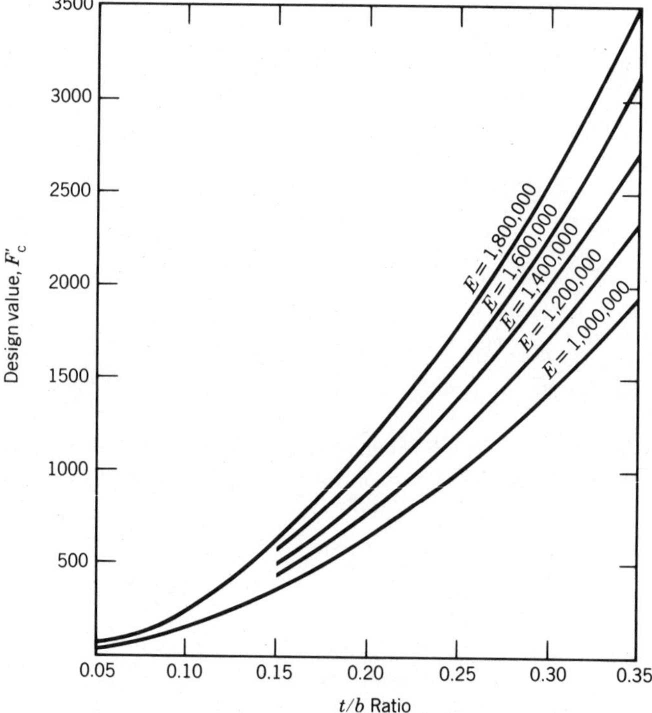

FIGURE 5.5 Design values for columns with outstanding flanges.

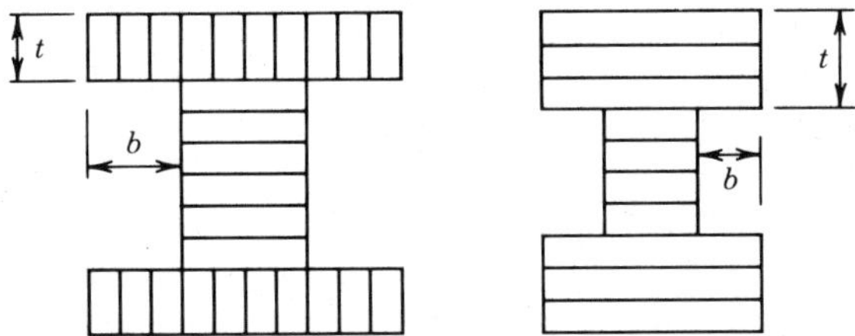

FIGURE 5.6 Glued laminated columns with outstanding flanges.

where E' = tabular modulus of elasticity (psi) adjusted by applicable modifiers,
t = thickness of the flange (in.), and
b = width of outstanding flange (in.).

The design value for the entire column cross section must not be exceeded after application of the appropriate column formula.

Figure 5.5 may be used to determine design values F'_c, for flanged columns for various E and thickness–breadth, t/b, ratios. The reduction of design values varies with modulus of elasticity, and in general, no reduction is necessary for t/b ratios of more than 0.35.

The column formula used must be modified to reflect the radius of gyration of the shape being designed. The tabular design value for compression parallel to grain, F_c, to use in determining F'_c is that shown for four or more laminations of the combination being used when the arrangements shown in Figure 5.6 are used. When the flanges contain less than four laminations and the t/b is one or more, the tabular design value of F_c for four or more laminations may be used. When t/b is less than 1 and the flanges contain less than four laminations, the tabular design value to use is that for two or three laminations.

COMBINED AXIAL COMPRESSION AND BENDING MEMBERS

Columns or compression members may be subject to bending stresses as well as axial compressive stresses. Beams, girders, or other members that normally support loads that cause bending may also be subject to axial compression loads. These members are discussed on page 5-199. Combined axial compression and bending loads may occur in any one or a combination of the following loading conditions:

1. Axial compression plus applied end moments.
2. Concentric end load and side loads.
3. Eccentric end load only.
4. Combined end loads, side loads, and eccentricity.
5. Loads applied at side brackets on columns.

The design of columns in each case is usually a trial-and-error process in which a tentative size is selected and analyzed. If the size is too large or too small, adjustments in size are made until the design criteria are satisfied. Where members are subjected to combined axial compression and bending moments, combined bending and compression caused by an eccentric load or side load, or an eccentric load in combination with other loads, an exact solution may be very time consuming and not justified in view of the variability of modulus of elasticity and other strength properties or exact knowledge of the loading condition. When a more accurate method is required, the secant formula is recommended. However, for the uses most often encountered in design, the simplified formulas from *NDS* (3) appear to be appropriate. These formulas are based on the following assumptions:

1. The stresses that cause a given deflection as a sinusoidal curve are the same as those for a beam with a uniform side load.
2. For a single concentrated side load, the stress under the load can be used, regardless of the position of load with reference to the length of the column.
3. The stress to use with a system of side loads is the maximum stress due to the system. (With large side loads near each end, some slight error on the side of overload will occur.)
4. For columns with a l_e/d ratio of 11 or less, stress due to deflection of the column may be neglected.

When a member is subjected to a combination of axial compression and bending about both axes, the following generalized interaction equation must be satisfied. See Figure 5.7 for an illustration of the axes.

$$\frac{f_c}{F'_c} + \frac{f_{bx}}{F'_{bx} - J_x f_c} + \frac{f_{by}}{F'_{by} - J_y f_c} \leqslant 1 \qquad (5\text{-}18)$$

where f_c = actual compression parallel to grain stress induced by axial load, ($f_c = P/A$) (psi),

F'_c = design value in compression parallel to grain if compression alone existed (psi) (this value is based on the maximum l_e/d ratio of the member),

f_{bx}, f_{by} = bending stresses about the x–x and y–y axes, respectively (psi),

F'_{bx}, F'_{by} = design value in bending if flexural stress about that axis existed alone (psi),

J_x, J_y = 0 when $(l_e/d)_x, (l_e/d)_y \leqslant 11$ (short column),

J_x, J_y = $[(l_e/d)_x - 11]/(K_x - 11)$, $[(l_e/d)_y - 11]/(K_y - 11)$ when $11 < (l_e/d)_x, (l_e/d)_y < K_x, K_y$ (intermediate column),

J_x, J_y = 1 when $(l_e/d)_x, (l_e/d)_y \geqslant K_x, K_y$ (long column),

and where $(l_e/d)_x, (l_e/d)_y$ = maximum l_e/d ratio for buckling about the x–x and y–y axes, respectively.

$K_x, K_y = 0.671\sqrt{E'_x/F''_c}$, $0.671\sqrt{E'_y/F''_c}$, respectively. The Jf_c terms in the interaction equation account for the increase in bending stresses caused by the moment induced by the axial load acting away from the centroid of the deflected mem-

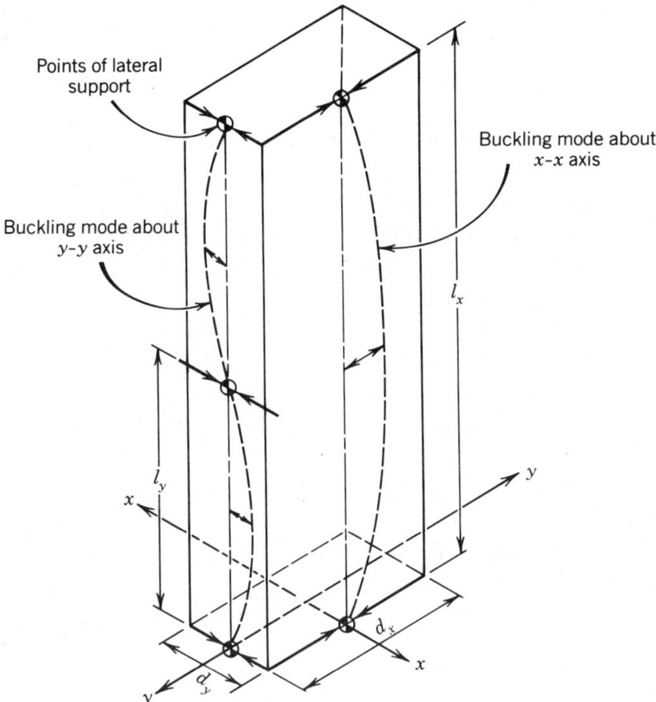

FIGURE 5.7 Column notations for combined bending and compression.

ber. This is sometimes referred to as the *P*–Δ effect. The *J* factor used should be for the plane of bending being considered to reduce the allowable bending value.

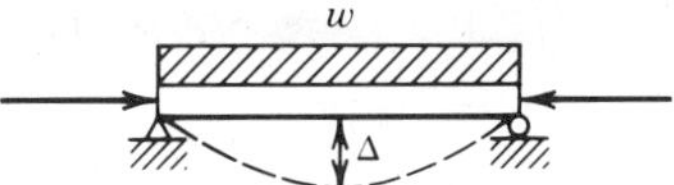

In the general equation, the value of F_c' is determined by considering the largest l_e/d value. In determining *J* for a bending moment about the *x*–*x* axis, the l_e/d ratio for column buckling about the *x*–*x* axis, $(l_e/d)_x$, is used; and for a bending moment about the *y*–*y* axis, the l_e/d ratio for column buckling about the *y*–*y* axis, $(l_e/d)_y$, is used.

The value of F_b' is determined by multiplying F_b by all applicable modifiers except that C_F and C_L are not applied cumulatively unless C_F is 1 or more. When axial compressive stress, f_c, equals or exceeds $F_b''(1 - C_F)$, C_F does not apply because the net stress on the face of the member where bending creates a tensile stress will not exceed $F_b''C_F$. For information on the application of C_L, see page 5-164.

There is some question about using the reduced value of F_b' based on lateral buckling in bending when column buckling is about a different axis. Most references use the more conservative approach of modifying F_b by the lateral stability of beams factor, C_L, regardless of the direction of column buckling. This procedure is used in the examples in this manual. When members are designed primarily as compression members, the depth–breadth ratio is usually close to 1 and the effect of using the more conservative approach is small. When the depth–breadth ratio is larger, in the range of 3–7, as is the case with members designed primarily as beams, the effect of including C_L in the formula is more significant, but the beam buckling and the column buckling are more likely to be about the same axis. C_L is not usually applied in the design of arches.

It is convenient to consider compressive stresses as negative and tension stresses caused by bending as positive in illustrating the combination of stresses.

An axial load will cause equal compressive stresses at each point *A*, *B*, *C*, and *D* as illustrated in Figure 5.8*a*.

A load creating bending about the *x–x* axis as illustrated in Figure 5.8*b* will result in negative stresses at points *D* and *C*. A load creating bending about the *y–y* axis as illustrated in Figure 5.8*c* will result in negative stresses at points *B* and *C*. The maximum combination of stresses as illustrated in Figure 5.8*d* will occur at point *C*.

When additional bending stresses are caused by eccentric loading of the member, the bending stress caused by the induced moment should be calculated by conventional methods and its effect in the interaction formula accounted for by modifying the design value in bending, F'_b, by the term Jf_c to account for the increased P–Δ effect. If the moment induced by the eccentricity is in the same di-

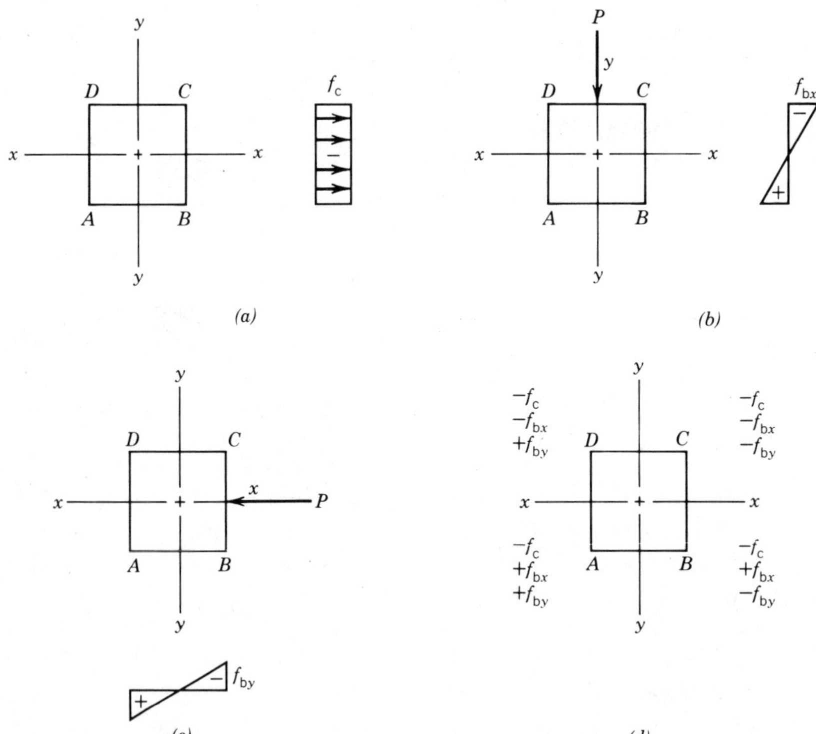

FIGURE 5.8 Combined bending and axial compression.

rection as the moment caused by the side loads, this stress in bending should be added to any other bending stresses caused by side loads. If the induced moment due to eccentricity is in the opposite direction, the stresses are subtracted. If the net resulting moment is negative, the absolute value should be used in the summation of the interaction equation. When the terms are rearranged and eccentric forces exist, the interaction formula becomes

$$\frac{f_c}{F'_c} + \frac{f_{bx} \pm f_c(6 + 1.5J_x)(e_x/d_x)}{F'_{bx} - J_x f_c} + \frac{f_{by} \pm f_c(6 + 1.5J_y)(e_y/d_y)}{F'_{by} - J_y f_c} \leqslant 1 \qquad (5\text{-}19)$$

where e_x, e_y = eccentricity of axial loads perpendicular to the x and y axes, respectively (in.) and

d_x, d_y = dimensions of the member perpendicular to the x and y axes respectively (in.).

If no bending stresses exist about the respective axes, the terms f_{bx} and f_{by} drop out of the equation and the bending stress due to eccentric loads, $f_c(6 + 1.5J_x)$ (e_x/d_x), is the only term in the numerator. In most practical design problems, many terms in the interaction equation drop out and a much simpler form of the expression can be applied. The following examples illustrate the application of the interaction formula to problems with varying degrees of complexity.

Example. Check the design of a 5 × 9-in. glued laminated timber column for the loading conditions illustrated. (*Note:* 5-in. width used for Southern Pine.) The column is braced at both ends in the x and y directions. Pinned ends are assumed in the design. For purposes of illustration, the dry-use stresses of Southern Pine combination 16F-V3 will be used.

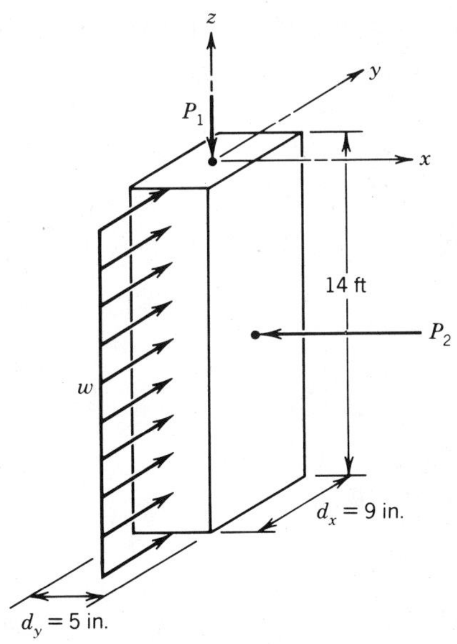

$F_c = 1450$ psi
$F_{bx} = 1600$ psi
$F_{by} = 1450$ psi
$E_x = 1{,}400{,}000$ psi
$E_y = 1{,}300{,}000$ psi
$P_1 = 10{,}000$ lb (dead load + snow load)
$P_2 = 1000$ lb (wind load)
$w = 150$ plf (wind load only)
Section properties: $A = 45$ in.^{2}, $S_x = 67.5$ in.^{3}, $S_y = 37.5$ in.^{3}

1. Determine stresses from applied loads:

$$f_c = \frac{P_1}{A} = \frac{10{,}000}{45} = 222 \text{ psi}$$

$$f_{bx} = \frac{M_x}{S_x} = \frac{wl^2}{8S_x} = \frac{(150)(14)^2(12)}{(8)(67.5)} = 653 \text{ psi}$$

$$f_{by} = \frac{M_y}{S_y} = \frac{P_2 l}{4S_y} = \frac{(1000)(14)(12)}{(4)(37.5)} = 1120 \text{ psi}$$

2. Determine design values:

$$F''_c = F_c C_D = (1450)(1.33) = 1930 \text{ psi} \quad (C_D \text{ for wind})$$

$$\left(\frac{l_e}{d}\right)_x = \frac{(14)(12)}{9} = 18.7 \qquad \left(\frac{l_e}{d}\right)_y = \frac{(14)(12)}{5} = 33.6.$$

$$K_x = 0.671\sqrt{\frac{E'_x}{F''_c}} = 0.671\sqrt{\frac{1{,}400{,}000}{1930}} = 18.1$$

$$K_y = 0.671\sqrt{\frac{E'_y}{F''_c}} = 0.671\sqrt{\frac{1{,}300{,}000}{1930}} = 17.4$$

Since $K_x < (l_e/d)_x < 50$, use the long-column formula, and since $K_y < (l_e/d)_y < 50$, use the long-column formula and largest l_e/d value.

$$F'_c = \frac{0.30E_y}{(l_e/d)_y^2} = \frac{(0.30)(1{,}300{,}000)}{(33.6)^2} = 345 \text{ psi}$$

$$F''_{bx} = F_{bx}C_D = (1600)(1.33) = 2130 \text{ psi} \quad (C_D \text{ for wind})$$

$$(l_e)_x = 1.92\,l_u = (1.92)(14)(12) = 322.6 \text{ in.}$$

$$(C_s)_x = \sqrt{\frac{l_e d}{b^2}} = \sqrt{\frac{(322.6)(9)}{(5)^2}} = 10.8$$

$$(C_k)_x = 0.956\sqrt{\frac{E'_y}{F''_{bx}}} = 0.956\sqrt{\frac{1{,}300{,}000}{2130}} = 23.6$$

Since $10 < C_s < C_k$, use the intermediate-beam formula.

$$(C_L)_x = \left[1 - \frac{1}{3}\left(\frac{C_s}{C_k}\right)_x^4\right] = \left[1 - \frac{1}{3}\left(\frac{10.8}{23.6}\right)^4\right] = 0.99$$

$$F'_{bx} = F''_{bx}C_L = (2130)(0.99) = 2100 \text{ psi}$$

From footnote 6, Table 2, AITC 117—Design (1), $C_F = 1.10$ (*Note:* C_F applies in this case because it is greater than 1.)

$$F''_{by} = F_{by}C_D C_F = (1450)(1.33)(1.10) = 2120 \text{ psi}$$

$$(l_e)_y = (l_e)_x = 322.6 \text{ in.}$$

$$(C_s)_y = \sqrt{\frac{(322.6)(5)}{(9)^2}} = 4.5 < 10$$

Therefore, use the short-beam formula.

$$F'_{by} = F''_{by} = 2120 \text{ psi}$$

3. Check the interaction formula. Because e_x and e_y are zero, the interaction equation becomes

$$\frac{f_c}{F'_c} + \frac{f_{bx}}{F'_{bx} - J_x f_c} + \frac{f_{by}}{F'_{by} - J_y f_c} \leqslant 1$$

For the long-column formula $J_x = 1$, $J_y = 1$.

$$\frac{222}{345} + \frac{643}{2100 - 222} + \frac{1120}{2120 - 222} = 0.64 + 0.35 + 0.59 = 1.58$$

Member is inadequate; increase both the width and depth and recheck. Try $6\frac{3}{4} \times 10\frac{1}{2}$-in. member with the same design values:

$$A = 70.88 \text{ in.}^2 \qquad S_x = 124.0 \text{ in.}^3 \qquad S_y = 79.73 \text{ in.}^3$$

1. Determine stresses from applied loads:

$$f_c = \frac{10{,}000}{70.88} = 141 \text{ psi}$$

$$f_{bx} = \frac{(150)(14)^2(12)}{(8)(124.0)} = 356 \text{ psi}$$

$$f_{by} = \frac{(1000)(14)(12)}{(4)(79.73)} = 527 \text{ psi}$$

2. Determine design values:

$$F''_c = 1930 \text{ psi}$$

$$\left(\frac{l_e}{d}\right)_x = \frac{(14)(12)}{10.5} = 16$$ Because $(l_e/d)_x < K_x$ use intermediate-column formula.

$$J_x = \frac{(l_e/d)_x - 11}{K_x - 11} = 0.70$$

$$\left(\frac{l_e}{d}\right)_y = \frac{(14)(12)}{6.75} = 24.9 > K_y, \text{ use the long-column formula.}$$

$$J_y = 1$$

$$F'_c = \frac{(0.30)(1{,}300{,}000)}{(24.9)^2} = 630 \text{ psi}$$

$$(C_s)_x = \sqrt{\frac{(322.6)(10.5)}{(6.75)^2}} = 8.6 \quad \text{(use short-beam formula)}$$

$$F'_{bx} = F''_{bx} = 2130 \text{ psi}$$

$$F''_{by} = (1450)(1.33)(1.07) = 2060 \text{ psi}$$

$$(C_s)_y = \sqrt{\frac{(322.6)(6.75)}{(10.5)^2}} = 4.4 \quad \text{(use short-beam formula)}$$

$$F'_{by} = F''_{by} = 2060 \text{ psi}$$

3. Check interaction equation:

$$\frac{141}{630} + \frac{356}{2130 - 0.70(141)} + \frac{527}{2060 - 141}$$

$$= 0.22 + 0.18 + 0.27 = 0.67 < 1 \quad \text{O.K.}$$

In most wood structures loaded with axial and bending loads, bending occurs about only one of the axes, and one of the bending terms in the general interaction equation drops out. The following example illustrates the design of a column subjected to bending about one axis.

Example. Check the design of a $6\frac{3}{4} \times 10\frac{1}{2}$-in. glued laminated timber column for the loading and condition illustrated. The column is braced in the x and y directions at both the top and bottom. Use Douglas Fir Combination 2 and wet conditions of use.

$F_c = 1900$ psi
$F_{bx} = 1700$ psi
$E_x = E_y = 1{,}700{,}000$ psi
$P = 17{,}000$ lb (including dead and snow loads)
$w = 200$ plf (wind load only)
Section properties: $A = 70.88$ in.^{2} $\quad S_x = 124.0$ in.^{3}

1. Check column for total combination of loads, C_D (wind) $= 1.33$, C_D (snow) $= 1.15$, C_M (bending) $= 0.80$, C_M (compression) $= 0.73$, C_M (modulus of elasticity) $= 0.833$

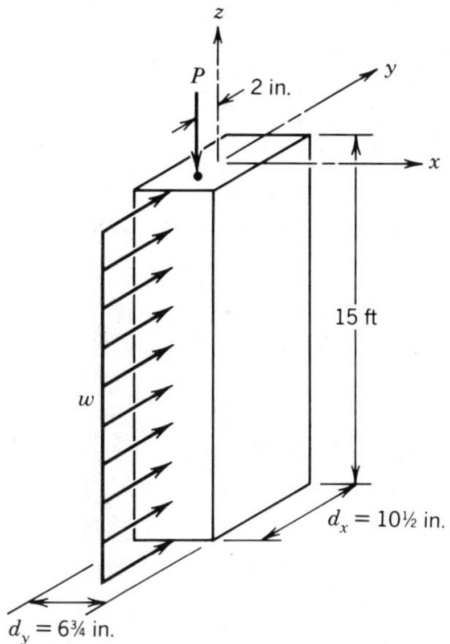

(a) Determine f_c and F'_c

$$f_c = \frac{P}{A} = \frac{17{,}000}{70.88} = 240 \text{ psi}$$

$$F''_c = F_c C_D C_M = (1900)(1.33)(0.73) = 1845 \text{ psi}$$

$$E' = EC_M = (1{,}700{,}000)(0.833) = 1{,}420{,}000 \text{ psi}$$

$$K_x = K_y = 0.671\sqrt{\frac{E'}{F''_c}} = 0.671\sqrt{\frac{1{,}420{,}000}{1845}} = 18.6$$

$$\left(\frac{l_e}{d}\right)_y = \frac{(15)(12)}{6.75} = 26.67 \quad \text{(use long-column formula)}$$

$$F'_c = \frac{0.30E}{(l_e/d)^2} = \frac{(0.30)(1{,}420{,}000)}{(26.67)^2} = 597 \text{ psi}$$

(b) Determine f_{bx} and F'_{bx}

$$f_{bx} = \frac{M}{S_x} = \frac{wl^2}{8S_x} = \frac{(200)(15)^2(12)}{(8)(124.0)} = 544 \text{ psi}$$

$$F''_{bx} = F_{bx} C_D C_M = (1700)(1.33)(0.8) = 1810 \text{ psi}$$

$$(l_e)_x = 1.92 l_u = (1.92)(15)(12) = 345.6 \text{ in.} \quad \text{(for uniform load)}$$

$$(C_s)_x = \sqrt{\frac{(l_e)_x d}{b^2}} = \sqrt{\frac{(345.6)(10.5)}{(6.75)^2}} = 8.92 < 10 \quad \text{(use short-beam formula)}$$

$$F'_{bx} = F''_{bx} = 1810 \text{ psi}$$

(c) Check combined loading by interaction formula: With no bending or eccentricity about the y–y axis, the interaction equation is

$$\frac{f_c}{F'_c} + \frac{f_{bx} + f_c(6 + 1.5J_x)(e_x/d_x)}{F'_{bx} - J_x f_c} \leqslant 1$$

$$\left(\frac{l_e}{d}\right)_x = \frac{(15)(12)}{(10.5)} = 17.1$$

$$J_x = \frac{(l_e/d)_x - 11}{K_x - 11} = \frac{17.1 - 11}{18.6 - 11} = 0.81$$

$$\frac{240}{597} + \frac{544 + (240)[6 + 1.5(0.81)](2/10.5)}{1810 - (0.81)(240)} = 0.94 \quad \text{O.K.}$$

2. Check column for eccentric load only, C_D for snow load = 1.15

$$F''_c = (1900)(1.15)(0.73) = 1595 \text{ psi}$$

$$K_x = K_y = 0.671\sqrt{\frac{E'}{F''_c}} = 0.671\sqrt{\frac{1{,}420{,}000}{1595}} = 20$$

$$\left(\frac{l_e}{d}\right)_y = 26.67 > 20 \quad \text{(use long-column formula)}$$

$$F'_c = 597 \text{ psi} \quad \text{(as determined in 1)}$$

$$F''_{bx} = (1700)(1.15)(0.8) = 1564 \text{ psi}$$

$$C_s = 8.92 \quad \text{(short-beam formula as determined in 1.)}$$

$$F'_{bx} = F''_b = 1564 \text{ psi}$$

$$J_x = \frac{17.1 - 11}{20 - 11} = 0.68$$

$$\frac{f_c}{F'_c} + \frac{f_c(6 + 1.5J_x)(e_x/d_x)}{F'_{bx} - J_x f_c} \leqslant 1$$

$$\frac{240}{597} + \frac{240[6 + (1.5)(0.68)](2/10.5)}{1564 - (0.68)(240)} = 0.63 \quad \text{O.K.}$$

Column is adequate.

Columns with Bracket Loads

Special design procedures are recommended when an eccentric load is applied through a bracket some distance from the top as shown in Figure 5.9 rather than at the top. It is assumed that the bracket load P applied at distance a from the center of the column is replaced by the same load P centrally applied at the top

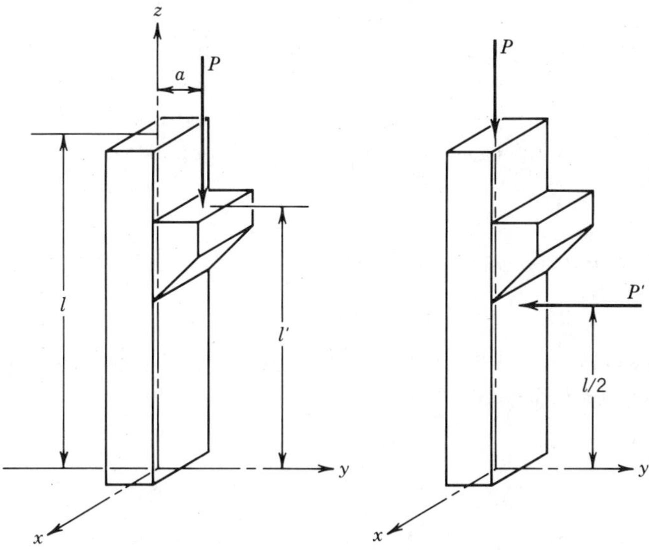

FIGURE 5.9 Column with load applied at side bracket.

of the columns plus a side load P' applied at midheight, $l/2$. This condition is shown in Figure 5.9.

The load P' is determined by the empirical formula

$$P' = \frac{3al'P}{l^2} \tag{5-20}$$

where P' = assumed horizontal side load, placed midheight of the column (lb),
P = actual load on bracket (lb),
a = horizontal distance from the bracket load to center of column (in.),
l' = distance from point of application of load on the bracket to further end of column (in.), and
l = length of column (in.).

The assumed axially applied load P should be added to other concentric column loads, and the calculated side load P' should be used to determine the flexural stress

$$f_b = \frac{3al'P}{4lS} \tag{5-21}$$

where S is the section modulus and the other terms are as defined previously. Substitution of this expression for f_{bx} in Eq. (5-18) results in

$$\frac{f_c}{F'_c} + \frac{3al'P/4lS}{F'_{bx} - J_x f_c} \leqslant 1 \tag{5-22}$$

This formula for combined axial compression and bending assumes that the column is of such a size that the combination of loads is critical. The designer is cautioned that this empirical method is to be used only in checking the combined axial and bending stress and should not be used to determine shear or horizontal reactions. These should be determined by the usual method of statics.

Example. Determine the size of a sawn wood column 12 ft long to carry a 14,000-lb concentric axial load and a permanent equipment load of 2000 lb mounted on a side bracket as shown. Use Douglas Fir-Larch, Posts and Timbers, No. 1 grade

$$F_b = 1200 \text{ psi} \qquad F_c = 1000 \text{ psi} \qquad E = 1{,}600{,}000 \text{ psi}$$

$$a = 24 \text{ in.} \qquad l = 144 \text{ in.}$$

The top of the bracket is 36 in. below the column; duration-of-load factor C_D = 1.15 for snow load; dry conditions of use C_M = 1.00.

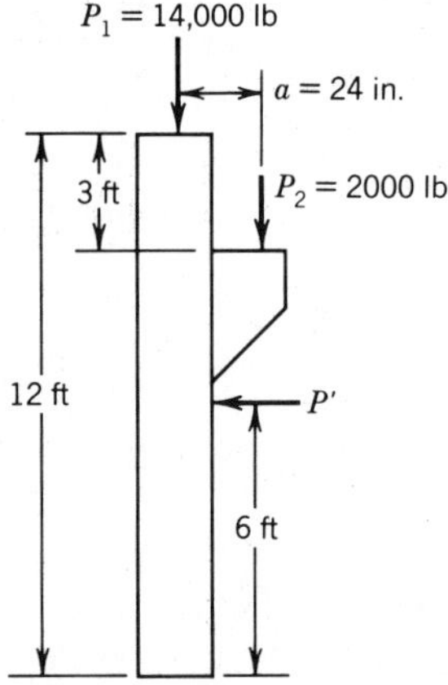

End fixity condition is assumed to be pin end at both ends (K_e = 1.0). Try a 6 × 8-in. column with the 8-in. direction in the plane of the side load:

$$A = 41.25 \text{ in.}^2 \qquad S = 51.56 \text{ in.}^3 \qquad F_c'' = (1000)(1.15) = 1150 \text{ psi}$$

$$F_b'' = (1200)(1.15) = 1380 \text{ psi} \qquad E' = E = 1{,}600{,}000 \text{ psi}$$

The maximum l_e/d ratio applies in determining F_c'.

$$\left(\frac{l_e}{d}\right)_x = \frac{144}{7.5} = 19.2 \qquad \left(\frac{l_e}{d}\right)_y = \frac{144}{5.5} = 26.18$$

$$K_x = K_y = 0.671\sqrt{\frac{E'}{F_c''}} = 0.671\sqrt{\frac{1{,}600{,}000}{1150}} = 25.0 < 26.18$$

(use long-column formula)

Because $(l_e/d)_x < K_x$, calculate J_x for buckling about the x axis for use in the interaction formula.

$$J_x = \frac{(l_e/d)_x - 11}{K_x - 11} = \frac{19.2 - 11}{25.0 - 11} = 0.58$$

The compression parallel to grain stress that would exist if the column were axially loaded only is

$$f_c = \frac{P}{A} = \frac{14{,}000 + 2000}{41.25} = 388 \text{ psi}$$

$$F'_c = \frac{0.30E'}{(l_e/d)^2} = \frac{(0.30)(1{,}600{,}000)}{(26.18)^2} = 700 \text{ psi}$$

$$l_e = 1.61l = (1.61)(144) = 232 \text{ in.} \quad \text{(from Table 5.11)}$$

$$C_s = \sqrt{\frac{l_e d}{b^2}} = \sqrt{\frac{(232)(7.5)}{(5.5)^2}} = 7.6 < 10 \quad \text{(use short-beam formula)}$$

$$F'_{bx} = F''_{bx}$$

$$\frac{f_c}{F'_c} + \frac{3al'P/4lS}{F'_{bx} - J_x f_c} \leqslant 1$$

$$\frac{388}{700} + \frac{(3)(24)(108)(2000)/(4)(144)\ (51.66)}{1380 - (0.58)(388)} = 1.01 \leqslant 1 \quad \text{(consider O.K.)}$$

Example. Determine the adequacy of an 8 × 8-in. pin-end column subjected to the following conditions:

Load P is a snow load of 10,000 lb on bracket, $C_D = 1.15$.
Assume dry conditions of use, $C_M = 1.0$.

$$a = 12 \text{ in} \qquad l_e = l = 10 \text{ ft} \qquad l' = 8 \text{ ft}$$

Douglas Fir-Larch, Dense No. 1, Post and Timber

$$F_b = 1400 \text{ psi} \qquad F_c = 1200 \text{ psi} \qquad E = 1{,}700{,}000 \text{ psi}$$

$$S = 70.31 \text{ in.}^3 \qquad A = 56.25 \text{ in.}^2$$

$$F''_c = 1200\,(1.15) = 1380 \text{ psi} \qquad F''_b = 1400\,(1.15) = 1610 \text{ psi}$$

Top of bracket at point of load application is located 2 ft below top of column. Analyze for combined stresses due to assumed concentric load P and side load P'.

Determine F'_c:

$$\left(\frac{l_e}{d}\right)_x = \left(\frac{l_e}{d}\right)_y = \frac{(10)(12)}{7.5} = 16.0$$

$$K_x = K_y = 0.671\sqrt{\frac{E'}{F''_c}} = 0.671\sqrt{\frac{1{,}700{,}000}{1380}} = 23.55$$

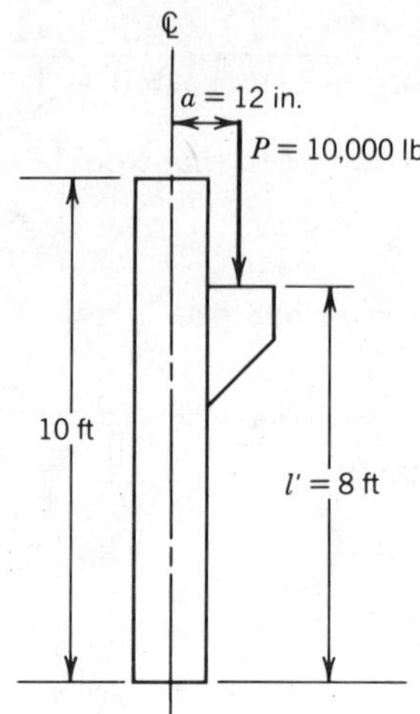

Since 11 < 16.0 < 23.55, use intermediate-column formula.

$$C_P = \left[1 - \frac{1}{3}\left(\frac{l_e/d}{K}\right)^4\right] = \left[1 - \frac{1}{3}\left(\frac{16.0}{23.55}\right)^4\right] = 0.93$$

$$F'_c = F''_c\ C_P = (1380)(0.93) = 1280 \text{ psi}$$

Determine J_x:

$$J_x = \frac{(l_e/d)_x - 11}{K_x - 11} = \frac{16 - 11}{23.55 - 11} = 0.398$$

Determine F'_b:

$$l_e = 1.61l = (1.61)(10)(12) = 193.2 \quad \text{(from Table 5.11).}$$

$$C_s = \sqrt{\frac{l_e d}{b^2}} = \sqrt{\frac{(193.2)(7.5)}{(7.5)^2}} = 5.08 < 10$$

(use short-beam formula, $C_L = 1$)

$$F'_b = F''_b = 1610 \text{ psi}$$

$$f_c = \frac{10{,}000}{56.25} = 178 \text{ psi}$$

$$\frac{f_c}{F'_c} + \frac{(3al'P/4lS)}{F'_b - Jf_c} \leqslant 1$$

$$\frac{178}{1280} + \frac{(3)(12)(96)(10{,}000)/(4)(120)(70.31)}{1610 - (0.398)(178)} = 0.80 < 1 \quad \text{O.K.}$$

A smaller size, a 6 × 8-in with the 8 in. side in the direction of the side bracket load, will be tried next. The 6 × 8-in. size is in the Post and Timber category, so the tabular design values are the same as for the 8 × 8-in. column. The K and F''_c are the same as 8 × 8-in. column.

Determine F'_c:

$$\left(\frac{l_e}{d}\right)_y = \frac{(10)(12)}{5.5} = 21.82, \qquad \left(\frac{l_e}{d}\right)_x = \frac{(10)(12)}{7.5} = 16$$

$$J_x = \frac{16 - 11}{23.55 - 11} = 0.398$$

Since 11 < 21.82 < 23.55, use intermediate-column formula.

$$C_P = \left[1 - \frac{1}{3}\left(\frac{21.82}{23.55}\right)^4\right] = 0.75$$

$$F'_c = C_P F''_c = (0.75)(1380) = 1040 \text{ psi}$$

Determine F'_b:

$$C_s = \sqrt{\frac{(1.61)(120)(7.5)}{(5.5)^2}} = 6.92 < 10$$

(use short-beam formula, $C_L = 1$)

$$F'_{bx} = F''_{bx} = 1610 \text{ psi}$$

$$S = 51.56 \text{ in.}^3 \qquad A = 41.25 \text{ in.}^2$$

$$f_c = \frac{P}{A} = \frac{10{,}000}{41.25} = 242 \text{ psi}$$

$$\frac{242}{1040} + \frac{(3)(12)(96)(10{,}000)/(4)(120)(51.56)}{1610 - 0.398(242)} = 1.16 > 1 \quad \text{N.G.}$$

Use the 8 × 8-in. column.

End Eccentricity

The column design formulas customarily used are based on pin-end conditions. In practice, most columns are square cut. This increases the capacity due to the extra amount of fixity at the ends.

In actual practice, the loads transmitted to columns from beams may be eccentric, especially when a column supports the end of a beam. Also, a slight error in fabrication can result in the load being eccentrically applied. Usually this causes no problem when the eccentricity is small because the partial fixity of square cut ends tends to compensate for the eccentricity. Where the design is critical, the designer may wish to check for some assumed eccentricity and make allowances for it in design. The following example illustrates the effect of 1 in. of eccentricity in a typical wood column.

Example. Column size is 6 × 6-in. sawn Southern Pine No. 1 Dense SR grade

$F_b = 1550$ psi $\quad F_c = 925$ psi $\quad E = 1{,}600{,}000$ psi

$A = 30.25$ in.^{2} $\quad S = 27.73$ in.^{3} $\quad$ Length $= 13$ ft

Axial load $P = 15{,}000$ lb

Normal duration of load $C_D = 1.0$

Dry condition of use $C_M = 1$

$F'_{bx} = F_{bx} = 1550$ psi $\quad F''_c = F_c = 925$ psi $\quad E' = E = 1{,}600{,}000$ psi

$K_e = 1 \quad l_e = (1)(13) = 13$ ft

$$\left(\frac{l_e}{d}\right)_x = \left(\frac{l_e}{d}\right)_y = \frac{(13)(12)}{5.5} = 28.36$$

$$K_x = K_y = 0.671\sqrt{\frac{E'}{F''_c}} = 0.671\sqrt{\frac{1{,}600{,}000}{925}} = 27.91$$

Since $l_e/d > K$, use long-column formula, $J_x = J_y = 1$.
Determine F'_c:

$$F'_c = \frac{0.30E'}{(l_e/d)_x^2} = \frac{(0.30)(1{,}600{,}000)}{(28.36)^2} = 597 \text{ psi}$$

$$f_c = \frac{P}{A} = \frac{15{,}000}{30.25} = 496 \text{ psi} < 597 \quad \text{O.K.}$$

Assume that the load is displaced 1 in. from the center of the column producing an eccentric load condition and bending about the x–x axis.

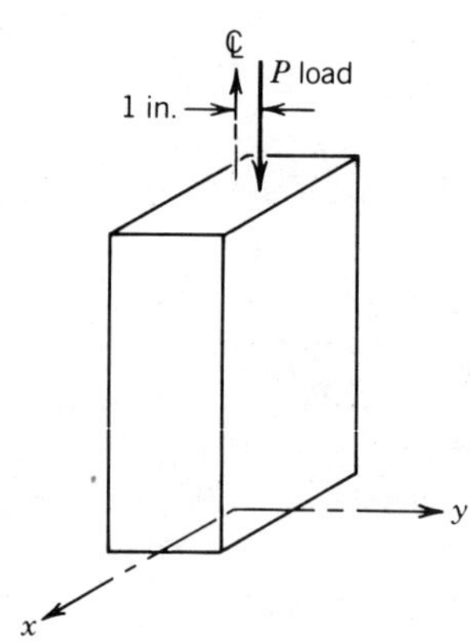

Determine F'_b:

$$C_s = \sqrt{\frac{l_e d}{b^2}} = \sqrt{\frac{(1.92)(13)(12)(5.5)}{(5.5)^2}} = 7.4 < 10$$

(use short-beam formula, from Table 5.11)

$$F'_{bx} = F''_{bx} = 1550 \text{ psi}$$

The interaction equation with $J_x = 1$ and $f_{by} = 0$ becomes

$$\frac{f_c}{F'_c} + \frac{f_c\,(7.5)\,(e/d)}{F'_{bx} - J_x f_c} \leqslant 1$$

$$\frac{496}{597} + \frac{(496)(7.5)(1/5.5)}{1{,}550 - (1)(496)} = 1.47 > 1 \quad \text{(overstressed)}$$

The column is now overloaded based on the pin-end assumption.

If it is determined that the design should include an eccentricity of 1 in., a trial size of 6 × 8-in. should be examined; $A = 41.25$ in.^{2}, $S_x = 51.56$ in.^{3}. The direction of the nominal 8-in. measurement is in the plane containing the eccentric load.

$$\left(\frac{l_e}{d}\right)_x = \frac{(13)(12)}{7.5} = 20.8$$

$$\left(\frac{l_e}{d}\right)_y = \frac{(13)(12)}{5.5} = 28.36 > 27.91 \quad \text{(use long-column formula)}$$

$$f_c = \frac{P}{A} = \frac{15{,}000}{41.25} = 364 \text{ psi}$$

$$F'_c = \frac{0.30E'}{(l_e/d)_y^2} = \frac{(0.30)(1{,}600{,}000)}{(28.36)^2} = 597 \text{ psi}$$

$$C_s = \sqrt{\frac{(1.92)(13)(12)(7.5)}{(5.5)^2}} = 8.6 < 10 \quad \text{(use short-beam formula)}$$

$$F'_b = F''_b = 1550 \text{ psi}$$

$$J_x = \frac{(l_e/d)_x - 11}{(K_x - 11)} = \frac{20.8 - 11}{27.91 - 11} = 0.58$$

The interaction equation with $J_x = 0.58$ and $F_{by} = 0$ becomes

$$\frac{f_c}{F'_c} + \frac{f_c(6 + 1.5J_x)(e_x/d_x)}{F'_{bx} - J_x f_c} =$$

$$\frac{364}{597} + \frac{(364)\,[6 + (1.5)(0.58)](1/7.5)}{1550 - (0.58)(364)} = 0.86 \quad \text{O.K.}$$

BEAMS AND BENDING MEMBERS

Wood members subjected to bending as the principle method of loading are usually designated as joists, purlins, beams, or girders. The term *beam* is often used to designate any of these members. Wood planks and decking are also subjected to bending. Beams are classified according to structural use: simple beams, cantilever beams, or continuous beams. They are also classified by shape: straight, single tapered, double tapered, curved, and double-tapered pitched and curved beams. Straight beams and tapered beams may be cambered.

Most sawn lumber or timber beams are used in simple-span applications. They may also be used as continuous or cantilevered beams.

Glued laminated timber beams are readily obtainable in long lengths, and for many applications beams cantilevered over a support are economical. In some cases, continuous beams are more suitable. Simple-span beams can provide long clear spans that are limited only by manufacturing and shipping restrictions.

The design of both sawn and glued laminated timber beams should be in accordance with established engineering practice. The following structural analysis should be taken into account in design:

1. Bending.
2. Deflection.
3. Horizontal shear.
4. Compression perpendicular to grain.
5. Lateral stability.
6. Radial stress in curved members.

Bending Stress

Bending stresses are determined by standard engineering formulas usually found in handbooks for the loading conditions most frequently encountered in design. Section 7 contains formulas for various loading and support conditions for simple, continuous, or cantilevered beam design. Tables 7.1 and 7.2 contain section properties for sawn lumber and glued laminated timber, respectively. Where the moment M and the section modulus S of a straight or relatively straight prismatic member are known, the bending stress f_b can be calculated by the flexure formula

$$f_b = \frac{M}{S}$$

The bending stress obtained from this formula must be modified for tapered, curved, and pitched and tapered curved beams as explained in the appropriate portion of this section. The tabular design values for bending as well as horizontal shear, compression perpendicular to grain, and modulus of elasticity are given in Tables 7.3 and 7.4 for sawn lumber and in AITC 117—Design (1) for glued laminated timber. Modification of these tabular design values are necessary when the conditions of use are other than stated in the tables.

Deflection and Camber

The following deflection and camber recommendations apply only to beams.

Deflection

A member of given span and stiffness will deflect, on application of load, in direct proportion to the load applied and by an amount approximating that computed by standard engineering formulas. Table 5.6 may be used to simplify the calculation of the deflection of beams under more than one type of loading by combining the formulas as given in the table for the various types of loading (see page 4-90 for the effects of long-time loading). The following example illustrates the use of Table 5.6.

Example. Determine deflection (1) at free end of overhang and (2) at centerline of supported span.

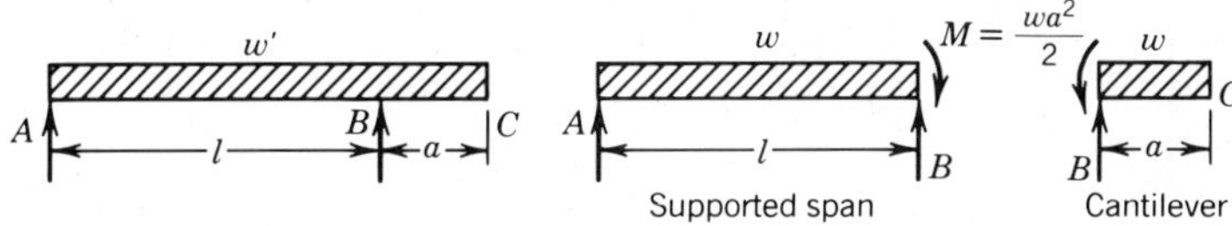

1. The cantilever portion has an initial slope θ_B at its left end equal to the slope at the right end of the supported portion, or

$$\theta_B = \theta_M + \theta_w = \frac{Ml}{3EI} + \frac{wl^3}{24EI} = \frac{wa^2 l}{6EI} - \frac{wl^3}{24EI}$$

$$\theta_B = \frac{wl}{24EI}(4a^2 - l^2)$$

Deflection at the free end: $\Delta_c = \Delta_w + a\theta_B$ or

$$\Delta_c = \frac{wa^4}{8EI} + \frac{wal}{24EI}(4a^2 - l^2) = \frac{wa}{24EI}(3a^3 + 4a^2 l - l^3)$$

2. Deflection at centerline of supported span:

$$\Delta_{℄} = \Delta_w + \Delta_M = \frac{5wl^4}{384EI} - \frac{Ml^2}{16EI}$$

$$= \frac{5wl^4}{384EI} - \frac{wa^2 l^2}{32EI} = \frac{wl^2}{384EI}(5l^2 - 12a^2)$$

Deflection Limitations

Conditions often require that a beam not exceed certain deflection limitations. Tables 5.7 and 5.8 give deflection limits for timber beams as recommended by AITC.

Member size as determined by deflection is usually limited by either applied load only or by applied load plus dead load, whichever governs, in accordance with Table 5.7. Applied load is live load, snow load, wind load, and so on.

TABLE 5.6

Beam Deflection Formulas for Various Types of Loading

	Point on Beam	Slope of Tangent to Elastic Curve		Deflection	
	A ℄ B	θ_A	θ_B	Δ_x	Δ_{max} and $\Delta_{℄}$
1	P; a, b; l	$\frac{Pb(l^2-b^2)}{6lEI}$	$\frac{Pab(2l-b)}{6lEI}$	When $x < a$: $\frac{Pbx}{6lEI}(l^2-x^2-b^2)$ When $x > a$: $\frac{Pb}{6lEI}\left[\frac{l}{b}(x-a)^3+(l^2-b^2)x-x^3\right]$	$\Delta_{max}\left(\text{at } x=\sqrt{\frac{l^2-b^2}{3}}\right)$: $\frac{Pb(l^2-b^2)^{3/2}}{9\sqrt{3}lEI}$ $\Delta_{℄}$ (if $a > b$): $\frac{Pb}{48EI}(3l^2-4b^2)$
2	P; l/2, l/2; l	$\frac{Pl^2}{16EI}$		When $x < l/2$: $\frac{Px}{48EI}(3l^2-4x^2)$	$\frac{Pl^3}{48EI}$
3	P, P; a, a; l	$\frac{Pa(l-a)}{2EI}$		When $x < a$: $\frac{Px}{6EI}(3la-3a^2-x^2)$ When $(l-a) > x > a$: $\frac{Pa}{6EI}(3lx-3x^2-a^2)$	$\frac{Pa}{24EI}(3l^2-4a^2)$

4	P P P; l/4 l/4 l/4 l/4; l	$\frac{5Pl^2}{32EI}$		When $x < l/4$: $\frac{Px}{32EI}(5l^2 - 8x^2)$ When $l/2 > x > l/4$: $\frac{P}{384EI}(72l^2x - 48lx^2 - 32x^3 - l^3)$	$\frac{19Pl^3}{384EI}$
5	w; l	$\frac{wl^3}{24EI}$		$\frac{wx}{24EI}(l^3 - 2lx^2 + x^3)$	$\frac{5wl^4}{384EI}$
6	w; a; b; l	$\frac{wa^2(l+b)^2}{24lEI}$	$\frac{wa^2(2l^2 - a^2)}{24lEI}$	When $x < a$: $\frac{wx}{24lEI}\left[a^2(l+b)^2 - 2ax^2(l+b) + lx^3\right]$ When $x > a$: $\frac{wa^2(l-x)}{24lEI}(4xl - 2x^2 - a^2)$	$\Delta_{max}\left(\text{at } x = l - \sqrt{\frac{2l^2 + a^2}{6}}\right)$: $\frac{wa^2(2l^2 - a^2)}{36lEI}\sqrt{\frac{2l^2 + a^2}{6}}$ $\Delta_{℄}$ (if $a < b$): $\frac{wa^2}{96EI}(3l^2 - 2a^2)$
7	w w; a a; l	$\frac{wa^2}{12EI}(3l - 2a)$		When $x < a$: $\frac{wx}{24EI}\left[2a^2(3l - 2a) + x^2(x - 4a)\right]$ When $x > a$: $\frac{wa^2}{24EI}(6lx - 6x^2 - a^2)$	$\frac{wa^2}{48EI}(3l^2 - 2a^2)$

TABLE 5.6 (*Continued*)

	Point on Beam	Slope of Tangent to Elastic Curve		Deflection	
	A ℄ B	θ_A	θ_B	Δ_x	Δ_{max} and $\Delta_{℄}$
8	a, w, a, l	$\frac{w}{24EI}\left[l^3 - 2a^2(3l - 2a)\right]$		When $x < a$: $\frac{wx}{24EI}\left[l^3 - 2x^2(l - 2a) - 2a^2(3l - 2a)\right]$ When $x > a$: $\frac{w}{24EI}\left[a^4 + l^3x - x^3(2l - x) - 6a^2x(l - x)\right]$	$\frac{w}{384EI}(l^2 - 4a^2)(5l^2 - 4a^2)$
9	M, l	$\frac{Ml}{3EI}$	$\frac{Ml}{6EI}$	$\frac{Mx}{6lEI}(l - x)(2l - x)$	$\Delta_{max}\left(\text{at } x = l - \frac{l}{\sqrt{3}}\right)$: $\frac{Ml^2}{9\sqrt{3}EI}$ $\Delta_{℄} = \frac{Ml^2}{16EI}$
10	M, l	$\frac{Ml}{6EI}$	$\frac{Ml}{3EI}$	$\frac{Mlx}{6EI}\left(1 - \frac{x^2}{l^2}\right)$	$\Delta_{max}\left(\text{at } x = \frac{l}{\sqrt{3}}\right)$: $\frac{Ml^2}{9\sqrt{3}EI}$ $\Delta_{℄} = \frac{Ml^2}{16EI}$

	$A \quad ℄ \quad B$	θ_B (at free end)	Δ_x	Δ_{max} (at free end)
11	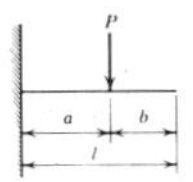 P; a; b; l	$\frac{Pa^2}{2EI}$	When $x < a$: $\frac{Px^2}{6EI}(3a - x)$ When $x > a$: $\frac{Pa^2}{6EI}(3x - a)$	$\frac{Pa^2}{6EI}(3l - a)$
12	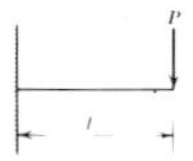 P; l	$\frac{Pl^2}{2EI}$	$\frac{Px^2}{6EI}(3l - x)$	$\frac{Pl^3}{3EI}$
13	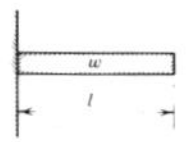 w; l	$\frac{wl^3}{6EI}$	$\frac{wx^2}{24EI}(x^2 + 6l^2 - 4lx)$	$\frac{wl^4}{8EI}$
14	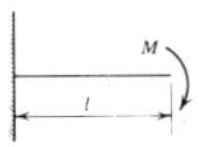 M; l	$\frac{Ml}{EI}$	$\frac{Mx^2}{2EI}$	$\frac{Ml^2}{2EI}$

TABLE 5.7

Recommended Deflection Limitations

Use Classification	Applied Load Only	Applied Load + Dead Load
Roof beams		
Industrial	l/180	l/120
Commerical and institutional		
Without plaster ceiling	l/240	l/180
With plaster ceiling	l/360	l/240
Floor beams		
Ordinary usage[a]	l/360	l/240
Highway bridge stringers	l/200 to l/300	
Railway bridge stringers	l/300 to l/400	

[a]Ordinary usage classification for floors is intended for construction in which walking comfort and minimized plaster cracking are the main considerations. These recommended deflection limits may not eliminate all objections to vibrations such as in long spans approaching the maximum limits or for some office and institutional applications where increased floor stiffness is desired. For these usages, the deflection limitations in Table 5.8 have been found to provide additional stiffness.

TABLE 5.8

Deflection Limitations for Uses Where Increased Floor Stiffness is Desired

Use Classification	Applied Load Only	Applied Load + K(Dead Load)[a]
Floor Beams		
Commercial, Office and Institutional		
Floor joists, spans to 26 ft[b]		
LL $\leqslant$ 60 psf	l/480	l/360
60 psf < LL < 80 psf	l/480	l/360
LL $\geqslant$80 psf	l/420	l/300
Girders, spans to 36 ft[b]		
LL $\leqslant$ 60 psf	l/480[c]	l/360
60 psf < LL < 80 psf	l/420[c]	l/300
LL $\leqslant$80 psf	l/360[c]	l/240

[a]K = 1 except for seasoned members where $K = 0.5$. Seasoned members for this usage are defined as having a moisture content of less than 16% at the time of installation.

[b]For girder spans greater than 36 ft and joist spans greater than 26 ft, special design considerations may be required such as more restrictive deflection limits and vibration considerations that include the total mass of the floor.

[c]Based on reduction of live load as permitted by the code.

For special uses, such as beams supporting vibrating machinery or carrying moving loads, more severe limitations may be required.

Camber

Camber is built into a structural member by introducing a curvature, either circular or parabolic, opposite to the anticipated deflection movement. Camber recommendations (see Table 5.9) vary with design criteria for various conditions of use. In addition, recommendations for camber depend on whether the member is of simple, continuous, or cantilever span, whether roof drainage is to be provided by the camber, and other factors. Reverse camber may be required in continuous and cantilever spans to permit adequate drainage.

Ponding

The roof loads caused by ponding are discussed in Section 4. It is strongly recommended that roofs be provided with adequate slope and drainage to prevent ponding. However, if such slope and drainage cannot be provided, the roofs must be designed to carry the added load. The complete analysis and design of a roof system subject to ponding is very time consuming because the deflection of all elements of the roof must be considered. When the deflection of purlins, stiffeners, and roof sheathing is significant, the analysis and design is best accomplished by use of a computer. The examples shown on page 5-186 illustrate the application of the magnification factor in the analysis of beams intended to support ponding loads.

TABLE 5.9

Recommended Minimum Camber for Glued Laminated Timber Beams

Roof beams[a]	$1\frac{1}{2}$ times dead load deflection
Floor beams[b]	$1\frac{1}{2}$ times dead load deflection
Bridge beams[c]	
Long span	2 times dead load deflection
Short span	2 times dead load + $\frac{1}{2}$ of applied load deflection

[a]Roof beams—the minimum camber of $1\frac{1}{2}$ times dead-load deflection will produce a nearly level member under dead load alone after plastic deformation has occurred. Additional camber is usually provided to improve appearance and/or provide necessary roof drainage. Roof beams should have a positive slope or camber equivalent to $\frac{1}{4}$ in. per foot of horizontal distance between the level of the drain and the high point of the roof, in addition to the minimum camber, to avoid the ponding of water. In addition, on long spans, level roof beams may not be desirable because of the optical illusion that the ceiling sags. This condition may also apply to floor beams in multistory buildings.

[b]Floor beams—the minimum camber of $1\frac{1}{2}$ times dead-load deflection will produce a nearly level member under dead load alone after plastic deformation has occurred. For warehouse or similar floors where live load may remain for long periods, additional camber should be provided to give a level floor under the permanently applied load.

[c]Bridge beams—bridge members are normally cambered for dead load only on multiple spans to obtain acceptable riding qualities.

Shear in Beams

Vertical Shear

It is ordinarily not necessary to compute or check the strength of wood beams in cross-grain (vertical) shear.

Horizontal shear

The general formula for calculating horizontal shear stress, f_v in a beam is

$$f_v = \frac{VQ}{Ib} \qquad (5\text{-}23)$$

where V = vertical shear force (lb),
Q = statical moment of area above or below neutral axis about the neutral axis (in.^{3}),
I = moment of inertia of section (in.^{4}), and
b = width of beam at neutral axis (in.).

For a rectangular beam, the formula becomes

$$f_v = \frac{3V}{2bd} \qquad (5\text{-}24)$$

where d = depth of beam (in.) and
b = width of beam (in.).

The horizontal shear stress as determined by these formulas, except in the case of checked sawn beams, may not exceed the design value in horizontal shear, F_v'. The formula should not be used when the beam end is notched, when the beam is supported by fastenings at the end such as bolts, or when the beam supports hanging loads near the end (see Section 6 for further information). Note the following when calculating the reaction, V:

(a) All loads within a distance from either support equal to the depth of the beam may be neglected when the loads are applied to one surface of the beam and the beam is fully supported by the opposite surface (see Fig. 5.10).

(b) When there is a single moving load, or one moving load that is considerably greater than the others, place that load at a distance from the support equal to the depth of the beam; place any other loads in their normal relation. When

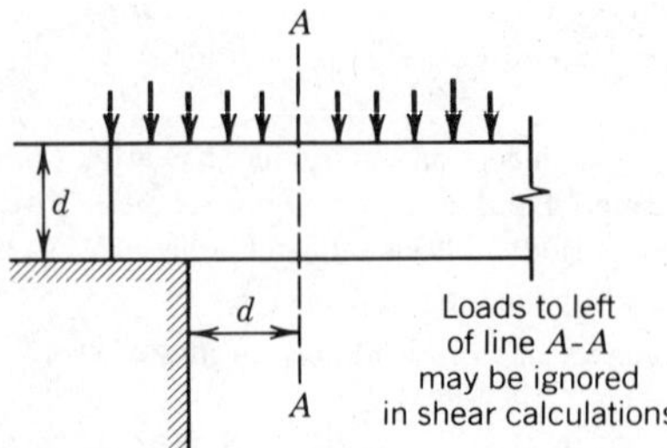

FIGURE 5.10 Loads within depth, d, from support.

there are two or more moving loads of approximately equal weight and in proximity, place the loads in the position that produces the maximum value for V, neglecting any loads within a distance from the support equal to the depth of the beam.

Horizontal Shear in Checked Beams

The design values in shear for sawn lumber beams contained in Tables 7.3 and 7.4 include allowances for checks, end splits, and shakes. These design values are recommended for use with Eq. (5-24). *NDS* (3) contains an alternate procedure for shear design that considers a checked sawn beam as if the shear were resisted by two beams, that is, the upper and lower halves of the checked beam.

End-Notched Members

Normally beams should not be notched or tapered on the tension side. If it becomes necessary to notch a bending member at its end at a support on the tension side, the horizontal shear f_v should be computed by the equation

$$f_v = \frac{3R_v}{2bd_e}\left(\frac{d}{d_e}\right) \tag{5-25}$$

where R_v = vertical reaction (lb),
b = width of beam (in.),
d = depth of beam (in.), and
d_e = depth of beam less the depth of the notch (in.) (see Fig. 5.11).

In calculating the vertical reaction R_v for use in Eq. (5-25), the loads within a depth d_e from the face of the support may be neglected. The notching of a bending member on the tension side results in a decrease in strength caused by stress concentrations around the notch as well as a reduction of the area resisting the shear forces. The notch induces tension perpendicular to grain stresses, which interact with the horizontal shear creating a splitting tendency. For these reasons, the notching of large glued laminated timbers is not recommended and it should be limited to smaller wood members. The equation given above is an empirical equation developed for the condition of a square-cornered end notch, and the ratio of the depth of the notch to the depth of the beam should be limited to 1 : 10. The designer should also consider reducing the stress concentration that occurs when a member is notched by using a gradual tapered notch configuration in lieu of

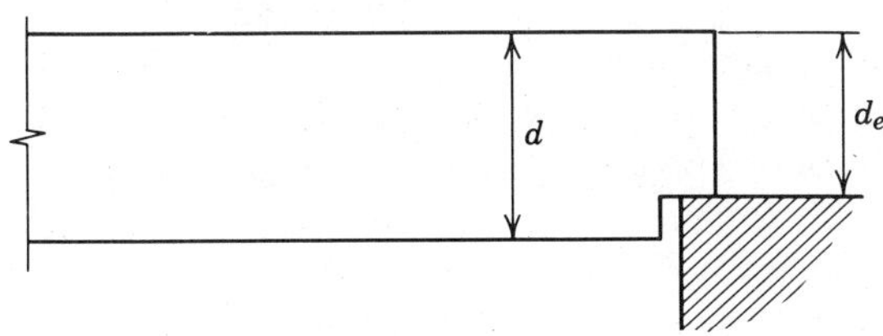

FIGURE 5.11 End-notched beam (notched on tension side).

a square-cornered notch. Also, the designer should consider the use of reinforcement such as full threaded lag bolts to resist the tendency to split at the notch. Notching on the tension side of simple beams in the center of the span is not recommended.

When a beam is notched or beveled on its upper (compression) side at the ends, a less severe condition from the standpoint of stress concentrations is realized. If such a notch is square cornered, as illustrated in Figure 5.12, the shear should be checked by the formula

$$f_v = \frac{3R_v}{2b\left[d - \left(\dfrac{d - d_e}{d_e}\right)e\right]}$$

where e is the distance the notch extends inside the inner edge of the support (in.). If e exceeds d_e, the preceding formula is not used; instead, the shear strength is computed by using only the depth of beam below the notch, d_e. In no case should a notch on the upper side of a beam exceed 40% of the total depth of a beam. If the end of a beam is beveled (as shown by the dotted line in Fig. 5.12), d_e is measured from the inner edge of the support to the bevel. The same formula and considerations as for beams notched on the upper side apply.

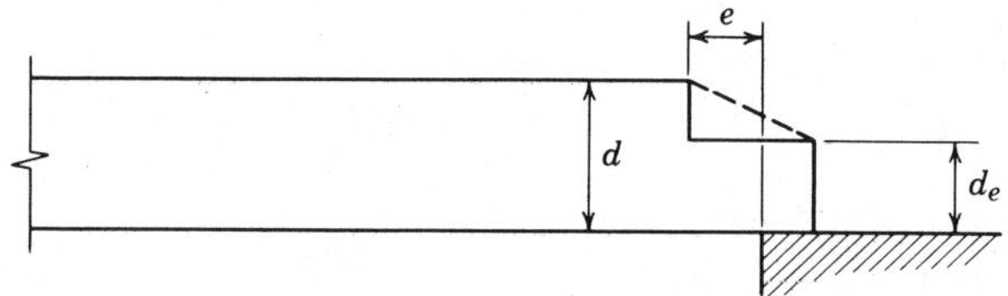

FIGURE 5.12 End-notched beam (notched on compression side).

Compression Perpendicular to Grain

The stress induced in compression perpendicular to grain, $f_{c\perp}$, at beam reactions or from loads applied to members (except as modified in the next paragraph) should not exceed the tabulated compression perpendicular to grain design values, $F_{c\perp}$. These tabulated design values apply to bearings of any length at the ends of a member and to all bearings 6 in. or more in length at any other location. When calculating the bearing area at the ends of members, it is traditional to make no allowance for the fact that as the member bends, the end rotation will result in increased pressure on the inner edge of the bearing.

For bearings less than 6 in. in length and not nearer than 3 in. to the end of a member, the maximum load per square inch is obtained by multiplying the tabular design values in compression perpendicular to grain by the following factor

$$\frac{l_b + 0.375}{l_b} \tag{5-27}$$

where l_b is the length of bearing along the grain of the wood (in.). This formula gives the following multiplying factors for the indicated lengths of bearing on such small areas as plates and washers.

Length of bearing, in.	$\frac{1}{2}$	1	$1\frac{1}{2}$	2	3	4	6 or more
Factor	1.75	1.38	1.25	1.19	1.13	1.10	1.00

In using the preceding formula and table for round bearing areas, such as washers, use a length of bearing equal to the diameter.

The tabular design values for compression perpendicular to grain in Tables 7.3 and 7.4 for sawn lumber and in AITC 117—Design (1) for glued laminated timber are based on a deformation limit of 0.04 in. obtained when tested in accordance with the standard ASTM D 143 (6) for compression perpendicular to grain. In special applications where deformation may be critical, use of a reduced compression perpendicular to grain design value may be appropriate.

For glued laminated timber, the following equation may be used for a deformation of 0.02 in., which is 50% of that associated with the values tabulated in AITC 117—Design (1).

$$F_{c\perp(0.02)} = 0.73F_{c\perp} \tag{5-28}$$

where $F_{c\perp(0.02)}$ = compression perpendicular to grain at 50% of deformation limit associated with tabulated $F_{c\perp}$ values (0.02 in.) (psi) and

$F_{c\perp}$ = compression perpendicular to grain at 0.04 in. deformation limit (psi).

For sawn lumber *NDS* (3) recommends that 5.60 psi be added to the value calculated by Eq. (5.28).

Prior to 1982 for sawn lumber and 1983 for glued laminated timber, the compression perpendicular to grain design values were based on the observed proportional limit. These design values were lower than the new values but were subject to modification for duration of load. The design values for compression perpendicular to grain in Tables 7.3 and 7.4 and in AITC 117—Design (1) should not be modified for duration of load.

The design values for compression perpendicular to grain for glued laminated timber are generally lower than those for sawn lumber for the same deformation limit. This is based, in part, on the consideration of the larger sizes of glued laminated timber, the length of bearing, and the method used to derive the design values.

Angle of Load to Grain (Hankinson Formula)

The angle of load to grain is the angle between the direction of the resultant load acting on a member and the longitudinal axis of the member. In this manual, tabular design values in compression are given for end grain in bearing, F_g, and

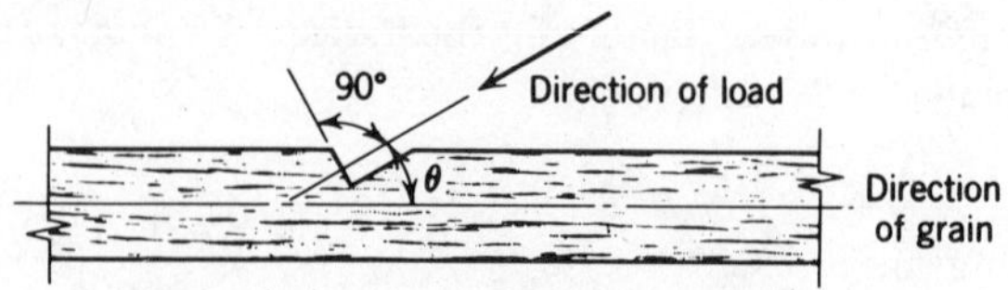

FIGURE 5.13 Angle of load to grain.

compresssion perpendicular to grain, $F_{c\perp}$. Design values for F_g may be obtained in Table 3.1 for sawn lumber and in Annex A to AITC 117—Design (1) for glued laminated timber. To obtain design values in compression at any angle of load to grain, the Hankinson formula may be used when the loaded surface is perpendicular to the direction of load (see Fig. 5.13). The Hankinson formula is

$$F'_n = \frac{F'_g F'_{c\perp}}{F'_g \sin^2\theta + F'_{c\perp} \cos^2\theta} \tag{5-29}$$

where F'_n = design value in compression acting perpendicular to the inclined surface (psi),

F'_g = design value in end grain in bearing, including all applicable modifiers (psi),

$F'_{c\perp}$ = design value in compression perpendicular to grain (psi) including all applicable modifiers (*Note:* $F_{c\perp}$ is not modified for duration of load C_D), and

θ = angle between the direction of load and the direction of grain (degrees).

When the resultant load is at an angle other than 90° with the surface being considered, the angle θ is the angle between the direction of grain and the direction of the load component.

The Hankinson formula may also be solved graphically through the use of nomographs in Figure 6.3. In this nomograph, set $F'_g = P'$, $F'_{c\perp} = Q'$, and $F'_n = N'$.

Lateral Support

To prevent sidewise buckling in bending and compression members and to permit the members to carry maximum loads, it is usually necessary to provide lateral support at intervals. The length of the intervals will depend on the dimensions of the member.

Sawn Beams

For sawn rectangular beams and joists, the approximate lateral support rules, based on nominal dimensions, listed in Table 5.10 may be applied by the designer. When adequately stabilized purlins or joists are set between bending members, the depth of the member below the bottom of the purlins or joist is used as the dimension d in determining depth–breadth ratio (d/b). For more exact engineering analysis of the lateral stability of sawn bending members, the slenderness factor considerations used for glued laminated members may be applied.

When joists have depth–breadth ratios of 8 or 9 (such joists have been in use for many years) and the compression edge is continuously supported by a deck,

TABLE 5.10

Approximate Lateral Support Rules for Sawn Beams

Depth–Breadth Ratio[a]	Rule
2:1 or less	No lateral support is required
3:1 to 4:1	The ends shall be held in position, as by full-depth solid blocking, bridging, hangers, nailing or bolting to other framing members, or other acceptable means
5:1	One edge shall be held in line for its entire length
6:1	Bridging, full-depth solid blocking, or cross bracing shall be installed at intervals not exceeding 8 ft unless both edges are held in line or unless the compression edge of the member is supported throughout its length to prevent lateral displacement, as by adequate sheathing or subflooring, and the ends at points of bearing have lateral support to prevent rotation
7:1	Both edges shall be held in line for their entire length

[a]Based on nominal dimensions. If a bending member is subject to both flexure and compression parallel to grain, the depth–breath ratio may be as much as 5:1 if one edge is held firmly in line. If under all combinations of load the unbraced edge of the member is in tension, the ratio for the beam may be 6:1.

the end bridging prevents torsional rotation. In such cases, intermediate bridging is provided to distribute concentrated loads to adjacent joists.

Glued Laminated Timber Beams

Economy in glued laminated timber beam design usually favors a deep and narrow section. Such a section increases the likelihood of lateral buckling of the compression edge. In the past, many building codes contained rules governing depth–breadth ratios to minimize such lateral buckling tendencies. The limitation of depth–breadth ratio in a beam to provide for lateral stability is believed to be an inefficient method of preventing lateral buckling in glued laminated beams. The following design procedure takes into account the unsupported length of the compression side of the bending member as well as the stability of the compression side of the member.

Slenderness Factor

The slenderness factor of a beam is calculated by the formula

$$C_s = \sqrt{\frac{l_e d}{b^2}} \tag{5-30}$$

where C_s = slenderness factor (in no case shall C_s exceed 50),
l_e = effective length of beam (in.) (see Table 5.11),
d = depth of beam (in.), and
b = breadth of beam (in.).

TABLE 5.11

Effective Length of Glued Laminated Beams

Type of Beam	Figure 5.14[a]	Nature of Load	Kind of Lateral Support and Unsupported Lengths[b]	Value of Effective Length (l_e)[c]
Simple beam	*a*	Uniformly distributed load	II	1.92 l_u
	b	Load concentrated at center	II	1.61 l_u
	c	Equal-end moments	II	1.84 l_u
	d	Any concentrated loads and spacing	I	1.92 l_u
	e	Load concentrated at center	I	1.11 l_u
	f	Two equal concentrated loads at $\frac{1}{3}$ points	I	1.68 l_u
	g	Three equal concentrated loads at $\frac{1}{4}$ points	I	1.54 l_u
	h	Four equal concentrated loads at $\frac{1}{5}$ points	I	1.68 l_u
	i	Five equal concentrated loads at $\frac{1}{6}$ points	I	1.73 l_u
	j	Six equal concentrated loads at $\frac{1}{7}$ points	I	1.78 l_u

	k	Seven or more equal concentrated loads equally spaced	I	$1.84\ l_u$
	l	Any loading	II	$1.92\ l_u$
Cantilever beam	*m*	Uniformly distributed load	II	$1.23\ l_u$
	n	Load concentrated at unsupported end	II	$1.69\ l_u$
	o	Uniformly distributed load with concentrated load at unsupported end	II	$1.69\ l_u$
	p	Any loading	II	$1.92\ l_u$

[a]Letters in this column refer to individual diagrams in Figure 5.14.

[b]Lateral support conditions and unsupported lengths:

I. When simply supported beams are provided with lateral support at bearing points as required in III and loaded with concentrated forces from secondary framing members such as purlins, the unsupported length, l_u, shall be taken as the maximum purlin spacing, provided the purlins are so connected to the beam that they prevent lateral displacement of the compression edge of the beam at the purlin locations.

II. When a beam is provided with lateral support as required in III and no other support to prevent lateral displacement is provided throughout the length of the beam, the unsupported length, l_u, shall be the distance between points of bearing or the length of the cantilever.

III. When the depth exceeds the width, lateral support shall be provided at the points of bearing, and shall be so arranged to prevent rotation and lateral displacement of the beam at those points in a plane perpendicular to its longitudinal axis.

[c]Where l_u is the unsupported length of beam with lateral support as defined in footnote *b*. Appendix O, *NDS*(3), provides for adjustment of l_e for span–depth ratios other than 17.

Source: *Code for Engineering Design in Wood*, CAN-086 (7).

Unsupported Length

When the compression edge of a beam is supported throughout its length to prevent its lateral displacement and the ends at points of bearing have lateral support to prevent lateral rotation, the unsupported length may be taken as zero.

When lateral support is provided to prevent lateral rotation at the points of bearing, but no other support to prevent lateral rotation or lateral displacement is provided throughout the length of a beam, the unsupported length is the distance between such points of bearing. See Figure 5.14 for other conditions of support. Figure 5.15 illustrates the unsupported length for cantilever beam conditions.

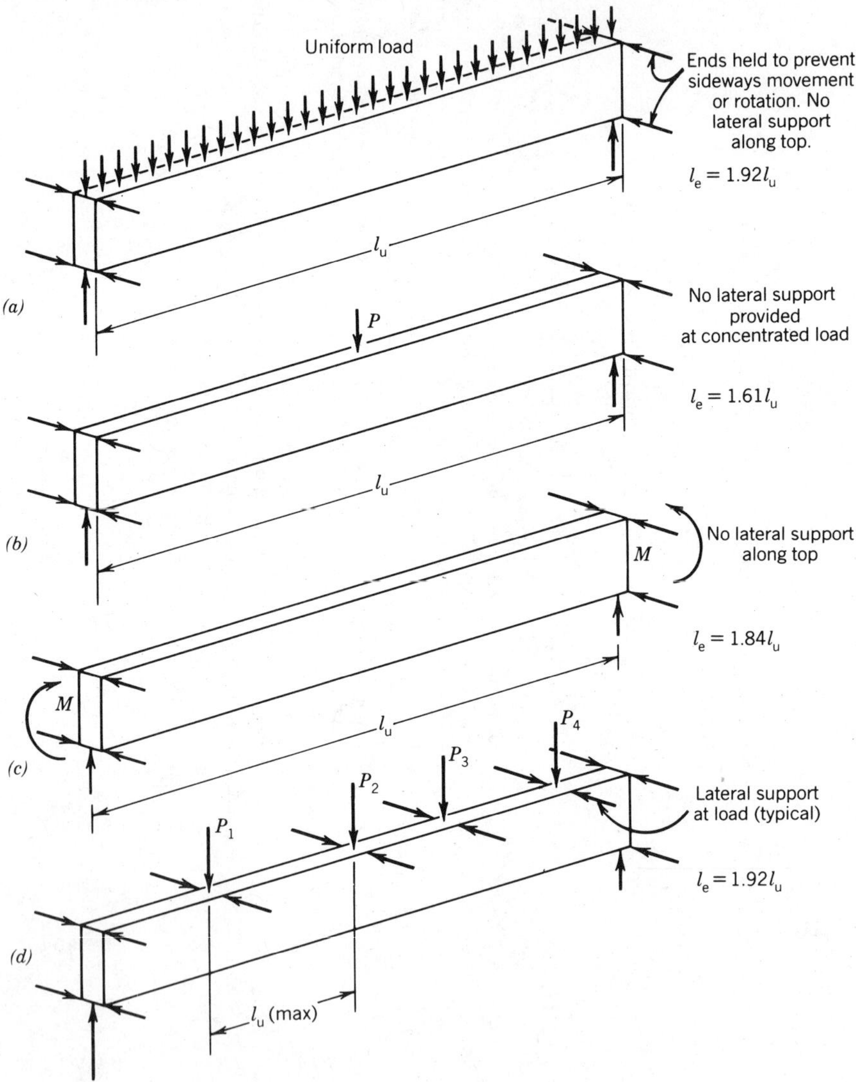

FIGURE 5.14*a–d* Various conditions of lateral support.

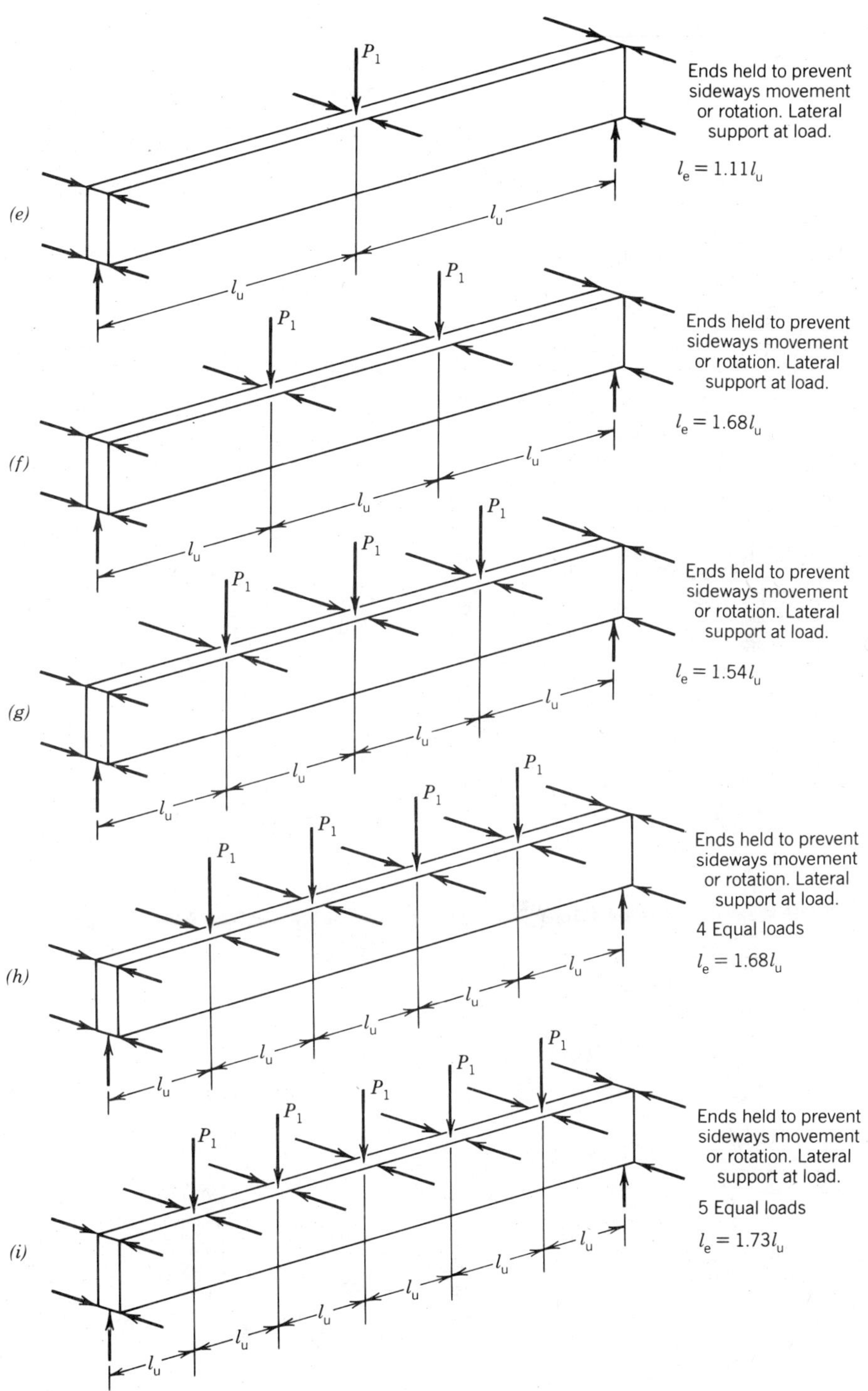

FIGURE 5.14*e–i* Various conditions of lateral support.

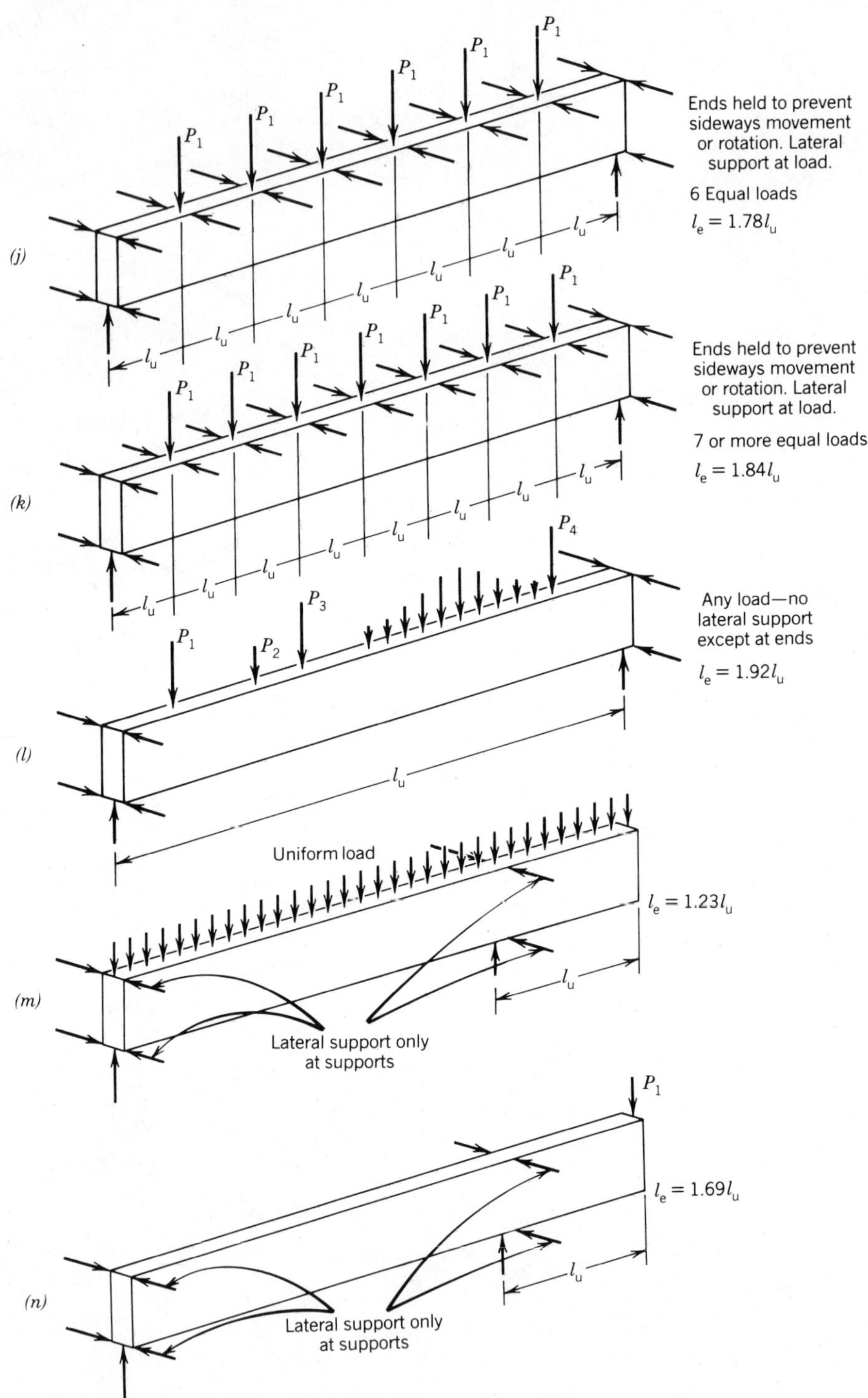

FIGURE 5.14*j–n* Various conditions of lateral support.

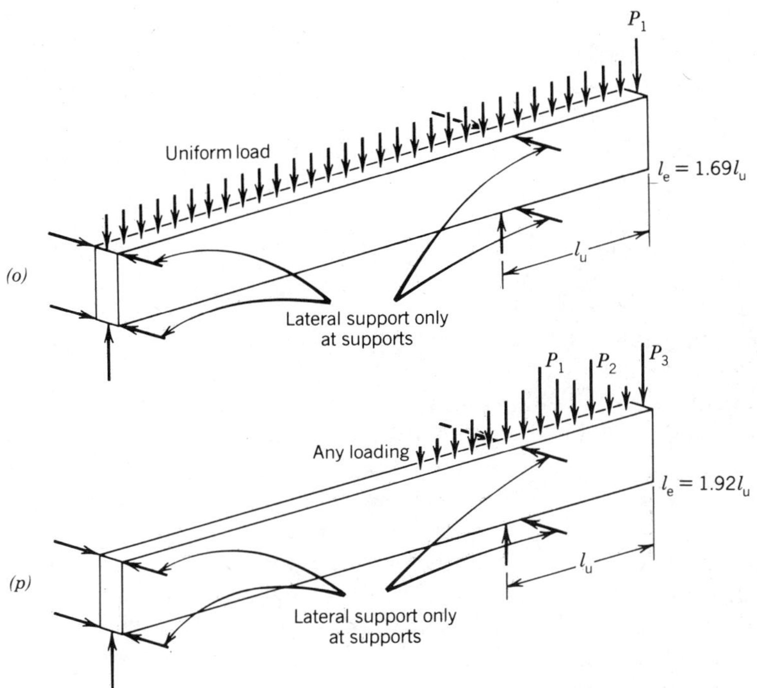

FIGURE 5.14o–p Various conditions of lateral support.

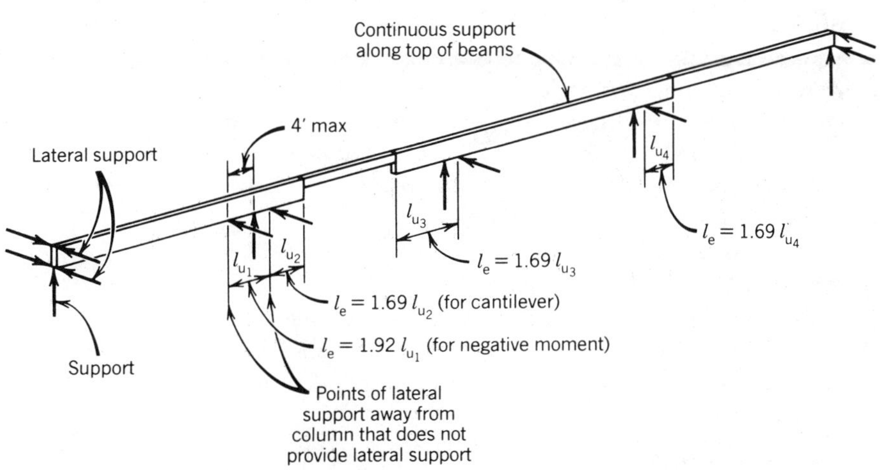

FIGURE 5.15 Effective lengths for various cantilevered support conditions.

When the beams are provided with lateral support to prevent both rotational and lateral displacement at intermediate points as well as at the ends, the unsupported length may be taken as the distance between such points of intermediate lateral support. If lateral displacement is not prevented at these points of intermediate support, the unsupported length must be defined as the full distance between points of bearing with adequate provision against rotation or as the full length of the cantilever.

Some of the means available for preventing lateral rotation of a beam at its points of bearing are to anchor the bottom of the beam to the wall or pilaster and the top of the beam to the parapet; to attach the roof diaphragm to the wall; and to provide a girt at the top of the wall and diagonal rod bracing for beams on wood columns with open sidewalls.

Connection of the roof sheathing to the compression side of the member is a good method of achieving lateral support for a beam. A plywood diaphragm nailed to the beam is an example of such support. When plank decking is used, nailing patterns are most important so that nail couples will be created. A nailing pattern that contains only one nail per deck plank and no nails between planks does not provide a system with adequate lateral support. If a wood deck is to supply such support, each piece must be securely nailed directly to the beam and to adjacent pieces to provide adequate diaphragm action. If adjacent deck planks are nailed to each other so that little or no differential movement can occur between planks, the decking will act as a diaphragm. If other kinds of decks are to provide such support, they must supply equivalent rigidity as a diaphragm.

When lateral support is provided, lateral movement of the compression edge is prevented by a continuous support, and the ends at points of bearing are held to prevent lateral rotation, there is no danger of lateral buckling and the design values require no reduction. Also, there is no need to limit the depth–breadth ratio to 5 or 6. When the depth of a beam does not exceed its breadth, no lateral support is required and the design value in bending is determined by applying the other applicable modifying factors to the tabular design values. When the depth of a beam exceeds the breadth, bracing must be provided at the points of bearing, and it must be so arranged as to prevent lateral rotation of the beam at those points in a plane perpendicular to its longitudinal axis. The design value in bending is calculated by the following procedure.

Design Values for Slender Beams

The tabular design value in bending F_b must be adjusted by all applicable modifiers as shown in Table 5.1 including the lateral stability of beams factor C_L when applicable. The lateral stability of beams factor C_L is determined as follows for short beams and intermediate beams:

Short Beams. When the slenderness factor, C_s does not exceed 10, C_L is unity and F_b' is determined by multiplying F_b by all other applicable modifiers.

Intermediate Beams. When the slenderness factor C_s is greater than 10 but does not exceed C_k, the lateral stability of beams factor C_L is determined as follows:

$$C_L = \left[1 - \frac{1}{3}\left(\frac{C_s}{C_k}\right)^4\right] \tag{5-31}$$

$$C_{kx} \text{ (bending about the } x\text{–}x \text{ axis)} = 0.956 \sqrt{E'_y / F''_{bx}} \text{ and} \tag{5-32}$$

$$C_{ky} \text{ (bending about the } y\text{–}y \text{ axis)} = 0.956 \sqrt{E'_x / F''_{by}}$$

where F''_{bx}, F''_{by} = tabular design value in bending (psi) adjusted by applicable modifiers (except for size factor C_F and lateral stability of beams factor C_L associated with the axis of bending) and

E'_x, E'_y = modulus of elasticity (psi) adjusted by applicable modifiers associated with the axis of lateral torsional buckling.

The design value F'_b is equal to either $F''_b C_L$ or $F''_b C_F$; C_L is not cumulative with C_F; and the smaller factor controls.

For sawn lumber, $F''_{bx} = F''_{by} = F_b$ and $E'_x = E'_y = E_y$ because these values do not generally vary with regard to axis.

The formula for the design value for an intermediate beam can also be written in the same form as the formula for an intermediate column.

$$F'_b = F''_b \left[1 - \frac{1}{3}\left(\frac{C_s}{C_k}\right)^4\right] \tag{5-33}$$

Long Beams. When the slenderness factor C_s is greater than C_k but less than 50, the design value in bending is determined directly by the long beam formula

$$F'_b = \frac{0.609 E'}{C_s^2} \tag{5-34}$$

where E' is modulus of elasticity E (psi) adjusted by applicable modifiers.

Figure 5.16 graphically illustrates the relationship between C_s and F'_b for a tabular design value in bending of 2400 psi and a modulus of elasticity of 1,800,000 psi. Similar plots can be developed for any combination of tabular bending stress F_b and modulus of elasticity E.

Equations (5-32) and (5-34) apply to glued laminated timber and MSR lumber. For products of more variability such as visually graded lumber, the formula for C_k is

$$C_k = 0.811 \sqrt{\frac{E'}{F''_b}} \tag{5-35}$$

and the long beam formula is

$$F'_b = \frac{0.438 E'}{C_s^2} \tag{5-36}$$

Note that the modifying factor for duration of load, C_D, is not applicable to members in the long beam classification because it does not apply to E. However,

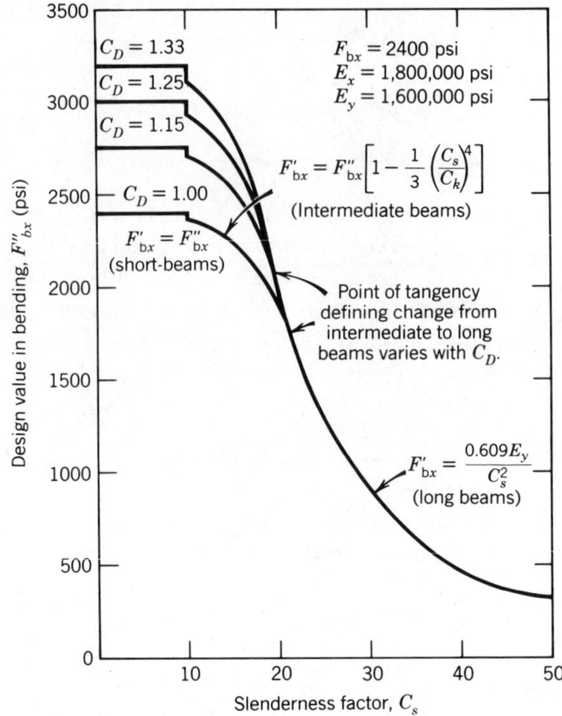

FIGURE 5.16 F'_b versus C_s.

F'_b cannot exceed the tabular design value F_b multiplied by all other applicable modifying factors.

To illustrate the preceding concepts, the following three examples representing each of the possible slenderness categories have been selected. For each example, assume that the design situation requires a simple-span glued laminated timber beam having a clear span of 60 ft. A preliminary analysis of the loads indicates that the bending moment required by this design is

$$M = 1{,}500{,}000 \text{ in.-lb}$$

For these examples, it has been assumed that the roof sheathing is not attached directly to the roof beams in a manner that will provide lateral support for the compression side of the beam. Futhermore, assume that a 24*F* combination has been specified with moduli of elasticity $E_x = 1{,}800{,}000$ psi and $E_y = 1{,}600{,}000$ psi. The roof load is primarily snow and the duration-of-load factor $C_D = 1.15$.

Determine trial size based on bending stress; assume a short beam:

$$S_n \text{ (required)} = \frac{M}{F_b C_D} = \frac{1{,}500{,}000}{(2400)(1.15)} = 543 \text{ in.}^3$$

where S_n is section modulus S times size factor C_F. From Table 7.2, a $5\frac{1}{8} \times$ 27-in. member with $S_n = 569$ in.^{3} is selected; then, $S = 622.7$ in.^{3} and $C_F = 0.91$.

Also, size factor C_F is not applicable to members classified as long beams. For purposes of comparing the reduction in design values for long beams, C_L can be assumed to be F'_b/F''_b. This ratio cannot exceed 0.667, which is the dividing point between a long beam and an intermediate beam. A member would have to be over 35 ft deep for C_F to be less than C_L. Therefore, the size factor is not applicable to members classified as long beams. See page 5-197 for additional information when axial loads are combined with bending.

Check for slenderness assuming the following cases of lateral bracing:

Example a. Assume the 60-ft main beam is laterally braced by joists placed 24 in. on centers with the only mode of lateral buckling taking place between the joists. For this case, the effective length l_e can be determined by use of Table 5.11 for uniformly distributed load because this condition closely approximates uniform loading. $C_D = 1.15$.

$$l_e = 1.92\, l_u = (1.92)(24) = 46.08 \text{ in.}$$

$$C_s = \sqrt{\frac{l_e d}{b^2}} = \sqrt{\frac{(46.08)(27)}{(5.125)^2}} = 6.9 < 10 \quad \text{(short beam)}$$

Thus, because $C_s < 10$, there is no reduction in bending stress resulting from slenderness, $C_L = 1$. Use $C_F = 0.91$:

$$F'_b = F_b C_D C_F = (2400)(1.15)(0.91) = 2512 \text{ psi}$$

$$S = \frac{M}{F'_b} = \frac{1{,}500{,}000}{2512} = 597 \text{ in.}^3 < 622.7 \text{ in.}^3$$

The $5\frac{1}{8} \times 27$-in. member is satisfactory.

Example b. Assume the 60-ft main beam is laterally braced by purlins placed 60 in. on centers with the only mode of lateral buckling taking place between the purlins.

$$F''_b = (2400)(1.15) = 2760 \text{ psi}$$

From Table 5.11 for seven or more concentrated loads,

$$l_e = (1.84)(60) = 110.4 \text{ in.}$$

Note that for 12 concentrated loads, it might also be reasonable to consider this loading condition as uniform and set $l_e = 1.92\, l_u$.

$$C_s = \sqrt{\frac{l_e d}{b^2}} = \sqrt{\frac{(110.4)(27)}{(5.125)^2}} = 10.7 > 10$$

$$C_k = 0.956 \sqrt{\frac{E'_y}{F''_{bx}}} = 0.956 \sqrt{\frac{1{,}600{,}000}{2760}} = 23.0$$

Since $10 < 10.7 < 23.0$, use the intermediate-beam formula.

$$C_L = 1 - \frac{1}{3}\left(\frac{C_s}{C_k}\right)^4 = 1 - \frac{1}{3}\left(\frac{10.7}{23.0}\right)^4 = 0.98$$

Since 0.98 > 0.91, size factor controls and $F_b' = 2512$ psi as determined initially based on application of size factor is applicable. The $5\frac{1}{8} \times 27$-in. section originally analyzed is satisfactory for this condition of lateral bracing.

Example c. Assume the 60-ft main beam is laterally braced by a secondary framing member 25 ft from one end with the buckling mode taking place between the end supports and this bracing member. From Table 5.11 use case *a* with an $l_u = 35$ ft and $l_e = 1.92\,l_u$ as being the closest condition to this:

$$l_e = (1.92)(420) = 806 \text{ in.}$$

$$C_s = \sqrt{\frac{l_e d}{b_2}} = \sqrt{\frac{(806)(27)}{(5.125)^2}} = 28.8$$

Since 50 > 28.8 > 23.0, use the long-beam formula.

$$F_b' = \frac{0.609\,E_y'}{C_s^2} = \frac{(0.609)(1{,}600{,}000)}{(28.8)^2} = 1175 \text{ psi}$$

The section modulus required is

$$S = \frac{1{,}500{,}000}{1175} = 1275 \text{ in.}^3$$

The $5\frac{1}{8} \times 27$-in. beam is inadequate and a larger beam must be used. A deeper $5\frac{1}{8}$-in.-wide beam could be used, but the *d/b* ratio becomes quite high. Try a $6\frac{3}{4} \times 25\frac{1}{2}$-in. beam, $S = 731.5$ in.^{3}:

$$C_s = \sqrt{\frac{(806)(25.5)}{(6.75)^2}} = 21.2 < 23.0 \quad \text{(intermediate beam)}$$

$$C_L = \left[1 - \frac{1}{3}\left(\frac{21.2}{23.0}\right)^4\right] = 0.76$$

Since 0.76 < 0.92, the lateral stability of beams factor C_L controls.

$$F_b' = (2400)(1.15)(0.76) = 2100 \text{ psi}$$

The section modulus required is

$$S = \frac{1{,}500{,}000}{2100} = 716.9 \text{ in.}^3 < 731.5 \text{ in.}^3 \quad \text{O.K.}$$

Torsion

Torsional stresses are sometimes induced in timber members. Torsional loads are forces that are offset from the shear center of the section under consideration, which cause the members to twist. For square and rectangular members, the shear center is the same as the centroid of the cross section. The torsional stress in a member is the stress that resists the tendency of a torsional load to twist the member.

Some examples of designs where torsion might be significant are beams with heavy eccentric side loads, bridge stringers resisting guard rail loads, and utility towers.

The torsional stress f_s in a rectangular section is

$$f_s = \frac{T(3a + 1.8b)}{8a^2b^2} \tag{5-37}$$

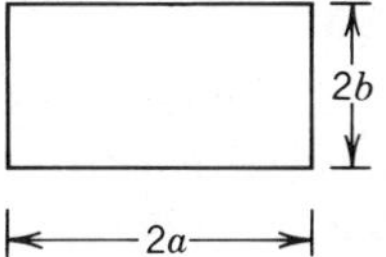

where f_s = maximum torsional stress at midpoint of each long side (psi),
T = applied torque (in.-lb),
a = one-half long side dimension (in.), and
b = one-half short side dimension (in.).

For a square section, the torsional stress is

$$f_s = \frac{4.8\,T}{s^3} \tag{5-38}$$

where s = dimension of square (in.).

There is limited information available on design values to be used when analyzing torsion in timber members. Until further research is done in this area, it is recommended that the design value in torsion, F_s, be limited to the same values recommended for radial stress.

If high torsional stresses are anticipated, the structural system should be modified to resist rotation and torsional stresses. Torsional twisting of girders may be resisted by fastening supported beams to the girders with connections that will resist the twist of the girder. These types of connections should be designed to account for shrinkage of the wood while providing the necessary rigidity.

Example. A $6\frac{3}{4} \times 30$-in. Douglas Fir-Larch glued laminated timber member is subjected to a torque T of 5000 in.-lb. The design value for radial tension stress F_{rt} is 15 psi.

$$2a = 30 \text{ in.} \qquad a = 15 \text{ in.}$$

$$2b = 6.75 \text{ in.} \qquad b = 3.375 \text{ in.}$$

$$f_s = \frac{T(3a + 1.8b)}{8a^2 b^2}$$

$$= \frac{(5000)\,[(3)(15) + (1.8)(3.375)]}{(8)(15)^2(3.375)^2}$$

$$= 12.5 \text{ psi} < F_{rt} \text{ (15 psi)} \quad \text{O.K.}$$

Design of Glued Laminated Timber Beams

Straight or Slightly Cambered Beams

Glued laminated timber beams should be designed with the ends held in place and the compression side supported so that no reduction for lack of lateral stability is required for simple-span beams. Cantilevered beams should be designed, proportioned, and braced so that the reduction for lateral stability is minimized. When it is not possible to provide this stability, the lateral stability of beams factor C_L should be applied. Also, beams should be provided with adequate roof slope to drain the roof and prevent ponding.

For cost-efficient glued laminated timber beams, the stress values determined by design should be specified rather than a particular species or stress combination. It is recommended that the tabular bending stress be selected first and a trial size be determined based on bending. Other tabular design values should be checked based on this trial size. If the required tabular design value for any of these is higher than that which can be reasonably obtained, the size of the member should be increased. Sometimes this increased size will permit the use of a lower tabular design value in bending than was originally assumed. Table 1, AITC 117—Design (1), contains design values for laminating combinations for members intended to be used to resist bending loads.

Design Procedure for Glued Laminated Timber Roof Beams

The following steps are recommended in the design of a glued laminated timber beam:

1. *Select the most practical framing system.* Simple-span beams of up to 50 ft in length are sufficiently economical to permit framing without interior columns or bearing walls. Where desired, long spans are competitive in cost with other construction materials. However, the cost is usually greater than that of systems using interior supports. For spans over 50 ft, systems using cantilever beams are often the most practical. Spacing of beams usually varies from 15 to 25 ft center to center with greater spacing possible with small glued laminated timber purlins or light trusses in place of conventional sawn purlins or joist framing. Figure 5.17 shows a few examples of cantilever patterns that can be used in determining preliminary framing systems. However, the actual length of the cantilevered portion of the beam will depend on the ratio of the live load to dead load because most building

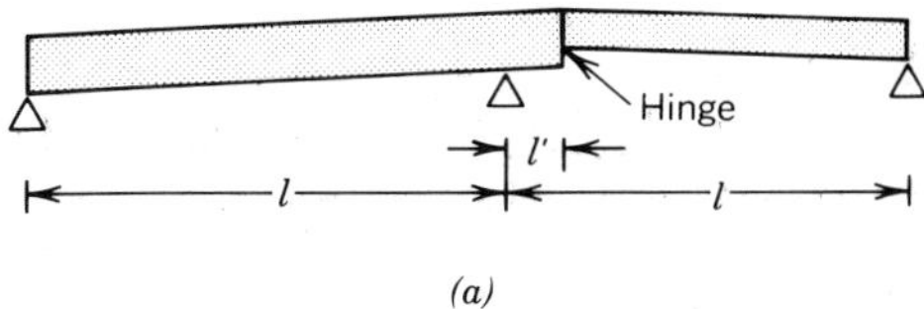

(a)

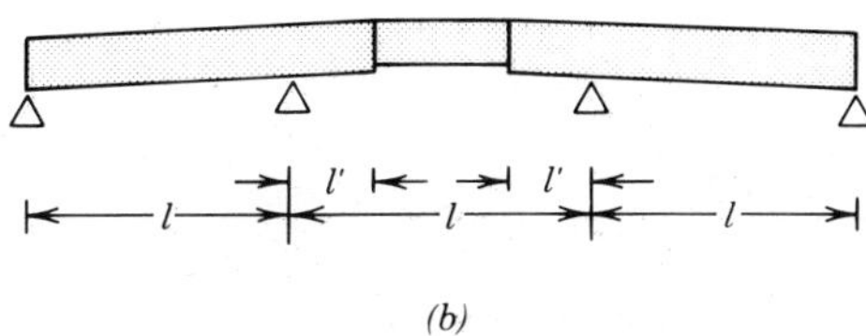

(b)

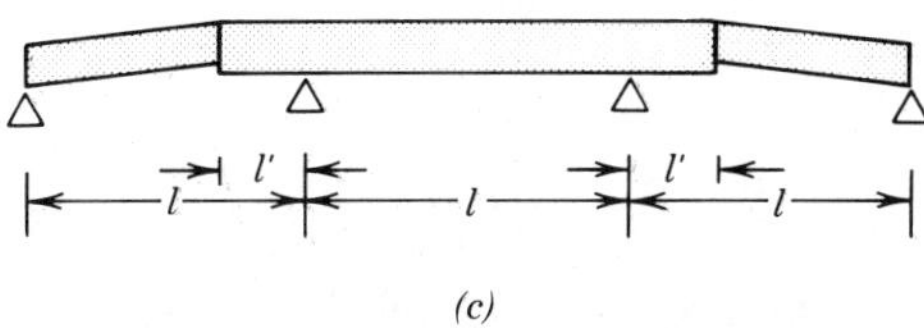

(c)

FIGURE 5.17 Cantilever beam system. (*a*) Two-span cantilever system: Cantilevered beam extends over center support with the length of cantilever l' equal to approximately 0.20 × main support spacing l, (*b*) Three-span cantilever system: End members cantilevered over intermediate column supports and carrying the suspended beam. Length of cantilevers, l' equal to approximately 0.25 × main support spacing l, (*c*) Three-span cantilever system: Center member double-cantilevered over intermediate column supports and carrying the suspended wall beams. Length of cantilevers, l' equal to approximately 0.17 × main support spacing l.

codes require the investigation of maximum bending stresses that occur both when all spans are fully loaded with dead load plus live, snow, or ponding loads and also when only alternate spans are fully loaded and dead load alone is applied on unloaded spans.

2. *Determine the design loads.*

(a) Live load includes all applied loads superimposed by the occupancy and the use of the structure, except wind, snow, ponding, earthquake, or dead loads. If applied loads act continuously, they should be treated as permanent loads and considered to be the same as a dead load. Minimum live loads are usually established by the building codes.

Under the provisions of many building codes, certain reductions in live loads are allowed based on the tributary area of loading supported by the glued laminated timber. This is an important consideration that must be recognized during the preliminary design stage.

(b) Dead load includes the weight of the glued laminated timber plus the weight of the appropriate portion of the structure that it supports as well as permanently applied loads including fixed service equipment such as heating, ventilating, and air-conditioning equipment.

(c) Snow loads (where applicable) are those designated by the governing building code.

(d) Wind loads (where applicable) are those designated by the governing building code.

For additional information on loads, see Section 4.

3. *Select design values in bending.* Glued laminated timber beams are manufactured from several species of lumber and consequently vary in availability regionally throughout the United States. As a general rule, availability can be increased by specifying the stresses required unless there are specific reasons for selecting a given species or combination. Table 1, AITC 117—Design (1), contains a list of combinations and corresponding design values that may be available. In general, glued laminated timber can be readily obtained with tabular design values in bending up to 2400 psi. The design values in Table 1 are based on members subjected to loads of normal duration, dry conditions of use, normal temperatures (see page 1-18), members 12 in. in depth, members laterally supported, and straight prismatic members.

In most cases, the only stress modifiers necessary are duration of load C_D and size factor C_F. The most direct design approach is to use S_n for the section modulus, which includes C_F. Modifiers for other conditions are given on page 5-107.

When glued laminated timbers are used under other conditions, adjustments to the design values by applicable modifiers must be made.

4. *Determine size required by bending.* Use normal engineering procedures to determine the bending moment. Then, using the selected tabular design value in bending adjusted by applicable modifiers, determine the required section modulus. The actual size used depends on several factors. Members with the larger depth–width ratios provide greater efficiency. However, practical considerations, such as lateral stability, size factor, excessive wall height, handling, and architectural considerations limit the actual depth used. A normal relationship of depth to width for beams is between 4:1 and 5:1 with an upper maximum of 7:1 when a review of construction details permit such use. Thus, $5\frac{1}{8}$-in.-wide beams with depths up to $25\frac{1}{2}$ in. are very common and depths up to a maximum of $34\frac{1}{2}$ in. may be used where adequate bracing is provided. Occasionally, beam sizes are determined by architectural considerations rather than structural requirements. In those cases, where the bending stresses are very low, a larger depth–width ratio may be possible. Length limitations vary with the manufacturer's facilities, shipping conditions between the plant and job site, and site construction limitations. Beams up to 100 ft are common. When longer beams are needed, manufacturing and shipping conditions should be investigated.

5. *Determine the required tabular design value for horizontal shear F_v.* Determine the end reaction associated with shear V and the required tabular design value for shear, F_v.

6. *Determine the required tabular design value for compression perpendicular to grain $F_{c\perp}$.* Determine the area in bearing and then calculate the required tabular design

value in compression perpendicular to grain. If the design value is higher than what may be reasonably available, the bearing area needs to be increased. Note that the duration-of-load modification factor C_D does not apply to compression perpendicular to grain.

7. *Determine the required tabular design value for modulus of elasticity E.* Recommended deflection limitations for glued laminated timber roof beams are included in Table 5.7. The recommended deflection limitations are independent of the required camber. They are minimum recommendations, and it may be necessary under some circumstances to provide greater stiffness by increasing the member size.

Once the deflection limitation has been determined, the required tabular design value for E is determined by use of the appropriate deflection formula.

8. *Determine camber requirements.* Glued laminated beams can be manufactured with camber to compensate for the deflection due to the load carried by the beam. For roof beams, this camber is in addition to any curvature that may be used to provide a slope for drainage. Due to creep under long-term applied loads, the usual practice is to provide a minimum camber equal to 1.5 times dead-load deflection.

9. *Consider detailing requirements including roof drainage.* The plan for roof drainage should be detailed showing drainage paths. Downspouts and/or scuppers should be designed and detailed to handle the anticipated flow of water. In order to provide positive effective drainage and prevent ponding, a minimum slope of $\frac{1}{4}$ in./ft after dead-load deflection has taken place should be designed into the roof system. This slope must be considered when proposed camber and anticipated deflection are calculated. Additional slope or camber may be required where camber or deflection problems may occur. One method of improving drainage through scuppers and downspouts is to slope the ends of the beams with a tapered cut on the top face. Varying beam seat elevations is another effective means to improve drainage. See page 4-88 for more information on ponding and the procedures necessary to prevent ponding.

10. *Specification.* The specification for glued laminated timber should include the following:

Size (width, depth, length)
Camber
Appearance grade
Required design values
  Bending F_b
  Shear F_v
  Compression perpendicular to grain
    $F_{c\perp}$ tension face
    $F_{c\perp}$ compression face
  Modulus of elasticity E

Example. Design a glued laminated timber beam to support the roof for an industrial building 45 × 60 ft as shown for dry conditions of use.

1. A simple joist framing system is selected with glued laminated timber beams

spaced 20 ft apart spanning the 45-ft direction. Beams will be laterally braced at the ends and held in place along the top (compression side) by the roof decking. Therefore, no modification for lateral stability is required.

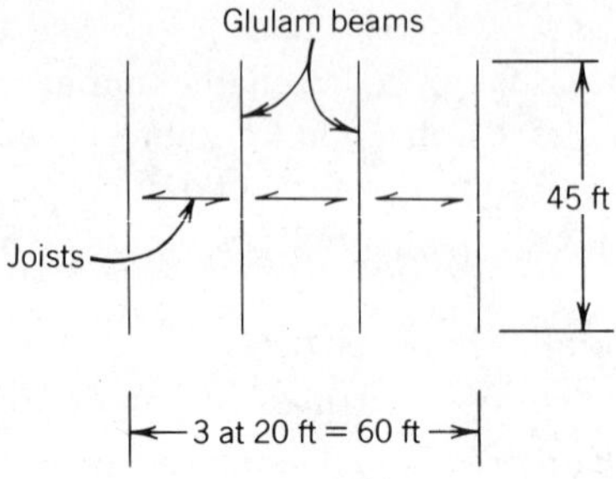

2. Determine design loads. The controlling building code requires a minimum roof live load of 20 psf. A 40% reduction in live load is permitted when the tributary area of a member exceeds 600 ft^2.

(a) Assumed loads:

LL on joists = 20 psf (no snow)
Beam tributary area (45)(20) = 900 ft^2 > 600 ft^2
LL on glulam beam, use (20)(1 − 0.40) = 12 psf (code reduction in load for tributary area)

DL = 6.0 psf (roofing)
= 1.6 psf ($\frac{1}{2}$ in. plywood)
= 1.7 psf (2 × 10 joists at 24 in. on center)
= 4.0 psf (mechanical system)
= 1.7 psf (glulam beam)

Total DL = 15.0 psf

Design load = LL + DL
= 12 + 15 = 27 psf
w_{TL} = (27)(20) = 540 plf
w_{DL} = (15)(20) = 300 plf

Note that some designers prefer to include additional dead load as a precaution to cover unanticipated loads or reroofing.

3. Select design value in bending. Try F_b = 2400 psi; duration of load modifying factor C_D = 1.25

$$F''_{bx} = (2400)\,(1.25) = 3000 \text{ psi}$$

4. Determine size required by bending:

$$M = \frac{wL^2}{8} = \frac{(540)(45)^2\,(12)}{8} = 1{,}640{,}000 \text{ in.-lb}$$

$$S_n \text{ (required)} = \frac{M}{F''_{bx}} = \frac{1{,}640{,}000}{3000} = 547 \text{ in.}^3$$

From Table 7.2, assuming a beam width of $5\frac{1}{8}$ in., try a $5\frac{1}{8} \times 27$ in. beam

$$S_n = 569 \text{ in.}^3 \qquad S = 622.7 \text{ in.}^3 \qquad I = 8406 \text{ in.}^4 \qquad A = 138.4 \text{ in.}^2$$

5. Determine required tabular design value for horizontal shear F_v. The effective length L_e for determining horizontal shear is

$$L_e = L - \frac{2d}{12} = 45 - \frac{(2)(27)}{12} = 40.5 \text{ ft}$$

$$V = \frac{wL_e}{2} = \frac{(540)(40.5)}{2} = 10{,}900 \text{ lb}$$

$$f_v = \frac{3V}{2A} \qquad F_v \text{ (required)} = \frac{f_v}{C_D}$$

$$F_v\text{(required)} = \frac{(3)(10{,}900)}{(2)(138.4)(1.25)} = 94.8 \text{ psi}$$

6. Determine the required tabular design value for compression perpendicular to grain, $F_{c\perp}$. Tension face: a typical commercial beam seat provides a length of bearing of 5 in.

$$\text{Area in bearing} = A = (5.125)(5) = 25.625 \text{ in.}$$

$$\text{Load on bearing plate} = P = \frac{wL}{2} = \frac{(540)(45)}{2} = 12{,}150 \text{ lb}$$

$$F_{c\perp} \text{ (required)} = \frac{P}{A} = \frac{12{,}150}{25.625} = 474 \text{ psi} \quad \text{(tension face)}$$

Compression face: the loads from the joists are transferred through joist hangers. A commercial joist hanger with a $1\frac{5}{16} \times 3\frac{1}{2}$-in. top plate to transfer the load will be used.

$$\text{Area in bearing} = A = (1.3125)(3.5) = 4.59 \text{ in.}^2$$

$$\text{Load on plate} = P = \frac{wL}{2} = \frac{(2)(27 - 1.7)(20)}{2} = 506 \text{ lb}$$

$$F_{c\perp} \text{ (required)} = \frac{P}{A} = \frac{506}{4.59} = 110 \text{ psi} \quad \text{(compression face)}$$

7. Determine the required tabular design value for modulus of elasticity E. The recommended deflection limitations for an industrial building from Table 5.7 are:

For live load plus dead load,

$$\Delta_{(LL+DL)} = \frac{l}{120} = \frac{(45)(12)}{120} = 4.5 \text{ in.}$$

For live load only,

$$\Delta_{LL} = \frac{l}{180} = \frac{(45)(12)}{180} = 3 \text{ in.}$$

$$\frac{LL}{TL} = \frac{12}{27} = 0.444 \qquad \frac{LL}{TL}(\Delta_{(LL + DL)}) = (0.444)(4.5) = 2 \text{ in.} < 3 \text{ in.}$$

Live load plus dead load controls.

For a uniform load,

$$\Delta = \frac{5wL^4}{384EI} \qquad E(\text{required}) = \frac{5wL^4}{384I\Delta}$$

$$E\,(\text{required}) = \frac{(5)(540)(45)^4\,(12)^3}{(384)(8406)(4.5)} = 1{,}320{,}000 \text{ psi}$$

The lowest tabular design value for a readily available 24*F* visual combination is 1,700,000 psi. Therefore, an *E* of 1,700,000 will be specified and used to calculate dead-load deflection for camber.

8. Determine camber requirements:

$$\Delta_{DL} = \frac{5w_{DL}L^4}{384EI} = \frac{(5)(300)(45)^4\,(12)^3}{(384)(1{,}700{,}000)(8406)} = 1.94 \text{ in.}$$

$$\text{Camber} = 1.5\Delta_{DL} = (1.5)(1.94) = 2.9 \text{ in.} \quad (\text{use 3 in.})$$

9. Roof drainage—The simplest method of roof drainage with straight beams in a building of this type is to drain the water to one side by varying the elevations of the beam seats so that a slope of at least $\frac{1}{4}$ in./ft is obtained. If the slope of the roof drainage is in the direction of the longitudinal axis of the glued laminated timbers, the difference in elevation is $(45)(\frac{1}{4}) = 11.25$ in.

If drainage to each side of the beam is required, the required slope for drainage can be obtained by adding additional camber or curvature to the member.

$$\text{Total camber} = 3 + (45/2)(\tfrac{1}{4}) = 8.625 \text{ in.} \quad (\text{use 9 in.})$$

10. Specification. Two $5\frac{1}{8}$ in. × 27 in. × 45 ft glulam beams (note that length should be the exact out to out dimensions required)

Camber = 3 in. (slope for drainage)

Appearance grade—Industrial.
The members shall be made from a laminating combination that provides tabular design values equal to or exceeding the following:

Bending F_{bx} = 2400 psi
Horizontal shear F_v = 95 psi
Compression perpendicular to grain
  Tension face $F_{c\perp}$ = 475 psi
  Compression face $F_{c\perp}$ = 110 psi
Modulus of elasticity E = 1,700,000 psi

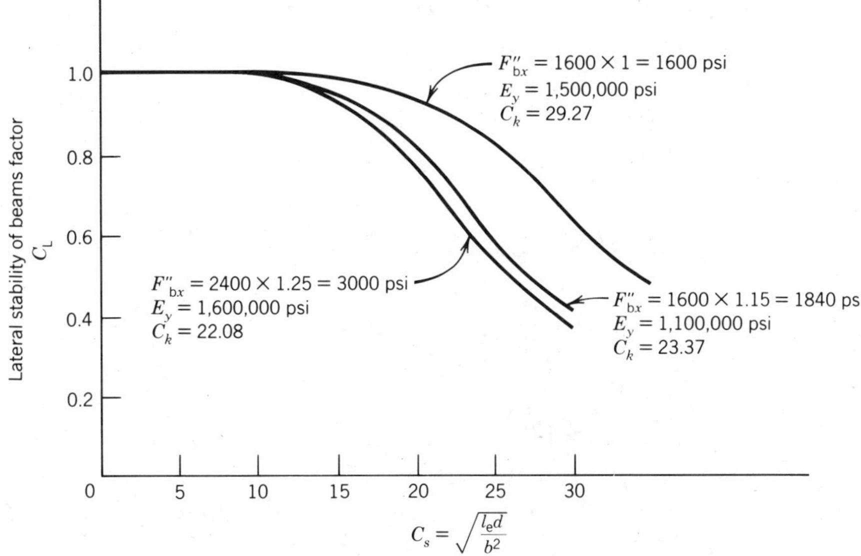

FIGURE 5.18 Relationship of lateral stability of beam factor, C_L to C_s.

Designing for Lateral Stability

Usually the most efficient method of roof framing is to provide continuous support to the glued laminated timbers with sheathing, decking, or joists and purlins spaced closely enough to provide lateral support. When this cannot be done, such as when purlins are installed over the top of the beam without direct connection between beam and roof sheathing, the lateral stability of the beam must be investigated.

See pages 5-164 to 5-168 for information on use of lateral stability of beam factor C_L. Figure 5.18 shows the relationship between C_s and C_L for various values of F''_{bx} and E_y. The curves for other relationships will in most cases, fall between the upper and lower curves. Note that the factors C_s and C_L are not cumulative and the smaller of the two controls for bending about the x axis. In many cases C_F is less than C_L and no reduction for increased distances between points of support is required.

Example. Assume that all conditions and requirements of the previous example are the same except that the roof system consists of nominal 4-in. purlins spaced 8 ft apart on top of the beams with a short space on one end. Assume no change

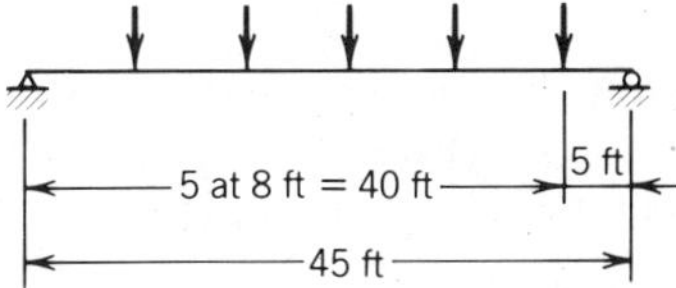

in dead loads. From Table 5.11, use condition *i* as being closest to design application (five equal concentrated loads).

$$l_e = 1.73\,l_u = (1.73)(8)(12) = 166.1 \text{ in.}$$

$$C_s = \sqrt{\frac{l_e d}{b^2}} = \sqrt{\frac{(166.1)(27)}{(5.125)^2}} = 13.07$$

$$F_b'' = (2400)(1.25) = 3000 \text{ psi} \qquad E_y' = 1{,}600{,}000 \text{ psi}$$

$$C_k = 0.956\sqrt{\frac{E_y'}{F_{bx}''}} = 0.956\sqrt{\frac{1{,}600{,}000}{3000}} = 22.08$$

Since $10 < 13.07 < 22.08$, use the intermediate-beam formula.

$$F_b' = F_{bx}'' \text{ multiplied by } C_F \text{ or } C_L, \text{ whichever is lower.}$$

$$C_L = 1 - \frac{1}{3}\left(\frac{C_s}{C_k}\right)^4 = 1 - \frac{1}{3}\left(\frac{13.07}{22.07}\right)^4 = 0.96$$

From Table 7.3, $C_F = 0.91 < 0.96$, and the size factor controls

$$F_b' = F_b''\,C_F = (3000)(0.91) = 2730 \text{ psi}$$

$$S\text{ (required)} = \frac{M}{F_b'} = \frac{1{,}640{,}000}{2730} = 600.8 \text{ in.}^3$$

A $5\frac{1}{8} \times 27$-in. member has a section modulus of 622.7 in.^{3} In this case, changing the roof-framing system resulted in the same-size member because the size factor controlled design.

Roof Beams with Tapered End Cuts

Taper cuts on the top at the end of beams are sometimes used to improve drainage, to provide extra head for downspouts and scuppers, to facilitate discharge of water, and to reduce the height of the wall. This sloping cut is commonly made

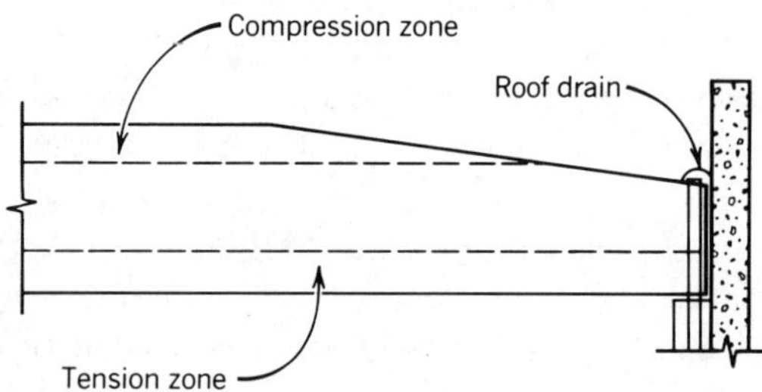

through the compression zone laminations exposing the lower strength core laminations and results in a member with a lower design value in bending than listed for the combination in Table 1, AITC 117–Design (1). The design values of F_{bx}, E_x, and $F_{c\perp}$ for taper cuts up to one-half the depth of the member can be obtained from Table 5.12. The variation in design values for combinations with the same design value in bending is caused by the use of various grades or species in the

TABLE 5.12

Design Values for Members with Taper Cuts on Compression Side

Combination Symbol	F_{bx}^{a}(psi)	E_{x}^{a}(psi × 10^{-6})	$F_{c\perp}^{b}$(psi)
VISUALLY GRADED WESTERN SPECIES			
16F-V1	1350	1.3	255
16F-V2	1600	1.3	375
16F-V3	1600	1.6	560
16F-V4	850	1.3	255
16F-V5	1250	1.6	470
16F-V6	1600	1.6	560
16F-V7	1600	1.3	375
16F-V8	1600	1.2	500
20F-V1	1400	1.3	255
20F-V2	1900	1.5	375
20F-V3	2000	1.6	560
20F-V4	2000	1.6	560
20F-V5	850	1.3	255
20F-V6	1250	1.6	470
20F-V7	2000	1.6	560
20F-V8	2000	1.6	560
20F-V9	1900	1.5	375
20F-V10	1950	1.5	375
20F-V11	1850	1.3	500
22F-V1	1400	1.3	255
22F-V2	1900	1.5	375
22F-V3	2200	1.7	560
22F-V4	2200	1.7	560
22F-V5	700	1.3	255
22F-V6	1300	1.7	470
22F-V7	2200	1.7	560
22F-V8	2200	1.7	560
22F-V9	1900	1.5	375
22F-V10	1900	1.5	500
24F-V1	1450	1.3	255
24F-V2	1900	1.5	375
24F-V3	2200	1.7	560
24F-V4	2200	1.7	560
24F-V5	2100	1.6	375
24F-V6	1000	1.5	255
24F-V7	1300	1.7	470
24F-V8	2200	1.7	560
24F-V9	1900	1.5	375
24F-V10	2000	1.6	375
24F-V11	2100	1.5	500

(*continued*)

TABLE 5.12 (*Continued*)

Combination Symbol	F_{bx}^{a}(psi)	E_{x}^{a}(psi × 10^{-6})	$F_{c\perp}^{b}$(psi)
E-Rated Western Species			
16F-E1	1400	1.3	255
16F-E2	1500	1.3	375
16F-E3	850	1.6	560
16F-E4	1250	1.3	255
16F-E5	1600	1.6	470
16F-E6	1500	1.6	560
16F-E7	1500	1.3	375
20F-E1	1400	1.3	255
20F-E2	1900	1.5	375
20F-E3	2000	1.6	560
20F-E4	850	1.3	255
20F-E5	1350	1.6	470
20F-E6	2000	1.6	560
20F-E7	1850	1.5	375
22F-E1	2200	1.7	560
22F-E2	2000	1.6	375
22F-E3	850	1.3	255
22F-E4	1350	1.7	470
22F-E5	2200	1.7	560
22F-E6	2000	1.5	375
24F-E1	2300	1.8	560
24F-E2	2000	1.6	375
24F-E3	2000	1.6	375
24F-E4	2300	1.7	560
24F-E5	2300	1.8	560
24F-E6	1450	1.3	255
24F-E7	1000	1.4	255
24F-E8	1350	1.9	470
24F-E9	850	1.3	375
24F-E10	2300	1.8	560
24F-E11	2000	1.6	375
24F-E12	2000	1.6	375
24F-E13	2300	1.7	560
24F-E14	2300	1.8	560
Visually Graded Southern Pine			
16F-V1	1550	1.4	560
16F-V2	1500	1.4	560
16F-V3	1550	1.4	560
16F-V4	1050	1.3	470
16F-V5	1500	1.4	560
20F-V1	1600	1.5	560

TABLE 5.12 *(Continued)*

Combination Symbol	F_{bx}^{a}(psi)	E_{x}^{a}(psi × 10^{-6})	$F_{c\perp}^{b}$(psi)
VISUALLY GRADED SOUTHERN PINE (cont)			
20F-V2	1600	1.5	560
20F-V3	2000	1.4	560
20F-V4	1150	1.4	470
20F-V5	1550	1.5	560
22F-V1	2100	1.5	560
22F-V2	2100	1.4	560
22F-V3	1600	1.5	560
22F-V4	1200	1.5	470
22F-V5	2100	1.5	560
24F-V1	1650	1.6	560
24F-V2	2300	1.6	560
24F-V3	2300	1.6	560
24F-V4	1200	1.5	560
24F-V5	2300	1.6	470
24F-V6	2300	1.6	560
E-RATED SOUTHERN PINE			
16F-E1	1600	1.6	560
16F-E2	1100	1.4	470
16F-E3	1600	1.6	560
20F-E1	2000	1.6	560
20F-E2	1200	1.5	470
20F-E3	2000	1.6	560
22F-E1	2200	1.7	560
22F-E2	1200	1.5	470
22F-E3	2100	1.6	560
24F-E1	2200	1.7	560
24F-E2	2400	1.7	560
24F-E3	1200	1.5	470
24F-E4	2200	1.7	560
CALIFORNIA REDWOOD			
B-16F-V1	1600	1.1	315

[a] Design value applicable to members that have up to one-half the depth on the compression side removed by taper cutting. Values are for normal duration of load, dry conditions of use, and 12 in. or less depth

[b] Design value in compression perpendicular to grain for the core laminations of the combinations.

core of the combination. The value of F_v remains unchanged. The values of E and $F_{c\perp}$ are usually less because of the removal of some of the higher-grade lumber in the compression zone.

The laminated beam is usually designed as a straight prismatic member assuming full depth throughout its length, and the design values specified are based on this assumption. The effect of the sloping end cut is then checked, and required design values are increased if necessary. The length of taper cuts vary depending on the length of span, the roof-framing systems, or other requirements. Sloping cuts from 4 to 8 ft long are commonly used. The depth of the end cut usually depends on the drainage requirements. However, the remaining cross section at the end of the member must be able to resist the shear stresses. The sloping end cut is then checked by application of the interaction stress factor C_I or use of the interaction formula [see Eq. (5-44)].

Example. Redesign the beam selected in the previous example of the joist roof system for a tapered end cut 6 ft long with a depth of cut $7\frac{1}{2}$ in. on the end as shown:

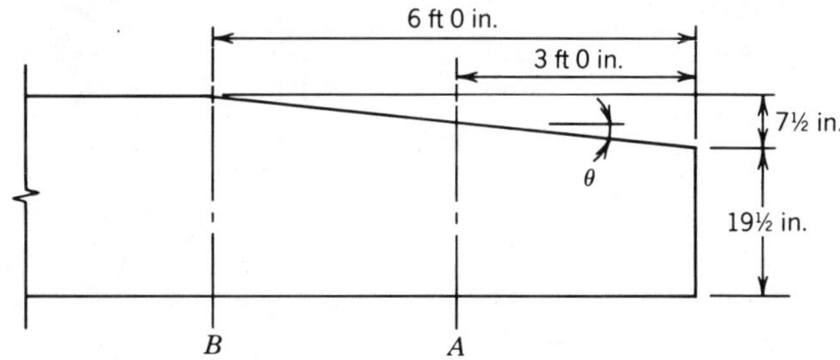

1. Determine required design value in shear

$$L_e = 45 - \frac{(2)(19.5)}{12} = 41.75 \text{ ft}$$

$$V = \frac{(540)(41.75)}{2} = 11{,}300 \text{ lb}$$

$$F_{v(\text{required})} = \frac{3V}{2AC_D} = \frac{(3)(11{,}300)}{(2)(5.125)(19.5)(1.25)} = 135 \text{ psi}$$

2. Determine the interaction factor C_I.

$$\tan\theta = \frac{7.5}{(6)(12)} = 0.1042$$

Assume trial design values in bending, shear, and compression perpendicular to grain for the tapered surface. Assume $F_v = 155$ psi (which will permit Hem-Fir, Douglas Fir-Larch, and Southern Pine to be used); $F_b = 1600$ psi; and $F_{c\perp} = 375$ psi for the tapered slope. This will permit most of the 24F visually graded combinations to be used. From Table 5.13, $C_I = 0.693$. If Table 5.13 does not contain the combination of design values selected, C_I can be calculated by use of the interaction formula (5.44).

3. Check bending stresses at Sections A and B.

Section A:

$$M_A = (12{,}150)(3)(12) - \frac{(540)(3)^2(12)}{2} = 408{,}000 \text{ in.-lb}$$

$$\text{Depth at A} = d = 27 - \frac{7.5}{2} = 23.25 \text{ in.}$$

$$S_A = \frac{bd^2}{6} = \frac{(5.125)(23.25)^2}{6} = 461.7 \text{ in.}^3$$

$$f_{bA} = \frac{M_A}{S_A} = \frac{408{,}000}{461.7} = 884 \text{ psi}$$

$$C_F \text{ for } d = 23.25 \text{ in.} = 0.93 \quad \text{(Table 7.2)}$$

Since $0.93 > 0.693$, C_I controls (C_I and C_F are not cumulative)

$$F'_{bx} = F_b C_D C_I = (1600)(1.25)(0.693) = 1386 \text{ psi} > 884 \text{ psi} \quad \text{O.K.}$$

Section B:

$$M_B = (12{,}150)(6)(12) - \frac{(540)(6)^2(12)}{2} = 758{,}000 \text{ in.-lb}$$

$$S_B = 622.7 \text{ in.}^3 \quad \text{(from Table 7.2)}$$

$$f_{bB} = \frac{M_B}{S_B} = \frac{758{,}000}{622.7} = 1220 \text{ psi}$$

The size factor at B for $d = 27$ in. is 0.91 (Table 7.2), C_I controls,

$$F'_{bx} = 1386 \text{ psi} > 1220 \text{ psi} \quad \text{O.K.}$$

4. Review assumed design values. If the assumed design values along the tapered cut are too high or too low, other assumptions can be made and a new size determined. The required compression perpendicular to grain under the joist hangers is 110 psi and the assumed design value for $F_{c\perp}$ of 375 psi is adequate. The modulus of elasticity is slightly reduced in the tapered section, but because the length of taper is small, the effect is negligible and can be ignored (when a fully tapered beam is used, a reduced E value may be necessary).

5. Check specification. Tapering the beam at the ends requires a change in the specification of F_v and the additional provision for design values along the sloping cut as follows:

Specification—Two $5\frac{1}{8}$ in. × 27 in. × 45 ft glulam beams taper cut on one end as shown on drawing on page 5-182.

TABLE 5.13

Interaction Stress Factor C_I as a Function of tan θ[a]

tan θ	C_I^t	C_I^c	C_I^d	C_I^e	C_I^f	C_I^g	C_I^h	C_I^i	C_I^j	C_I^k	C_I^l
0.02	0.960	0.972	0.962	0.966	0.977	0.972	0.968	0.981	0.982	0.979	0.987
0.025	0.940	0.958	0.942	0.949	0.964	0.957	0.952	0.970	0.972	0.968	0.981
0.03	0.910	0.941	0.920	0.928	0.950	0.940	0.933	0.958	0.960	0.955	0.972
0.035	0.891	0.922	0.896	0.906	0.933	0.921	0.911	0.944	0.947	0.940	0.963
0.04	0.864	0.902	0.870	0.882	0.915	0.900	0.889	0.928	0.932	0.924	0.952
0.045	0.837	0.880	0.843	0.857	0.896	0.878	0.865	0.912	0.917	0.907	0.941
0.05	0.809	0.857	0.815	0.832	0.876	0.855	0.840	0.894	0.900	0.889	0.928
0.055	0.781	0.835	0.788	0.806	0.856	0.832	0.815	0.876	0.882	0.870	0.915
0.06	0.753	0.811	0.761	0.781	0.835	0.809	0.791	0.857	0.864	0.850	0.901
0.065	0.727	0.788	0.735	0.756	0.813	0.785	0.766	0.838	0.846	0.830	0.887
0.07	0.701	0.766	0.709	0.731	0.792	0.762	0.742	0.819	0.827	0.810	0.872
0.075	0.676	0.743	0.685	0.707	0.771	0.740	0.718	0.800	0.809	0.790	0.857
0.08	0.652	0.721	0.661	0.684	0.751	0.718	0.696	0.781	0.790	0.771	0.842
0.085	0.629	0.700	0.638	0.662	0.730	0.696	0.674	0.762	0.772	0.751	0.827
0.09	0.607	0.679	0.616	0.640	0.710	0.676	0.652	0.743	0.753	0.732	0.811
0.095	0.586	0.659	0.595	0.619	0.691	0.656	0.632	0.725	0.735	0.714	0.796

0.100	0.566	0.640	0.576	0.600	0.672	0.636	0.612	0.707	0.718	0.695	0.781
0.105	0.548	0.621	0.557	0.581	0.654	0.618	0.593	0.689	0.701	0.678	0.765
0.110	0.530	0.604	0.539	0.563	0.637	0.600	0.576	0.672	0.684	0.661	0.750
0.115	0.513	0.586	0.522	0.546	0.620	0.583	0.558	0.656	0.667	0.644	0.736
0.120	0.497	0.570	0.506	0.530	0.604	0.566	0.542	0.640	0.652	0.628	0.721
0.125	0.482	0.554	0.491	0.514	0.588	0.551	0.526	0.624	0.636	0.612	0.707
0.150	0.416	0.485	0.424	0.447	0.518	0.482	0.458	0.554	0.566	0.542	0.640
0.200	0.325	0.384	0.331	0.351	0.413	0.381	0.360	0.446	0.458	0.435	0.529

[a]Applicable when taper cut is on compression side only.

[b]F_b = 2400 psi; F_v = 165 psi; $F_{c\perp}$ = 560 psi.

[c]F_b = 2400 psi; F_v = 200 psi; $F_{c\perp}$ = 560 psi.

[d]F_b = 2200 psi; F_v = 155 psi; $F_{c\perp}$ = 375 psi.

[e]F_b = 2200 psi; F_v = 165 psi; $F_{c\perp}$ = 560 psi.

[f]F_b = 2200 psi; F_v = 200 psi; $F_{c\perp}$ = 560 psi.

[g]F_b = 2000 psi; F_v = 165 psi; $F_{c\perp}$ = 560 psi.

[h]F_b = 2000 psi; F_v = 155 psi; $F_{c\perp}$ = 375 psi.

[i]F_b = 2000 psi; F_v = 200 psi; $F_{c\perp}$ = 560 psi.

[j]F_b = 1600 psi; F_v = 165 psi; $F_{c\perp}$ = 560 psi.

[k]F_b = 1600 psi; F_v = 155 psi; $F_{c\perp}$ = 375 psi.

[l]F_b = 1600 psi; F_v = 200 psi; $F_{c\perp}$ = 560 psi.

Camber = 3 in.
Appearance grade = Industrial
Bending F_{bx} = 2400 psi
Horizontal shear F_v = 155 psi
Compression perpendicular to grain
Tension face $F_{c\perp}$ = 475 psi
Compression face $F_{c\perp}$ = 375 psi
Modules of Elasticity E = 1,500,000 psi

In addition, the interior core laminations shall provide the following minimum design values along the sloping cut.

Bending F_{bx} = 1600 psi
Compression perpendicular to grain $F_{c\perp}$ = 375 psi

Design Considerations for Ponding

The subject of ponding of water on roofs was introduced on page 4-88. The following examples illustrate the application of ponding analysis. They are based on a simplified approach for ponding calculations that can be used when the effect of the deflection of the secondary roof-framing members on ponding is small.

To illustrate the effect of the magnification factor on flat roof design, two examples have been selected: (1) a beam designed for dead load plus snow load and (2) a beam with a 2-in.-high gravel stop designed for dead load plus live load.

Example 1. Check an $8\frac{3}{4} \times 36$-in. glued laminated timber simple-span beam under the following conditions:

$L = 62$ ft $F'_b = (2400)(1.15)(0.88) = 2460$ psi

Spacing $= s = 16$ ft $F'_v = (165)(1.15) = 190$ psi

$A = 315$ in.^{2} $E' = E = 1{,}800{,}000$ psi

$S = 1890$ in.^{3} Allowable total-load deflection

$I = 34{,}020$ in.^{4} $= l/120 = (62)(12)/120 = 6.20$ in.

$C_F = 0.88$

Duration of load factor $C_D = 1.15$.

SL = 25 psf $w_{SL} = 400$ plf $f_b = 2343$ psi (dead load + snow load)

DL = 23 psf $w_{DL} = 368$ plf $f_v = 102$ psi (dead load + snow load)

TL = 48 psf $w_{TL} = 768$ plf $\Delta_{TL} = 4.17$ in.

Check for the effects of ponding:

$$\Delta \text{ (for 5-psf load)} = (4.17)\frac{5}{48} = 0.434 \text{ in.} < 0.5 \quad \text{(spring constant)} \quad \text{O.K.}$$

This simplified check does not take into account that snow load may be acting at the same time ponding occurs. If maximum snow load is assumed to be acting at the same time, a further check should be made by use of the magnification factor C_p:

$$C_p = \frac{1}{1 - W'L^3/\pi^4 EI}$$

$$W' = 5.2Ls = (5.2)(62)(16) = 5160 \text{ lb}$$

$$C_p = \left[1 - \frac{(5160)(62)^3(1728)}{\pi^4(1{,}800{,}000)(34{,}020)}\right]^{-1} = 1.55$$

Multiply the calculated stresses for the member under consideration by C_p to determine the effects of ponding in combination with dead and snow loads:

$$f_b C_p = (2343)(1.55) = 3640 \text{ psi} > 2460 \quad \text{N.G.}$$

$$f_v C_p = (102)(1.55) = 158 \text{ psi} < 190 \quad \text{O.K.}$$

$$\Delta_{TL} C_p = (4.17)(1.55) = 6.48 \text{ in.} > 6.20 \quad \text{N.G.}$$

In this example, it can be seen that the beam is overstressed in bending and that the deflection limit is exceeded as the result of ponding in combination with the design loads; therefore, if it is not possible to eliminate the cause of ponding; a larger section is required.

Try an $8\frac{3}{4} \times 40\frac{1}{2}$-in. section:

$$A = 354.4 \text{ in.}^2 \qquad S = 2392 \text{ in.}^3$$

$$C_F = 0.87 \qquad F_b' = (2400)(1.15)(0.87) = 2400 \text{ psi}$$

$$f_b = 1850 \text{ psi} \qquad I = 48{,}440 \text{ in.}^4$$

$$f_v = 89.8 \text{ psi}$$

$$\Delta_{TL} = 2.93 \text{ in.}$$

$$C_p = \left[1 - \frac{(5160)(62)^3(1728)}{\pi^4(1{,}800{,}000)(48{,}440)}\right]^{-1} = 1.33$$

Effects of ponding:

$$f_b C_p = (1850)(1.33) = 2467 \text{ psi} > 2400 \quad \text{(consider O.K.)}$$

$$f_v C_p = (89.8)(1.33) = 120 \text{ psi} < 190 \quad \text{O.K.}$$

$$\Delta_{TL} C_p = (2.93)(1.33) = 3.90 \text{ in.} < 6.20 \text{ in.} \quad \text{O.K.}$$

The larger section is adequate.

Example 2. A $6\frac{3}{4} \times 31\frac{1}{2}$-in. glued laminated timber simple-span beam with a 2-in. high gravel stop is used under the following conditions:

$$L = 58 \text{ ft} \qquad F'_b = (2400)(1.25)(0.90) = 2700 \text{ psi}$$

$$\text{Spacing} = 16 \text{ ft} \qquad F'_v = (200)(1.25) = 250 \text{ psi}$$

$$A = 212.6 \text{ in.}^2 \qquad E' = E = 1{,}800{,}000 \text{ psi}$$

$$S = 1117 \text{ in.}^3 \qquad \text{Allowable live-load deflection}$$

$$I = 17{,}580 \text{ in.}^4 \qquad = l/240 = (58)(12)/240 = 2.90 \text{ in.}$$

$$C_F = 0.90$$

Duration-of-load factor $C_D = 1.25$:

$$\text{LL} = 12 \text{ psf} \qquad w_{LL} = 192 \text{ plf}$$

$$\text{DL} = \underline{15 \text{ psf}} \qquad w_{DL} = \underline{240 \text{ plf}}$$

$$\text{TL} = 27 \text{ psf} \qquad w_{TL} = 432 \text{ plf}$$

For the loading conditions shown, the following stresses are calculated.

$$f_b = 1950 \text{ psi} \qquad f_v = 80.4 \text{ psi} \qquad \Delta_{LL} = 1.54 \text{ in.}$$

$$\Delta \text{ (for 5-psf load)} = (1.54)\frac{5}{12} = 0.64 \text{ in.} > 0.5$$

Since the $6\frac{3}{4} \times 31\frac{1}{2}$-in. section has insufficient stiffness to meet the criterion that a 5-psf load will cause not more than $\frac{1}{2}$ in. deflection, a stiffer section must be selected.

Try a $6\frac{3}{4} \times 34\frac{1}{2}$-in. section:

$$A = 232.9 \text{ in.}^2 \qquad S = 1339 \text{ in.}^3$$

$$C_F = 0.89 \qquad I = 23{,}100 \text{ in.}^4$$

$$f_b = 1628 \text{ psi} \qquad F'_b = (2400)(1.25)(0.89) = 2670 \text{ psi}$$

$$\Delta_{LL} = 1.17 \text{ in.} \qquad f_v = 72.7 \text{ psi}$$

$$\Delta \text{ (for 5-psf load)} = (1.17)\frac{5}{12} = 0.488 \text{ in.} < 0.5 \quad \text{O.K.}$$

This simplified check works satisfactorily when only a small amount of water accumulates to start the deflection process. When a larger amount of water may be present due to inadequate drainage, high gravel stops, and so on, a further check by use of the magnification factor C_p should be made.

A 2-in. depth of water caused by the gravel stop will be assumed. It will also be assumed that the design live load (12 psf) is not acting simultaneously with the ponding load.

A 2-in. depth of water over the roof will cause a load of (2/12)(62.4)(16) = 166 plf. The total load on the roof beam is equal to the dead load plus the water load, or

$$w_{\mathrm{TL}} = 240 + 166 = 406 \text{ plf}$$

$$f_b = (1628)\frac{406}{432} = 1530 \text{ psi}$$

$$f_v = (72.7)\frac{406}{432} = 68.3 \text{ psi}$$

$$\Delta_{\mathrm{LL}} = (1.17)\frac{406}{432} = 1.10 \text{ in.}$$

$$W' = 5.2Ls = (5.2)(58)(16) = 4825 \text{ lb}$$

$$C_p = \left[1 - \frac{(4825)(58)^3(1728)}{\pi^4(1{,}800{,}000)(23{,}100)}\right]^{-1} = 1.67$$

Effects of ponding:

$$f_bC_p = (1530)(1.67) = 2555 \text{ psi} < 2670 \quad \text{O.K.}$$

$$f_vC_p = (68.3)(1.67) = 114 \text{ psi} < 250 \quad \text{O.K.}$$

$$\Delta_{\mathrm{LL}}C_p = (1.10)(1.67) = 1.84 \text{ in.} < 2.90 \quad \text{O.K.}$$

The $6\frac{3}{4} \times 34\frac{1}{2}$-in. section is adequate. (In some cases the live load may occur simultaneously with the ponding. If so, the live load should also be included in determining the total design load.)

Design Procedure for Glued Laminated Timber Floor Beams

The following steps are recommended for the design of floor beams. They are similar to those for the design of roof beams except that the order of calculations is different. This is because deflection is more critical in the design of floor beams and the check of modulus of elasticity is recommended after the size required by bending is determined. Table 5.7 contains the traditional deflection limitations for floor beams that meet most code requirements. When the length of spans increase and when a more rigid floor system is desired for certain uses such as commercial, office, and institutional buildings, these limits may not provide the desired stiffness and freedom from vibration. The deflection limitations for floors in Table 5.8 are suggested for such uses. Requirements for determination of floor performance vary with the intended use, and the design requirements are usually subjective. The deflection limitations included in Table 5.8 have been found by experience to provide floors that meet the more restrictive requirements. In addition to considering the deflection limitations of the glued laminated timber girders and joists, attention must also be given to the material spanning between the joists. The recommended $1\frac{1}{8}$-in. plywood on joists spaced 2 ft on centers as shown in the example is more than required by codes but is recommended in order to provide additional stiffness.

1. Select the most practical framing system: Simple-span girders with spans of up to 35 or 40 ft are practical for floor-framing systems that provide a reasonably stiff floor for use in commercial, office, or institutional applications. Girders may be spaced up to about 26 ft on center with glued laminated timber joists spanning between them. These joists may be spaced 2 ft on centers with $1\frac{1}{8}$-in.-thick plywood and lightweight concrete topping over the joists. Added stiffness is obtained by gluing the plywood to both the joists and girders. This provides extra stiffness; however, the deflection limitations shown in Table 5.8 were based on excluding the stiffness provided by the plywood from the computations. Cantilevered girder or joist systems are generally not recommended for floor systems because of difficulties in cambering the members as well as avoiding transmission of floor vibrations between sections of the building.

2. Determine the design loads. Minimum floor live loads are generally specified by the applicable building code. In some cases, it may be advisable to design the floor for live loads higher than the minimums recommended by the code in order to obtain a more rigid floor. Many building codes provide for reductions in live loads based on the tributary area and dead load/live load ratio.

3. Select design value in bending. For glued laminated timber girders, design value in bending, F_b, of 2400 or 2200 psi is recommended. Because of the many glued laminated timber joists used in floor systems, it is recommended that they be designed at the 1600 psi bending stress level.

The design values in Table 1, AITC 117—Design (1), are based on members subjected to loads of normal duration, dry conditions of use, and normal temperatures and members that are laterally supported. All of these conditions usually apply to floor beams and no adjustments for other conditions will be needed. The design values are also based on members 12 in. deep. For deeper members, the size factor C_F should be applied.

4. Determine size required by bending. An adjusted section modulus that includes the size factor can be calculated by dividing the bending moment by the design value in bending selected in step 3.

5. Determine the required tabular design value for modulus of elasticity E. Recommended limitations for deflection of floor beams are included in Tables 5.7 and 5.8. These recommendations are minimums, and it may be necessary to apply more restrictive limitations where stiffer floors are required.

6. Determine the required tabular design value for horizontal shear F_v.

7. Determine the required tabular design value for compression perpendicular to grain, $F_{c\perp}$.

8. Determine camber requirements.

9. Specification.

Example. Design a glued laminated timber joist and girder system for the second floor bay of an office building 48 × 35 ft as shown:

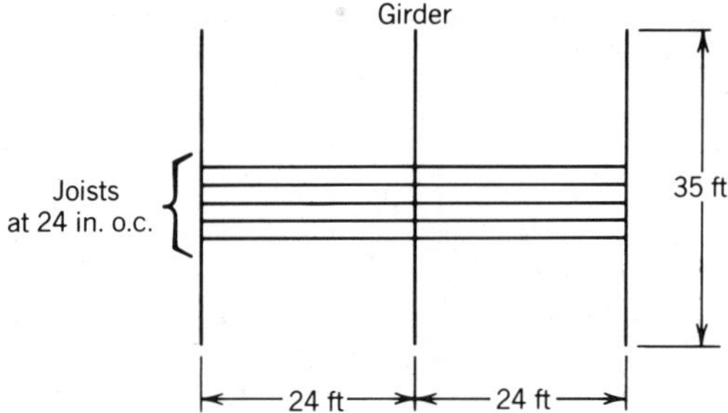

1. The framing system selected is for girders spanning 35 ft and joists 24 in. on centers framing into each side of the girder. The structural floor consists of $1\frac{1}{8}$-in. plywood topped with lightweight concrete.

2. Determine design loads.

Dead Loads on joists (see Table 7.11)

Carpet	1 psf
Lightweight concrete topping ($\frac{3}{4}$ in.)	$6\frac{1}{2}$ psf
$1\frac{1}{8}$-in. plywood	$3\frac{1}{2}$ psf
Joist (glued laminated timber)	6 psf
Mechanical allowance	4 psf
Suspended ceiling	2 psf
Partition allowance	20 psf
	43 psf
Add 2 psf for girder weight	
Live load on joists	50 psf

The applicable building code permits the following live load-reduction, R, for the girder:

$$R = r(A - 150) = (0.08)\,[(35)(24) - 150] = 55.2\%$$

$$R = 23.1\left(1 + \frac{\text{DL}}{\text{LL}}\right) = (23.1)\left(1 + \frac{45}{50}\right) = 43.9\%$$

$$R_{\text{max}} = 40\%$$

$$\text{LL (girder)} = (50)(1 - 0.40) = 30 \text{ psf}$$

where r = rate of reduction (0.08 for floors), and A = area supported by member (ft^2).

3. Select design value in bending for joists. Because of the close spacing and for economy in design, try 1600 psi.

$$F''_{bx} = F_{bx} = 1600 \text{ psi}$$

4. Determine size required by bending:

$$M = \frac{wL^2}{8} = \frac{(93)(2)(24)^2(12)}{8} = 161{,}000 \text{ in.-lb}$$

$$S_{n(\text{required})} = \frac{M}{F''_{bx}} = \frac{161{,}000}{1600} = 100.4 \text{ in.}^3$$

From Table 7.2, assuming a $3\frac{1}{8}$-in. width, try a $3\frac{1}{8} \times 15$-in. joist:

$S_n = 114.3 \text{ in.}^3 \qquad S = 117.2 \text{ in.}^3 \qquad I = 878.9 \text{ in.}^4 \qquad A = 46.88 \text{ in.}^2$

5. Check the stiffness of the joists without considering the composite behavior between the joist and the plywood floor sheathing due to gluing of the plywood to the joists. The recommended deflection limitations from Table 5.8 are: for live load + 1/2 dead load, $\Delta = l/360 = (24)(12)/360 = 0.8$ in.; for live load only, $\Delta = l/480 = (24)(12)/480 = 0.6$ in.

$$\frac{\text{LL}}{\text{LL} + \frac{1}{2}\text{DL}} = \frac{50}{50 + (\frac{1}{2})(23)} = 0.8$$

Since (0.8)(0.8 in.) = 0.64 in. > 0.6 in., live load only controls.

$$E_{(\text{required})} = \frac{5wL^4}{384EI\Delta_{\text{LL}}} = \frac{(5)(50)(2)(24)^4(12)^3}{(384)(878.9)(0.6)}$$

$$= 1{,}420{,}000 \text{ psi} \quad (\text{use } 1{,}500{,}000 \text{ psi})$$

6. Determine required tabular design value for horizontal shear F_v.

$$L_e = L - \frac{2d}{12} = 24 - \frac{(2)(15)}{12} = 21.5 \text{ ft}$$

$$V = \frac{wL_e}{2} = \frac{(93)(2)(21.5)}{2} = 2000 \text{ lb}$$

$$f_v = \frac{3V}{2A} = \frac{(3)(2000)}{(2)(46.88)} = 64 \text{ psi}$$

$$F_{v(\text{required})} = f_v = 64 \text{ psi}$$

7. Determine required tabular design value for compression perpendicular to grain, $F_{c\perp}$. A typical commercial joist hanger has a $2\frac{1}{2}$-in. bearing length.

$$\text{Joist reaction} = \frac{wL}{2} = \frac{(93)(2)(24)}{2} = 2230 \text{ lb}$$

$$F_{c\perp(\text{required})} = \frac{2230}{(3.125)(2.5)} = 290 \text{ psi}$$

There is no critical bearing stress on the compression face of the joist.

8. Determine camber requirements. Camber recommendations given in Table 5.9 for floor beams are for ordinary usage and in some cases result in excessive

camber. For the type of floor system in this example the camber recommendation is $1\frac{1}{4}$ times the dead load deflection, which does not include the 20-psf partition load.

$$\Delta_{DL} = \frac{(5)(23)(2)(24)^4(12)^3}{(384)(1{,}500{,}000)(878.9)} = 0.26 \text{ in.}$$

$$\text{Camber} = (0.26)(1.25) = 0.33 \text{ in.} \quad (\text{use } \tfrac{3}{8} \text{ in.})$$

9. Specification for joist:

$$\text{Size} = 3\tfrac{1}{8} \times 15 \text{ in.}, \qquad \text{Camber} = \tfrac{3}{8} \text{ in.}$$

$$F_{bx} = 1600 \text{ psi}, \qquad F_v = 90 \text{ psi}$$

$$F_{c\perp(\text{tension face})} = 375 \text{ psi}, \qquad F_{c\perp(\text{compression face})} = 375 \text{ psi}$$

$$E_x = 1{,}500{,}000 \text{ psi}$$

Start at step 3 of the procedure for the design of the girder.

3. Select design value in bending for girder. Try 2400 psi.

$$F''_{bx} = F_{bx} = 2400 \text{ psi}$$

4. Determine size required by bending. (Use reduced live load allowed by code.)

$$w = \text{DL} + \text{LL} = 45 + 30 = 75 \text{ psf}$$

$$M = \frac{wL^2}{8} = \frac{(75)(24)(35)^2(12)}{8} = 3{,}310{,}000 \text{ in.-lb}$$

$$S_{n(\text{required})} = \frac{M}{F''_{bx}} = \frac{3{,}310{,}000}{2400} = 1378 \text{ in.}^3$$

From Table 7.2, assuming a $6\frac{3}{4}$-in. width, try a $6\frac{3}{4} \times 37\frac{1}{2}$-in. girder.

$$S_n = 1394 \text{ in.}^3 \qquad S = 1582 \text{ in.}^3 \qquad I = 29{,}660 \text{ in.}^4 \qquad A = 253.1 \text{ in.}^2$$

5. Determine the required tabular design value for modulus of elasticity E. The recommended deflection limitations from Table 5.8 are: for live load + 1/2(dead load), $\Delta = l/360 = (35)(12)/360 = 1.17$ in.; for live load only, $\Delta = l/480 = 35(12)/480 = 0.875$ in.

$$\frac{\text{LL}}{\text{LL} + \frac{1}{2}\text{DL}} = \frac{30}{30 + (0.5)(45)} = 0.57$$

$$(0.57)(1.17) = 0.63 \text{ in.} < 0.875 \text{ in.}$$

Live load + $\frac{1}{2}$ dead load controls.

$$E_{(\text{required})} = \frac{5[30 + (0.5)(45)](24)(35)^4(12)^3}{(384)(29{,}660)(1.17)} = 1{,}230{,}000 \text{ psi}$$

Use $E = 1{,}700{,}000$ psi, because this is the lowest E value readily available for 24F combinations.

6. Determine the required tabular design value for horizontal shear, F_v.

$$L_e = L - \frac{2d}{12} = 35 - \frac{(2)(37.5)}{12} = 28.75 \text{ ft}$$

$$V = \frac{(75)(24)(28.75)}{2} = 25{,}900 \text{ lb}$$

$$f_v = \frac{3V}{2A} = \frac{(3)(25{,}900)}{(2)(253.1)} = 153 \text{ psi}$$

$$F_{v(\text{required})} = f_v = 153 \text{ psi}$$

7. Determine the required tabular design value for compression perpendicular to grain, $F_{c\perp}$. A typical commercial beam hanger has a 8-in. bearing length.

$$\text{Reaction} = \frac{(75)(24)(35)}{2} = 31{,}500 \text{ lb}$$

$$F_{c\perp(\text{required})} = f_{c\perp} = \frac{(31{,}500)}{(6.75)(8)} = 583 \text{ psi}$$

8. Determine camber requirements. (Do not camber for the 20-psf partition load.)

$$\Delta_{DL} = \frac{(5)(25)(24)(35)^4(12)^3}{(384)(29{,}660)(1{,}700{,}000)} = 0.40 \text{ in.}$$

$$\text{Camber} = (1.25)(0.40) = 0.50 \text{ in.} \quad (\text{use } \tfrac{1}{2} \text{ in.})$$

9. Specification for girder:

$$\text{Size} = 6\tfrac{3}{4} \times 37\tfrac{1}{2} \text{ in.}, \qquad \text{Camber} = \tfrac{1}{2} \text{ in.}$$

$$F_{bx} = 2400 \text{ psi}, \qquad F_v = 155 \text{ psi}$$

$$F_{c\perp(\text{tension face})} = 650 \text{ psi}, \qquad F_{c\perp(\text{compression face})} = 500 \text{ psi}$$

$$E_x = 1{,}700{,}000 \text{ psi}$$

Design of Sawn Beams

Sawn lumber and timber beams are usually used for simple spans consisting of straight members. Section properties are given in Table 7.1 and tabular design values are in Tables 7.3 and 7.4. The same general design considerations used for straight glued laminated timber beams are also applicable to sawn timber except that sawn timbers cannot be cambered. Bending values are tabulated for single member and repetitive member use (see NDS). As a general rule, however, sawn members may contain a small amount of curvature prior to erection and the crown of the member should be turned up.

Size effect has already been taken into account in establishing the design values for sawn lumber except for pieces thicker than 4 in. and deeper than 12 in. when loaded in bending about the strong axis. Therefore, the size factor C_F applies only to sawn members that are wider than 4 in. and deeper than 12 in. Lateral support can also be taken into account by the application of approximate rules to prevent lateral rotation or lateral displacement as shown in Table 5.10.

The designer should check on the availability of the larger sizes of sawn timbers prior to design.

Examples. Analyze a 4 × 16-in. sawn timber roof beam under the following conditions. The roof deck is applied directly to the beams, the beams are seated in hangers with 3.5 × 4-in. bearing area, the ends are restrained to prevent lateral rotation, and the slope of the roof is $\frac{3}{8}$ in./ft with adequate downspouts.

$$
\begin{aligned}
L &= 20 \text{ ft} & F'_b &= (1500)(1.15) = 1725 \text{ psi} \\
\text{Spacing} &= 10 \text{ ft} & F'_v &= (95)(1.15) = 109 \text{ psi} \\
\text{SL} &= 25 \text{ psf} & F'_{c\perp} &= 625 \text{ psi} \\
\text{DL} &= 10 \text{ psf} & E' &= 1{,}800{,}000 \text{ psi} \\
C_D &= 1.15 \quad \text{(snow load)} & \Delta_{SL} &= \frac{l}{240} = 1 \text{ in.} \\
 & & \Delta_{TL} &= \frac{l}{180} = 1.33 \text{ in.}
\end{aligned}
$$

1. Determine actual bending stress f_b:

$$w_{DL} = (\text{spacing})(\text{DL}) = (10)(10) = 100 \text{ plf}$$

$$w_{SL} = (\text{spacing})(\text{SL}) = (10)(25) = \underline{250 \text{ plf}}$$

$$w_{TL} = w_{DL} + w_{SL} \qquad = 350 \text{ plf}$$

For a 4 × 16-in. member: $A = 53.38$ in., $S = 135.7$ in.^{3}, and $I = 1034$ in.^{4} from Table 7.1.

$$M = \frac{wL^2}{8} = \frac{(350)(20)^2(12)}{8} = 210{,}000 \text{ in.-lb}$$

$$f_b = \frac{M}{S} = \frac{210{,}000}{135.7} = 1548 \text{ psi} < F'_b = 1725 \text{ psi} \quad \text{O.K.}$$

2. Determine actual horizontal shear stress f_v:

$$\text{Effective span } L_e = L - 2d = 20 - \frac{(2)(15.25)}{12} = 17.46 \text{ ft}$$

$$\text{Design shear at end } V = \frac{(17.46)(350)}{2} = 3055 \text{ lb}$$

$$f_v = \frac{3V}{2A} = \frac{(3)(3055)}{(2)(53.38)} = 86 \text{ psi} < F'_v = 109 \text{ psi} \quad \text{O.K.}$$

3. Determine the actual deflection of the beam:

$$\text{Deflection for snow load, } \Delta_{SL} = \frac{5wL^4}{384EI} = \frac{(5)(250)(20)^4(12)^3}{(384)(1{,}800{,}000)(1034)}$$

$$= 0.48 \text{ in.} < 1 \text{ in.} \quad \text{O.K.}$$

$$\text{Deflection for total load} = (0.48)\frac{350}{250} = 0.68 \text{ in.} < 1.33 \text{ in.} \quad \text{O.K.}$$

4. Determine actual compression perpendicular to grain stress, $f_{c\perp}$, at the ends of the beam. Beam reaction:

$$R_v = \frac{(20)(350)}{2} = 3500 \text{ lb}$$

$$f_{c\perp} = \frac{3500}{(3.5)(4)} = 250 \text{ psi} < 625 \text{ psi} \quad \text{O.K.}$$

5. Check lateral stability. Ratio of depth to width (nominal size) = 16/4 or 4:1. This complies with criteria in Table 5.10 for lateral stability of sawn members and no other support is necessary. It also complies with lateral stability requirements because unsupported length is zero and ends are restrained from rotation.

6. Ponding. Roof slope of $\frac{3}{8}$ in./ft with adequate roof drainage should prevent accumulation of water, and ponding analysis is not required.

Designing for Fire Resistance

The fire resistance of large timber members has been recognized for years by building code and fire officials by allowing larger building areas and more uses than for light-frame or ordinary construction. For certain uses, however, the building codes require specific hourly fire resistance ratings for structural components. The fire rating of large sawn members and glued laminated timber members can be calculated as shown on pages 1-22 to 1-24. The fire rating of a timber member is a function of the size of the member and the percent of full design load carried by the member.

Example. Determine the dimensions of a glued laminated timber beam to carry an 800-plf load (normal duration of load), and to provide a fire rating of 1 hr for the following conditions:

Simple beam span, $L = 40$ ft
Beam laterally supported at ends and along the top
Use a $24F$ combination
Determine size required to carry a full load:

$$M = \frac{wL^2}{8} = \frac{(800)(40)^2(12)}{8} = 1{,}920{,}000 \text{ in.-lb}$$

$$S_n = \frac{M}{F'_b} = \frac{1{,}920{,}000}{2400} = 800 \text{ in.}^3$$

Try a $6\frac{3}{4}$-in. wide member
From Table 7.2, a $6\frac{3}{4} \times 28\frac{1}{2}$-in. beam provides an $S_n = 830$ in.^{3}
For a beam exposed to fire on 3 sides,

$$t = 2.54Zb\left(4 - \frac{b}{d}\right) \quad \text{(Eq. 1.2)}$$

For $6\frac{3}{4} \times 28\frac{1}{2}$-in. beam, ratio of stress = 800/830 =0.96
From Figure 1.2, $Z = 1.0$:

$$b = 6.75 \text{ in.} \qquad d = 28.5 \text{ in.}$$

$$t = (2.54)(1.0)(6.75)\left(4 - \frac{6.75}{28.5}\right) = 64 \text{ min.} > 60 \text{ min.} \quad \text{O.K.}$$

Use a $6\frac{3}{4} \times 28\frac{1}{2}$-in. beam.

COMBINED BENDING AND AXIAL STRESSES

A combination of axial stresses and bending stresses must be taken into account in design by use of the interaction formula.

Bending and Axial Tension

For a combination of axial stresses and bending stresses, the formula for checking strength becomes

$$\frac{f_t}{F'_t} + \frac{f_b}{F''_b C_F} \leqslant 1 \tag{5-39}$$

where f_t = applied axial stress in tension (psi),
F'_t = tabular design values in axial tension (psi) adjusted by applicable modifiers,
f_b = applied bending stress (psi), and
F''_b = tabular design value in bending (psi) adjusted by applicable modifiers except C_F and C_L

The size factor C_F applies to the tension side of a bending member so it always applies as a modifier in determining the design value in bending adjusted by all applicable modifiers to F'_b for use in interaction formula (5-39) for combined bending and tension. When bending is in a plane perpendicular to the wide faces of the laminations of glued laminated timber, F_{bx} should be used. When bending is in a plane parallel to the wide faces of the laminations, F_{by} should be used.

The lateral stability of beams factor C_L affects the compression side and is offset by the axial tension stresses in the member. When the lateral support of a beam is such that it is classified as an intermediate or long beam and the design value in bending, F'_b, is determined by use of the lateral stability of beams factor C_L rather than the application of the size factor C_F, the following stability check equation should be used:

$$\frac{f_b - f_t}{F_b'' C_L} \leqslant 1 \tag{5-40}$$

where terms are as defined in Eq. (5.39).

See page 5-164 for a discussion of C_L for intermediate and long beams. The designer should design a tension member by determining the required tabular design value for both tension and bending. However, the specified design values should be realistic in that combinations that are reasonably available can fulfill the stress requirements. If bending is the predominant stress, the combinations in Table 1, AITC 117—Design (1), can be used as guides. If tension is the predominant stress, one of the combinations in Table 2, AITC 117—Design (1), may be more appropriate than those in Table 1 because they were developed primarily for axial stresses.

Example. Select a 10-ft-long glued laminated timber tension member to resist a 30,000-lb axial tension load and a moment of 75,000 in.-lb. No lateral support is provided between the ends of the member. Design for normal duration of load and dry condition of use. Ends of member are laterally supported to prevent lateral rotation.

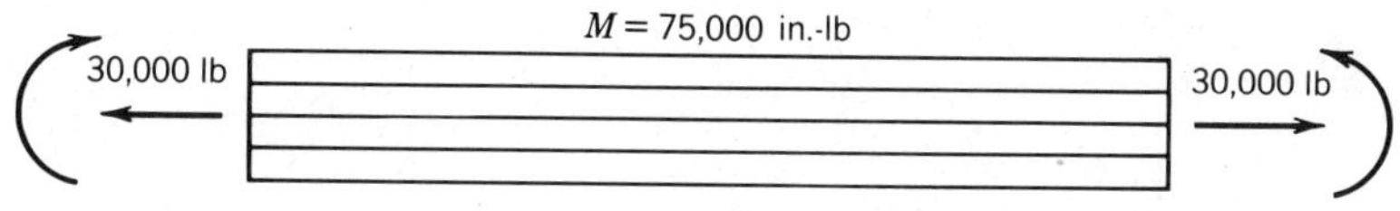

Use combination 3 (L2D Douglas Fir)

$$F_{by} = 2100 \text{ psi} \qquad F_{bx} = 2000 \text{ psi}$$

$$F_t = 1450 \text{ psi} \qquad E_x = E_y = 1{,}800{,}000 \text{ psi}$$

Assume all modifying factors are unity for the determination of a trial size

$$F''_{bx} = F_{bx} = 2000 \text{ psi}$$

$$F''_{by} = F_{by} = 2100 \text{ psi}$$

$$F'_t = F_t = 1450 \text{ psi}$$

$$E' = E = 1{,}800{,}000 \text{ psi} \qquad \text{(either } E_x \text{ or } E_y\text{)}$$

Assume a trial size of $5\frac{1}{8} \times 10\frac{1}{2}$ in.

$$A = 53.8 \text{ in.}^2 \qquad S = 94.17 \text{ in.}^3 \qquad C_F = 1$$

$$f_t = \frac{30{,}000}{53.81} = 557 \text{ psi}$$

$$f_{bx} = \frac{75{,}000}{94.17} = 796 \text{ psi}$$

Check lateral support. From Figure 5.14*c* for equal end moments,

$$l_e = 1.84 l_u = (1.84)(10)(12) = 220.8$$

$$C_s = \sqrt{\frac{l_e d}{b^2}} = \sqrt{\frac{(220.8)(10.5)}{(5.125)^2}} = 9.4$$

Since 9.4 < 10, use short-beam formula; $C_L = 1$:

$$F'_{bx} = F''_{bx} = 2000 \text{ psi}$$

$$\frac{f_t}{F'_t} + \frac{f_{bx}}{F''_{bx} C_F} = \frac{557}{1450} + \frac{796}{(2000)(1)} = 0.78 < 1 \quad \text{O.K.}$$

$$\frac{f_{bx} - f_t}{F''_{bx} C_L} = \frac{796 - 557}{(2000)(1)} = 0.12 < 1 \quad \text{O.K.}$$

A $5\frac{1}{8} \times 10\frac{1}{2}$-in. member is satisfactory.

If the length of the member had been 20 ft instead of 10 ft,

$$l_e = (1.84)(20)(12) = 441.6$$

$$C_s = \sqrt{\frac{(441.6)(10.5)}{(5.125)^2}} = 13.3 > 10$$

$$C_{kx} = 0.956 \sqrt{\frac{E'_y}{F''_{bx}}} = 0.956 \sqrt{\frac{1{,}800{,}000}{2000}} = 28.7$$

Since 10 < 13.3 < 28.7, use intermediate-beam formula.

$$C_L = \left[1 - \frac{1}{3}\left(\frac{13.3}{28.7}\right)^4\right] = 0.98$$

$$\frac{f_t}{F'_t} + \frac{f_{bx}}{F''_{bx} C_F} = \frac{557}{1450} + \frac{796}{(2000)(1)} = 0.78 \leqslant 1 \quad \text{O.K.}$$

$$\frac{f_b - f_t}{F''_b C_L} \leqslant 1 \qquad \frac{796 - 557}{(2000)(0.98)} = 0.12 \leqslant 1 \quad \text{O.K.}$$

Bending and Axial Compression

When a member is subjected to both bending and axial compression, the combination of the compressive stresses and the tensile stresses caused by bending results in a smaller absolute value of stress on the tension side than on the compression side. This affects the manner in which the size factor C_F and the lateral stability of beams factor C_L are applied. The size of member affects the bending strength of the member because the ability to resist tension decreases as the size increases. This is accounted for in design by the application of the size factor C_F. When a member is subjected to both axial compressive and bending stresses, the compressive stressses reduce the net stress on the tension side. When $C_F < 1$ and the axial compressive stress f_c is equal to or larger than $F''_b(1 - C_F)$. Note $F''_b = F_b$ adjusted by all applicable modifiers except C_L and C_F. The application of the size factor is inappropriate because of the decrease in tension stress caused by the compressive force. For this condition, C_F does not apply. When the axial compressive stress f_c is less than $F''_b(1 - C_F)$, the design value in bending, F'_b, should not exceed that determined as follows:

$$F'_b = F''_b C_F + f_c$$

where F''_b = tabular design value in bending (psi) adjusted by all applicable modifiers except C_L and other terms are as previously noted.

The lateral stability of beams factor C_L is applied to prevent buckling of the compresssion side of a member subjected to bending. Thus, reductions for lateral stability (compression effect) and size (tension effect) are not cumulative. In beams with combined axial compression and bending stresses, the design value in bending, F'_b, should not exceed $F''_b C_L$ or $F''_b C_F + f_c$, whichever is smaller. When the interaction factor C_I is involved, it is cumulative when the taper cut is on the compression side. In this case, the design value in bending, F'_b, is calculated as follows:

$$F'_b = F''_b C_L C_I$$

Example. Determine F'_b when size factor $C_F = 0.8$ and the lateral stability of beams factor $C_L = 0.9$. Loads are primarily snow ($C_D = 1.15$). Assume a 24*F* combination is being used; $F_b = 2400$ psi, $f_c = 400$ psi:

$$F''_b = F_b C_D = (2400)(1.15) = 2760 \text{ psi}$$

$$F''_b(1 - C_F) = (2760)(1 - 0.8) = 552 > f_c = 400 \text{ psi}$$

Therefore

$$F'_b = F''_b C_F + f_c = (2760)(0.8) + 400 = 2608 \text{ psi}$$

or

$$F'_b = F''_b C_L = (2760)(0.9) = 2484 \text{ psi}$$

Use $F'_b = 2484$ psi.

Glued laminated arches, discussed beginning on page 5-249, are another example of members subjected to a combination of axial compression and bending stresses. In most cases, the value of f_c is large enough so that no size factor C_F need be applied, but the preceding check should be made.

The application of an axial compressive load to a bending member causes a P–Δ effect that is similar to that in a compression member subjected to a bending stress. Therefore, a member subjected to both bending about the x–x axis and axial compression should be designed so that Eq. (5-19) is satisfied as follows:

$$\frac{f_c}{F'_c} + \frac{f_{bx} \pm (6 + 1.5J_x)(e_x/d_x)}{F'_{bx} - J_x f_c} \leqslant 1$$

See pages 5-128 and 5-131 for definition of terms. A similar interaction formula would be used for bending about the y–y axis.

When the axial load is not eccentric, the interaction equation becomes

$$\frac{f_c}{F'_c} + \frac{f_{bx}}{F'_{bx} - J_x f_c} \leq 1$$

Example. Select a 20-ft-long glued laminated timber beam to resist a 200-plf uniform load acting perpendicular to the longitudinal axis of the member and an axial compressive load of 20,000 lb. The member is braced perpendicular to the plane of the y–y axis at the ends and throughout its length at the top. Assume normal duration of load and dry conditions of use. For a trial design, use $F_{bx} = 2000$ psi, $F_c = 1350$ psi, $F_v = 90$ psi, $E_x = 1{,}500{,}000$ psi, $E_y = 1{,}400{,}000$ psi. Try a $5\frac{1}{8} \times 12$-in. member: $A = 61.5$ in.^{2}, $S = 123$ in.^{3}, $C_F = 1$, $F''_{bx} = F_{bx}$, $F''_c = F_c$, $E'_x = E_x$, and $E'_y = E_y$.

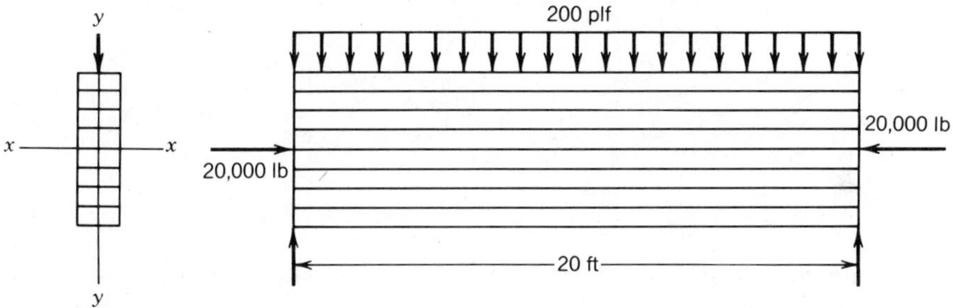

Determine F'_{bx}: Beam is braced along the top and at the ends (short beam) and $C_L = 1$.

$$F'_{bx} = F_{bx} = 2000 \text{ psi}$$

Determine f_{bx}:

$$f_{bx} = \frac{M}{S} = \frac{(200)(20)^2(12)/8}{128} = 976 \text{ psi}$$

Determine F'_c:

$$\left(\frac{l_e}{d}\right)_x = \frac{(20)(12)}{12} = 20$$

E_x will be used in the calculation of K_x because buckling takes place about the x–x axis.

$$K_x = 0.671\sqrt{\frac{E'_x}{F''_c}} = 0.671\sqrt{\frac{1{,}500{,}000}{1350}} = 22.4$$

Since $11 < 20 < 22.4$, use intermediate-column formula.

$$f_c = P/A = 20{,}000/61.5 = 325 \text{ psi}$$

$$C_P = \left[1 - \frac{1}{3}\left(\frac{(l_e/d)_x}{K_x}\right)^4\right] = \left[1 - \frac{1}{3}\left(\frac{20}{22.4}\right)^4\right] = 0.787$$

$$F'_c = F''_c C_P = (1350)(0.787) = 1060 \text{ psi}$$

Determine J about the axis of bending:

$$J_x = \frac{(l_e/d)_x - 11}{K_x - 11} = \frac{20 - 11}{22.4 - 11} = 0.792$$

$$\frac{f_c}{F'_c} + \frac{f_{bx}}{F'_{bx} - J_x f_c} = \frac{325}{1064} + \frac{976}{2000 - (0.792)(325)} = 0.87 < 1 \quad \text{O.K.}$$

Using the same procedure, a $5\frac{1}{8} \times 10\frac{1}{2}$-in. beam was investigated and found to be too small.

Check shear:

$$L_e = L - 2d = 20 - (2)(1) = 18 \text{ ft}$$

$$V = \frac{(18)(200)}{2} = 1800 \text{ lb}$$

$$f_v = \frac{3V}{2A} = \frac{(3)(1800)}{(2)(61.5)} = 44 \text{ psi} < 90 \text{ psi} \quad \text{O.K.}$$

Any combination from Table 1, AITC 117–Design (1), that provides the following design values or larger is satisfactory.

$$F_{bx} = 2000 \text{ psi} \qquad F_c = 1350 \text{ psi} \qquad F_v = 90 \text{ psi}$$

$$E_x = 1{,}500{,}000 \text{ psi} \qquad E_y = 1{,}400{,}000 \text{ psi}$$

Note that E_y was not used in this example, but it is shown for use in the next example.

Example. Assume that the member in the preceding example was not braced along the top and all other conditions are the same. The lack of top bracing creates a tendency for the member to buckle in bending and compression about the y–y axis and a wider beam ($6\frac{3}{4}$ in.) with the same depth (12 in.) will be assumed as a trial size.

$$A = 81 \text{ in.}^2 \qquad S = 162 \text{ in.}^3 \qquad C_F = 1.0$$

Determine F'_{bx}:

$$l_e = 1.92 l_u = (1.92)(20)(12) = 460.8 \text{ in.} \quad \text{(from Table 5.11)}$$

$$C_s = \sqrt{\frac{l_e d}{b^2}} = \sqrt{\frac{(460.8)(12)}{(6.75)^2}} = 11.0$$

$$C_{kx} = 0.956\sqrt{\frac{E'_y}{F''_{bx}}} = 0.956\sqrt{\frac{1{,}400{,}000}{2000}} = 25.3$$

Since 10 < 11 < 25.3, use intermediate-beam formula.

$$F'_{bx} = F''_{bx}C_L$$

$$C_L = \left[1 - \frac{1}{3}\left(\frac{C_s}{C_k}\right)^4\right] = \left[1 - \frac{1}{3}\left(\frac{11.0}{25.3}\right)^4\right] = 0.988$$

$$F'_{bx} = (2000)(0.988) = 1976 \text{ psi}$$

Determine F'_c and f_c:

$$l_e = K_e l = 1l = (1)(20)(12) = 240 \text{ in.}$$

$$\left(\frac{l_e}{d}\right)_x = \frac{240}{12} = 20$$

$$\left(\frac{l_e}{d}\right)_y = \frac{240}{6.75} = 35.6$$

Because column buckling is governed by $(l_e/d)_y$, E'_y will be used to determine K_y and F'_c.

$$K_y = 0.671\sqrt{\frac{E'_y}{F''_c}} = 0.671\sqrt{\frac{1{,}400{,}000}{1350}} = 21.6$$

From previous example, $K_x = 22.4$ and K_y governs. Since 21.6 < 35.6 < 50, use long-column formula.

$$F'_c = \frac{0.30E_y}{(l_e/d)_y^2} = \frac{(0.30)(1{,}400{,}000)}{(35.6)^2} = 332 \text{ psi}$$

$$f_c = \frac{20{,}000}{81} = 247 \text{ psi}$$

$$f_{bx} = \frac{M}{S} = \frac{(200)(20)^2(12)}{(8)(162)} = 741 \text{ psi}$$

Calculate J_x for bending about the x–x axis:

$$J_x = \frac{20 - 11}{22.4 - 11} = 0.792$$

$$\frac{f_c}{F'_c} + \frac{f_{bx}}{F'_{bx} - J_x f_c} \leqslant 1 \qquad \frac{247}{332} + \frac{741}{1976 - (0.792)(247)} = 1.16 > 1 \quad \text{N.G.}$$

The $6\frac{3}{4} \times 12$-in. member is not adequate. Try an $8\frac{3}{4} \times 12$-in. member:

$$A = 105 \text{ in.}^2 \qquad S = 210 \text{ in.}^3 \qquad C_F = 1.0$$

Determine F'_c:

$$\left(\frac{l_e}{d}\right)_y = \frac{240}{8.75} = 27.4$$

From previous example, $K_y = 21.6$, and since $21.6 < 27.4 < 50$, use long-column formula.

$$F'_c = \frac{(0.30)(1{,}400{,}000)}{(27.4)^2} = 558 \text{ psi}$$

$$f_c = \frac{20{,}000}{105} = 190 \text{ psi}$$

Determine F'_{bx}:

$$C_s = \sqrt{\frac{(460.8)(12)}{(8.75)^2}} = 8.5 < 10 \quad \text{(use short-beam formula, } C_L = 1\text{)}$$

$$F'_{bx} = F''_{bx} = 2000 \text{ psi}$$

$$f_{bx} = \frac{(200)(20)^2(12)}{(8)(210)} = 571 \text{ psi}$$

$(l_e/d)_x$, K_x and J_x remain unchanged.

$$\frac{f_c}{F'_c} + \frac{f_b}{F'_{bx} - J_x f_c} = \frac{190}{558} + \frac{571}{2000 - (0.792)(190)} = 0.65 < 1 \quad \text{O.K.}$$

The $8\frac{3}{4} \times 12$-in. member is adequate. A shallower member with an $8\frac{3}{4}$-in. width might also prove to be adequate. Any combination from Table 1, AITC 117—Design (1), that provides the following design values or larger is satisfactory for the condition checked.

$$F_b = 2000 \text{ psi} \qquad F_c = 1350 \text{ psi}$$

$$E_x = 1{,}500{,}000 \text{ psi} \qquad E_y = 1{,}400{,}000 \text{ psi}$$

MEMBERS SUBJECTED TO BENDING ABOUT BOTH AXES

When members are subjected to bending about both axes, the interaction formulas must also be used. For glued laminated timber, the design values in bending about the x–x axis, F_{bx}, are usually different from those about the y–y axis, F_{by}. Therefore, the bending stresses cannot be added directly at the corner of the cross section where the stresses are additive and the interaction formula must be applied.

In Figure 5.19, the tension stresses caused by bending are additive along the line marked A–A'. The compressive stresses are additive along line D–D'. The interaction of these stresses should be checked by the formula

$$\frac{f_{bx}}{F'_{bx}} + \frac{f_{by}}{F'_{by}} \leqslant 1 \tag{5-41}$$

When members are subjected to bending about both axes and an axial tension

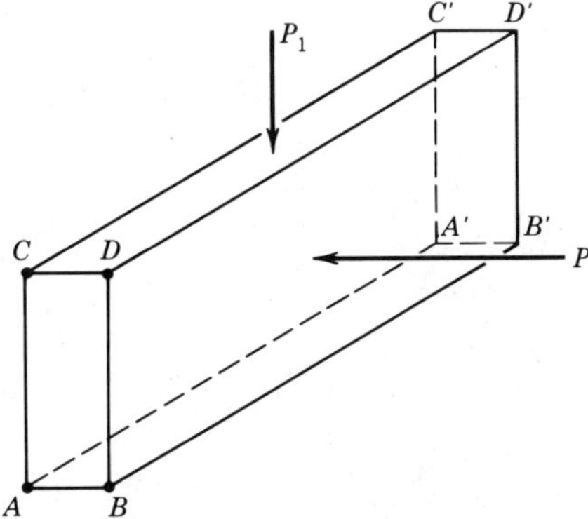

FIGURE 5.19 Bending about both axes.

load is applied, the following formula applies:

$$\frac{f_t}{F'_t} + \frac{f_{bx}}{F'_{bx}} + \frac{f_{by}}{F'_{by}} \leqslant 1 \qquad (5\text{-}42)$$

where terms are as defined on page 5-128.

When members are subjected to bending about both axes and an axial compression load is applied, the following formula applies:

$$\frac{f_c}{F'_c} + \frac{f_{bx}}{F'_{bx} - J_x f_c} + \frac{f_{by}}{F'_{by} - J_y f_c} \leqslant 1$$

where terms are as defined on page 5-128.

When the size factor C_F is less than 1 and when the compressive stress f_c exceeds $F''_{bx}(1 - C_{Fx})$ in the x–x plane, F_{bx} is not modified by C_F. If the compressive stress exceeds $F''_{by}(1 - C_{Fy})$ in the y–y plane, F_{by} is not modified by C_F. When the compressive stress f_c is less than $F''_{bx}(1 - C_{Fx})$, the design bending value is $F'_{bx} = F''_{bx} C_{Fx} + f_c$ but shall not exceed $F''_{bx} C_L$. When the compressive stress is less than $F''_{by}(1 - C_{Fy})$, the design bending value $F'_{by} = F''_{by} C_{Fy} + f_c$ but shall not exceed $F''_{by} C_L$. Note that size factor about the x–x axis and y–y axis will be different due to differences in depth.

Example: A 16-ft-long member is subjected to bending about both axes and a compression load as follows. Assume normal duration load and dry conditions of use. The member is laterally unsupported except at ends.

$$P = 3000 \text{ lb} \qquad M_x = 50{,}000 \text{ in.-lb}$$

$$M_y = 12{,}000 \text{ in.-lb}$$

x
y
3000 lb
M_x = 50,000 in. lb
3000 lb
M_y = 12,000 in. lb
y
x

The following interaction formula must be satisfied:

$$\frac{f_c}{F'_c} + \frac{f_{bx}}{F'_{bx} - J_x f_c} + \frac{f_{by}}{F'_{by} - J_y f_c} \leqslant 1$$

Table 1, AITC 117—Design, is used as a guide for trial design values. Assume $F_{bx} = 2000$ psi; $F_{by} = 1200$ psi; $F_c = 1300$ psi; $E_y = 1{,}400{,}000$ psi; and $E_x = 1{,}500{,}000$ psi. Try a $5\frac{1}{8} \times 10\frac{1}{2}$-in. member.

$$A = 53.81 \text{ in.}^2 \qquad S_x = 94.17 \text{ in.}^3 \qquad S_y = 45.96 \text{ in.}^3$$

$$C_{Fx} = 1.0 \qquad C_{Fy} = 1.10 \quad \text{(see footnote } f\text{, Table 2, AITC 117—Design)}$$

$$l_e \text{ for compression} = (16)(12) = 192 \text{ in.}$$

$$f_c = \frac{3000}{53.81} = 55.8 \text{ psi} \qquad f_{bx} = \frac{50{,}000}{94.17} = 531 \text{ psi}$$

$$f_{by} = \frac{12{,}000}{45.96} = 261 \text{ psi}$$

$$\left(\frac{l_e}{d}\right)_y = \frac{192}{5.125} = 37.46 < 50 \quad \text{O.K.}$$

$$\left(\frac{l_e}{d}\right)_x = \frac{192}{10.5} = 18.3$$

Determine F''_c, F''_{bx}, F''_{by}, and E':

$$F''_c = F_c = 1300 \text{ psi}$$

$$F''_{bx} = F_{bx} = 2000 \text{ psi} \qquad F''_{by} = F_{by} = 1200 \text{ psi}$$

$$E'_y = E_y = 1{,}400{,}000 \text{ psi}$$

Determine F'_c (buckling about y–y axis).

$$K_y = 0.671\sqrt{\frac{E'_y}{F''_c}} = 0.671\sqrt{\frac{1{,}400{,}000}{1300}} = 22 < 37.46$$

(use long-column formula)

$$F'_c = \frac{0.30E'_y}{(l_e/d)_y^2} = \frac{(0.30)(1{,}400{,}000)}{(37.46)^2} = 299 \text{ psi}$$

Determine F'_{bx}:

$$C_{kx} = 0.956\sqrt{\frac{E'_y}{F''_{bx}}} = 0.956\sqrt{\frac{1{,}400{,}000}{2000}} = 25.3$$

$$l_e = 1.92\,l_u$$

for bending with any loading condition, from Figure 5.14*l*.

$$C_{sx} = \sqrt{\frac{l_e d}{b^2}} = \sqrt{\frac{(1.92)(192)(10.5)}{(5.125)^2)}} = 12.1$$

Since 10 < 12.1 < 25.3, use intermediate-beam formula.

$$C_L = \left[1 - \frac{1}{3}\left(\frac{12.1}{25.3}\right)^4\right] = 0.98$$

Note C_{Lx} is 0.98 and $C_{Fx} = 1$; C_{Lx} applies.

$$F'_{bx} = F''_{bx} C_L = (2000)(0.98) = 1965 \text{ psi}$$

Determine F'_{by}:

$$C_{ky} = 0.956 \sqrt{\frac{E'_x}{F''_{by}}} = 0.956 \sqrt{\frac{1,500,000}{1200}} = 33.8$$

$$C_{sy} = \sqrt{\frac{l_e d}{b^2}} = \sqrt{\frac{(1.92)(192)(5.125)}{(10.5)^2}} = 4.1 < 10 \quad \text{(use short-beam formula)}$$

Note that when $C_F > 1$, C_F is cumulative with C_L; $C_L = 1$.

$$F'_{by} = F''_{by} C_{Fy} C_L = (1200)(1.10)(1) = 1320 \text{ psi}$$

$$K_x = 0.671 \sqrt{\frac{E'_x}{F''_c}} = 0.671 \sqrt{\frac{1,500,000}{1300}} = 22.8$$

$$J_x = \frac{(l_e/d)_x - 11}{K_x - 11} = \frac{18.3 - 11}{22.8 - 11} = 0.62$$

$$J_y = 1 \quad \text{(for long-columns)}$$

$$\frac{f_c}{F_c} = \frac{f_{bx}}{F'_{bx} - J_x f_c} + \frac{f_{by}}{F'_{by} - J_y f_c} \leqslant 1$$

$$\frac{55.8}{299} + \frac{531}{1965 - (0.62)(55.8)} + \frac{261}{1320 - (1)(55.8)} = 0.67 < 1 \quad \text{O.K.}$$

A $5\frac{1}{8} \times 9$-in. beam was checked by the same procedure and found to be satisfactory.

A $5\frac{1}{8}$-in. × 9-in. × 16-ft member made from a laminating combination with the tabular design values equal to or exceeding the following should be specified.

$$F_{bx} = 2000 \text{ psi} \qquad F_{by} = 1200 \text{ psi} \qquad F_c = 1300 \text{ psi}$$

$$E_x = 1,500,000 \text{ psi} \qquad E_y = 1,400,000 \text{ psi}$$

The final design should include a check of shear and deflection.

TAPERED STRAIGHT BEAMS OF CONSTANT WIDTH

Glued laminated beams are often tapered to meet architectural requirements, to provide pitched roofs, to facilitate drainage, and to lower wall height requirements at the end supports. Figure 5.20 illustrates the most common forms of these beams.

It is recommended that any sawn taper cuts be made only on the compression face of tapered beams. The laminations should be parallel to the tension face of the beam.

Camber may be provided in beams that are taper cut on the compression face by providing camber on the tension face similar to that in nontapered beams and by sawing camber into the compression face. For single-tapered straight beams, the minimum camber based on dead-load deflection may be built into the tension face as well as sawn into the tapered face. For double-tapered straight beams, the minimum camber is usually built into the tension face, and if this camber is over 2 in., one-fourth the centerline camber is sawn into the tapered face. The designer should specify the required camber both for the tension face and the sawn compression face. Camber is usually not provided by the fabricator unless specified by the designer.

The design methods for tapered beams are based on the procedures contained in *Deflection and Stresses of Tapered Wood Beams* (9). Those procedures presented are based on the Bernoulli–Euler theory of bending and beams of isotropic material. This results in an approximate solution for wood beams, which is considered satisfactory for design purposes. Additional research is being conducted on this subject using the finite-element technique and testing wood as an orthotropic material. Preliminary results indicate that the procedures for tapered beams now being used may be somewhat conservative. Until such time as that research can be fully evaluated, they are the best procedures available and continued use is recommended.

Stresses in Tapered Beams of Constant Width

When the top and bottom of a beam are not parallel to each other (a variable depth along the length of the beam is created by a sawn surface), consideration

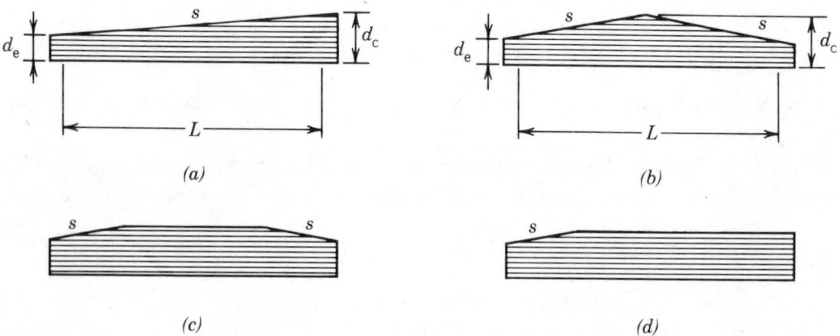

FIGURE 5.20 Simple-span single-tapered straight beams. (*a*) Single-tapered straight; (*b*) double-tapered straight; (*c*) tapered both ends straight; (*d*) tapered one end straight.

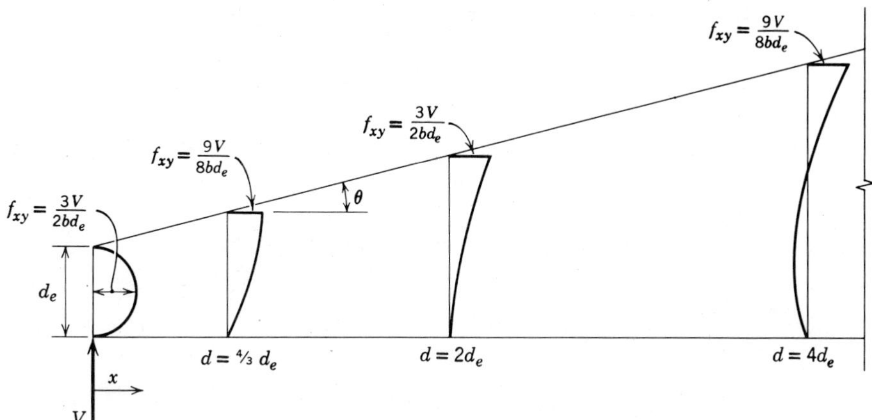

FIGURE 5.21 Distribution of shear stresses in a simply supported beam under concentrated loads.

must be given to the combined effects of bending, compression, tension, and shear parallel to grain and also to compression or tension perpendicular to grain. An illustration of the distribution of the shear stresses existing in a tapered beam is shown in Figure 5.21.

This analysis involves an interaction formula when stresses f_x, f_y, and f_{xy} occur simultaneously to reduce the beam capacity below the capacity that would result from considering each stress separately.

In (9), the basic theory was expanded to verify the applicability of an interaction formula in predicting the strength of tapered timber bending members. Such a formula, expressing the effects of combined stresses is also used for materials other than wood and takes the form

$$\frac{f_x^2}{F_x^2} + \frac{f_y^2}{F_y^2} + \frac{f_{xy}^2}{F_{xy}^2} \leqslant 1 \tag{5-43}$$

where f_x = actual bending stress (psi),

f_y = actual compression or tension stress perpendicular to grain (psi) (compression if the taper cut is on the compression face and tension if the taper cut is on the tension face) (note that taper cutting on the tension face is not recommended),

f_{xy} = actual shear stress (psi),

F_x = tabular design value in bending (psi), F_b, adjusted by all applicable modifiers except C_F, C_I, and C_L,

F_y = tabular design value in compression perpendicular to grain (psi), $F_{c\perp}$, adjusted by all applicable modifiers (when the taper cut is on the tension side (not recommended), $F_y = F_{rt}$), and

F_{xy} = tabular design value in horizontal shear (psi), F_v, adjusted by all applicable modifiers.

For tapered beams, the actual stresses at the tapered edge can be expressed as a function of the bending stress f_x as follows:

$$f_{xy} = f_x \tan\theta \qquad f_y = f_x \tan^2\theta \qquad \text{or} \qquad f_y = f_{xy}\tan\theta$$

where $\tan\theta$ = slope of tapered face.

By assuming an optimum design (i.e., the interaction stress equation set equal to 1), it is possible to write this equation as

$$\frac{f_x^2}{F_b^2} + \frac{f_x^2}{F_v^2}\tan^2\theta + \frac{f_x^2}{F_{c\perp}^2}\tan^4\theta = 1$$

Solving this equation for f_x,

$$f_x = F_b\, C_I$$

where C_I = interaction stress factor:

$$C_I = \left[\frac{1}{1 + (F_b \tan\theta / F_v)^2 + (F_b \tan^2\theta / F_{c\perp})^2}\right]^{1/2} \qquad (5\text{-}44)$$

To account for interaction stresses in design, the design value in bending, F_b, is modified by C_I:

$$F_b' = F_b''\, C_I$$

where F_b'' = tabular design value in bending (psi) adjusted by all applicable modifiers except C_L and C_F.

For a taper cut on the compression side of a beam, C_I is cumulative with adjustments for lateral stability C_L but is not cumulative with C_F except when C_F is greater than 1. Although taper cutting on the tension side of beams is not recommended, C_I in such cases is cumulative with C_F but not C_L. When the taper cut is on the tension side, the design value obtained by the use of C_I and other applicable modifiers is then compared to the actual calculated bending stress f_b to determine if the section is adequate. Because C_I is dependent only on the slope of the tapered face and the material properties, tabular values of C_I as a function of $\tan\theta$ have been developed for commonly used species-bending combinations and are given in Table 5.13.

In the calculation of stresses and the interaction of stresses, it is necessary to know where the maximum stress conditions will occur. The shear stress f_{xy} and the perpendicular to grain stress f_y are functions of the bending stress f_x. Determination of the location of the highest bending stress also gives the location of the highest values of f_{xy} and f_y. The location of the point of maximum stress can be obtained by simple calculation or by inspection in some members. In other members, it may be necessary to divide the members into finite sections and calculate the maximum bending stress at each cross section.

Unless otherwise specified, tapered glued laminated timber beams will have the same grades on the taper cut as required for the compression side of the combination being used. The tabular design values required for calculating C_I (F_{bx}, F_{vx}, and $F_{c\perp}$) can be obtained from Table 1, AITC 117—Design (1), included in

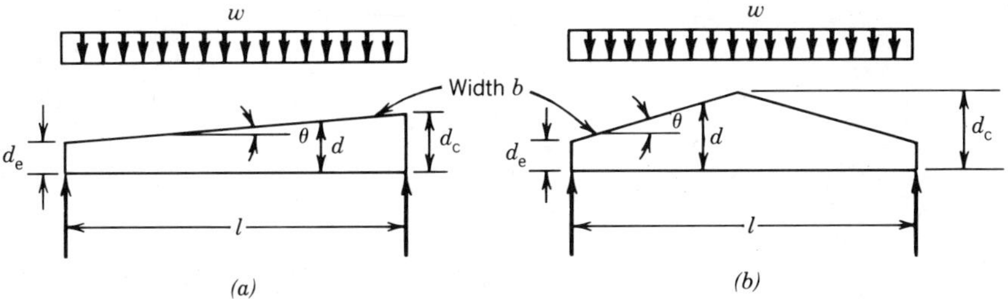

FIGURE 5.22 Tapered beams: (*a*) single-tapered straight beam; (*b*) double-tapered straight beam.

Section 7 of this manual. Tapered beams can also be manufactured by cutting through the compression zone as illustrated on page 5-178 for straight beams with end cuts. If this is done, the tabular design values in Table 5.12 for F_{bx}, $F_{c\perp}$, and E_x should be used to calculate C_I. Other design values in Table 1, AITC 117—Design (1), remain unchanged. There must be a clear understanding between the designer and the manufacturer when this process is used.

The depth d of tapered beams where maximum bending stress occurs for uniformly loaded single-tapered straight beams (a in Fig. 5.22) or symmetrical double-tapered beams (b in Fig. 5.22) can be determined by the formula

$$d = 2d_e \frac{d_e + l \tan \theta}{2d_e + l \tan \theta}$$

Depth at Maximum Stress, Uniform Load

The above formula can be converted to the following forms for easier computations.

For single-tapered straight beams (Fig. 5.22*a*):

$$d = 2d_e \frac{d_c}{d_e + d_c} \tag{5-45}$$

For symmetrical double-tapered straight beams (Fig. 5.22*b*):

$$d = \frac{d_e}{d_c}(2d_c - d_e) \tag{5-46}$$

Bending Stress, Uniform Load

The bending stress f_x at the point of maximum stress for both types of beams shown in Figure 5.22 can be determined by the following formula for a uniform load:

$$f_x = \frac{3\,Wl}{4\,bd_e(d_e + l \tan \theta)} \tag{5-47}$$

where W = total uniform load (lb) and
l = span (in.).

For single-tapered straight beams (Fig. 5.22*a*), the formula reduces to

$$f_x = \frac{3\,Wl}{4\,bd_c d_e} \tag{5-48}$$

For symmetrical double-tapered straight beams (Fig. 5.22*b*), the formula converts to

$$f_x = \frac{3\,Wl}{4\,bd_e(2d_c - d_e)} \tag{5-49}$$

Depth at Maximum Stress, Concentrated Load

For a beam loaded with a single concentrated load as shown in Figure 5.23, the depth at which maximum stress occurs can be determined by the formula

$$d = 2d_e$$

when d lies within the region 0 to z.

When d as determined from this formula lies outside of the region 0 to z, the depth at which maximum bending stress occurs is at the load point. If the depth d under the load point is such that $\frac{4}{3}d_e \leqslant d < 2d_e$, the maximum shear at the section also occurs on the tapered surface. If the depth d under the load point is such that $d_e < d < \frac{4}{3}d_e$, the maximum value of shear stress lies within the beam, but because the stresses along the taper cut are those considered in the interaction equation, the value of shear stress as determined by the formula $f_{xy} = f_x \tan\theta$ is still applicable for use in the interaction equation.

Stresses

The stresses at the tapered surface can be determined by the formulas

$$f_x = \frac{6M}{bd^2} \qquad f_{xy} = f_x \tan\theta \qquad f_y = f_x \tan^2\theta$$

where M = moment at point where depth d occurs (in.-lb),
b = width at point where depth d occurs (in.), and
d = depth where maximum bending stress occurs (in.).

When a tapered member contains loads or combinations of loads other than a uniform load or a single concentrated load or if a nonsymmetrical double-tapered straight member is used, the simplified formulas for determining the depth at

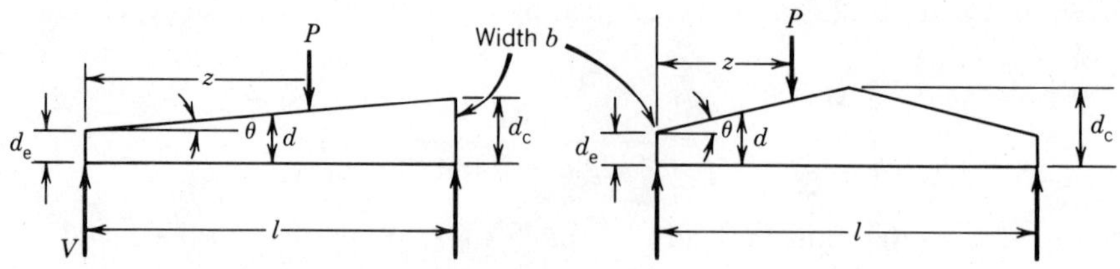

FIGURE 5.23 Depth at which maximum bending stress occurs.

which the maximum stress occurs do not apply. In these cases it is generally necessary to determine f_b (f_x) at various points along the member to determine the maximum bending stress. Once the maximum bending stress f_x has been obtained, f_y and f_{xy} can be calculated for that cross section and the effect of their interaction determined.

Deflection

The shear deflection in tapered members is larger than in prismatic members and the resulting total deflection including both bending deflection and shear deflection is slightly larger than that obtained by the customary methods of calculating deflection in prismatic members. (See (9) for further information.) Acceptable accuracy in determining the deflection of tapered members can be obtained by determining the deflection of an equivalent prismatic member. The depth of an equivalent member of constant cross section of the same width that will have the same deflection as a tapered beam can be determined by the formula

$$d = C_{dt}\, d_e \tag{5-50}$$

where d_e = depth of ends of symmetrical double-tapered beam or smaller end depth of single-tapered beam (in.) and
C_{dt} = constant derived from relationship of formulas for deflection of tapered beams and straight prismatic beams.

For a uniformly loaded symmetrical double-tapered beam as shown in Figure 5.24*b*,

$$C_{dt} = 1 + 0.66C_y \quad \text{when} \quad 0 < C_y \leqslant 1$$

$$C_{dt} = 1 + 0.62C_y \quad \text{when} \quad 1 < C_y \leqslant 3$$

$$C_y = \frac{d_c - d_e}{d_e} \tag{5-51}$$

For a uniformly loaded single tapered straight beam as shown in Figure 5.24*a*,

$$C_{dt} = 1 + 0.46\,C_y \quad \text{when} \quad 0 < C_y < 1.1$$

$$C_{dt} = 1 + 0.43\,C_y \quad \text{when} \quad 1.1 < C_y \leqslant 2$$

where the terms are as previously defined.

The calculation of the deflection of tapered beams with other than uniform loads or single loads at the midpoint becomes somewhat complicated. In many cases,

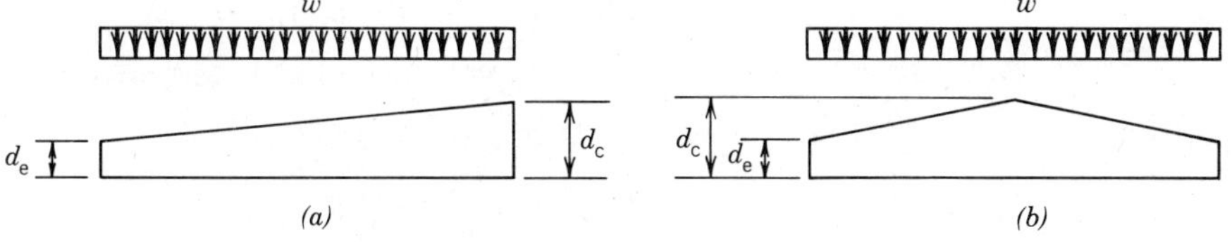

FIGURE 5.24 Notations for end and center depths

the resulting moment diagram approximates the moment diagram for a uniform load and a close approximation can be obtained by assuming the load to be uniform. When this cannot be done, it is recommended that one of the other methods of determining deflection such as the virtual work (dummy load) method be used.

Theoretically, both the shear deflection Δ_s and the bending deflection Δ_b should be determined. The total deflection Δ is the summation of the two.

$$\Delta = \Delta_s + \Delta_b$$

The tabular design values for modulus of elasticity already contain small elements of shear deflection because they are based on the deflection of a member with an l/d of 21. In most cases, sufficient accuracy in deflection can be obtained by using only the tabular modulus of elasticity in deflection calculations for bending deflection Δ_b and ignoring the shear deflection Δ_s because it has been largely compensated for by the lower tabular design values for E.

If it is desired to calculate both Δ_b and Δ_s, the modulus of elasticity E in the formula for bending deflection Δ_b should be 5% larger than the tabular design value of E. The shear deflection can be approximated by the formula

$$\Delta_s = \frac{3Wl}{20Gbd_e} \tag{5-52}$$

where G = shear modulus (psi) (approximately equal to $0.06E$)
and other terms are as previously defined. See page 4-89 for a detailed discussion on the variability of modulus of elasticity.

Example. Design a double-tapered straight glued laminated timber roof beam to meet the following requirements. Roof decking is applied directly to beams to provide lateral stability and ends are held in place. Assume dry conditions of use and the following parameters:

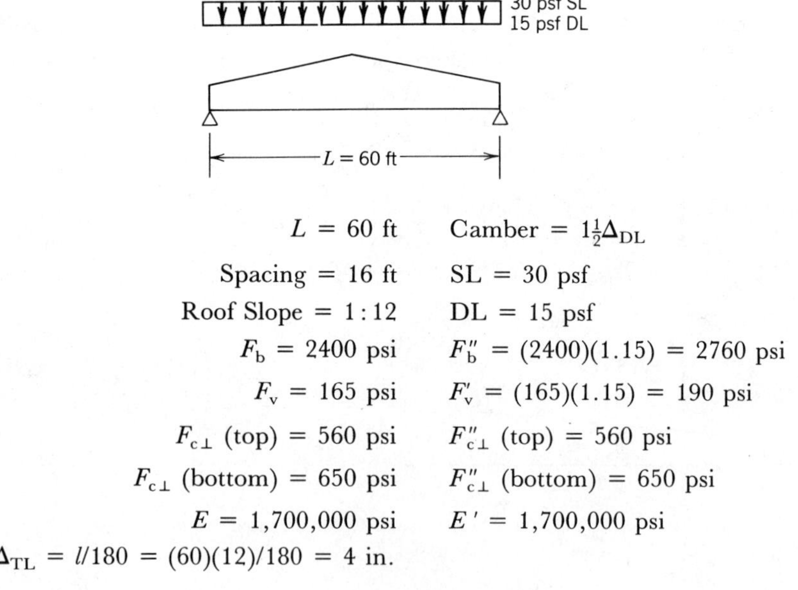

$L = 60$ ft   Camber $= 1\frac{1}{2}\Delta_{DL}$

Spacing $= 16$ ft   SL $= 30$ psf

Roof Slope $= 1:12$ DL $= 15$ psf

$F_b = 2400$ psi $F_b'' = (2400)(1.15) = 2760$ psi

$F_v = 165$ psi $F_v' = (165)(1.15) = 190$ psi

$F_{c\perp}$ (top) $= 560$ psi $F_{c\perp}''$ (top) $= 560$ psi

$F_{c\perp}$ (bottom) $= 650$ psi $F_{c\perp}''$ (bottom) $= 650$ psi

$E = 1{,}700{,}000$ psi $E' = 1{,}700{,}000$ psi

$\Delta_{TL} = l/180 = (60)(12)/180 = 4$ in.

1. Determine end depth: Assume effective span for shear $L_e = 55$ ft and width $b = 5\frac{1}{8}$ in.

$$V = \frac{wL_e}{2} = \frac{(30 + 15)(16)(55)}{2} = 19{,}800 \text{ lb}$$

$$\text{End depth} = d_e = \frac{3V}{2bF'_v} = \frac{(3)(19{,}800)}{(2)(5.125)(190)} = 30.5 \text{ in.} \quad \text{(use 30 in.)}$$

Check L_e:

$$L_e = 60 - \frac{(2)(30)}{12} = 55 \text{ ft} \quad \text{O.K.}$$

2. Determine trial centerline depth: Centerline depth $= d_c = d_e +$ (roof slope)(span/2)

$$d_c = 30 + \left(\frac{1}{12}\right)\left(\frac{60}{2}\right)(12) = 60 \text{ in.}$$

3. Check maximum deflection: Determine depth d of equivalent prismatic beam that has the same deflection as the tapered beam:

$$d = d_e C_{dt}$$

where d_e = depth at end (in.),
C_{dt} = multiplying factor obtained from the following formula for a uniformly loaded double-tapered straight beam,
$C_{dt} = 1 + 0.66C_y$

$$C_y = \frac{d_c - d_e}{d_e} = \frac{60 - 30}{30} = 1$$

$$C_{dt} = 1 + (0.66)(1) = 1.66$$

$$d = d_e C_{dt} = (30)(1.66) = 49.8 \text{ in.}$$

$$I \text{ of equivalent beam} = \frac{(5.125)(49.8)^3}{12} = 52{,}700 \text{ in.}^4$$

$$\Delta_{TL} = \frac{5wL^4}{384EI} = \frac{(5)(45)(16)(60)^4(12)^3}{(384)(1{,}700{,}000)(52{,}700)} = 2.34 \text{ in.} < 4 \text{ in.} \quad \text{O.K.}$$

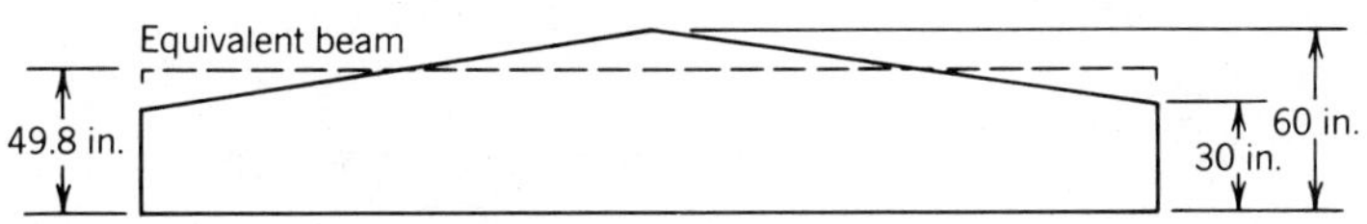

4. Check the actual stresses: Determine depth at which maximum stresses occur:

$$d = \frac{d_e}{d_c}(2d_c - d_e) = \frac{30}{60}(120 - 30) = 45 \text{ in.}$$

$$C_F = 0.86 \quad \text{(from Table 7.2)}$$

Assuming size factor controls:

$$F_b' = F_b'' C_F = (2760)(0.86) = 2370 \text{ psi}$$

$$f_x = \frac{3Wl}{4bd_e(2d_c - d_e)}$$

$$= \frac{(3)(720)(60)(60)(12)}{(4)(5.125)(30)(120 - 30)} = 1690 \text{ psi} < 2370 \text{ psi} \quad \text{O.K.}$$

Actual shear stress at point where $d = 45$ in.:

$$f_{xy} = f_x \tan\theta = 1690\left(\frac{1}{12}\right) = 140 \text{ psi} < 190 \quad \text{O.K.}$$

Actual compression perpendicular to grain stress:

$$f_y = f_x \tan^2\theta = 1690\left(\frac{1}{12}\right)^2 = 11.7 \text{ psi} < 560 \quad \text{O.K.}$$

5. Check the effects of combined stresses at the point of maximum stress ($d = 45$ in.) by either the interaction formula or use of the interaction stress factor C_I. (Both methods are shown for purposes of illustration.)

Using interaction formula:

$$F_x = F_b' \qquad F_y = F_{c\perp}' \qquad F_{xy} = F_v'$$

$$\frac{f_x^2}{F_x^2} + \frac{f_y^2}{F_y^2} + \frac{f_{xy}^2}{F_{xy}^2} \leq 1$$

$$\frac{(1690)^2}{(2760)^2} + \frac{(11.7)^2}{(560)^2} + \frac{(140)^2}{(190)^2} = 0.92 < 1 \quad \text{O.K.}$$

Using interaction stress factor C_I:

$$\tan\theta = \frac{1}{12} = 0.083$$

$$C_I = 0.638 \quad \text{(from Table 5.12)}$$

$$C_F = 0.86 \quad \text{(from Table 7.2)}$$

$C_I < C_F$, and C_I controls.

$$F_b' = F_b'' C_I = (2760)(0.638) = 1760 \text{ psi} > 1690 \text{ psi} \quad \text{O.K.}$$

6. Determine camber:

$$\text{Dead load deflection} = \Delta_{DL} = \Delta_{TL}\frac{w_{DL}}{w_{TL}}$$

$$\Delta_{DL} = 2.34\left(\frac{240}{720}\right) = 0.78 \text{ in.}$$

$$\text{Centerline camber} = 1.5\Delta_{DL} = (1.5)(0.78) = 1.17 \text{ in.}$$

No compression face camber is required because it is less than 2 in. The assumed design values are satisfactory. For greater choices of available laminating combinations, the designer should recheck the design using lower design values for compression perpendicular to grain and shear.

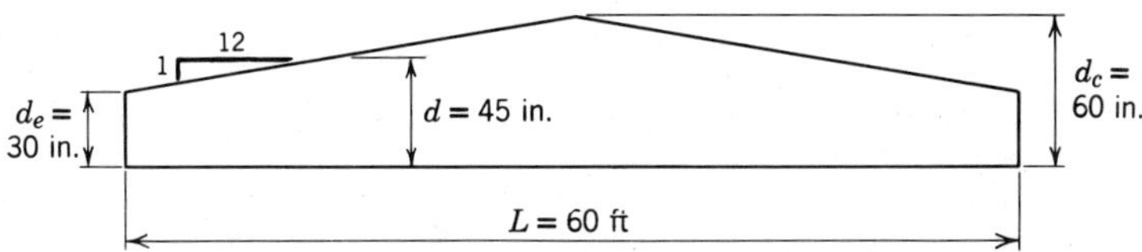

CURVED MEMBERS

Minimum Bending Radii

The recommended minimum radii of curvature R for curved structural glued laminated timbers are 9 ft 4 in. for Western Species and 7 ft 0 in. for Southern Pine for a lamination thickness t of $\frac{3}{4}$ in.; and 27 ft 6 in. for a lamination thickness of $1\frac{1}{2}$ in. for all species. If required for architectural or other design considerations, other radii of curvature may be used with these thicknesses and other radius–thickness combinations may be used provided that the ratio t/R does not exceed 1/100 for hardwoods and Southern Pine, or 1/125 for other softwoods.

Curvature Factor

Because of stresses induced in bending laminations to a required curvature, the allowable flexural stress in curved laminated members is less than in straight members. Therefore, the allowable unit stress in bending for a curved member must be modified by multiplying by the curvature factor C_C given in Eq. (3-2) as follows:

$$C_C = 1 - 2000\left(\frac{t}{R}\right)^2$$

where t = thickness of lamination (in.) and
R = radius of curvature (bending radius) of the lamination (in.).

The curvature factor C_C should not be applied to stresses in straight portions of a member, regardless of curvature elsewhere in the member. See Figure 3.2 for a graphical solution of the curvature factor equation.

Radial Tension and Compression

When curved members are subjected to a bending moment M, radial stresses are set up in a direction parallel to the radius of curvature R of the centerline of the member (perpendicular to grain). If the moment increases the radius of curvature (causes the member to become straighter), the stress is tension; if it decreases the radius (causes the member to become more sharply curved), the stress is compression.

The formula for computing the actual stress f_r in members of constant cross section is:

$$f_r = \frac{3M}{2R_m bd} \tag{5-53}$$

where M = bending moment (in.-lb),
b = width of rectangular member (in.),
d = depth of rectangular member (in.), and
R_m = radius of curvature of centerline of member (in.).

For these members, when M causes stress in tension across the grain (radial tension), the tensile stress shall be limited to one-third the design value in horizontal shear for Southern Pine (67 psi) and California Redwood (42 psi) for all load conditions and for Douglas Fir-Larch and Douglas Fir South (55 psi), Hem-Fir (52 psi), and Western Woods (47 psi) for wind and earthquake loadings. The limit shall be 15 psi for Douglas Fir-Larch, Douglas Fir South, Hem-Fir, and Western Woods for other conditions of loading. These values are subject to modification for duration of load.

If the calculated stress exceeds the applicable design value indicated, the design should be reevaluated by changing the geometry of the section by increasing the radius of curvature or by changing the pitch to increase the centerline depth of the member. As an alternate procedure for Douglas Fir-Larch, Douglas Fir South, Hem-Fir, and Western Woods, mechanical reinforcement may be designed sufficient to resist the full magnitude of the calculated radial tension stress in accordance with the recommendations given on pages 5-229 to 5-234 of this manual. When radial reinforcement is used, it resists shrinkage that may occur between the time of installation of the reinforcement and the time the member reaches equilibrium. If the shrinkage is excessive, this restraint may cause cracking in the member. To minimize this condition where members are to be used in dry conditions, the member should be manufactured from laminations with a maximum moisture content of 12%. With radial reinforcement, the calculated radial tension stress f_{rt} is limited to 55 psi for Douglas Fir-Larch and Douglas Fir South, 52 psi for Hem-Fir, and 47 psi for Western Woods subject to modification for duration of load.

When the moment is in a direction causing a stress in compression across the grain, this stress shall be limited to the design value in compression perpendicular to the grain, $F_{c\perp}$, for the species involved.

When designing a curved bending member of variable cross section such as a double-tapered curved beam, the radial stress f_r is computed by the equation

$$f_r = K_r C_r \frac{6M}{bd_c^2} \tag{5-54}$$

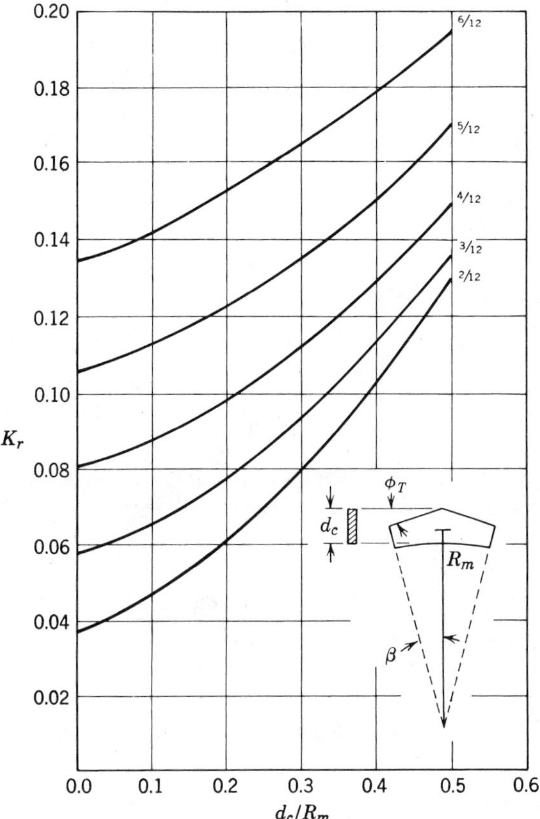

FIGURE 5.25 Determination of K_r.

where K_r = radial stress factor obtained from Figure 5.25 or calculated from Table 5.14,

M = bending moment at midspan (in.-lb),

b = width of cross section (in.),

d_c = depth of cross section at centerline (in.), and

C_r = reduction factor, a function of the shape of the member obtained from Figures 5.32–5.35.

The design values in radial tension, F_{rt}, are the same as the design values given for members of constant cross section.

The radial stress factor K_r varies with the ratio of the slope of the bottom face to the slope of the top face, ϕ_B/ϕ_T. To reduce the design complexity, K_r has been selected for one ϕ_B/ϕ_T ratio. This results in a slightly conservative design for other ϕ_B/ϕ_T ratios.

$$K_r = A + B\left(\frac{d_c}{R_m}\right) + C\left(\frac{d_c}{R_m}\right)^2 \tag{5-55}$$

TABLE 5.14

Polynomial Approximation to K_r[a]

Angle of Upper Tapered Surface, ϕ_T (degrees)	Factors		
	A	*B*	*C*
2.5	0.0079	0.1747	0.1284
5.0	0.0174	0.1251	0.1939
7.5	0.0279	0.0937	0.2162
10.0	0.0391	0.0754	0.2119
15.0	0.0629	0.0619	0.1722
20.0	0.0893	0.0608	0.1393
25.0	0.1214	0.0605	0.1238
30.0	0.1649	0.0603	0.1115

[a]For intermediate values of ϕ_T, use straight-line interpolation between calculated values of K_r as illustrated for K_θ on page 5-237.

where A, B, C = are dimensionless factors from Table 5.14,
d_c = depth of centerline of member (in.), and
R_m = radius of curvature of centerline of member (in.).

The values of factors A, B, and C are shown graphically in Figure 5.26.

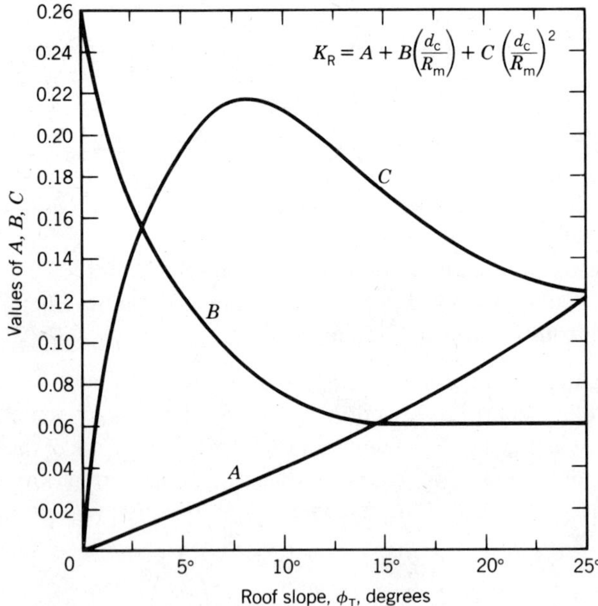

FIGURE 5.26 Roof slope versus *A*, *B*, and *C*. From *Timber Design Manual*, Laminated Timber Institute of Canada.

CURVED BEAMS

A number of different types of curved beams can be manufactured from glued laminated timbers. The most commonly used are shown in Figure 5.27:

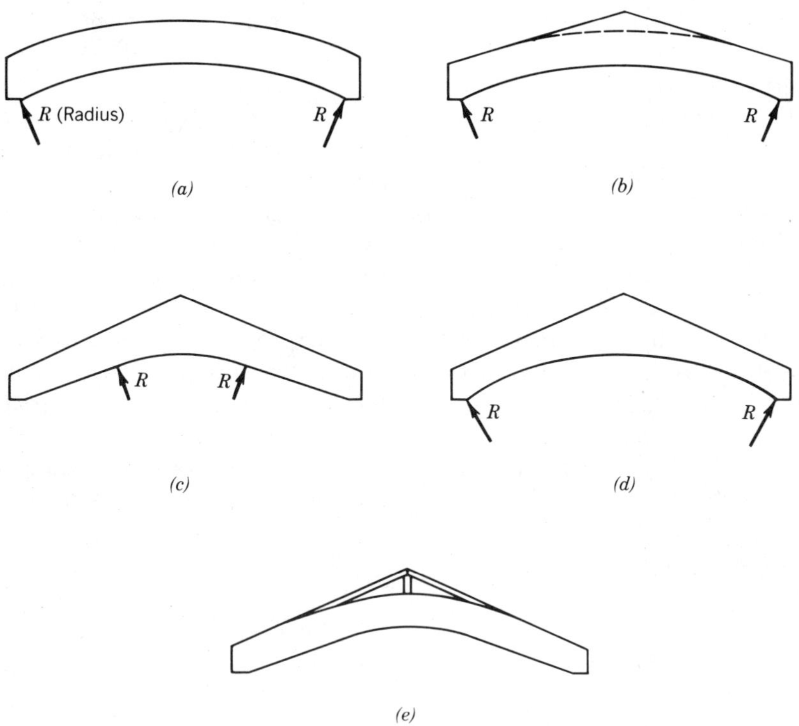

FIGURE 5.27 Simple span curved beam forms. (*a*) Curved beam with constant cross section; (*b*) Double-tapered pitched, with constant cross section and mechanically attached haunch: (*c*) Double-tapered pitched (tangent ends); (*d*) Double-tapered pitched (constant curvature); (*e*) Double-tapered pitched, with constant cross section and framed haunch.

Double-tapered pitched and curved glued laminated beams are among the most popular types of structural roof members where a sloping roof and maximum interior clearance are desired. In these beams, as shown in Figure 5.28, the top edge slopes from the apex at the centerline toward the supports at an angle to the horizontal ϕ_T and the lower edge is curved between the tangent points P.T. and slopes at an angle ϕ_B between the tangent point and the supports. The end portion of the member is usually tapered but may be of constant cross section between the end and the tangent point. The tangent points are usually located near the quarter points, but the location can vary from a point located as close to the center as the minimum radius of curvature permits to a point at the end support. The procedure illustrated applies only to symmetrical double-tapered pitched and curved beams which are uniformly loaded and laterally supported along the top edge.

The design of double-tapered pitched and curved beams is similar to that of straight prismatic beams, with the exception that radial tension stress f_{rt} is induced and the distribution of bending stresses about the assumed neutral axis is different from a prismatic member. Radial tension stresses were recognized by T. R. C. Wilson in Technical Bulletin No. 691 (10). The equation $f_r = 3M/2Rbd$ was derived on the basis of an analogy to a pressure vessel. It is applicable to members of constant cross section only. The procedure included in this manual is based on subsequent research.

The bending and radial stresses in double-tapered pitched and curved beams are affected by the variable shape of the section, and their exact determination is complex. The procedures presented in this manual are based on a simplification of those contained in *Behavior and Design of Double-Tapered Pitched and Curved Glulam Beams* (11). The effect of interaction of stresses on the top tapered cut of these types of members is usually very small and can be ignored in beams of the usual configurations. Because the factors for determining bending and radial tension stresses are a function of the geometric configuration of the member, they cannot be accurately determined until the final size and shape are known. Therefore, the procedure presented in this manual is a trial-and-error method whereby a trial size is determined and adjusted until the correct size is obtained. Other methods may be used provided the design criteria are satisfied. The stress in bending, f_b, is increased by the bending stress factor K_θ due to the shape of the member.

$$f_b = K_\theta \frac{6M}{bd_c^2} \tag{5-56}$$

where M = moment at midspan (in.-lb),
b = width (in.),
d_c = depth at midspan (in.)
$K_\theta = D + E(d_c/R_m) + F(d_c/R_m)^2$,
D, E, F = dimensionless factors from Table 5.15,

and the other terms are as defined previously.

TABLE 5.15

Coefficients for Determining K_θ[a]

ϕ_T (degrees)	D	E	F
2.5	1.042	4.247	−6.201
5	1.149	2.036	−1.825
10	1.330	0.0	0.927
15	1.738	0.0	0.0
20	1.961	0.0	0.0
25	2.625	−2.829	3.538
30	3.062	−2.594	2.440

[a]For intermediate values of K_θ, use straight-line interpolation between the values of θ_T (see examples).

The deflection is also affected by the shape of the member and the following equation represents a close approximation based on test data:

$$\Delta_c = \frac{5\,Wl^3}{32\,E'bd_{eb}^3} \tag{5-57}$$

where Δ_c = deflection at midspan (in.),
W = total uniform load (lb),
l = span (in.),
E' = modulus of elasticity (psi) adjusted by appropriate modifiers,
b = width (in.),
d_{eb} = $(d_e + d_c)(0.5 + 0.735 \tan \phi_T) - 1.41(d_c) \tan \phi_B$,
ϕ_B = slope of bottom (soffit), at ends (degrees),
ϕ_T = slope of top (degrees)

and d_c and d_e are as shown in fig. 5.28. Other methods of determining deflection such as the virtual work method may also be used.

Design Procedures for Double-Tapered Pitched and Curved Beams

The design procedure is based on a mathematical solution of the beam geometry (see Fig. 5.28) and eliminates the need for a graphical solution. Other methods may be used provided the requirements in steps 7 and 8 are met. When reinforcement is needed, step 10 must be followed.

1. Determine minimum end depth d_e:

$$d_e = \frac{3\,V}{2\,bF'_v} \tag{5-58}$$

where V = shear based on effective span (lb),
b = width (in.), (assumed) and
F'_v = tabular design value in shear (psi) adjusted by the applicable modifiers.

2. Determine approximate centerline depth d_{cb} required to meet bending

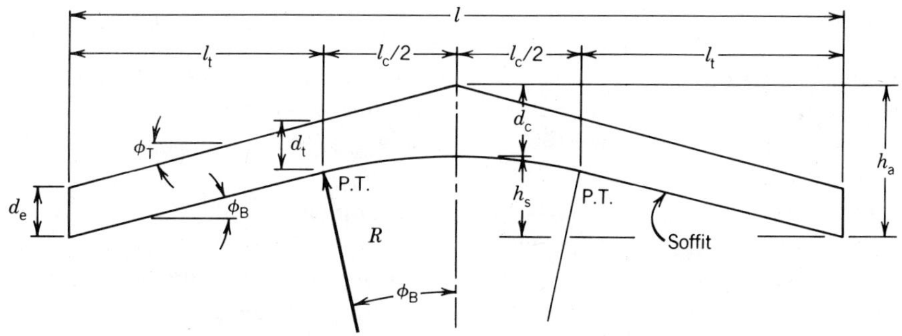

FIGURE 5.28 Pitched and tapered curved beam.

stress limitations:

$$d_{cb} = \sqrt{\frac{6MD}{bF'_b}} \qquad (5\text{-}59)$$

where D = coefficient obtained from Table 5.15.
M = maximum bending moment (in.-lb), and
F'_b = tabular design value in bending (psi) adjusted by applicable modifiers (including C_F *or* C_L).

3. Determine approximate centerline depth $d_{c\Delta}$ to meet deflection limitations (Δ_{max}) as shown in Table 5.7. Determine approximate effective centerline depth d_{eff} for calculating $d_{c\Delta}$:

$$d_{eff} = \sqrt[3]{\frac{Wl^3}{6.4E'b\Delta_{max}(\cos\phi_T)^3}} \qquad (5\text{-}60)$$

where W = total uniform load (lb),
l = span (in.),
E' = tabular design value of E (psi), adjusted by applicable modifiers
b = width (in.), and
ϕ_T = slope of top face (degrees).

The approximate centerline depth is

$$d_{c\Delta} = 2d_{eff} - d_e \qquad (5\text{-}61)$$

4. Determine the trial minimum centerline depth d_{crt} due to the radial tension stress limitation.

$$d_{crt} = \sqrt{\frac{6MK_r}{bF'_{rt}}} \qquad (5\text{-}62)$$

where K_r = value obtained from Figure 5.25,
R_m = radius of curvature at neutral axis (in.),
M = maximum bending moment (in.-lb), and
F'_{rt} = design value for radial tension (psi) adjusted by applicable modifiers,
b = width (in.).

The determination of K_r requires an estimate of a value for d_c/R_m. It is recommended that the first trial value of K_r be determined by taking d_c as the greater of the trial values of d_{cb} (from step 2) or $d_{c\Delta}$ (from step 3). A trial value for R_m can be taken as the span in inches.

If the design value for Douglas Fir-Larch, Douglas Fir South, Hem-Fir, or Western Woods is exceeded (see page 5-218), the member should be reproportioned or mechanical reinforcing sufficient to resist all of the radial tension load should be used. When radial reinforcement is used, the design value for radial stress, F'_{rt}, may not exceed one-third the tabular design value in shear adjusted by applicable modifiers for the species (55 psi for Douglas Fir-Larch and Douglas

Fir South, 52 psi for Hem-Fir, and 47 psi for Western Woods). As an alternative to mechanical reinforcement, the beam may be manufactured with the peaked portion (haunch) not glued to the remainder of the beam as a nonstructural part or the peaked area may be replaced by open framing. In these cases, the member may be designed as a curved beam (see page 5-243). If d_{crt} is greater than the height of the apex point, h_a (see step 5), or is impracticably large for unreinforced Douglas Fir-Larch, Douglas Fir South, Hem-Fir, or Western Woods recalculate d_{crt} based on F_{rt} for radial reinforcement. In no case should F_{rt} exceed one-third the design value in horizontal shear, with or without mechanical reinforcement.

5. Determine the height of the apex point h_a the height of the soffit at midspan, h_s, a trial bottom slope and soffit radius R as follows:

$$h_a = d_e + \frac{l}{2} \tan \phi_T \tag{5-63}$$

where l = span length (in.), and
ϕ_T = top slope (degrees).

$$h_s = h_a - d_c \tag{5-64}$$

where d_c = largest of d_{cb}, $d_{c\Delta}$, or d_{crt}.

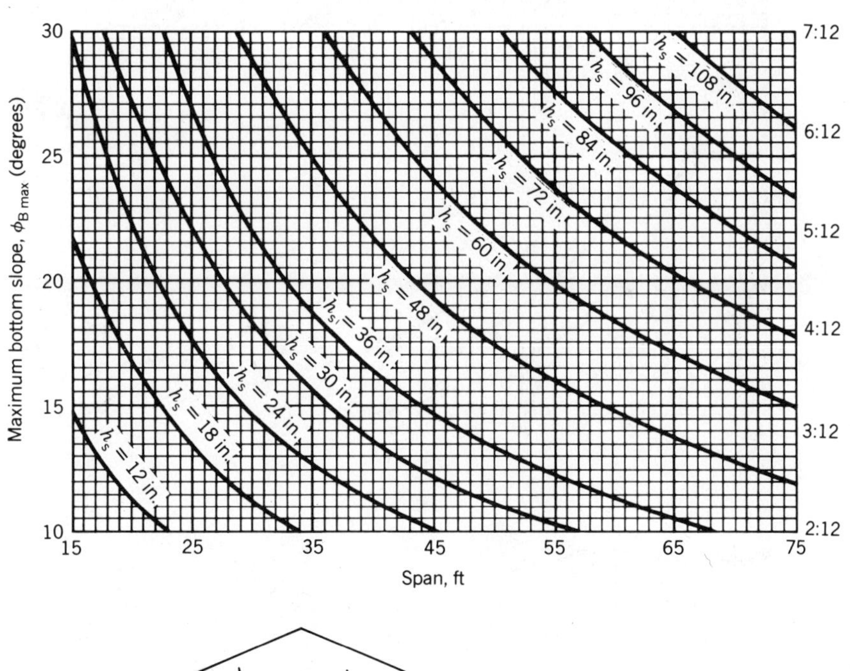

FIGURE 5.29 Graph for determining the largest possible bottom slope (ϕB_{max}).

From Figure 5.29, determine the maximum bottom slope, $\phi_{B\,max}$. If $\phi_{B\,max}$ is greater than the top slope, ϕ_T, set $\phi_{B\,max} = \phi_T$ (in which case the beam becomes a double-tapered curved beam as shown in Figure 5.27*d*). From Figure 5.30, determine the smallest possible value of the bottom slope, $\phi_{B\,min}$. If the value for $\phi_{B\,min}$ cannot be obtained from Figure 5.30 (because of the curves not extending far enough) or if the value of $\phi_{B\,min}$ obtained from Figure 5.30 is greater than ϕ_T, then use $\frac{3}{4}$-in. laminations and determine $\phi_{B\,min}$ from Figure 5.31.

Choose the trial bottom slope that is between the maximum and minimum permitted as:

$$\phi_B = 0.45(\phi_{B\,max} + \phi_{B\,min}) \quad \text{(but not less than } \phi_{B\,min}) \tag{5-65}$$

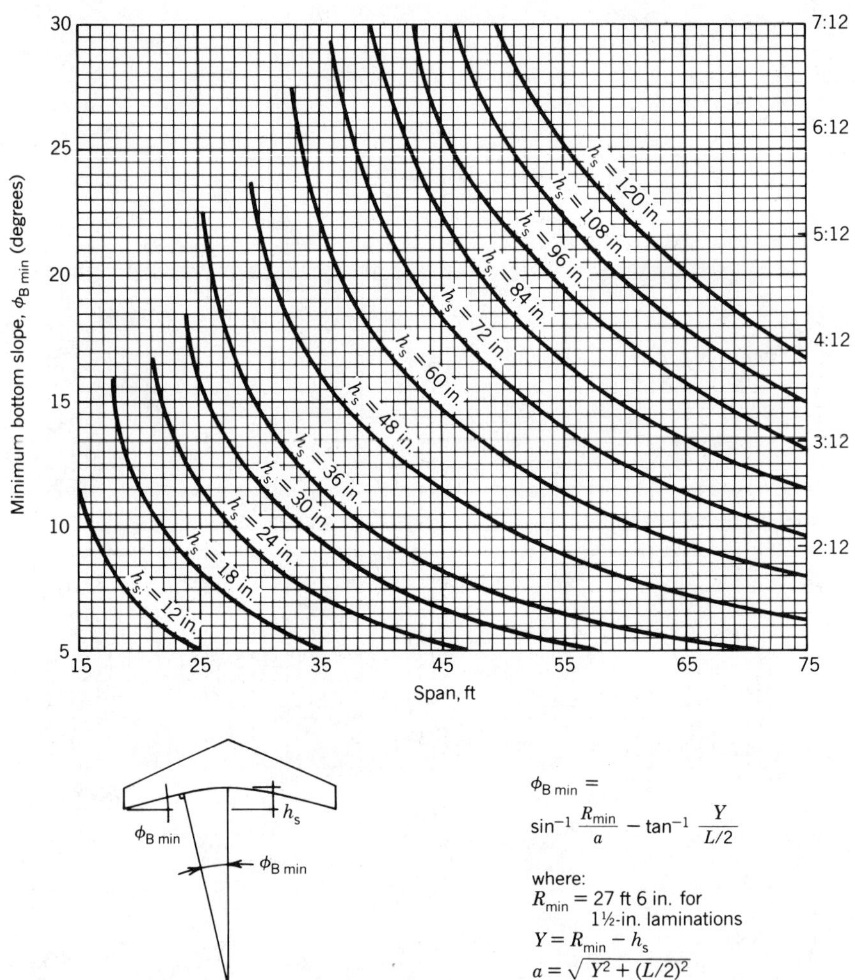

FIGURE 5.30 Graph for determining the smallest possible bottom slope ($\phi_{B_{min}}$) when using $1\frac{1}{2}$-in. laminations (Western Species).

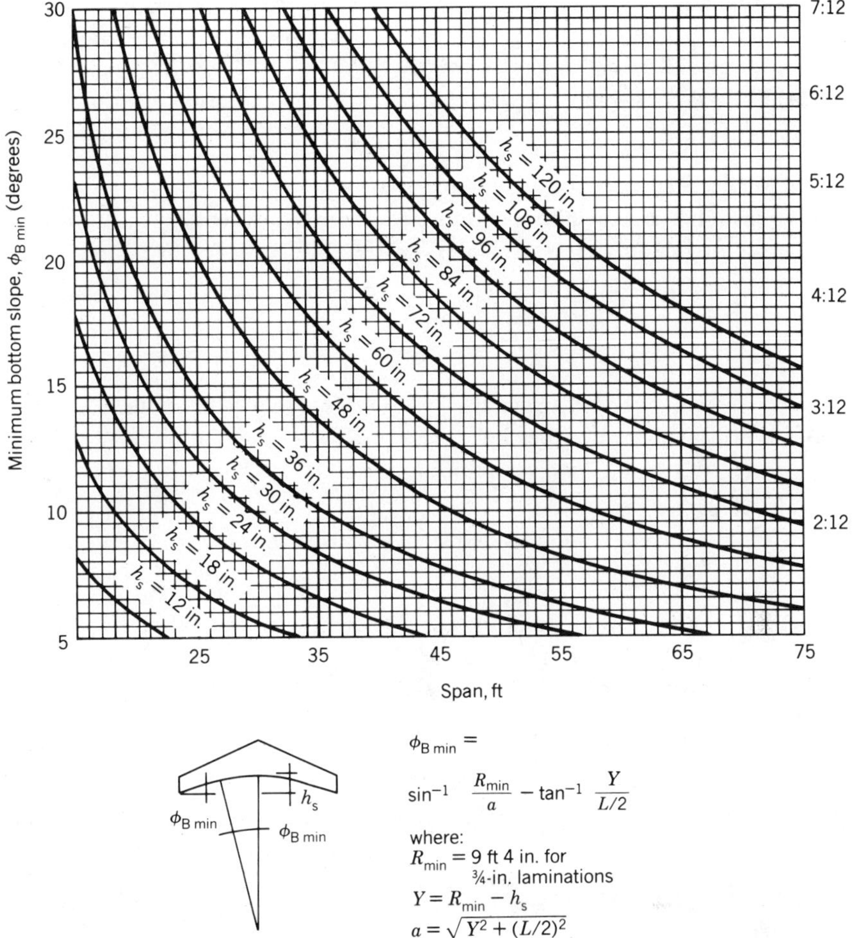

FIGURE 5.31 Graph for determining the smallest possible bottom slope ($\phi_{B_{min}}$) when using $\frac{3}{4}$ in. laminations (Western Species).

$$R = \frac{h_s - \dfrac{l}{2}\tan\phi_B}{1 - \cos\phi_B - \sin\phi_B \tan\phi_B} \tag{5-66}$$

6. Determine the following values using the trial bottom slope ϕ_B and soffit radius R determined in step 5. The following notations are used in the calculations:

(a) $l_t = l/2 - R\sin\phi_B$ = length of tapered leg (in.),

(b) $l/l_c = l/(l - 2l_t)$ = ratio of span length to distance between tangent points,

(c) $d_t = d_e + l_t(\tan\phi_T - \tan\phi_B)$ = depth of beam at tangent point (in.),

(d) $d_c/R_m = d_c/(R + d_c/2)$, and

(e) $f_o = 6M/bd_c^2$ = reference stress at centerline used for convenience in the calculations where M = maximum bending moment at centerline (in.-lb).

7. Check deflection at the centerline Δ_c.

$$d_{eb} = (d_e + d_c)(0.5 + 0.735 \tan \phi_T) - 1.41 d_c \tan \phi_B \qquad (5\text{-}67)$$

$$\Delta_c = \frac{5Wl^3}{32E'db_{eb}^3} \qquad (5\text{-}68)$$

8. Check stresses.

(a) Determine extreme fiber bending stress at the apex centerline. From Table 5.15, determine bending stress factor K_θ corresponding to ϕ_T and d_c/R_m:

$$K_\theta = D + E(d_c/R_m) + F(d_c/R_m)^2 \qquad (5\text{-}69)$$

$$f_b = K_\theta f_o \qquad (5\text{-}70)$$

The stress f_b must be no larger than the design value in bending, F_b'.

(b) Determine extreme fiber bending stress at the tangent point:

$$f_{bt} = 6M_t / bd_t^2 \qquad (5\text{-}71)$$

where M_t = bending moment at a distance L_t from the end (in.-lb); f_{bt} must be no larger than F_b'.

(c) Check the maximum radial stress. Determine the radial stress factor K_r corresponding to d_c/R_m for the given ϕ_T from Figure 5.25. Determine the reduction factor C_r corresponding to l/l_c for the given ϕ_T from Figures 5.32, 5.33, 5.34, or 5.35.

$$f_{rt} = K_r C_r f_o$$

For values of l/l_c between those shown in the figures, use straight-line interpolation.

The applied radial tension stress f_{rt} must be no larger than the design value in radial tension F_{rt}'.

Note: If either the deflection or the radial stress limitations cannot be met, redesign by increasing depth d_c. Redesign requires going through steps 5–8 again. Steps 8a and 8b can be skipped in the redesign if the bending stress limitations were satisfied in the earlier trial geometry. If only the bending stress at the tangent point does not meet the limitations, revise the geometry by decreasing the angle ϕ_B (maintain $\phi_B \geq \phi_{B\,min}$). Redesign requires going through steps 5, 6, 8b, and 8c only (because the magnitude of the maximum deflection and the bending stresses at the centerline decrease due to this revision in geometry). If the revised geometry is not adequate to satisfy the limitation, increase the depth (d_c) and redesign.

9. Determine the horizontal movement at the supports, Δ_H, using the formula

$$\Delta_H = \frac{2h\Delta_c}{l} \qquad (5\text{-}72)$$

where h = rise in centroidal axis from end to center (in.) ($h = h_a - d_c/2$),
l = span length (in.), and
Δ_c = centerline deflection as determined in step 7.

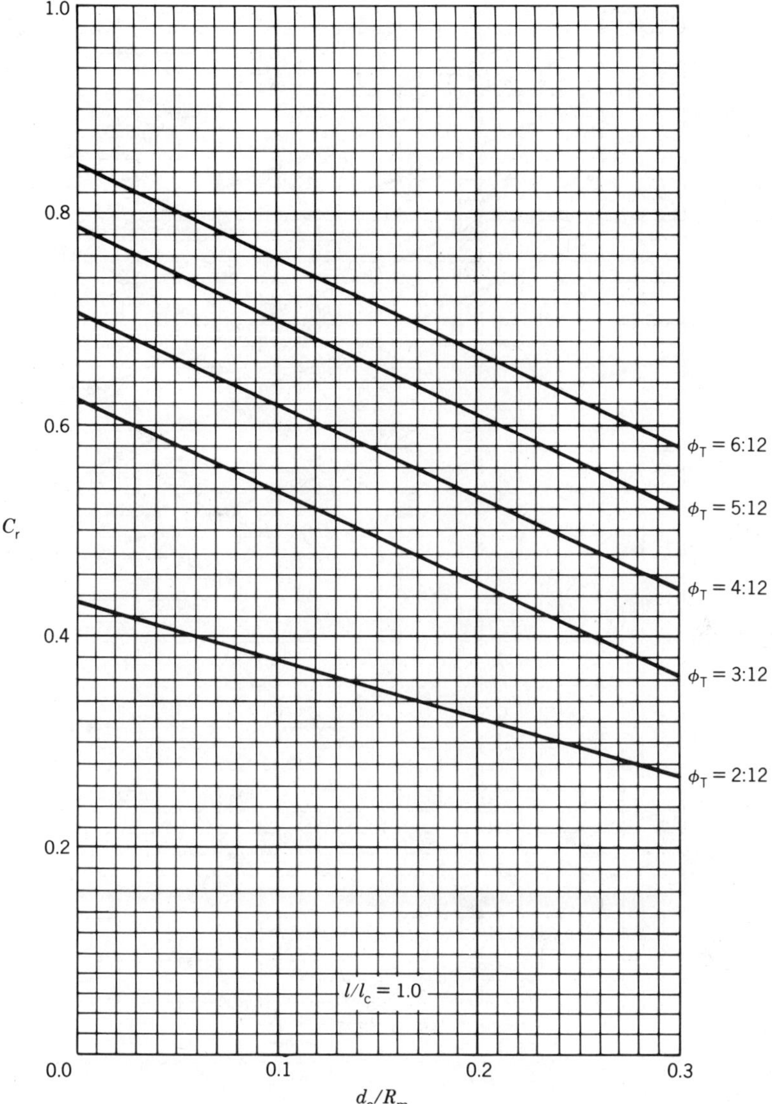

FIGURE 5.32 Reduction factor C_r to obtain radial stress for uniformly distributed loading case.

Note that horizontal movement should be provided for by the use of a slotted connection and an antifriction pad.

10. Design the radial reinforcement in accordance with the recommended procedures given in the next section if required by the analysis of radial stresses in step 8c.

Radial Reinforcement

If the radial tension stress in step 8c exceeds 15 psi adjusted by applicable modifiers for Douglas Fir-Larch, Douglas Fir South, Hem-Fir, or Western Woods, the

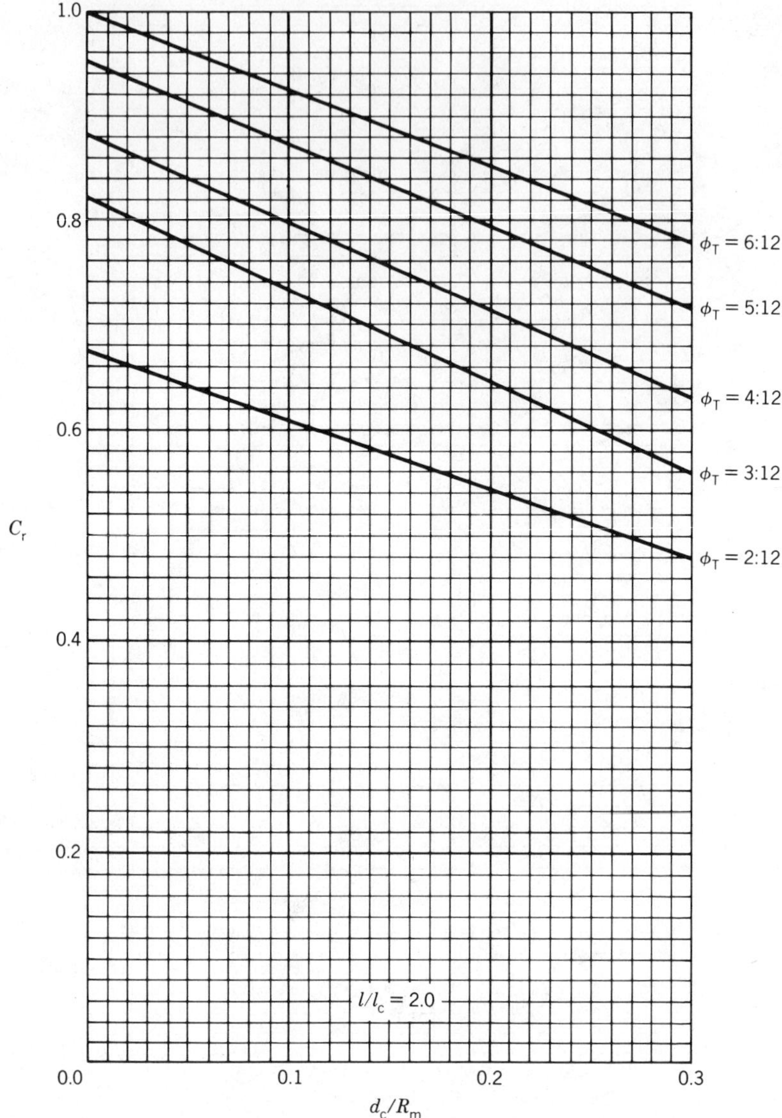

FIGURE 5.33 Reduction factor C_r to obtain radial stress for uniformly distributed loading case.

design procedure is to use radial reinforcement sufficient to resist the full magnitude of radial tension force. This reinforcement should be designed on the basis of sound engineering principles. Any type of reinforcement such as lag bolts mechanically attached to the wood or deformed bars bonded by adhesive that will effectively transfer the radial tension stresses between the wood and the reinforcement throughout the entire depth of embedment may be used. The method of bonding or attaching the reinforcement to the wood should be of a durable quality and capable of developing the required tensile strength of the reinforcement.

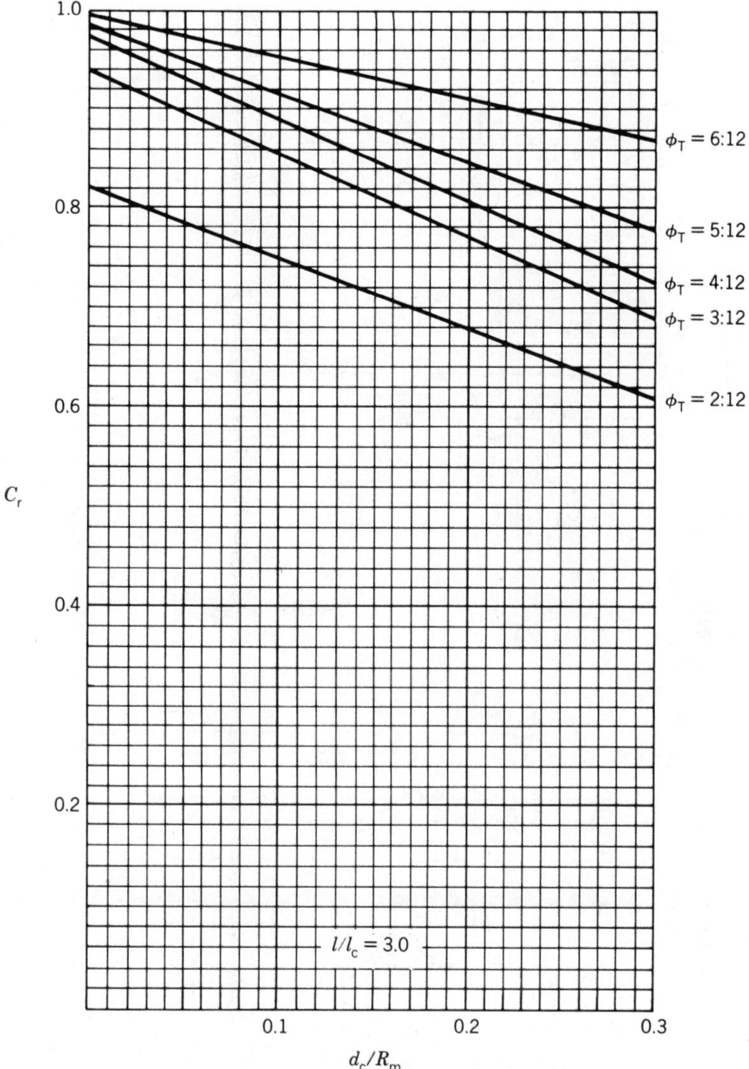

FIGURE 5.34 Reduction factor C_r to obtain radial stress for uniformly distributed loading case.

When radial reinforcement is used, it tends to restrain shrinkage due to moisture loss. Because of this, it is recommended that the moisture content of the laminations prior to gluing not exceed 12% for dry conditions of use.

A typical design is to utilize fully threaded lag bolts as radial tension reinforcement. The following comments provide general installation recommendations and design guidelines for the use of lag bolt reinforcement.

(a) The shop-installed radial tension reinforcing to be used is lag bolts threaded full length. (Although the lag bolts are specified to be threaded full length, manufacturing the lag bolts requires a small length of unthreaded shank.) These lag

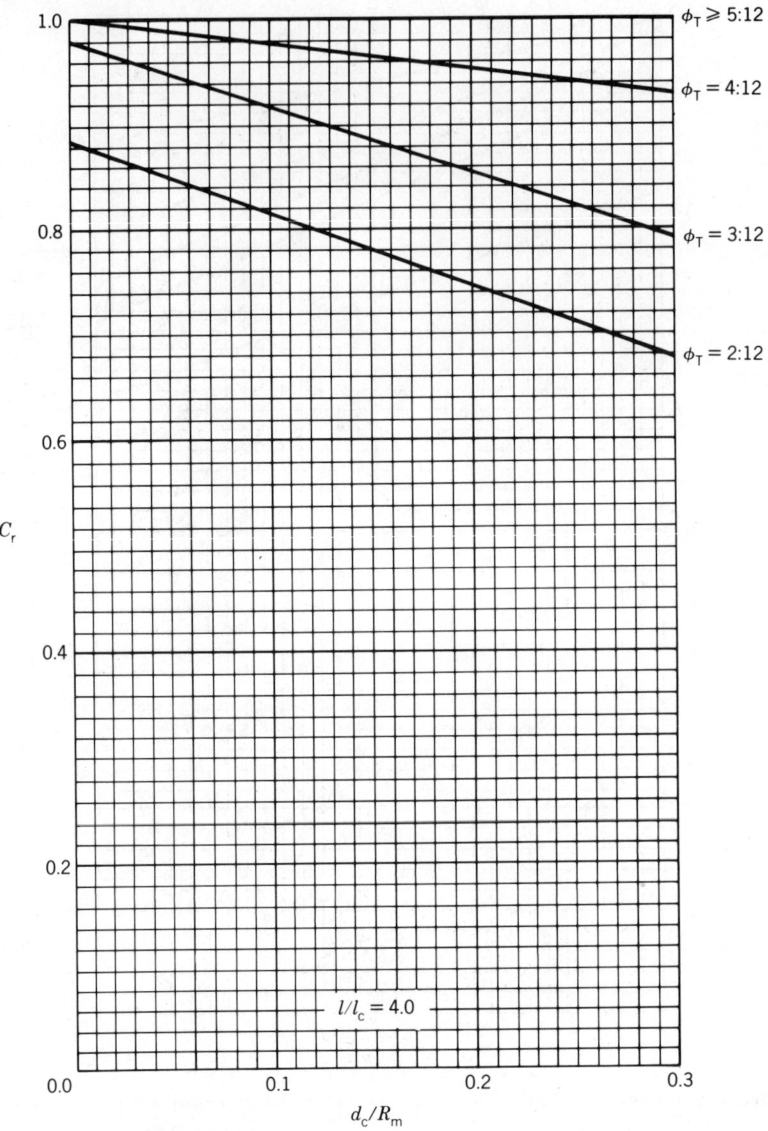

FIGURE 5.35 Reduction factor C_r to obtain radial stress for uniformly distributed loading case.

bolts should be shop installed in a prebored hole from the top of the member on the width centerline of the member. They should be used only in the curved or radially stressed portion of the member. The lag bolts should be installed normal to the axis of the lamination (90° to the direction of the glue line) at the section where the lag bolt is located. No washer should be used under the head of the lag bolt. It is desirable, for structural reasons, not to countersink the head of the lag bolt into the top of the beam. If it is necessary to install the lag bolt in such a way that the top surface is smooth following installation; the head of the lag bolt can be sawn off flush with the member or it can be countersunk, but the countersink should only be large enough to install the lag bolt head flush with the top of the beam.

(b) The prebored lead hole for the threaded portion of the lag bolt should have a diameter not greater than 85% of the shank diameter for Group II species and 80% for Group III species (see Table 6.1 for species groupings) and a depth equal to the length of the threaded portion.

The design values for lag bolts in Table 5.16 for Douglas Fir-Larch are 85% of the design values for lag bolts in Table 6.28 to account for the larger hole, which is necessary for the longer lag bolts. For other species, use 85% of the design value in Table 6.28 for lag bolts used for radial reinforcement.

(c) The length of the full thread of the lag bolt should extend from the top (head end of the lag bolt) of the beam to not less than 2 in. or not more than 3 in. from the soffit of the beam. Lag bolt lengths should be in multiples of 1 in.

(d) In general, lag bolts from $\frac{5}{8}$ to 1 in. in diameter have been found satisfactory to cover most reinforcing requirements. The 1-in.-diameter lag bolts should

TABLE 5.16

Lag Bolt Reinforcement Data for Douglas Fir-Larch

	Steel		Wood		
			Allowable Withdrawal Load (lb/in.) of Threaded Portion[b]		
Lag Bolt Shank Diameter, D (in.)	Net Area at Root of Thread (in.^{2})	Allowable Tension Load (lb) for a Unit Stress of 20,000 psi[a]	Normal 1.00[c]	Snow 1.15[c]	7-Day 1.25[c]
$\frac{1}{4}$	0.0235	470	197	227	246
$\frac{5}{16}$	0.0405	810	233	268	291
$\frac{3}{8}$	0.0552	1,105	267	307	334
$\frac{7}{16}$	0.0845	1,690	300	345	375
$\frac{1}{2}$	0.108	2,160	332	381	414
$\frac{9}{16}$	0.149	2,970	362	416	453
$\frac{5}{8}$	0.174	3,485	392	451	490
$\frac{3}{4}$	0.263	5,265	449	516	561
$\frac{7}{8}$	0.366	7,330	504	580	630
1	0.478	9,555	558	641	697
$1\frac{1}{8}$	0.618	12,360	609	700	761
$1\frac{1}{4}$	0.804	16,090	659	758	823

[a]Rounded to nearest 5 lb.

[b]Based on Douglas Fir–Larch; specific gravity of 0.51 based on weight and volume when oven dry. For other species, see Table 6.28. These values are 85% of the values in Table 6.28 because of the larger lead hole used for long lag bolts used for radial reinforcement.

[c]Duration-of-load factors.

be limited to a maximum length of 60 in. and the $\frac{5}{8}$-in.-diameter lag bolts to a maximum length of 30 in. If greater lengths of reinforcement are required, larger-diameter lag bolts should be specified.

(e) The fully threaded lag bolts should be designed to take the entire radial tension stress developed in the member with no radial tension stress carried by the wood. The lag bolts should be so spaced along the width centerline throughout the curved portion of the member as to carry the entire radial tension force.

(f) The magnitude of the radial tension force to be carried by each lag bolt should not exceed either of the following:

1. Maximum allowable tension in lag bolts on net area at root of thread—The maximum tension load to be carried by each lag bolt is limited to a unit stress of 20,000 psi on the net area at the root of the thread. Table 5.16 lists the net area at the root of threads and the allowable tension at a unit stress of 20,000 psi.
2. Maximum allowable tension in lag bolts on allowable thread holding in wood (allowable withdrawal)—The maximum tension to be carried by each lag bolt should not exceed the allowable withdrawal thread holding of the lag bolt threads in the wood. The effective embedded thread length to be used in this determination is the embedded thread length from the neutral axis of the member to the point end of the lag bolt or the thread length from the neutral axis of the member to the head end of the lag bolt, whichever is the lesser. The allowable unit withdrawal load for the lag bolt threads in the wood should be in accordance with the values given in Table 5.16.

(g) The section modulus is reduced somewhat by the hole bored for the radial reinforcement. In many cases the effect is small but the reduction should be considered in design. The reduced section modulus is determined by subtracting only that portion of the lag bolt hole below the neutral axis, because the lag bolt completely fills the hole on the compression side. Consider the diameter of the hole as being equal to the shank diameter of the lag bolt.

Example: Douglas Fir-Larch Beam. Design a double-tapered pitched and curved glued laminated timber roof beam to meet the following requirements. Assume decking is applied to the top of the beam, providing adequate lateral support.

Species: Douglas Fir-Larch

$L = 60$ ft	$F''_b = (2000)(1.15) = 2300$ psi
Spacing = 16 ft	$F'_v = (165)(1.15) = 190$ psi
Roof slope = 3:12 (14.0°)	$F'_{rt} = (15)(1.15) = 17.2$ psi
Snow load = 30 psf ($C_D = 1.15$)	$E' = 1{,}600{,}000$ psi
Dead load = 15 psf	$\Delta_{TL(max)} = \dfrac{l}{180}$ for total load

1. Determine end depth d_e. Assume $b = 6\frac{3}{4}$ in. Calculate end reaction R_v:

$$R_v = \frac{wL}{2} = \frac{(45)(16)(60)}{2} = 21{,}600 \text{ lb}$$

Calculate trial end depth d:

$$d = \frac{3R_v}{2bF'_v} = \frac{(3)(21{,}600)}{(2)(6.75)(190)} = 25.3 \text{ in.}$$

$$L_e = L - 2d = 60 - 2\left(\frac{25.3}{12}\right) = 55.8 \text{ ft} \quad \text{(use 56 ft)}$$

$$V = \frac{wL_e}{2} = \frac{(45)(16)(56)}{2} = 20{,}200 \text{ lb}$$

$$d_e = \frac{3V}{2bF'_v} = \frac{(3)(20{,}200)}{(2)(6.75)(190)} = 23.6 \text{ in.} \quad \text{(use 24 in.)}$$

2. Determine approximate trial centerline depth d_{cb} from the bending stress limitation:

$$M = \frac{wL^2}{8} = \frac{(45)(16)(60)^2(12)}{8} = 3{,}890{,}000 \text{ in.-lb}$$

Assume $C_F = 0.85$:

$$F'_b = (2000)(1.15)(0.85) = 1955 \text{ psi}$$

From Table 5.15, D for 10° = 1.330 and D for 15° = 1.738.

$$D\,(14.0°) = 1.330 + \frac{(1.738 - 1.330)(4.0)}{5} = 1.66$$

$$d_{cb} = \sqrt{\frac{6MD}{bF'_b}} = \sqrt{\frac{(6)(3{,}890{,}000)(1.66)}{(6.75)(1955)}} = 54.1 \text{ in.}$$

3. Determine trial minimum centerline depth $d_{c\Delta}$ from the deflection limitation:

$$\Delta_{max} = \frac{l}{180} = \frac{(60)(12)}{180} = 4 \text{ in.}$$

$$d_{eff} = \sqrt[3]{\frac{Wl^3}{6.4E'b\Delta_{max}(\cos\phi_T)^3}}$$

$$= \sqrt[3]{\frac{(45)(16)(60)(720)^3}{(6.4)(1{,}600{,}000)(6.75)(4)(0.1970)^3}} = 40.0 \text{ in.}$$

$$d_{c\Delta} = 2d_{eff} - d_e = (2)(40.0) - 24 = 56.0 \text{ in.}$$

4. Determine trial minimum centerline depth d_{crt} from radial stress limitation. Assume $d_c = 56$ in. and $R_m = L = 60$ ft; $d_c/R_m = 56/(60)(12) = 0.08$. From Figure 5.25, $K_r = 0.064$

$$d_{crt} = \sqrt{\frac{6MK_r}{bF'_{rt}}} = \sqrt{\frac{(6)(3{,}890{,}000)(0.064)}{(6.75)(17.2)}} = 113 \text{ in.}$$

This is a very large depth. Therefore, to reduce the depth, the beam will be designed for radial reinforcement, using a maximum radial tension stress for Douglas Fir–Larch of $F_v/3 = 165/3 = 55$ psi; $F'_{rt} = (55)(1.15) = 63.25$ psi.

$$d_{crt} = \sqrt{\frac{(6)(3{,}890{,}000)(0.064)}{(6.75)(63.25)}} = 59.1 \text{ in.} \quad \text{(use 60 in.)}$$

5. Determine height of apex h_a, trial bottom slope ϕ_B, and soffit radius R:

$$h_a = d_e + \left(\frac{l}{2}\right) \tan \phi_T = 24 + \left[\frac{(60)(12)}{2}\right]\left(\frac{3}{12}\right) = 114 \text{ in.}$$

$$d_c = 60 \text{ in.}$$

$$h_s = h_a - d_c = 114 - 60 = 54 \text{ in.}$$

From Figure 5.29, by interpolation,

$$\phi_{B\max} = 17.1° > \phi_T$$

Therefore, set $\phi_{B\max} = 14.0°$. From Figure 5.30, by interpolation, $\phi_{B\min} = 9.2°$

Choose trial bottom slope

$$\phi_B = 0.45(\theta_{B\max} + \phi_{B\min}) = (0.45)(14.0 + 9.2) = 10.4°$$

$$R = \frac{h_s - (l/2) \tan \phi_B}{1 - \cos \phi_B - \sin \phi_B \tan \phi_B} = \frac{54 - (360)(0.184)}{1 - 0.984 - (0.181)(0.184)} = 707 \text{ in.}$$

6. Find the following values based on the values of ϕ_B and R determined in step 5:

(a) $l_t = l/2 - R \sin \phi_B = (30)(12) - 707 \sin 10.4 = 233$ in.

(b) $l_c = l - 2l_t = (60)(12) - (2)(233) = 257$ in.
$l/l_c = 720/257 = 2.82$.

(c) $d_t = d_e + l_t(\tan \phi_T - \tan \phi_B) = 24 + (19.4)(12)(0.250 - 0.184)$
$= 39.3$ in.

(d) $R_m = d_c/2 + R = 60/2 + 707 = 737$ in.
$d_c/R_m = 60/737 = 0.081$.

(e) $f_0 = 6M/bd_c^2 = (6)(3{,}890{,}000)/(6.75)(60)^2 = 960$ psi (reference stress at centerline).

7. Check maximum deflection:

$$d_{eb} = (d_e + d_c)(0.5 + 0.735 \tan \phi_T) - 1.41 d_c \tan \phi_B$$

$$= (24 + 60)[0.5 + (0.735)(0.250)] - (1.41)(60)(0.184) = 41.9 \text{ in.}$$

$$\Delta_c = \frac{5Wl^3}{32E'bd_{eb}^3}$$

$$\frac{(5)(45)(16)(60)(720)^3}{(32)(1,6000,000)(6.75)(41.9)^3} = 3.17 \text{ in.} < 4.0 \text{ in.} \quad \text{O.K.}$$

8. Check stresses:

(a) Bending stress at centerline: From Table 5.15, determine K_θ (for ϕ_T of 10° and d_c/R_m of 0.081) = 1.33 + 0.0(0.081) + (0.927)(0.081) = 1.41 and K_θ (for ϕ_T of 15° and d_c/R_m of 0.081) = 1.738. Then, by straight-line interpolation, K_θ for ϕ_T of 14.0° and d_c/R_m of 0.081:

$$K_\theta = 1.41 + \frac{(1.738 - 1.41)(4.0)}{5} = 1.67$$

$$f_b = K_\theta f_0 = (1.67)(960) = 1600 \text{ psi}$$

$$F'_b = F_b C_D C_F = (2000)(1.15)(0.84) = 1930 \text{ psi} > 1600 \text{ psi} \quad \text{O.K.}$$

(b) Bending stress at tangent point:

$$M_t = \frac{wl(l_t)}{2} - \frac{wl_t^2}{2} = \frac{(45)(16)(60)(19.4)(12)}{2} - \frac{(45)(16)(19.4)^2(12)}{2}$$

$$= 3,403,000 \text{ in.-lb}$$

$$f_{bt} = \frac{6M_t}{bd_t^2} = \frac{(6)(3,403,000)}{(6.75)(39.3)^2} = 1960 \text{ psi}$$

$$C_F = 0.88$$

$$F'_b = (2000)(1.15)(0.88) = 2020 \text{ psi} \quad \text{O.K.}$$

(c) Radial stress at centerline: From Figure 5.25, $K_r = 0.064$. For $l/l_c = 2.82$, from Figures 5.33 and 5.34 by interpolation,

$$C_r = 0.85$$

$$f_{rt} = K_r C_r f_0 = (0.064)(0.85)(960) = 52.2 \text{ psi}$$

$$F'_{rt} = (55)(1.15) = 63.2 \text{ psi} \quad \text{O.K.}$$

9. Determine horizontal movement at supports:

$$h = h_a - \frac{d_c}{2} = 114 - \frac{60}{2} = 84 \text{ in.}$$

$$\Delta_H = \frac{2h\Delta_c}{l} = \frac{(2)(84)(3.17)}{720} = 0.74 \text{ in.}$$

10. Design radial reinforcement:

$$\text{Radial force per inch} = (52.2)(6.75) = 352 \text{ lb/in.}$$

Assume $\frac{3}{4}$-in lag bolts used for radial reinforcement. From Table 5.16,

$$\text{Allowable steel tension load} = 5265 \text{ lb}$$

$$\text{Allowable withdrawal load per in. of thread} = 516 \text{ lb}$$

$$\text{Effective thread penetration} = \frac{d_c}{2} - 2 = \frac{60}{2} - 2 = 28 \text{ in.}$$

$$\begin{aligned}\text{Allowable withdrawal load} &= (516)(28) \\ &= 14{,}450 \text{ lb} > 5256 \text{ lb} \quad (\text{use } 5265 \text{ lb})\end{aligned}$$

$$\text{Maximum spacing (controlled by steel)} = \frac{5625}{352} = 15 \text{ in.}$$

Use $\frac{3}{4}$-in. lag bolts at 15 in. on centers, symmetric about centerline; nine lag bolts on each side, starting $7\frac{1}{2}$ in. on each side of centerline; lag bolts to extend to within 2 in. of the soffit.

Recheck bending stresses at reduced section caused by lag bolt hole at the centerline and tangent points. Locate the neutral axis assuming that the portion of the reinforcement in the tension zone of the beam does not carry stress:

$$c = \frac{bd^2 - b'c^2 + b'a^2}{2(bd - b'c + b'a)}$$

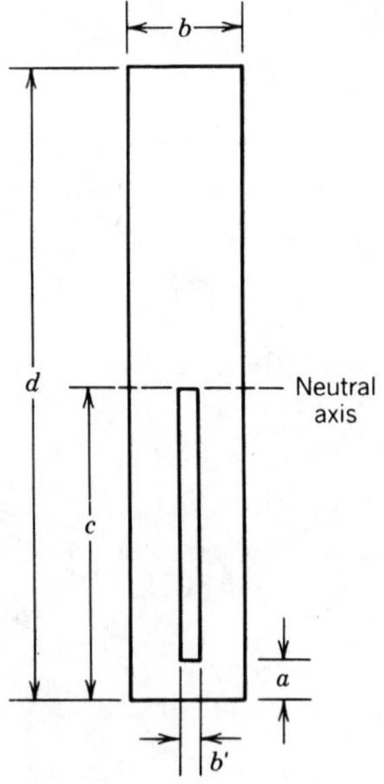

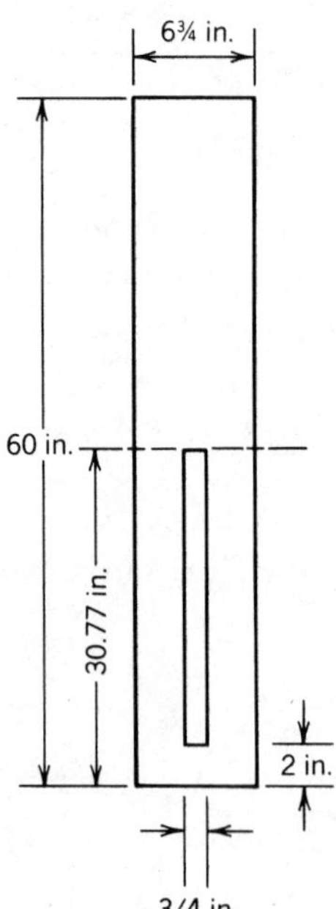

Using the actual dimensions at the centerline and tangent point depths and solving the quadratic equation,

$$c = 30.77 \text{ in.} \quad \text{(centerline)}$$

$$c = 20.11 \text{ in.} \quad \text{(tangent point)}$$

$$I_{CL} = \frac{(6.75)(60)^3}{12} + (6.75)(60)(0.77)^2 - \frac{(0.75)(28.77)^3}{3} = 115{,}787 \text{ in.}^4$$

$$F'_b = (2000)(1.15)(0.84) = 1930 \text{ psi}$$

$$f_0 = \frac{(3{,}890{,}000)(30.77)}{115{,}787} = 1030 \text{ psi} \quad \text{O.K.}$$

$$f_b = 1.67 f_0 = (1.67)(1030) = 1730 \text{ psi} \quad \text{O.K.}$$

$$I_T = \frac{(6.75)(39.3)^3}{12} + (6.75)(39.3)(0.46) - \frac{(0.75)(18.11)^3}{3} = 32{,}714 \text{ in.}^4$$

$$F'_b = (2000)(1.15)(0.88) = 2020 \text{ psi}$$

$$f_b = \frac{Mc}{I} = \frac{(3{,}403{,}000)(20.11)}{32{,}714} = 2090 \text{ psi} \quad \text{consider O.K.}$$

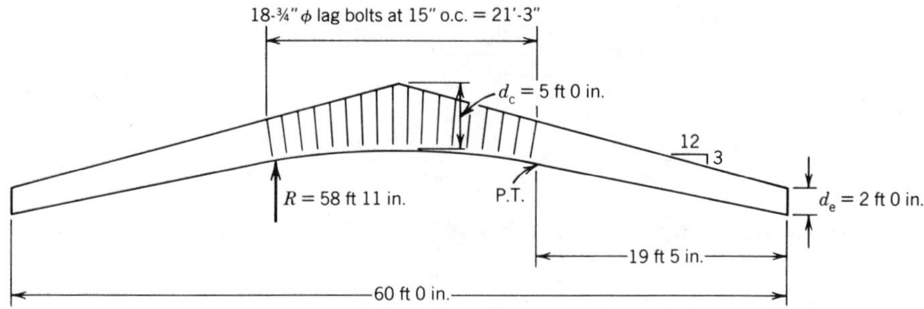

Example: Southern Pine Beam. Design a double-tapered pitched and curved glued laminated timber roof beam to meet the following requirements (assume decking is applied to the top of the beam providing adequate lateral support):

Species: Southern Pine	
$L = 45$ ft	$F''_b = (2400)(1.15) = 2760$ psi
Spacing = 14 ft	$F'_v = (200)(1.15) = 230$ psi
Roof slope = 4:12 (18.4°)	$F'_{rt} = (200/3)(1.15) = 77$ psi
Snow load = 30 psf ($C_D = 1.15$)	$E = 1{,}800{,}000$ psi
Dead load = 15 psf	$\Delta_{max} = l/180$ for total load

1. Determine end depth d_e. Assume $b = 5\frac{1}{8}$ in. Calculate end reaction R_v:

$$R_v = \frac{wL}{2} = \frac{(45)(14)(45)}{2} = 14{,}200 \text{ lb}$$

Calculate trial end depth d:

$$d = \frac{3R_v}{2bF'_v} = \frac{(3)(14{,}200)}{(2)(5.125)(230)} = 18.0 \text{ in.}$$

$$L_e = L - 2d = 45 - \frac{(2)(18)}{12} = 42 \text{ ft}$$

$$V = \frac{wL_e}{2} = \frac{(45)(14)(42)}{2} = 13{,}200 \text{ lb}$$

$$d_e = \frac{3V}{2bF'_v} = \frac{(3)(13{,}200)}{(2)(5.125)(230)} = 16.8 \text{ in.} \quad \text{(use 17 in.)}$$

2. Determine approximate centerline depth d_{cb} from the bending stress limitation:

$$M = \frac{wL^2}{8} = \frac{(45)(14)(45)^2(12)}{8} = 1{,}910{,}000 \text{ in.-lb}$$

Assume $C_F = 0.85$.

$$F'_b = (2400)(1.15)(0.85) = 2350 \text{ psi}$$

From Table 5.15, D for 15° = 1.738 and D for 20° = 1.961.

$$D\,(18.4°) = 1.738 + \frac{(1.961 - 1.738)(3.4)}{5} = 1.89$$

$$d_{cb} = \sqrt{\frac{6MD}{bF'_b}} = \sqrt{\frac{(6)(1{,}910{,}000)(1.89)}{(5.125)(2350)}} = 42.5 \text{ in.}$$

3. Determine minimum centerline depth $d_{c\Delta}$ from the deflection limitation:

$$\Delta_{max} = \frac{l}{180} = \frac{(45)(12)}{180} = 3 \text{ in.}$$

$$d_{eff} = \sqrt[3]{\frac{Wl^3}{6.4E'b\Delta_{max}(\cos\phi_T)^3}}$$

$$= \sqrt[3]{\frac{(45)(14)(45)(540)^3}{(6.4)(1{,}800{,}000)(5.125)(3)(0.949)^3}} = 30.9 \text{ in.}$$

$$d_{c\Delta} = 2d_{eff} - d_e = (2)(30.9) - 17 = 44.8 \text{ in.}$$

4. Determine minimum centerline depth d_{crt} from the radial stress limitation. Assume $d_c = 45$ in. and $R_m = L = 45$ ft; $d_c/R_m = 45/(45)(12) = 0.08$. From Figure 5.25, $K_r = 0.086$.

$$d_{crt} = \sqrt{\frac{6MK_r}{bF'_{rt}}} = \sqrt{\frac{(6)(1{,}910{,}000)(0.086)}{(5.125)(77)}} = 50.0 \text{ in.}$$

5. Determine height of apex h_a, trial bottom slope ϕ_B, and soffit radius R:

$$h_a = d_e + \frac{l}{2} \tan \phi_T = 17 + \frac{(45)(12)(0.333)}{12} = 107 \text{ in.}$$

$$d_c = 50 \text{ in.}$$

$$h_s = h_a - d_c = 107 - 50 = 57 \text{ in.}$$

From Figure 5.29, by interpolation, $\phi_{B\max} = 23.7° > \phi_T$; therefore, set $\phi_{B\max} = \phi_T = 18.4°$. From Figure 5.30, by interpolation, $\phi_{B\min} = 13.9°$. Note that the graphs in Figures 5.30 and 5.31 are based on R_{min} values for Douglas Fir-Larch. For Southern Pine, the R_{min} value for $1\frac{1}{2}$-in. laminations is 18 ft and for $\frac{3}{4}$-in. laminations is 7 ft. These values may be used in the formulas on the graphs to determine values for $\phi_{B\min}$ for Southern Pine. For this problem, the value of $\phi_{B\min}$ is determined as follows:

$$Y = R_{min} - h_s = (18)(12) - 57 = 159 \text{ in.}$$

$$a = \sqrt{Y^2 + (l/2)^2} = \sqrt{(159)^2 + (270)^2} = 313 \text{ in.}$$

$$\phi_{B\min} = \sin^{-1}\left(\frac{R_{min}}{a}\right) - tan^{-1}\left(\frac{Y}{l/2}\right) = \sin^{-1}\left(\frac{216}{313}\right) - \tan^{-1}\left(\frac{159}{270}\right) = 13.1°$$

Choose the trial bottom slope:

$$\phi_B = 0.45(\phi_{B\max} + \phi_{B\min}) = (0.45)(18.4 + 13.1) = 14.2°$$

$$R = \frac{h_s - l/2 \tan \phi_B}{1 - \cos \phi_B - \sin \phi_B \tan \phi_B} = \frac{57 - (270)(0.253)}{1 - 0.969 - (0.245)(0.253)} = 365 \text{ in.}$$

6. Find the following values based on the values of ϕ_B and R determined in step 5:

(a) $l_t = l/2 - R \sin \phi_B = (22.5)(12) - 365 \sin 14.2° = 180$ in.
(b) $l_c = l - 2l_t = (45)(12) - (2)(15)(12) = 225$ in.; $l/l_c = 540/225 = 3.00$.
(c) $d_t = d_e + l_t(\tan \phi_T - \tan \phi_B) = 17 + (15)(12)(0.333 - 0.253) = 31.4$ in.
(d) $R_m = d_c/2 + R = 50/2 + 365 = 390$ in.; $d_c/R_m = 50/390 = 0.13$.
(e) $f_0 = 6M/bd_c^2 = (6)(1{,}910{,}000)/(5.125)(50)^2 = 896$ psi (reference stress at centerline.

7. Check maximum deflection:

$$d_{eb} = (d_e + d_c)(0.5 + 0.735 \tan \phi_T) - 1.41 d_c \tan \phi_B$$

$$= (17 + 50)[0.5 + (0.735)(0.333)] - (1.41)(50)(0.253) = 32.1 \text{ in.}$$

$$\Delta_c = \frac{5Wl^3}{32E'bd_{eb}^3}$$

$$= \frac{(5)(45)(14)(45)(540)^3}{(32)(1{,}800{,}000)(5.125)(32.1)^3} = 2.29 \text{ in.}$$

8. Check stresses:

(a) Bending stress at centerline: Determine K_θ using Eq. (5-69) and Table 5.15.

$$K_\theta \text{ (for } \phi_T \text{ of } 15° \text{ and } \frac{d_c}{R_m} \text{ of } 0.13) = 1.738$$

$$K_\theta \text{ (for } \phi_T \text{ of } 20° \text{ and } \frac{d_c}{R_m} \text{ of } 0.13) = 1.961$$

$$K_\theta \text{ (for } \phi_T \text{ of } 18.4° \text{ and } \frac{d_c}{R_m} \text{ of } 0.13) = 1.738 + \frac{(1.961 - 1.738)(3.4)}{5} = 1.89$$

$$f'_b = K_\theta f_0 = (1.89)(896) = 1960 \text{ psi}$$

$$C_F = 0.85$$

$$F'_b = F_b C_D C_F = (2400)(1.15)(0.85) = 2350 \text{ psi} \quad \text{O.K.}$$

(b) Check bending stress at tangent point:

$$M_t = \frac{wLl_t}{2} - \frac{wl_t^2}{2}$$

$$= \frac{(45)(14)(45)(15)(12)}{2} - \frac{(45)(14)(15)^2(12)^2}{(2)(12)}$$

$$= 1{,}700{,}000 \text{ in.-lb}$$

$$f_{bt} = \frac{6M}{bd_t^2} = \frac{(6)(1{,}700{,}000)}{(5.125)(31.4)^2} = 2020 \text{ psi}$$

$$C_F = 0.90$$

$$F'_b = F_b C_D C_F = (2400)(1.15)(0.90) = 2480 \text{ psi} \quad \text{O.K.}$$

(c) Check radial stress at centerline. From Figure 5.25, $K_r = 0.09$. From Figure 5.34, $C_r = 0.86$.

$$f_{rt} = K_r C_r f_0 = (0.09)(0.86)(896) = 69 \text{ psi} < 77 \text{ psi} \quad \text{O.K.}$$

9. Determine horizontal movement at supports:

$$h = h_a - \frac{d_c}{2} = 107 - \frac{50}{2} = 82 \text{ in.}$$

$$\Delta_H = \frac{2h}{l}\Delta_c = \frac{(2)(82)(2.28)}{540} = 0.69 \text{ in.}$$

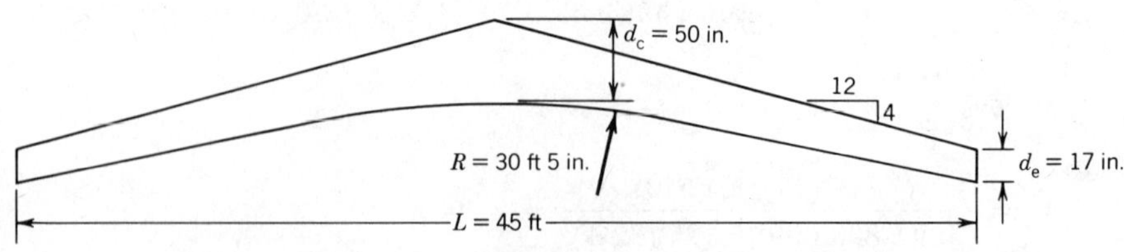

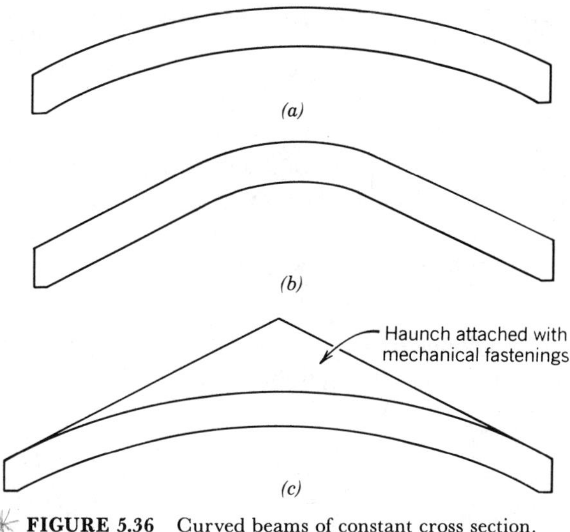

FIGURE 5.36 Curved beams of constant cross section.

Curved Beams of Constant Cross Section

For curved beams of constant cross section, the radial tension is calculated by Eq. (5-53). These beams may be of constant cross section and constant curvature as shown in Figure 5.36*a*, of constant cross section but with tangent ends (Fig. 5.36*b*), or of constant cross section with tangent ends with the haunch attached using mechanical fastenings (Fig. 5.36*c*). The configuration shown in Fig. 5.36*c* is sometimes used to reduce radial tension stresses in Western species beams in order to eliminate the need for mechanical reinforcement.

One of the more common shapes is shown in Figure 5.37. The member is manufactured with a loose haunch that is attached mechanically. The curved portion is of constant cross section and the tangent ends are tapered.

Radial tension in the curved portion can be determined by the following formula (which is different from the formula for a member with an integral haunch):

$$f_{rt} = \frac{3M}{2R_m b d_c}$$

where d_c = depth at centerline and at tangent points (in.)

and other terms are as defined for Eq. (5-53).

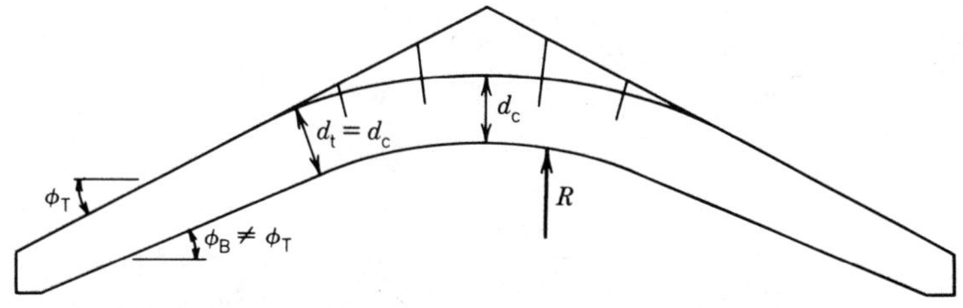

FIGURE 5.37 Double-tapered pitched beam with attached haunch.

The fiber stress in bending in the curved portion is

$$f_b = \frac{M}{S} = \frac{6M}{bd_c^2}$$

The tabular design value in bending, F_b, is adjusted by the curvature factor C_C in addition to other applicable modifiers. The tapered portion should be considered in the same manner as a straight tapered beam where $\theta = \phi_T - \phi_B$. The tabular design value for bending in this portion of the member is further adjusted by the interaction factor C_I in addition to other applicable modifiers.

Example. Design a double-tapered pitched and curved glued laminated timber roof beam to meet the following requirements:

Species: Douglas Fir-Larch	$F_b = 2400$ psi, $F_b'' = (2400)(1.15) = 2760$ psi
$L = 60$ ft	$F_v = 165$, $F_v' = (165)(1.15) = 190$ psi
Spacing $= 16$ ft	$E_x = 1{,}800{,}000$, $E_x' = 1{,}800{,}000$ psi
Roof slope $= 3:12$	$E_y = 1{,}600{,}000$ psi, $E_y' = 1{,}600{,}000$ psi
Camber $= 1\frac{1}{2}\Delta_{DL}$	$G = 0.06E = 108{,}000$ psi, $G' = 108{,}000$ psi
Snow load $= 30$ psf	$\Delta_{TL} = l/180 = (60)(12)/180 = 4$ in.
Dead Load $= 15$ psf	$F_{rt} = 15$ psi, $F_{rt}' = (15)(1.15) = 17.2$ psi

The haunch is to be mechanically attached and no radial reinforcement is to be used.

1. Determine end depth d_e. Assume $b = 6\frac{3}{4}$ in.

Calculate end reaction R_v:

$$R_v = \frac{wL}{2} = \frac{(45)(16)(60)}{2} = 21{,}600 \text{ lb}$$

Calculate trial end depth d:

$$d = \frac{3R_v}{2bF_v'} = \frac{(3)(21{,}600)}{(2)(6.75)(190)} = 25.26 \text{ in.}$$

$$L_e = 60 - 2\left(\frac{25.26}{12}\right) = 55.8 \text{ ft}$$

$$V = \frac{wL_e}{2} = \frac{(45)(16)(55.8)}{2} = 20{,}100 \text{ lb}$$

$$d_e = \frac{3V}{2bF_v'} = \frac{(3)(20{,}100)}{(2)(6.75)(190)} = 23.5 \text{ in.} \quad \text{(use 24 in.)}$$

2. Determine the approximate trial centerline depth d_{cb} from the bending stress limitation:

$$M = \frac{wL^2}{8} = \frac{(45)(16)(60)^2(12)}{8} = 3{,}980{,}000 \text{ in.-lb}$$

$$F'_b = F_b C_D C_F C_C$$

$$C_D = 1.15 \quad \text{(snow load)}$$

$$C_F = \text{Assume } 0.85$$

$$C_C = 1 - 2000\left[\frac{(1.5)}{(60)(12)}\right]^2 = 0.99 \quad (\text{assume radius } R = \text{span} = 60 \text{ ft})$$

$$F'_b = (2400)(1.15)(0.85)(0.99) = 2320 \text{ psi}$$

$$d_{cb} = \sqrt{\frac{6M}{bF'_b}} = \sqrt{\frac{(6)(3{,}890{,}000)}{(6.75)(2320)}} = 38.6 \text{ in.}$$

3. Determine trial minimum centerline depth $d_{c\Delta}$ from deflection limitation. For preliminary design, assume deflection is 10% larger than for a straight beam with the same depth as the constant cross section.

$$\Delta = \frac{(1.1)(5)wl^4}{384E'(bd_{c\Delta}^3/12)}$$

$$\Delta_{max} = 4 \text{ in.}$$

$$d_{c\Delta} = \sqrt[3]{\frac{5.5wl^4}{384\Delta_{max}E'b}} = \sqrt[3]{\frac{(5.5)(45)(16)(720)^4}{(384)(4)(1{,}800{,}000)(6.75)}} = 38.5 \text{ in.}$$

4. Determine the trial minimum centerline depth d_{crt} due to radial tension stress. Assume radius = span = 60 ft.

$$d_{crt} = \frac{3M}{2R_m bF'_{rt}} = \frac{(3)(3{,}890{,}000)}{(2)(60)(12)(6.75)(17.2)} = 69.80 \text{ in.}$$

At this point, a decision needs to be made as to whether the depth is acceptable or whether to increase the radius or the width to reduce the depth. A direct solution to the problem is difficult and a trial-and-error approach that gives due consideration to the geometry involved is suggested. The depth d_{crt} is inversely proportional to both the radius and the width. Usually, changing the radius should be investigated first. Consideration should also be given to the final shape of the member for aesthetic as well as structural considerations. The minimum depth required to satisfy both bending stress and deflection requirements is 38.6 in.

From an appearance standpoint, the tangent point is usually located near the quarter point. For convenience of calculations, the point of tangency is assumed to be 15 ft from the support measured along the bottom face of the member. The

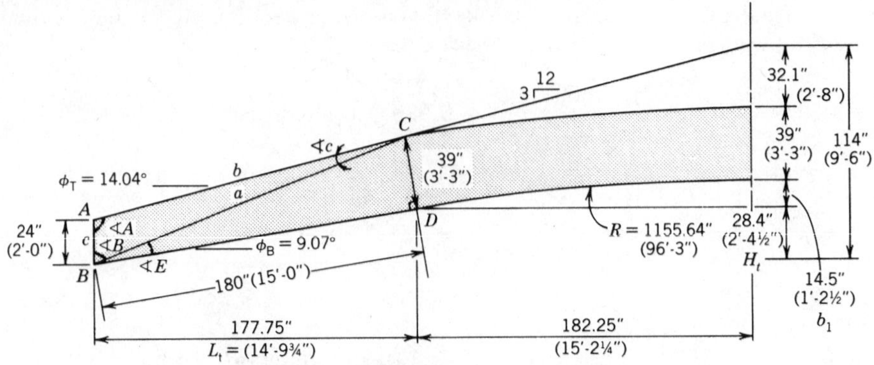

FIGURE 5.38 Geometry for mechanically attached haunch (trial size).

depth of the member at the tangent points and curved portion of the member, d_z, is assumed to be 39 in. From these assumptions, the critical dimensions and radius for this configuration are determined. Figure 5.38 gives critical dimensions calculated from these assumed dimensions.

1. Calculate angle A:

$$A = 90° + \phi_T = 104.04°$$

2. Calculate the length of side a:

$$a = \sqrt{(15)^2 + (3.25)^2} = 15.3 \text{ ft}$$

3. From triangle solution (side, side, angle), calculate angle B for triangle ABC:

$$B = 68.7°$$

4. Calculate angle E:

$$E = \tan^{-1}\left(\frac{3.25}{15}\right) = 12.2°$$

5. Calculate ϕ_B:

$$\phi_B = 90° - B - E = 90° - 68.7° - 12.2° = 9.1°$$

6. Calculate the horizontal distance from the support to the tangent point, l_t:

$$l_t = (15)(12)\cos\phi_B = (15)(12)\cos 9.1° = 178 \text{ in.}$$

7. Calculate the vertical distance from the support to the tangent point, H_t:

$$H_t = (15)(12)\sin\phi_B = (15)(12)\sin 9.1° = 28.4 \text{ in.}$$

8. Calculate the chord distance between tangent points, l_c:

$$l_c = l - 2l_t = (60)(12) - (2)(178) = 364.5 \text{ in.}$$

9. Calculate rise, b_1:

$$b_1 = \frac{l_c}{2} \tan \frac{\phi_B}{2} = \frac{364.5°}{2} \tan \frac{9.1°}{2} = 14.5 \text{ in.}$$

10. Calculate the radius of the soffit, R:

$$R = \frac{4b_1^2 + l_c^2}{8b} = \frac{(4)(14.5)^2 + (364.5)^2}{(8)(14.5)} = 1153 \text{ in.} = 96 \text{ ft}$$

The radius determined for this assumed location of tangent point and depth is 96.0 ft. The required d_{crt} for a 96.0-ft radius is

$$R_m = R + \frac{d_c}{2} = 96.0 + \frac{3.25}{2} = 97.6 \text{ ft}$$

$$d_{crt} = \frac{(3)(3{,}890{,}000)}{(2)(97.6)(12)(6.75)(17.2)} = 42.9 \text{ in.}$$

Changing the depth to 42.9 in. will decrease ϕ_B and change the location of point of tangency. After several trials, a tangent depth of 41 in. and a radius of 103 ft is determined to satisfy the radial stress limit.

5. Check deflection: Assume deflection is 10% higher than that determined for a beam of constant cross section.

$$\Delta_{TL} = \frac{(1.1)(5)wl^4}{384E'bd^3/12} = \frac{(1.1)(5)(45)(16)(720)^4}{(384)(1{,}800{,}000)(6.75)(41)^3} = 3.31 \text{ in} < 4 \text{ in.} \quad \text{O.K.}$$

6. Check stresses:

$$f_b = \frac{6M}{bd^2} = \frac{(6)(3{,}890{,}000)}{(6.75)(41)^2} = 2060 \text{ psi}$$

$$C_C = 1 - 2000\left[\frac{1.15}{(103)(12)}\right]^2 = 0.997$$

$$C_D = 1.15 \quad \text{(snow load)}$$

$$C_F = 0.87 \text{ for 41-in. depth} \quad \text{(see Table 7.2)}$$

$$F_b' = (2400)(1.15)(0.997)(0.87) = 2390 \text{ psi} > 2060 \text{ psi} \quad \text{O.K.}$$

Check lateral support between points of connection of the haunch to the beam. The lateral stability equations used in this example are those intended for use with straight members. However, they will be applied for the lateral stability check of pitched and tapered curved beams with a detached haunch. Assume connection is made 10 ft on either side of the centerline.

$$l_u = 20 \text{ ft} \qquad l_e = 1.92l_u = (1.92)(20) = 38.4 \text{ ft}$$

$$C_s = \sqrt{\frac{l_e d}{b^2}} = \sqrt{\frac{(38.4)(12)(41)}{(6.75)^2}} = 20.4$$

$$F_b'' = (2400)(0.997)(1.15) = 2750 \text{ psi}$$

$$C_{kx} = 0.956\sqrt{\frac{E_y'}{F_{bx}''}} = 0.956\sqrt{\frac{1{,}600{,}000}{2750}} = 23.1$$

Since $C_s < C_k$, use intermediate-beam formula.

$$C_L = 1 - \frac{1}{3}\left(\frac{C_s}{C_k}\right)^4 = 1 - \frac{1}{3}\left(\frac{20.4}{23.1}\right)^4 = 0.80 < 0.87 \quad (C_L \text{ controls})$$

$$F_b' = (2400)(0.997)(1.15)(0.80) = 2310 \text{ psi} > 2060 \text{ psi} \quad \text{O.K.}$$

Check interaction of stresses in the straight tapered portion of the beam. Determine the slope of tapered top surface θ:

$$\theta = \phi_T - \phi_B = 14.0 - 9.1 = 4.9°$$

$$\tan\theta = 0.087$$

From Table 5.13, $C_I = 0.62$. Check stress at tangent point.

$$F_b' = F_b C_D C_I \qquad C_I < C_F$$

$$F_b' = (2400)(1.15)(0.62) = 1710 \text{ psi}$$

$$M = (21{,}600)(14.8)(12) - \frac{(720)(14.8)^2(12)}{2} = 2{,}890{,}000 \text{ in.-lb}$$

$$S = \frac{(6.75)(41)^2}{6} = 1891 \text{ in.}^3$$

$$f_b = \frac{M}{S} = \frac{2{,}890{,}000}{1891} = 1530 < 1710 \text{ psi} \quad \text{O.K.}$$

$$R_m = R + \frac{d_c}{2} = 103 + \frac{41}{(12)(2)} = 104.7 \text{ ft}$$

7. $$f_{rt} = \frac{3M}{2R_m bd} = \frac{(3)(3{,}890{,}000)}{(2)(104.7)(12)(6.75)(41)} = 16.8 < 17.2 \text{ psi} \quad \text{O.K.}$$

8. $$h = 114 - 33 - 41/2 = 60.5 \text{ in.}$$

$$\Delta_H = \frac{2h\Delta_c}{l} = \frac{(2)(60.5)(3.31)}{(60)(12)} = 0.56 \text{ in.} \quad (\text{use } 0.6 \text{ in.})$$

The haunch should be designed to be mechanically attached to the beam as shown in Figure 5.37. This portion carries no bending load and may be of all L3 grade Douglas Fir-Larch (Western Species combination 1), which is sufficient to transfer the load from the roof decking to the beam. The top is attached by $\frac{1}{2}$-in. lag bolts with depth of penetration of lag bolts into the main member of at least 6 in. to develop the full strength of the lag bolt in tension as shown in the following calculations.

Allowable load in tension = 2160 lb (see Table 5.16)
Withdrawal load for a $\frac{1}{2}$-in. lag bolt = 390 lb/in. of penetration of point (see Table 6.28)
Depth of penetration = 2160/390 = 5.5 in.; use 6 in.

Spacing of lag bolts is arbitrary. A spacing of approximately 4 ft should be satisfactory. Heads should be countersunk as they may interfere with the placement of decking, purlins, or roof joists.

Note: This member contains more wood and is deeper at the tangent points and the centerline than the beam designed with radial reinforcement. The tangent depth is 41 in. compared to 39.3 in. and the centerline depth as shown in Figure 5.37 is 74.1 in. compared to 60 in. This results in approximately 4 in. less head room at the centerline. However, in some cases, it may be more economical and more desirable without reinforcement.

ARCHES

Three-Hinged Arches

Three-hinged arches, as the name implies, are hinged at each support and at the crown or peak. They may take the shapes shown in Figure 5.39.

Design Procedure

The following procedures may be used to design simple three-hinged arches under usual loading conditions. These procedures are essentially trial-and-error methods, and revision of original values may be necessary. These procedures are given for a three-hinged tudor arch. The design of other three-hinged arches is similar.

Either a graphical or mathematical procedure can be used in the design process. The graphical solution works very well with a trial-and-error procedure, whereas the mathematical procedure is useful where advantage can be taken of modern calculating methods including computer technology. Many building codes require a more realistic application of wind loads than is shown in the wind load diagram in Figure 5.41. The graphical solution to wind loads applied perpendicular to the roof surface is laborious. One method for a graphical solution is to divide the outside surface of the arch into short segments and to assume that the wind load over each segment is applied as a point load perpendicular to the arch surface. This load is then resolved into its vertical and horizontal components. An equilibrium polygon for the series of vertical loads is constructed and another is constructed for the horizontal loads. The reactions and moments of the vertical and horizontal loads are determined separately and then combined. The solution to wind loads applied perpendicular to the outside surface of an arch usually can be determined best by mathematical rather than graphical methods.

Graphical Solution

1. Determine the dead, live, or snow loads and wind loads and the combinations of loads that should be considered. The roof loads may be assumed to be uniform if roof decking is applied directly to the arch or they may be concentrated

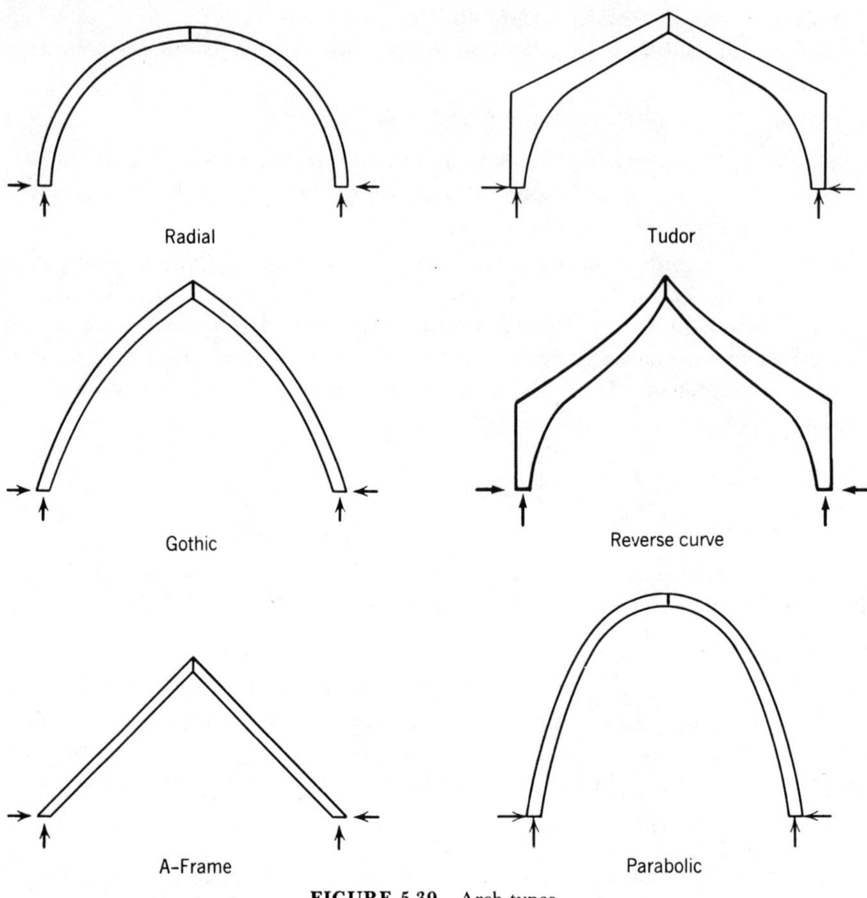

FIGURE 5.39 Arch types.

at purlin points. See Section 4 for dead, live, snow, and wind load recommendations.

2. Lay out to a convenient scale the arch outline indicating external architectural dimensions (see Fig. 5.40*a*). Radius of curvature is also an architectural dimension, and it should not be less than the minimums recommended for various lamination thicknesses on page 5-217. In the case of arches with concentrated loads applied at purlin points, note the locations of the load.

3. Calculate and tabulate the approximate reactions for various combinations of loading conditions. Figure 5.41 indicates some of the more common loading conditions. Other loading conditions to be checked are given in Table 4.1.

4. Determine minimum section size at base, bd_b, by the formula

$$bd_b = \frac{3R_H}{2F_v'} \tag{5-73}$$

where b = arch width (in.),
d_b = arch depth at base (in.),
R_H = horizontal reaction (lb), as determined in step 3, and
F_v' = design value in horizontal shear (psi) adjusted by applicable modifiers.

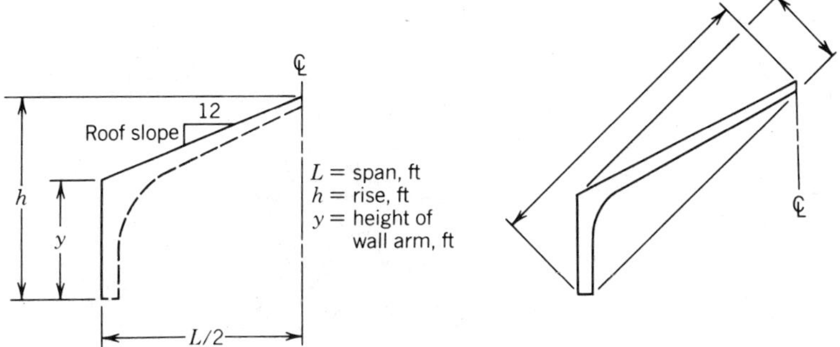

FIGURE 5.40 (*a*) Arch dimensions; (*b*) critical dimensions for shipping.

5. As a first trial in the trial-and-error process, determine approximate depths of arch at crown, d_c, and tangent points, d_t, by the formulas

$$d_c = 1.5b \qquad \text{and} \qquad d_t = 1.5d_b$$

The section at the crown is often proportioned for architectural appearance or to match the depth of purlins that frame into the arch at this point. In no case should the crown depth be less than the arch width. For design simplicity, the upper and lower tangent point depths are often assumed to be equal.

Note the the shipping heights or widths for the tudor arch configuration may be a controlling factor, and the designer should check with the manufacturer to determine the applicable limitations (see Fig. 5.40*b*).

6. If a graphical solution of the arch is used to determine the design moments and section properties along the arch axis, the following procedures are recommended.

Lay out the arch to scale, using the approximate dimensions determined in steps 4 and 5 and the assumed radius of curvature R. Locate the arch axis at the midpoint of the arch depth. Determine the effective span L' and effective rise h' (see Fig. 5.42). Finally, determine the design moments through the use of equilibrium polygons.

Equilibrium polygons are moment diagrams for a specified loading condition drawn to scale and in a position that they pass through the hinges. Equilibrium polygons for each loading condition are illustrated in Figure 5.41. For loading condition (*a*) in Figure 5.41, uniform load on full span, the equilibrium polygon is a parabola with vertex at the crown hinge 0. For condition (*b*), uniform load on right-half span, the equilibrium polygon is a straight line from base hinge A to 0 and a parabola from 0 to base hinge B with its vertex $L/8$ to the right and $h/8$ above

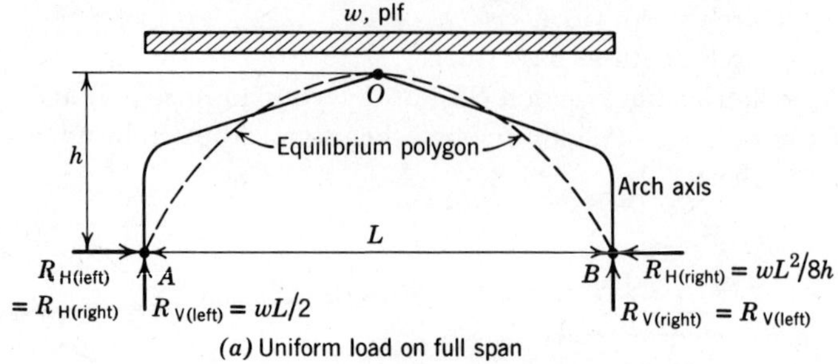

(a) Uniform load on full span

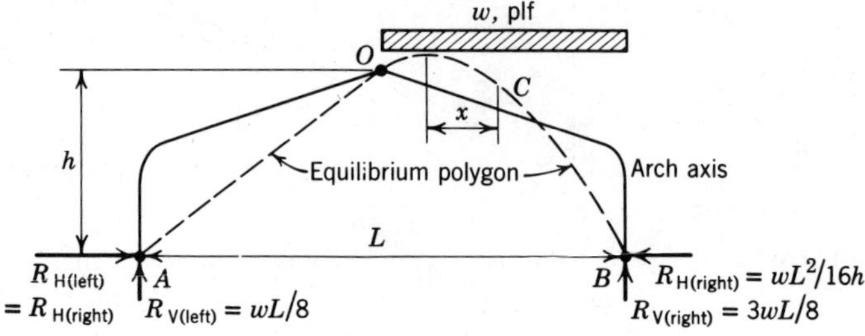

(b) Uniform load on right half span

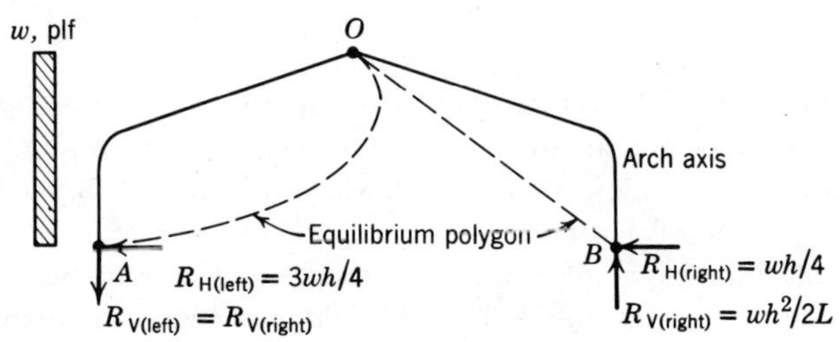

(c) Wind load on left side

FIGURE 5.41 Three-hinged arch reactions and equilibrium polygons.

0. For condition (*c*), uniform wind load on left vertical projection, the equilibrium polygon is a parabola from *A* to 0 with its vertex $L/16$ to the right and $h/4$ below 0 and a straight line from 0 to *B*.

Note: When the governing building code requires that wind loads be applied perpendicular to the roof surface, the graphical solution described on page 5-249 should be used or else a mathematical solution is required.

Parabolas may easily be constructed by applying the fact that the perpendicular offset from a tangent to a parabola varies directly as the square of the tangent distances. For example, the vertical offset y to point *C* from the horizontal tangent through the vertex of the equilibrium polygon for loading conditions (*b*) is equal to $(h/8)[x/(L/8)]^2$.

The advantages of the equilibrium polygons are that they provide a means of determining the moment at any point on the arch axis and a means of determining

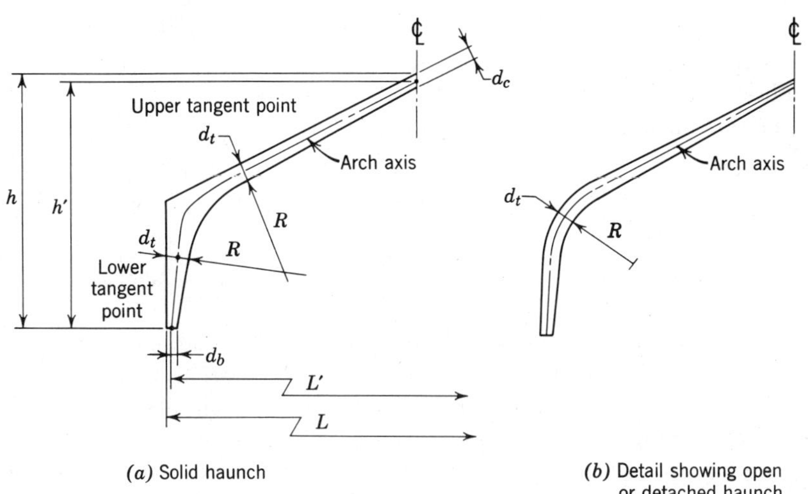

FIGURE 5.42 Arch details.

by inspection the points of maximum moment and the signs of the moments; however, the maximum moment point may not be at the point of maximum stress in a tapered roof arm or vertical leg. When stresses approach the allowable design values, several points along the roof arm or vertical leg must be checked to determine the location of maximum stress.

For vertical loads, loading conditions (*a*) and (*b*) in Figure 5.41, the moment at any point may be determined by multiplying the scaled length of the vertical line between the arch axis and the equilibrium polygon by the horizontal reaction R_H as determined in step 3. Moments are positive if the equilibrium polygon lies above the arch axis and negative if it lies below.

For horizontal loads, loading condition (*c*) in Figure 5.41, the moment at any point is equal to the vertical reaction (R_v from step 3) times the scaled length of the horizontal line between the arch axis and the equilibrium polygon. Moments are positive at a point on the arch axis if the point is between the equilibrium polygon and the load and negative if the equilibrium polygon and the load are to the same side of the point. Thus, for the wind from the left as illustrated in Figure 5.41, loading condition (*c*), moments in the left half of the arch are positive and those in the right half are negative.

7. Recalculate horizontal reactions R_H for vertical loads and vertical reactions R_v for horizontal loads using the effective span and rise from step 6 in the appropriate formula from step 3.

8. Check the arch section at several points along the arch axis for combined bending and axial compression. Points to be checked include the upper and lower tangent points and at least two points on the roof arm; and when the haunch is not glued integrally, additional points between the upper and lower tangent points should be checked. Manufacturing and shipping requirements that would determine whether or not the haunch will be integrally glued should be coordinated

with potential manufacturers. More points on the roof arm should be checked if the two points investigated have stresses approaching the allowable values. One or more points on the vertical leg should also be checked.

The combined stress formula to be applied is

$$\frac{f_c}{F'_c} + \frac{f_{bx}}{F'_{bx}} \leq 1$$

where f_c = actual axial compressive stress at the point under consideration (psi) ($f_c = P'/A$),
A = cross-sectional area at point under consideration (in.^{2}),
P' = axial compression at point under consideration (lb) (determined by laying out loads and reactions to scale, as in Figure 5.43),
F'_c = design value in compression parallel to grain if compression alone existed (psi) (this value is based on the maximum l_e/d ratio of the member),
f_{bx} = actual bending stress (psi), and
F'_{bx} = design value in bending if flexural stress about the x–x axis alone existed (psi) (note that $F'_{bx} - J_x f_c$ is not used for arch design.).

When $f_c > F''_{bx}(1 - C_F)$, the size factor C_F is not applicable in the interaction equation and $F'_{bx} = F''_{bx}$.

When $f_c < F''_b\ (1 - C_F)$, $F'_{bx} = F''_{bx}\ C_F + f_c$.

See page 5-199 for a discussion of members subjected to bending and axial compression.

If the combined stress formula value exceeds 1, it will be necessary to revise

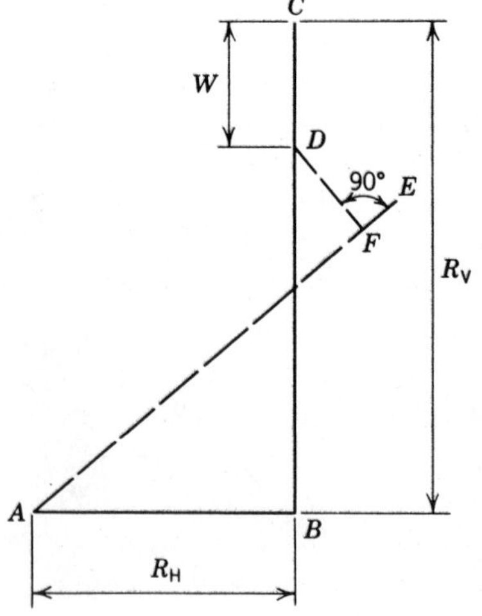

FIGURE 5.43 Force diagram. (1) Lay out lines to scale: AB = horizontal reaction (steps 3 and 7); BC = vertical reaction (steps 3 and 7); CD = sum of vertical loads to left of point under consideration. (2) Draw line AE parallel to the tangent to the arch axis at the point under consideration. (3) Draw line through D perpendicular to line AE = axial compression at point under consideration = P', and DF = shear perpendicular to arch axis at point under consideration.

the original trial sections approximated in step 5 and to repeat the subsequent steps.

9. Lateral stability considerations. When one edge of the arch is braced by decking fastened directly to the arch or braced at frequent intervals, as by girts or roof purlins, the ratio of tangent point depth d_t to breadth b of the arch based on actual dimensions should not exceed 6. When such lateral bracing is lacking, the ratio should not exceed 5, and the arch should be checked for column action in the lateral (width) direction.

The least dimension d in the slenderness ratio l_e/d for lateral column action is the breadth b of the arch. For the roof arm of the arch, the column length l_e in the slenderness ratio is the distance between points of lateral support to the roof. For the wall arm, the unsupported length is either the height of the wall arm (y in Fig. 5.39*a*) or the distance between points of lateral support to the wall. The lateral stability of beams factor, C_L, is not usually applied to arches.

10. Check radial stress. Radial stresses in arches seldom control the design because for most loading conditions the radial stress is compressive. However, for some arch configurations such as tudor arches with steep roof slopes, the wind loads may cause high radial tension stresses. Note that the design value in radial tension for wind loads for all species including Douglas Fir-Larch, Douglas Fir South, Hem-Fir, and Western species is one-third the design value in shear adjusted by the duration-of-load factor of 1.33. When the arch is of constant cross section, the radial stress can be checked by Eq. (5-53):

$$f_r = \frac{3M}{2R_m bd}$$

where f_r = actual radial stress (psi), tension f_{rt}, or compression f_{rc},
M = maximum bending moment (in.-lb),
R_m = radius of curvature at arch axis (in.),
b = arch width (in.), and
d = arch depth at point of maximum moment (in.).

Limitations on allowable radial stress values are given on page 5-218.

When the arch is of variable cross section such as tudor-type arches, the radial stresses can be approximated by modifying the procedure used for double-tapered pitched and curved beams of variable cross section as shown in Eq. (5-54). Assume that the wall arm of the arch represents one-half of a symmetrical beam with a moment at the centerline equal to the moment at the haunch of the arch as shown in the following drawing:

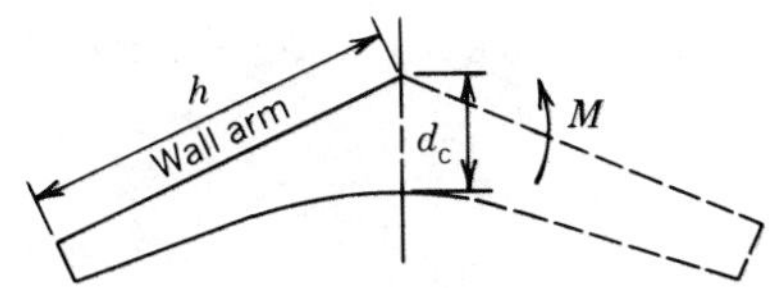

11. Deflection of arches. Specific deflection limitations for arches have not been generally established because this is a function of the intended use. Vertical as well as horizontal deflection of arches must be considered by the designer of the building. This is especially important when adjoining construction will not tolerate the anticipated movement of the arch.

Determine deflection due to bending and change in moisture content. The deflection at any point in an arch due to bending may be determined by the formula

$$\Delta_a = \frac{s}{E} \sum \frac{Mm}{I} \tag{5-74}$$

where Δ_a = actual deflection (in.),
s = length of segments along arch axis (in.) (each arch half should be divided into equal segments of length along the arch axis),
E = modulus of elasticity (psi),
M = moment at midpoint of each segment s (in.-lb) under the loading condition for which the deflection is desired,
m = moment at midpoint of each segment s (in.) under a unit load applied at the point where the deflection is desired and in the direction in which the magnitude of the deflection is desired, and
I = moment of inertia of a section perpendicular to the arch axis at midpoint of each segment s (in.^{4}).

Deflection in curved members caused by changes in moisture content can be significant and should be considered by the designer when it is critical. An explanation of this deflection is contained in *Fabrication and Design of Glued Laminated Wood Structural Members* (8) and AITC Technical Note No. 2 (12). Most arches are manufactured at a slightly higher moisture content than the equilibrium moisture content in service. This change in moisture content, although small, can have a significant effect on deflection. The determination of deflection due to moisture content is complex because the exact moisture content of members at the time of fabrication is difficult to determine. The moisture content at the time of fabrication can vary as much as 5%. The equilibrium moisture content of the member in service will also vary depending on a number of factors. An approximation of the deflection can be obtained by assuming a reasonable decrease in moisture content such as 3 or 4%. The deflection is caused by a change in the thickness in the curved portion of a member. When the moisture content changes, the thickness t changes and the lengths L and l remain relatively unchanged. Therefore the angle α must change to accommodate the change in thickness (Fig. 5.44). The change in angle α can be approximated as $q\alpha$ where q is equal to $-k$ and k is equal to the change in thickness. Slightly greater accuracy can be obtained by using the formula

$$q = \frac{-k\alpha}{l + k} \tag{5-75}$$

Calculation of deflection due to a change in moisture content involves a number

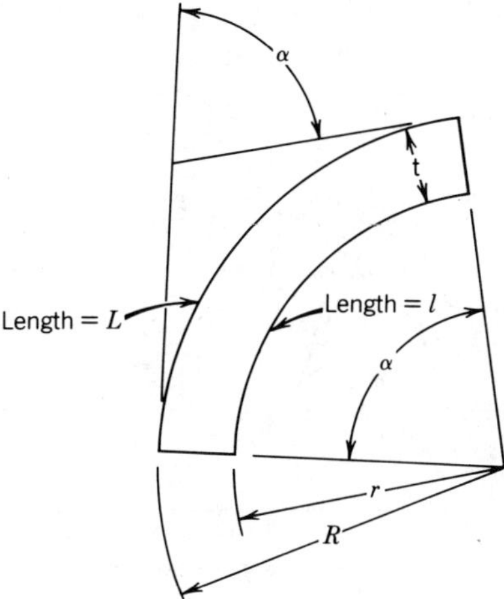

FIGURE 5.44 Notations for deflection.

of approximations, and unless accurate data is available, the value of k for Douglas Fir-Larch and Southern Pine can be taken as 0.2% shrinkage for each 1% change in moisture content.

Example. Determine the change in the angle for a 4% decrease in moisture content when α is 60°.

$$k = (-4)(0.2) = -0.8$$

$$q = -k = -(-0.8) = +0.8\%$$

$$q\alpha = (60)(0.008) = 0.48°$$

When the change in α is known, the deflection at the crown of tudor arches, Δ_m, can be determined by a combination of graphics and calculations or can be approximated very closely by the following formula

$$\Delta_m = \tan(q\alpha)\left[\frac{L'}{2}\right]\left(1 - \frac{h-y}{h}\right) \tag{5-76}$$

where $q\alpha$ = change in interior angle α due to shrinkage (degrees),
L' = span of arch measured from hinge to hinge (ft),
h = height of crown of arch (ft), and
y = wall height (ft).

When the crown of the arch deflects due to a decrease in moisture content, the sides of the arch deflect outward (Fig. 5.45). The deflection at the top of the wall, Δ_h, can be closely approximated by the formula

$$\Delta_h = \tan q\alpha \frac{(h-y)(y)}{h} \tag{5-77}$$

where terms are as defined previously.

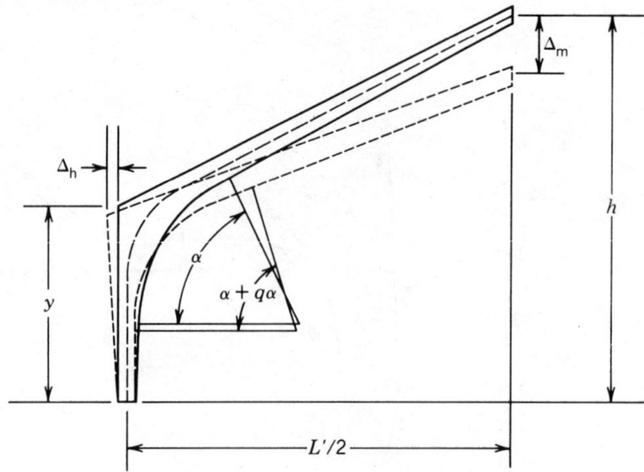

FIGURE 5.45 Deflection due to drying.

Example. Design a three-hinged glued laminated tudor arch to meet the following conditions using a graphical solution.

Spacing = 15 ft
Radius of curvature = R = 10.5 ft
Thickness of lamination = $t = \frac{3}{4}$ in.
Curvature factor = $C_C = 1 - 2000(t/R)^2$

$$C_C = 1 - 2000\left[\frac{0.75}{(10.5)(12)}\right]^2 = 0.929$$

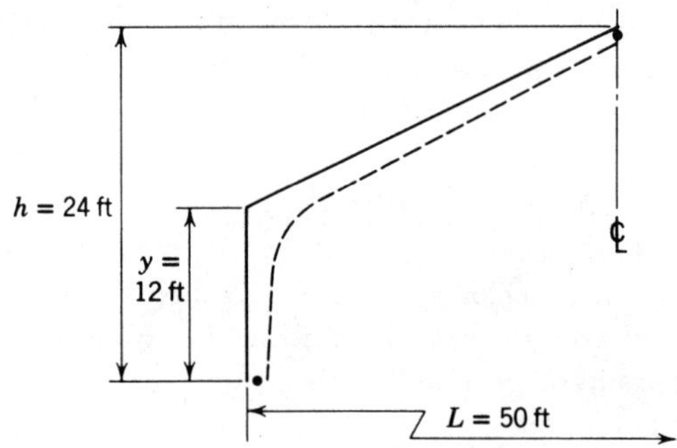

Dry conditions of use, $C_M = 1.0$

F_b = 2400 psi	Wall is laterally supported at midheight and at top.
F_c = 1500 psi	Decking is applied directly to roof arm
$F_{c\perp}$ = 560 psi	DL = 13.5 psf (estimated)
F_v = 200 psi	DL converted to psf of horizontal projection = 15 psf
E_x = 1,800,000 psi	SL = 25 psf (C_D = 1.15)

E_y = 1,600,000 psi   WL = 20 psf of vertical projection. (When governing code requires wind load acting perpendicular to the arch surface, see discussion on page 5-249.)

1. Application of loads:

$$w_{DL} = \text{(spacing)(DL)} = (15)(15) = 225 \text{ plf}$$

$$w_{SL} = \text{(spacing)(SL)} = (15)(25) = \underline{375} \text{ plf}$$

$$\text{Total vertical load} = 600 \text{ plf}$$

2. Arch layout, see sketch.
3. Calculate reactions.

Determine reactions for all loading conditions by use of equations of statics (Table 5.17). The initial calculations for reactions are based on the outside dimensions of the arch and will be recalculated when effective dimensions are determined.

TABLE 5.17

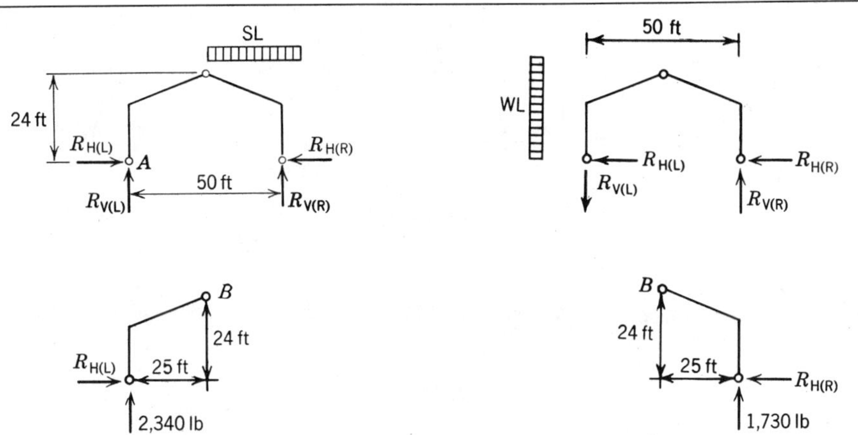

Load	Vertical Reactions (lb) (Upward Reactions Positive)		Horizontal Reactions (lb) (Reactions Acting Right Positive)	
	$R_{V(\text{left})}$	$R_{V(\text{right})}$	$R_{H(\text{left})}$	$R_{H(\text{right})}$
DL	+5,620	+5,620	+2,930	−2,930
SL (full span)	+9,380	+9,380	+4,880	−4,880
DL + SL (full span)	+15,000	+15,000	+7,810	−7,810
SL (right half)	+2,340	+7,030	+2,440	−2,440
DL + SL (right half)	+7,960	+12,650	+5,370	−5,370
WL (from left)	−1,730	+1,730	−5,400	−1,800
DL + WL (from left)	+3,890	+7,350	−2,470	−4,730

4. Determine section at base.

$$\text{Maximum } R_{\text{H}} = 7810 \text{ lb (DL + full SL)}$$

$$bd_b = \frac{3R_H}{2F_v} = \frac{(3)(7810)}{(2)(230)} = 50.9 \text{ in.}^2$$

$$\text{Try } 5\tfrac{1}{8} \times 10\tfrac{1}{2} \text{ section, } A = 53.8 \text{ in.}^2$$

5. Approximate crown and tangent point depths.

$d_c = 1.5b = (1.5)(5.125) = 7.7$ in. (use 7.75 in.)
$d_t = 1.5d_b = (1.5)(10.5) = 15.75$ in.

6. Lay out arch to scale and determine effective dimensions. Construct equilibrium polygons. (See Fig. 5.46.)

$L' = 49.2$ ft
$h' = 23.6$ ft

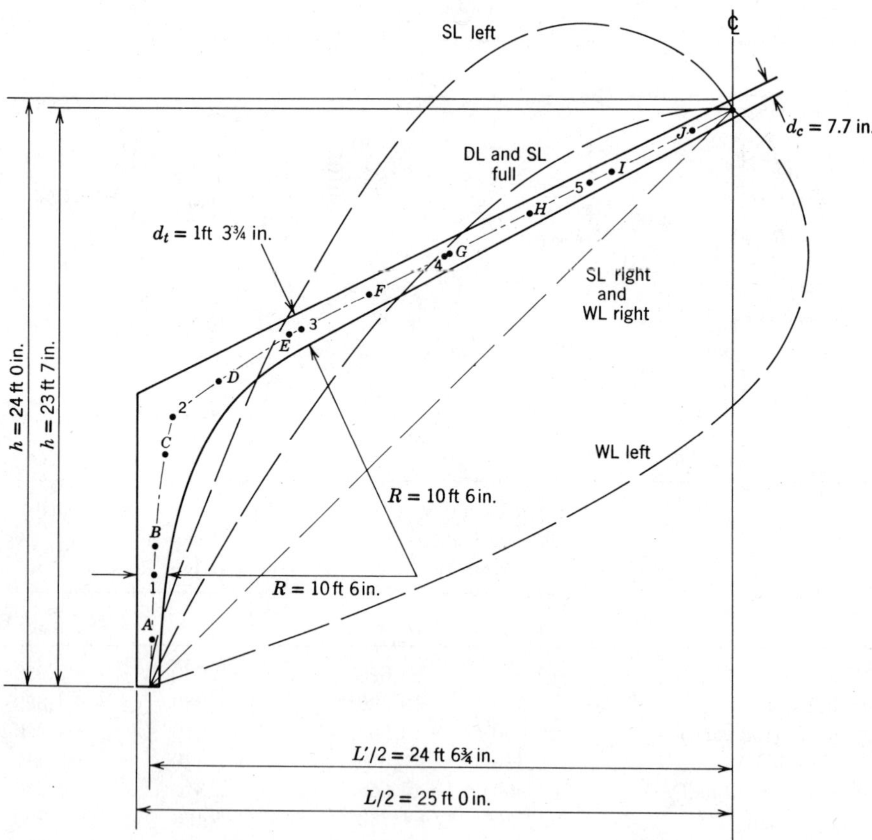

FIGURE 5.46

TABLE 5.18

Load	Vertical Reactions (lb) (Upward Reactions Positive)		Horizontal Reactions (lb) (Reactions Acting Right Positive)	
	$R_{V(left)}$	$R_{V(right)}$	$R_{H(left)}$	$R_{H(right)}$
DL	+5,620	+5,620	+2,880	−2,880
SL (full span)	+9,380	+9,380	+4,800	−4,800
DL + SL (full span)	+15,000	+15,000	+7,680	−7,680
SL (right half)	+2,340	+7,030	+2,520	−2,520
DL + SL (right half)	+7,960	+12,650	+5,400	−5,400
WL (from left)	−1,700	+1,700	−5,400	−1,800
DL + WL (from left)	+3,920	+7,320	−2,520	−4,680

7. Recalculate reactions (Table 5.18).
8. Check arch sections. Using the equilibrium polygons, determine moments at points indicated in Figure 5.46 and also at corresponding points on right half of arch (Table 5.19).

Tabulate maximum moments (Table 5.20) for each point along with the corresponding thrust and shear as determined from Figures 5.47 and 5.48. Properties of sections with depths scaled from Figure 5.46 are given in Table 5.21.

Check each point using the combined stress formula

$$\frac{f_c}{F'_c} + \frac{f_{bx}}{F'_{bx}} \leqslant 1$$

Points 1 and 1′—DL + full SL: Lateral unbraced length = l = 6 ft—0 in. = 72 in.; assume K_e = 1, $l_e = K_e l$ = 72 in.

$$\frac{l_e}{d} = \frac{72}{5.125} = 14.05$$

$$F''_c = F_c C_D = (1500)(1.15) = 1725 \text{ psi}$$

$$E'_y = E_y = 1{,}600{,}000 \text{ psi}$$

$$K_y = 0.671\sqrt{\frac{E'_y}{F''_c}} = 0.671\sqrt{\frac{1{,}600{,}000}{1725}} = 20.44$$

Since 11 < 14.05 < 20.44, use intermediate-column formula.

$$C_P = \left[1 - \frac{1}{3}\left(\frac{l_e/d}{K}\right)^4\right] = 1 - \frac{1}{3}\left(\frac{14.05}{20.44}\right)^4 = 0.93$$

$$F'_c = F''_c C_P = (1725)(0.93) = 1600 \text{ psi}$$

TABLE 5.19

(*a*) Dead Load Only and Dead Load Plus Snow Load on Full Span

Point	Vertical Distance between Arch Axis and Equilibrium Polygon	Horizontal Reaction(lb)		Moment (in.-lb)	
		DL	DL + SL	DL	DL + SL
1 and 1′	4 ft 5 in. = 53 in.	2,880	7,680	−152,600	−407,000
2 and 2′	9 ft 3 in. = 111 in.	2,880	7,680	−319,700	−852,500
3 and 3′	3 ft 8 in. = 44 in.	2,880	7,680	−126,700	−337,900
4 and 4′	3 in. = 3 in.	2,880	7,680	+860	+23,000
5 and 5′	1 ft 7 in. = 19 in.	2,880	7,680	+54,700	+145,900

(*b*) Snow Load on Right-Half Span

Point	Vertical Distance between Arch Axis and Equilibrium Polygon	Horizontal Reaction (lb)	Moment (in.-lb)	
			SL only	DL + SL
Left				
1	4 ft 6 in. = 54 in.	2,520	−136,000	−288,600
2	10 ft 1 in. = 121 in.	2,520	−304,900	−624,600
3	8 ft 5 in. = 101 in.	2,520	−254,500	−381,200
4	5 ft 7 in. = 67 in.	2,520	−168,800	−177,400
5	2 ft 9 in. = 33 in.	2,520	−83,200	−137,900
Right				
5′	6 ft 5 in. = 77 in.	2,520	+194,000	+248,700
4′	6 ft 9 in. = 81 in.	2,520	+204,100	+212,700
3′	1 ft 7 in. = 19 in.	2,520	+47,900	−78,800
2′	7 ft 2 in. = 86 in.	2,520	−216,700	−536,400
1′	3 ft 9 in. = 45 in.	2,520	−113,400	−266,000

(*c*) Wind Load from Left

Point	Horizontal Distance between Arch Axis and Equilibrium Polygon	Vertical Reaction (lb)	Moment (in.-lb)	
			WL only	DL + WL
Left				
1	12 ft 5 in. = 149 in.	1,700	+253,300	+100,700
2	22 ft 10 in. = 274 in.	1,700	+465,800	+146,100
3	20 ft 5 in. = 245 in.	1,700	+416,500	+289,800
4	15 ft 2 in. = 182 in.	1,700	+309,400	+318,000
5	8 ft 4 in. = 100 in.	1,700	+170,800	+224,700
Right				
5′	2 ft 10 in. = 34 in.	1,700	−57,800	−3,100
4′	5 ft 9 in. = 69 in.	1,700	−117,300	−108,700
3′	8 ft 8 in. = 104 in.	1,700	−176,800	−303,500
2′	10 ft 5 in. = 125 in.	1,700	−212,500	−532,200
1′	4 ft 7 in. = 55 in.	1,700	−93,500	−246,100

TABLE 5.20

Point	Loading Condition	Maximum Moment M (in.-lb)	Thrust P' (lb)	Shear Perpendicular to Arch Axis (lb)
Left				
1	DL + full SL	−407,000	15,240	6,480
2	DL + full SL	−852,500	16,000	1,140
3	DL + SL right	−381,200	7,640	3,240
4	DL + WL left	+318,000	2,680	220
5	DL + WL left	+224,700	2,100	1,500
Right				
5′	DL + SL right	+248,700	5,360	1,360
4′	DL + SL right	+212,700	6,960	1,860
3′	DL + full SL	−337,900	11,640	6,160
2′	DL + full SL	−852,500	16,000	1,140
1′	DL + full SL	−407,000	15,240	6,480

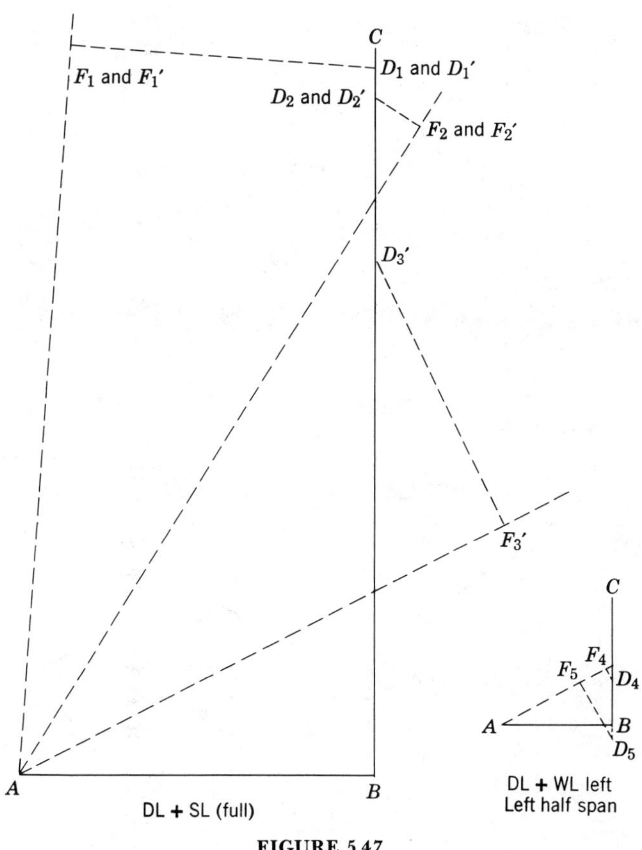

FIGURE 5.47

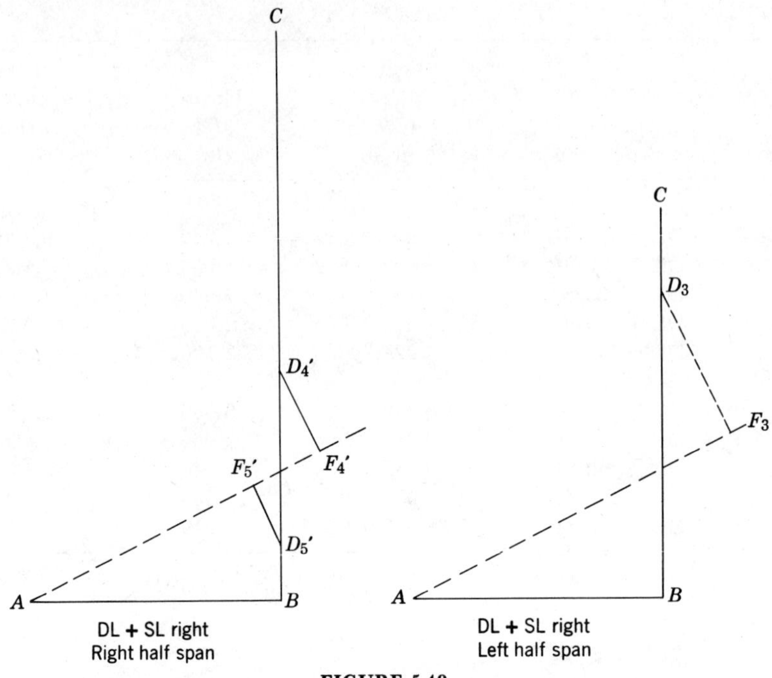

FIGURE 5.48

$$f_c = \frac{P'}{A} = \frac{15{,}240}{80.7} = 189 \text{ psi}$$

$$F_b'' = F_b C_D C_C = (2400)(1.15)(0.93) = 2570 \text{ psi}$$

Check for applicability of size factor:

$$F_b''(1 - C_F) = (2570)(1 - 0.97) = 77 \text{ psi} < f_c \quad (189 \text{ psi})$$

Size factor is not applicable.

TABLE 5.21

PROPERTIES OF SECTIONS

Point	Depth d (in.)[a]	Area A (in.^{2})	Section Modulus S (in.^{3})	Size Factor C_F
1 and 1′	15.75	80.7	212	0.97
2 and 2′	43.0	220.4	1579	0.87
3 and 3′	15.75	80.7	212	0.97
4 and 4′	14.0	71.8	167	0.98
5 and 5′	11.0	56.4	103	1.00

[a]Scaled from Figure 5.46.

$$F'_b = F''_b = 2570 \text{ psi}$$

$$f_b = \frac{M}{S} = \frac{407{,}000}{212} = 1920 \text{ psi}$$

$$\frac{189}{1600} + \frac{1920}{2570} = 0.87 < 1 \quad \text{O.K.}$$

Points 2 and 2′—laterally braced $l_e/d < 11$:

$$F'_c = F_c C_D = (1500)(1.15) = 1725 \text{ psi}$$

$$f_c = \frac{P'}{A} = \frac{16{,}000}{80.7} = 198 \text{ psi}$$

Check for applicability of size factor:

$$F''_b = 2570 \text{ psi} \quad \text{(same as point 1)}$$

$$C_F = 0.87$$

$$F''_{bx}(1 - C_F) = (2570)(1 - 0.87) = 334 \text{ psi} > f_c \quad (198 \text{ psi})$$

$$F'_{bx} = F''_{bx} C_F + f_c = (2570)(0.87) + 198 = 2430 \text{ psi}$$

$$f_{bx} = \frac{M}{S} = \frac{852{,}500}{1579} = 540 \text{ psi}$$

$$\frac{198}{1725} + \frac{540}{2430} = 0.337$$

Point 3—DL + SL right:

$$\frac{l_e}{d} < 11 \qquad F'_c = 1725 \text{ psi}$$

$$f_c = \frac{P'}{A} = \frac{7640}{80.7} = 94.7 \text{ psi}$$

$$F''_b = 2570 \quad \text{(same as point 1)}$$

$$C_F = 0.97 \quad \text{(same as point 1)}$$

Check for applicability of size factor:

$$F''_b(1 - C_F) = (2570)(1 - 0.97) = 77 \text{ psi} < f_c \quad (94.7 \text{ psi})$$

Size factor is not applicable.

$$F'_b = F''_b = 2570 \text{ psi}$$

$$f_b = \frac{M}{S} = \frac{381{,}200}{212} = 1800 \text{ psi}$$

$$\frac{94.7}{1725} + \frac{1800}{2570} = 0.755 < 1 \quad \text{O.K.}$$

Point 3′—DL + full SL:

$$\frac{l_e}{d} < 11 \qquad F'_c = 1725 \text{ psi}$$

$$f_c = \frac{P'}{A} = \frac{11{,}610}{80.7} = 144 \text{ psi}$$

$$F''_b = 2570 \text{ psi} \quad \text{(same as point 1)}$$

$$C_F = 0.97 \quad \text{(same as point 1)}$$

Check for applicability of size factor:

$$F''_b(1 - C_F) = (2570)(1 - 0.97) = 77 \text{ psi} < f_c \quad (144 \text{ psi})$$

Size factor is not applicable.

$$F'_b = 2570 \text{ psi} \quad \text{(same as point 3)}$$

$$f_b = \frac{M}{S} = \frac{337{,}900}{212} = 1590 \text{ psi}$$

$$\frac{144}{1720} + \frac{1590}{2570} = 0.704 < 1 \quad \text{O.K.}$$

Point 4—DL and WL from left:

$$\frac{l_e}{d} < 11 \qquad F'_c = F_c C_D = (1500)(1.33) = 1990 \text{ psi}$$

$$f_c = \frac{P'}{A} = \frac{2680}{71.8} = 37.3 \text{ psi}$$

Check for applicability of size factor:

$$C_F = 0.98$$

$$F''_b = F_b C_D = (2400)(1.33) = 3190 \text{ psi}$$

$$F''_b(1 - C_F) = (3190)(1 - 0.98) = 63.8 \text{ psi} > f_c \quad (37.3 \text{ psi})$$

Size factor is applicable when $f_c < F''_b(1 - C_F)$.

$$F'_b = F''_b C_F + f_c$$

$$F'_b = (3190)(0.98) + 37.3 = 3160 \text{ psi}$$

$$f_b = \frac{M}{S} = \frac{318{,}000}{167} = 1900 \text{ psi}$$

$$\frac{37.3}{1990} + \frac{1900}{3160} = 0.62 < 1 \quad \text{O.K.}$$

Point 4′—DL and SL right:

$$\frac{l_e}{d} < 11 \qquad F'_c = F_c C_D = (1500)(1.15) = 1720 \text{ psi}$$

$$f_c = \frac{P'}{A} = \frac{6960}{71.8} = 97 \text{ psi}$$

Check for applicability of size factor:

$$C_F = 0.98 \quad \text{(same as point 4)}$$

$$F_b'' = F_b C_D = (2400)(1.15) = 2760 \text{ psi}$$

$$F_b''(1 - C_F) = (2760)(1 - 0.98) = 55.2 \text{ psi} < 97 \text{ psi}$$

Size factor is not applicable.

$$F_b' = F_b'' = 2760 \text{ psi}$$

$$f_b = \frac{M}{S} = \frac{212{,}700}{167} = 1270 \text{ psi}$$

$$\frac{97}{1720} + \frac{1270}{2760} = 0.517 < 1 \quad \text{O.K.}$$

Point 5—DL + WL left:

$$\frac{l_e}{d} < 11 \qquad F_c' = 1990 \text{ psi} \quad \text{(same as point 4)}$$

$$f_c = \frac{P'}{A} = \frac{2100}{56.4} = 37.2 \text{ psi}$$

$$d = 11 \text{ in.} \qquad C_F = 1$$

$$F_b'' = 3190 \text{ psi} \quad \text{(same as point 4)}$$

$$f_b = \frac{M}{S} = \frac{224{,}700}{103} = 2180 \text{ psi}$$

$$\frac{37.2}{1990} + \frac{2180}{3190} = 0.702 < 1 \quad \text{O.K.}$$

Point 5′—DL and SL right:

$$\frac{l_e}{d} < 11 \qquad F_c' = F_c C_D = 1720 \quad \text{(same as point 4')}$$

$$f_c = \frac{5360}{56.4} = 95 \text{ psi}$$

$$d = 11 \text{ in.} \qquad C_F = 1$$

$$F_b' = F_b'' = 2760 \text{ psi}$$

$$f_b = \frac{248{,}700}{103} = 2410 \text{ psi}$$

$$\frac{95}{1720} + \frac{2410}{2760} = 0.93 < 1 \quad \text{O.K.}$$

The depths of the arch at the crown and base are initially set based on shear requirements. Shear will seldom govern along the length of the arch. A sample calculation is shown for point 1.

$$F'_v = 230 \text{ psi}$$

$$f_v = \frac{3V}{2A} = \frac{(3)(6480)}{(2)(80.7)} = 120 \text{ psi} < 230 \text{ psi} \quad \text{O.K.}$$

9. Check arch for lateral stability in regard to tangent point depth–width ratio

$$\frac{d_t}{b} = \frac{15.75}{5.125} = 3.07 < 6 \quad \text{O.K.}$$

10. Check radial stress (see page 5-289 for example of calculation of radial stress).

11. Check deflection (see page 5-289 for example of calculation of deflection of three-hinged arch).

Summary of actual section sizes:

Base depth = $10\frac{1}{2}$ in.
Crown depth = $7\frac{3}{4}$ in.
Tangent depths = $15\frac{3}{4}$ in.
Arch width = $5\frac{1}{8}$ in.

Mathematical Solution. The mathematical or algebraic solution of arch design allows advantage to be taken of modern calculating methods including computer technology. It is a very useful procedure when a computer program can be developed for use with a number of designs. However, when only one or two designs are being considered, the graphical method is usually faster. The mathematical solution also can be readily used when wind loads are applied perpendicular to the outer surfaces of the arch.

The following recommended design procedure is based on common industry practices, for example, in regard to location of tangent points. Other similar methods may be used that can produce arch shapes that vary slightly from this method. The first five steps in the design are the same as for the graphical solution except that the drawing in step 2 need not be an exact scale. The suggested design procedure is as follows:

Step 6. Calculate the arch dimensions using geometrical relationships and the forces in the arch from statics.

Steps 7, 8, 9, and 10 are the same as shown for the graphical solution.

The arch shape and design requirements used to illustrate the mathematical solution of a tudor arch in the following example are the same as those for the graphical solution except the wind load is applied perpendicular to the outer surfaces of the arch.

Example. Design a three-hinged glued laminated tudor arch using a mathematical solution to meet the following conditions.

Spacing = 15 ft
Radius of curvature = R = 10.5 ft
Thickness of lamination = t = $\frac{3}{4}$ in.
Curvature factor = $C_C = 1 - 2000\,(t/R)^2$:

$$C_C = 1 - 2000\left[\frac{0.75}{(10.5)(12)}\right]^2 = 0.93$$

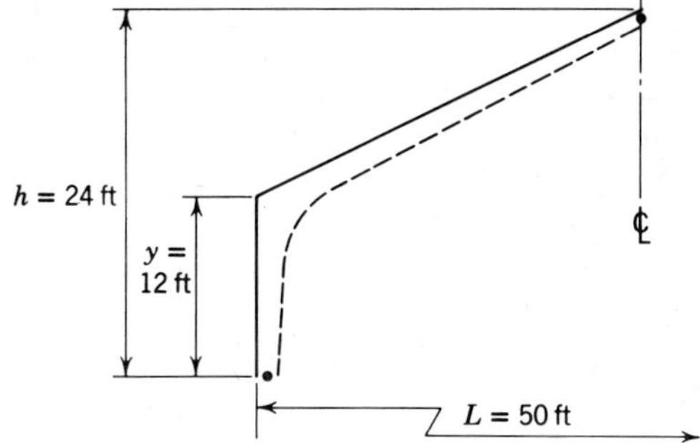

Dry conditions of use, $C_M = 1.0$

$F_b = 2400$ psi	Wall leg is laterally supported at midheight and at top, decking is applied directly to roof arm.
$F_c = 1500$ psi	DL = 13.5 psf (estimated)
$F_{c\perp} = 560$ psi	DL converted to psf of horizontal projection = 15 psf
$F_v = 200$ psi	SL = 25 psf ($C_D = 1.15$)
$E_x = 1{,}800{,}000$ psi	Wind load—the governing code requires that the structure be designed to resist a 90-mph wind for which the basic wind load pressure is 16 psf.
$E_y = 1{,}600{,}000$ psi	

The loads are to be applied as shown in Figure 5.49.

1. Application of loads:

$$w_{DL} = (\text{spacing})(\text{DL}) = (15)(15) = 225 \text{ plf}$$

$$w_{SL} = (\text{spacing})(\text{SL}) = (15)(25) = \underline{375} \text{ plf}$$

$$\text{Total vertical load} = 600 \text{ plf}$$

2. Arch layout, see Figure 5.49

3. Calculate reactions. Determine reactions (see Table 5.22) for all loading conditions by use of equations of statics. The initial calculations for reactions are based on the outside dimensions of the arch and will be recalculated when effective dimensions are determined. See Figure 5.49 for loading conditions.

4. Determine section at base:

$$\text{Maximum } R_H = 7810 \text{ lb (DL + full SL)}$$

$$F'_v = F_v C_D = (200)(1.15) = 230 \text{ psi}$$

$$bd_b = \frac{3R_H}{2F'_v} = \frac{(3)(7810)}{(2)(230)} = 50.9 \text{ in.}^2$$

Try $5\frac{1}{8}$-in.-wide arch:

$$d_b = \frac{50.9}{5.125} = 9.9 \text{ in.} \quad (\text{use } 10\tfrac{1}{2} \text{ in.})$$

5. Approximate crown and tangent point depths:

$$d_c = 1.5b = (1.5)(5.125) = 7.7 \text{ in.} \quad (\text{use } 7\tfrac{3}{4} \text{ in.})$$

$$d_t = 1.5d_b = (1.5)(10.5) = 15.75 \text{ in.}$$

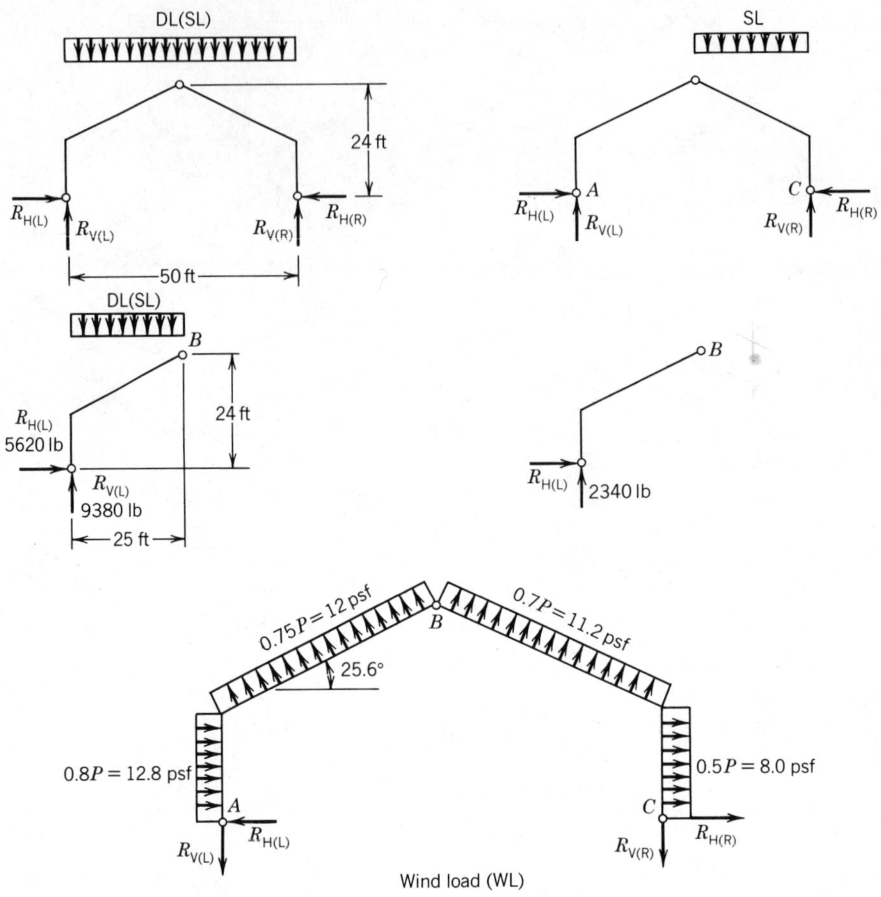

FIGURE 5.49 Arch loading conditions.

TABLE 5.22

Summary of Reactions

	Vertical Reactions (lb) (Upward Reactions +)		Horizontal Reactions (lb) (Reactions Acting Right +)	
Load	$R_{V(L)}$	$R_{V'(R)}$	$R_{H(L)}$	$R_{H(R)}$
DL	+5,620	+5,620	+2,930	−2,930
SL (full span)	+9,380	+9,380	+4,880	−4,880
DL + SL (full span)	+15,000	+15,000	+7,810	−7,810
SL (right half)	+2,340	+7,030	+2,440	−2,440
DL + SL (right half)	+7,960	+12,650	+5,370	−5,370
WL (from left)	−4,820	−3,880	−3,870	+270
DL + WL (from left)	+800	+1,740	−940	−2,660

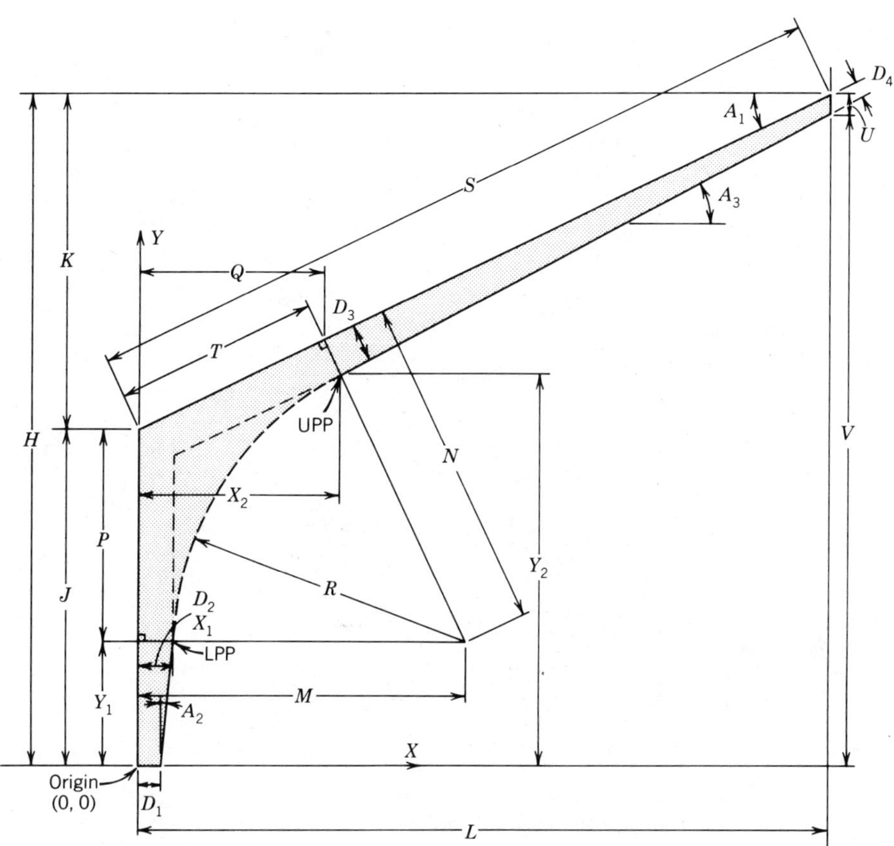

FIGURE 5.50 Location of pivot points.

6. Calculate section properties, moments, shears, and axial forces. Calculate dimensions shown in Figure 5.50 (dimensions H, J, and L are usually set by other than structural considerations).

Using $D_2 = d_t$,

$$M = R + \frac{D_2}{12} = 10.5 + \frac{15.75}{12} = 11.8 \text{ ft}$$

Using $D_3 = d_t$,

$$N = R + \frac{D_3}{12} = 10.5 + \frac{15.75}{12} = 11.8 \text{ ft}$$

$$A_1 = \arctan \frac{K}{L} = \arctan\left(\frac{12}{25}\right) = 25.6°$$

$$Q = M - N \sin A_1 = 11.8 - 11.8 \sin 25.6° = 6.7 \text{ ft}$$

$$P = N \cos A_1 - Q \tan A_1 = 11.8 \cos 25.6° - 6.7 \tan 25.6° = 7.4 \text{ ft.}$$

$$Y_1 = J - P = 12 - 7.4 = 4.6 \text{ ft}$$

$$T = \frac{Q}{\cos A_1} = \frac{6.7}{\cos 25.6°} = 7.4 \text{ ft}$$

$$S = \frac{K}{\sin A_1} = \frac{12}{\sin 25.6°} = 27.7 \text{ ft}$$

Upper and lower pivot points as shown in Figure 5.50 are used as an initial step in determining the arch geometry. Locate the lower pivot point (LPP) from the origin:

$$X_1 = D_2 = \frac{15.75}{12} = 1.3 \text{ ft} \qquad Y_1 = 4.6 \text{ ft}$$

Locate the upper pivot point (UPP) from the origin:

$$X_2 = M - R \sin A_1 = 11.8 - 10.5 \sin 25.6° = 7.3 \text{ ft}$$

$$Y_2 = Y_1 + R \cos A_1 = 4.6 + 10.5 \cos 25.6° = 14.0 \text{ ft.}$$

Determine soffit angles A_2 and A_3

$$A_2 = \arctan\left(\frac{D_2 - D_1}{Y_1}\right) = \arctan\left(\frac{15.75 - 10.5}{(4.6)(12)}\right) = 5.5°$$

$$U = \frac{D_4}{12 \cos A_1} = \frac{7.75}{12 \cos 25.6°} = 0.7 \text{ ft}$$

$$V = H - U = 24 - 0.7 = 23.3 \text{ ft}$$

$$A_3 = \arctan\left(\frac{V - Y_2}{L - X_2}\right) = \arctan\left(\frac{23.3 - 14.0}{25 - 7.3}\right) = 27.6°$$

Locate intersection point of leg soffit and rafter soffit (X_3, Y_3) (see Figure 5.51). The equation for R_2 is derived as follows. From Figure 5.51,

$$R_1 \cos A_2 + R_2 \sin A_3 = V \quad \text{(Eq. A)}$$

$$R_1 \sin A_2 + R_2 \cos A_3 = L - \frac{D_1}{12} \quad \text{(Eq. B)}$$

Solve Eq. A for R_1:

$$R_1 = \frac{V}{\cos A_2} - \frac{R_2 \sin A_3}{\cos A_2}$$

Substitute into Eq. B:

$$\left[\frac{V}{\cos A_2} - \frac{R_2 \sin A_3}{\cos A_2}\right] \sin A_2 + R_2 \cos A_3 = L - \frac{D_1}{12}$$

$$\frac{V \sin A_2}{\cos A_2} - \frac{R_2 \sin A_3 \sin A_2}{\cos A_2} + R_2 \cos A_3 = L - \frac{D_1}{12}$$

$$V \tan A_2 - R_2 \tan A_2 \sin A_3 + R_2 \cos A_3 = L - \frac{D_1}{12}$$

$$R_2 \cos A_3 - R_2 \tan A_2 \sin A_3 = L - \frac{D_1}{12} - V \tan A_2$$

$$R_2 (\cos A_3 - \tan A_2 \sin A_3) = L - \frac{D_1}{12} - V \tan A_2$$

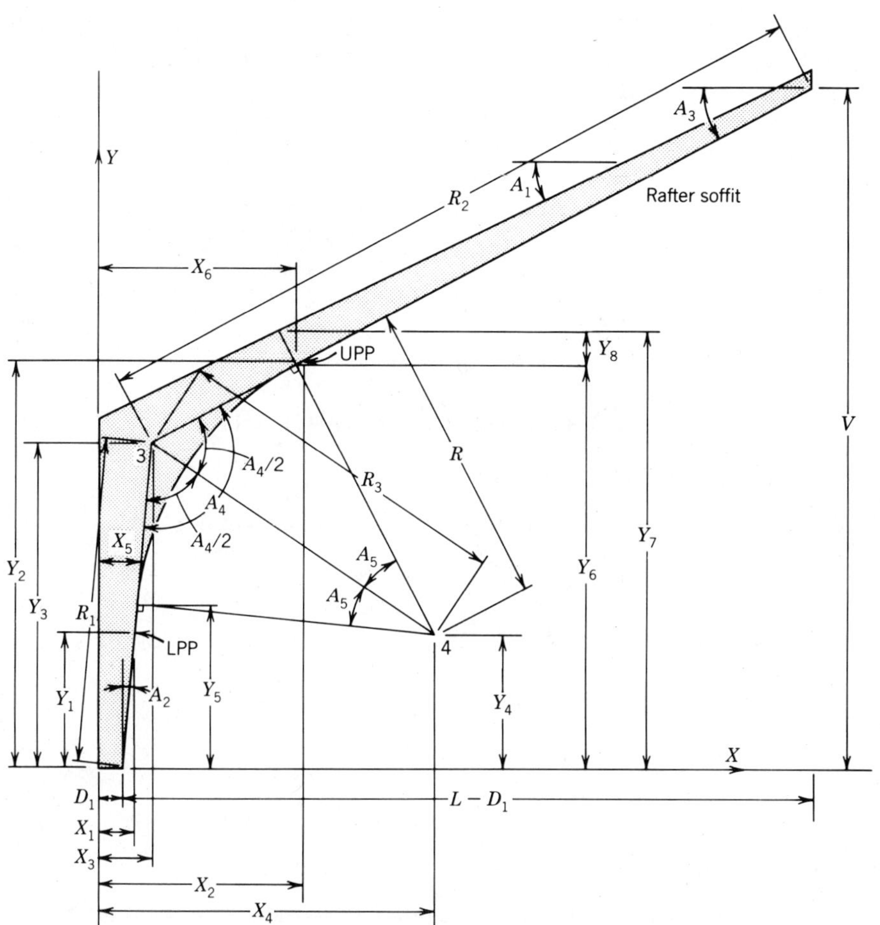

FIGURE 5.51 Location of center of radius and pivot points.

$$R_2 = \frac{L - (D_1/12) - V \tan A_2}{\cos A_3 - \tan A_2 \sin A_3}$$

$$= \frac{25 - (10.5/12) - 23.3 \tan 5.5°}{\cos 27.6° - \tan 5.5 \sin 27.6°} = 26.0 \text{ ft}$$

$$R_1 = \frac{V}{\cos A_2} - \frac{R_2 \sin A_3}{\cos A_2}$$

$$= \frac{23.3}{\cos 5.5°} - \frac{26.0 \sin 27.6°}{\cos 5.5°} = 11.3 \text{ ft}$$

$$X_3 = R_1 \sin A_2 + D_1 = 11.3 \sin 5.5° + \frac{10.5}{12} = 2.0 \text{ ft}$$

$$Y_3 = R_1 \cos A_2 = 11.3 \cos 5.5° = 11.2 \text{ ft}$$

Determine location of center of curvature (X_4, Y_4). From Figure 5.51, angle $A_4 = A_2 + 90° + A_3 = 5.5° + 90° + 27.6° = 123.1°$. The line between points 3 and 4 bisects angle A_4.

$$A_4/2 = 61.5°$$

$$A_5 = 90° - \frac{A_4}{2} = 90° - 61.5° = 28.4°$$

$$R_3 = \frac{R}{\cos A_5} = \frac{10.5}{\cos 28.4°} = 11.9 \text{ ft}$$

$$X_4 = X_3 + R_3 \cos (A_2 + A_5)$$

$$= 2.0 + 11.9 \cos (5.5° + 28.4°) = 11.8 \text{ ft}$$

$$Y_4 = Y_3 - R_3 \sin (A_2 + A_5)$$

$$= 11.2 - 11.9 \sin (5.5° + 28.4°) = 4.6 \text{ ft}$$

Locate lower tangent point (X_5, Y_5):

$$Y_5 = Y_4 + R \sin A_2$$

$$= 4.6 + 10.5 \sin 5.5° = 5.6 \text{ ft}$$

$$X_5 = D_1 + Y_5 \tan A_2$$

$$= \frac{10.5}{12} + 5.6 \tan 5.5° = 1.4 \text{ ft}$$

Lower tangent point depth, $(X_5)(12) = (1.4)(12) = 16.9$ in.
Locate upper tangent point (X_6, Y_6):

$$X_6 = X_4 - R \sin A_3$$

$$= 11.8 - 10.5 \sin 27.6° = 7.0 \text{ ft}$$

$$Y_6 = Y_4 + R \cos A_3$$

$$= 4.6 + 10.5 \cos 27.6° = 13.9 \text{ ft}$$

Determine haunch depth (D_5) (see Figure 5.52):

$$P = J - Y_1 = 12 - 4.6 = 7.4 \text{ ft (see Figure 5.50)}$$

$$A_6 = \arctan \frac{P}{X_4} = \arctan \frac{7.4}{11.8} = 32.1°$$

$$R_4 = \frac{P}{\sin A_6} = \frac{7.4}{\sin 32.1°} = 14.0 \text{ ft}$$

$$D_5 = R_4 - R = 14.0 - 10.5 = 3.5 \text{ ft}$$

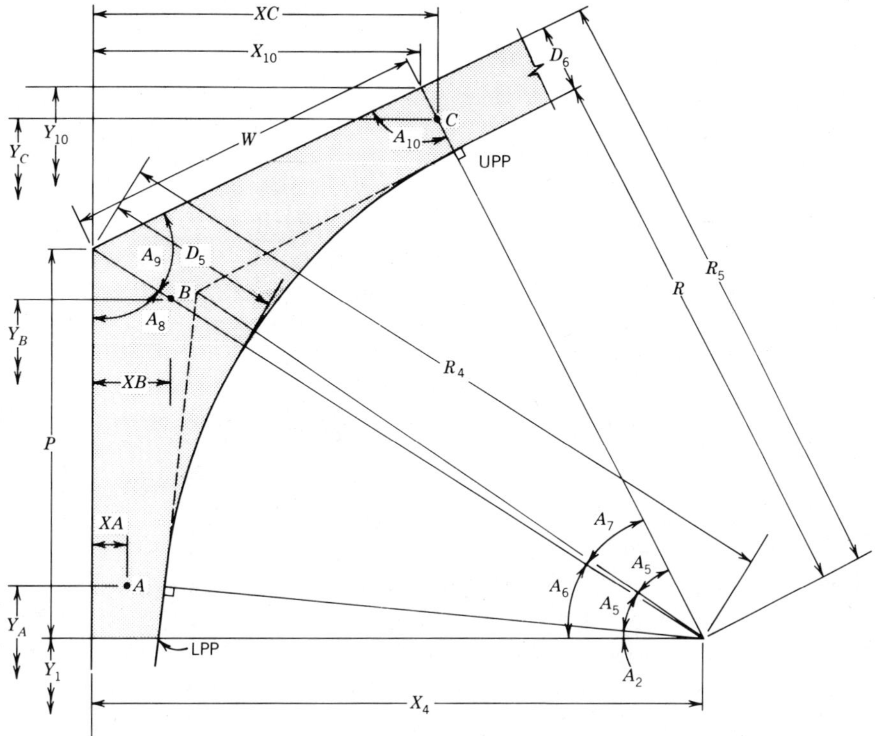

FIGURE 5.52 Haunch detail.

Determine coordinates of points A and B:

$$XA = \frac{X_5}{2} = \frac{1.4}{2} = 0.7 \text{ ft}$$

$$YA = Y_5 = 5.6 \text{ ft}$$

$$XB = \frac{D_5}{2} \cos A_6 = \frac{3.5}{2} \cos 32.1° = 1.5 \text{ ft}$$

$$YB = J - \frac{D_5}{2} \sin A_6 = 12 - \frac{3.5}{2} \sin 32.1° = 11.1 \text{ ft}$$

Determine upper tangent point depth (D_6):

$$A_7 = 2A_5 + A_2 - A_6 = 2(28.4°) + 5.5° - 32.1° = 30.3°$$

$$A_8 = 90° - A_6 = 90° - 32.1° = 57.9°$$

$$A_9 = 90° + A_1 - A_8 = 90° + 25.6° - 57.9° = 57.7°$$

$$A_{10} = 180° - A_7 - A_9 = 92.0°$$

$$R_5 = \left(\frac{\sin A_9}{\sin A_{10}}\right) R_4 = \left(\frac{\sin 57.7°}{\sin 92.0°}\right) 14 = 11.8 \text{ ft}$$

$$D_6 = R_5 - R = 11.8 - 10.5 = 1.3 \text{ ft}$$

Determine coordinates of point C:

$$W = R_4 \left(\frac{\sin A_7}{\sin A_{10}}\right) = 14\left(\frac{\sin 30.3°}{\sin 92.0°}\right) = 7.1 \text{ ft}$$

$$Y_{10} = J + W \sin A_1 = 12 + 7.1 \sin 25.6° = 15.1 \text{ ft}$$

$$X_{10} = W \cos A_1 = 7.1 \cos 25.6° = 6.4 \text{ ft}$$

$$XC = X_{10} + \frac{D_6}{2} \sin A_3 = 6.4 + \frac{1.3}{2} \sin 27.6° = 6.7 \text{ ft}$$

$$YC = Y_{10} - \frac{D_6}{2} \cos A_3 = 15.1 - \frac{1.3}{2} \cos 27.6° = 14.5 \text{ ft}$$

Determine coordinates of points D and E (See Figure 5.53)

$$Z = \frac{L - X_6}{\cos A_3} = \frac{25 - 7}{\cos 27.6°} = 20.3 \text{ ft}$$

$$Z/3 = 6.8 \text{ ft}$$

$$X_{11} = X_6 + \frac{Z}{3} \cos A_3 = 7.0 + 6.8 \cos 27.6° = 13.0 \text{ ft}$$

$$Y_{11} = Y_6 + \frac{Z}{3} \sin A_3 = 13.9 + 6.8 \sin 27.6° = 17.0 \text{ ft}$$

$$D_7 = \frac{D_4}{\cos (A_3 - A_1)} = \frac{7.75}{12} \cos (27.6° - 25.6°) = 0.6 \text{ ft}$$

$$D_8 = D_6 - \frac{D_6 - D_7}{3} = 1.3 - \frac{1.3 - 0.6}{3} = 1.1 \text{ ft}$$

$$XD = X_{11} - \frac{D_8}{2} \sin A_3 = 13.0 - \frac{1.1}{2} \sin 27.6° = 12.7 \text{ ft}$$

$$YD = Y_{11} + \frac{D_8}{2} \cos A_3 = 17.0 + \frac{1.1}{2} \cos 27.6° = 17.5 \text{ ft}$$

$$X_{12} = X_6 + 2\frac{Z}{3} \cos A_3 = 7.0 + (2)(6.8) \cos 27.6° = 19.0 \text{ ft}$$

$$Y_{12} = Y_6 + 2\frac{Z}{3} \sin A_3 = 13.9 + (2)(6.8) \sin 27.6° = 20.1 \text{ ft}$$

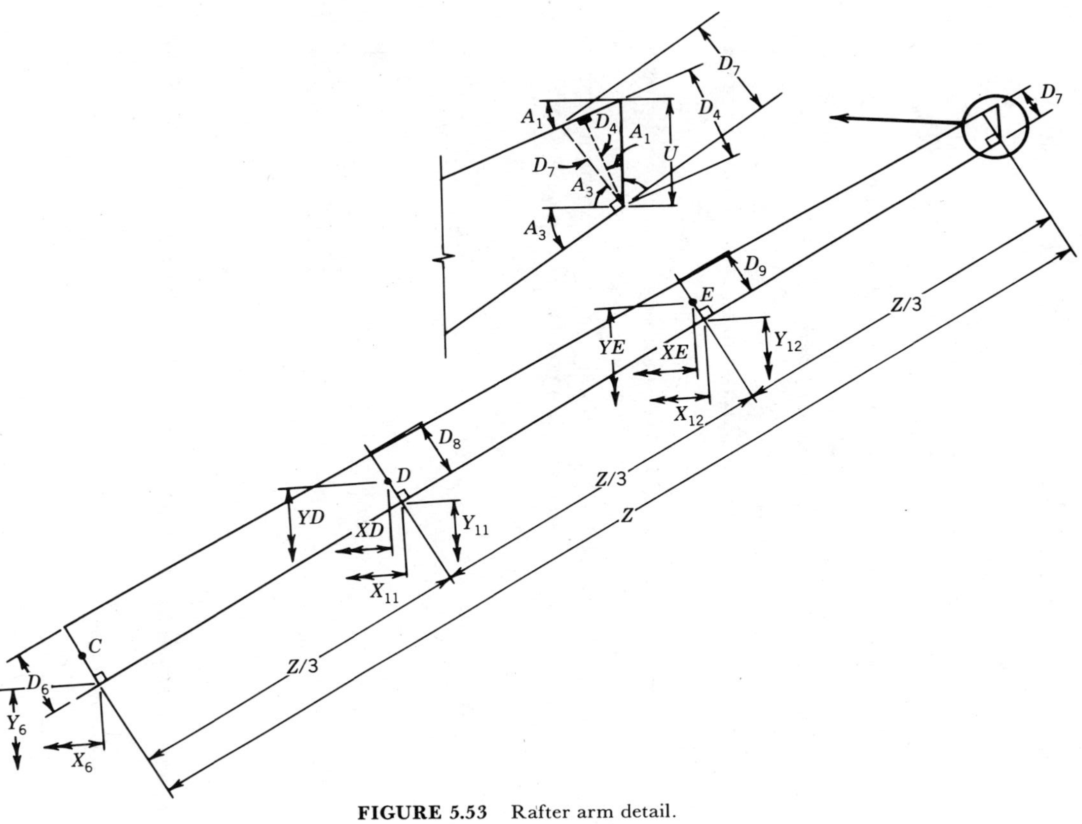

FIGURE 5.53 Rafter arm detail.

$$D_9 = D_6 - \frac{2(D_6 - D_7)}{3} = 1.3 - \frac{2(1.3 - 0.6)}{3} = 0.9 \text{ ft}$$

$$XE = X_{12} - \frac{D_9}{2} \sin A_3 = 19.0 - \frac{0.9}{2} \sin 27.6° = 18.8 \text{ ft}$$

$$YE = Y_{12} + \frac{D_9}{2} \cos A_3 = 20.1 + \frac{0.9}{2} \cos 27.6° = 20.5 \text{ ft}$$

See Figure 5.54 for summary of coordinate locations.

7. Determine effective dimensions using Figure 5.55.

8. Recalculate reactions based on effective dimensions to center of reactions (see Table 5.23).

Calculate thrust, shear, and moment at each section based on the equations of statics shown on the following pages. For convenience of calculation, the cross sections are not perpendicular to the arch axis but are as shown in Figures 5.56 and 5.57, which results in only an insignificant error. The equations for P, V, and

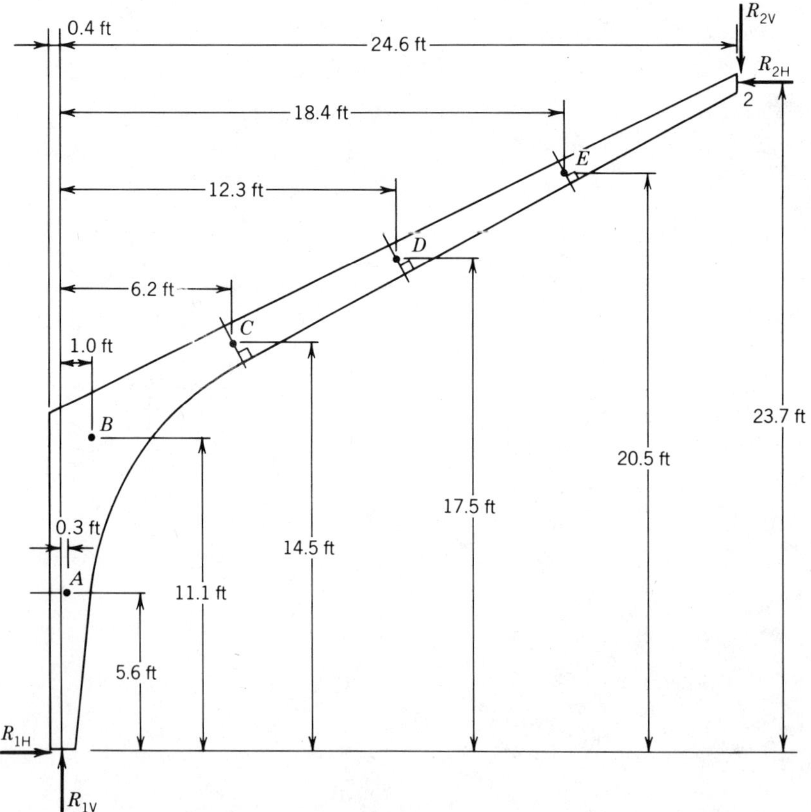

FIGURE 5.54 Summary of geometry for stress calculations.

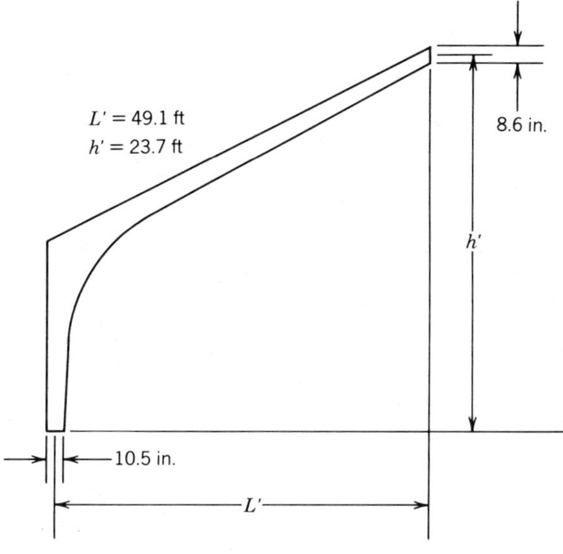

FIGURE 5.55

M include all possible loading conditions. When calculating values for only one loading condition, some terms in the equations drop out.

From Figure 5.56*a*,

$$P_A = R_{V(L)}$$

$$V_A = W_w Y_A + R_{H(L)}$$

$$M_A = + R_{V(L)} X_A - R_{H(L)} Y_A - \frac{W_w Y_A^2}{2}$$

TABLE 5.23

Summary of Reactions

	Vertical Reactions (lb) (Upward Reactions +)		Horizontal Reactions (lb) (Reactions Acting Right +)	
Load	$R_{V(L)}$	$R_{V(R)}$	$R_{H(L)}$	$R_{H(R)}$
DL	+5,620	+5,620	+2,870	−2,870
SL (full span)	+9,380	+9,380	+4,780	−4,780
DL + SL (full span)	+15,000	+15,000	+7,650	−7,650
SL (right half)	+2,300	+7,070	+2,390	−2,390
DL + SL (right half)	+7,920	+12,690	+5,260	−5,260
WL (from left)	−4,810	−3,850	−3,820	+220
DL +WL (from left)	+810	+1,770	−950	−2,650

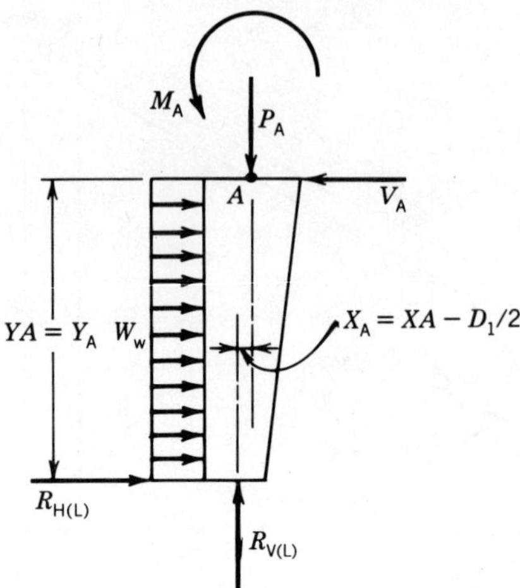

FIGURE 5.56*a*

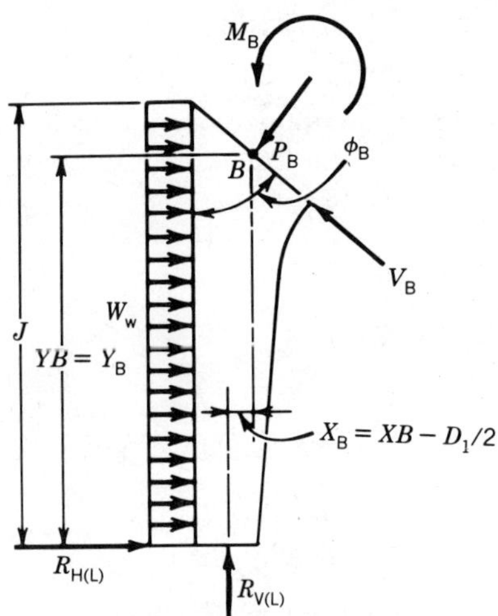

FIGURE 5.56*b*

From Figure 5.56*b*,

$$P_B = (R_{H(L)} + W_w J) \cos \phi_B + R_{V(L)} \sin \phi_B$$

$$V_B = (R_{H(L)} + W_w J) \sin \phi_B - R_{V(L)} \cos \phi_B$$

$$M_B = +R_{V(L)} X_B - R_{H(L)} Y_B - W_w J (Y_B - J/2)$$

Note in Figure 5.56*c* that W_R is applied perpendicular to the roof surface such as for wind and W_H is applied on the horizontal projection such as for snow.

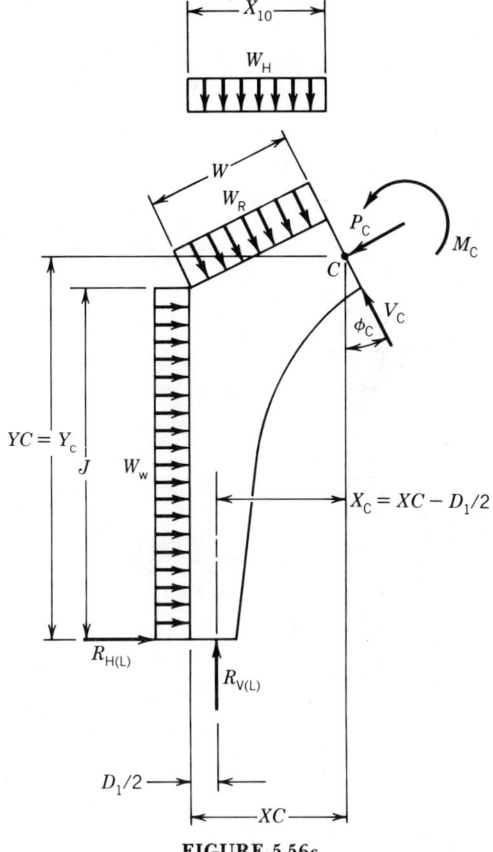

FIGURE 5.56*c*

From Figure 5.56*c*,

$$P_C = (R_{H(L)} + W_w J) \cos \phi_C + (R_{v(L)} - W_H X_{10}) \sin \phi_C$$

$$V_c = (R_{H(L)} + W_w J) \sin \phi_C + (W_H X_{10} - R_{v(L)}) \cos \phi_C + W_R W$$

$$M_C = + R_{v(L)} X_C - R_{H(L)} Y_C - W_w J \, [Y_C - J/2] - \frac{W_R W^2}{2}$$

$$- W_H X_{10} \, [XC - X_{10}/2]$$

Note that for points C, D, and E, it is assumed that the loads acting perpendicular to the roof surface are parallel to the surface of the cross section, which introduces a slight error in the calculation of P, V, and M that is conservative.

Example. Calculate P_c for wind loading ($W_H = 0$).

$$P_C = [-3820 + (12.8)(15)(12)] \cos 27.6° + (-4810) \sin 27.6°$$

$$= -3570 \text{ lb}$$

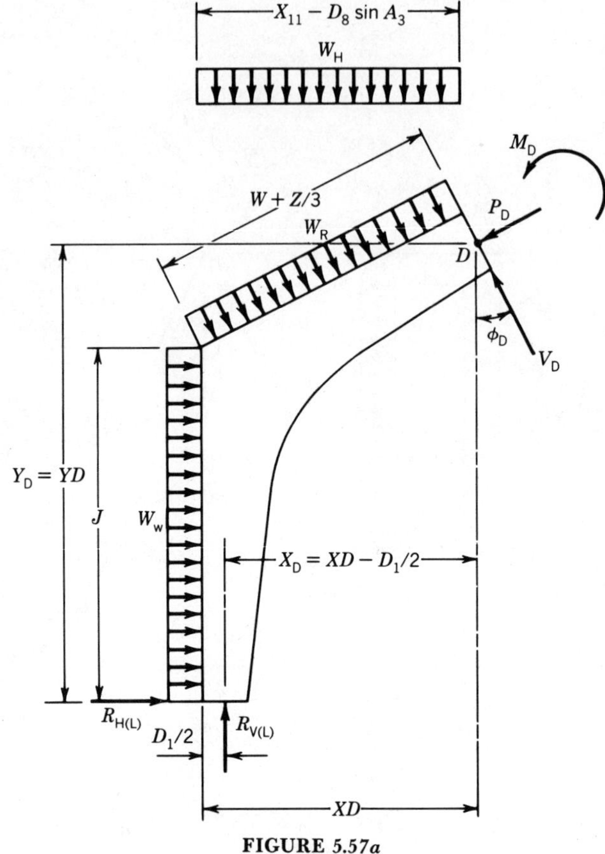

FIGURE 5.57*a*

From Figure 5.57*a*,

$$P_D = (R_{H(L)} + W_w J)\cos\phi_D + [R_{v(L)} - W_H(X_{11} - D_8 \sin A_3)]\sin\phi_D$$

$$V_D = (R_{H(L)} + W_w J)\sin\phi_D$$
$$+ [W_H(X_{11} - D_8\sin A_3) - R_{V(L)}]\cos\phi_D + W_R(W + Z/3)$$

$$M_D = +R_{V(L)}X_D - R_{H(L)}Y_D - W_w J\,[Y_D - J/2] - \frac{W_R}{2}(W + Z/3)^2$$
$$- W_H\,[X_{11} - D_8\sin A_3]\left[XD - \frac{(X_{11} - D_8\sin A_3)}{2}\right]$$

From Figure 5.57*b*,

$$P_E = (R_{H(L)} + W_w J)\cos\phi_E + [R_{V(L)} - W_H(X_{12} - D_9\sin A_3)]\sin\phi_E$$

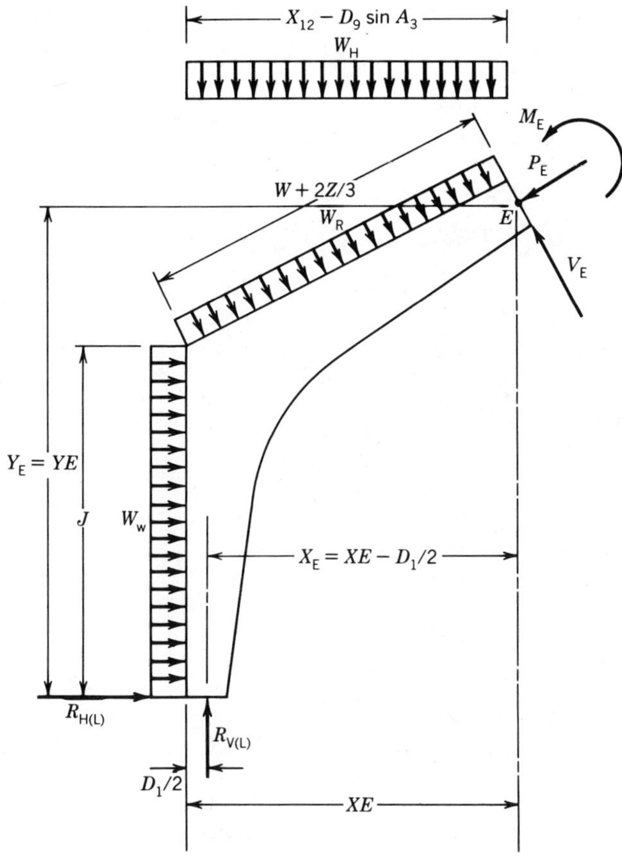

FIGURE 5.57b

$$V_E = (R_{H(L)} + W_w J) \sin \phi_E + [W_H(X_{12} - D_9 \sin A_3)] \cos \phi_E + W_R(W + 2Z/3)$$

$$M_E = +R_{V(L)} X_E - R_{H(L)} Y_E - W_w J \,[Y_E - J/2] - \frac{W_R}{2}(W + 2Z/3)^2 - W_H\,[X_{12} - D_9 \sin A_3]\left[XE - \frac{(X_{12} - D_9 \sin A_3)}{2}\right]$$

The summary in Table 5.26 is based on the assumption that the f_{bx}/F'_{bx} ratios are more significant than the ratios for compression. If compression ratios (f_c/F'_c) are significant in the following interaction checks, a table similar to Table 5.26 could be developed for the load combinations that summarizes the maximum axial thrust and the moment associated with the maximum thrust.

TABLE 5.24

Summary of Moment, Thrust, and Shear

Point	Moment M (compression outside +) (in.-lb)	Thrust Perpendicular to Plane through Point, P (lb) (compression +)	Shear Parallel to Plane through Point, V (lb) (inward +)
DEAD LOAD			
A & *A'*	− 173,200	5,620	−2,870
B & *B'*	−310,200	6,290	+ 560
C & *C'*	−136,900	4,450	+2,330
D & *D'*	+ 9,600	3,820	+1,110
E & *E'*	+ 56,200	3,190	− 90
FULL SNOW LOAD			
A & *A'*	−288,700	9,380	−4,780
B & *B'*	−517,100	10,480	+ 930
C & *C'*	−228,100	7,420	+3,880
D & *D'*	+ 16,100	6,360	+1,850
E & *E'*	+ 93,700	5,310	− 150
SNOW LOAD ON RIGHT SIDE			
A	−152,000	2,300	−2,390
B	−288,400	3,220	− 800
C	−242,200	3,180	+ 930
D	−161,500	3,180	+ 930
E	− 81,100	3,180	+ 930
E'	+175,000	2,130	−1,080
D'	+177,700	3,180	+ 930
C'	+ 14,200	4,240	+2,950
B'	−228,600	7,260	+1,730
A'	−136,700	7,070	−2,390
WIND LOAD FROM LEFT			
A	+203,900	−4,810	+2,750
B	+307,000	−4,810	−1,270
C	+123,500	−3,570	−2,290
D	− 19,100	−3,460	−2,070
E	− 61,500	−3,450	+ 150
E'	−114,700	−3,990	− 820
D'	−130,600	−3,180	− 320
C'	− 53,100	−3,210	−1,460
B'	+ 68,500	−4,140	− 640
A'	+ 24,600	−3,850	+ 890

TABLE 5.25

Properties of Sections at Points Shown in Figure 5.54

Point		Depth d (in.)	Area A (in.^{2})	Section Modulus S (in.^{3})
Base (D_1)		10.5	53.8	—
A & A'	(X_5)	16.9	86.6	244.0
B & B'	(D_5)	42.0	215.2	1,506.8
C & C'	(D_6)	16.1	82.5	221.4
D & D'		13.3	68.2	151.1
E & E'		10.4	53.3	92.4
Crown (D_7)		7.8	44.1	—

9. Check each point (Table 5.29) using the interaction formula

$$f_c/F'_c + f_b/F'_b \leq 1.$$

Points A and A'—lateral unbraced length $= l = 6$ ft $= 72$ in.:

$$\left(\frac{l_e}{d}\right)_y = \frac{72}{5.125} = 14.05$$

$$F''_c = F_c C_D = (1500)(1.15) = 1725 \text{ psi}$$

$$K_y = 0.671\sqrt{E'_y/F''_c} = 0.671\sqrt{\frac{1,600,000}{1725}} = 20.4$$

TABLE 5.26

Summary of Maximum Moments at Points Shown in Figure 5.54

Point	Loading Condition	Maximum Moment M (in.-lb)	Thrust Associated with Maximum Moment, P (lb)
A	DL + full SL	−461,900	15,000
B	DL + full SL	−827,300	16,770
C	DL + SL right	−379,100	7,630
D	DL + SL right	−151,900	7,000
E	DL + full SL	+149,900	8,500
E'	DL + SL right	+231,200	5,320
D'	DL + SL right	+187,300	7,000
C'	DL + full SL	−365,000	11,870
B'	DL + full SL	−827,300	16,770
A'	DL + full SL	−461,900	15,000

Since 11 < 14.05 < 20.4, use intermediate-column formula.

$$C_P = \left[1 - \frac{1}{3}\left(\frac{(l_e/d)_y}{K_y}\right)^4\right] = F_b'' \left[1 - \frac{1}{3}\left(\frac{14.05}{20.4}\right)^4\right] = 0.93$$

$$F_c' = F_c'' C_P = (1725)(0.93) = 1600 \text{ psi}$$

$$F_b' = F_b C_D C_C = (2400)(1.15)(0.93) = 2570 \text{ psi}$$

$$f_c = \frac{15{,}000}{86.6} = 173 \text{ psi}$$

$$\frac{f_c}{F_c'} + \frac{f_b}{F_b'} = \frac{173}{1600} + \frac{1890}{2570} = 0.108 + 0.737 = 0.845$$

Points B and B'—$l_e/d = 0$; $F_c' = F_c C_D = (1500)(1.15) = 1725$:

$$\frac{f_c}{F_c'} + \frac{f_b}{F_b'} = \frac{78}{1725} + \frac{549}{2310} = 0.045 + 0.238 = 0.283$$

Point C:

$$\frac{f_c}{F_c'} + \frac{f_b}{F_b'} = \frac{92}{1725} + \frac{1712}{2570} = 0.054 + 0.668 = 0.722$$

Point D—C_C not applicable in straight portion:

$$\frac{f_c}{F_c'} + \frac{f_b}{F_b'} = \frac{103}{1725} + \frac{1005}{2760} = 0.041 + 0.364 = 0.405$$

Point E:

$$\frac{f_c}{F_c'} + \frac{f_b}{F_b'} = \frac{159}{1725} + \frac{1622}{2760} = 0.092 + 0.588 = 0.680$$

TABLE 5.27

Summary of Bending and Compressive Stresses

Point	Bending Stress f_b (psi)	Compressive Stress f_c (psi)
A	1893	173
B	549	78
C	1712	92
D	1005	103
E	1622	159
E'	2502	100
D'	1240	103
C'	1649	144
B'	549	78
A'	1893	173

TABLE 5.28

Applicability of Size Factor

Point	$F_b''^{a}$	C_F	$F_b''(1 - C_F)$	f_c	C_F Applicable
A	2567	0.963	95	173	No
B	2567	0.870	334	78	Yes[b]
C	2567	0.968	82	92	No
D	2760	0.989	30	103	No
E	2760	1.000	0	159	No
E'	2760	1.000	0	100	No
D'	2760	0.989	30	103	No
C'	2567	0.968	82	144	No
B'	2567	0.870	334	78	Yes[b]
A'	2567	0.963	95	173	No

[a] $F_b'' = F_b$ multiplied by all applicable modifiers except C_F.
[b] For points *B* and *B'*, $F_b' = F_b'' C_F + f_c = (2567)(0.87) + 78 = 2311$ psi.

Point *E'*:

$$\frac{f_c}{F_c'} + \frac{f_b}{F_b'} = \frac{100}{1725} + \frac{2502}{2760} = 0.058 + 0.907 = 0.965$$

Point *D'*:

$$\frac{f_c}{F_c'} + \frac{f_b}{F_b'} = \frac{103}{1725} + \frac{1240}{2760} = 0.041 + 0.449 = 0.490$$

Point *C'*:

$$\frac{f_c}{F_c'} + \frac{f_b}{F_b'} = \frac{144}{1725} + \frac{1649}{2570} = 0.083 + 0.643 = 0.726$$

As shown in Table 5.29, all values of $f_c/F_c' + f_b/F_b'$ are less than 1. The assumed sizes are adequate for the combined bending and compression check.

TABLE 5.29

Point	f_c/F_c'	f_b/F_b'	$f_c/F_c' + f_b/F_b'$
A	0.108	0.737	0.845
B	0.045	0.238	0.283
C	0.054	0.668	0.722
D	0.041	0.364	0.405
E	0.092	0.588	0.680
E'	0.058	0.907	0.965
D'	0.041	0.449	0.490
C'	0.083	0.643	0.726
B'	0.045	0.238	0.283
A'	0.105	0.737	0.845

TABLE 5.30

Point	Loading Condition	Maximum Shear V (lb)	$3V/2A$ (psi)
Base	DL + full SL	−7650	213
A	DL + full SL	−7650	133
B	DL + full SL	+1490	10
C	DL + full SL	+6210	113
D	DL + full SL	+2960	65
E	DL + SL right	+ 840	24
Crown	DL + SL right	+2340	80
E'	DL + SL right	+1170	33
D'	DL + full SL	+2960	65
C'	DL + full SL	+6210	113
B'	DL + SL right	+2290	16
A'	DL + full SL	−7650	133
Base	DL + full SL	−7650	213

Check shear at each point as summarized in Table 5.30 and at the base and crown using the formula $f_v = 3V/2A$.

All values are less than $F_v' = F_v C_D = (200)(1.15) = 230$ psi.

The largest moments in a tudor-type arch normally occur at the haunch. The section provided by the haunch is usually more than adequate to resist this moment when the full depth of the haunch is considered to be effective. The designer, however, should calculate the required structural depth of the arch and include this in the drawings or specifications.

The approximate depth required at point B is

$$d = \sqrt{\frac{6M}{F_b C_D C_C b}} = \sqrt{\frac{(6)(827{,}300)}{(2400)(1.15)(0.93)(5.125)}} = 19.4 \text{ in.}$$

Check application of size factor:

$$f_c = \frac{16{,}770}{(5.125)(19.4)} = 169 \text{ psi}$$

$$F_b''(1 - C_F) = (2400)(1.15)(0.93)(1 - 0.95) = 128 \text{ psi} < f_c \quad (169 \text{ psi})$$

Size factor is not applicable, and depth required is 19.4 in.

In large arches, it is sometimes necessary to provide detachable haunches to meet shipping clearance requirements. In these cases, the structural depth of the arch does not include the attached haunch. The section remaining after the detachable haunch is removed should not be less than that calculated at point B.

Summary of actual section sizes:

Base depth = $10\frac{1}{2}$ in.
Crown depth = $7\frac{3}{4}$ in.
Tangent depth = $15\frac{3}{4}$ in.

10. Check lateral stability at the tangent point ($d_t/b < 6$):

$$\frac{15.75}{5.125} = 3.07 < 6 \quad \text{O.K.}$$

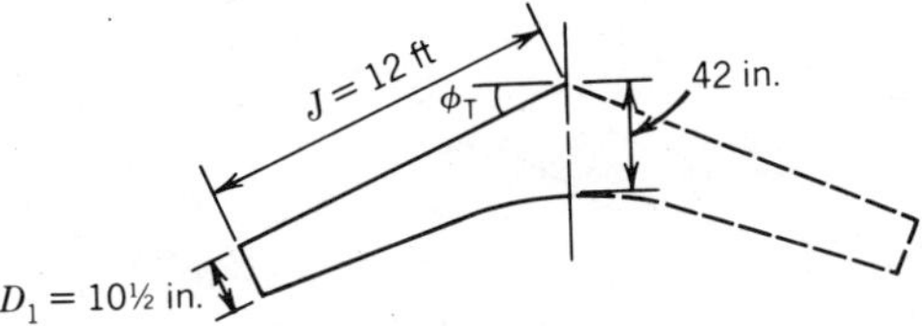

11. Check radial stress:

$$d_c = D_5 = 3.5 \text{ ft} = 42 \text{ in.}$$

$$R_m = R + \frac{d_c}{2} = (10.5)(12) + \frac{42}{2} = 147 \text{ in.}$$

$$\frac{d_c}{R_m} = \frac{42}{147} = 0.29$$

$$\phi_T = 90° - A_8 = 90° - 57.9° = 32.1°$$

Use values from Figure 5.25 for a ϕ_T of 6:12; $K_r = 0.165$. Because the member being analyzed is not uniformly loaded, assume $C_R = 1$. The maximum moment at the haunch is $M = 827{,}300$ in.-lb, which results in radial compression. From Eq. (5-54)

$$f_{rc} = K_r C_r \frac{6M}{bd_c^2} = \frac{(0.165)(1)(6)(827{,}300)}{(5.125)(42)^2} = 91 \text{ psi}$$

$$F_{rc} = F_{c\perp} \quad \text{(for the core laminations)} = 560 \text{ psi}$$

$$F'_{rc} = \mathrm{F}_{rc} = 560 \text{ psi} > 91 \text{ psi} \quad \text{O.K.}$$

In this example, the maximum moment from wind load when combined with the dead-load moment does not result in tension perpendicular to grain stresses in the haunch. On arches with steeply sloped rafter arms, the wind load may result in radial tension stresses in the haunch.

12. Determine the vertical deflection at the peak and the horizontal deflection at the haunch under full snow load plus dead load. Also estimate the deflection due to shrinkage. See discussion on page 5-256.

To determine the deflection due to loads, the arch is divided into a number of segments. For convenience, in this example the arch leg is divided into two segments, the haunch into four segments by the bisectors of angles A_6 and A_7, and the rafter portion is divided into six segments. These divisions are shown in Figure 5.58.

The coordinates of the centroid of each segment, the depth at the centroid, and the segment length are calculated. Those values are summarized in Table 5.31.

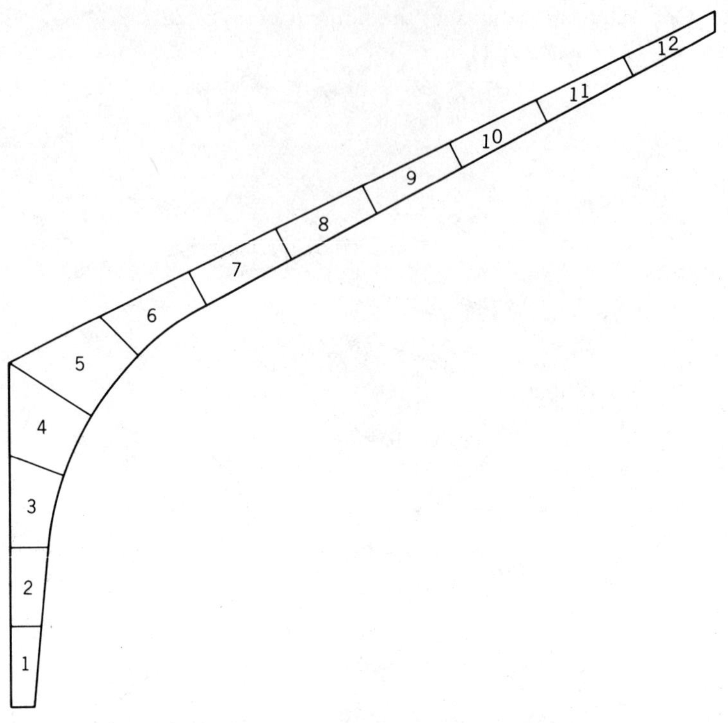

FIGURE 5.58 Segments for deflection calculations.

TABLE 5.31

Segment N	Coordinates XN (ft)	Coordinates YN (ft)	Depth, DN (ft)	Length, SN (ft)
1	0.504	1.40	1.01	2.80
2	0.638	4.20	1.28	2.80
3	0.797	6.81	1.63	2.68
4	1.19	9.66	2.63	2.72
5	2.81	12.10	2.49	3.24
6	5.39	13.78	1.47	2.95
7	8.17	15.21	1.22	3.39
8	11.22	16.73	1.16	3.39
9	14.25	18.25	1.05	3.39
10	17.28	19.77	0.93	3.39
11	20.31	21.29	0.81	3.39
12	23.34	22.81	0.69	3.55

To calculate the deflection due to dead load plus snow load, the moment at the midpoint of each segment due to these loads is calculated. Also, the moment at the centroid of each segment is calculated due to a unit load placed at the point where deflection is desired and in the direction the deflection is to be estimated. The deflection is calculated by the formula

$$\Delta_c = \frac{1}{E} \sum \frac{Mms}{I} \tag{5-78}$$

where Δ_c = deflection of the peak (in.),
E = modulus of elasticity (psi),
M = moment at the midpoint of each segment due to applied loads (in.-lb),
m = moment at the midpoint of each segment due to the unit load at the peak (in.),
s = length of each segment (in.), and
I = moment of inertia at the centroid of each segment (in.^{4}).

Table 5.32 summarizes the calculation for peak deflection and the calculation of horizontal movement at the haunch.

To calculate the horizontal deflection, a unit load is applied as shown in the following sketch:

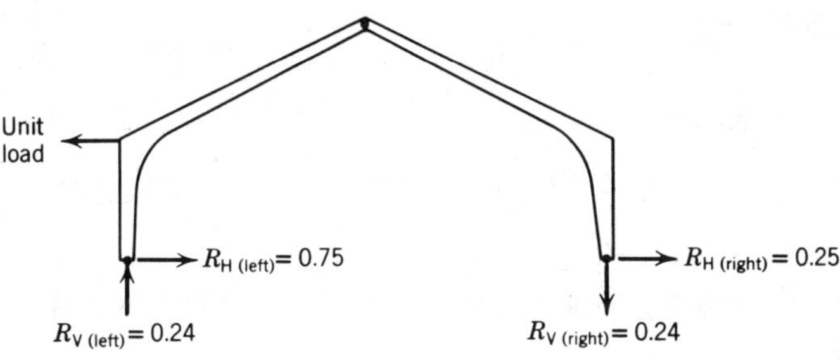

$$\Delta_h = \frac{1}{E} \sum \frac{Mms}{I} = \frac{212{,}273}{1{,}800{,}000} = 0.12 \text{ in.}$$

The outward movement at the haunch, Δ_h, caused by a change in moisture content of -3% is calculated as follows: (see page 5-256)

$$q = -k = -(-3)(0.2) = 0.6\%$$

$$\alpha = (2)(A_5) = (2)(28.4°) = 56.8°$$

$$\Delta_h = \tan q\alpha \frac{(h - y)(y)}{h}$$

$$= \tan [(0.006)(56.8°)] \left[\frac{(24 - 12)(12)(12)}{24}\right] = 0.43 \text{ in.}$$

Total outward deflection = 0.12 + 0.43 = 0.55 in.

TABLE 5.32

Peak Deflection

Point Left	Unit Load Moment m (in.)	Moment (DL + SL) M (in.-lb)	Moment of Inertia I (in.^{4})	Segment Length s (in.)	$\frac{Mms}{I}$
1	−5.70	−37,800	760	33.6	9,526
2	−22.33	−270,700	1,548	33.6	131,203
3	−37.63	−481,700	3,196	32.2	182,625
4	−53.02	−672,600	13,425	32.2	85,534
5	−58.50	−630,100	11,393	38.9	125,857
6	−53.48	−398,300	2,344	35.4	321,697
7	−45.71	−165,700	1,340	40.7	230,051
8	−36.87	−30,800	1,152	40.7	40,120
9	−28.16	+158,800	854	40.7	−230,118
10	−19.45	+220,700	594	40.7	−294,124
11	−10.73	+216,500	392	40.7	−241,194
12	−2.02	+146,200	242	42.6	−51,987
				Σ *Mms/I* =	326,200

Horizontal Movement at Haunch

Point	Unit Load Moment m (in.) (Compression Outside +)	Moment (DL + SL) M (in.-lb) (Compression Outside +)	Moment of Inertia I (in.^{4})	Segment Length s (in.)	$\frac{Mms}{I}$
Left					
1	−11.06	−37,800	760	33.6	18,483
2	−35.73	−270,700	1,548	33.6	209,937
3	−58.63	−481,700	3,196	32.2	284,542
4	−82.99	−672,600	13,425	32.2	133,883
5	−98.89	−630,100	11,393	38.9	212,752
6	−86.22	−398,300	2,344	35.4	518,638
7	−73.72	−165,717	1,340	40.7	371,058
8	−60.15	−30,831	1,152	40.7	65,519
9	−46.65	+158,808	854	40.7	−353,070
10	−33.14	+220,689	594	40.7	−501,119
11	−19.64	+216,545	392	40.7	−441,568
12	−6.14	+146,239	242	42.6	−158,061
Right					
12′	+1.18	+146,239	242	42.6	30,377
11′	+5.42	+216,545	392	40.7	121,858
10′	+9.66	+220,689	594	40.7	146,072
9′	+13.90	+158,808	854	40.7	105,202
8′	+18.14	−30,831	1,152	40.7	−19,759
7′	+22.44	−165,717	1,340	40.7	−112,948
6′	+26.22	−398,300	2,344	35.4	−157,721
5′	+28.65	−630,100	11,393	38.9	−61,643
4′	+25.96	−672,600	13,425	32.2	−41,880
3′	+18.42	−481,700	3,196	32.2	− 89,395
2′	+10.93	−270,700	1,548	33.6	−64,221
1′	+2.79	−37,800	760	33.6	−4,663
				Σ *Mms/I* =	212,273

Circular Arches

Three-hinged circular or parabolic arches are designed in a manner similar to that for three-hinged tudor arches. They will usually require slightly more wood than two-hinged arches but are frequently more economical because of the elimination of the moment splice required for the two-hinged arches.

Two-Hinged Arches

Two-hinged arches are hinged at each base and are statically indeterminate. They may have a profile of any shape with any combination of straight or curved sections with constant or variable section depth. When shipping restrictions limit the size of member that can be transported, it is common to utilize moment splices to reduce component dimensions. The horizontal thrust at the base must be resisted by some adequate means, such as tie rods, abutments, or foundations. After the reactions, moments, shears, and axial forces have been determined, the part of the design for determining the required arch section is similar to that for the three-hinged arch.

Design Procedure

Two-hinged arches are statically indeterminate to the first degree. The usual design procedure is to determine the horizontal reactions by one of the energy methods and complete the solution by statics. Usually, the left hinge of the arch is considered the origin of a coordinate system and horizontal distances x and vertical distances y are measured from this point. Sometimes the configuration is such that an algebraic solution of the location of x and y coordinates of various points along the arch axis is difficult and a graphical solution may be preferred.

When arches are of constant cross section, constant E, and symmetrical about the centerline, both E and I drop out of the energy equations and symmetry reduces the number of calculations. The design is further simplified when the shape of the arch axis is an arc of a circle or a parabola because an algebraic solution is readily obtainable. The effect of tie rod elongation, differential settlement, or spread of abutments is usually small and is commonly neglected in design. Tie rod elongation can change values in larger arches unless the elongation is compensated for by initial shortening of the rod with a provision for sliding bearings to permit movement of the arch base. The following procedure shows the steps required for the design of a two-hinged arch.

1. Lay out one-half of the arch axis to a convenient scale, and divide it into any number of equal divisions (the more divisions used, the more precise the results will be). Determine and tabulate the x and y distances for each midpoint of a division (see Fig. 5.59). Tabulate the values for y^2.

2. Determine vertical reactions R_v for the various loading conditions by the summation of vertical forces equal to zero or by taking moments about one hinge. For balanced loading, R_v = one-half the total live or snow and dead loads on the arch.

3. The arch is made statically determinate by assuming that the left horizontal reaction $R_{H\,(\text{left})}$ has been removed and the left support is free to move. For vertical loads, the entire arch can be considered as a simple-span beam. Compute and

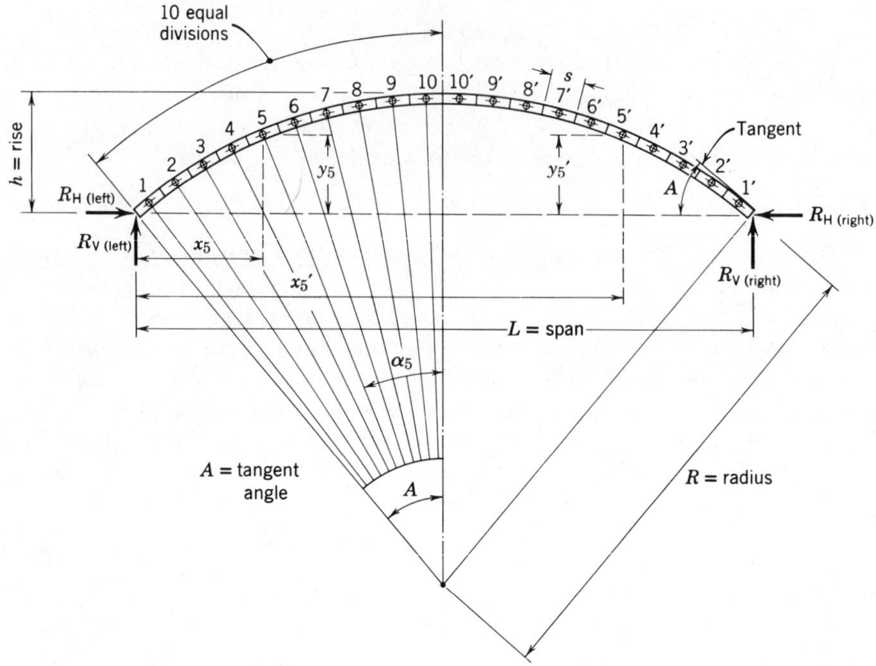

FIGURE 5.59 Two-hinged arch notations.

tabulate the bending moment M_s at each division point for dead loads and balanced and unbalanced live or snow loads. For wind loads, M_s is determined by use of the principles of statics.

4. Multiply the M_s values by the corresponding y values and tabulate.

5. Compute the horizontal reactions R_H for the various loadings by the formula

$$R_H = \frac{\Sigma\,(M_s y)}{\Sigma\,(y^2)} \tag{5-79}$$

This formula neglects the effect of tie rod elongation or spread in abutments because these movements are usually small.

6. Determine and tabulate bending moment M on the arch at each division point using the following formula. For dead and vertical loads:

$$M = M_s - R_H y \tag{5-80}$$

The procedure for calculating wind loads depends on the code requirements for distribution of wind load. When the projected area method is used, moments are determined by Eq. (5-80). When the wind is from the left and with the sign convention for moments used, M_s is negative and $R_H y$ is positive. Equation (5-80) is then written as $M = R_H y - M_s$.

In the absence of a governing building code, ANSI A58.1 is recommended for determination of wind loads as discussed in Section 4.

7. Determine the axial thrust P' at each division point.

8. On the basis of the assumption of full lateral support and using the following formulas, determine a trial size required at each division point.

$$A = \frac{P'}{F''_c} \qquad S = \frac{M}{F''_b}$$

where A = cross-sectional area (in.^{2}),
P' = axial thrust as determined in step 7 (lb),
F''_c = design value in compression parallel to grain (psi) adjusted by applicable modifiers except C_p,
S = section modulus (in.^{3}),
M = bending moment as determined in step 6, (in.-lb), and
F''_b = design value in bending (psi) adjusted by applicable modifiers except C_L.

On the basis of these requirements and industry standard depths and widths (see Table 7.2), determine the required arch depths d and widths b at each division point. When one edge of the arch is braced by decking fastened directly to the arch or braced at frequent intervals, as by girts or purlins, the depth–breadth ratio d/b of the arch based on actual dimensions should not exceed 6. When such lateral bracing is lacking, the ratio should not exceed 5.

9. Check the section for combined bending and axial compression as follows:

$$\frac{f_c}{F'_c} + \frac{f_{bx}}{F'_{bx}} \leqslant 1$$

where f_c = actual compression parallel to grain stress induced by the axial loads (psi),
f_{bx} = actual bending stress about the x–x axis (psi),
F'_c = design value in compression parallel to grain (psi) adjusted by applicable modifiers, and
F'_{bx} = design value in bending about the x–x axis (psi) adjusted by applicable modifiers.

Depending on the loading condition, the length for determining the slenderness ratio, $(l_e/d)_x$, may be either of the following straight-line distances and both should be checked:

(a) hinge to point of contraflexure on the same half of the arch or
(b) point of contraflexure to point of contraflexure.

In either case, d is the arch depth. For determining the length for calculating the slenderness ratio, $(l_e/d)_y$, the distance between purlins should be used for l_u. If the combined stress formula exceeds 1, it will be necessary to revise the chosen section size and repeat this step.

10. Check the section for shear stress at the base and at the point of maximum shear, using Eq. (5-24):

$$f_v = \frac{3V}{2A}$$

where f_v = actual shear stress (psi),
V = shear perpendicular to arch axis (lb), and
A = cross-sectional area (in.^{2}).

This value may not exceed the design value in shear, F_v', where $F_v' = F_v$ adjusted by all applicable modifiers.

11. Check radial stress f_r by Eq. (5-53):

$$f_r = \frac{3M}{2R_m A}$$

where f_r = actual radial stress (psi) tension f_{rt} or compression f_{rc},
M = maximum bending moment (in.-lb),
R_m = radius of curvature at arch axis (in.), and
A = cross-sectional area (in.^{2}).

Limitations on radial design values are given on page 5-218.

12. Determine the deflection due to bending. The deflection at any point in the arch may be determined by the formula

$$\Delta_a = \frac{s}{EI} \sum Mm \tag{5-81}$$

where Δ_a = actual deflection (in.),
s = length of segments along arch axis (in.) (any number of equal-length segments may be used; however, the greater the number of segments, the greater the accuracy),
E = modulus of elasticity (psi),
I = moment of inertia of a section perpendicular to the arch axis (in.^{4}),
M = moment at midpoint of each segment s (in.-lb) under the loading conditions for which deflection is desired, and
m = moment at midpoint of each segment s (in.) under a unit load applied at point where deflection is desired and in the direction in which magnitude of deflection is desired.

To facilitate deflection calculations, tabulate the values.

Constant-Radius Arches

The following formulas will assist in the design of two-hinged, constant-section, constant-radius arches.

Step 1:

$$x_n = R(\sin A - \sin \alpha_n) \tag{5-82}$$

$$y_n = R(\cos \alpha_n - \cos A) \tag{5-83}$$

where α_n = the central angle to point under consideration (degrees)
x_n, y_n = x and y coordinates corresponding with central angle α_n (see Fig. 5.59)

and other terms are as defined in Figure 5.59.

Step 3:
Dead load:

$$M_s = R_{V(\text{left})}x - w_{DL}Ry + w_{DL}R^2 \sin \alpha_n(A - \alpha_n) \quad \text{(for left half span)} \tag{5-84}$$

where w_{DL} = uniform dead load along arch axis (plf) and
$A - \alpha_n$ = difference between tangent angle and central angle to point under consideration (radians).

Balanced live or snow load:

$$M_s = R_{V(\text{left})}x - \tfrac{1}{2}w_{SL}x^2 \quad \text{(for left half span)} \tag{5-85}$$

where w_{SL} = uniform snow load or live load along horizontal projection (plf).
Unbalanced snow load on right half span:

$$M_{s(\text{left})} = R_{V(\text{left})}x \quad \text{(simple moments in left or unloaded half span)} \tag{5-86}$$

$$M_{s(\text{right})} = R_{V(\text{right})}(L - x) - \tfrac{1}{2}w_{SL}(L - x)^2 \tag{5-87}$$

(simple moments in right or loaded half span)

Wind load from the left (projected area method):

$$M_{s(\text{left})} = -R_{V(\text{left})}x - \tfrac{1}{2}w_{WL}y^2 \quad \text{(simple moments in left half span)} \tag{5-88}$$

$$M_{s(\text{right half span})} = R_{V(\text{right})}(L - x) - R_{H(\text{right})}y \quad \text{(simple moments in right half span)} \tag{5-89}$$

where w_{WL} = uniform wind load along vertical projection (plf)

and other terms are as shown in Figure 5.59.

Step 7: At the point of maximum moment (which may be located from the tabulation in step 6), the axial thrust P' is determined by the formula

$$P' = R_H \cos \alpha_n + (R_V - \textstyle\sum wx) \sin \alpha_n \tag{5-90}$$

where R_H = horizontal reaction for loading causing maximum moment (lb) (for maximum moment occurring in left half span, use $R_{H(\text{left})}$; in right half span, use $R_{H(\text{right})}$)
R_V = vertical reaction for the loading causing maximum moment (lb),
α_n = central angle to point of maximum moment (degrees), and
Σwx = sum of loads to left (for point on left half span) or to right (for point on right half span) of point of maximum moment (lb).

Step 10: Check shear. The maximum horizontal and vertical reactions occur under dead load plus full snow load; therefore, check shear at base for this condition.

$$V = (R_{\text{VDL}} + R_{\text{VSL(left)}}) \cos A - (R_{\text{HDL}} + R_{\text{HSL(left)}}) \sin A \qquad (5\text{-}91a)$$

$$f_v = \frac{3V}{2A}$$

The maximum shear at the crown occurs under dead load (DL) and unbalanced snow load (UL).

$$V = R_{\text{VDL}} + R_{\text{VUL(left)}} - \sum w_{\text{DL}}x \quad \text{(snow load on right)} \qquad (5\text{-}91b)$$

Example. Design a constant-section, constant-radius glued laminated two-hinged arch to meet the following conditions:

L = 86 ft

Spacing = B = 17 ft

R = 64 ft (radius of centerline of arch)

Loads:

w_{DL} = 10 psf

w_{SL} = 20 psf

Wind loading for 70-mph wind using ANSI A58.1 procedure (13)

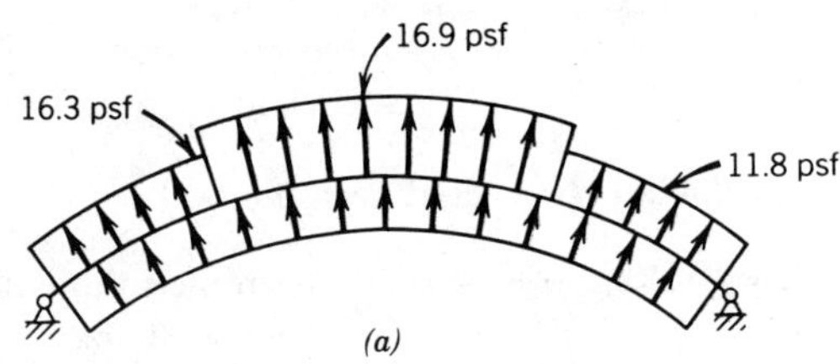

(a)

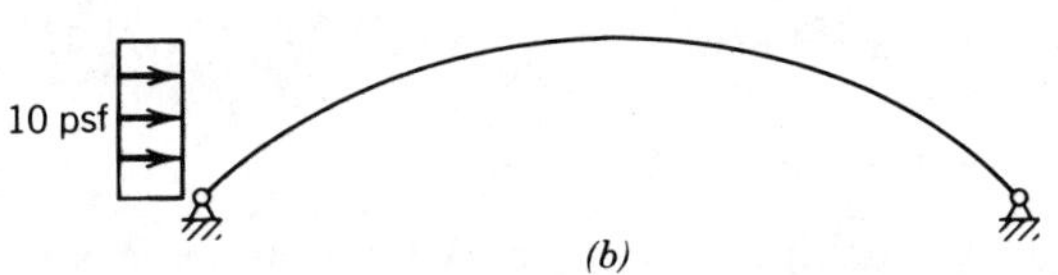

(b)

Assuming the arch is on buttresses 15 ft high, the wind loading is shown in (a). By inspection, it is obvious that when wind load stresses are added to dead load stresses, this loading will not control. Therefore, the minimum horizontal load of 10 psf acting on the vertical projection is used in design as shown in (b).

Lamination thickness $t = 1\frac{1}{2}$ in.
Curvature factor $C_C = 1 - 2000(t/R)^2 = 1 - 2000[(1.5)/(64)(12)]^2 = 0.992$
Duration of load factor C_D = 1.15(DL + SL), 1.33(DL + WL)
$F_b'' = F_bC_DC_C = (2400)(1.15)(0.992) = 2740$ psi
$F_c'' = F_cC_D = (1650)(1.15) = 1900$ psi
$F_{c\perp}' = F_{c\perp} = 560$ psi
$F_{rt}' = F_{rt}C_D = (15)(1.15) = 17.25$ psi
$F_v' = F_vC_D = (165)(1.15) = 190$ psi
$E_x' = E_x = 1{,}800{,}000$ psi, $E_y' = E_y = 1{,}600{,}000$ psi

Lateral support provided by 3 × 12 purlins spaced at intervals of 9.57 ft measured along the arch.

1. Determine A = tangent angle:

$$\sin A = \frac{L/2}{R} = \frac{43}{64} \qquad A = 42.21°$$

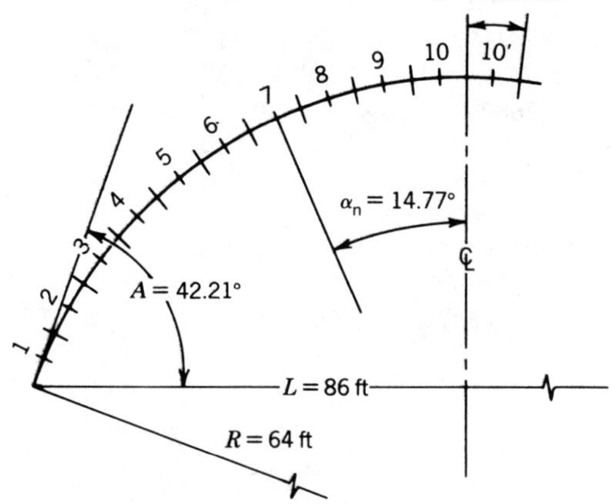

Divide arch half into 10 equal parts. Determine length of arc, s.

$$s = \frac{A2\pi R}{(360)(10)} = \frac{(42.21°)(2)(\pi)(64)(12)}{(360)(10)} = 56.58 \text{ in.}$$

Determine and tabulate α_n, x, y, y^2. (See Table 5.33.)

2. Determine vertical reactions.

Dead load:

$$R_{\text{V(left)}} = R_{\text{V(right)}} = \frac{w_{\text{DL}}\, B(\text{length of arch axis})}{2}$$

$$= \frac{(10)(17)(94.3)}{2} = 8015 \text{ lb}$$

Balanced snow load:

$$R_{\text{V(left)}} = R_{\text{V(right)}} = w_{\text{SL}}\frac{BL}{2} = \frac{(20)(17)(86)}{2} = 14{,}620 \text{ lb}$$

Unbalanced snow load (full snow load on right half span, no snow load on left half span): Note that some codes may require the combination of full snow load on onehalf with one-half snow load on the other half, which may be more critical than full unbalanced loading on onehalf.

By the sum of the moments about left hinge = 0,

$$R_{\text{V(right)}} = \frac{3\, w_{\text{SL}}\, BL}{8} = \frac{(3)(20)(17)(86)}{8} = 10{,}960 \text{ lb}$$

$$R_{\text{V(left)}} = \frac{w_{\text{SL}}\, BL}{2} - R_{\text{V(right)}} = \frac{(20)(17)(86)}{2} - 10{,}960 = 3660 \text{ lb}$$

TABLE 5.33[a]

Point	α_n (degrees)	α_n (rads)	sin α_n	cos α_n	x (ft)	y (ft)	y^2 (ft^2)
Left							
1	40.10	0.69990050	0.64414	0.76491	1.77	1.55	2.41
2	35.88	0.62622679	0.58609	0.81024	5.49	4.45	19.83
3	31.66	0.55255305	0.52486	0.85119	9.41	7.07	50.03
4	27.44	0.47887931	0.46078	0.88751	13.51	9.40	88.33
5	23.22	0.40520557	0.39421	0.91902	17.77	11.41	130.3
6	19.00	0.33153183	0.32549	0.94554	22.17	13.11	171.9
7	14.77	0.25789809	0.25501	0.96694	26.68	14.48	209.7
8	10.55	0.18418435	0.18314	0.98309	31.28	15.51	240.7
9	6.33	0.11051061	0.11029	0.99390	35.94	16.21	262.7
10	2.11	0.03683687	0.03683	0.99932	40.64	16.55	274.0
Right							
10′					45.36	16.55	274.0
9′					50.06	16.21	262.7
8′					54.72	15.51	240.7
7′					59.32	14.48	209.7
6′					63.83	13.11	171.9
5′					68.23	11.41	130.3
4′					72.49	9.40	88.33
3′					76.59	7.07	50.03
2′					80.51	4.45	19.83
1′					84.23	1.55	2.41
						Σy^2 =	2899.8 ft^2

[a] $A = 42.21° = 0.73674$ rad; $\sin A = 0.671875$; $\cos A = 0.740665$.

Wind load acting from the left:

$$\text{Arch rise} = \frac{L}{2}\tan\frac{A}{2} = 43\tan\frac{42.21°}{2} = 16.60 \text{ ft}$$

Assume arch is 20 in. deep. Add 10 in. for one-half depth of arch and 6 in. for decking and roofing:

$$h = 16.60 + \frac{16}{12} = 17.93 \text{ ft}$$

By the sum of the moments about left hinge = 0,

$$R_{\text{V(right)}} = \frac{w_{\text{WL}}\,Bh^2}{2L} = \frac{(10)(17)(17.93)^2}{(2)(86)} = 318 \text{ lb}$$

$$R_{\text{V(left)}} = -R_{\text{V(right)}} = -318 \text{ lb} \quad \text{(acts downward)}$$

Steps 3 and 4. Calculate and tabulate values of M_s and $M_s y$ (see Table 5.34).

TABLE 5.34

	Dead Load		Simple Beam Moments (ft-lb)				Wind Load	
			Balanced SL		Unbalanced SL			
Point	M_s	$M_s y$	M_s	$M_s y$	M_s	$M_s y$	M_s	$M_s y$
Left								
1	13,850	21,460	25,340	39,280	6,480	10,040	767	1,190
2	40,690	181,060	75,140	334,370	20,090	89,420	3,430	15,260
3	65,820	465,320	122,520	866,220	34,440	243,500	7,240	51,190
4	88,750	834,270	166,490	1,564,990	49,450	464,800	11,810	110,980
5	109,300	1,247,100	206,120	2,351,780	65,040	742,090	16,720	190,740
6	126,920	1,663,870	240,570	3,153,860	81,140	1,063,770	21,660	283,950
7	141,300	2,046,060	269,050	3,895,870	97,650	1,413,950	26,310	380,910
8	152,410	2,363,830	290,980	4,513,080	114,480	1,775,660	30,400	471,420
9	159,770	2,589,900	305,860	4,957,940	131,540	2,132,270	33,760	547,310
10	163,610	2,707,730	313,380	5,186,490	148,740	2,461,690	36,200	599,200
Right								
10′					164,640	2,724,800	37,520	621,000
9′					174,320	2,825,660	37,980	615,670
8′					176,490	2,737,430	37,330	578,970
7′					171,400	2,481,910	35,650	516,240
6′					159,430	2,090,080	32,910	431,460
5′					141,080	1,609,700	29,130	332,350
4′					117,040	1,100,190	24,360	228,950
3′					88,080	622,730	18,560	131,200
2′					55,050	244,960	11,820	52,590
1′					18,870	29,240	4,160	6,450
	$\Sigma M_s y = 14{,}120{,}600$ $\times 2 = 28{,}241{,}200$		$\Sigma M_s y = 26{,}863{,}800$ $\times 2 = 53{,}727{,}760$		$\Sigma M_s y = 26{,}863{,}890$		$\Sigma M_s y = 6{,}167{,}030$	

By sum of the horizontal forces = 0,

$$R_{\mathrm{H(right)}} = w_{\mathrm{WL}}\,Bh = (10)(17)(17.93) = 3050\ \mathrm{lb}$$

5. Compute horizontal reactions

$$R_{\mathrm{H}} = \frac{\Sigma M_s y}{\Sigma y^2}$$

$$R_{\mathrm{HDL}} = \frac{28{,}241{,}200}{2899.8} = 9740\ \mathrm{lb}$$

$$R_{\mathrm{HSL}} = \frac{53{,}727{,}760}{2899.8} = 18{,}530\ \mathrm{lb}$$

$$R_{\mathrm{HUL}} = \frac{26{,}863{,}890}{2899.8} = 9260\ \mathrm{lb}$$

$$R_{\mathrm{HWL(left)}} = \frac{6{,}164{,}390}{2899.8} = 2{,}130\ \mathrm{lb}\quad \text{(acting to left)}$$

$$R_{\mathrm{HWL(right)}} = w_{\mathrm{WL}}h - R_{\mathrm{HWL(left)}}$$

$$= (10)(17)(17.93) - 2130 = 920\ \mathrm{lb}\ \text{(acting to left)}$$

6. Actual moment values are tabulated in Table 5.35.

7. Determine axial thrust. From step 6, the point of maximum moment occurs at point 5 under dead load on full span and full snow load on the right half span (unbalanced snow load). At point 5,

$$P' = [R_{\mathrm{HDL(left)}} + R_{\mathrm{HUL(left)}}]\cos 23.22^\circ$$

$$+ [R_{\mathrm{VDL(left)}} + R_{\mathrm{VUL(left)}} - \Sigma wx]\sin 23.22^\circ$$

$$\Sigma wx = Rw_{\mathrm{DL}}(\mathrm{A} - \alpha_n)$$

$$= (64)(10)(17)(0.73674 - 0.40520) = 3610\ \mathrm{lb}$$

$$P' = (9740 + 9260)(0.91902)$$

$$+ (8015 + 3660 - 3610)(0.39421) = 20{,}640\ \mathrm{lb}$$

8. Determine section size

$$A = \frac{P'}{F''_c} = \frac{20{,}640}{1900} = 10.87\ \mathrm{in.}^2$$

$$S = \frac{M}{F''_b} = \frac{(42{,}490)(12)}{(2740)} = 186\ \mathrm{in.}^3$$

Try a $5\frac{1}{8} \times 16\frac{1}{2}$-in. section, $A = 84.56\ \mathrm{in.}^2$, $S = 232.5\ \mathrm{in.}^3$

9. Check for combined stresses

(a) Compression: Consider buckling about the x–x axis. From step 6, point

TABLE 5.35

Actual Moment Values

	Actual Moments (ft-lb)						
Point	DL	SL	$\frac{1}{2}$SL	WL	DL + SL	DL + $\frac{1}{2}$ SL	DL + WL
Left							
1	−1,250	−3,380	−7,880	+2,530	−4,630	−9,130	+1,280
2	−2,650	−7,320	−21,130	+6,040	−9,970	−23,780	+3,390
3	−3,040	−8,490	−31,060	+7,800	−11,530	−34,100	+4,760
4	−2,810	−7,690	−37,630	+8,190	−10,500	−40,440	+5,380
5	−1,830	−5,310	−40,660	+7,550	−7,140	−42,490	+5,720
6	−770	−2,360	−40,310	+6,230	−3,130	−41,080	+5,460
7	+260	+740	−36,490	+4,490	+1,000	−36,230	+4,750
8	+1,340	+3,580	−29,200	+2,590	+4,920	−27,860	+3,930
9	+1,880	+5,490	−18,630	+710	+7,370	−16,750	+2,590
10	+2,410	+6,710	−4,580	−1,000	+9,120	−2,170	+1,410
Right							
10′	+2,410	+6,710	+11,320	−2,320	+9,120	+13,730	+90
9′	+1,880	+5,490	+24,150	−3,500	+7,370	+26,030	−1,620
8′	+1,340	+3,580	+32,800	−4,340	+4,920	+34,140	−3,000
7′	+260	+740	+37,260	−4,850	+1,000	+37,520	−4,590
6′	−770	−2,360	+37,980	−5,020	−3,130	+37,210	−5,790
5′	−1,830	−5,310	+35,380	−4,860	−7,140	+33,550	−6,690
4′	−2,810	−7,690	+29,960	−4,360	−10,500	+27,150	−7,170
3′	−3,040	−8,490	+22,580	−3,520	−11,530	+19,540	−6,560
2′	−2,650	−7,320	+13,830	−2,350	−9,970	+11,170	−5,000
1′	−1,250	−3,380	+4,510	−860	−4,630	+3,260	−2,110

of contraflexure for this loading condition occurs near the centerline of the arch. Chord distance between base of arch to center of arch is

$$\frac{L/2}{\cos(A/2)} = 46.1 \text{ ft} = 553 \text{ in.}$$

$$\left(\frac{l_e}{d}\right)_x = \frac{553}{16.5} = 33.5$$

Consider buckling between purlins equally spaced at 9.57 ft = 114.8 in. (l_u).

$$\left(\frac{l_e}{d}\right)_y = \frac{114.8}{5.125} = 22.4 \quad (l_e/d)_x \text{ governs}$$

$$K_x = 0.671\sqrt{\frac{E_x}{F_c''}} = 0.671\sqrt{\frac{1{,}800{,}000}{1900}} = 20.7$$

Since 20.7 < 33.5 < 50, use long-column formula.

$$F'_c = \frac{0.30E}{(l_e/d)_x^2} = \frac{(0.30)(1{,}800{,}000)}{(33.5)^2} = 481 \text{ psi}$$

$$f_c = \frac{P'}{A} = \frac{20{,}640}{84.56} = 244 \text{ psi}$$

(6) Bending (check for buckling about y–y axis):

$$C_F = 0.97 \qquad F''_{bx}(1 - C_F) = 2740(1 - 0.97) = 82 \text{ psi} < f_c$$

Size factor is not applicable.

$$l_e = 1.68 l_u = (1.68)(114.8) = 192.9 \text{ in.}$$

that is, a beam with four equal loads (Table 5.10).

$$C_s = \sqrt{\frac{l_e d}{b^2}} = \sqrt{\frac{(192.9)(16.5)}{(5.125)^2}} = 11.0 > 10$$

$$C_k = 0.956\sqrt{\frac{E_y}{F''_{bx}}} = 0.956\sqrt{\frac{1{,}600{,}000}{2740}} = 23.1$$

Since 10 < 11 < 23.1, use intermediate-beam formula.

$$C_L = \left[1 - \frac{1}{3}\left(\frac{11.0}{23.1}\right)^4\right] = 0.98$$

$$F'_{bx} = F''_{bx} C_L = (2740)(0.98) = 2690 \text{ psi}$$

$$f_{bx} = \frac{M}{S} = \frac{(42{,}490)(12)}{232.5} = 2190 \text{ psi}$$

$$\frac{f_c}{F'_c} + \frac{f_{bx}}{F'_{bx}} \le 1$$

$$\frac{244}{481} + \frac{2{,}190}{2690} = 0.51 + 0.81 = 1.32 \quad \text{(overstressed)}$$

Repeating the same procedure using a $5\frac{1}{8} \times 19\frac{1}{2}$-in. section,

$$A = 99.98 \text{ in.}^2 \qquad S = 324.8 \text{ in.}^3$$

$$C_F = 0.95 \qquad I = 3167 \text{ in.}^4$$

$$\left(\frac{l_e}{d}\right)_x = \frac{553}{19.5} = 28.4 \qquad \left(\frac{l_e}{d}\right)_y = 22.4 \text{ in.} \qquad \left(\frac{l_e}{d}\right)_x \text{ governs}$$

Since 20.7 < 28.4 < 50, use long-column formula.

$$F'_c = \frac{(0.30)(1{,}800{,}000)}{(28.4)^2} = 670 \text{ psi}$$

$$f_c = \frac{20{,}640}{99.98} = 207 \text{ psi}$$

$$F''_{bx}(1 - C_F) = 2740(1 - 0.95) = 137 \text{ psi} < f_c$$

Size factor is not applicable.

$$C_s = \sqrt{\frac{(192.9)(19.5)}{(5.125)^2}} = 12.0 > 10$$

Since 10 < 12 < 23.1, use intermediate-beam formula.

$$C_L = \left[1 - \frac{1}{3}\left(\frac{12.0}{23.1}\right)^4\right] = 0.98$$

$$F'_{bx} = (2740)(0.98) = 2680 \text{ psi}$$

$$f_{bx} = \frac{(42{,}490)(12)}{324.8} = 1570 \text{ psi}$$

$$\frac{207}{670} + \frac{1{,}570}{2680} = 0.31 + 0.59 = 0.90 \quad \text{O.K.}$$

10. Check for shear. Maximum horizontal and vertical reactions occur under dead load plus full snow load; therefore, check shear at the base for this condition.

$$\begin{aligned} V &= (R_{VDL} + R_{VSL})\cos A - (R_{HDL} + R_{HSL})\sin A \\ &= (8015 + 14{,}620)(0.740665) - (9740 + 18{,}530)(0.671875) \\ &= 2230 \text{ lb} \end{aligned}$$

$$f_v = \frac{3V}{2A} = \frac{(3)(2230)}{(2)(99.98)} = 33.5 \text{ psi} < 190 \text{ psi} \quad \text{O.K.}$$

At the crown of the arch, the point of maximum shear occurs under dead load plus unbalanced snow load.

$$V = R_{VDL} + R_{VUL(left)} - \Sigma w_{DL}x \quad \text{(snow load on right)}$$

$$\Sigma w_{DL}x = (64)(10)(17)(0.73674) = 8015 \text{ lb}$$

$$V = 8015 + 3660 - 8015 = 3660 \text{ lb}$$

$$f_v = \frac{3V}{2\text{A}} = \frac{(3)(3660)}{(2)(99.98)} = 55.0 \text{ psi} < 190 \text{ psi} \quad \text{O.K.}$$

Check d/b ratios with new sizes.

$$\frac{d}{b} = \frac{19.5 - 11.25}{5.125} = 1.61 < 5 \quad \text{O.K.}$$

Purlins are equally spaced at 9.57 ft = 114.8 in.

$$d = b = 5.125 \text{ in.} \qquad (l_e/d)_y = \frac{114.8}{5.125} = 22.4$$

Since 21.7 < 22.4 < 50, use long-column formula.

$$F'_c = \frac{(0.30)(1{,}800{,}000)}{(22.4)^2} = 1080 > 660 \text{ psi} \quad \text{O.K.}$$

11. Check radial stress.

$$R = \text{radius of curvature} = (64)(12) = 768 \text{ in.}$$

Maximum moment is negative; therefore, radial stress is compressive.

$$F'_{rc} = F_{c\perp} = 560 \text{ psi}$$

$$f'_{rc} = \frac{3M}{2RA} = \frac{(3)(42{,}490)(12)}{(2)(768)(99.98)} = 10 \text{ psi} < 560 \quad \text{O.K.}$$

Maximum positive moment occurs at point 6′ = 37,210 ft-lb (DL + $\frac{1}{2}$ SL)

$$F'_{rt} = (15)(1.15) = 17.25 \text{ psi}$$

$$f_{rt} = \frac{3M}{2RA} = \frac{(3)(37{,}210)(12)}{(2)(768)(99.98)} = 8.7 \text{ psi} < 17.25 \text{ psi} \quad \text{O.K.}$$

12. Determine vertical deflection at the centerline under unbalanced snow load. Using the layout from step 1 of this example, s = 56.58 in. Moments due to unbalanced snow loads were determined in step 6 and are retabulated. Moments due to the unit load are calculated in the same manner. They are also tabulated as in Table 5.36.

$$R_{H(\text{left})} = \frac{\Sigma M_s y}{\Sigma y^2} = \frac{2840.8}{2899.8} = 0.97965$$

$$m = m_s - R_{H(\text{left})} y$$

$$\Delta_a = \frac{s}{EI} \Sigma Mm = \frac{(56.58)(184{,}900)(144)}{(1{,}800{,}000)(3167)} = 0.26 \text{ in.}$$

MOMENT SPLICES

The use of glued laminated timber has reduced the need for moment splices in timber beams and girders, but such splices are sometimes used in arches and rigid frames where shipping requirements make the shipping of full-size structural frames impractical or uneconomical. Moment splices should be located at points of minimum moment whenever possible. They must be designed to resist bending moment, shear, stress reversal, uplift, and axial forces by various means, such as tension or compression straps located on the sides or edges and compression

TABLE 5.36

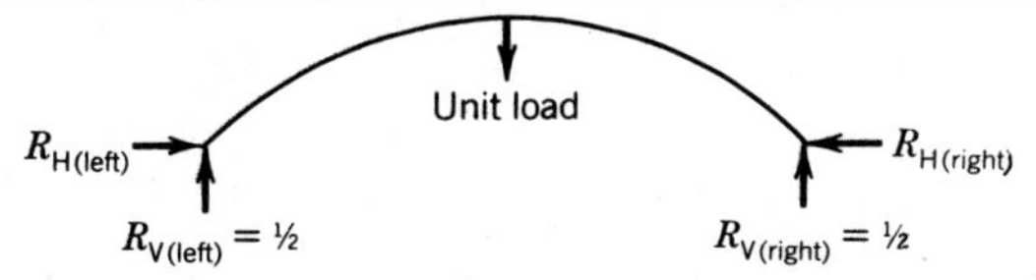

Point	Unit Load m_s	$m_s y$	m	Moments M DL + $\frac{1}{2}$ SL (ft-lb)	Mm
Left					
1	0.885	1.3718	−0.63344	− 9,130	+5,780
2	2.745	12.215	−1.61440	−23,780	+38,390
3	4.705	33.264	−2.22105	−34,100	+75,740
4	6.755	63.497	−2.4536	−40,440	+99,220
5	8.885	101.38	−2.29270	−42,490	+97,420
6	11.085	145.32	−1.75808	−41,080	+72,220
7	13.335	193.16	−0.84518	−36,230	+30,620
8	15.64	242.58	+0.44579	−27,860	−12,420
9	17.97	291.29	+2.0900	−16,750	−35,010
10	20.32	336.30	+4.1070	−2,170	−8,910
		1420.38			
Right					
10′			+4.1070	+13,730	+56,390
9′			+2.0900	+26,030	+54,400
8′			+0.44579	+34,140	+15,220
7′			+0.84518	+37,520	−31,710
6′			−1.75808	+37,210	−65,420
5′			−2.2927	+33,550	−76,920
4′			−2.4536	+27,150	−66,620
3′			−2.22105	+19,540	−43,400
2′			−1.61440	+11,170	−18,030
1′			−0.63344	+3,260	−2,060
	$\Sigma m_s y$ = (2)(1420.38) = 2840.8				ΣMm = +184,900

plates between members. A means of holding the two sections in alignment must also be provided.

Close tolerances are required in the fabrication of moment splices. Consideration should be given to inelastic as well as elastic deformation in the joint. For sawn lumber sections, and for many smaller glued laminated timbers, the simplest form of moment connection is a plywood splice plate (or pair of plates). Such plates can be glued or nailed but are also frequently used in conjunction with bolts and shear connectors.

A typical moment splice for glued laminated timber is illustrated in the construction details in AITC 104 in Part III of this manual. In this splice, compression stress is taken in end bearing between the two sections. If the unit stress for end grain in bearing parallel to grain f_g as determined by the loading conditions and splice location, exceeds 75% of the design value in end grain bearing F_g', a snug-fitting metal bearing plate not thinner than 20 gage and not deeper than one-half the depth of the member should be installed between the abutting ends. See Annex A, AITC 117—Design (1) for design values for end grain in bearing.

Tension stress is taken across the splice by means of steel straps and shear plates as illustrated or by wood splice plates and split rings. Connector capacity must be adequate to carry the tension stress. Additional side straps and shear plates are required to keep the sides and top of the members in position and to take the stress reversals from erection loads or wind uplift. Shear connections must be provided to transfer the maximum design shear through the splice connection.

Consideration should be given to the use of separate side plates for each row of connectors in order to minimize secondary stresses due to shrinkage effects in the member. See Section 6, "Fastenings and Connections," for spacing of rows.

The design of moment-resisting splices of this type is affected by the inelastic slip in the connections that can modify the usual assumptions on which elastic design with materials of different moduli of elasticity are based. The recommendations in this section consider this inelastic slip. If the assumption is made that the force on the compression side of the member is resisted entirely by the wood and the tension force is resisted by the steel splice plates and no inelastic slip exists between the wood and steel, a transformed section can be assumed as shown.

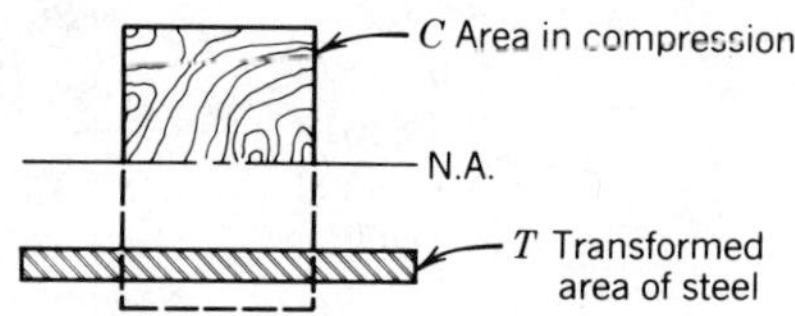

When the moment is very small compared to the compressive force (less than $Pd/6$) where P is the axial thrust and d is the depth of the member at the splice), the entire cross section is in compression (see Fig. 5.60*a*). This condition may exist on circular or parabolic shaped arches which are subjected to uniform loads only. When wind loads or unbalanced loads are taken into account, the moment increases and the resultant of the compressive forces moves away from the neutral axis. The area under compression starts to decrease when the resultant has moved more than one-sixth of the depth of the member from the neutral axis. The moment and the axial thrust can be shown as an equivalent loading of $P'e$ and P' where P' is equal to the axial thrust and $P'e$ is equivalent to the moment (see Fig. 5.60*b*).

When the ratio of the moment to axial thrust is high, e may be greater than $d/2$ and P' is shown acting outside of the cross section (see Fig. 5.60*c*). Each loading condition should be examined so that the different ratios of bending moment to axial thrust can be checked.

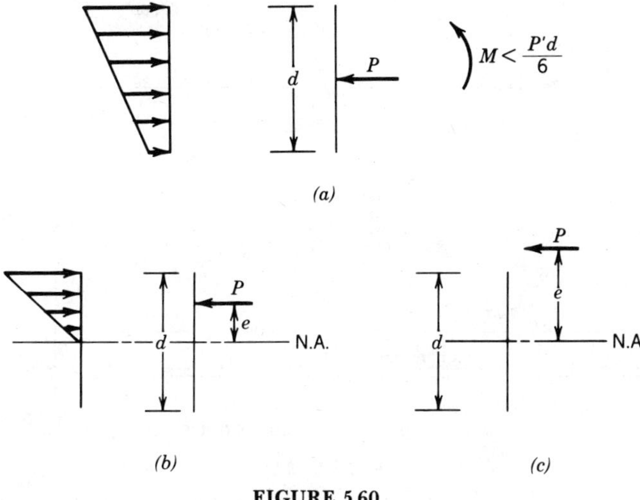

FIGURE 5.60

In case (a) and some case (b) conditions, the structure would be stable provided that its members are held in line, a method for transferring shear forces is provided and end grain bearing is not exceeded. However, in both cases, a splice is needed to provide continuity so that the assumption of a continuous member in a two-hinged arch design is validated. In case (c), the structure is unstable and a moment splice is required for stability as well as continuity.

The inelastic slip in mechanical fastenings, crushing of wood fibers in end grain bearing, and inaccuracies in fabrication result in errors in location of the neutral axis and some designers use arbitrary methods of locating the area under compression such as using the entire area above the neutral axis. Others prefer to use the transformed section approach.

A conservative approach is to design the splice so that it can resist the entire applied moment. The tension force in the splice is taken by a steel strap and the compressive force by end bearing of the wood. As shown in Figure 5.61, the maximum end grain in bearing, f_g, is the summation of the axial compressive stress f_c and the compressive stress in bending caused by the splice f_b.

Moment-resisting splices can also be designed so that the compression force is carried by splice plates, in which case the compression splice plates must be designed as compression members. These types require more material and fab-

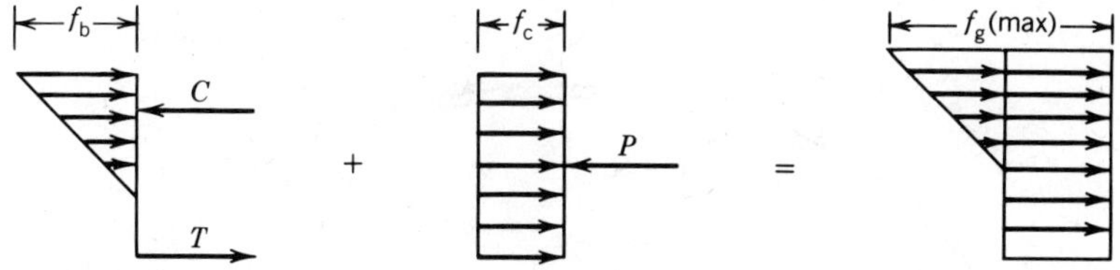

f_b C T + f_c P = f_g(max)

FIGURE 5.61

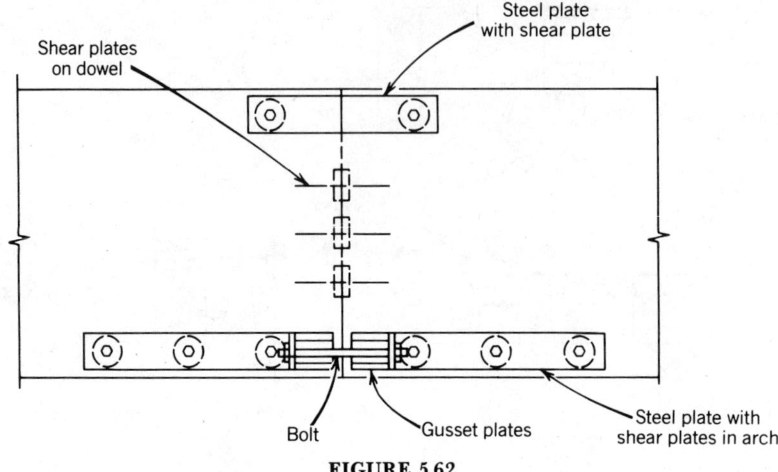

FIGURE 5.62

rication than those where the ends of the member abut in compression. However, the location of the resultant of the compressive force is more precise.

Inelastic slip in the tension splice plates can be offset by using a tension-type splice with bolts that can be tightened to eliminate the slip, as shown in Figure 5.62. These also require extra materials and fabrication, but the location of forces to resist the moment can be determined more accurately with this system and the slippage that occurs in the tension connection can be compensated for with the bolt.

Example. Design a moment splice of the type illustrated in *Typical Construction Details*, AITC 104, for a two-hinged arch with a radius of 60 ft. At the location of the splice, positive moment is 120,500 in.-lb caused by DL + SL. The maximum axial compression is 29,400 lb caused by DL + SL and the maximum shear is 5230 lb caused by DL + SL (on half the span). The maximum negative moment is 15,000 in.-lb caused by DL + WL.

$$C_D = 1.15 \text{ (snow load)} \qquad C = 1.33 \text{ (wind load)}$$

Calculate design value, F'_g; for snow loading:

$$E_x = 1{,}800{,}000 \text{ psi}$$

$$F'_g = F_g C_D = (2350)(1.15) = 2700 \text{ psi}$$

The section size is $5\frac{1}{8} \times 21$ in.

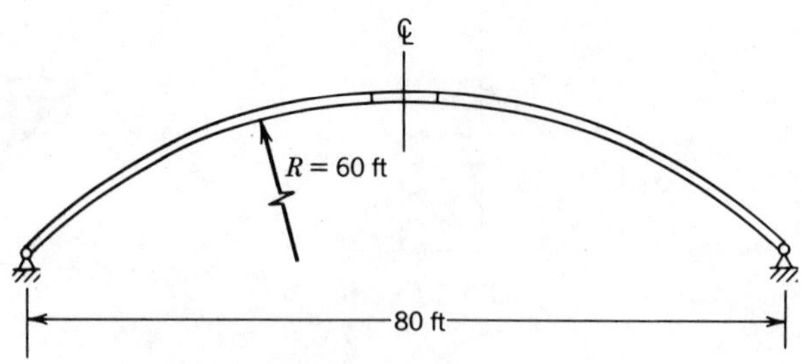

1. Determine location of resultant compression force.

$$P'e = M \qquad e = \frac{M}{P'} = \frac{120{,}500}{29{,}400} = 4.1 \text{ in.}$$

$$\frac{Y}{3} = \frac{21}{2} - 4.1$$

$$Y = 19.2 \text{ in.}$$

The volume of the stress block equals the force P'. Therefore

$$f_g = \frac{2P'}{Yb} = \frac{(2)(29{,}400)}{(19.2)(5.125)} = 598 \text{ psi} < 2700 \text{ O.K.}$$

$$f_g < (0.75)(2700)$$

No metal bearing plate is needed.

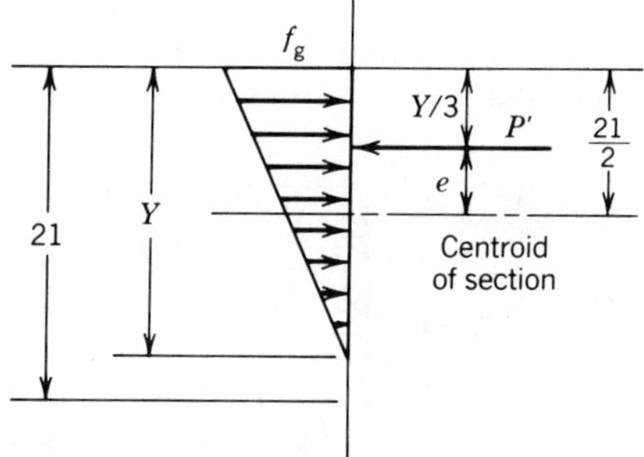

2. Determine tension force required. In this example, the resultant of the axial compressive force P' is close enough to the neutral axis that the joint is theoretically stable provided it is tied together and fastenings are used to resist the shear force. However, the splice should be designed to resist all of the moment (neglecting the effect of the axial force P). When the moment alone is considered, the location of the neutral axis changes from that shown in step 3. For a trial design, the neutral axis will be assumed to be at middepth and the location of the tension force is determined by assuming the use of $2\frac{5}{8}$-in. shear plates. For a $2\frac{5}{8}$-in. shear plate, minimum edge distance = $1\frac{3}{4}$ in. (see Table 6.13).

$$X = 8.75 + \frac{(2)(10.5)}{3} = 15.75 \text{ in.}$$

$$T = \frac{M}{X} = \frac{120{,}500}{15.75} = 7650 \text{ lb}$$

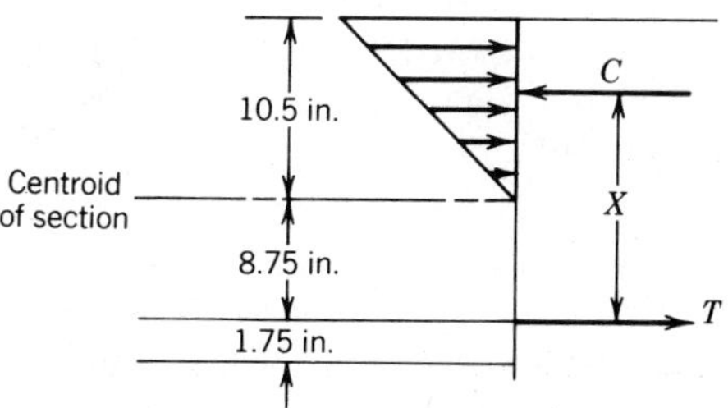

3. Determine the neutral axis by the transformed section method. Tension is to be resisted by two steel straps.

Area of the steel required: (see AISC *Manual of Steel Construction* (14))

$$A_S = \frac{T}{F_{t(\text{steel})}} = \frac{7650}{22{,}000} = 0.35 \text{ in.}^2$$

$$\text{Net area for one strap} = \frac{0.35}{2} = 0.17 \text{ in.}^2$$

Try a 3 × $\frac{1}{4}$-in. steel bar:

$$\text{Net area} = 0.75 - (0.25)\,(13/16) = 0.55 \text{ in.}^2$$

$$\text{Transformed net tension area} = (2)(0.55)\,\frac{29{,}000{,}000}{1{,}800{,}000} = 17.72 \text{ in.}^2$$

The neutral axis has equal moments of areas above and below.

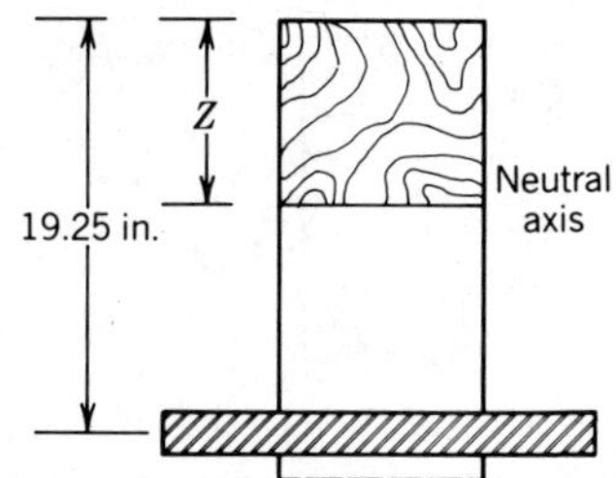

$$5.125Z\left(\frac{Z}{2}\right) = (17.72)(19.25 - Z)$$

$$Z = 8.59 \text{ in.}$$

4. Determine T and C. The steel selected for side straps is 3 in. wide to cover the shear plates. Therefore, the neutral axis will remain as calculated. (Theoretically a smaller strap could be used.)

$$T = C = \frac{M}{X} = \frac{120{,}500}{19.25 - (8.59/3)} = 7355 \text{ lb}$$

5. Determine tension resistance requirements: Angle of load to grain = 0°. Tabular value of one $2\frac{5}{8}$-in. shear plate for Group B species ($\frac{3}{4}$-in. bolt): P = 2860 lb with metal side plates (see Table 6.13); C_D = 1.15, P' = PC_D = (2860)(1.15) = 3290 lb > 2900 lb maximum allowable (see footnote to Table 6.13). Therefore, use 2900 lb.

$$\text{Number of shear plates required} = T/2900 = 7355/2900 = 2.54$$

Use four on each side of splice with two in a row, C_g = 1.00; therefore, there is no reduction in design load of connector for fastenings in a row.

$$\text{Percentage of design load value used} = \frac{2.54}{4} = 63.5\%$$

Minimum spacing parallel to grain required is $4\frac{3}{8}$ in. (by interpolation) (see Fig. 6.14 or Table 6.15). Minimum end distance is approximately $2\frac{3}{4}$ in. Minimum edge distance is $1\frac{3}{4}$ in. Use two shear plates on each face of member on each side of splice and one tension strap each side of member (see Fig. 5.63). The end distance and spacing calculated are the minimum. Unless space is limited, end distance and spacing for full load are recommended as follows:

End distance = $5\frac{1}{2}$ in.
Edge distance = $1\frac{3}{4}$ in.
Spacing = $6\frac{3}{4}$ in.

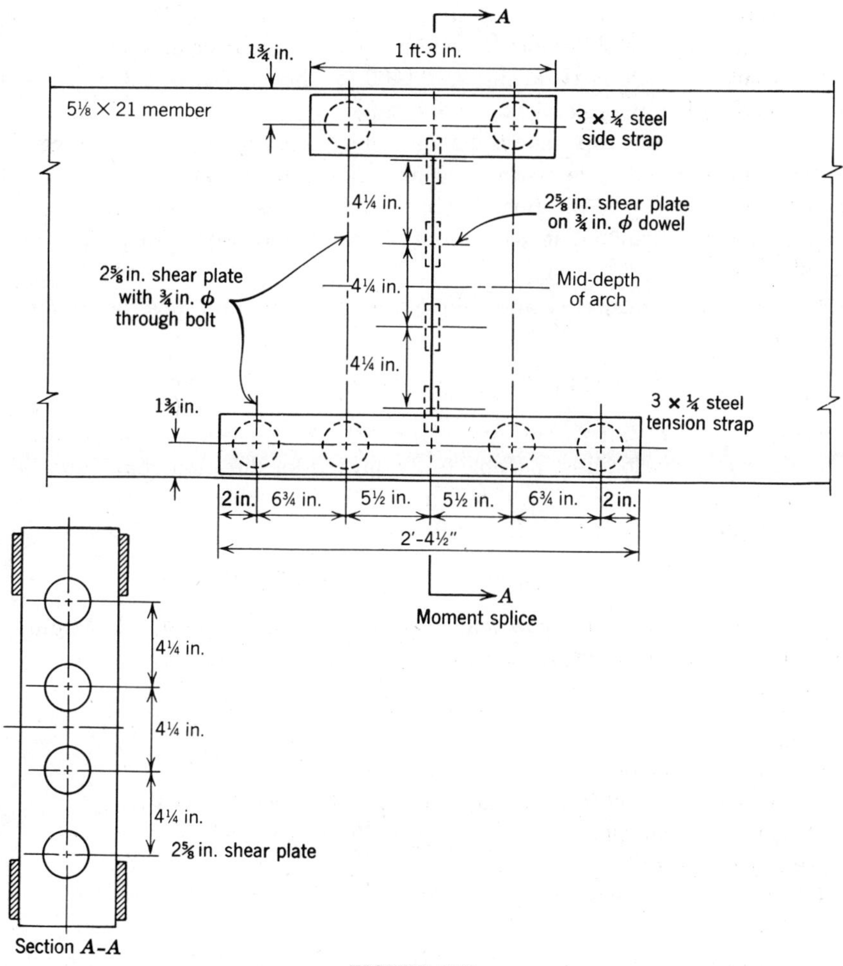

FIGURE 5.63

$$\text{Load resisted by one tension strap} = \frac{T}{2} = \frac{7355}{2} = 3678 \text{ lb}$$

$$\text{Area of steel required} = \frac{3678}{22{,}000} = 0.17 \text{ in.}^2$$

Try a 3 × $\frac{1}{4}$-in. steel bar. Check stress at the net section with a $\frac{13}{16}$-in. hole in strap, and the stress on the gross cross section.

$$\text{Stress at net section} = \frac{3678}{0.55} = 6690 \text{ psi} < 29{,}000 \text{ psi}$$

$$\text{Stress on gross area} = \frac{3678}{0.75} = 4900 < 22{,}000 \text{ psi}$$

See AISC *Manual of Steel Construction* (14).

Check capacity of bolt in steel side strap (see AISC *Manual of Steel Construction*). Load resisted by one bolt = 3678/2 = 1840 lb. Three-quarter-in. diameter A307 bolt capacity is 16,300 lb (bearing) and 4400 lb (shear). Use one 3 × $\frac{1}{4}$-in steel bar on each side of member.

The splice at the top should be of sufficient strength to resist negative moment due to handling and stress reversal. The negative moment is very small and by inspection two $2\frac{5}{8}$-in. shear plates on each side as shown in Figure 5.63 are adequate for the calculated negative moment and should also be adequate to resist handling stresses.

6. Determine shear resistance requirements. Maximum shear occurs under DL + $\frac{1}{2}$ SL.

$$V = 5230 \text{ lb} \qquad \text{Angle of load to grain} = 90°$$

For a $2\frac{5}{8}$-in. shear plate in end grain, use 60% of the value tabulated for perpendicular to grain loading in Table 6.12 adjusted for duration of loading. Edge distance provided is 2.56 in., which is greater than the $1\frac{3}{4}$-in. minimum edge distance for full load; therefore, $C_e = 1.0$.

$$P' = (1990)(1.15)(0.60) = 1370 \text{ lb}$$

Try four shear plates in a row loaded perpendicular to the grain. Minimum spacing for full allowable load perpendicular to grain stress is $4\frac{1}{4}$ in.

$A_1 = A_2 = (4.25)(5.125) = 21.8 \text{ in.}^2$
$A_1/A_2 = 1.0$
Modification factor $C_g = 0.97$ (from Table 6.7)
(1370)(0.97) = 1330 lb (average allowable load per connector)
Number of shear plates required = V/1330 = 5230/1330 = 3.9
Number used = 4
Percentage = 3.9/4 = 98%
Use $4\frac{1}{4}$-in. spacing
Dowel length—Use standard penetration required for lag bolt used with a shear plate as shown in Table 6.11 and add 1 in. on each end.

Standard penetration = 5 × lag bolt diameter
Length of dowel = [(5)(0.75) + 1)] 2 = 9.5 in.

TRUSSES

The subject of timber truss design is quite broad. This discussion is limited to basic design procedures and highlighting of features unique to timber truss construction.

Truss Types

The types of timber trusses most commonly built are (see Fig. 5.64): (a) parallel chord; (b) pitched or triangular; (c) bowstring; (d) camelback and (e) special, such as crescent, scissors, sawtooth, lenticular, and king or queen post trusses.

Some roof shapes are dictated by architectural considerations, such as a sawtooth roof. Generally, however, the designer selects the truss type for best overall economy along with the necessary clear span, clearances, and aesthetic appearance desired for the structure. The maximum economical span for any given type of timber truss will vary with the materials available, loading conditions, ratio of labor to material cost, and fabrication methods. Table 1.1 gives span ranges for various primary truss-framing systems. Also, span ranges for various secondary framing systems are given, which can be used to help determine truss spacings.

Truss members are designated in three categories: top chord, bottom chord, and webs (all interior vertical or diagonal members between the top and bottom chords are called webs). Joints at which members intersect and connect are called panel points. If flat or parallel chord trusses are used as roof trusses, adequate pitch must be provided for drainage. (See 4-88.)

Pratt trusses have the advantage that for normal loading the longer (diagonal) webs are in tension whereas the shorter (vertical) webs are in compression. Flat or low-pitched Pratt roof trusses can be used for clear spans up to 120 ft with the most economy in spans 80 ft and under.

Pitched Pratt and Belgian trusses can be used for clear spans up to 100 ft. They become more economical than Fink trusses when the pitch is less than 25° and the span is over 50 ft. The roof pitch should be a minimum of 3 : 12.

Fink trusses can be used for clear spans up to 80 ft with roof pitches 25° or more from the horizontal. Optimum span range is 40–60 ft.

Bowstring trusses are most economical for spans from 80 to 150 ft but may range from 50 to 200 ft. Usually the radius of the top chord is equal to the span. Modified bowstring trusses, where the center portion of the top chord is straight instead of curved, can be used for spans up to 300 ft. The chords and heel connections take the major stresses; the web stresses under uniform loading conditions are negligible, and for unbalanced loading, web stresses are comparatively light. Bowstring trusses, because of low web stresses, have light webs and web connections. Chord stresses are nearly equal throughout their length and, with constant cross-section, chords have full-length economy.

Bowstring trusses usually have the upper chord shaped to the form of a circular arc or parabola. With the normal depth–span ratios used, the variation between the

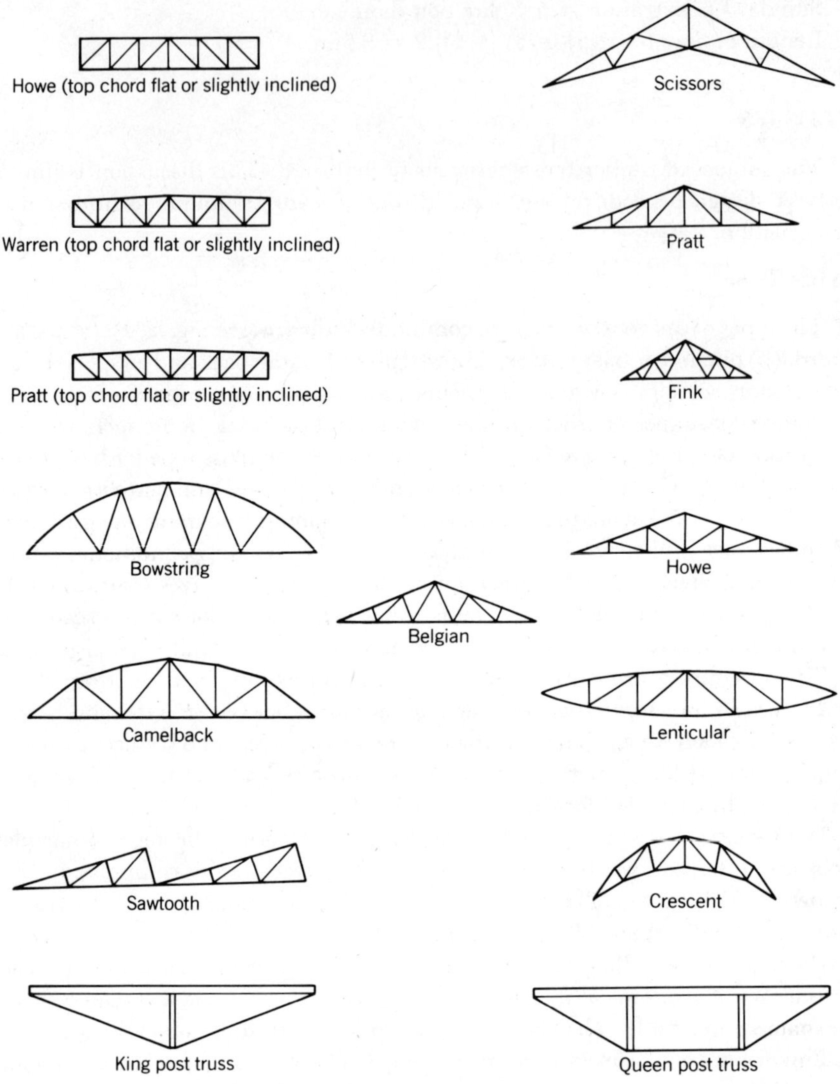

FIGURE 5.64 Truss types.

parabola and the circular arc is very slight and is not sufficient to change stresses materially. Circular curves are customarily used. In addition to normal axial and bending stresses, upper chords of bowstring trusses are subject to moments due to the eccentricity of their curved shape.

Scissor trusses can be efficiently designed for clear spans up to 80 ft and are used to provide more height clearance toward the midspan. They are primarily used in churches, gymnasiums, and various types of assembly halls.

Bridge trusses can be designed for economy with clear spans up to 200 ft.

Table 5.37 lists aids for determining economical truss dimensions without a

TABLE 5.37

Recommended Timber Truss Depth–Span Ratios

Flat or parallel chord	1/8 to 1/10
Triangular or pitched	1/6 or deeper
Bowstring	1/6 to 1/8

great deal of preliminary calculations. When these ratios of depth to span are lower, larger forces and members usually result. Also, these trusses may deflect excessively under applied loads.

Chords and webs in all truss types may be constructed as single-leaf, double-leaf or multileaf members. Single-leaf chords are also known as monochords. The most common arrangements of trusses are monochord trusses with single-leaf chords and webs and those with double-leaf chords having single-leaf webs located between the chord leaves. Webs may be attached to sides of chords or may be in the same plane and attached thereto with steel straps or gussets. Truss members may be sawn or glued laminated timber. The use of steel rods or other steel shapes for members in timber trusses is acceptable if they fulfill all conditions of design and service.

Timber truss web systems should be selected for convenience of connection and economy. Web locations and panel point spacings may be dictated by selection of secondary purlin framing so as to minimize chord bending stresses. In parallel chord trusses, diagonal webs should be sloped between 45° and 60° from the horizontal for greatest economy. Bowstring trusses use panel lengths from 8 to 14 feet depending on the truss span.

Monochord trusses may require steel gusset plate connections with through bolts alone or in combination with shear plates. In multileaf trusses, connections can often be made more easily by using through bolts and shear plates or split rings. When split-ring or shear plate connectors are required, it is usually best to have the wood trusses fabricated in a shop.

Glued laminated timber provides many features desirable for truss designs. Glued laminated timber can be made in almost any shape, size, or length and provides higher design values than do sawn timbers. Sawn timbers are limited in maximum length and cross-sectional size and are prone to checking. However, sawn members may provide cost savings where they will serve. Thus, depending on truss span and loading requirements, trusses can be made of all sawn timber, all glued laminated timber, or a combination of both and may include some steel tension members. If sawn members are used in conjunction with glued laminated timber, care must be taken to match widths at connections and provide for differential shrinkage.

Deflection

The deflection Δ_a at any point in a truss may be determined by the method of virtual work. In this method, a unit load that acts in the direction of the desired component of movement is applied at the point being investigated. The force caused in each truss member by this unit load is determined. Deflection is then determined by the summation of the effects in each member of the unit load and actual design loads, using the formula

$$\Delta_a = \sum \frac{Pul}{AE} \qquad (5\text{-}92)$$

where P = axial force load in a truss member caused by design loads (lb),
u = force in a truss member caused by a unit load (dimensionless),
l = length of truss member (in.),
A = cross-sectional area of truss member (in.^{2}), and
E = modulus of elasticity (psi).

The effects of truss deflection should be considered in the design of columns and walls and in the working of truss-supported doors or other building fixtures.

Camber

Truss camber (Fig. 5.65) should be such that total load deflection does not produce a sag below a straight line between points of support. Camber may be determined by using Eq. (5-92) to determine the dead-load elastic deformation to which the inelastic deformation (or slip) is added, test data, the recommendations given in CSA Standard 086 (7), or the following empirical formula recommended by the Timber Engineering Company (15) for multileaf, timber-connected trusses:

$$\text{Camber} = K_1 \frac{L^3}{h} + K_2 \frac{L^2}{h} \qquad (5\text{-}93)$$

where L = span (ft),
h = rise (ft), and
K_1, K_2 = coefficients depending on the truss configuration (see Fig. 5.65)

and camber is in inches.

Both lower and upper chords may be cambered, but this is generally done only in flat or parallel chord trusses and in longer triangular trusses. When only the lower chord is cambered, as is usual for bowstring and some triangular or pitched trusses, the effective depth of the truss should be reduced accordingly in stress computations.

The following camber recommendations by the Canadian Standards Association, CSA Standard 086, (7) are provided as a guide and should be modified where unusual conditions exist. The amount of camber built into a truss varies

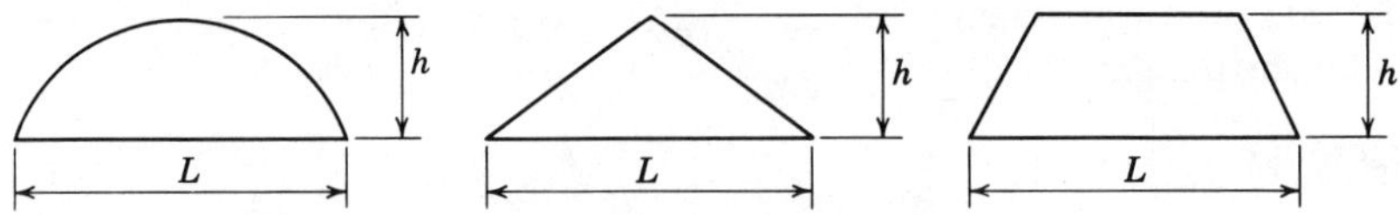

FIGURE 5.65 Truss camber. Coefficients vary with truss geometry and roof height. See applicable building codes or ANSI (13) for specific information on application of wind loads.

with the use to which the truss is put and the expectancy of full design load. The following camber limitations are suggestions only and are for simple-span trusses only:

(a) Bowstring trusses: (i) with continuous glued laminated top chord, camber in bottom chord only, $\frac{1}{4}$ in. per 10 ft of span; (ii) segmental overlapping sawn chord trusses, camber in bottom chord only, $\frac{3}{8}$ in. per 10 ft of span.

(b) Triangular trusses—where possible, camber should be built into both top and bottom chords: (i) with sawn timber or rod and sawn timber trusses, $\frac{1}{2}$ in. per 10 ft of span on bottom chord; at midpoint of half span on top chord, $\frac{3}{8}$ in. per 10 ft of span; (ii) with glued laminated timber or rod and glued laminated timber trusses, $\frac{3}{8}$ in. per 10 ft of span on bottom chord; at mid-point of half span on top chord, $\frac{1}{4}$ in. per 10 ft of span.

(c) Pratt and Howe trusses—camber both top and bottom chords similarly: (i) for sawn timber or rod and sawn trusses, $\frac{1}{2}$ in. per 10 ft of span; (ii) for glued laminated timber or rod and glued laminated timber trusses, $\frac{3}{8}$ in. per 10 ft of span.

General Design Procedure

Given the truss type, span, depth, spacing, and species and grade of lumber to be used, the following general procedure may be used for all truss types. Special design features for each particular truss type must also be considered.

1. Determine dead loads acting on the truss. Dead load should include weights of all roof and ceiling construction, mechanical and electrical equipment, the estimated truss weight, and other permanently applied loads that will act throughout the service life of the truss.

2. Determine snow or live loads, wind loads, and any special loads required by applicable building codes and/or good engineering judgment. Apply loading magnification factors such as for snow drifting, water ponding, and impact as may be required. Most building codes and ordinances specify loading combinations to be used. Unbalanced loading conditions frequently control truss designs. If net uplift is possible, special considerations for stress reversals and bottom chord buckling will be required.

Preliminary Design. Most timber trusses have chords that are continuous across several panel points and also have distributed loads on the chords. Member forces and architectural considerations generally determine the type of connections to be used and can result in pinned, fixed, or semifixed-type connections. These fixed or semifixed conditions affect the distribution of forces throughout the truss and result in the truss being internally indeterminate. If the overall height of a truss must be held to a given dimension, the geometry of the centerlines of the members will be dependent on the truss member sizes. Consequently, it is usually necessary to conduct a preliminary truss design to determine approximate member sizes and connection types.

For the purposes of a preliminary design, loads are placed at panel points and all joints are assumed pinned. Member forces can then be determined graphically or analytically. Preliminary chord sizes can be selected based on these axial forces plus approximate moments due to any distributed loads or concentrated loads that do not occur at panel points. Preliminary web sizes can also be determined using the axial forces. Refer to design of members for axial forces or bending plus axial forces, which are covered on page 5-197 of this manual. Locate splices in chords considering available lengths and minimal axial forces. Design the joint connections considering first the joint or joints carrying the greatest load. Connections should be concentric and hinged wherever possible. Consideration should be given to the effect of possible wood shrinkage on the connection. Member sizes may have to be increased to compensate for loss of section at connections.

Final Design. Draw truss to scale using member sizes from the preliminary design, locate centerlines of members, and chord splices, and indicate locations of hinged and fixed connections. Indicate loading exactly as it will be applied to the truss. The designer must now decide the method of analysis to be used. Table 5.38 lists typical truss parameters along with methods of analysis and results possible.

If no members are continuous but are all pinned at each panel point, a simple mathematical or graphical analysis will give a correct distribution of member forces (Table 5.38, Type A). If chords are continuous or any connections are not pinned, the distribution of forces will be affected and a more complex method of analysis should be considered. In reality, most truss top and bottom chords have continuity and may be pinned only at the middle of the truss span and at the ends, although webs are usually pinned to the chords (Table 5.38, Type B).

If the truss span is relatively short, the depth–span ratio is in accordance with Table 5.37 and loads are fairly low, the designer may decide that a manual analytical

TABLE 5.38

Truss Parameters and Methods of Analysis

	Truss Parameters	Method of Analysis	Design Results
A	All members discontinuous with pins at every panel point	Manual analytical, graphical, or computer	Exact[a]
B	Chords continuous through panel points; chords pinned at ends and splices and all webs pinned to chords	Same as above Stiffness analysis by computer or complex manual method	Approximate Exact[a]
C	Continuous chords with fixed connections at all panel points	Stiffness analysis by computer or complex manual method	Exact[a]

[a]Based on the assumption of average modulus of elasticity in each member and the accuracy of the assumed joint fixity.

or graphical analysis will be satisfactory, although not exact. In such cases, the chords should be analyzed as multispan beams in order to determine support reactions at panel points (assume rigid supports and apply loading exactly as it will be transmitted from secondary framing). Apply these reactions to the truss as panel point loads and analyze for axial member forces. Member sizes and connections can then be designed. Connections should be concentric, pinned (in accordance with design assumption), and given consideration as to possible wood shrinkage. The designer must keep in mind that this analysis is not exact. Deflection of the truss under load will cause a redistribution of member forces and possible higher forces and moments in some locations.

As a general rule, if truss members are continuous or any joint is not hinged, a more complex and accurate method of analysis must take into account the relative stiffness of members, deflection of the truss under load, and resulting redistribution of member forces and moments. Likewise, if there are knee braces or other structural elements attached to the truss that can lend support or influence load distribution, they must be included as integral with the truss in the overall analysis.

The design of truss chord members will be governed primarily by axial tension or compression if chords are not continuous and if all secondary framing transmits loads into panel points only. However, if distributed loads are applied to chord members or chord members are continuous over panel points or panel point connections are fixed, then significant bending stresses along with axial stresses could result and a combined stress unity check must be made. Combined loading in beams and lateral stability considerations are discussed on page 5-197. If chords are curved, moments resulting from eccentricity and axial forces must be included in the design.

Truss web member designs are governed by axial tension or compression loads for all pinned webs. If web connections offer moment-resisting capacity, the webs must be sized to resist the resulting combined loading, the same as for chords. Slenderness ratios of compression webs should be checked as columns using the width as the least dimension and the distance between panel points as the length. Unbalanced, wind, wind uplift, or other special loadings can cause increased stresses or reversal of stresses and thus control portions of a total design. Net member sections must be used where drilling and dapping for connection hardware is required. The required truss member sizes are determined for appropriate conditions of loading based on the above-stated procedures utilizing the latest available design values for structural glued laminated timber. For members stressed principally in axial tension or axial compression, refer to Table 2, AITC 117—Design (1). If chord members are highly stressed in bending, refer to Table 1, AITC 117—Design (1).

Connections are to be designed in accordance with Section 6 of this manual. Truss connections should be concentric; that is, the centerlines of members at each joint should intersect at a common point so that moments are not induced into the members being connected.

Connections should be of the pinned type whenever possible. Truss members tend to rotate in relation to each other as a result of distortion and deflection of the truss under loading. If rotation is restrained by the joints, moments and additional forces in members and connections will result.

Determine the size and type of fastening to be used. When possible, use fastenings of the same size and type throughout the truss. Check spacing, end distance, and edge distance requirements for each fastener. See Section 6 on design considerations for fastenings. Table 3, AITC 117—Design (1), contains information on the connector and fastener groups to be used. Possible shrinkage of wood members must be considered. For splices in deep members where steel plates are used, it may be necessary to utilize two rows of splice plates. Checking due to a decrease in moisture content is a possibility in all wood members and should be considered, especially if members are large or if rapid or large changes in moisture content may take place. It is generally advisable to stagger connectors.

Monochord Trusses

Design of steel straps in compression must be based on actual support conditions for the plates. For plates connecting webs to chords, generally the web does not provide lateral stability for the plate but instead the plate is an extension of the web. Therefore, the end of the plate fastened to the web cannot be considered fixed in determining the plate effective length. When the web plates are pinned at the chord, the spacing between bolts in the web should be large enough to provide stability and prevent splitting of the web. Normal fabrication tolerances must be considered. If steel rods are used as tension elements in a truss, written instruction should be given to the truss assembler regarding tightening of the rods. Otherwise, overtightening could occur, which might overstress truss elements when design loads are applied.

Truss Bracing

In structures employing trusses, a system of bracing is required to provide resistance to lateral forces, to hold the trusses true and plumb, and to hold compression elements in line. Both permanent bracing and temporary erection bracing should be designed according to accepted engineering principles to resist all loads that will normally act on the system. Erection bracing is that which is installed during erection to hold the trusses in a safe position until sufficient permanent construction is in place to provide full stability. Permanent bracing is that which forms an integral part of the completed structure. Part or all of the permanent bracing may also act as erection bracing.

Permanent bracing must be provided to resist both transverse and longitudinal forces and may consist of either of the following systems:

1. A structural diaphragm in the plane of the top chord. The diaphragm, acting as a plate girder, transmits forces to end and side walls (see "Structural Diaphragms," beginning on page 5-385 of this manual). This system is the preferred method of providing truss bracing and is usually most economical.

2. Horizontal bracing between trusses and in the plane of the bottom and/or top chords. In effect, this is a horizontal truss. It is a positive method of bracing,

but it is more costly and should be used only where the strength of system 1 is insufficient. If horizontal bracing is used, it should be designed to resist all the transverse and longitudinal loads that will normally act on the system.

If the roof construction does not provide proper top chord strut action, separate additional members should be provided.

If bottom chords can go into compression under wind uplift or some other load application, additional bracing may be required to satisfy l/d requirements. To provide lateral support to the compression chord of a truss, the bracing system should be designed to withstand a horizontal force equal to at least 2% of the compressive force in the truss chord if the members are in perfect alignment; or, if the members are not aligned, the calculation of the force should be based on the eccentricity of the members due to the misalignment.

In all cases, vertical X-bracing between trusses should be installed in each third or fourth bay at intervals of approximately 35 ft measured parallel to trusses. Bottom chord lateral bracing should be installed perpendicular to the trusses and in line with vertical X-bracing and extend from end wall to end wall.

Bracing is also required to provide overall building stability and transmit forces from the roof to the ground. This bracing typically consists of shear walls, diagonal bracing, buttresses, cantilevered columns, or knee braces.

Use of knee braces between the trusses and columns is a method of bracing particularly adaptable to buildings that have a large length–width ratio. The stresses induced in the trusses and columns can be critical. Lateral loads acting in the long direction of the building are usually resisted by means of diagonal bracing and struts in the end bays.

Erection Truss Bracing

Erection truss bracing is that installed to hold trusses true and plumb and in a safe condition until permanent truss bracing and other permanent components, such as joists and sheathing contributing to the rigidity of the complete roof structure, are in place.

Erection truss bracing may consist of struts, ties, cables, guys, shores, or similar items. Joists, purlins and other permanent elements may be used as part of the erection bracing.

Example. Design a triangular Howe truss using glued laminated wood members for the following conditions:

Span = 48 ft center to center of supports
Truss spacing = 12 ft center to center
Panel point spacing = 8 ft
Pitch = 6 : 12

Heavy timber decking is attached directly to the top chord. Bottom chord is braced laterally at 16-ft intervals.

Loading: DL = 15 psf on horizontal projection (no ceiling load)
SL = 30 psf on horizontal projection (2-month duration)
WL = 25 psf applied as shown in the sketch

Use the following tabular design values:

Top chord	Bottom chord and webs
$F_{bx} = 2400$ psi	$F_{bx} = 1600$ psi
$E_x = E_y = 1{,}700{,}000$ psi	$F_t = 1250$ psi
$F_c = 1600$ psi	$F_c = 1900$ psi
$F_{c\perp} = 560$ psi	$E_x = E_y = 1{,}700{,}000$ psi
	$F_{c\perp} = 560$ psi

Steel plates ASTM A36, bolts ASTM A307.

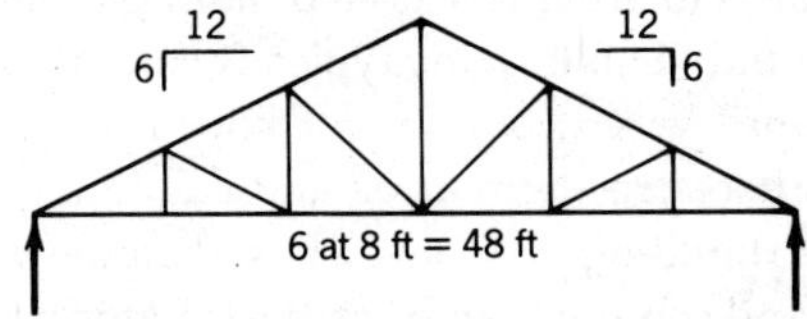

Preliminary size of top chord:

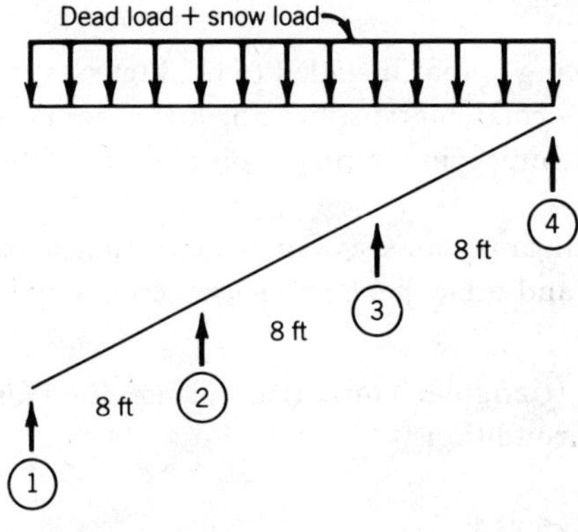

$$w_{DL} = (15)(12) = 180 \text{ lb/ft}$$
$$w_{SL} = (30)(12) = \underline{360}$$
$$540 \text{ lb/ft}$$

Calculate support reactions and moments using standard engineering formulas for a three-span continuous beam.

$$R_1 = R_4 = 0.4wl = (0.4)(540)(8) = 1730 \text{ lb}$$

$$R_2 = R_3 = 1.1wl = (1.1)(540)(8) = 4750 \text{ lb}$$

$$M_{1-2} = M_{3-4} = 0.08wL^2 = (0.08)(540)(8)^2(12) = 33{,}180 \text{ in.-lb}$$

$$M_2 = M_3 = -0.10wL^2 = -(0.10)(540)(8)^2(12) = -41{,}470 \text{ in.-lb}$$

$$M_{2-3} = 0.025wL^2 = (0.025)(540)(8)^2(12) = 10{,}370 \text{ in.-lb}$$

where M_{1-2} = maximum moment between points 1 and 2
M_2 = moment at point 2
M_{2-3} = maximum moment between points 2 and 3
M_3 = moment at point 3
M_{3-4} = maximum moment between points 3 and 4

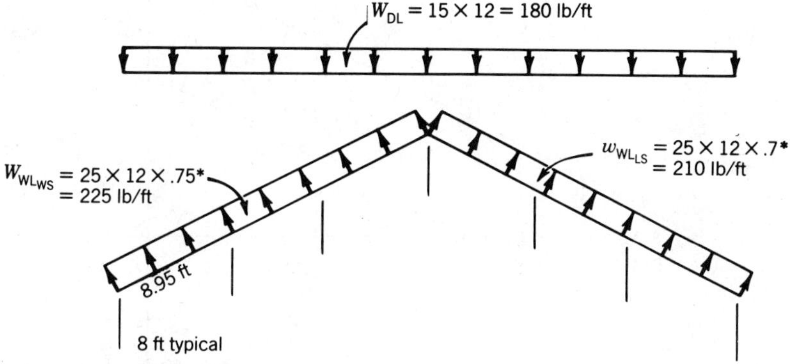

*Coefficients vary with truss geometry and roof height. See applicable building code or ANSI A58.1(13) for specific information on application of wind loads.

Maximum moments—windward side:

$$M_{DL} = -0.10wL^2 = -(0.10)(180)(8)^2(12) = -13{,}820 \text{ in.-lb}$$

$$M_{WL} = 0.10wL^2 = (0.10)(225)(8.95)^2(12) = \underline{21{,}630} \text{ in.-lb}$$

$$7810 \text{ in.-lb}$$

Maximum moments—leeward side:

$$M_{DL} = \quad -13{,}820 \text{ in.-lb}$$

$$M_{WL} = 0.10wL^2 = (0.10)(210)(8.95)^2(12) = \underline{20{,}190} \text{ in.-lb}$$

$$6370 \text{ in.-lb}$$

Determine preliminary axial member forces graphically as shown in Figure 5.66 for DL + SL, Figure 5.67 for DL + unbalanced snow load, and Figure 5.68 for DL + WL. Results are summarized in Table 5.39.

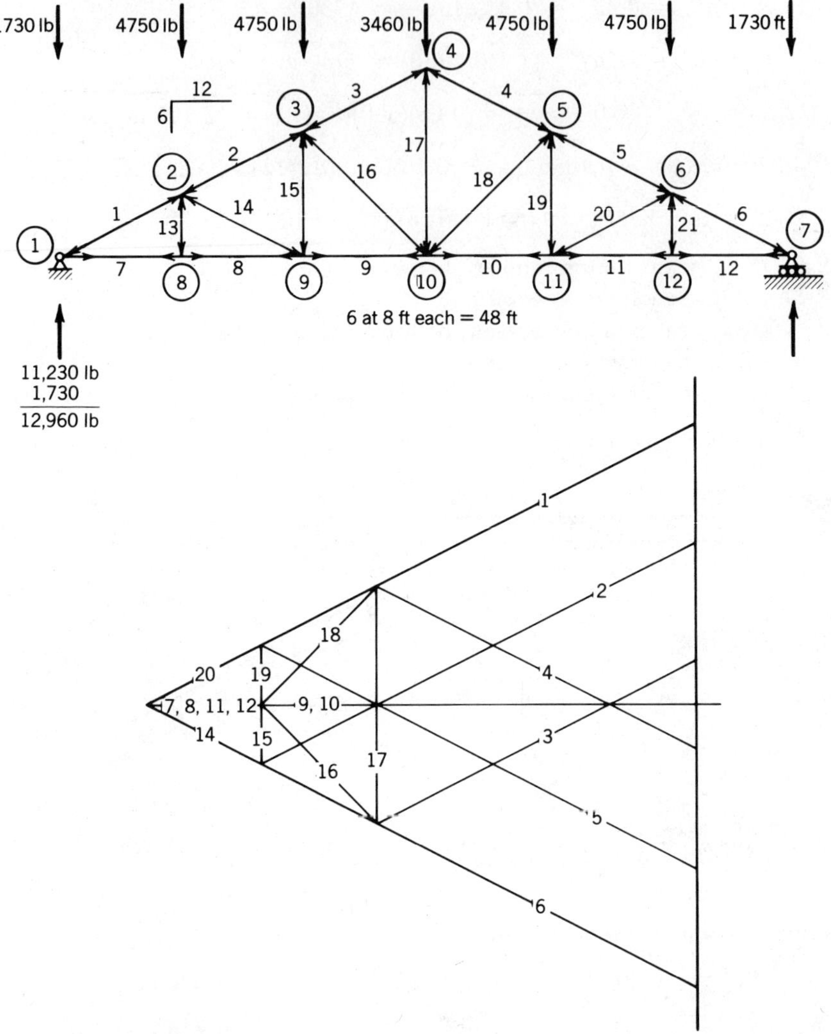

FIGURE 5.66 Graphical solution for dead load plus snow load. DL + SL; symmetrical about centerline; 1 in. = 5000 lb.

Member	Force	Member	Force
1	−25,100	13	0
2	−19,900	14	−5,300
3	−14,700	15	+2,300
7	+22,500	16	−6,700
8	+22,500	17	+9,400
9	+17,800		

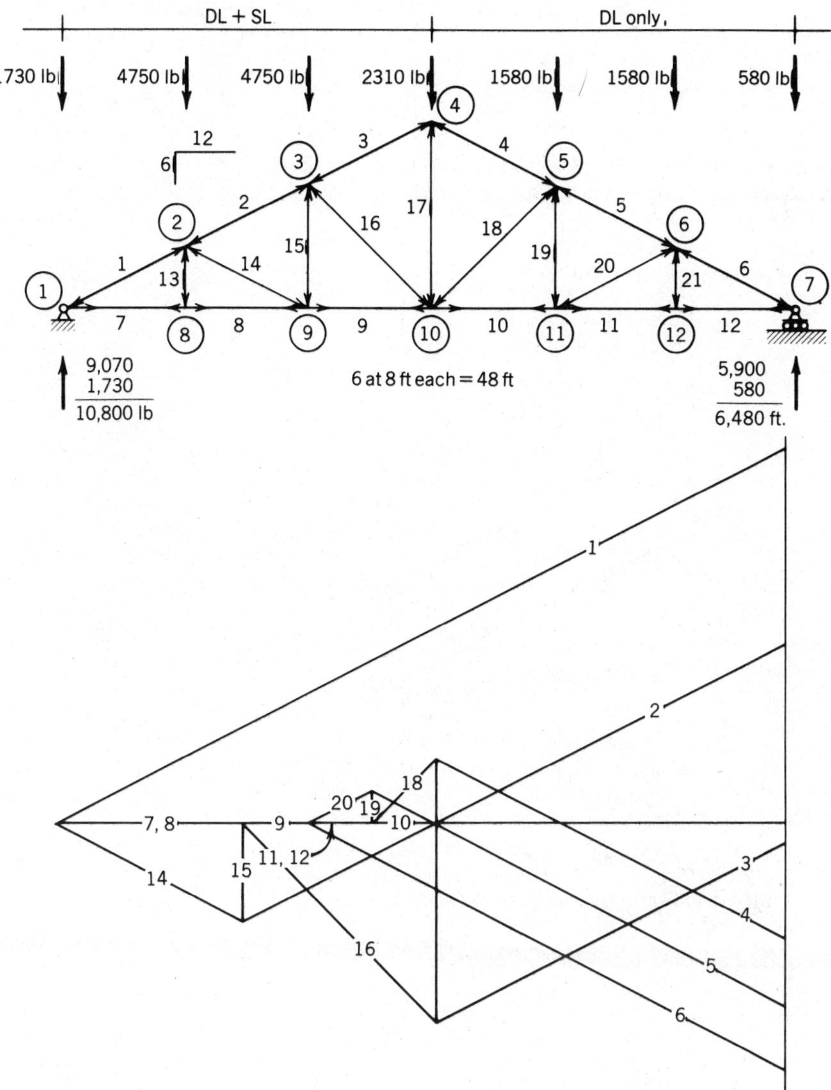

FIGURE 5.67 Graphical solution for dead load plus unbalanced snow load. DL + USL; 1 in. = 3000 lb.

Member	Force	Member	Force
1	−20,300	13	0
2	−15,000	14	−5,300
3	−9,700	15	+2,400
4	−9,700	16	−6,700
5	−11,400	17	+6,300
6	−13,200	18	−2,200
7,8	+18,100	19	+800
9	+13,400	20	0
10	+10,200	21	0
11,12	+11,800		

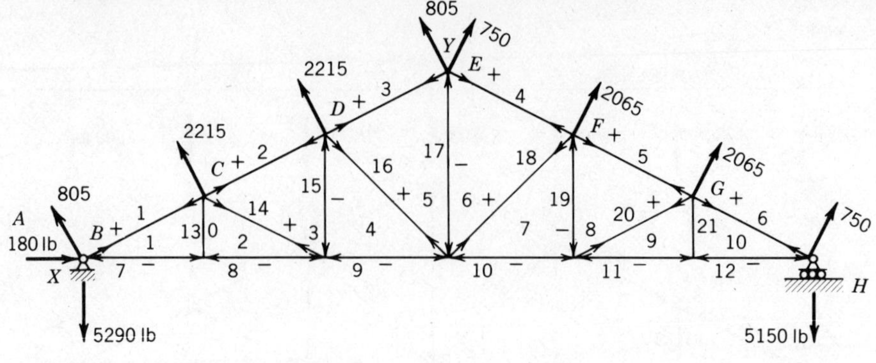

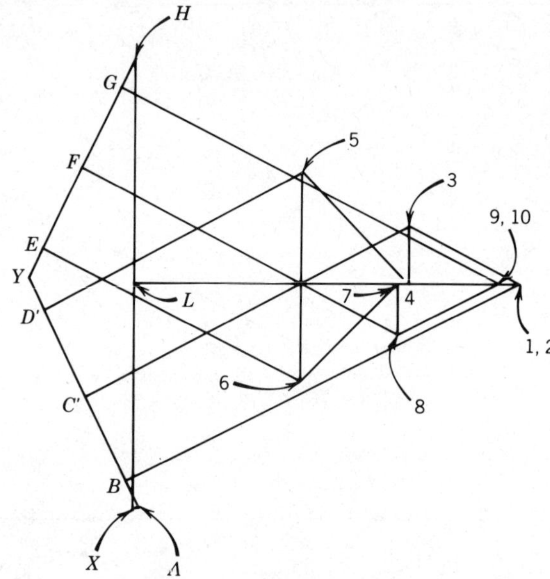

FIGURE 5.68 Graphical solution for dead load plus wind loads. DL + WL.

Forces (lb)

Member	WL	DL	WL + DL
1	+10,300	−8,500	+1,800
2	+8,500	−6,700	+1,800
3	+6,800	−4,900	+1,900
4	+6,800	−4,900	+1,900
5	+8,300	−6,700	+1,600
6	+9,900	−8,500	+1,400
7	−9,000	+7,600	−1,400
8	−9,000	+7,600	−1,400
9	−6,400	+6,000	−400
10	−6,100	+6,000	−100
11	−8,500	+7,600	−900
12	−8,500	+7,600	−900
13	0	0	0
14	+2,900	−1,800	+1,100
15	−1,300	+800	−500
16	+3,500	−2,200	+1,300
17	−4,800	+3,100	−1,700
18	+3,200	−2,200	+1,000
19	−1,200	+800	−400
20	+2,700	−1,800	+900
21	0	0	0

From the preliminary axial forces, it appears that DL + SL will control. Check compression P and bending M in the top chord.

$$P_{TC} = 25{,}100 \text{ lb} \quad (\text{DL} + \text{SL})$$

$$M_{TC} = 41{,}470 \text{ in.-lb} \quad (\text{DL} + \text{SL})$$

Try $5\frac{1}{8} \times 6$ in., $A = 30.75$ in.^{2}, $S = 30.75$ in.^{3}

$$E_x' = E_x = 1{,}700{,}000 \text{ psi}$$

$$F_c'' = F_c C_D = (1600)(1.15) = 1840 \text{ psi}$$

$$F_b' = F_b C_D = (2400)(1.15) = 2760 \text{ psi}$$

$$(l_e/d)_x = \frac{(8.95)(12)}{6} = 17.9$$

$$K_x = 0.671\sqrt{\frac{E_x'}{F_c''}} = 0.671\sqrt{\frac{1{,}700{,}000}{1840}} = 20.4$$

$$J_x = \frac{(l_e/d)_x - 11}{K_x - 11} = \frac{17.9 - 11}{20.4 - 11} = 0.73$$

Since $11 < 17.9 < 20.4$, use intermediate-column formula.

$$F_c' = F_c''\left[1 - \frac{1}{3}\left(\frac{l/d}{K_x}\right)^4\right] = 1840\left[1 - \frac{1}{3}\left(\frac{17.9}{20.4}\right)^4\right] = 1480 \text{ psi}$$

$$f_c = \frac{P_{TC}}{A} = \frac{25{,}100}{30.75} = 815 \text{ psi}$$

$$f_b = \frac{M_{TC}}{S} = \frac{41{,}470}{30.75} = 1350 \text{ psi}$$

$$\frac{f_c}{F_c'} + \frac{f_b}{F_{bx}' - J_x f_c} = \frac{815}{1480} + \frac{1350}{2760 - (0.73)(815)} = 0.55 + 0.62 = 1.17$$

For a preliminary trial size, use $5\frac{1}{8} \times 7\frac{1}{2}$ in. in order to make allowance for a reduction in net section by connectors and to reduce the combined stress to less than one. Note that DL + SL + WL does not control, by observation, for this particular truss design because the wind uplift reduces the DL + SL stresses.

Determine the preliminary size of the bottom chord based on tension :

$$T_{BC} = 22{,}500 \text{ lb } (\text{DL} + \text{SL})$$

$$F_t' = (1250)(1.15) = 1440 \text{ psi}$$

$$A = \frac{T_{BC}}{F_t'} = \frac{22{,}500}{1440} = 16 \text{ in.}^2$$

Try a $5\frac{1}{8} \times 7\frac{1}{2}$-in. section (same size as top chord), which is larger than that cal-

TABLE 5.39

Summary of Preliminary Member Loads Based on Graphical Analysis of Truss with Pinned Joints[a]

	Loading	DL + SL		DL + USL[b]		DL + WL	
	Member	F	M_{max}	F	M_{max}	F	M_{max}
Top chord	1	−25.1	41.47	−20.3	41.47	+1.8	7.8
	2	−19.9	41.47	−15.0	41.47	+1.8	7.8
	3	−14.7	41.47	−9.7	41.47	+1.9	7.8
	4	−14.7	41.47	−9.7	13.82	+1.9	6.4
	5	−19.9	41.47	−11.4	13.82	+1.6	6.4
	6	−25.1	41.47	−13.2	13.82	+1.4	6.4
Bottom chord	7	+22.5	0	+18.1	0	−1.4	0
	8	+22.5	0	+18.1	0	−1.4	0
	9	+17.8	0	+13.4	0	−0.4	0
	10	+17.8	0	+10.2	0	−0.1	0
	11	+22.5	0	+11.8	0	−0.9	0
	12	+22.5	0	+11.8	0	−0.9	0
Webs	13	0	0	0	0	0	0
	14	−5.3	0	−5.3	0	+1.1	0
	15	+2.3	0	+2.4	0	−0.5	0
	16	−6.7	0	−6.7	0	+1.3	0
	17	+9.4	0	+6.3	0	−1.7	0
	18	−6.7	0	−2.2	0	+1.0	0
	19	+2.3	0	+0.8	0	−0.4	0
	20	−5.3	0	−1.8	0	+0.9	0
	21	0	0	0	0	0	0

[a]Units—kips; (+) tension, (−) compression, in.-kips.
[b]USL—unbalanced snow load.

culated because of considerations of net section at connectors and moment that will be induced into the bottom chord due to truss deflection.

Determine the preliminary size of web members:

$$P_{max} = 6700 \text{ lb} \quad (\text{DL} + \text{SL})$$
$$T_{max} = 9400 \text{ lb} \quad (\text{DL} + \text{SL})$$

Try $5\frac{1}{8} \times 6$ in., $A = 30.75$ in.^{2} Check member 16, Figure 5.66.

$$\left(\frac{l_e}{d}\right)y = \frac{(11.3)(12)}{5.125} = 26.5$$

$$F_c'' = (1900)(1.15) = 2190 \text{ psi}$$

$$K_y = 0.671\sqrt{\frac{E_y}{F_c''}} = 0.671\sqrt{\frac{1,700,000}{2190}}$$

$$K_y = 18.7 < 26.5 \quad \text{(use long-column formula)}$$

$$F_c' = \frac{(0.3)(1{,}700{,}000)}{(26.5)^2} = 730 \text{ psi}$$

$$f_c = \frac{6700}{30.75} = 220 \text{ psi} \leqslant 730 \text{ psi} \quad \text{O.K.}$$

$$F_t' = F_t C_D = (1250)(1.15) = 1440 \text{ psi}$$

$$f_t = \frac{9400}{30.75} = 310 \text{ psi} \leqslant F_t' \quad \text{O.K.}$$

Use $5\frac{1}{8} \times 6$ in. as a trial size for all web members considering connections will reduce net sections and a smaller size could present problems in making connections.

Based on these preliminary member sizes, the member forces shown in Table 5.40 were determined using computer analysis.

Design connections on the basis that they will be pinned and will not significantly resist rotations.

Design heel connection. Note that this connection should be as shown in the bottom detail of Figure 5.69 so as to be pinned and have concentric force lines. The upper detail of Figure 5.69 does not have concentric force lines and is not a pinned connection. The restraint caused by the four bolts in each chord will tend to cause splitting in the chord.

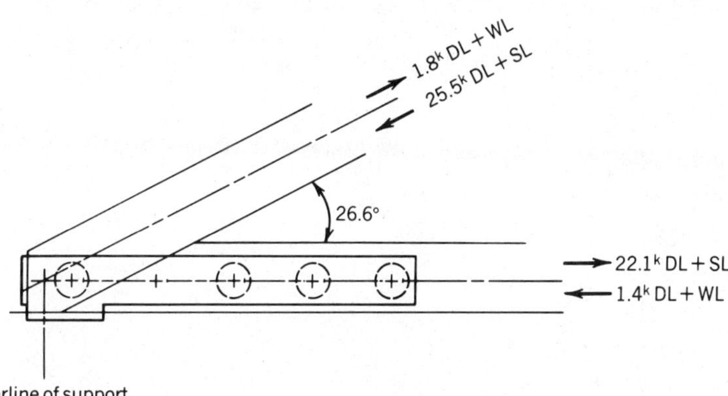

Determine number of connectors in bottom chord. Use 4-in. diameter shear plates. Members are Group B species. From Table 6.12, the design value is 4320 lb for one shear plate. When used with metal side plates this value is increased 11% for Group B species. Value could also be obtained directly from Table 6.13.

Number of shear plates required $= T_{BC}/(4320)(1.11)(1.15)$

$= 22{,}100/(4320)(1.11)(1.15) = 4.0.$

Try six shear plates on three $\frac{3}{4}$-in. bolts.

Try $\frac{1}{4} \times 4\frac{1}{2}$ in. side plate. See (14) for design of steel plates and definition of terms.

TABLE 5.40

Summary of Forces and Moments[a]

Joint Condition		Pinned				Fixed			
Loading		DL + SL		DL + USL[b]		DL + SL		DL + USL[b]	
	Member	*F*	*M*	*F*	*M*	*F*	*M*	*F*	*M*
Top chord	1	−25.5	29.6	−20.7	31.6	−25.6	29.2	−20.9	31.5
	2	−20.7	39.9	−15.9	40.5	−20.6	33.0	−15.8	35.4
	3	−15.6	39.9	−10.8	40.5	−15.7	39.8	−10.9	33.4
	4	−15.6	39.9	−10.0	12.6	−15.7	39.8	−10.1	21.7
	5	−20.7	39.9	−11.7	12.6	−20.6	33.0	−11.7	8.7
	6	−25.5	29.6	−13.3	7.8	−25.6	29.2	−13.3	7.5
Bottom chord	7	+22.1	6.9	+17.8	5.7	+22.1	12.0	+17.9	12.3
	8	+22.1	6.9	+17.8	5.7	+22.0	5.4	+17.7	4.4
	9	+17.7	0.5	+13.4	0.4	+17.6	2.8	+13.3	2.4
	10	+17.7	0.5	+10.2	0.3	+17.6	2.8	+10.1	1.3
	11	+22.1	6.9	+11.6	3.5	+22.0	5.4	+11.6	2.8
	12	+22.1	6.9	+11.6	3.5	+22.1	12.0	+11.6	3.8
Webs	13	−0.14	0	−0.11	0	−0.04	4.6	−0.06	5.1
	14	−4.9	0	−5.0	0	−4.9	2.7	−4.9	1.8
	15	+2.2	0	+2.3	0	+2.2	3.8	+2.2	2.2
	16	−6.7	0	−6.7	0	−6.1	1.2	−6.3	0.4
	17	+9.5	0	+6.3	0	+8.7	0	+5.8	4.3
	18	−6.7	0	−2.2	0	−6.1	1.2	−1.8	1.3
	19	+2.2	0	+0.73	0	+2.2	3.8	+0.7	2.9
	20	−4.9	0	−1.6	0	−4.9	2.7	−1.6	1.8
	21	−0.14	0	−0.07	0	−0.04	4.6	−0.02	0.9

[a]Units—kips; (+) tension, (−) compression, in.-kips.

[b]USL—unbalanced snow load.

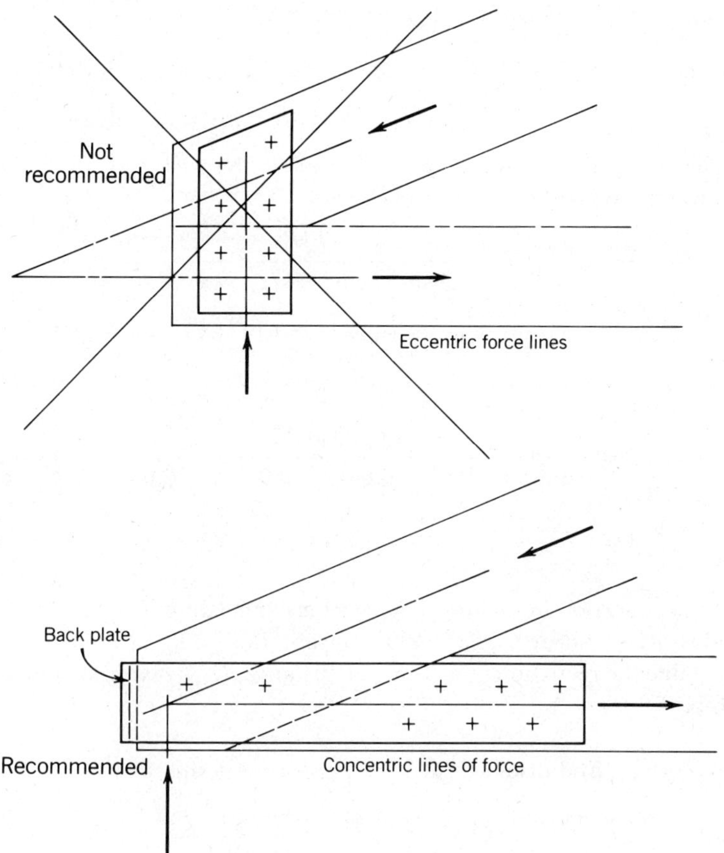

FIGURE 5.69 Truss heel connection.

$$A_{net} = (0.25)(2)(4.5 - 0.81) = 1.85 \text{ in.}^2,$$

$$A_{gross} = (0.25)(2)(4.5) = 2.25 \text{ in.}^2$$

$$\text{Capacity} = 0.50\ F_u A_{net} = (29{,}000)(1.85)$$

$$= 53{,}650 \text{ lb} > 22{,}100 \text{ lb} \quad \text{O.K.}$$

$$= 0.60\ F_y A_{gross} = (22{,}000)(2.25)$$

$$= 49{,}500 \text{ lb} > 22{,}100 \text{ lb} \quad \text{O.K.}$$

Check the capacity of the connection considering reduction for fasteners in a row.

$$A_1 = (5.125)(7.5) = 38.4 \text{ in.}^2$$

$$A_2 = (0.25)(4.5)(2) = 2.25 \text{ in.}^2$$

$$\frac{A_1}{A_2} = \frac{38.4}{2.25} = 17.1$$

From Table 6.8, group action factor, C_g, is 0.96.

$$\text{Capacity of six 4-in. shear plates} = (6)(4320)(1.11)(1.15)(0.96)$$
$$= 31{,}800 \text{ lb} > 22{,}100 \text{ lb} \quad \text{O.K.}$$

Check the end grain bearing of the top chord on the back plate of the heel connection using the Hankinson formula

$$F'_n = F'_{c-26.6°} = \frac{F'_g F'_{c\perp}}{F'_g \sin^2 26.6° + F'_{c\perp} \cos^2 26.6°}$$

$$F_g = 2300 \text{ psi} \quad \text{(from Table A-1 AITC 117—Design)}$$

$$F_{c\perp} = 560 \text{ psi}$$

$$\text{F}_{c-26.6°} = \frac{(2300)(1.15)(560)}{(2300)(1.15)\sin^2 26.6° + (560)\cos^2 26.6°} = 1540 \text{ psi}$$

$$f_c = \frac{22{,}100}{(5.125)(4.5)} = 960 \text{ psi} < 1540 \text{ psi} \quad \text{O.K.}$$

Check the connection in the top chord for tension loading. Capacity of one $\frac{3}{4}$-in. bolt and two 4-in. shear plates is 4800 lb from Table 6.13.

Determine the reduction for a 5-in. end distance, C_n, by interpolation of values from Table 6.15.

End distance (in.)	Percent of design load
7	100
$3\frac{1}{2}$	62.5

$$C_n = \left[\frac{(1.5)}{(3.5)}\right] 0.375 + 0.625 = 0.79$$

$$P = (4800)(1.33)(0.79)(2) = 10{,}100 \text{ lb} > 1800 \text{ lb} \quad \text{(DL + WL)}$$

The bottom connection plate should be checked for buckling. If critical, a tie bolt can be added near the end of the bottom plate.

The truss chord to the web connections should be designed as shown in the bottom detail of Figure 5.70. Web members should be fastened to the side plates with a minimum of two bolts. Determine the number and size of bolts in the ends of web members.

Maximum web forces:

$$\text{Tension} = +9500 \text{ lb} \quad \text{(DL + SL)}$$
$$\text{Compression} = -6700 \text{ lb} \quad \text{(DL + SL)}$$

The capacity of one $\frac{3}{4}$-in. bolt parallel to grain is 3580 lb from Table 6.21. This value can be increased by 1.15 for duration of load for snow.

$$P' = (3580)(1.15) = 4120 \text{ lb}$$

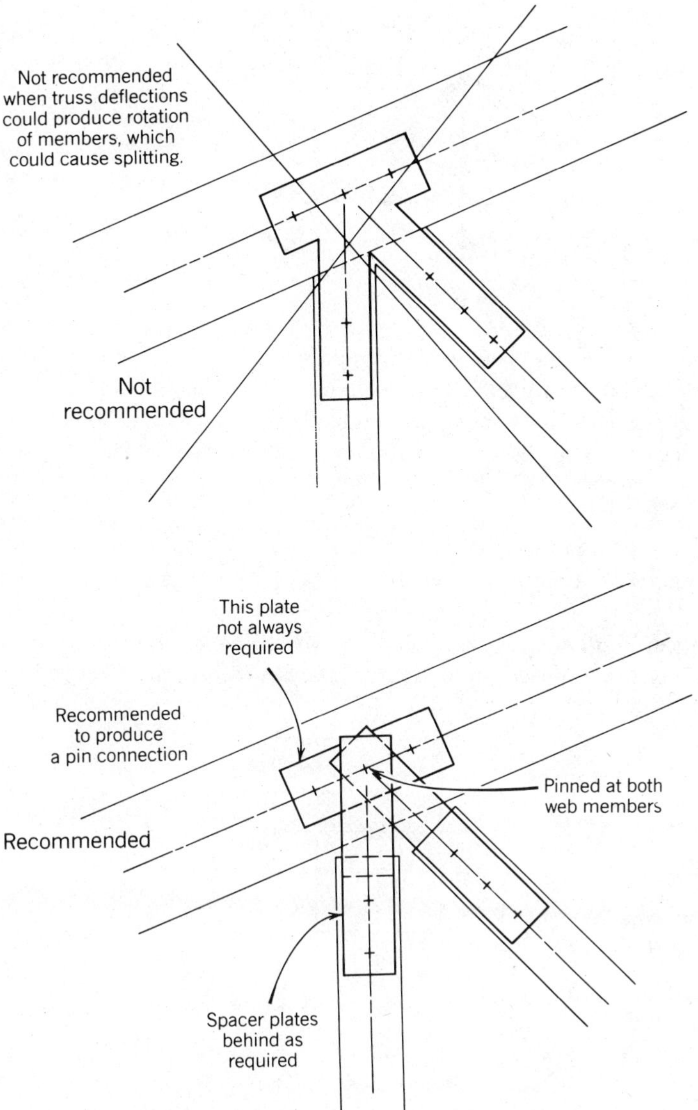

FIGURE 5.70 Truss web and chord connection.

Two $\frac{3}{4}$-in. bolts are adequate for the maximum forces except for the ends of member 17, which requires three bolts. Use two $\frac{3}{4}$-in. bolts in the end of each web member (three $\frac{3}{4}$-in. bolts for member 17).

Spacing should be 4 times the bolt diameter (3 in.) and minimum end distance should be 7 times the bolt diameter ($5\frac{1}{4}$ in.). Check the connection between webs 15 and 16 and the top chord at point 3.

$$\text{Perpendicular to grain load} = 6360 - 2060 = 4300 \text{ lb}$$

$$\text{Parallel to grain load} = 1030 + 2120 = 3150 \text{ lb}$$

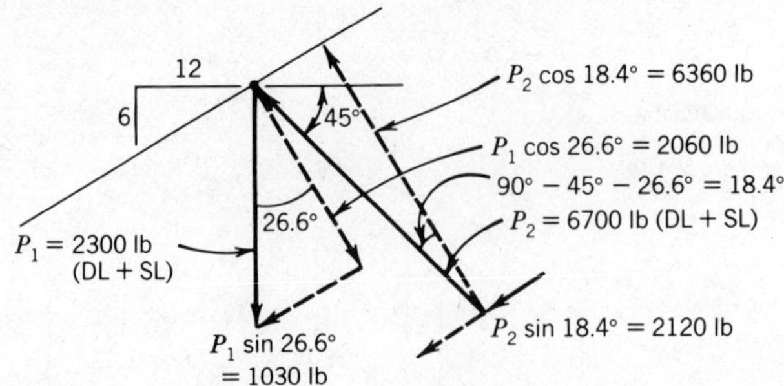

The capacity of one $\frac{3}{4}$-in. bolt perpendicular to the grain is 1780 lb. Use three $\frac{3}{4}$-in. bolts in the top chord as shown in the bottom detail of Figure 5.70. The capacity of one 4-in. shear plate perpendicular to the grain is 3000 lb. Two 4-in. shear plates on a $\frac{3}{4}$-in. bolt = (2)(3000) = 6000. Increase 15% for duration of load,

$$P' = (2)(3000)(1.15) = 6900 > 4300 \text{ lb}$$

An alternate would be to use two 4-in. shear plates on a $\frac{3}{4}$-in. bolt for this connection.

Check the member sizes based on reduced member cross section at connectors. Use member forces based on computer analysis with pinned joints.

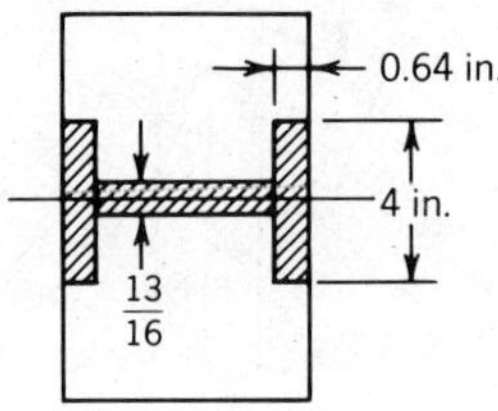

Check member 1:

$$P = 25{,}500 \text{ lb} \quad (\text{DL} + \text{SL})$$

$$M = 29{,}600 \text{ in.-lb} \quad (\text{DL} + \text{SL})$$

$$A_{\text{net}} = 38.44 - 8.2 = 30.2 \text{ in.}^2$$

$$S_{\text{net}} = \tfrac{1}{6}\,[(5.125)(7.5)^2 - (2)(0.64)(4)^2 - (3.84)(0.81)^2] = 44.2 \text{ in.}^3$$

$$\left(\frac{l}{d}\right)_x = \frac{(8.95)(12)}{(7.5)} = 14.3$$

$$K_x = 0.671\sqrt{\frac{1{,}700{,}000}{1840}} = 20.4$$

Since 11 < 14.3 < 20.4, use intermediate-column formula.

$$F_c' = 1840\left[1 - \frac{1}{3}\left(\frac{14.3}{20.4}\right)^4\right] = 1690 \text{ psi}$$

$$f_c = \frac{25{,}500}{30.2} = 844 \text{ psi}$$

$$f_{bx} = \frac{29{,}600}{44.2} = 670 \text{ psi}$$

$$J_x = \frac{14.3 - 11}{20.4 - 11} = 0.35$$

$$\frac{f_c}{F_c'} + \frac{f_b}{F_b' - J_x F_c} = \frac{844}{1690} + \frac{670}{2760 - (0.35)(844)}$$

$$= 0.50 + 0.27 = 0.77 \quad \text{O.K.}$$

Check member 2:

$$f_c = \frac{20{,}700}{30.2} = 685 \text{ psi}$$

$$f_b = \frac{39{,}900}{44.2} = 903 \text{ psi}$$

$$\frac{685}{1690} + \frac{903}{2760 - (0.35)(685)} = 0.41 + 0.36 = 0.77 \quad \text{O.K.}$$

Check member 7:

$$T = 22{,}100 \text{ lb} \quad (\text{DL} + \text{SL})$$

$$M = 6900 \text{ in.-lb} \quad (\text{DL} + \text{SL})$$

$$f_t = \frac{22{,}100}{30.2} = 732 \text{ psi}$$

$$f_b = \frac{6900}{44.2} = 156 \text{ psi}$$

$$\frac{732}{(1250)(1.15)} + \frac{156}{(1600)(1.15)} = 0.51 + 0.08 = 0.59 \quad \text{O.K.}$$

Check compression in bottom chord taking member forces from graphical analysis of DL + WL:

$$P = 1400 \text{ lb} \quad (\text{DL} + \text{WL})$$

$$\left(\frac{l}{d}\right)_y = \frac{(16)(12)}{(5.125)} = 37.5$$

$$F_c'' = (1900)(1.33) = 2530.$$

$$K_y = 0.671\sqrt{\frac{1{,}700{,}000}{2530}} = 17.4$$

Since 37.5 > 17.4, use long-column formula.

$$F_c' = \frac{(0.3)(1{,}700{,}000)}{(37.5)^2} = 363 \text{ psi}$$

$$f_c = \frac{1400}{30.2} = 46 \text{ psi} < 363 \text{ psi} \quad \text{O.K.}$$

Check members 16 and 18:

$$P = 6700 \text{ lb} \quad (\text{DL} + \text{SL})$$

From preliminary size check, web members of $5\frac{1}{8} \times 6$ in. are O.K.

Check member 17:

$$T = 9500 \text{ lb} \quad (\text{DL} + \text{SL})$$

For a $5\frac{1}{8} \times 6$ in. member, $A_{net} = 30.75 - 8.2 = 22.6 \text{ in.}^2$

$$F_t' = 1250(1.15) = 1440 \text{ psi}$$

$$f_t = \frac{9500}{22.6} = 420 \text{ psi} < 1440 \text{ psi} \quad \text{O.K.}$$

All other web members of $5\frac{1}{8} \times 6$ in. are O.K. by comparison with above check.

At this point, adjustment of sizes for minimum design should be considered. If adjustments are made, the stiffness analysis should be recalculated since some redistribution of forces and moments will likely occur. Also, after connection designs are completed, member net sections should be reviewed and compared to those used in member designs. If pinned connections are not used, the forces and moments from the "fixed-joint" analysis must be used in both the design of members and conncctions.

POLE-TYPE FRAMING

Pole-type frame structures generally consist of preservatively treated round timber poles set in the ground as the main upright supporting members. Preservatively treated sawn timber or glued laminated timber posts may also be used. The hole in which the pole or post is set furnishes both vertical and horizontal support for the structure. Setting the pole in the ground tends to prevent rotation of the bottom of the pole, thereby providing some or all of the required bracing. The roof portion of the structure is usually framed with lumber or glued laminated timber. Preservative treatment of the poles or posts that are set in the ground should be in accordance with the *American Wood-Preservers' Association Standards* (16). Use AWPA C1 for all timber products, C4 for round poles, C2 for sawn timber posts, and C28 for glued laminated timber posts.

General Considerations

General considerations applicable to all pole-type frame structures include the following.

1. A bracing system can be provided at the top of a pole in order to reduce bending moments at the base of the pole and to distribute loads. The design of

buildings supported by poles without bracing requires good knowledge of soil conditions in order to eliminate excessive deflection or sidesway.

2. Bearing values under butt ends of poles should be checked. It is common practice to backfill holes around poles with well-tamped native soil, sand, or gravel. Backfilling with concrete or soil-cement can develop a more effective pole diameter; consequently, it can be used as a means of reducing required depth of embedment. Concrete backfill also increases the area of the pole for skin friction and this increases the bearing capacity. Skin friction is also effective where uplift due to wind may act on a pole through its connections to the roof framing.

3. In order to increase bearing capacity under pole butts, concrete footings may be used. If they are used, they should be designed to withstand the punching shear of the pole and load. Concrete footings should be considered even in firm soils such as dry hard clay, coarse firm sand, or gravel.

The intended use of the structure largely determines such general features as height, overall length and width, spacing of poles, height at eaves, type of roof framing, and the kind of flooring to be used, as well as any special features such as wide bays, unsymmetrical layouts, or the possible suspending of particular loads from the roof framing. These general design features having been determined, the following procedure may be used.

Design Values for Poles

In the absence of other information, the design values for timber piles based on single-member use are suggested for use in pole design. Pole frame buildings may be used for purposes (such as farm buildings) where the importance of the structure is such that higher design values are justified. The designer should check the applicable building code for the appropriate design values.

The design values in Table 5.41 are consistent with other wood design values in regard to safety factors. They are based on the provisions of ASTM D 3200, *Round Timber Construction Poles* (17) and include reduction factors for conditioning prior to treatment in accordance with ASTM D 1760-76 as well as safety factors for single-member use. When air- or kiln-dried conditioning only is used prior to treatment, the seasoning conditioning modification factor, C_{co}, is applied to all design values except modulus of elasticity. For Southern Pine, the value of C_{co} is 1.18 and for other species 1.11. The design values for compression perpendicular to grain, $F_{c\perp}$, are calculated by the same method used for sawn lumber based on wet conditions of use and further modified for conditioning prior to treatment. The design values for shear F_v are based on wet conditions of use with the reduction for conditioning prior to treatment applied. The design values for fastenings are the same as for sawn lumber. For bolts with l/d values other than those shown in Table 5.41, the design values may be determined by interpolation.

Design Procedures

1. Determine loads. The principal load on a pole-type frame is generally the horizontal wind load; therefore, the wind load value required by the governing building code or, in the absence of a governing code, as determined from ANSI

TABLE 5.41

Design Values for Preservatively Treated Poles[a,b], Wet and Dry Conditions of Use

Species	Bending F_b (psi)	Compression Parallel to Grain F_c (psi)	Modulus of Elasticity E (psi)	Compression Perpendicular to Grain $F_{c\perp}$ (psi)	Horizontal Shear F_v (psi)	End Grain in Bearing[c] F_g (psi)
Douglas Fir, Coast	1,850	1,000[d]	1,500,000	375	115	1,200
Jack Pine	1,500	800	1,100,000	280	95	870
Lodgepole Pine	1,350	700	1,100,000	240	85	870
Northern White Cedar	1,050	525	600,000	225	80	670
Ponderosa Pine	1,300	650	1,000,000	320	90	820
Red or Norway Pine	1,450	725	1,300,000	265	85	790

Southern Pine	1,700	900[d]	1,500,000	320	105	1,120
Western Red Cedar	1,350	750	900,000	255	95	940
Western Hemlock	1,650	900	1,300,000	245	115	1,120
Western Larch	2,050	1,075	1,500,000	375	120	1,210

[a]The design values are based on ASTM D 2899-74, *Method for Establishing Design Stresses for Round Timber Piles* (22). The values are for single-member uses of poles and assume that the conditioning prior to treatment was in accordance with ASTM D 1760-76. If the poles are conditioned by air or kiln drying only prior to treatment, the design values, with the exception of modulus of elasticity, may be increased by application of the modifying factor for seasoning conditioning, C_{co}, which is 1.18 for Southern Pine and 1.11 for other species.

[b]The compression perpendicular to grain design values $F_{c\perp}$ are based on the design values for sawn lumber under wet conditions of use, which have been reduced for conditioning. For dry conditions of use, multiply by 1.5.

[c]End grain in bearing design values, F_g, are based on the design values for sawn lumber used in a wet location and reduced for conditioning. When used in a dry location, they may be increased by 10% ($C_M = 1.10$).

[d]Design values for compression parallel to grain, F_c, in Douglas Fir and Southern Pine may be increased 0.2% for each foot of length from the tip of the pole to the critical section. This increase shall not exceed 10%. The modification factor for location of critical section, $C_{cs} = 1 + 0.002L$, where L = distance (ft) from the tip to the critical section. C_{cs} shall not exceed 1.10.

A58.1 (13), should be used. In addition to dead load and other loads, roofs will transmit vertical components of wind load to poles. When wind loads are involved, design values may be increased one-third for duration of loading.

Determine the resultant of the horizontal forces acting on the poles and the height above the ground surface at which it acts. ANSI A58.1 requires determination of windward loading and a leeward loading of the structure. When glued laminated timber or other roofing systems of similar rigidity are used, the tops of the poles deflect approximately the same amount. Because the leeward forces may be less than the windward forces, the leeward pole or poles tend to support the windward pole. When only two poles are used in a bent, the following method can be used to determine the forces acting on the poles. In this method, it is assumed that the wind loads are applied through girts that approximate a uniform load on the poles. If not, the actual loading condition should be used.

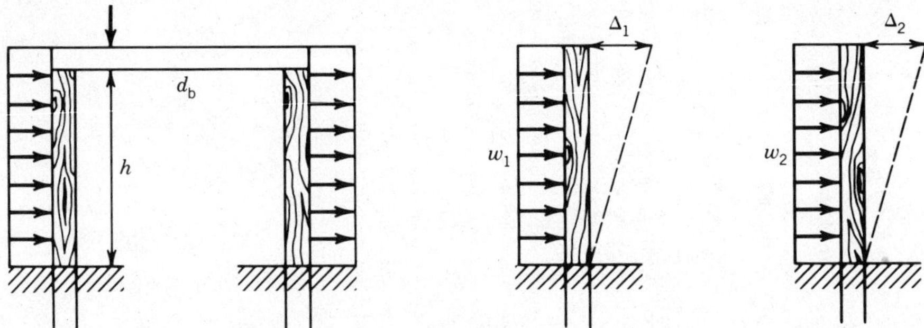

(a) The top of both poles can be assumed to deflect the same amount. First assume that the poles are not connected by the roof structure and each is allowed to deflect separately. Let the deflection of the windward pole $= \Delta_1$, and the deflection of the leeward pole $= \Delta_2$. Next calculate the force P_3 that will move the windward pole deflection Δ_3 to the left and the leeward pole deflection Δ_3 to the right so that $\Delta_A = \Delta_B$. Next determine the resultant P on the windward pole of the wind load and its distance h from the ground line.

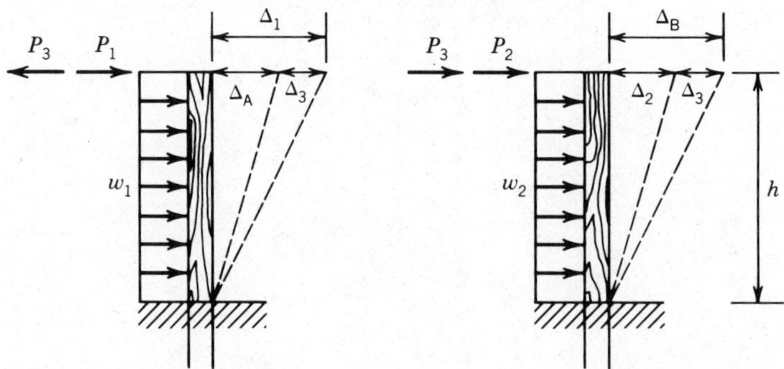

Determine the wind load acting on the structure above the top of the poles, P_1 and P_2:

$$P_1 = w_1 d_b, \qquad P_2 = w_2 d_b$$

Determine P_3:

$$\Delta_1 = \frac{w_1 h^4}{8EI} + \frac{w_1 d_b h^3}{3EI}$$

$$\Delta_2 = \frac{w_2 h^4}{8EI} + \frac{w_2 d_b h^3}{3EI}$$

$$\Delta_3 = \frac{P_3 h^3}{3EI}$$

where w_1 = wind load on windward side (plf),
w_2 = wind load on leeward side (plf),
d_b = depth of glued laminated beam, plus any additional roof structure above beam (ft),
EI = the stiffness modulus, which is assumed to be the same for each pole, and
h = length of pole below beam (ft).

$$\Delta_A = \Delta_1 - \Delta_3 \qquad \Delta_B = \Delta_2 + \Delta_3 \qquad \Delta_A = \Delta_B$$

therefore,

$$\Delta_1 - \Delta_3 = \Delta_2 + \Delta_3, \quad \frac{w_1 h^4}{8EI} + \frac{w_1 d_b h^3}{3EI} - \frac{P_3 h^3}{3EI} = \frac{w_2 h^4}{8EI} + \frac{w_2 d_b h^3}{3EI} + \frac{P_3 h^3}{3EI}$$

Rearranging terms and solving for P_3,

$$P_3 = \frac{3h(w_1 - w_2)}{16} + d_b\left(\frac{w_1 - w_2}{2}\right) \qquad (5\text{-}94)$$

Determine the resultant of the forces on the windward pole, P, and the height above ground line, h', at which it acts.

$$P = w_1 h + P_1 - P_3$$

$$h' = \frac{(w_1 h^2/2) + (P_1 - P_3)h}{P} \qquad (5\text{-}95)$$

(b) When knee braces are used, the deflection at the points of contraflexure can be assumed to be equal with only a small error. The forces acting on the windward pole at and below the point of contraflexure are calculated in a manner similar to that for poles without knee braces.

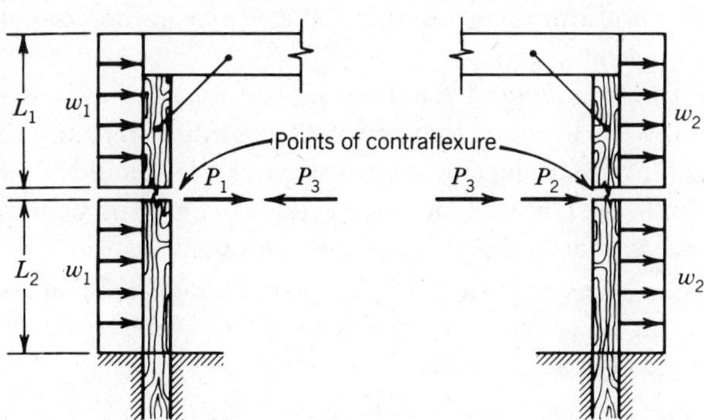

$$P_1 = w_1 L_1$$

$$P_3 = \frac{L_1(w_1 - w_2)}{2} + \frac{3L_2(w_1 - w_2)}{16}$$

$$P = w_1(L_2 + L_1) - P_3$$

$$h' = \frac{(P_1 - P_3)\,L_2 + (w_1 L_2^2)}{P}$$

The following combinations of loads are recommended for roofs unless specified otherwise by the applicable building code: dead load plus snow load, dead load plus wind load, and deal load plus wind load plus one-half snow load.

2. Determine soil values. If soil tests are not available, visual inspection and a careful estimate of the bearing value of samples of the soil should be made. Building codes generally contain allowable design values for direct bearing assigned to various soil classifications.

When building codes do not give established allowable lateral passive soil pressure values, the values may be determined by using Rankin's formula:

$$p = \frac{1 + \sin\phi}{1 - \sin\phi}\,wd$$

where p = allowable passive soil pressure (psf per ft of depth),
ϕ = angle of internal friction of soil (degrees),
w = weight of soil (pcf), and
d = depth below grade (ft).

Allowable soil pressure values for various classes of soil are also given in Table 5.42.

3. Estimate the size of pole required. The American National Standards Institute (ANSI) has established certain pole classes. Table 5.43 tabulates the size requirements corresponding to ANSI classes H-6 through H-10 for Douglas Fir and Southern Pine poles. The size requirements for other species may be obtained from ANSI 05.1 (13).

Poles of a given class and length are selected to have approximately the same load-carrying capacity regardless of species. The minimum circumferences specified at 6 ft from the butt in Table 5.43 are based on the maximum fiber stress in bending that will occur at the groundline due to a given horizontal load applied 2 ft from the top of the pole.

The size of a pole is generally influenced more by bending stresses than by compressive stresses. However, these stresses are interactive and the interaction formula for combined bending and compression stresses should be used in the final design. To estimate the pole size, determine the size based on bending alone, then increase that size based on the effect of the compressive stress.

Circumferences at points other than those tabulated in Table 5.43 may be determined assuming an average taper of 0.25 in. circumference per foot of length of pole for Douglas Fir and Southern Pine.

4. Determine required embedment of pole. Equation (5-96) may be used in

TABLE 5.42

Allowable Foundation and Lateral Pressure[a]

Class of Materials[b]	Allowable Foundation Pressure S_B (psf)[c]	Lateral Bearing of Depth below Natural Grade S_0 (psf)[d]
1. Massive crystalline bedrock	4000	1,200
2. Sedimentary and foliated rock	2000	400
3. Sandy gravel and/or gravel (GW and GP)	2000	200
4. Sand, silty sand, clayey sand, silty gravel, and clayey gravel, (SW, SP, SM, SC, GM, and GC)	1,500	150
5. Clay, sandy clay, silty clay, and clayey silt (CM, ML, MH, and CH)	1000[e]	100

[a]Source: Reproduced from *The Uniform Building Code*, 1982 ed., Copyright 1982, with permission of the International Conference of Building Officials.

[b]For soil classifications such as organic clays and peat, a foundation investigation shall be required.

[c]All values of allowable foundation pressure are for footings having a minimum width of 12 in. and a minimum depth of 12 in. into natural grade. Except as in footnote *e* below, an increase of 20% is allowed for each additional foot of width and/or depth to a maximum value of 3 times the designated value.

[d]May be increased by the amount of the designated value for each additional foot of depth to a maximum of 15 times the designated value. Isolated poles for uses such as flagpoles or signs and poles used to support buildings that are not adversely affected by a $\frac{1}{2}$-in. motion at ground surface due to short-term lateral loads may be designed using lateral bearing values equal to 2 times the tabulated values.

[e]No increase for width is allowed.

determining required embedment depth where no constraint (such as a rigid floor or surface pavement) is provided at the ground surface. Note that this procedure requires the assumption of an initial embedment depth for determining lateral soil bearing pressure.

$$d = \frac{A}{2}\left(1 + \sqrt{\frac{4.36h}{A}}\right) \quad \text{(see Fig. 5.71)} \tag{5-96}$$

where d = depth of embedment in earth (ft),

$A = 2.34P/S_1B$,

P = applied horizontal force on pole (lb),

S_1 = allowable lateral soil-bearing pressure as shown in Table 5.42 based on a depth of one-third the depth of embedment but not over 12 ft. For purposes of determining lateral pressure (psf), $S_1 = S_0 d/3$,

h = height above groundline at which force P is applied (ft), and

B = butt diameter of pole, diagonal of square pole, or diameter of concrete casing (ft).

TABLE 5.43

Dimensions of Douglas Fir (Both Types) and Southern Pine Poles[a,b]

Class		H-6	H-5	H-4	H-3	H-2	H-1	1	2	3	4	5	6	7	9	10
Minimum Circumference at Top (in.)		39	37	35	33	31	29	27	25	23	21	19	17	15	15	12
Length of Pole (ft)	Groundline Distance from Butt[c] (ft)	Minimum Circumference at 6 ft from Butt (in.)														
20	4.0							31.0	29.0	27.0	25.0	23.0	21.0	19.5	17.5	14.0
25	5.0							33.5	31.5	29.5	27.5	25.5	23.0	21.5	19.5	15.0
30	5.5							36.5	34.0	32.0	29.5	27.5	25.0	23.5	20.5	
35	6.0					43.5	41.5	39.0	36.5	34.0	31.5	29.0	27.0	25.0		
40	6.0			51.0	48.5	46.0	43.5	41.0	38.5	36.0	33.5	31.0	28.5			
45	6.5	58.5	56.0	53.5	51.0	48.5	45.5	43.0	40.5	37.5	35.0	32.5	30.0			
50	7.0	61.0	58.5	55.5	53.0	50.5	47.5	45.0	42.0	39.0	36.5	34.0				
55	7.5	63.5	60.5	58.0	55.0	52.0	49.5	46.5	43.5	40.5	38.0					

60	8.0	65.5	62.5	59.5	57.0	54.0	51.0	48.0	45.0	42.0	39.0
65	8.5	67.5	64.5	61.5	58.5	55.5	52.5	49.5	46.5	43.5	40.5
70	9.0	69.0	66.5	63.5	60.5	57.0	54.0	51.0	48.0	45.0	41.5
75	9.5	71.0	68.0	65.0	62.0	59.0	55.5	52.5	49.0	46.0	
80	10.0	72.5	69.5	66.5	63.5	60.0	57.0	54.0	50.5	47.0	
85	10.5	74.5	71.5	68.0	65.0	61.5	58.5	55.0	51.5	48.0	
90	11.0	76.0	73.0	69.5	66.5	63.0	59.5	56.0	53.0	49.0	
95	11.0	77.5	74.5	71.0	67.5	64.5	61.0	57.0	54.0		
100	11.0	79.0	76.0	72.5	69.0	65.5	62.0	58.5	55.0		
105	12.0	80.5	77.0	74.0	70.5	67.0	63.0	59.5	56.0		
110	12.0	82.0	78.5	75.0	71.5	68.0	64.5	60.5	57.0		
115	12.0	83.5	80.0	76.5	72.5	69.0	65.5	61.5	58.0		
120	12.0	85.0	81.0	77.5	74.0	70.0	66.5	62.5	59.0		
125	12.0	86.0	82.5	78.5	75.0	71.0	67.5	63.5	59.5		

[a]This material is reproduced with permission from American National Standard *Specifications and Dimensions for Wood Poles*, ANSI 05.1, copyright 1979 by the American National Standards Institute. Copies of this standard may be purchased from the American National Standards Institute at 1430 Broadway, New York, N.Y. 10018.

[b]Douglas Fir includes Douglas Fir, Interior North, and Douglas Fir, Coastal.

[c]The figures in this column are intended for use only when a definition of groundline is necessary in order to apply requirements relating to scars, straightness, etc.

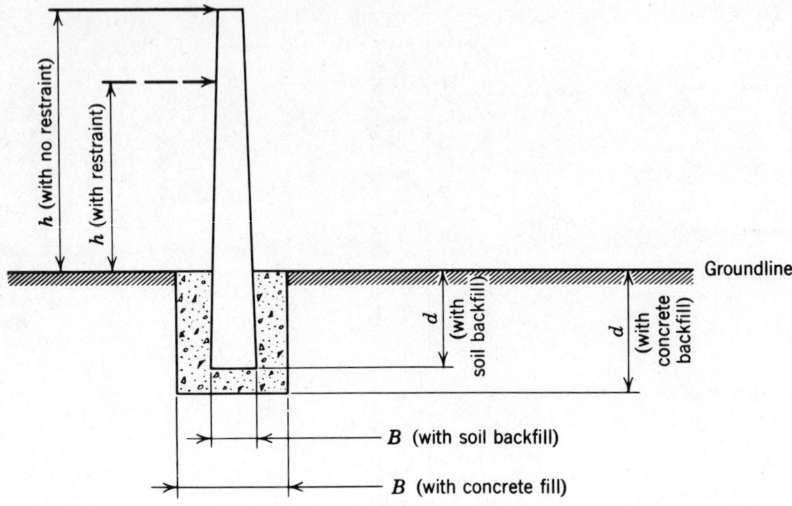

FIGURE 5.71 Depth of embedment.

If the pole is restrained at the groundline, such as by a rigid concrete floor, the following formula is used:

$$d = \sqrt{\frac{4.25\,Ph}{S_3 B}} \tag{5-97}$$

where S_3 = allowable lateral soil-bearing pressure as set forth in Table 5.42 based on a depth equal to the depth of embedment (psf)

and other terms are as previously defined ($S_3 = S_0 d$).

It is sometimes more convenient to use the formulas in the form

$$d = 1.97\sqrt[3]{\frac{P(h + 0.93\,d)}{BS_4}} \quad \text{(for unrestrained footings)} \tag{5-98}$$

and

$$d = 1.63\sqrt[3]{\frac{Ph}{BS_4}} \quad \text{(for restrained footings)} \tag{5-99}$$

where S_4 = tabular value in Table 5.42 for lateral soil pressure S_0 adjusted for duration of load and allowable movement (psf) and other terms are as previously defined.

Note that when buildings are not adversely affected by a $\frac{1}{2}$-in. movement at the ground surface due to short-term lateral loads, the tabular design value of S_0 may be doubled.

For poles having some degree of fixity or restraint at the eave line, such as provided by knee braces, the point at which the applied horizontal force P acts is at the point of contraflexure of the pole. For round tapered poles, the point of contraflexure is assumed to be at two-thirds the distance from the groundline to the point of restraint. For poles with no restraint, h is calculated as shown in Figure 5.71.

It may be desirable to repeat the procedure for determining embedment with the first determination of embedment as the basis for recalculating the value of S_1 or S_3.

5. Check pole bending stress. The actual bending stress in the pole at the point of maximum moment may be determined by the formula

$$f_b = \frac{32\pi^2 M}{C^3} = \frac{32M}{\pi D^3} \tag{5-100}$$

where f_b = actual bending stress (psi),
C = pole circumference at point of maximum moment (in.),
D = pole diameter at point of maximum moment (in.), and
M = maximum moment (in.-lb), which, for a pole embedded in earth, is assumed to occur at one-fourth the depth of embedment below groundline; it may be determined if P and h, as previously defined, are known. For a pole embedded in concrete, the maximum moment can be assumed to act at the top of the concrete.

The actual bending stress cannot exceed design values in bending in Table 5.41. These tabular design values may be increased one-third for wind duration of loading.

When knee braces are used, the bending moment above the pole's point of inflection is maximum at the bottom of the roof bracing. The bending stress at this point should be checked to make sure that it does not exceed the allowable bending stress.

6. Check the pole compression stress. The vertical loads on one pole include, in addition to applied live loads and/or snow loads and dead load of the supported structure, the vertical component of wind load and the weight of the pole above the critical section.

The design procedures for short, intermediate, and long round tapered columns are shown on page

The ratio l_e/D should not exceed 44 for round columns. The unsupported length of long poles may be reduced by providing bracing in both directions at approximately the inflection point of the pole. The actual compression parallel to grain stress at the top, or small end, of the pole should be checked, and it should not exceed design values for a short column. Actual compressive stress f_c may be determined by the formula

$$f_c = \frac{P}{A}$$

where P = total vertical load on the pole (lb) and
A = cross-sectional area at the top of the pole (in.^{2}).

7. Check interaction of stresses by use of Eq. (5-18). Assume that only half of the snow load is acting when maximum wind occurs and check interaction of stresses at the point of maximum bending stress, which can be assumed to occur at the groundline when concrete backfill is used or one-fourth the depth of embedment below the groundline when soil is used for the backfill. The design should also be checked by the interaction equation along the length of the pole. The exact solution is difficult because of the taper of the member, variation of moment along the length, and difficulty in determining the effective length l_e of the column. Usually, conservative assumptions can be made to determine whether or not a problem exists.

8. Determine footing requirements. The pole footing requirements can be determined if the kind of soil, the vertical load on the pole, and the size of pole required are known. Soil-bearing pressure (psf) can be determined by the formula

$$\text{Soil bearing pressure} = \frac{P}{A}$$

where P = total vertical load on pole (lb) and
A = area of butt of pole (ft^2).

If this load exceeds the allowable bearing capacity S_B for the type of soil in question, a concrete footing is required.

The required area A_c in square feet of a concrete footing may be determined by the formula

$$A_c = \frac{P}{S_B}$$

where S_B = allowable bearing capacity (psf)
and other terms are as previously defined.

Concrete footings should be designed in accordance with the recommendations of the American Concrete Institute.

Example. Design the pole for the illustrated typical bent of a pole-type building to meet the stated conditions.

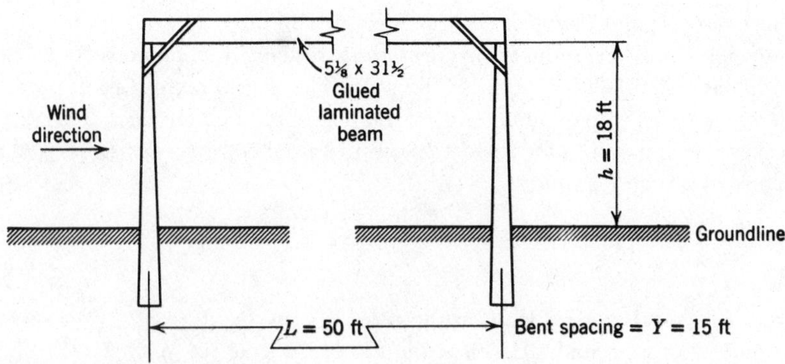

Loads: DL = 10 psf Pole design values: F_b = 1850 psi
SL = 20 psf (C_D = 1.15) F_c = 1000 psi
WL = 10 psf on windward side plus E = 1,500,000 psi
5 psf on leeward side
(C_D = 1.33)

1. Determine loads:

$$\text{Vertical load on one pole} = V = \frac{YL(DL + SL)}{2}$$

$$V = \frac{(15)(50)(10 + 20)}{2} = 11{,}250 \text{ lb} \quad \text{(full snow load)}$$

$$V' = \frac{(15)(50)(10 + 10)}{2} = 7500 \text{ lb} \quad \text{(one-half snow load)}$$

Wind loads:

$$\text{windward side, } w_1 = (10)(15) = 150 \text{ plf}$$

$$\text{leeward side, } w_2 = (5)(15) = 75 \text{ plf}$$

Determine the wind load acting on structure above top of the poles. Assume roofing extends as much as $2\frac{1}{2}$ in. above the beam,

$$d_b = 31\tfrac{1}{2} + 2\tfrac{1}{2} = 34 \text{ in.}$$

$$P_1 = (34/12)(150) = 425 \text{ lb} \quad \text{(windward side)}$$

$$P_2 = (34/12)(75) = 212.5 \text{ lb} \quad \text{(leeward side)}$$

Determine the force P_3 from Eq. (5-94):

$$P_3 = \frac{3h(w_1 - w_2)}{16} + d_b \frac{(w_1 - w_2)}{2}$$

$$P_3 = \frac{(3)(15)(150 - 75)}{16} + \frac{(34/12)(150 - 75)}{2} = 317 \text{ lb}$$

Determine resultant of the wind forces on the windward pole, P, and the height above ground at which it acts.

$$P = w_1 h + P_1 - P_3$$

$$P = (150)(15) + 425 - 317 = 2360 \text{ lb}$$

$$h' = \frac{[w_1 h^2/2] + (P_2 - P_3)h}{P}$$

$$h' = \frac{[(150)(15)^2/2] + (425 - 317)(15)}{2360} = 7.84 \text{ ft}$$

2. Determine soil values; soil is well-graded sand (Type 3 soil):

$$S_0 = 200 \text{ psf/ft} \quad \text{from Table 5.42}$$

Increase by 1.33 for wind and also multiply by 2 when $\frac{1}{2}$-in. movement at base permitted.

$$S_4 = (200)(1.33)(2) = 532 \text{ psf}$$

3. Estimate size of pole required.

Determine size of pole required to resist wind load in bending:

$$M = (2360)(7.84)(12) = 221{,}840 \text{ in.-lb}$$

$$F'_b = (1850)(1.33) = 2460 \text{ psi}$$

$$C = \sqrt[3]{\frac{32\pi^2 M}{F'_b}} = \sqrt[3]{\frac{32\pi^2(221{,}840)}{2460}} = 30.5 \text{ in.}$$

where C = circumference (in.) and
M = moment (in.-lb).

Assume that dead load plus $\frac{1}{2}$ snow load is acting at the same time as the maximum wind load.

$$V' = 7500 \text{ lb}$$

Assume critical diameter of pole = 10 in.

$$f_c = \frac{V'}{A} = \frac{7500}{(\pi)(10)^2/4} = 95 \text{ psi}$$

Assume a slightly larger pole than required to resist bending alone. Assume depth of embedment d_1 = 8 ft. A 25-ft class 1 pole has a circumference of 33.5 in. six ft from butt (Table 5.43).

4. Determine required embedment. Douglas Fir and Southern Pine poles have a taper of approximately 0.25 in. per ft in circumference.

$$\text{Butt diameter} = B = \frac{33.5 + (6)(0.25)}{\pi} = 11.14 \text{ in.} = 0.928 \text{ ft}$$

Assumed depth of embedment is 8 ft.

$$\text{Calculated depth of embedment } d = 1.97 \sqrt[3]{\frac{P(h + 0.93d)}{BS_4}}$$

$$= 1.97 \sqrt[3]{\frac{2360\,[7.84 + (0.93)(8)]}{(0.928)(532)}} = 8.23 \text{ ft}$$

Try $d = 8.25$ ft:

$$d = 1.97 \sqrt[3]{\frac{2360\,[7.84 + (0.93)(8.25)]}{(0.928)(532)}} = 8.27 \text{ ft} \quad (\text{use } 8.3 \text{ ft})$$

$$h + d = 15 + 8.3 = 23.3 \text{ ft}$$

A 25-ft-long pole is required. A 25-ft Class 1 Douglas Fir pole has a minimum circumference at the top of 27 in. and a minimum circumference 6 ft from the butt of 33.5 in. (Table 5.43).

5. Check bending stress. Assume backfill conditions meet the requirements for maximum bending moment to occur at the groundline. Circumference, C, at the groundline is calculated as

$$C = 33.5 - (8.3 - 6)(0.25) = 32.93 \text{ in.}$$

$$M = 221{,}840 \text{ in.-lb} \quad (\text{from step 3})$$

$$f_b = \frac{32\pi^2 M}{C^3} = \frac{32\pi^2\,(221{,}840)}{32.93^3} = 1960 \text{ psi} < 2460 \text{ psi} \quad \text{O.K.}$$

6. Check compression stress for full snow load on one pole, $V = 11{,}250$ lb; the weight of a 25-ft Class 1 pole is 625 lb (approximate, based on assumed weight of 50 lb/ft^3 and minimum sizes). Assume critical stress for compression occurs one-third the height of the pole from the top. Determine weight of pole above critical section:

$$W = (5/25)(625) = 125 \text{ lb}$$

Determine circumference at critical section:

$$C = 27 + (5 + 1.7)(0.25) = 28.68 \text{ in.}$$

Note that 1.7 ft was cut off the top of the pole.

$$P = V + W = 11{,}250 + 125 = 11{,}375 \text{ lb}$$

Determine f'_c. Assume structure unbraced for sidesway; from Table 5.4, $K_e =$ 2.1.

$$l_e = lK_e = (15)(2.1) = 31.5 \text{ ft}$$

$$D = \frac{C}{\pi} = \frac{28.68}{\pi} = 9.13 \text{ in.}$$

$$\frac{l_e}{D} = \frac{(31.5)(12)}{9.13} = 41.4$$

$$F''_c = F_c C_D C_{cs} = (1000)(1.15)[1 + (0.002)(6.7)] = 1165 \text{ psi}$$

Note that C_{cs} is obtained from the footnotes to Table 5.41.

$$K_R = 0.58\sqrt{\frac{E}{F''_c}} = 0.58\sqrt{\frac{1,500,000}{1165}} = 20.8$$

Since 20.8 < 41.4 < 44, use long-column formula.

$$F'_c = \frac{0.225E'}{(l_e/D)^2} = \frac{(0.225)(1,500,000)}{(41.4)^2} = 197 \text{ psi}$$

$$f_c = \frac{P}{A} = \frac{11,375}{(28.68)^2/4\pi} = 174 \text{ psi} < 197 \text{ psi} \quad \text{O.K.}$$

7. Check interaction of stresses at point of maximum bending stress (assumed to be at the ground line). Assume only one-half of snow load is acting when maximum wind load occurs.

$$V' = 7500 \text{ lb}$$

Calculate weight of pole above ground line.

$$W' = (15/25)(625) = 375 \text{ lb}$$

$$P = V' + W' = 7500 + 375 = 7875 \text{ lb}$$

$$f_c = \frac{P}{C^2/4\pi} = \frac{7875}{(32.93)^2/4\pi} = 91.2 \text{ psi}$$

$$F'_c = (1000)(1.33)[1 + (0.002)(25 - 8.25)] = 1375 \text{ psi}$$

$$\frac{f_c}{F'_c} + \frac{f_b}{F'_b} = \frac{91.2}{1375} + \frac{1960}{2460} = 0.863 < 1 \quad \text{O.K.}$$

In the above equation, F'_c was taken as F''_c, the tabular design value modified by all applicable modifiers except for slenderness, because the critical section as determined by maximum moment is at the ground surface.

8. Determine footing requirements.

$$\text{Area of pole butt} = A = \frac{[33.5 + (6)(0.25)]^2}{4\pi} = 97.5 \text{ in.}^2 = 0.677 \text{ ft}^2$$

$$P = 11,250 + 625 = 11,875 \text{ lb}$$

From Table 5.42, allowable soil bearing is

$$S_B = (2000)[1 + (0.2)(8.25 - 1)] = 4900 \text{ psf}$$

$$\text{Soil bearing load} = \frac{P}{A} = \frac{11,875}{0.677} = 17,540 \text{ psf} > 4900 \text{ psf}$$

Therefore, use concrete footing. Required area of concrete footing is calculated as

$$A_c = \frac{P'}{S_B} = \frac{11,875}{4900} = 2.42 \text{ ft}^2$$

Minimum diameter of hole

$$D_H = \sqrt{\frac{4A_c}{\pi}} = \sqrt{\frac{4(2.42)}{\pi}} = 1.76 \text{ ft}$$

The hole should be backfilled with concrete so that the maximum moment acts at groundline. For practical reasons, the hole should be large enough to place the concrete backfill. The diameter of the butt of the pole calculates to be slightly less than 1 ft but may actually be larger. Using a 2-ft-diameter hole will allow approximately 6 in. between the side of the hole and the pole, which should be adequate. Another advantage of concrete backfill is that it increases the effective diameter of the pole below the groundline, and by using this in Eq. (5-94), the depth of embedment can be reduced.

Step 4 will be repeated using 2-ft-diameter concrete backfill as the diameter of the pole butt B. Assume depth $d = 6.0$ ft.

$$d = 1.97\sqrt[3]{\frac{P[h + 0.93d]}{BS_4}} = 1.97\sqrt[3]{\frac{2360[7.84 + (0.93)(6)]}{(2)(532)}} = 6.1 \text{ ft}$$

Length of pole required = 15 + 6.1 = 21.1 ft > 20 ft

A 25-ft-long pole is still required, but digging the hole will be less costly.

The use of knee braces cuts down the effective length of the column. However, care should be used in designing connections for knee braces because forces in the knee braces may become quite large. The effect of sidesway can also be reduced by use of a roof diaphragm and shear walls. A typical pole frame structure has light roofing and end walls and depends on the poles for lateral stability.

TIMBER PILES

Recommendations for the use of timber piles in foundations may be found in *Pile Foundations: Know-How* (19). The ASTM D 25 *Standard Specifications for Round Timber Piles* (20) classifies round timber piles according to specified butt circumferences with mimimum tip circumferences or specified tip circumferences with minimum butt circumferences. Piles are generally designed at a critical section governed by the tip size, butt size, or an intermediate section. The ASTM classification allows the designer to specify a pile with adequate dimensions at the critical section. For example, a pile depending on frictional forces along the side of the pile to support the vertical load will generally have a critical section located away from the tip. On the other hand, an end-bearing pile may have the critical section located at the tip.

Tables 5.44 and 5.45 list size requirements for piles in accordance with ASTM D 25.

A number of species are used for piles with Southern Pine, Douglas Fir, and Oak being the most commonly used. Piles must be relatively straight and possess the strength to resist driving stresses and carry the imposed loads. When piles are

TABLE 5.44

Specified Butt Circumferences with Minimum Tip Circumferences[a]

Required Minimum Circumference 3 ft from Butt (in.)	22	25	28	31	35	38	41	44	47	50	57
Length (ft)	Minimum Tip Circumferences (in.)										
20	16.0	16.0	16.0	18.0	22.0	25.0	28.0				
30	16.0	16.0	16.0	16.0	19.0	22.0	25.0	28.0			
40				16.0	17.0	20.0	23.0	26.0	29.0		
50					16.0	17.0	19.0	22.0	25.0	28.0	
60						16.0	16.0	18.6	21.6	24.6	31.6
70						16.0	16.0	16.0	16.2	19.2	26.2
80							16.0	16.0	16.0	16.0	21.8
90							16.0	16.0	16.0	16.0	19.5
100							16.0	16.0	16.0	16.0	18.0
110										16.0	16.0
120											16.0

[a]Source: ASTM D 25.

TABLE 5.45

Specified Tip Circumferences with Minimum Butt Circumferences[a]

Required Minimum Tip Circumference (in.)	16	19	22	25	28	31	35	38
Length (ft)	Minimum Circumferences 3 ft from Butt (in.)							
20	22.0	24.0	27.0	30.0	33.0	36.0	40.0	43.0
30	23.5	26.5	29.5	32.5	35.5	38.5	42.5	45.5
40	26.0	29.0	32.0	35.0	38.0	41.0	45.0	48.0
50	28.5	31.5	34.5	37.5	40.5	43.5	47.5	50.5
60	31.0	34.0	37.0	40.0	43.0	46.0	50.0	53.0
70	33.5	36.5	39.5	42.5	45.5	48.5	52.5	55.5
80	36.0	39.0	42.0	45.0	48.0	51.0	55.0	58.0
90	38.6	41.6	44.6	47.6	50.6	53.6	57.6	60.5
100	41.0	44.0	47.0	50.0	53.0	56.0	60.0	
110	43.6	46.6	49.6	52.6	55.6	61.0		
120	46.0	49.0	52.0	55.0	58.0			

[a]Source: ASTM D 25.

used below the permanent water table or completely submerged in fresh water, preservative treatment is not necessary. However, for most permanent structures where these conditions do not exist, preservative treatment is necessary.

Preservative Treatment of Piles

Foundation piles are most commonly used where some part of the piles are exposed above the permanent water table. Therefore, preservative treatment is required. The preservative treatment should conform to recognized specifications such as Federal Specification TT-W-571 (21) and American Wood-Preservers' Association (AWPA) Standards C1 and C3 (16). Cutoffs at the tops of piles exposing untreated wood should be field treated in accordance with AWPA Standard M4 (16).

Piles used in salt water are subject to attack by marine borers, and special treatment techniques must be used to minimize the problem. The treatment of piles for use in saltwater is covered in Federal Specification TT-W-571 and AWPA Standard C18 (16).

Pile Driving

Equipment used for driving timber piles is of special importance. The energy used to drive the piles must be sufficient to drive the pile but must not impart excessive forces. Pile butts and tips may be damaged severely by sharp blows. For this reason, it is not desirable to drive timber piles with a drop hammer unless a suitable block is employed to dampen the impact. Air, steam, and diesel hammers are commonly used.

Generally, it is desirable to band the butt or driven end of a pile to minimize damage during driving. In some cases where tip damage may occur, a special shoe or fitting is used to protect the tip.

Before selecting a pile foundation, the soil conditions should be investigated. Group action, effective unsupported length, lateral loads, rebound, and other design and construction features should be considered.

Design Values for Piles

Design values for piles are determined in accordance with ASTM D 2899-74 (22). Table 5.46 contains design values for commonly used species.

These design values apply to both wet and dry conditions of use. They are subject to modification for duration of load C_D, conditioning prior to treatment C_{co}, and lateral stability C_L. The form factor for circular cross section C_{sf} was applied in the development of the design values and it should not be applied again by the designer. The size factor C_F is applied at the critical section in bending when the pile circumference exceeds 43 in. (diameter = 13.5 in.). The design values in Table 5.46 include an adjustment for strength reductions due to conditioning prior to treatment done in accordance with ASTM D 1760 (31). Where piles are conditioned prior to treatment by air or kiln drying only, the design values, except modulus of elasticity, can be increased by multiplying by the modifier C_{co}. For Southern Pine, C_{co} = 1.18, and for Pacific Coast Douglas Fir, Red Oak, and Red Pine, C_{co} = 1.11. When untreated piles are used for temporary structures, the design values can also be increased by application of C_{co}.

The design values shown in Table 5.46 are based on the strength at the tip.

TABLE 5.46

Design Values for Treated Round Timber Piles[a]

Species	Compression Parallel to Grain F_c (psi)	Extreme Fiber in Bending F_b (psi)	Horizontal Shear F_v (psi)	Compression Perpendicular to Grain $F_{c\perp}$ (psi)	Modulus of Elasticity E (psi)
Pacific Coast Douglas Fir[b]	1250	2450	115	230	1,500,000
Southern Pine[c]	1200	2400	110	250	1,500,000
Red Oak[d]	1100	2450	135	350	1,250,000
Red Pine[e]	900	1900	85	155	1,280,000

[a]Design values for normal load duration and wet conditions of use. Source: *National Design Specification* (3).

[b]Pacific Coast Douglas Fir values apply to this species as defined in ASTM D 1760-76, *Standard Specification for Pressure Treatment of Timber Products*. For fastener design, use Douglas Fir-Larch design values.

[c]Southern Pine values apply to Longleaf, Slash, Loblolly, and Shortleaf Pines.

[d]Red Oak values apply to Northern and Southern Red Oak.

[e]Red Pine values apply to Red Pine grown in the United States. For fastener design, use Northern Pine design values.

TABLE 5.47

Applicable Modifiers for Piles

	Modifier								
Design Value	C_D	C_M	C_F	C_L	C_P	C_R	C_f	C_{co}	C_{sf}
Compression parallel to grain F_c	Yes	No	No	No	Yes	Yes	No	Yes	Yes
Compression perpendicular to grain $F_{c\perp}$	No	No	No	No	No	Yes	No	Yes	No
Bending F_b	Yes	No	Yes	Yes	No	Yes	No	Yes	Yes
Shear F_v	Yes	No	No	No	No	Yes	No	Yes	No
End grain in bearing F_g	Yes	No	No	No	No	Yes	No	Yes	No
Modulus of elasticity E	No	No	No	No	No	Yes	No	No	No

Strength increases toward the butt. The design values for compression parallel to grain, F_c, may be increased by 0.2% for each foot of length from the tip to the critical section except that the total increase should not exceed 10%.

When piles or portions of piles are standing unbraced in air, water, or material not capable of lateral support, they should be designed in accordance with the applicable column formula. When piles are subjected to lateral loads, the applicable formulas for flexure also apply.

The tabular design values in compression parallel to grain, F_c, and bending, F_b, are based on piles used in a group so that load sharing takes place. When piles are used in such a manner that each pile must support a specific load, an additional safety factor of 1.25 should be applied to the tabular design value in compression parallel to grain, F_c, and a factor of 1.3 to tabular design values in bending, F_b, which results in a modification factor, C_{sf}, of 0.80 for compression parallel to grain, F_c, and 0.77 for bending, F_b. See Table 5.47 for application of modifiers for piles.

Example. Application of modifiers: Determine the design values in bending and compression parallel to grain for a single Southern Pine pile 10 ft from the tip that was conditioned prior to treatment by air drying. The critical section has a 17-in. diameter. The critical stresses are caused by wind. Each pile must support a specific load under wet conditions of use. The unsupported length of the pile is such that it is classified in compression as an intermediate column with a C_P of 0.91 and in bending as a short beam.

Compression parallel to grain:

$$F_c' = F_c C_D C_{co} C_{sf} C_P[1 + (0.002)(10)]$$
$$= (1200)(1.33)(1.18)(0.8)(0.91)[1 + (0.002)(10)] = 1400 \text{ psi}$$

Bending:

$$F_b' = F_b C_D C_{co} C_F C_{sf}$$
$$= (2400)(1.33)(1.18)(0.96)(0.77)$$
$$= 2780 \text{ psi}$$

TIMBER BRIDGES

Timber bridges consist of several basic types, including trestles, girder bridges, truss bridges, and arch bridges. In the design of any of these types, the considerations given under "Moving Loads" in Section 4 of this manual should be taken into account. For highway bridges, design loads and their application should be in accordance with the recommendations of the American Association of State Highway and Transportation Officials (AASHTO) (23). For railway bridges, the recommendations of the American Railway Engineering Association (AREA) (24) should be followed.

Trestles

The trestle is probably the simplest type of timber bridge. Timber trestles consist of stringers supported by pile or frame bents. The bridge deck is applied to the stringers. Pile and frame bents are capped by timbers 12 × 12 or larger, adequately fastened to the tops of the piles or posts.

If pile penetration or height of bent is such that piles longer than those commercially available are required, or if pile bearing values are low and a large number of piles must be driven, posts may be used on top of the pile bents. Frame bents must rest on some type of foundation structure, such as concrete footings or piles. Sway bracing and longitudinal tower bracing, appropriate to the height of the bent, must be provided.

Spacing of bents is determined, in part, by the commercially available lengths of stringers, which are fabricated in even-foot increments. The ends of interior stringers are usually lapped and fastened at the bearing on the caps, whereas exterior stringers are butted at the ends and spliced over the bent caps.

Stringers are designed as simple-span beams under the loadings recommended by AASHTO or AREA. Sizes and spacing are determined by the span and loading conditions. Standard sizes of glued laminated or sawn timbers as given in Tables 7.1 and 7.2 should be used. Solid bridging should be provided at the ends of stringers to hold them in line and also to serve as a fire stop. Bridging should also be placed between stringers at midspan or, on long spans, at third-points.

Girder Bridges

Girder bridges, consisting of glued laminated or sawn timber girders supporting a bridge deck, may be used for spans that exceed the practical limits of timber trestles, for spans less than those economical for truss bridges, or where a truss bridge is not desirable. Substructures similar to those used for timber trestles can be used for girder bridges, the girder being fastened to the bent caps by means of a fabricated steel girder seat.

Timber girders are designed as beams in accordance with the recommendations in this section, including those for lateral support (see Table 5.10). Lateral forces acting on the girder will enter into the design of the lateral bracing system. Construction economies will usually result if the standard sizes for glued laminated or sawn timber as given in Tables 7.1 and 7.2 of this manual are used.

Truss Bridges

Truss bridges may be of either of two types: deck-truss bridges, in which trusses support the bridge deck and roadway; or through-truss bridges, in which the roadway passes between two parallel trusses forming the bridge structure. The deck-truss type is the more economical, since substructures and lateral bracing are narrower. The use of deck trusses may be limited, however, by under-clearance.

Deck trusses may be of the parallel chord type or of the bowstring type with the truss built up to the level of the floor beams. Through-trusses may be of either of these truss types also but the bowstring is usually more economical. Bowstring trusses particularly are often used as pony truss bridges, which are through-trusses with no overhead bracing above the roadway. Through-trusses have the disadvantage of potential damage from vehicles.

Substructures for truss bridges may be similar to those for timber trestles; however, because the vertical loads are greater and are concentrated at the ends of the trusses, the bents must be capable of carrying a greater load and a system of cribbing is required for bent caps. For longer span and heavier loads, timber, stone, or concrete piers may be required. Lateral forces are greater on truss bridges, and a carefully designed substructure sway bracing system is necessary.

The design of trusses for bridges is similar to that for roof trusses, the length of truss panel being determined by economical spacing of floor beams, a minimum number of joints, and commercially available lengths of timber. As in roof truss design, the joint design is an important consideration. Bridge truss joints should be designed to eliminate or minimize pockets, which may tend to collect moisture.

Arch Bridges

When site conditions are such that considerable height is required between foundation and roadway or a relatively long clear span is required, an arch bridge may be most economical because of the lesser necessity for substructure framing. Arch bridges may be of the two-hinged or three-hinged type, two-hinged designs being more frequently used on short spans and three hinged designs on long spans.

Glued laminated timber arches may be fabricated to the desired shape and the ends built up to the level of the roadway by means of post bents. Post bents may be connected to the arch by means of steel gusset plates, which should be designed for erection loads, possible stress reversals, and lateral forces as well as for the anticipated bridge loads.

Bridge Decks

The selection of decks for timber bridges is determined by density of traffic and economics. Plank decks may be used for light traffic or for temporary bridges. Laminated decks can be used for heavier traffic conditions. Asphaltic wearing surfaces may be applied on the decking, although this is not usually done for plank decks.

Composite timber-concrete decks are sometimes used in timber bridge construction. Composite timber-concrete construction combines timber and concrete in such a manner that the wood is in tension and the concrete is in compression (except at the supports of continuous spans, where negative bending occurs and these stresses are reversed). Composite timber-concrete construction is of two basic types: T-beams and "slab" decks.

T-beams consist of timber stringers, which form the stems, and concrete slabs, which form the flanges of a series of T-shapes. Composite beams of this type are usually simple-span bridges. Slab decks use, as a base for the concrete, a mechanically laminated wooden deck made up of planks set on edge, with alternate planks raised 2 in. to form longitudinal grooves. This grooved surface is usually obtained by using planks of two different widths and alternating them in assembly. This composite type has been used for continuous-span bridges and trestles, but is not very common today.

In both types, a means of horizontal shear resistance and a means of preventing separations are needed at the joint between the two materials. In T-beams, resistance to horizontal shear is generally provided by a series of notches $\frac{1}{2}$ to $\frac{3}{4}$ in. deep cut into the top of the sawn timber stringer and about $1\frac{1}{2}$ in. deep for a glued laminated timber stringer, while nails and spikes partially driven into the top prevent vertical separation of the concrete and timber. Other adequate methods may be used. In slab decks, shear resistance is accomplished either by means of notches $\frac{1}{2}$-in. deep cut into the tops of all laminations, by triangular steel plate "shear developers" driven into precut slots in the channels formed by the raised laminations, or other suitable shear connectors. When the $\frac{1}{2}$-in. notches are used, grooves are milled the full length of both faces of each raised lamination to resist the uplift and separation of the wood and concrete. When the steel shear developers are used, nails or spikes are partially driven into the tops of raised laminations to resist separation.

In T-beam design, secondary shearing stresses due to temperature must be considered in designing for horizontal shear resistance. These stresses are induced by the thermal expansion or contraction of the concrete, both of which are resisted by the wood, which is assumed to be unaffected by normal temperature changes. Shear connections for temperature change are neglected in slab deck-type composite construction; however, expansion joints should be provided in the concrete slab.

The concrete slab should be reinforced for temperature stresses. In continuous spans, steel sufficient to develop negative bending stresses is necessary over interior supports.

The dead load of the composite structure is considered to be carried entirely by the timber section. The composite structure carries positive bending moment and, over interior supports in continuous spans, steel reinforcing and the wood act to resist negative bending moment.

In designing a composite structure, if it is assumed that the junction between the two materials is without inelastic deformation and has elastic characteristics in keeping with the materials, the structure can be designed by the transformed-area method, that is, by transforming the composite section into an equivalent homogeneous section. This is accomplished by multiplying the concrete width or depth by the ratio of the moduli of elasticity of the materials.

Glued laminated timber bridges are commonly used for highway bridges and design examples are included. Because nail-laminated decks and composite timber-concrete decks are not common, design examples for these decks are not included in this manual. For bridges utilizing longitudinal stringers and transverse decks, the following design procedures illustrate the superstructure design with interconnecting steel dowels between the panels and without the dowels. Note that this procedure and examples are for vertical loading from vehicles as described in *Standard Specifications for Highway Bridges* (23). The reader should refer to this standard for additional loading requirements on the superstructure components.

Longitudinal Stringers and Doweled Deck

The design examples do not show camber for the glued laminated stringers because camber may vary with the particular use. Generally, camber equal to dead-load deflection plus additional amounts for drainage and appearance is sufficient. Glued laminated timber design values are based on AITC 117—Design (1). Wet-use design values were used for the design of deck panels because tests in the field indicate the moisture content is likely to exceed 16% in service. Dry-use design values are used for the design of the glued laminated timber stringers, with the exception of compression perpendicular to grain stresses, $F_{c\perp}$. The glued laminated timber deck panel, when properly treated and installed, should provide a roof for the stringers, preventing them from exceeding an in-service moisture content of 16%.

Wind-driven moisture can cause a surface wetting on the exposed face of the edge stringers, but this superficial wetting should not affect the design. Localized moisture accumulations can develop and preservative treatment is recommended for the stringers to protect against possible decay hazards such as at areas of steel connections and at bearing locations.

Example I.

Criteria

HS 15-44 live load, two lanes, follow AASHTO Standards
3-in. asphalt wearing surface
Span: 40 ft 0 in.
Width 34 ft 0 in.

Materials

Decking: Glued laminated L2 Douglas Fir (Combination 2) or No. 2 MG Southern Pine (Combination 47); dry condition of use

Stringers: Douglas fir or Southern Pine; 24F-V4 DF/DF or 24F-V3 SP/SP

Stringer Design

Assume: Six girders at 6 ft 7 in. center-to-center
$5\frac{1}{8}$-in.-thick deck panels, 35 ft long
Weight of wood, 50 pcf
Weight of asphalt, 150 pcf

Dead load (on one stringer): Deck $\dfrac{(5.125)(50)(6.58)}{12} = 140$ plf

$$\text{Wearing surface: } \frac{(3)(150)(6.58)}{12} = 247 \text{ plf}$$

$$\text{Stringer (assumed): } = 100 \text{ plf}$$

$$\text{Total (preliminary): } = 487 \text{ plf}$$

Live load (on one stringer): HS 15-44 (AASHTO, Appendix A)

$$\text{Distribution factor} = \frac{\text{spacing}}{5.0} = \frac{6.58}{5} = 1.32$$

$$\text{Impact factor} = 0 \text{ (none required for timber)}$$

$$\text{Moment} = \frac{\text{(ft-kips/lane) (distribution factor)}}{\text{wheel lines/lane}}$$

$$= \frac{(337.4)(1.32)}{2} = 222 \text{ ft-kips}$$

Try $6\frac{3}{4} \times 43\frac{1}{2}$-in. girder
Area $= 293.6$ in.^{2}
Section modulus $= 2129$ in.^{3}
Moment of inertia $= 46{,}300$ in.^{4}

$$\text{Size factor } C_F = \left(\frac{12}{d}\right)^{1/9} = \left(\frac{12}{43.5}\right)^{1/9} = 0.87$$

$$\text{Weight} = \frac{(50)(293.6)}{144} = 102 \text{ plf}$$

Actual dead load $= 140 + 247 + 102 = 489$ plf
Bending:

$$\text{Dead load moment} = \frac{wL^2}{8} = \frac{(0.489)(40)^2}{8} = 98 \text{ ft-kips}$$

$$\text{Total load moment} = 98 + 222 = 320 \text{ ft-kips}$$

$$\text{Actual bending stress } f_b = \frac{M}{S} = \frac{(320)(12)(1{,}000)}{2129} = 1800 \text{ psi}$$

The deck is fastened to the stringers to provide lateral bracing; therefore $C_L = 1$.
Design value in bending, $F_b' = F_b C_F = (2400)(0.87) = 2088$ psi
Horizontal shear: points considered
Three times depth from end:

$$\frac{(3)(43.5)}{12} = 10.875 \text{ ft}$$

Span quarter point:

$$\frac{40.0}{4} = 10.0 \quad \text{(controls)}$$

$$\text{Dead load} = 9.8 - (0.489)(10) = 4.9 \text{ kips}$$

$$\text{Live load} = 0.5\,[(0.6)(13.95) + (1.32)(13.95)]$$

$$= 13.4 \text{ kips}$$

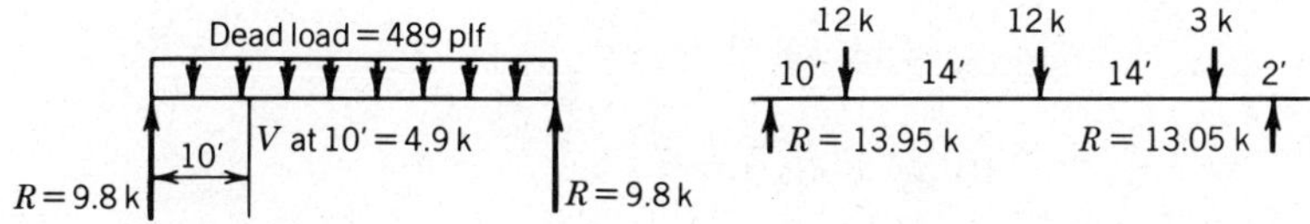

Actual horizontal shear stress:

$$f_v = \frac{3V}{2bd} = \frac{(3)(13.4 + 4.9)(1000)}{(2)(6.75)(43.5)} = 93.5 \text{ psi}$$

A check of the shear at one stringer depth from the end for undistributed live load indicates that the shear stress is

$$f_v = \frac{(3)(18.3 + 8.0)(1000)}{(2)(6.75)(43.5)} = 134 \text{ psi}$$

The design value in horizontal shear is 165 psi for Douglas Fir or 200 psi for Southern Pine.

Overload provisions: apply for single lane; distribution factor = 6.58/6 = 1.10. Bending—100% increase in truck or lane load:

$$\text{Moment} = \left(\frac{1.10}{1.32}\right)(222)(2) + 98 = 468 \text{ ft-kips}$$

$$\text{Actual bending stress} = \frac{M}{S} = \frac{(468)(12)(1000)}{2129} = 2638 \text{ psi}$$

Increase allowable stress by 50%.

$$\text{Allowable bending stress} = 1.5\,F_bC_F = (1.5)(2400)(0.87) = 3130 \text{ psi} \quad \text{O.K.}$$

Horizontal shear with 100% increase in truck or lane load:

$$\text{Live load shear at L/4} = (13.95)(2) = 27.9 \text{ kips}$$

$$\text{Distributed live load shear} = 0.5\,[(0.6)(27.9) + (1.10)(27.9)]$$

$$= 23.7 \text{ kips}$$

$$\text{Actual horizontal shear stress } f_v = \frac{3V}{2bd} = \frac{(3)(23.7 + 4.9)(1000)}{(2)(6.75)(43.5)} = 146 \text{ psi}$$

Check shear with overload at one d from end:

$$f_v = \frac{(3)(36.6 + 8.0)(1000)}{(2)(6.75)(43.5)} = 228 \text{ psi}$$

Increase allowable stress by 50%; then allowable horizontal shear stress = (1.5)(165) = 248 psi for Douglas fir or (1.5)(200) = 300 psi for Southern Pine.

Deck Design

Basic design equations: (see AASHTO 3.25)

$$M_x = P\,[(0.51 \log_{10} s) - K]$$

$$R_x = 0.034\,P$$

$$t = \sqrt{\frac{6M_x}{F'_b}} \quad \text{or} \quad t = \frac{3R_x}{2F'_v} \quad \text{(whichever is greater)}$$

where M_x = primary bending moment (in.-lb/in.),
R_x = primary shear (lb/in.),
P = design wheel load = 12,000 lb,
s = effective deck span = clear span + one-half girder width (75.625 in. or clear span + assumed deck thickness, 77.375 in., whichever is less),
t = deck thickness (assumed to be $5\frac{1}{8}$ in.)
K = 0.44 for H 10 loading, 0.47 for H 15 loading, 0.51 for H 20 loading.
F_b = 1750 psi, Combination 2 Douglas Fir or Combination No. 46 Southern Pine,
F_v = 145 psi (Douglas Fir) or 175 psi (Southern Pine)

Example (continued)

C_F = 1.10 for $5\frac{1}{8}$ in. thickness from footnote f, Table 2, AITC—Design (1),
$F'_b = F_b C_F C_M$ = (1750)(1.10)(0.8) = 1540 psi,
$F'_v = F_v C_M$ = (145)(0.875) = 127 psi (Douglas Fir)
M_x = 12,000[0.51 (log 75.625) − 0.47] = 12,000[(0.51)(1.878665) − 0.47] = 5860 in.-lb/in.,
R_x = (0.034)(12,000) = 408 lb/in., and

$$t = \sqrt{6M/F'_b} = \sqrt{(6)(5860)/1540} = 4.78 \text{ in.}$$

or

$$t = 3R/2F'_v = (3)(408)/(2)(127) = 4.82 \text{ in.}$$

Use $5\frac{1}{8}$-in. glued laminated deck.

Dowel Design, (for basic design equations see AASHTO, 3.25)

$$n = \frac{1000}{\sigma_{PL}}\left[\frac{\overline{R_y}}{R_D} + \frac{\overline{M_y}}{M_D}\right]$$

$$\overline{R_y} = \frac{6Ps}{1000} \quad \text{for } s \leqslant 50 \text{ in.}$$

$$\overline{R_y} = \frac{P}{2s}(s - 20) \quad \text{for } s > 50 \text{ in.}$$

$$\overline{M_y} = \frac{Ps}{1600}(s - 10) \quad \text{for } s \leqslant 50 \text{ in.}$$

$$\overline{M_y} = \frac{Ps(s - 30)}{20(s - 10)} \quad \text{for } s > 50 \text{ in.}$$

where R_D, M_D = constants dependent on dowel diameter and length, and
σ_{PL} = 1000 psi for Douglas Fir or Southern Pine

Try $1\frac{1}{2}$-in. diameter dowels:

$$R_D = 2770 \text{ lb} \qquad M_D = 8990 \text{ in.-lb}$$

$$\overline{R_y} = \frac{(12{,}000)(75.625-20)}{(2)(75.625)} = 4410 \text{ lb}$$

$$\overline{M_y} = \frac{(12{,}000)(75.625)(75.625-30)}{(20)(75.625-10)} = 31{,}550 \text{ in.-lb}$$

$$n = \frac{1000}{1000}\left[\frac{4410}{2770} + \frac{31{,}550}{8990}\right] = 5.10$$

Use five $1\frac{1}{2}$-in. diameter by $19\frac{1}{2}$-in. long dowels in each span. Check stress in dowels:

$$\sigma = \frac{1}{n}(C_R\overline{R_y} + C_M\overline{M_y})$$

$$C_R = 3.11 \text{ in.}^2$$

$$C_M = 3.02 \text{ in.}^3$$

$$\sigma = (1/5)[(3.11)(4410) + (3.02)(31{,}550)] = 21{,}800 \text{ psi} \quad \text{O.K.}$$

because this is less than the elastic limit of the dowels.

Longitudinal Stringers and Nondoweled Deck

For nondoweled decks carrying HS-20-44 or HS-15-44 loading, the following chart shows the maximum clear span between stringers and the maximum deck overhang beyond the outside stringer.

Deck Thickness (in.)	Maximum Clear Span	Maximum Clear Overhang
$3\frac{1}{8}$	2 ft 6 in.	1 ft 3 in.
$5\frac{1}{8}$	4 ft 6 in.	2 ft 3 in.
$6\frac{3}{4}$	6 ft 0 in.	3 ft 0 in.

These values are based on experience in constructing nondoweled decks that are fastened to the stringers using the bracket system shown in Figure 5.72 or similar deck connection systems.

Example II. The following design example illustrates the procedures to follow in the design of nondoweled decks on longitudinal stringers.

Criteria

HS 20 loading, two lanes
Asphalt wearing surface, 3 in.
Span = 50 ft 0 in. center to center of bearings
Width = 34 ft 0 in.

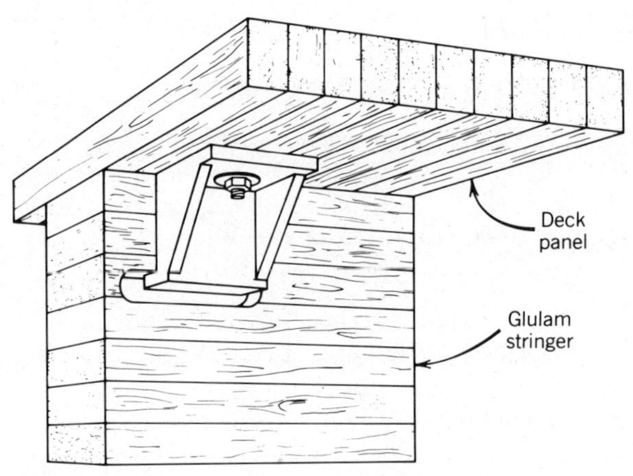

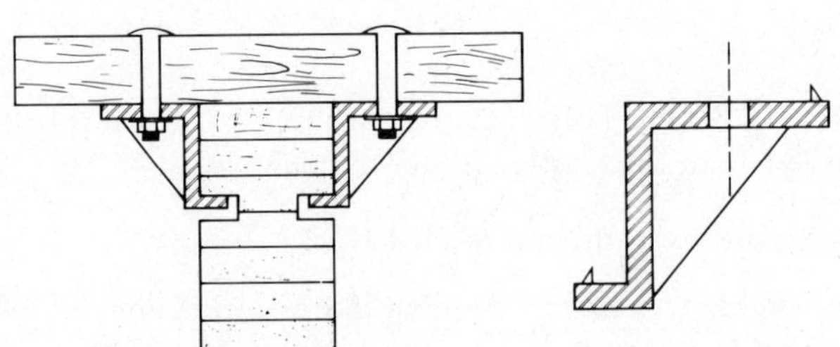

FIGURE 5.72 Bracket for fastening deck to glulam stringer.

Stringer Design

Determine number of stringers, n, assuming $8\frac{3}{4}$-in. wide stringers:

$$n = \frac{34 - 2(2.25)}{4.5 + (8.75/12)} = 5.64 \quad \text{(use six stringers)}$$

$$\text{Stringer spacing} = \frac{34 - (2)(2.25)}{5} = 5.9 \text{ ft} \quad \text{(use 6 ft)}$$

$F_b = 2400$ psi $\quad F_v = 165$ psi $\quad F_{c\perp} = 650$ psi $\quad E = 1{,}800{,}000$ psi

$$\text{Dead loads: 3 in. asphalt} \quad \frac{(3)(150)(6)}{12} = 225 \text{ plf}$$

$$6\tfrac{3}{4} \text{ in. deck} \quad \frac{(6.75)(50)(6)}{12} = 169 \text{ plf}$$

Railing, posts, and curb = 75 plf

Stringer (assume $8\frac{3}{4} \times 43\frac{1}{2}$ in.) = 130 plf

599 plf

Live load moment per wheel lane (from AASHTO, Appendix A):

$$M_{LL} = \frac{627{,}900}{2} = 314 \text{ ft-kips}$$

$$\text{Distribution factor} = \frac{\text{spacing}}{5} = \frac{6}{5} = 1.2$$

$$M_{DL} = \frac{wL^2}{8} = \frac{(0.599)(50)^2}{8} = 187 \text{ ft-kips}$$

$$M_{DL+LL} = 314 + 187 = 501 \text{ ft-kips}$$

Try an $8\frac{3}{4} \times 45$-in. stringer, $S = 2953$ in.^{3}:

$$f_b = \frac{M}{S} = \frac{(501)(12)(1000)}{2953} = 2030 \text{ psi}$$

Check lateral stability. The top of the stringer is laterally supported by brackets at 12 in. centers therefore no reduction for lateral stability is required.

$$F_b' = F_b C_F = (2400)(0.86) = 2060 \text{ psi} > 2030 \quad \text{O.K.}$$

Check shear:

$$\frac{\text{Span}}{4} = \frac{50}{4} = 12.5 \text{ ft}$$

$$3(\text{stringer depth}) = (3)(45/12) = 11.25 \text{ ft}$$

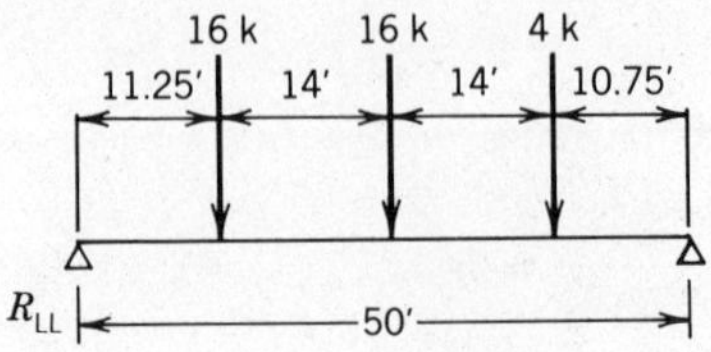

$$R_{LL} = (16)\left(\frac{38.75}{50}\right) + (16)\left(\frac{24.75}{50}\right) + (4)\left(\frac{10.75}{50}\right) = 21.2 \text{ kips}$$

$$V_{LL} = (0.5)[(0.6)(21.2) + (1.2)(21.2)] = 19.1 \text{ kips}$$

$$V_{DL} = \left(\frac{L}{2} - 11.25\right) w_{DL} = (25 - 11.25)(0.599) = 8.2 \text{ kips}$$

$$V_{LL} + V_{DL} = 19{,}100 + 8200 = 27{,}300 \text{ lb}$$

$$f_v = \frac{(3)(27{,}300)}{(2)(8.75)(45)} = 100 \text{ psi} < 165 \text{ psi} \quad \text{O.K.}$$

Check shear 1 times the depth from end:

$$R_{LL} = 26.6 \text{ kips} \qquad R_{DL} = 12.5 \text{ kips}$$

$$f_v = \frac{(3)(39{,}100)}{(2)(8.75)(45)} = 150 \text{ psi} < 165 \text{ psi} \quad \text{O.K.}$$

The overload provisions should also be checked as in Example 1.

Deck Design

$F_b = 1450$ psi $\qquad F_v = 145$ psi

The wheel load is distributed over a deck length of 15 in. plus the deck thickness 15 + 6.75 = 21.75 in.

Deck span = 72 − 8.75 + 8.75/2 = 67.62 in.

Deck span (maximum) = 72 − 8.75 + 6.75 = 70 in.

The tire imprint is distributed over a width b:

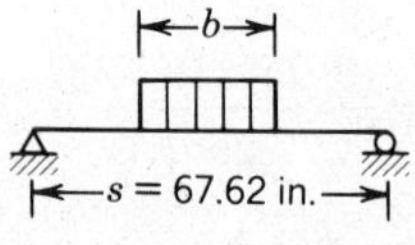

$$b = \sqrt{(0.01P)(2.5)} = \sqrt{(0.01)(12{,}000)(2.5)} = 17.3 \text{ in.}$$

$$M_{LL} = \left[\frac{Ps}{4} - \frac{Pb}{8}\right] = (12{,}000)\left[\frac{67.62}{4} - \frac{17.3}{8}\right] = 176{,}900 \text{ in.-lb}$$

$$w_{DL} = \left[\left(\frac{3}{12}\right)(150) + \left(\frac{6.75}{12}\right)(50)\right]\left(\frac{1}{12}\right)\left(\frac{21.75}{12}\right) = 9.9 \text{ lb/in.}$$

$$M_{DL} = \frac{wl^2}{8} = \frac{(9.9)(67.62)^2}{8} = 5660 \text{ in.-lb}$$

Because the deck is continuous over two or more stringers, the moment is 80% of the simple-span moment.

$$M = (0.8)(176{,}900 + 5660) = 146{,}000 \text{ in.-lb}$$

$$S = \frac{(21.75)(6.75)^2}{6} = 165.2 \text{ in.}^3$$

$$F_b' = F_b C_F C_M = (1450)(1.07)(0.8) = 1240 \text{ psi}$$

$$f_b = \frac{146{,}000}{165.2} = 884 \text{ psi}$$

Check shear by placing wheel load 15 in. from centerline of stringer.

$$V_{LL} = \left(\frac{52.62}{67.62}\right)(12{,}000) = 9340 \text{ lb}$$

$$V_{DL} = (9.9)\left(\frac{67.62}{2}\right) = 335 \text{ lb}$$

$$F_v' = F_v C_M = (145)(0.875) = 127 \text{ psi}$$

$$f_v = \frac{3V}{2bd} = \frac{(3)(9675)}{(2)(48)(6.75)} = 45 \text{ psi} < 127 \text{ psi} \quad \text{O.K.}$$

Post and Rail Design

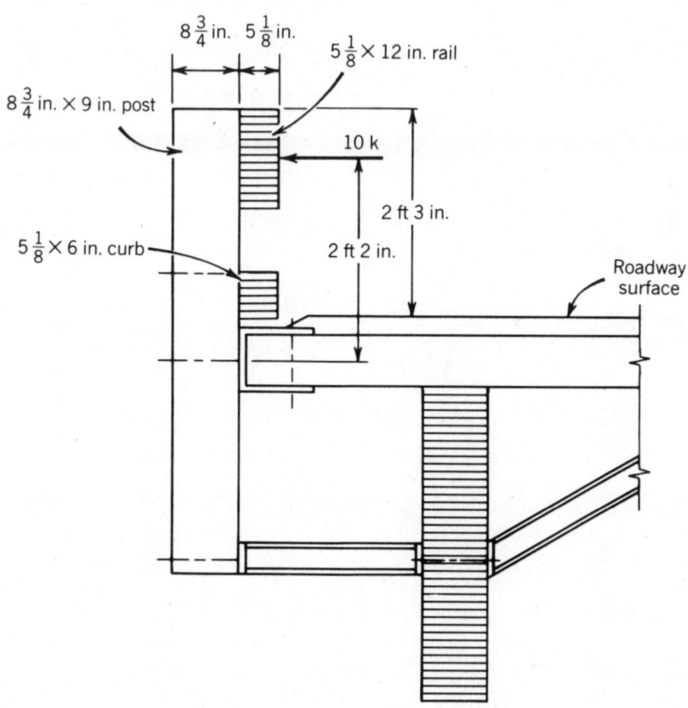

Load requirements are in accordance with AASHTO. Load $P = 10$ kips applied with the top of the rail 2 ft 3 in. above roadway surface (see sketch). The load is applied in an outward direction at post for post design and at midspan for rail design. For post design, a longitudinal load of 5 kips distributed over four posts must be applied simultaneously with the outward load. Each post shall be designed to resist an inward load of 10 kips/4. The attachment of the rail shall be designed to resist a vertical load of 10 kips/4 either upward or downward. The curb loading is 500 lb/ft.

Analyze the rail for bending using the AASHTO recommended formula for bending of $Pl/6$ and for shear by locating the load a distance from the face of the post equal to the rail width.

Bending:

$$M = \frac{Pl}{6} = \frac{(10{,}000)(8)(12)}{6} = 160{,}000 \text{ in.-lb}$$

$$S_y = 52.5 \text{ in.}^3 \quad \text{(from Table 7.2)}$$

$$f_b = \frac{M}{S_y} = \frac{160{,}000}{52.5} = 3045 \text{ psi}$$

For Combination 2, $F_b = 1800$ psi; for rail loading, $C_D = 1.65$. From footnote *f*, Table 2, AITC 117—Design, $C_F = 1.10$.

$$F_b' = F_b C_F C_D = (1800)(1.10)(1.65) = 3270 \text{ psi} > 3045 \text{ psi} \quad \text{O.K.}$$

Shear:

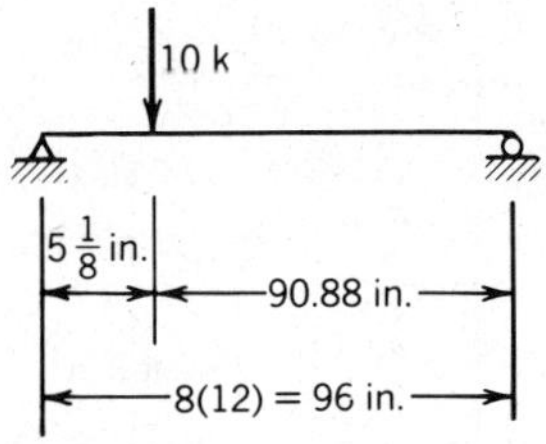

$$V_{max} = \left(\frac{90.88}{96}\right)(10{,}000) = 9470 \text{ lb}$$

$$f_v = \frac{3V}{2bd} = \frac{(3)(9470)}{(2)(12)(5.125)} = 230 \text{ psi}$$

$$F_v' = F_v C_D = (145)(1.65) = 240 \text{ psi} > 230 \text{ psi} \quad \text{O.K.}$$

Bearing: Check compression perpendicular to grain between rail and post (assume post is 9 in. wide).

$$f_{c\perp} = \frac{P}{A} = \frac{(10{,}000)}{(12)(9)} = 93 \text{ psi}$$

$$F_{c\perp}' = F_{c\perp} = 560 \text{ psi} > 93 \text{ psi} \quad \text{O.K.}$$

Check tension in bolts connecting rail to post for a load of one-fourth of the 10-kip load. Two $\frac{3}{4}$-in.-diameter A307 bolts provide a tensile capacity of (8.8)(2) = 17.6 kips, more than enough capacity for this load.

Check the shear in the bolts connecting the rail to the post.
From Table 6.22, one $\frac{3}{4}$-in.-diameter bolt capacity in single shear is 1435 lb. Two bolts are good for (2)(1435)(1.65) = 4735 lb. Vertical load on rail is 10 kips/4 = 2500 lb.

Check curb for bending and shear:

$$M = \frac{wL^2}{8} = \frac{(500)(8)^2(12)}{8} = 48{,}000 \text{ in.-lb}$$

$$S_y = 26.3 \text{ in.}^3$$

$$f_b = \frac{\text{M}}{\text{S}_y} = \frac{48{,}000}{26.3} = 1825 \text{ psi} < 3270 \text{ psi} \quad \text{O.K.}$$

$$V = \frac{wL}{2} = \frac{(500)(8)}{2} = 2000 \text{ lb}$$

$$f_v = \frac{3V}{2bd} = \frac{(3)(2000)}{(2)(5.125)(6)} = 98 \text{ psi} < 230 \text{ psi} \quad \text{O.K.}$$

Check bending in post with loads applied 26 in. above centerline of deck.

$$M_y = (10{,}000)(26) = 260{,}000 \text{ in.-lb}$$

$$M_x = (10{,}000/4)(26) = 65{,}000 \text{ in.-lb}$$

Assume an $8\frac{3}{4} \times 9$-in. post, Combination 5, $F_{by} = 2400$ psi, $F_{bx} = 2200$ psi. Assume two $1\text{-}\frac{1}{16}$-in. holes for bolts.

$$S_{y\,(\text{net})} = 87.7 \text{ in.}^3 \qquad S_{x\,(\text{net})} = 80.9 \text{ in.}^3$$

$$C_{F_y} = 1.04 \quad \text{(from footnote } f\text{, Table 2, AITC 117—Design)}$$

$$C_D = 1.65$$

$$F'_{by} = F_{by}C_{F_y}C_D = (2400)(1.04)(1.65) = 4120 \text{ psi}$$

$$F'_{bx} = F_{bx}C_D = (2200)(1.65) = 3630 \text{ psi}$$

$$f_{by} = \frac{M_y}{S_y} = \frac{260{,}000}{87.7} = 2960 \text{ psi}$$

$$f_{bx} = \frac{M_x}{S_x} = \frac{65{,}000}{80.9} = 800 \text{ psi}$$

$$\frac{f_{by}}{F_{by}} + \frac{f_{bx}}{F'_{bx}} = \frac{2960}{4120} + \frac{800}{3630} = 0.95 < 1 \quad \text{O.K.}$$

Check shear on net area:

$$A_{net} = (8.75)(9) - (2)(8.75)\,(1\text{-}\tfrac{1}{16}) = 60.2 \text{ in.}^2$$

$$f_v = \frac{3V}{2A} = \frac{(3)(10{,}000)}{(2)(60.2)} = 250 \text{ psi}$$

$$F_{vy} = 145 \text{ psi} \qquad F'_{vy} = F_{vy}C_D = (145)(1.65) = 240 \text{ psi} \approx 250 \text{ psi} \quad \text{O.K.}$$

Check post to deck connection:

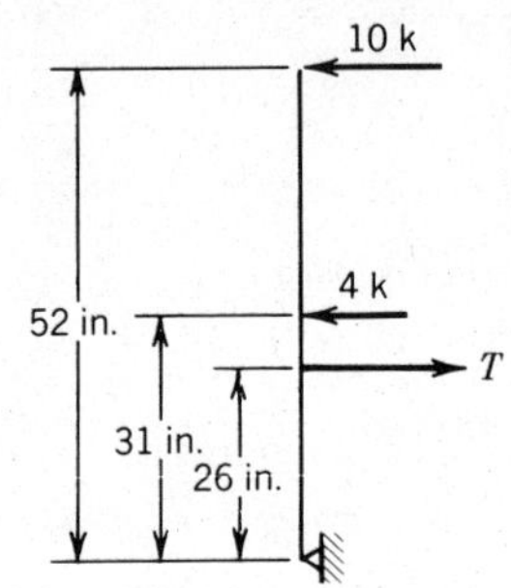

$$(4)(31) + (10)(52) = 26T$$

$$T = \frac{(10)(52) + (4)(31)}{26} = 24.8 \text{ kips}$$

Two 1-in.-diameter A307 bolts provide (15.7)(2) = 31.4 kips for post to bracket connection. For bracket to deck connection, determine number n of 1-in.-diameter bolts required. From Table 6.21, bolt value is 6370 lb, which may be increased by $C_D = 1.65$ for rail connection.

$$n = \frac{24{,}800}{(6370)(1.65)} = 2.4 \quad \text{(use three bolts)}$$

The reaction at the bottom of the post must be transferred to the stringer and, in turn, through a diagonal member to the deck.

Longitudinal Deck with Transverse Stiffeners

The following example illustrates the procedure for designing a simple-span longitudinal deck bridge using glued laminated timber panels.

Example III.

Criteria

HS 20 loading
Asphalt wearing surface, 3 in.
Span = 20 ft
Width = 32 ft

$$F_b = 1800 \text{ psi} \qquad F'_b = F_b C_F C_M = (1800)(1.01)(0.8)$$

$$= 1455 \text{ psi} \quad (10\tfrac{3}{4} \text{ in. deck assumed})$$

$$F_v = 145 \text{ psi} \qquad F'_v = F_v C_M = (145)(0.875) = 127 \text{ psi}$$

Width of deck over which wheel load is distributed, W, is width of tire plus twice the deck thickness.

$$\text{Width of tire, } b = \sqrt{(0.01P)(2.5)}$$

$$= \sqrt{(0.01)(16{,}000)(2.5)} = 20 \text{ in.}$$

$$W = 20 + (2)(10.75) = 41.5 \text{ in.}$$

The maximum moment of this span and loading condition occurs when the 16,000 lb-wheel load is placed at midspan.

$$M_{LL} = \frac{PL}{4} = \frac{(16{,}000)(20)(12)}{4} = 960{,}000 \text{ in.-lb}$$

Dead loads:			
asphalt	(3/12)(150)(38.8/12)	=	121 plf
deck	(10.75/12)(50)(38.8/12)	=	145 plf
			266 plf

$$M_{DL} = \frac{(266)(20)^2(12)}{8} = 159{,}600 \text{ in.-lb}$$

$$M_{DL+LL} = 159{,}600 + 960{,}000 = 1{,}120{,}000 \text{ in.-lb}$$

$$S = \frac{(41.5)(10.75)^2}{6} = 800 \text{ in.}^3$$

$$f_b = \frac{1{,}120{,}000}{800} = 1400 \text{ psi} \quad \text{O.K.}$$

Check shear:

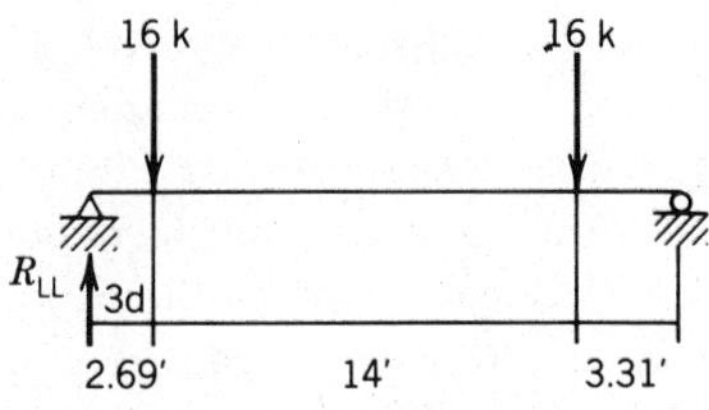

$$R_{LL} = 16\left(\frac{3.31}{20}\right) + 16\left(\frac{17.31}{20}\right) = 16.5 \text{ kips}$$

$$V_{LL} = R_{LL} = 16.5 \text{ kips}$$

$$V_{DL} = (L/2 - 2.69)w_{DL} = (10 - 2.69)(266) = 1.9 \text{ kips}$$

$$V_{LL} + V_{DL} = 16{,}500 + 1900 = 18{,}400 \text{ lb}$$

$$f_v = \frac{(3)(18{,}400)}{(2)(48)(10.75)} = 53 \text{ psi} < 127 \text{ psi} \quad \text{O.K.}$$

SHEATHING, FLOORING, AND DECKING

Sheathing and Flooring

Lumber

Sheathing consisting of lumber up to 1 in. in nominal thickness nailed transversely or diagonally at about 45° to studs or joists is sometimes used in wood frame construction. When subjected to lateral forces such as wind or earthquakes, lumber sheathing and its supporting framework may act as a diaphragm, when properly designed as such, serving to brace the building against the lateral forces and transmitting these forces to the foundations. Diaphragm design procedures for lumber sheathing 1 in. in nominal thickness are given in *Western Woods Use Book* (28).

The edges of lumber sheathing may be square, shiplapped, splined or tongue-and-groove. Sheathing runs should always be spliced over supports unless end matched or scarfed and glued end joints or splice blocks are used. Sheathing used as subflooring in wood frame construction has the effect of a shallow beam, and, therefore, its design procedure is similar to that for beams with certain modifications. Figure 5.73 presents span-load curves for 1 in. nominal thickness lumber sheathing for various types of spans.

Panels

Panels for construction and industrial applications can be manufactured in a variety of ways: as plywood (cross-laminated wood veneer), as composites (veneer faces bonded to reconstituted wood cores), or as nonveneered panels (including waferboard, oriented strand board, and certain specific classes of particleboard).

Some grades of veneered panels are manufactured under the provisions of U.S. Product Standard PS1, *Construction and Industrial Plywood* (25). These panels are referred to in this manual as plywood. Other veneered panels, however, as well as other composite and nonveneered panels are manufactured under provisions of American Plywood Association performance standards to establish their performance for specific construction applications. The panels meeting these APA standards are referred to in this manual as performance-rated panels.

Plywood is divided into two basic types by the exposure durability category. Exterior type is manufactured with a waterproof glue line and C grade or better veneers. Interior types may contain D grade veneers and may use a moisture-resistant glue line. However, most Interior-type plywood manufactured today is made with waterproof glue and designated Exposure 1. Performance-rated panels can be manufactured in three exposure durability classifications: Exterior, Exposure 1, and Exposure 2. Panels marked Exterior are for applications subject to continuous exposure to the weather or moisture and are comparable to Exterior-

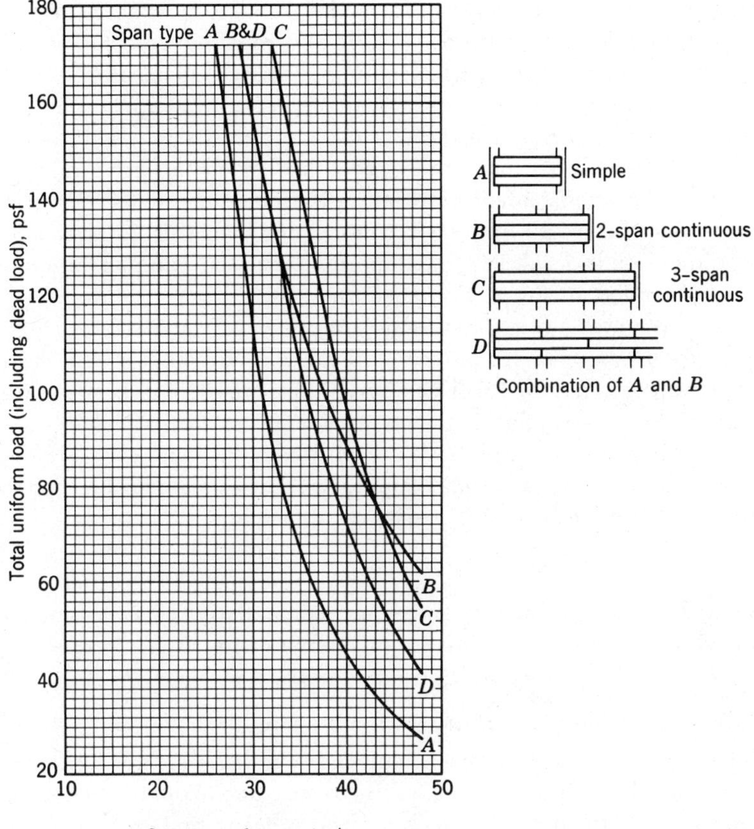

FIGURE 5.73 Span-load curves. For 1-in. nominal thickness lumber sheathing. Conditions: Deflection = $l/240$; Modulus of elasticity = 1,800,000 psi; design value in bending = 1200 psi, span types A–D illustrated at right of figure.

type plywood. Panels with an Exposure 1 designation are intended for protected construction applications where ability to resist moisture during long construction delays or similar exposure conditions is needed. Exposure 1 panels are comparable to Interior-type plywood with exterior glue. Panels with an Exposure 2 designation are intended for protected construction applications where moderate delays in providing protection from moisture and other similar exposure may be expected. Exposure 2 panels are comparable to Interior-type plywood with intermediate glue.

Plywood can be manufactured from over 70 species of wood. These species are divided into five groups by strength and stiffness categories. Group 1, containing Douglas Fir, Western Larch, and Southern Pine, is the strongest group. Group 2 contains the next strongest species, and so on.

Panel products are also categorized by span ratings, which denote the maximum recommended center-to-center spacing in inches of supports over which the panels should be placed in construction applications. A typical span rating for sheathing consists of two numbers separated by a slash, for example, 32/16. The left-hand number (32 in this example) indicates the maximum recommended spacing of supports when used as roof sheathing. The right-hand number indicates the maximum spacing for subflooring. Both of these ratings apply when the long dimension of the panel is across three or more supports.

Table 5.48 contains recommended roof live loads for performance-rated sheathing in conventional applications where the long dimension is perpendicular to the supports. Table 5.49 contains recommended live loads for plywood where the long dimension is parallel to the supports. Table 5.50 contains minimum fastener recommendations for performance-rated panels.

Panel edge support is required to resist concentrated loads in some cases, such as single-layer floors, or where the span–thickness ratio is high in roof sheathing. Such edge supports may be furnished by solid blocking cut in between framing or by tongue-and-groove joints in the plywood. For roof sheathing, specially manufactured H-shaped metal clips may be used. When the panel edges are required to transmit lateral shear, as in some diaphragms, they should be attached to solid blocking or otherwise fastened to resist lateral loading.

Decking

Timber decking is commonly used for floor and roof construction in conjunction with timber joists, purlins, beams, arches and trusses. Timber decking may also be used with other structural materials.

Two-, Three-, and Four-Inch Heavy Timber Decking

For information on species, sizes, patterns, lengths, moisture content, application, specifications, applicable allowable stresses, and allowable loads for 2-, 3-, and 4-in.-nominal-thickness tongue-and-groove heavy timber decking used as roof decking, see *Standard for Tongue-and-Groove Heavy Timber Roof Decking*, AITC 112(26), in Part III of this manual. If decking is to be used for floors or other purposes, special care should be given to end joint locations and to deflection limitations.

Mechanically Laminated Decking

Mechanically laminated decks consist of square-edged dimension lumber set on edge, wide face-to-wide face, with the pieces connected by nails or other fasteners. If side nails are used, they should be long enough to penetrate approximately two and one-half lamination thicknesses for load transfer. Where deck supports are 4 ft center to center or less, side nails should be spaced not more than 30 in. on centers and staggered one third of the spacing in adjacent laminations. When supports are spaced more than 4 ft. center to center, side nails should be spaced approximately 18 in. on centers alternately near top and bottom edges and also staggered one-third of the spacing in adjacent laminations. Two side nails should be used at each end of the butt-joined pieces.

Laminations should be toe-nailed to supports using 20*d* or larger common nails. When the supports are 4 ft center to center or less, alternate laminations should be toe-nailed to alternate supports; when supports are spaced more than 4 ft center to center, alternate laminations should be toe-nailed to every support.

TABLE 5.48

Recommended Uniform Roof Live Load for Performance-rated Panel Sheathing with Long Dimension Perpendicular to Supports[a,b,c]

Panel Span Rating	Panel Thickness (in.)	Maximum Span (in.)		Allowable Live Loads (psf),[d] Spacing of Supports Center-to-Center (in.)							
		With Edge Support[e]	Without Edge Support	12	16	20	24	32	40	48	60
12/0	5/16	12	12	30							
16/0	5/16, 3/8	16	16	55	30						
20/0	5/16, 3/8	20	20	70	50	30					
24/0	3/8, 7/16, 1/2	24	20[f]	90	65	55	30				
24/16	7/16, 1/2	24	24	135	100	75	40				
32/16	15/32, 1/2, 5/8	32	28	135	100	75	55	30			
40/20	19/32, 5/8, 3/4, 7/8	40	32	165	120	100	75	55	30		
48/24	23/32, 3/4, 7/8	48	36	210	155	130	100	65	50	35	
48 oc[g]	1-1/8	60	48				375	205	100	65	40

[a] Source: American Plywood Association.

[b] Rated sheathing and Structural I and II rated sheathing.

[c] When roofing is to be guaranteed by a performance bond, check with roofing manufacturer for minimum thickness, span, and edge support requirements.

[d] 10-psf dead load assumed.

[e] Tongue-and-groove edges, panel edge clips (one between each support, except two between supports 48 in. on center), lumber blocking, or other.

[f] 24 in. for 1/2 in. panels.

[g] Span rating applies to APA-rated Sturd-I-Floor "2-4-1."

TABLE 5.49

Recommended Uniform Loads (psf) for Plywood Roof Sheathing with Long Dimension Parallel to Supports[a, b]

					Number and Length of Spans							
					Four at 12 in.		Three at 16 in.		Two at 24 in.		One at 48 in.	
Panel Grade	Thickness (in.)	No. of Plies[c]	Span Rating	Maximum Span (in.)	Live Load	Total Load	Live Load	Total Load	Live Load	Total Load	Live Load	Total Load
APA Structural I Rated Sheathing	3/8	3	24/0	12	35	45						
	15/32	4	32/16	24	155	200	60	85	20	30		
	15/32	5	32/16	24	230	235	95	130	35	45		
	1/2	4	32/16	24	185	230	75	100	25	35		
	1/2	5	32/16	24	260	265	115	150	40	55		
	19/32 and 5/8	5	40/20	24			190	215	70	75		
	23/32 and 3/4	5 and 6	48/24	24			260	265	90	95		
	1-1/8	7	—	48							45	55

PS 1 Rated Sheathing other than Structural 1	15/32	3 and 4	32/16	12	50	65						
	15/32	5	32/16	24[d]	145	160	55	75	20	25		
	1/2	3 and 4	24/0, 32/16	12	45	60						
	1/2	5	24/0, 32/16	24[d]	160	180	65	85	25	30		
	19/32	4	40/20	24	165	210	70	90	25	35		
	19/32	5	40/20	24	255	260	115	145	40	50		
	5/8	4	32/16, 40/20	24	185	230	80	105	30	40		
	5/8	5	32/16, 40/20	24	280	285	120	160	45	55		
	23/32 and 3/4	4	40/20, 48/24	24			130	175	50	65		
	23/32 and 3/4	5 and 6	40/20, 48/24	24			160	185	60	65		
	7/8	5 and 7	40/20, 48/24	24			345	350	120	125		
	1-1/8, (2-4-1)	7	—	48							20[e]	30[e]

[a] Source: American Plywood Association.

[b] When roofing is to be guaranteed by a performance bond, check with roofing manufacturer for minimum panel thickness, span, and edge support requirements.

[c] Number of layers equal to number of piles, except 4 ply is 3 layer and 6 ply is 5 layer.

[d] Solid blocking recommended at 24-in. span.

[e] 25-psf live and 35-psf total load with solid blocking at panel ends.

TABLE 5.50

Recommended Minimum Fastening Schedule for Performance-rated Panel Roof Sheathing[a, b]

	Nailing[c]			Stapling[d, e]		
		Spacing (in.)		Leg Length	Spacing (in.)	
Panel Thickness (in.)	Size	Panel Edges	Intermediate	(in.)	Panel Edges	Intermediate
5/16	6d	6	12	$1\frac{1}{4}$	4	8
3/8	6d	6	12	$1\frac{3}{8}$	4	8
7/16, 15/32, 1/2	6d	6	12	$1\frac{1}{2}$	4	8
19/32, 5/8, 23/32, 3/4, 7/8	8d	6	12[f]	—	—	—
1-1/8, 1-1/4	8d or 10d	6	12[f]	—	—	—

[a] Source: American Plywood Association.

[b] Closer nail spacing may be required to obtain higher diaphragm shear values.

[c] Use common smooth or deformed shank nails with panels to 1 in. thick. For 1-1/8 and 1-1/4 in. panels, use 8d deformed shank or 10d common smooth-shank nails.

[d] Values are for 16-gage galvanized wire staples with a minimum crown width of 3/8 in.

[e] For stapling asphalt shingles to 5/16 in. and thicker panels, use staples with a 3/4 in. minimum crown width and a 3/4 in. leg length. Space according to shingle manufacturer's recommendations.

[f] For spans 48 in. or greater, space nails 6 in. at all supports.

The five-span arrangements in *Standard for Tongue-and-Groove Heavy Timber Roof Decking*, AITC 112 (26), in Part III of this manual are also applicable for mechanically laminated decks. The design formulas are as follows:

Simple span (Type I):

$$w_b = \frac{8F'_bI}{l^2c} \quad \text{and} \quad w_\Delta = \frac{100\ \Delta E'I}{5l^4}$$

Controlled random layup (Type 2):

$$w_b = \frac{20F'_bI}{3l^2c} \quad \text{and} \quad w\Delta = \frac{100\ \Delta E'I}{l^4}$$

Cantilevered pieces intermixed (Type 3):

$$w_b = \frac{20F'_bI}{3l^2c} \quad \text{and} \quad w\Delta = \frac{105\ \Delta E'I}{l^4}$$

Combination simple- and two-span continuous (Type 4):

$$w_b = \frac{8F'_bI}{l^2c} \quad \text{and} \quad w_\Delta = \frac{109\ \Delta E'I}{l^4}$$

Two-span continuous (Type 5):

$$w_b = \frac{8F'_bI}{l^2c} \quad \text{and} \quad w_\Delta = \frac{185\ \Delta E'I}{l^4}$$

where w_b = allowable load limited by bending (plf),
w_Δ = allowable load limited by deflection (plf),
F'_b = design value in bending (psi) modified by applicable modifiers,
E' = design value for modulus of elasticity (psi) modified by applicable modifiers,
I = moment of inertia (in.^{4}),
l = span (in.),
c = half the depth of the decking (in.), and
Δ = deflection limitation (in.).

Table 5.51 gives allowable uniformly distributed loads for mechanically laminated decks for various spans as limited by bending and deflection. The table is based on the use of seasoned lumber, normal duration of loading, and loads applied normal to the decking surface. The allowable loads include dead load.

Other Decking

Other special decking products are available. They include panelized decking, glued laminated decking, and heavy plywood decking such as "2-4-1." Panelized decking is a decking component made up of splined panels, usually about 2 ft wide and of any specified length up to a maximum, which may vary among manufacturers. Glued laminated decking is manufactured by laminating two or more pieces of lumber into single decking members. Common nominal lumber thick-

TABLE 5.51

Design Loads for Mechanically Laminated Decks[a]

Span (ft)	Nominal Thickness of Deck (in.)	Uniform Load (psf)						
		Types 1, 2, and 3[b]	Types 4 and 5[b]	Type 1[c]	Type 2[c]	Type 3[c]	Type 4[c]	Type 5[c]
4	4	1021	851	1488	3587	2114	2036	1939
5	4	653	544	762	1836	1082	1042	993
	6	1613	1344	2958	7126	4199	4045	3852
	4	454	378	441	1063	626	603	574
6	6	1120	934	1712	4124	2430	2341	2229
	8	1946	1622	3920	9446	5566	5362	5106
	4	333	278	278	669	394	380	362
7	6	823	686	1078	2597	1530	1474	1404
	8	1430	1192	2469	5948	3505	3376	3215
	4	255	213	186	448	264	254	242
8	6	630	525	722	1740	1025	988	940
	8	1095	912	1654	3985	2348	2262	2154
	10	1782	1485	3435	8276	4877	4698	4474
	4	201	168	131	315	186	179	170
9	6	498	415	507	1222	720	694	660
	8	865	721	1162	2799	1649	1589	1513
	10	1408	1174	2412	5813	3425	3299	3142
	6	403	336	370	891	525	506	481
10	8	701	584	847	2040	1202	1158	1103
	10	1141	951	1759	4237	2497	2405	2290
11	8	579	483	636	1533	903	870	829
	10	943	786	1321	3184	1876	1807	1721
12	8	487	406	490	1181	696	670	638
	10	792	660	1018	2452	1445	1392	1326
13	8	415	346	385	929	547	527	502
	10	675	562	800	1929	1137	1095	1043
14	8	357	298	309	744	438	422	402
	10	582	485	641	1544	910	877	835
15	8	311	260	251	605	356	343	327
	10	507	422	521	1256	740	713	679
16	8	274	228	207	498	294	283	269
	10	446	371	429	1034	610	587	559
17	8	242	202	172	415	245	236	224
	10	395	329	358	862	508	490	466

TABLE 5.51 (*Continued*)

Span (ft)	Nominal Thickness of Deck (in.)	Uniform Load (psf)						
		Types 1, 2, and 3[b]	Types 4 and 5[b]	Type 1[c]	Type 2[c]	Type 3[c]	Type 4[c]	Type 5[c]
18	8	216	180	145	350	206	199	189
	10	352	293	302	727	428	412	393
19	10	316	263	256	618	364	351	334
20	10	285	238	220	530	312	301	286

[a]For seasoned lumber, normal conditions of loading, and load applied normal to decking surface. Loads for other stress and deflection values can be determined by proportion. Load includes weight of decking, which should be subtracted to determine allowable superimposed load.

[b]Limited by bending $F_b = 1000$ psi.

[c]Limited by deflection. $E = 1{,}000{,}000$ psi, $\Delta_A = l/240$.

nesses of the decking are 2, 3, or 4 in., although exact finished sizes may differ between manufacturers. Plywood panels known as "2-4-1" plywood are $1\frac{1}{8}$ in. thick, for 32- or 48-in. floor spans and may also be used as roof decking in "heavy timber" construction. They are usually supplied with tongue-and-groove joints on the long edge. Manufacturers of these special decking products should be consulted for information concerning their products. Table 5.48 gives design loads for "2-4-1" and other plywood decking systems.

Cantilevered Overhangs

Under uniformly distributed loads, cantilevered overhangs act to reduce the deflections in the remaining areas of a deck structure. Thus, span increases are often justified where overhangs are used. The effects of cantilevered overhangs are varied. The span arrangement and the length of overhang affect support reactions, bending stress, and span and overhang deflections. Charts that permit the determination of these factors for simple-span, two-span continuous, and up to five-span continuous arrangements are given in Section 7 of this manual.

The designer should also consider the effects of heating on snow loads for roofs with cantilevered overhangs. Unbalanced loading may be created where melting occurs over heated areas but not over unheated overhangs.

STRUCTURAL DIAPHRAGMS

Structural diaphragms are relatively thin, usually rectangular, structural elements capable of resisting shear parallel to their edges. A conventional frame roof, wall, or floor will normally function as a structural diaphragm with only slight design modification. They may be used as walls in a vertical position, as roofs or floors in a horizontal position, or as roofs pitched or curved to conform with common truss shapes.

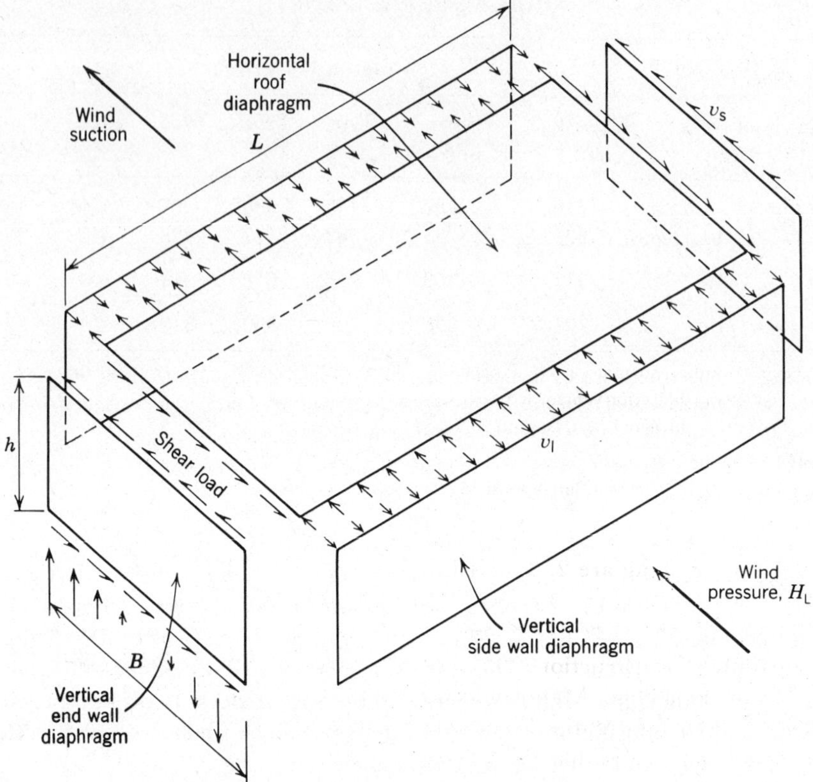

Figure 5.74 Lateral forces on a simple structure.

The function of the diaphragm is to brace a structure against lateral forces, such as wind or earthquake loads, and to transmit these forces to the other resisting elements of the structure. Figure 5.74 illustrates the distribution of lateral forces acting on a simple structure. The lateral loads produced act on the side walls spanning from foundation to roof. The top of the side wall thereby produces horizontal loads against the roof framing. When designed as a diaphragm, the roof framing system acts as a large plate girder, generally with continuous chords resisting bending moment as flanges and with the sheathing itself resisting shear forces as the web. The roof framing system carries the side wall reactions as horizontal loads and spans to the end walls, which acts as a cantilever system extending up from the foundation to provide the necessary horizontal support.

A bibliography on the design and performance of lumber and wood panel diaphragms can be found in (30).

Common Types

Common types of wood diaphragms are the following:

(a) Plywood-consists of sheets of plywood or other panel products fastened to cross members usually by means of nails, although sometimes by adhesives. For additional information see *Plywood Diaphragm Construction* (27).

(b) Heavy timber decking-consists of 2-, 3-, or 4-in. heavy timber decking and may be covered with plywood sheathing to improve diaphragm values. Available test data on this system indicate that when sheathed with plywood, it can be designed as equivalent to a blocked plywood diaphragm.

(c) Diagonal sheathing-consists either of 1-in.-nominal-thickness boards or 2-in.-nominal-thickness lumber, nailed at a 45° angle in a single layer to cross members. Considerably greater strength and stiffness result from this placement than when transverse sheathing is used. Tests verify that although there may be considerable bending in the sheathing, the primary load resistance in an efficient diaphragm is due to the axial stress in the sheathing. The axial stress may be either direct tension or compression. Moment forces are resisted by the continuous chords.

(d) Double diagonal sheathing-consists of two layers of diagonal sheathing, one on top of the other, with the sheathing in one layer at a 90° angle with the sheathing in the outer layer. This type is considerably stiffer and stronger than the types (e) and (c). One layer of sheathing is in axial tension and is counteracted by the other layer, which is in compression; thus, the effects counteract and cancel each other.

(e) Transverse sheathing-consists of either 1-in. nominal boards or 2-in.-nominal-thickness lumber, nailed in a single layer at right angles to the direction of cross members, such as joists or studs. This type is suitable when loads are light and when deflection is not important. When used vertically as a wall diaphragm, the load capacity of transverse sheathing is quite low compared to that of other types, and it is recommended that cross bracing be used to increase the strength and stiffness.

General Design Procedure

All diaphragms are essentially deep beams with a shear-resistant web. As the web is not considered effective in bending, ordinary beam theory rather than deep beam theory normally is used. The sheathing material acts as the beam web to carry the shear forces. Because the individual pieces of the sheathing are discontinuous, the sheathing connections are one of the most critical items in diaphragm design. The beams, girts, studs, columns, joists, and purlins act as stiffeners and cross ties for the web. Chord members must be provided to carry the flange forces of the beam. Either additional members, a portion of the walls, or some other continuous framing must be provided. For frame buildings with shear-resistant walls, some engineers consider that the chord forces are directly transferred to the foundation by the walls in shear, so no separate chord members are required.

Openings in diaphragms must have the shear force on their two sides distributed in proportion to their stiffeners (size). The bending stiffness is calculated from the main diaphragm chords and the secondary chords required at the edges of the opening. The resulting chord forces are the total of the force due to bending on the whole diaphragm with the opening removed plus the secondary bending on each side of the hole loaded with the moment due to the shear on the side times one-half the length of the opening. This secondary bending is of opposite sign at the two ends so it is additive to the main chords. The secondary chords must extend far enough into the diaphragm to anchor the chord forces into the diaphragm. At both ends of the opening, drag struts are required to transmit the shear force on the cut section to each side. Drag struts or collectors are the members used to collect shear forces from the diaphragm at openings or other discontinuities. Most frequently, they are anchored to shear walls.

For design of lumber diaphragms, see *Western Woods Use Book* (28). For plywood diaphragms, see *Plywood Diaphragm Construction* (27) or *Design of Wood Structures* (29).

REFERENCES

1. American Institute of Timber Construction, *Standard Specifications for Structural Glued Laminated Timber of Softwood Species*, AITC 117—Design, Englewood, CO, 1984.
2. American Institute of Timber Construction, *Standard Specification for Hardwood Glued Laminated Timber*, AITC 119, Englewood, CO, 1985.
3. National Forest Products Association, *National Design Specification for Wood Construction*, Washington, D.C., 1982.
4. United States Department of Agriculture, Forest Service, Forest Products Laboratory, *Wood Handbook*, Agriculture Handbook No. 72, Madison, WI, 1955.
5. United States Department of Agriculture, Forest Service, Forest Products Laboratory, *Wood Handbook: Wood as an Engineering Material*, Agriculture Handbook No. 72, Madison, WI, 1974.
6. American Society for Testing and Materials, *Standard Methods of Testing Small Clear Specimens of Timber*, ASTM D 143, Philadelphia, PA, 1952 (reapproved 1978).
7. Canadian Standards Association, *Code for Engineering Design in Wood*, CAN3-086, Rexdale, Ontario, 1980.
8. United States Department of Agriculture, Forest Service, Forest Products Laboratory, *Fabrication and Design of Glued Laminated Wood Structural Members*, Technical Bulletin No. 1069, 1954 (available from AITC).
9. United States Department of Agriculture, Forest Service, Forest Products Laboratory, *Deflection and Stresses of Tapered Wood Beams*, Research Paper FPL 34, Madison, WI, 1965.
10. United States Department of Agriculture, Forest Service, Forest Products Laboratory, *The Glued Laminated Wooden Arch*, Technical Bulletin No. 691, Madison, WI, 1939.
11. Colorado State University, Civil Engineering Department, *Behavior and Design of Double-Tapered Pitched and Curved Glulam Beams*, Structural Research Report No. 16, Ft. Collins, CO, 1976.
12. American Institute of Timber Construction, *Deflection of Arches*, Technical Note No. 2, Englewood, CO, 1976.
13. American National Standards Institute, *Minimum Design Loads for Buildings and Other Structures*, ANSI A58.1, New York, 1982.
14. American Institute of Steel Construction, *Manual of Steel Construction*, Chicago, IL, 1980.
15. Timber Engineering Company (TECO), *Design Manual for TECO Timber Connector Construction*, Washington, D.C., 1973.
16. American Wood-Preservers' Association, *Book of Standards*, Stevensville, MD, 1984.
17. American Society for Testing and Materials, *Standard Specifications and Methods for Establishing Recommended Design Stresses for Round Timber Construction Poles*, ASTM D 3200, Philadelphia, PA, 1974.
18. American National Standards Institute, *Specifications and Dimensions for Wood Poles*, ANSI 05.1, New York, 1979.
19. American Wood-Preserver's Institute, *Pile Foundations Know-How*, Vienna, VA, 1969.

20. American Society for Testing and Materials, *Standard Specifications for Round Timber Piles*, ASTM D25, Philadelphia, PA, 1979.
21. United States Government, Federal Specification, *Wood Preservation, Treating Practice,* TT-W-571i (2), 1972.
22. American Society for Testing and Materials, *Standard Method for Establishing Design Stresses for Round Timber Piles*, ASTM D 2899, Philadelphia, PA, 1974.
23. American Association of State Highway and Transportation Officials, *Standard Specifications for Highway Bridges*, Washington, D.C., 1983.
24. American Railway Engineering Association, *Manual of Recommended Practice*, Chicago, IL, 1969.
25. United States Department of Commerce, *Construction and Industrial Plywood*, PS 1, 1983.
26. American Institute of Timber Construction, *Standard for Tongue-and-Groove Heavy Timber Roof Decking*, AITC 112, Englewood, CO, 1981.
27. American Plywood Association, *Plywood Diaphragm Construction*, U310, Tacoma, WA, 1978.
28. Western Wood Products Association, *Western Woods Use Book*, Portland, OR, 1983.
29. D. E. Breyer, *Design of Wood Structures*, McGraw-Hill, New York, 1980.
30. American Society of Civil Engineers, *Bibliography of Lumber and Wood Diaphragms, Journal of Structural Engineering,* **109**(12) Dec. 1983.
31. American Society for Testing and Materials, *Standard Specification for Pressure Treatment of Timber Products*, ASTM D 1760, Philadelphia, PA, 1983.

SECTION 6

FASTENERS AND CONNECTIONS

FASTENERS—GENERAL CONSIDERATIONS

The following considerations, in general, apply to all the mechanical fasteners for timber joints covered in this section. Considerations applicable to a particular mechanical fastener will be found under the appropriate headings herein. The factors that require consideration in determining design values for mechanically fastened joints are (1) lumber species (density), (2) critical section, (3) angle of load to grain, (4) spacing of mechanical fastenings (5) edge and end distances, (6) conditions of loading, (7) eccentricity, and (8) modifications to tabular design values.

Lumber Species

The design values for mechanical fasteners vary with the species of wood with which they are used. The species load groups for sawn lumber in Table 6.1 and for glued laminated timber in Table 6.2 apply to design values for mechanical fasteners.

The recommendations herein are based on the assumption that the mechanical fasteners are used in structural glued laminated timber or lumber that meets the requirements for stress-graded lumber, and that proper fabrication practices have been followed in the installation of the fasteners.

Modification Factors for Fasteners

The design values for fasteners in the tables in this section are shown for one fastener in wood when fabricated with full edge and end distance requirements and spacing and used under the stated moisture content conditions for a normal duration of load.

When other conditions exist, these design values must be adjusted by the appropriate modification factors.

These modifying factors for design values of fasteners are applied in the same manner as modifying factors for strength properties.

TABLE 6.1

Species Groups for Fastener Design for Sawn Lumber

Species	Bolt Group[a]	Timber Connector Load Group[b] (Shear Plates and Split Rings)	Grouping for Lag Bolts, Drift Bolts, Nails, Spikes, Wood Screws, Staples, and Metal Plate Connectors	
			Group	Specific Gravity(G)[c]
Ash, Commercial White	2	A	I	0.62
Aspen	12	D	IV	0.40
Aspen, Northern[e]	12	C	III	0.42
Beech	4	A	I	0.68
Birch, Sweet and Yellow	4	A	I	0.66
Cedar, Northern White	12	D	IV	0.31
Cedars, Western[d]	9	D	IV	0.35
Coast Species[e]	12	D	IV	0.39
Cottonwood, Black	12	D	IV	0.33
Cottonwood, Eastern	12	D	IV	0.41
Cypress, Southern	3	C	III	0.48
Douglas Fir-Larch[d]	3	B	II	0.51
Douglas Fir-Larch (dense)	1	A	II	—[f]
Douglas Fir, South	6	C	III	0.48
Eastern Woods	12	D	IV	0.38
Fir, Balsam	11	D	IV	0.38
Hem-Fir[d]	8	C	III	0.42
Hemlock				
Eastern-Tamarack[e]	8	C	III	0.45
Mountain	9	C	III	0.47
Western[d]	8	C	III	0.48
Hickory and Pecan	2	A	I	0.75
Maple, Black and Sugar	4	A	I	0.66
Northern Species[e]	12	D	IV	0.35
Oak, Red and White	5	A	I	0.67
Pine				
Eastern White[d]	11	D	IV	0.38
Idaho White	11	D	IV	0.40
Lodgepole	10	C	III	0.44
Northern	9	C	III	0.46
Ponderosa[e]	11	C	III	0.49
Ponderosa-Sugar	11	C	III	0.42
Red[e]	11	C	III	0.42
Southern	3	B	II	0.55
Southern (dense)	1	A	II	—[f]
Western White	11	D	IV	0.40
Poplar, Yellow	10	C	III	0.46
Redwood				
California	3	C	III	0.42
California (open grain)	8	D	IV	0.37

TABLE 6.1 ***(Continued)***

Species	Bolt Group[a]	Timber Connector Load Group[b] (Shear Plates and Split Rings)	Grouping for Lag Bolts, Drift Bolts, Nails, Spikes, Wood Screws, Staples, and Metal Plate Connectors	
			Group	Specific Gravity(G)[c]
Spruce				
Eastern	10	C	III	0.43
Engelmann-Alpine Fir	12	D	IV	0.36
Sitka	10	C	III	0.43
Sitka, Coast[e]	10	D	IV	0.39
Spruce-Pine-Fir[e]	10	C	III	0.42
West Coast Woods (mixed species)	12	D	IV	0.35
White Woods (Western Woods)	12	D	IV	0.35

[a] See Tables 6.20–6.22 for species and density groupings applicable to bolt design.

[b] When stress graded.

[c] Based on weight and volume when oven-dry. These specific-gravity values are to be used for the determination of withdrawal design values for lag bolts, nails, spikes, and wood screws.

[d] Also applies when species name includes the designation "North."

[e] Applies when graded in accordance with National Lumber Grades Authority *Standard Grading Rules for Canadian Lumber* (2).

[f] The specific gravity of dense lumber is slightly higher than for medium-grain lumber. However, the design values for this group are based on the average specific gravity of the species, which is 0.51 for Douglas Fir–Larch and 0.55 for Southern Pine.

The modifying factors for fasteners consist of

C_D, Duration of load factor
C_M, Moisture content factor
C_R, Fire retardant treatment factor
C_t, Temperature factor
C_e, Edge distance factor
C_n, End distance factor
C_s, Spacing factor
C_d, Depth of embedment factor
C_g, Group action factor
C_{st}, Steel sideplate factor
C_{lb}, Lag bolt modifying factor

Table 6.3 shows the applicability of these modifying factors to the various types of fasteners.

Duration of Load Factor C_D

The same duration of load factors applicable to the strength properties of wood are also applicable to fasteners where the strength of the fastener is controlled by the wood. An exception is for shear plates where the strength is limited by the

TABLE 6.2

Species Groups for Fastener Design for Softwood Glued Laminated Timber[a,b]

Species	Bolt Load Group[c]	Timber Connector Load Group (Shear Plates and Split Rings)	Grouping for Lag Bolts, Drift Bolts, Nails, Spikes, Wood Screws, Staples, and Metal Plate Connectors	
			Group	Specific Gravity(G)[d]
Douglas Fir-Larch (dense)	1	A	II	0.51[e]
Douglas Fir-Larch (close grain)	3	B	II	0.51
Douglas Fir-Larch (medium) grain)	3	B	II	0.51
Douglas Fir-Larch (coarse grain)	8	C	III	0.41[f]
Hem-Fir	8	C	III	0.42
Western Woods	12	D	IV	0.36
Southern Pine (dense)	1	A	I	0.55[e]
Southern Pine (medium grain)	3	B	II	0.55
Southern Pine (coarse grain)	8	C	III	0.41[f]
California Redwood (close grain)	3	C	III	0.42
Douglas Fir South (dense)	6	B	II	0.48[e]
Douglas Fir South (medium grain)	6	C	III	0.48

[a]See Table 3, AITC 117—Design (7) for species and location of species within lamination combinations.

[b]For hardwoods use Table 6.1

[c]The bolt load groups are based on the numbering system used in the *National Design Specification* (3). For bolt design values in sawn lumber see Table 6.20 and Tables 6.21 and 6.22 for bolts in glued laminated timber.

[d]Based on oven-dry weight and volume—for use in determining withdrawal loads for lag bolts, nails, spikes, and wood screws.

[e]The specific gravity for dense lumber is slightly higher than medium-grain lumber but the design values in withdrawal for fasteners are based on the average specific gravity of the species or species group.

[f]The specific gravities listed for coarse grain Southern Pine and Douglas Fir are consistent with the strength properties.

bearing of the shear plate on the bolt. Where this occurs in Tables 6.12 and 6.13 the exceptions are indicated. (*Note:* C_D also applies to load perpendicular to grain design values for fasteners but not to compression perpendicular to grain design values.)

Moisture Content Factor C_M

Sawn lumber can be fabricated and used under various moisture content conditions. Tables 6.4 and 6.5 contain the modification factors for the various moisture contents at time of fabrication and moisture content in use. Glued laminated tim-

TABLE 6.3

Applicability of Modification Factors for Fasteners

	Modification Factor										
Type of Fastener	C_D	C_M	C_t	C_R	C_e	C_n	C_s	C_d	C_g	C_{st}	C_{lb}
Split rings	Yes	Yes	Yes	Yes	Yes	Yes	Yes	No	Yes	No	Yes
Shear plates	Yes[a]	Yes	Yes	Yes	Yes	Yes	Yes	No	Yes	No[c]	Yes
Bolts	Yes	Yes	Yes	Yes	No	Yes	Yes	No	Yes	—[d]	No
Lag bolts	Yes	Yes	Yes	Yes	No	Yes	Yes	No	Yes	No[c]	No
Drift bolts or pins	Yes	Yes	Yes	Yes	Yes	Yes	Yes	Yes	Yes	Yes	No
Wire nails and spikes	Yes	Yes	Yes	Yes	No[b]	No[b]	No[b]	Yes	No	Yes	No
Threaded, hardened steel nails	Yes	Yes	Yes	Yes	No[b]	No[b]	No[b]	Yes	No	Yes	No
Wood screws	Yes	Yes	Yes	Yes	No[b]	No[b]	No[b]	Yes	No	Yes	No
Metal plate connectors	Yes	Yes	Yes	Yes	No	No	No	No	No	No	No
Spiral dowels	Yes	Yes	Yes	Yes	Yes	Yes	Yes	Yes	Yes	Yes	No

[a]Duration of load increases apply except under some conditions where the capacity of the shear plate is determined by the steel (see Tables 6.12 and 6.13).

[b]The end distance, edge distance, and spacing shall be such as to avoid unusual splitting of the wood.

[c]Tabular design values for shear plates and lag bolts with steel side plates are included in Tables 6.13 and 6.26, respectively, and no further modification should be made.

[d]Tabular design values for bolts in sawn lumber included in Table 6.20 should be modified by the factor C_{st} when steel side plates are used. Tabular design values for bolts in glued laminated timber with steel side plates may be obtained directly from Table 6.21 and used without further modification.

ber is dry when fabricated but may be used under either dry or wet conditions. Table 6.6 contains the modification factors for glued laminated timber. (Note values of C_M for fasteners may vary from the values used for wood.)

Temperature Factor C_t

The strength of fasteners in wood is generally controlled by the strength of the wood. In those unusual cases where a reduction is required for strength properties of wood due to temperature, the design value for fasteners should also be adjusted by the temperature factor.

Fire Retardant Treatment Factor C_R

The design values for fasteners in pressure-impregnated fire retardant treated wood must be multiplied by a factor C_R. The manufacturer of the treatment should be contacted for specific information on the value of C_R, which is the same as is used for the species of lumber being fastened. In addition, lumber must be dried after treatment in accordance with the *Structural Lumber, Fire-Retardant Treatment by*

TABLE 6.4

Fastener Moisture Content Factors—Sawn Lumber

Type of Fastener	Condition of Wood[b] At Time of Fabrication	Condition of Wood[b] In Service	Modification Factor C_M
Timber connectors[c] (shear plates and split rings)	Dry	Dry	1.0
	Partially seasoned[d]	Dry	—[d]
	Wet	Dry	0.8
	Dry or wet	Partially seasoned or wet	0.67
Bolts or lag bolts	Dry	Dry	1.0
	Partially seasoned or wet[d]	Dry	See Table 6.5
	Dry or wet	Exposed to weather	0.75
	Dry or wet	Wet	0.67
Drift bolts or pins—Laterally loaded	Dry or wet	Dry	1.0
	Dry or wet	Partially seasoned or wet, or subject to wetting and drying	0.70
Wire nails and spikes			
Withdrawal loads	Dry	Dry	1.0
	Partially seasoned or wet	Will remain wet	1.0
	Partially seasoned or wet	Dry	0.25
	Dry	Subject to wetting and drying	0.25
Lateral loads	Dry	Dry	1.0
	Partially seasoned or wet	Dry or wet	0.75
	Dry	Partially seasoned or wet	0.75
Threaded, hardened steel nails	Dry or wet	Dry or wet	1.0
Wood screws	Dry or wet	Dry	1.0
	Dry or wet	Exposed to weather	0.75
	Dry or wet	Wet	0.67
Metal plate connectors	Dry	Dry	1.0
	Partially seasoned or wet	Dry or wet	0.8

TABLE 6.5

Moisture Content Factors for Laterally Loaded Bolts and Lag Bolts in Timber Seasoned in Place—Sawn Lumber[a, b]

Arrangement of Bolts or Lag Bolts	Type of Splice Plate	Modification Factor C_M
One fastener only, or Two or more fasteners placed in a single line parallel to grain, or Fasteners placed in two or more lines parallel to grain with separate splice plates for each line	Wood or metal	1.0
All other arrangements	Wood or metal	0.4

[a] Factors apply when wood is at or above the fiber saturation point (wet) at time of fabrication but dries to a moisture content of 19% or less (dry) before full design load is applied. For wood partially seasoned when fabricated, adjusted intermediate values may be used.

[b] Source: *National Design Specification for Wood Construction* (3).

Pressure Process, AWPA C20 (1) or *Plywood Fire-Retardant Treatment by Pressure Process,* AWPA C27 (1).

Edge Distance Factor C_e

The tabular design values for bolts, lag bolts, connectors, drift bolts, and drift pins, are based on the full edge distance requirements. When full edge distance is not used in fabrication, the tabular design loads must be modified by the edge distance factor, which is listed for each fastener where required.

Footnotes to Table 6.4

[a] Source: *National Design Specification for Wood Construction* (3).

[b] Condition of wood definitions applicable to fasteners are:
"Dry" wood has a moisture content of 19% or less.
"Wet" wood has a moisture content at or above the fiber saturation point (approximately 30%).
"Partially seasoned" wood, for the purposes of this table, has a moisture content greater than 19% but less than the fiber saturation point (approximately 30%).
"Exposed to weather" implies that the wood may vary in moisture content from dry to partially seasoned, but is not expected to reach the fiber saturation point at times when the joint is under full design load.
"Subject to wetting and drying" implies that the wood may vary in moisture content from dry to partially seasoned or wet, or vice versa, with consequent effects on the tightness of the joint.

[c] For timber connectors, moisture content limitations apply to a depth of $\frac{3}{4}$ in. from the surface of the wood.

[d] When timber connectors, bolts, or laterally loaded lag bolts are installed in wood that is partially seasoned at the time of fabrication but that will be dry before full design load is applied, proportional intermediate values may be used.

TABLE 6.6

Fastener Moisture Content Factors—Glued Laminated Timber[a, b]

	Condition of Wood		
Type of Fastener	At Time of Fabrication	In Service	Modification Factor, C_M
Timber connectors (shear plates and split rings)	Dry	Dry	1.0
		Wet	0.67
Bolts or lag bolts	Dry	Dry	1.0
		Wet	0.67
Drift bolts or pins-laterally loaded	Dry	Dry	1.0
		Wet	0.7
Wire nails and spikes			
Withdrawal loads	Dry	Dry	1.0
		Wet	0.25
Lateral loads	Dry	Dry	1.0
		Wet	0.75
Threaded, hardened steel nails	Dry	Dry	1.0
		Wet	1.0
Wood screws	Dry	Dry	1.0
		Wet	0.67
Spiral dowels	Dry	Dry	1.0
		Wet	0.67

[a] Dry glued laminated timber has a moisture content of 16% or less.

[b] Wet glued laminated timber in service is, for purposes of this table, that with a moisture content greater than 16%.

End Distance Factor C_n

The tabular design values for bolts, lag bolts, connectors, drift bolts, and drift pins are based on full end distance requirements. When full end distance is not used in fabrication, the tabular design loads must be modified by the end distance factor, which is listed for each fastener.

Spacing Factor C_s

The tabular design values for bolts, lag bolts, connectors, drift bolts, and drift pins are based on the full spacing requirements. When full spacing is not used, the tabular design values must be modified by the spacing factor, which is listed for each fastener.

Depth of Embedment Factor C_d

The tabular design values of lag bolts, wood screws, nails, and spikes are based on a specific embedment of the fastening in the piece receiving the point. The design value at a smaller embedment is less and is usually determined by interpolation between the full design value and the lower design value at minimum penetration. It is sometimes advantageous to show this decreased value as a decimal fraction of the full load. In which case, it is used as the depth of embedment factor, C_d.

Group Action Factor C_g

The design values of bolts, lag bolts, connectors, drift bolts, and drift pins are decreased when used in long rows. The modification factors are shown in Tables 6.7 and 6.8.

The factor is applicable to the group of fasteners acting as a whole, but for convenience of calculation, it may be applied to the single fastener value along with other modifiers. The total load that can be carried by the joint is the number of fasteners multiplied by the design value of each fastener.

Steel Side Plate Factor C_{st}

The design values of some fasteners may be increased when steel rather than wood side plates are used (See Table 6.3).

Values in Table 6.13 for shear plates, Table 6.21 for bolts in glued laminated timbers, and Table 6.26 for lag bolts, have been increased for steel side plates and no further increase is needed.

TABLE 6.7

Group Action Factor C_g for Wood Side Plate for Connector, Bolt, and Laterally Loaded Lag Bolt Joints[a]

		Number of fasteners in a row										
A_1/A_2[b]	A_1 (in^2)[c]	2	3	4	5	6	7	8	9	10	11	12
0.5[d]	<12	1.00	0.92	0.84	0.76	0.68	0.61	0.55	0.49	0.43	0.38	0.34
	12 – 19	1.00	0.95	0.88	0.82	0.75	0.68	0.62	0.57	0.52	0.48	0.43
	>19 – 28	1.00	0.97	0.93	0.88	0.82	0.77	0.71	0.67	0.63	0.59	0.55
	>28 – 40	1.00	0.98	0.96	0.92	0.87	0.83	0.79	0.75	0.71	0.69	0.66
	>40 – 64	1.00	1.00	0.97	0.94	0.90	0.86	0.83	0.79	0.76	0.74	0.72
	>64	1.00	1.00	0.98	0.95	0.91	0.88	0.85	0.82	0.80	0.78	0.76
1.0[d]	<12	1.00	0.97	0.92	0.85	0.78	0.71	0.65	0.59	0.54	0.49	0.44
	12 – 19	1.00	0.98	0.94	0.89	0.84	0.78	0.72	0.66	0.61	0.56	0.51
	>19 – 28	1.00	1.00	0.97	0.93	0.89	0.85	0.80	0.76	0.72	0.68	0.64
	>28 – 40	1.00	1.00	0.99	0.96	0.92	0.89	0.86	0.83	0.80	0.78	0.75
	>40 – 64	1.00	1.00	1.00	0.97	0.94	0.91	0.88	0.85	0.84	0.82	0.80
	>64	1.00	1.00	1.00	0.99	0.96	0.93	0.91	0.88	0.87	0.86	0.85

[a] Source: *National Design Specification for Wood Construction* (3).

[b] A_1 = cross-sectional area of main member(s) before boring or grooving and A_2 = sum of the cross-sectional areas of side members before boring or grooving.

[c] When A_1/A_2 exceeds 1.0, use A_2 instead of A_1.

[d] For A_1/A_2 between 0 and 1.0, interpolate or extrapolate from the tabulated values. When A_1/A_2 exceeds 1.0, use A_2/A_1.

TABLE 6.8

Group Action Factor C_g for Metal Side Plate for Connector, Bolt, and Laterally Loaded Lag Bolt Joints[a]

		Number of fasteners in a row										
A_1/A_2[b]	A_1 (in^2)	2	3	4	5	6	7	8	9	10	11	12
2–12	5 – 8	1.00	0.78	0.64	0.54	0.46	0.40	0.35	0.30	0.25	0.20	0.15
	9 – 16	1.00	0.85	0.73	0.63	0.54	0.48	0.42	0.38	0.34	0.30	0.26
	17 – 24	1.00	0.91	0.83	0.74	0.66	0.59	0.53	0.48	0.43	0.38	0.33
	25 – 39	1.00	0.94	0.87	0.80	0.73	0.67	0.61	0.56	0.51	0.46	0.42
	40 – 64	1.00	0.96	0.92	0.87	0.81	0.75	0.70	0.66	0.62	0.58	0.55
	65 – 119	1.00	0.98	0.95	0.91	0.87	0.82	0.78	0.75	0.72	0.69	0.66
	120 – 199	1.00	0.99	0.97	0.95	0.92	0.89	0.86	0.84	0.81	0.79	0.78
12–18	17 – 24	1.00	0.94	0.88	0.81	0.74	0.67	0.61	0.55	0.49	0.43	0.37
	25 – 39	1.00	0.96	0.91	0.86	0.80	0.74	0.68	0.62	0.56	0.50	0.44
	40 – 64	1.00	0.98	0.94	0.90	0.85	0.80	0.75	0.70	0.67	0.62	0.58
	65 – 119	1.00	0.99	0.96	0.93	0.90	0.86	0.82	0.79	0.75	0.72	0.69
	120 – 199	1.00	1.00	0.98	0.96	0.94	0.92	0.89	0.86	0.83	0.80	0.78
	200 or more	1.00	1.00	1.00	0.98	0.97	0.95	0.93	0.91	0.90	0.88	0.87
18–24	40 – 64	1.00	1.00	0.96	0.93	0.89	0.84	0.79	0.74	0.69	0.64	0.59
	65 – 119	1.00	1.00	0.97	0.94	0.92	0.89	0.86	0.83	0.80	0.76	0.73
	120 – 199	1.00	1.00	0.99	0.98	0.96	0.94	0.92	0.90	0.88	0.86	0.85
	200 or more	1.00	1.00	1.00	1.00	0.98	0.96	0.95	0.93	0.92	0.92	0.91
24–30	40 – 64	1.00	0.98	0.94	0.90	0.85	0.80	0.74	0.69	0.65	0.61	0.58
	65 – 119	1.00	0.99	0.97	0.93	0.90	0.86	0.82	0.79	0.76	0.73	0.71
	120 – 199	1.00	1.00	0.98	0.96	0.94	0.92	0.89	0.87	0.85	0.83	0.81
	200 or more	1.00	1.00	0.99	0.98	0.97	0.95	0.93	0.92	0.90	0.89	0.89
30–35	40 – 64	1.00	0.96	0.92	0.86	0.80	0.74	0.68	0.64	0.60	0.57	0.55
	65 – 119	1.00	0.98	0.95	0.90	0.86	0.81	0.76	0.72	0.68	0.65	0.62
	120 – 199	1.00	0.99	0.97	0.95	0.92	0.88	0.85	0.82	0.80	0.78	0.77
	200 or more	1.00	1.00	0.98	0.97	0.95	0.93	0.90	0.89	0.87	0.86	0.85
35–42	40 – 64	1.00	0.95	0.89	0.82	0.75	0.69	0.63	0.58	0.53	0.49	0.46
	65 – 119	1.00	0.97	0.93	0.88	0.82	0.77	0.71	0.67	0.63	0.59	0.56
	120 – 199	1.00	0.98	0.96	0.93	0.89	0.85	0.81	0.78	0.76	0.73	0.71
	200 or more	1.00	0.99	0.98	0.96	0.93	0.90	0.87	0.84	0.82	0.80	0.78

[a] Source: *National Design Specification for Wood Construction* (3).

[b] A_1 = cross-sectional area of main member before boring or grooving and A_2 = sum of cross-sectional areas of metal side plates before drilling.

Lag Bolt Modifying Factor C_{lb}

When lag bolts are used in lieu of bolts in split ring and shear plate connections, the design values must be modified by the lag bolt modifying factor given in Table 6.11.

Design Values

The design values are calculated by first obtaining the tabular design value of the fastener for parallel-to-grain loading P, or perpendicular-to-grain loading Q. These values are then adjusted by all of the applicable modifying factors. The equation for calculation of either P' or Q' has the following form:

$$P' = PC_D C_M C_t C_R C_e C_n C_s C_d C_g C_{st} C_{lb}$$

$$Q' = QC_D C_M C_t C_R C_e (C_n \text{ or } C_s)\ C_d C_g C_{st} C_{lb}$$

The modifying factors C_e, C_n, and C_s are not cumulative; the smallest applies.

In most cases only a few of the modification factors are applicable and only these are shown in the calculations. For instance, when only the duration of load factor is applicable, the calculations are shown as $P' = PC_D$ or $Q' = QC_D$.

Table 6.3 contains a checklist of the applicability of modifying factors to the various types of fasteners.

Critical Section

The critical section of a wood member in a joint is that section, taken at a right angle to the longitudinal axis of the member, that gives the maximum stress based on the net area. The net area at this section is equal to the full cross-sectional area of the timber less the projected area of that portion of the mechanical fastener within the member, including the projected area of associated holes not within the fastener's projected area. See provisions herein for determining net section when a specific type of mechanical fastener is staggered.

Angle of Load to Grain

Angle of load to grain is a factor in the determination of the design value on certain types of mechanical fasteners because wood has a greater bearing value parallel to grain than perpendicular to grain. The angle of load to grain is the angle between the resultant of the load exerted by a mechanical fastener acting on a member and the longitudinal axis of the member (angle θ in Fig. 6.1).

The letter P is used to designate parallel-to-grain tabular design values, Q is used for perpendicular-to-grain tabular design values, and N is used for design values at angles of load to grain between 0° and 90°.

The Hankinson formula is used to determine the design value N of loads at an angle of load to grain between 0° and 90° as shown in Figure 6.1. P', Q' or N' are used to designate the actual values used in design after adjustment by applicable modifiers.

The modifiers for fasteners are applied to the tabular values of P and Q prior to solution of the Hankinson formula except for shear plates and split rings as

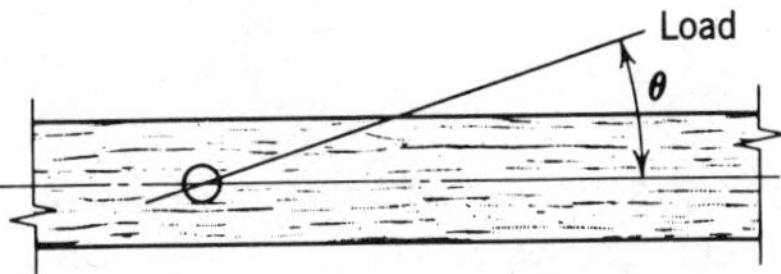

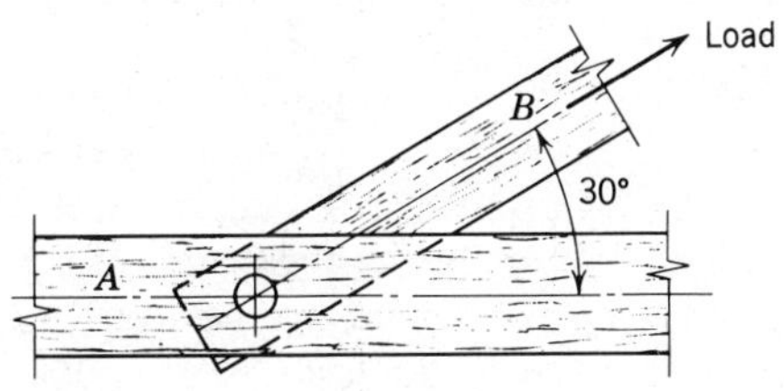

FIGURE 6.1 Angle of load to grain.

discussed in the following paragraph. The solution to the formula for N' then has all of the applicable modifiers included. The angle of load to grain applies only to the particular member under consideration; that is, the angle of load to grain may be different for the various members being connected by the same fastener. For example, in Figure 6.1, the angle of load to grain with respect to member A is 30°, whereas the angle of load to grain with respect to member B is zero.

For split ring and shear plate connectors the modifiers for edge distance, end distance, and spacing are based on the angle of load to grain and cannot be applied to the design values prior to solution by the Hankinson formula. Therefore for connectors, N, the load at an angle to grain, is calculated first and the modifiers are applied to N. For other fasteners the modifiers are applied to P and Q prior to use of the Hankinson formula as follows:

$$N' = \frac{P'Q'}{P' \sin^2 \theta + Q' \cos^2 \theta} \qquad (6\text{-}1)$$

where N' = design value for load acting at an angle θ with direction of grain (lb),
P' = design value for load acting parallel to grain (lb),
Q' = design value for load acting perpendicular to grain (lb), and
θ = angle between the direction of load and the direction of grain (degrees)

The design values P' and Q' have been adjusted by all applicable modifiers, therefore, N' needs no further adjustments and may be used directly in design.

The Hankinson formula may be solved graphically through use of the nomographs in Figure 6.2. The difference between the two nomographs is in their scale. The units on the vertical scales are shown in pounds for design values for fasteners and in pounds per square inch (psi) for design values for strength properties of wood such as bearing at an angle of load to grain between 0° and 90°. When the Hankinson formula is used to determine the design value in compression at an angle of load to grain, the modifiers must be applied prior to use of the formula.

Spacing of Mechanical Fasteners

The spacing of mechanical fasteners is the distance between centers of the fasteners measured on a straight line joining their centers. Spacing may also be measured parallel and perpendicular to grain. These measurements are illustrated in Figure 6.3. Spacing between fasteners in a group should be sufficient to develop the full strength of each fastener.

Edge Distance

Edge distance is the distance from the edge of a member to the center of the mechanical fastener closest to that edge measured perpendicular to the edge. The loaded edge is the edge toward which the load induced by the fastener acts. The

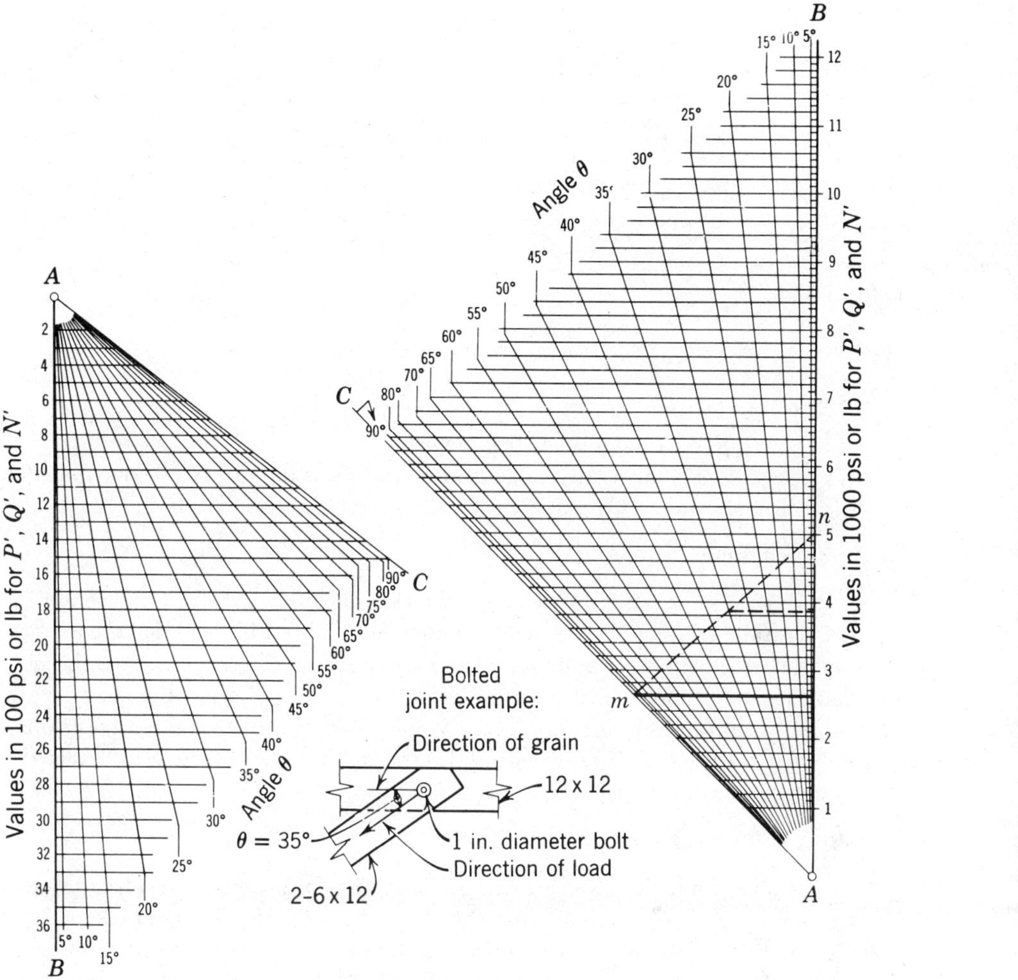

FIGURE 6.2 Scholten nomograph for solution of Hankinson formula. Example of use: Conditions—$P' = 5030$ lb, $Q' = 2620$ lb, $\theta = 35°$. Locate 5030 lb at n on line A–B on right-hand chart. Locate 2620 lb at m on line A–C, opposite that value on line A–B. Where line m–n intersects the 35° radial line, project to line A–B and read allowable load $N' = 3870$ lb. Source: *National Design Specification for Wood Construction* (3).

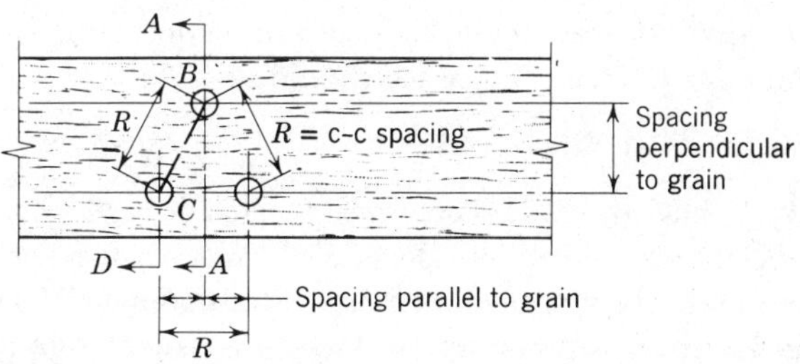

FIGURE 6.3 Spacing Measurements.

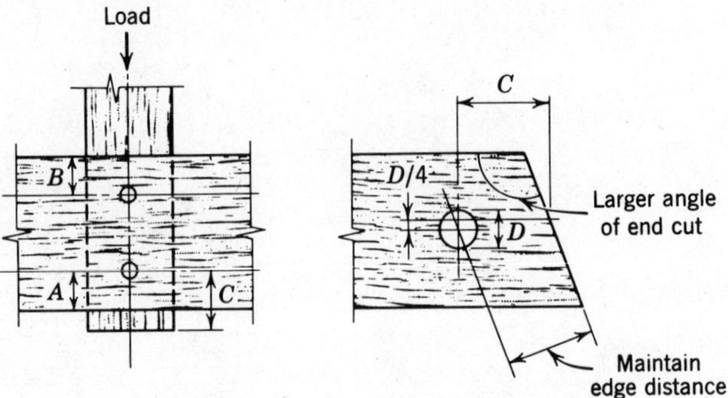

FIGURE 6.4 Edge and end distances.

unloaded edge is the edge away from which the load induced by the fastener acts. In Figure 6.4, A is the loaded edge distance and B is the unloaded edge distance. Edge distance should be sufficient to develop the required strength of fasteners.

End Distance

End distance is the distance, measured parallel to grain, from the center of a mechanical fastener to the square-cut end of a member. If the end of the member is cut at an angle, the end distance is measured parallel to the length of the piece on a line that is one-fourth of the fastener diameter, D, from the center of the connector and on the side of the larger angle of the end cut. The dimensions C in Figure 6.4 are end distances. End distances should be sufficient to develop the required strength of fasteners.

Effect of Treatment

No reduction in design values is recommended for preservatively treated wood except for fire retardant treatment. (See "Fire Retardant Treatment Factor C_R.") Also check with the manufacturer of the treatment to determine whether or not the fire retardant treatment has a corrosive effect on the fasteners being used.

Eccentricity

Eccentric timber connector, bolt, or lag screw joints as shown in Figure 6.5 should be avoided wherever possible, especially in heavily stressed members. If eccentric joints are used, the effect of the shear, moment, tension perpendicular to grain, and tension stresses should be taken into consideration in the design. Concentric joints as shown in Figure 6.6 should be used.

Shear Stress in Joints at Ends of Members

It is preferable to have the ends of beams supported by resting on another member, column cap, bolster, and so on. However, beams made of sawn lumber and smaller glued laminated beams have been successfully supported by fasteners such as illustrated in Figure 6.7. When this is done, the same design procedure used for beams notched on the end should be used.

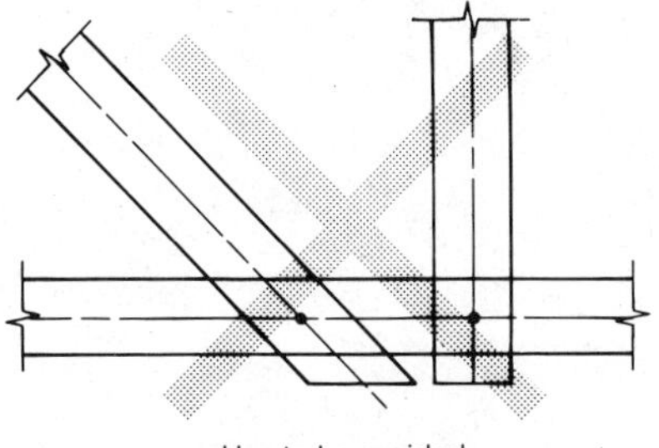

FIGURE 6.5 Truss connection—eccentric joint. When the centerlines of members do not intersect at common point in a truss, considerable shear, moment, and tension perpendicular to grain may result in the bottom chord. When these are combined with the presumably high-tension stress, the member may be over-stressed.

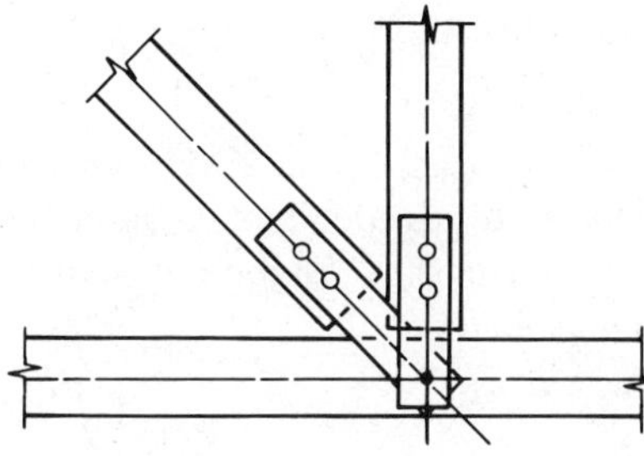

FIGURE 6.6 Truss connection—concentric joint. For trusses with single-piece upper and lower chords, such as bowstring trusses, the chords and web members should be in the same plane, with straps or gusset plates used for the connection.

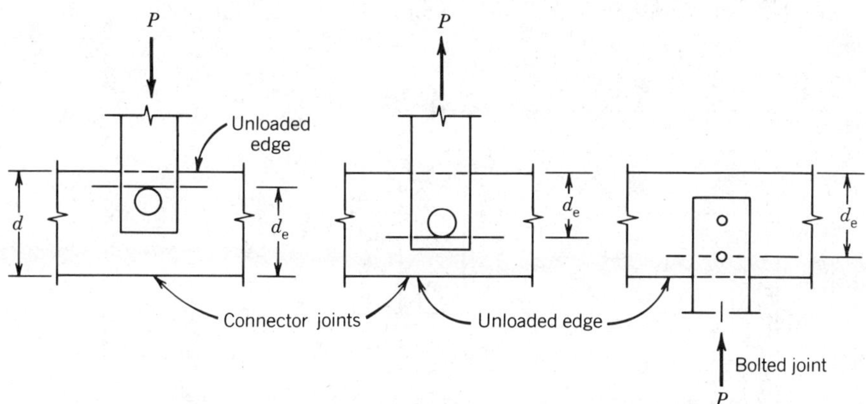

FIGURE 6.7 Depth, d_e for members with various fasteners.

The allowable shear V is determined by the following formula:

$$V = \frac{2F'_v b d_e^2}{3d} \tag{6-2}$$

where V = shear at the end of the member (lb) (V = vertical reaction, R_v, when member is horizontal),
F'_v = design value in shear (psi) adjusted by applicable modifiers,
b = width of beam (in.),
d = depth of beam (in.) (see Fig. 6.7), and
d_e = depth from fastener to the unloaded edge (in.).

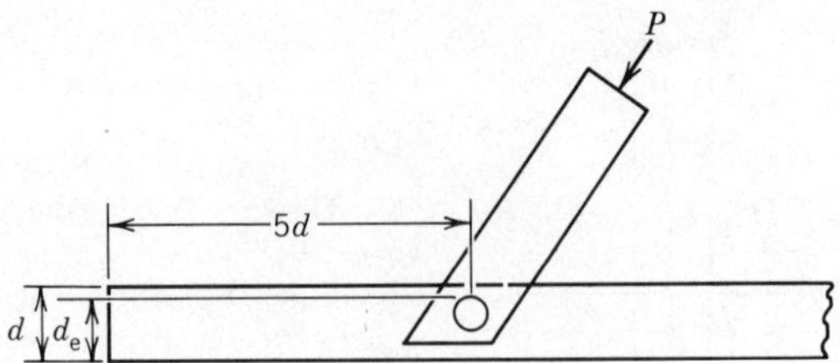

FIGURE 6.8 Shear in joint details.

The designer is cautioned against using fasteners to support the ends of large glued laminated beams. See Figure 6.7 for additional information.

Shear Stress in Joints away from Ends of Members

When the joint is located five times the depth or more from the end of a beam or when a load is applied to a member at least five times the depth of the member from the ends as shown in Figure 6.8, the design value in shear for joint detail may be increased 50% and the shear is determined as follows:

$$V = (2/3)(1.5F'_v bd_e) = F'_v bd_e \tag{6-3}$$

where terms are as described previously.

Example. Determine the maximum value of P that is limited by shear in joint detail.

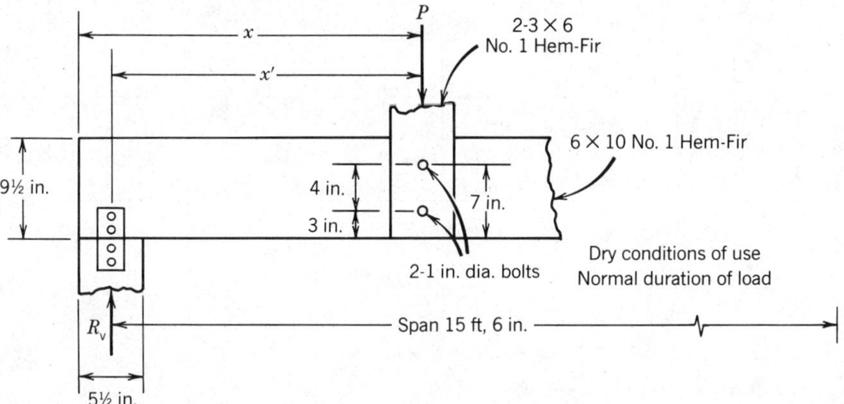

(a) When $x = 4$ ft, joint is located $\dfrac{(4)(12)}{9.5} = 5.05d$ from end;

therefore, shear in joint detail may be increased 50%. From Table 7.3, $F_v = 70$ psi, $F'_v = 70$ psi. For $x = 4$ ft,

$$x' = (12)(4) - \frac{5.5}{2} = 45.25 \text{ in.}$$

$$R_v = \frac{(15.5)(12) - (45.25)(P)}{(15.5)(12)} = 0.757P$$

$$V = (2/3)(1.5F'_v bd_e)$$

$$V = (2/3)(1.5)(70)(5.5)(7) = 2695 \text{ lb}$$

$$V = R_v = 0.757P$$

$$P = \frac{2695}{0.757} = 3560 \text{ lb}$$

However, P can be no larger than would be allowed based on use of full cross section with no increase for shear.

$$V = \tfrac{2}{3} F'_v bd = (\tfrac{2}{3})\ (70)(5.5)(9.5) = 2440 \text{ lb}$$

$$P = \frac{2440}{0.757} = 3220 \text{ lb} < 3560 \text{ lb}$$

Use $P = 3220$ lb.

(b) when $x = 3$ ft, joint is located $\frac{(3)(12)}{9.5} = 3.79d$ from end

Since $3.79d < 5d$, no increase for F_v.

$$F'_v = F_v = 70 \text{ psi}$$

$$V = (2/3)F'_v bd_e \left(\frac{d_e}{d}\right)$$

$$V = (2/3)(70)(5.5)(7)\left(\frac{7}{9.5}\right) = 1320 \text{ lb}$$

For $x = 3$ ft,

$$R_v = \frac{(15.5)(12) - [(3)(12) - 5.5/2]P}{(15.5)(12)} = 0.820P$$

$$R_v = V = 0.820P, \qquad P = \frac{1324}{0.820} = 1610 \text{ lb} < 3220 \text{ lb}$$

Use $P = 1610$ lb

Group Action of Fasteners

Research has indicated that a load carried by a row of fasteners is not equally divided among the fasteners; that is, end fasteners such as A and G in Figure 6.9 tend to carry a larger portion of the load than the intermediate fasteners. The distribution of load is determined by the relative stiffness of the main member and the side members. Tables 6.7 and 6.8 contain modification factors C_g to be applied to design values for 2 to 12 fasteners in a row. In the determination of the modification factor, the following principles apply:

1. A group of fasteners consists of one or more rows of fasteners.
2. A row of fasteners consists of either two or more bolts loaded in single or multiple shear or two or more split rings, shear plates, or lag bolts loaded in single shear. The row is aligned with the direction of the load.

   When fasteners in adjacent rows are staggered, and the distance a between the rows is less than $\frac{1}{4}$ of the spacing b between the closest fasteners

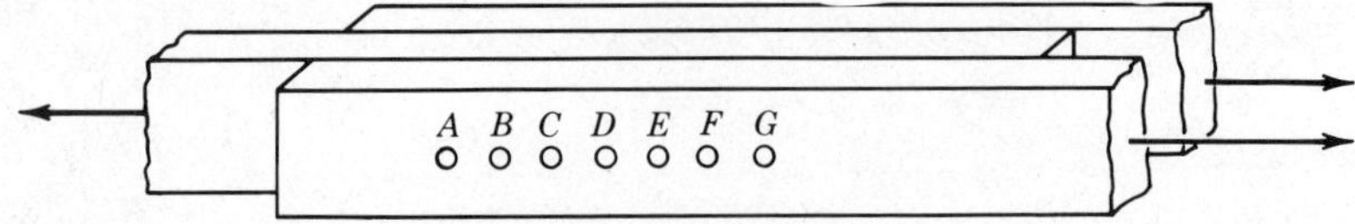

FIGURE 6.9 Row of fasteners.

in an adjacent row as shown in Figure 6.10*a*, the fasteners in adjacent rows should be considered as one row for the purpose of determining the modification factor C_g. The modification factor C_g is shown as K in some publications.

3. The load for each row of fasteners is determined by summing the individual loads for each fastener in the row and then multiplying this value by the modification factor C_g in Table 6.7 or 6.8.

   For convenience, C_g may be applied to individual fastener values prior to summation of values.

4. The design value for the group of fasteners is the sum of the design values of the rows in the group.
5. When a member is loaded perpendicular to grain, its equivalent cross-sectional area is the product of the thickness of the member and the overall width of the fastener group for calculating cross-sectional area ratios. When only one row of fasteners is used, the width is equal to the minimum spacing for full load for the type of fastening used. In general, long rows of fasteners perpendicular to grain should be avoided.

Uplift Loads

Where gravity loads are transferred by bearing perpendicular to grain, adequate fasteners must be provided to carry both horizontal and uplift loads. These loads are normally of a transient nature and are of short duration. The fastener is usually placed toward the lower edge of the member. The distance d_e, shown in Figure 6.11, must be no less than the distance required for the perpendicular-to-grain edge distance of the type of fastener used. It should also be deep enough so that the calculated horizontal shear f_v in the following formula (5-25) does not exceed the design value in shear, F'_v.

$$f_v = \frac{3Vd}{2bd_e^2}$$

where f_v = calculated horizontal shear (psi),
V = shear (lb), (V = vertical reaction R_v when member is horizontal),

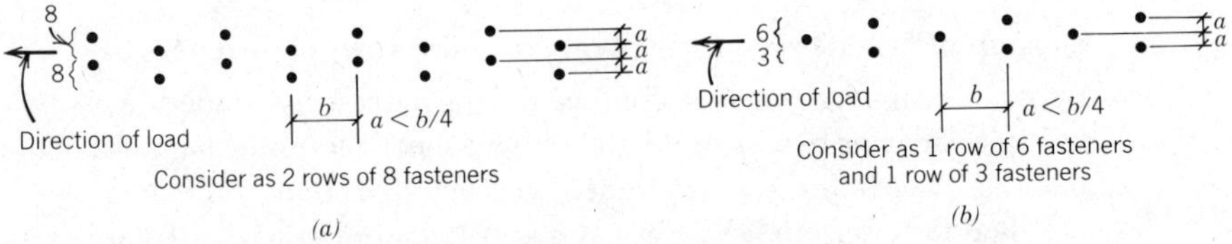

8 8 Direction of load b $a < b/4$ Consider as 2 rows of 8 fasteners

(*a*)

6 3 Direction of load b $a < b/4$ Consider as 1 row of 6 fasteners and 1 row of 3 fasteners

(*b*)

FIGURE 6.10 Group action of fasteners.

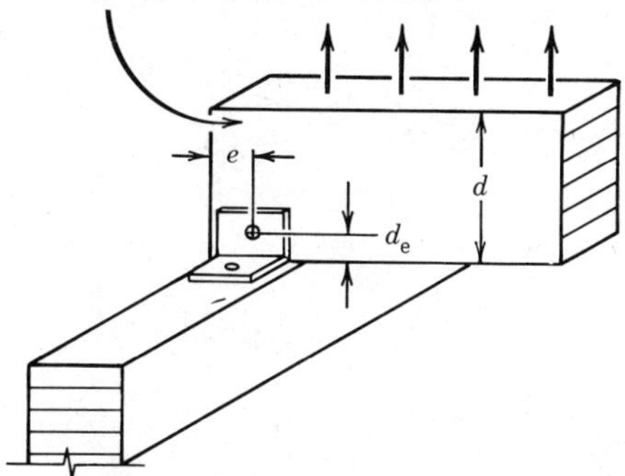

FIGURE 6.11 Uplift loading conditions.

b = width of member (in.),
d_e = effective depth of member (in.) (see Fig. 6.11), and
d = depth of member (in.) (see Fig. 6.11).

The end distance *e* must also be adequate for the type of fastener used. The uplift loads and horizontal loads are usually small compared to other loads, and frequently only one fastener is required. An adequate number of fasteners, however, should be used to resist the uplift load.

CONNECTIONS—GENERAL CONSIDERATIONS

In addition to being designed for strength to transfer loads, connections should be designed to avoid splitting the members and to permit swelling and shrinkage of the wood. Glued laminated timbers are often larger than sawn lumber and the loads transferred are also larger, therefore, the effect of increased size should be considered in the design of glued laminated timber connections.

Examples of good and poor detailing practices for glued laminated timber construction are included in *Typical Construction Details*, (4) AITC 104, in Part III of this manual.

TIMBER CONNECTORS—SHEAR PLATES AND SPLIT RINGS

The design values given in Table 6.12 are for two shear plates used back-to-back in the contact faces of a wood-to-wood joint with their bolt in single shear or for one shear plate with its bolt in single shear used in conjunction with a steel strap or shape in a wood-to-metal joint. Split rings and shear plates are illustrated in Figure 6.12. The design values given in Table 6.14 are for one split ring with its bolt in single shear. Projected areas of connectors and bolts for use in determining net sections are given in Table 6.9. Typical dimensions for timber connectors are given in Table 6.10.

In installing timber connectors and bolts, a nut must placed on each bolt. Washers, not smaller than the size given in Table 6.10, must be placed between the wood member and the bolt head and between the wood member and the nut. When a steel strap or shape is used in conjunction with shear plates, the washer may be omitted, except when desirable to extend the bolt length to prevent the metal from bearing on the threaded portion of the bolt.

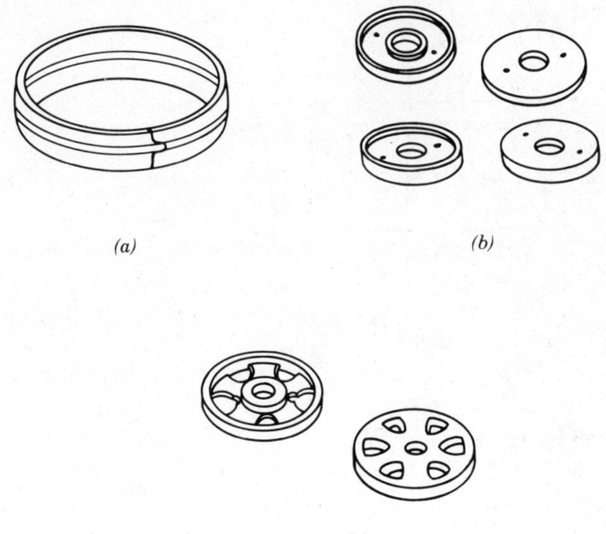

FIGURE 6.12 Typical timber connectors. (*a*) Split ring; (*b*) pressed steel shear plates; (*c*) malleable iron shear plates.

Lumber Species

The density of the wood affects the allowable loads for timber connectors. The species groupings for connectors in Tables 6.1 and 6.2 are based on density and related factors.

Connector Design Values

The connector design values in Tables 6.12–6.14 are for one connector unit installed in a seasoned timber joint under normal duration of loading and dry-use conditions. For connectors, lumber is considered to be seasoned if the moisture content is no higher than 19% for a depth of $\frac{3}{4}$ in. These design values must be modified if end distances, edge distances, or spacings are between the minimums shown and those required for maximum design value. The reductions due to reduced end distances, C_n, edge distances, C_e, and spacings, C_s, are not cumulative and the lowest value controls. However, any reduction due to moisture content factor C_M is cumulative with other modifiers. The reduction factor for connectors in a row, C_g, is cumulative with other modifiers.

The tabular design values in Tables 6.12 and 6.13 are based on the capacity of the shear plate in the wood. Values marked with an asterisk (*) may need to be reduced as indicated in footnote *d* when limited by the capacity of the steel. The higher values are included for use in calculating reduced spacing, edge, and end distances for the loads that are limited by the steel.

It is recommended that, wherever possible, a connector joint be designed with edge distance, end distance, and spacing for maximum design value as shown in Tables 6.12–6.15 or Figures 6.14 and 6.15. If space is not available, these dimensions

TABLE 6.9

Total Projected Area of Connectors and Bolts (in.^{2}) For Use in Determining Net Sections

Connector Type	Connector Size (in.)	Bolt Diameter (in.)	Placement of Connectors	Member Thickness (in.) 1$\frac{1}{2}$	2$\frac{1}{2}$	3$\frac{1}{8}$	3$\frac{1}{2}$	5$\frac{1}{8}$	5$\frac{1}{2}$	6$\frac{3}{4}$	8$\frac{3}{4}$	10$\frac{3}{4}$	12$\frac{1}{4}$
Split Rings													
1	2$\frac{1}{2}$	$\frac{1}{2}$	1 face	1.73	2.29	2.65	2.86	3.77	3.98	4.69	5.81	6.94	7.78
		$\frac{1}{2}$	2 faces	2.62	3.18	3.54	3.75	4.66	4.87	5.58	6.70	7.82	8.67
2	4	$\frac{3}{4}$	1 face	3.05	3.86	4.37	4.68	6.00	6.30	7.32	8.94	10.56	11.79
		$\frac{3}{4}$	2 faces	4.88	5.69	6.20	6.51	7.83	8.13	9.15	10.77	12.40	13.62
Shear Plates													
1	2$\frac{5}{8}$	$\frac{3}{4}$	1 face	2.03	2.85	3.35	3.66	4.98	5.28	6.30	7.92	9.55	10.77
		$\frac{3}{4}$	2 faces	2.84	3.66	4.16	4.47	5.79	6.09	7.11	8.73	10.36	11.58
1-LG[a]	2$\frac{5}{8}$	$\frac{3}{4}$	1 face	1.91	2.72	3.23	3.53	4.85	5.16	6.18	7.80	9.43	10.64
		$\frac{3}{4}$	2 faces	2.60	3.41	3.92	4.22	5.54	5.85	6.87	8.49	10.11	11.33
2	4	$\frac{3}{4}$	1 face	3.26	4.07	4.58	4.89	6.21	6.51	7.53	9.15	10.78	12.00
		$\frac{3}{4}$	2 faces	—	6.11	6.62	6.93	8.25	8.55	9.57	11.19	12.82	14.04
2-A	4	$\frac{7}{8}$	1 face	3.37	4.30	4.89	5.24	6.77	7.12	8.29	10.16	12.04	13.45
		$\frac{7}{8}$	2 faces	—	6.26	6.85	7.20	8.73	9.08	10.25	12.12	14.00	15.41

[a]Light gage.

TABLE 6.10

Dimensions and Specifications for Connectors[a]

Split Rings[b]

Size	$2\frac{1}{2}$ in.	4 in.		$2\frac{1}{2}$ in.	4 in.
Split ring			Washers, standard		
Inside diameter at center when closed	2.500	4.000	Round, cast or malleable iron, diameter	$2\frac{5}{8}$	3
Thickness of metal at center	0.163	0.193	Round, wrought iron (minimum):		
Depth of metal (width of ring)	0.750	1.000	Diameter	$1\frac{3}{8}$	2
Groove			Thickness	$\frac{3}{32}$	$\frac{5}{32}$
Inside diameter	2.56	4.08	Square plate		
Width	0.18	0.21	Length of side	2	3
Depth	0.375	0.50	Thickness	$\frac{1}{8}$	$\frac{3}{16}$
Bolt hole			Projected area		
Diameter	$\frac{9}{16}$	$\frac{13}{16}$	Portion of one ring within member (in.^{2})	1.10	2.24

Shear Plates[b, c]

Size	$2\frac{5}{8}$ in.	$2\frac{5}{8}$ in.	4 in.	4 in.
Shear plate material	Pressed steel	Light gage	Malleable iron	Malleable iron
Diameter of plate	2.62	2.62	4.03	4.03
Diameter of bolt hole	0.81	0.81	0.81	0.94
Thickness of plate	0.172	0.12	0.20	0.20
Depth of plate	0.42	0.35	0.64	0.64
Hole diameter in straps or shapes for bolts	$\frac{13}{16}$	$\frac{13}{16}$	$\frac{13}{16}$	$\frac{15}{16}$
Bolt hole—diameter in timber	$\frac{13}{16}$	$\frac{13}{16}$	$\frac{13}{16}$	$\frac{15}{16}$

Size	$2\frac{5}{8}$ in.	$2\frac{5}{8}$ in.	4 in.	4 in.
Washers, standard				
Round, cast or malleable iron, diameter	3	3	3	$3\frac{1}{2}$
Round, wrought iron, minimum				
Diameter	2	2	2	$2\frac{1}{4}$
Thickness	$\frac{5}{32}$	$\frac{5}{32}$	$\frac{5}{32}$	$\frac{11}{64}$
Square plate				
Length of side	3	3	3	3
Thickness	$\frac{1}{4}$	$\frac{1}{4}$	$\frac{1}{4}$	$\frac{1}{4}$
Projected area				
Portion of one shear plate within member (in.^{2})	1.18	1.00	2.58	2.58

[a] Source: *National Design Specification for Wood Construction* (3).

[b] Dimensions in inches.

[c] Steel straps or shapes, for use with shear plates, shall be designed in accordance with accepted engineering practices.

TABLE 6.11

Modification Factors C_{lb} for Timber Connectors Used with Lag Bolts[a]

Timber Connector Size and Type	Side Plates	Penetration	Penetration of Lag Bolt into Member Receiving Point (number of shank diameters) — Fastener Species Load Group[c] I	II	III	IV	Modification Factor[b] C_{lb}
$2\frac{1}{2}$-in. split ring, 4-in. split ring, or 4-in. shear plate	Wood or Metal	Standard	7	8	10	11	1.00
		Minimum	3	$3\frac{1}{2}$	4	$4\frac{1}{2}$	0.75
$2\frac{5}{8}$-in. shear plate	Wood	Standard	4	5	7	8	1.00
		Minimum	3	$3\frac{1}{2}$	4	$4\frac{1}{2}$	0.75
	Metal	Standard and Minimum	3	$3\frac{1}{2}$	4	$4\frac{1}{2}$	1.00

[a]Factors apply to design values tabulated for connector units used with bolts.

[b]Use straight line interpolation for intermediate penetrations.

[c]See Table 6.1.

can be reduced provided that the reduced design values of the connectors are capable of carrying the design load. Conversely, if the allowable load is reduced because of a reduced end distance, edge distance, or spacing, the other distances or spacings may be reduced to those resulting in the same allowable load. In determining the total load capacity of a joint, however, the design value for each individual connector is calculated, and the total load that can be carried by the joint is the lowest connector value obtained multiplied by the number of connectors.

The required end distance and spacing requirements may be reduced below those required for 100% of the tabulated load only when the full design load on the joint is reduced in proportion to the reduction in end distance or spacing. Conversely, if the end distance of one connector or the spacing of a pair of connectors of a group of connectors in a joint is less than that required for 100% of full load, the design load of each connector in the group shall not exceed that of the connector with the reduced end distance or spacing.

Modification Factors for Timber Connectors

Modification factors for timber connector design values for duration of load are given in Figure 3.1 or Table 3.2. Modification factors for various service and seasoning conditions are given in Tables 6.4–6.6. A summary of modification factors applicable to shear plates and split rings is contained in Table 6.3.

Lag bolts may be used instead of bolts in connector units, provided that they have the same shank diameter as the bolt specified for the connector and they meet all other provisions for lag bolts as given herein.

When lag bolts are used with timber connectors, the design values from Tables 6.12–6.14 are multiplied by the applicable modification factor from Table 6.11. Penetration of the lag bolt into the member receiving its point should not be less than the minimum specified in Table 6.11.

Critical or Net Section

The critical section of a timber connector joint will probably be at the centerline of the bolt and connector. The net area of this section is equal to the full cross-sectional area of the timber less the projected area of the connectors within the member and the projected bolt area (see Table 6.9 for projected areas). When connectors are staggered, adjacent connectors with parallel-to-grain spacing equal to or less than one connector diameter are considered to occur at the same critical section. In Figure 6.3, section *A-B-C-D* is the critical section when connectors *B* and *C* have parallel-to-grain spacing equal to or less than one connector diameter. For greater parallel-to-grain spacings, section *A-A* is the critical section.

Glued Laminated Timber

Knots occurring at, or near, the critical section are disregarded in determining net section. The net cross-sectional area, in square inches, required at the critical section is determined by dividing the total load, in pounds, transferred through the critical section of the member by the design value in tension, F'_t, for tension members and the design value in compression, F'_c, for compression members.

Sawn Lumber

The net cross-sectional area, in square inches, required at the critical section is determined by dividing the total load, in pounds, transferred through the critical section of the member by the design values in tension, F'_t, for tension members and the design value in compression, F'_c, for compression members. Values for F_t and F_c are contained in Table 7.3. These must be adjusted by the appropriate modifiers to obtain F'_c and F'_t. Conversely, the total load capacity in pounds may be determined by multiplying the net area in square inches by the appropriate design value.

When analyzing an existing structure, the required net area may be determined by dividing total load, in pounds, transferred through the critical section by the appropriate design value in Table 3.1, provided no knots occur in the critical section. This recommendation is based on the assumption that the area of the connector and bolt hole will be deducted from the net section. This procedure is not applicable to glued laminated timber.

Angle of Load to Grain

The design values of connectors for angles of load to grain other than 0° and 90° are determined by the application of the Hankinson formula (6-1) to the design values for parallel- and perpendicular-to-grain loading given in Tables 6.12–6.14.

TABLE 6.12

Shear Plate Design Values—Wood Side Plates[a,b]

Shear Plate diameter (in.)	Bolt Diameter (in.)	Numbers of Faces of Piece with Connectors on Same Bolt	Net Thickness of Piece (in.)	Loaded Parallel to Grain (0°): Minimum Edge Distance (in.)	Loaded Parallel to Grain (0°): Design Value per Connector Unit and Bolt (lb)[c]: Group A Woods	Group B Woods	Group C Woods	Group D Woods	Loaded Perpendicular to Grain (90°): Edge Distance (in.): Unloaded Edge, min	Loaded Edge[c]	Loaded Perpendicular to Grain (90°): Design Value per Connector Unit and Bolt (lb): Group A Woods	Group B Woods	Group C Woods	Group D Woods
$2\frac{5}{8}$	$\frac{3}{4}$	1	$1\frac{1}{2}$ min	$1\frac{3}{4}$	3110*	2670	2220	2010	$1\frac{3}{4}$	$1\frac{3}{4}$ min	1810	1550	1290	1110
										$2\frac{3}{4}$ or more	2170	1860	1550	1330
		2	$1\frac{1}{2}$ min	$1\frac{3}{4}$	2420	2080	1730	1500	$1\frac{3}{4}$	$1\frac{3}{4}$ min	1410	1210	1010	870
										$2\frac{3}{4}$ or more	1690	1450	1210	1040
			2	$1\frac{3}{4}$	3190*	2730	2270	1960	$1\frac{3}{4}$	$1\frac{3}{4}$ min	1850	1590	1320	1140
										$2\frac{3}{4}$ or more	2220	1910	1580	1370
			$2\frac{1}{2}$ or more	$1\frac{3}{4}$	3330*	2860	2380	2060	$1\frac{3}{4}$	$1\frac{3}{4}$ min	1940	1660	1380	1200
										$2\frac{3}{4}$ or more	2320	1990	1650	1440
4	$\frac{3}{4}$	1	$1\frac{1}{2}$ min	$2\frac{3}{4}$	4370	3750	3130	2700	$2\frac{3}{4}$	$2\frac{3}{4}$ min	2540	2180	1810	1550
										$3\frac{3}{4}$ or more	3040	2620	2170	1860
	or		$1\frac{3}{4}$ or more	$2\frac{3}{4}$	5090*	4360	3640	3140	$2\frac{3}{4}$	$2\frac{3}{4}$ min	2950	2530	2110	1810
										$3\frac{3}{4}$ or more	3540	3040	2530	2200
	$\frac{7}{8}$	2	$1\frac{3}{4}$ min.	$2\frac{3}{4}$	3390	2910	2420	2090	$2\frac{3}{4}$	$2\frac{3}{4}$ min	1970	1680	1400	1250
										$3\frac{3}{4}$ or more	2360	2020	1680	1410
			2	$2\frac{3}{4}$	3790	3240	2700	2330	$2\frac{3}{4}$	$2\frac{3}{4}$ min	2200	1880	1570	1360
										$3\frac{3}{4}$ or more	2640	2260	1880	1630
			$2\frac{1}{2}$	$2\frac{3}{4}$	4310	3690	3080	2660	$2\frac{3}{4}$	$2\frac{3}{4}$ min	2500	2140	1780	1540
										$3\frac{3}{4}$ or more	3000	2550	2140	1850
			3	$2\frac{3}{4}$	4830	4140	3450	2980	$2\frac{3}{4}$	$2\frac{3}{4}$ min	2800	2400	2000	1720
										$3\frac{3}{4}$ or more	3360	2880	2400	2060
			$3\frac{1}{2}$ or more	$2\frac{3}{4}$	5030*	4320	3600	3110	$2\frac{3}{4}$	$2\frac{3}{4}$ min	2920	2500	2090	1800
										$3\frac{3}{4}$ or more	3500	3000	2510	2160

[a]Source: *National Design Specification for Wood Construction* (3).

[b]Design values apply to one shear plate unit and bolt in single shear when installed between seasoned wood members that will remain dry in service and be subject to normal loading conditions.

[c]Loads followed by an asterisk (*) exceed those permitted by Note *d*, but are needed for proper determination of loads for other angles of load to grain. Note *d* limitations apply in all cases.

[d]Design loads (lb) on shear plates shall not exceed the following:

(1)	$2\frac{5}{8}$-in. shear plate	2900
(2)	4-in. shear plate with $\frac{3}{4}$ in. bolt	4400
(3)	4-in. shear plate with $\frac{7}{8}$ in. bolt	6000

[e]See Figure 6.4.

TABLE 6.13

Shear Plate Design Values—Steel Side Plates[a,b]

Shear Plate Diameter (in.)	Bolt Diameter (in.)	Number of Faces of Piece with Connectors on Same Bolt	Net Thickness of Piece (in.)	Minimum Edge Distance (in.)	Loaded Parallel to Grain (0°): Design Value per Connector Unit and Bolt (lb)[c] Group A Woods	Group B Woods	Group C Woods	Group D Woods	Loaded Perpendicular to Grain (90°): Edge Distance (in.) Unloaded Edge, min.	Loaded edge[c]	Design Value per Connector Unit and Bolt (lb) Group A Woods	Group B Woods	Group C Woods	Group D Woods
$2\frac{5}{8}$	$\frac{3}{4}$	1	$1\frac{1}{2}$ min	$1\frac{3}{4}$	3110*	2670	2220	2010	$1\frac{3}{4}$	$1\frac{3}{4}$ min	1810	1550	1290	1110
										$2\frac{3}{4}$ or more	2170	1860	1550	1330
		2	$1\frac{1}{2}$ min	$1\frac{3}{4}$	2420	2080	1730	1500	$1\frac{3}{4}$	$1\frac{3}{4}$ min	1410	1210	1010	870
										$2\frac{3}{4}$ or more	1690	1450	1210	1040
			2	$1\frac{3}{4}$	3190*	2730	2270	1960	$1\frac{3}{4}$	$1\frac{3}{4}$ min	1850	1590	1320	1140
										$2\frac{3}{4}$ or more	2220	1910	1580	1370
			$2\frac{1}{2}$ or more	$1\frac{3}{4}$	3330*	2860	2380	2060	$1\frac{3}{4}$	$1\frac{3}{4}$ min	1940	1660	1380	1200
										$2\frac{3}{4}$ or more	2320	1990	1650	1440
4	$\frac{3}{4}$	1	$1\frac{1}{2}$ min	$2\frac{3}{4}$	5160	4160	3290	2700	$2\frac{3}{4}$	$2\frac{3}{4}$ min	2540	2180	1810	1550
										$3\frac{3}{4}$ or more	3040	2620	2170	1860
	or	2	$1\frac{3}{4}$ or more	$2\frac{3}{4}$	6010*	4840*	3820	3140	$2\frac{3}{4}$	$2\frac{3}{4}$ min	2950	2530	2110	1810
										$3\frac{3}{4}$ or more	3540	3040	2530	2200
	$\frac{7}{8}$		$1\frac{3}{4}$ min	$2\frac{3}{4}$	4000	3230	2540	2090	$2\frac{3}{4}$	$2\frac{3}{4}$ min	1970	1680	1400	1250
										$3\frac{3}{4}$ or more	2360	2020	1680	1410
			2	$2\frac{3}{4}$	4470	3600	2830	2330	$2\frac{3}{4}$	$2\frac{3}{4}$ min	2200	1880	1570	1360
										$3\frac{3}{4}$ or more	2640	2260	1880	1630
			$2\frac{1}{2}$	$2\frac{3}{4}$	5090*	4100	3230	2660	$2\frac{3}{4}$	$2\frac{3}{4}$ min	2500	2140	1780	1540
										$3\frac{3}{4}$ or more	3000	2550	2140	1850
			3	$2\frac{3}{4}$	5700*	4600*	3620	2980	$2\frac{3}{4}$	$2\frac{3}{4}$ min	2800	2400	2000	1720
										$3\frac{3}{4}$ or more	3360	2880	2400	2060
			$3\frac{1}{2}$ or more	$2\frac{3}{4}$	5940*	4800*	3780	3110	$2\frac{3}{4}$	$2\frac{3}{4}$ min	2920	2500	2090	1800
										$3\frac{3}{4}$ or more	3500	3000	2510	2160

[a]Source: *National Design Specification for Wood Construction* (3).

[b]Design values apply to one shear plate unit and bolt in single shear when installed with seasoned wood members that will remain dry in service and be subject to normal loading conditions.

[c]Loads followed by an asterisk (*) exceed those permitted by Note *d*, but are needed for proper determination of loads for other angles of load to grain. Note *d* limitations apply in all cases.

[d]Design loads (lb) on shear plates shall not exceed the following:

(1)	$2\frac{5}{8}$-in. shear plate	2900
(2)	4-in. shear plate with $\frac{3}{4}$ in. bolt	4400
(3)	4-in. shear plate with $\frac{7}{8}$ in. bolt	6000

[e]See Figure 6.4.

TABLE 6.14

Split Ring Design Values[a,b]

Split Ring Diameter (in.)	Bolt Diameter (in.)	Number of Faces of Piece with Connectors on Same Bolt	Net Thickness of Piece (in.)	Loaded Parallel to Grain (0°): Minimum Edge Distance (in.)	Loaded Parallel to Grain (0°): Design Value per Connector Unit and Bolt (lb)				Loaded Perpendicular to Grain (90°): Edge Distance (in.)		Loaded Perpendicular to Grain (90°): Design Value per Connector Unit and Bolt (lb)			
					Group A Woods	Group B Woods	Group C Woods	Group D Woods	Unloaded Edge, min	Loaded Edge[c]	Group A Woods	Group B Woods	Group C Woods	Group D Woods
$2\frac{1}{2}$	$\frac{1}{2}$	1	1 min	$1\frac{3}{4}$	2630	2270	1900	1640	$1\frac{3}{4}$	$1\frac{3}{4}$ min	1580	1350	1130	970
										$2\frac{3}{4}$ or more	1900	1620	1350	1160
			$1\frac{1}{2}$ or more	$1\frac{3}{4}$	3160	2730	2290	1960	$1\frac{3}{4}$	$1\frac{3}{4}$ min	1900	1620	1350	1160
										$2\frac{3}{4}$ or more	2280	1940	1620	1390
		2	$1\frac{1}{2}$ min	$1\frac{3}{4}$	2430	2100	1760	1510	$1\frac{3}{4}$	$1\frac{3}{4}$ min.	1460	1250	1040	890
										$2\frac{3}{4}$ or more	1750	1500	1250	1070
			2 or more	$1\frac{3}{4}$	3160	2730	2290	1960	$1\frac{3}{4}$	$1\frac{3}{4}$ min	1900	1620	1350	1160
										$2\frac{3}{4}$ or more	2280	1940	1620	1390
4	$\frac{3}{4}$	1	1 min	$2\frac{3}{4}$	4090	3510	2920	2520	$2\frac{3}{4}$	$2\frac{3}{4}$ min	2370	2030	1700	1470
										$3\frac{3}{4}$ or more	2840	2440	2040	1760
			$1\frac{1}{2}$	$2\frac{3}{4}$	6020	5160	4280	3710	$2\frac{3}{4}$	$2\frac{3}{4}$ min	3490	2990	2490	2150
										$3\frac{3}{4}$ or more	4180	3590	2990	2580
			$1\frac{5}{8}$ or more	$2\frac{3}{4}$	6140	5260	4380	3790	$2\frac{3}{4}$	$2\frac{3}{4}$ min	3560	3050	2540	2190
										$3\frac{3}{4}$ or more	4270	3660	3050	2630
		2	$1\frac{1}{2}$ min	$2\frac{3}{4}$	4110	3520	2940	2540	$2\frac{3}{4}$	$2\frac{3}{4}$ min	2480	2040	1700	1470
										$3\frac{3}{4}$ or more	2980	2450	2040	1760
			2	$2\frac{3}{4}$	4950	4250	3540	3050	$2\frac{3}{4}$	$2\frac{3}{4}$ min	2870	2470	2050	1770
										$3\frac{3}{4}$ or more	3440	2960	2460	2120
			$2\frac{1}{2}$	$2\frac{3}{4}$	5830	5000	4160	3600	$2\frac{3}{4}$	$2\frac{3}{4}$ min	3380	2900	2410	2080
										$3\frac{3}{4}$ or more	4050	3480	2890	2500
			3 or more	$2\frac{3}{4}$	6140	5260	4380	3790	$2\frac{3}{4}$	$2\frac{3}{4}$ min	3560	3050	2540	2190
										$3\frac{3}{4}$ or more	4270	3660	3050	2630

[a]Source: *National Design Specification for Wood Construction* (3).

[b]Design values apply to one split ring and bolt in single shear when installed in seasoned wood that will remain dry in service and be subject to normal loading conditions.

[c]See Figure 6.4.

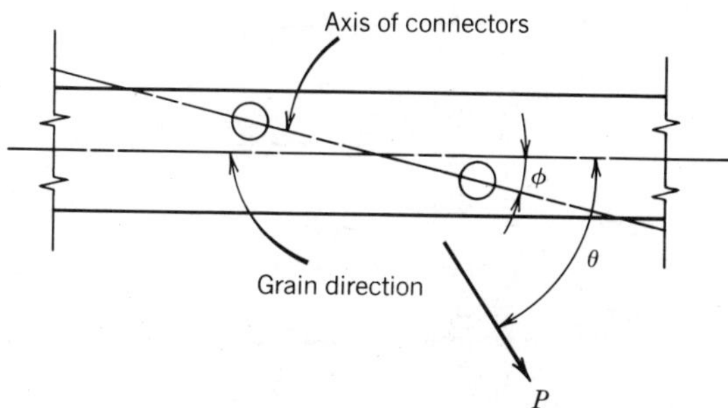

FIGURE 6.13 Angle of axis of connectors to grain, ϕ; angle of load to grain, θ.

Angle of Axis of Connectors to Grain

The connector axis is formed by a line joining the centers of any two adjacent connectors located in the same face of a member in a joint. The angle of axis of the connectors is the angle formed by the axis line of the connectors and the longitudinal axis of the member. This angle is a factor in the determination of the required spacing of connectors for a given load as illustrated by angle ϕ in Figure 6.13.

Spacing of Connectors

Spacing between connectors must be considered in determining connector loads because it controls the shearing area that develops the connector load. Factors that influence spacing are angle of load to grain and angle of axis of connectors to grain. Table 6.15 contains the spacing required for 100 and 50% of full load for parallel- and perpendicular-to-grain loading. The design values for intermediate spacings can be obtained by straight-line interpolation. The spacing in members loaded at an angle to grain with the connector axis at various angles to the grain can be calculated by Eq. (6-4) or obtained graphically from Figures 6.14 and 6.15. For tension and compression members loaded at an angle of grain, θ, other than 0° and 90° with connector axis and various angles with the grain, ϕ, other than 0° and 90°, the spacing, R, can be determined by the following formula:

$$R = \frac{AB}{\sqrt{A^2 \sin^2\phi + B^2 \cos^2\phi}} \tag{6-4}$$

where R = spacing along connector axis as shown in Figure 6.3 for the required design value (in.),

A = dimension from column 3, Table 6.16, opposite the connector in column 1 and angle of load to grain, θ, in column 2 (in.),

B = dimension from column 4, Table 6.16, opposite the connector in column 1 and angle of load to grain, θ, in column 2 (in.), and

ϕ = angle of axis of connectors to grain as shown in Figure 6.13 (degrees).

Dimension C in column 5, Table 6.16, is the minimum spacing along the connector axis and will permit 50% of the design load for the spacing R determined for full load. For spacings intermediate between R and C, the design load is determined by straight-line interpolation.

TABLE 6.15

Connector Spacings and End Distances[a,b]

		Spacing Parallel to Grain		Spacing Perpendicular to Grain		End Distance		
Split Ring Diameter (in.)	Shear Plate Diameter (in.)	Spacing (in.)	Percentage of Tabulated Load on Joint (%)	Minimum (in.)	Percentage of Tabulated Load on Joint (%)	Tension Member (in.)	Compression Member (in.)	Percentage of Tabulated Load on Joint (%)
Parallel to grain loading								
$2\frac{1}{2}$	$2\frac{5}{8}$	$6\frac{3}{4}$	100	$3\frac{1}{2}$ min	100	$5\frac{1}{2}$	4	100
$2\frac{1}{2}$	$2\frac{5}{8}$	$3\frac{1}{2}$ min	50	$3\frac{1}{2}$ min	100	$2\frac{3}{4}$ min	$2\frac{1}{2}$ min	62.5
4	4	9	100	5 min	100	7	$5\frac{1}{2}$	100
4	4	5 min	50	5 min	100	$3\frac{1}{2}$ min	$3\frac{1}{4}$ min	62.5
Perpendicular to grain loading								
$2\frac{1}{2}$	$2\frac{5}{8}$	$3\frac{1}{2}$ min	100	$4\frac{1}{4}$	100	$5\frac{1}{2}$	$5\frac{1}{2}$	100
$2\frac{1}{2}$	$2\frac{5}{8}$	$3\frac{1}{2}$ min	100	$3\frac{1}{2}$ min	50	$2\frac{3}{4}$ min	$2\frac{3}{4}$ min	62.5
4	4	5 min	100	6	100	7	7	100
4	4	5 min	100	5 min	50	$3\frac{1}{2}$ min	$3\frac{1}{2}$ min	62.5

[a]Source: *National Design Specification for Wood Construction* (3).

[b]With corresponding percentages of tabulated design values.

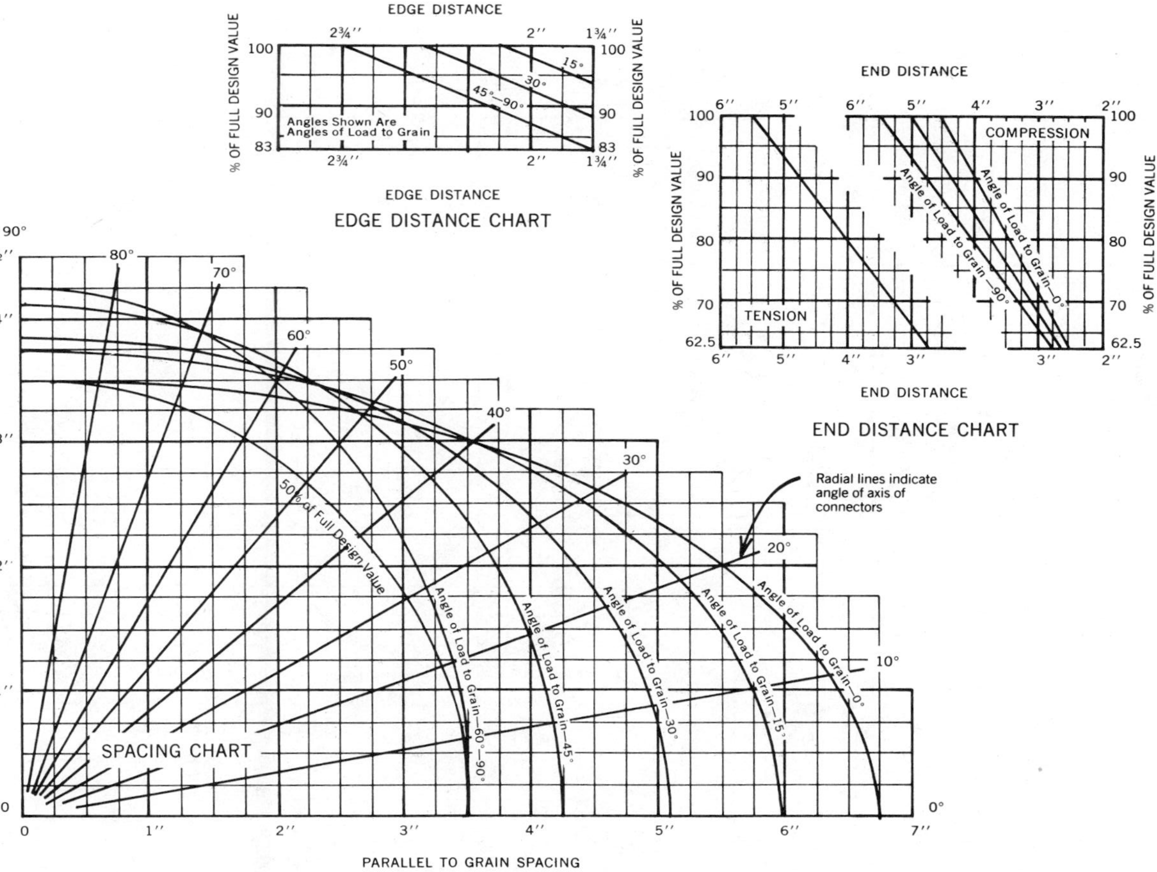

FIGURE 6.14 Design value data for connectors: $2\frac{5}{8}$-in. shear plates and $2\frac{1}{2}$-in. split rings. [Based on data from (10); copyright, TECO (Timber Engineering Co.).]

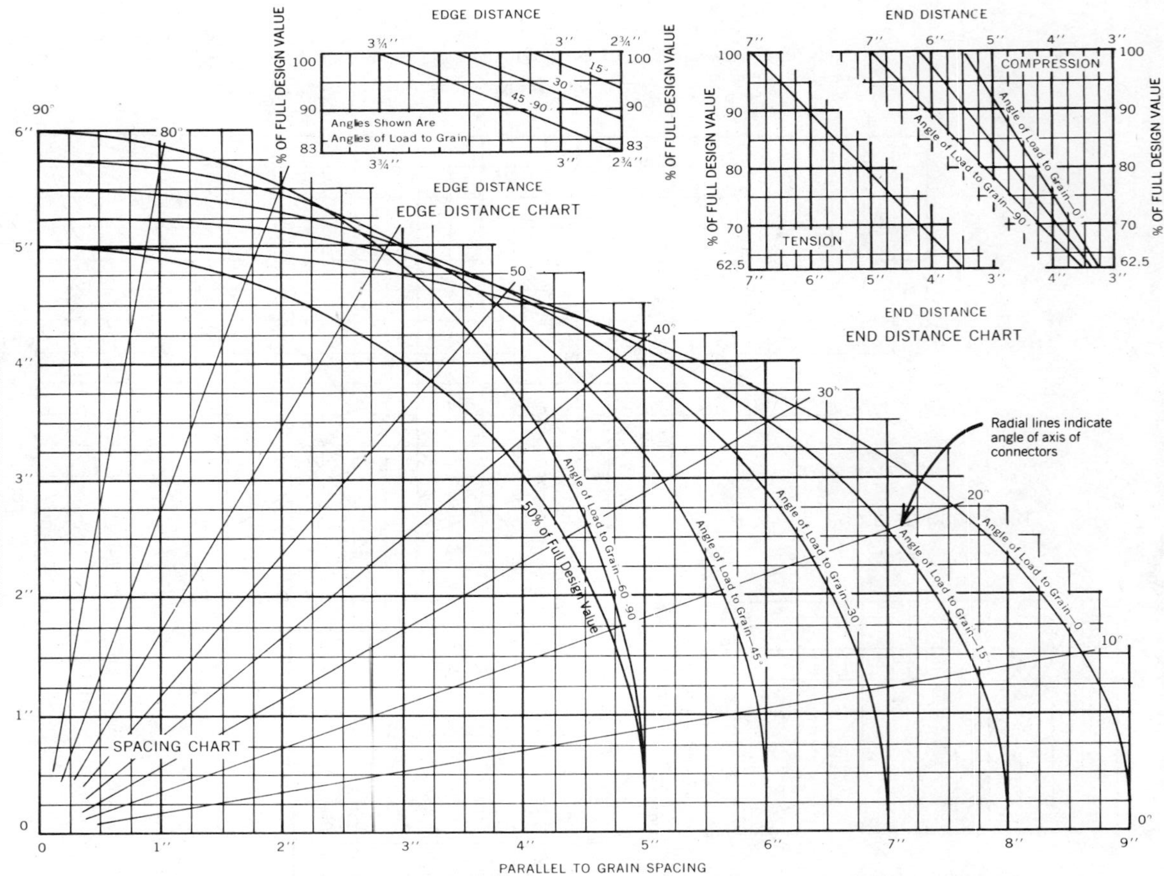

FIGURE 6.15 Design value data for connector: 4-in. shear plates and 4-in. split rings. [Based on data from (10); copyright TECO (Timber Engineering Co.).]

TABLE 6.16

Values to Use with Eq. (6-4)[a,b]

1	2	3	4	5
Type and Size of Connector	Angle of Load to Grain (θ) (deg)	A (in.)	B (in.)	C (50% value) (in.)
$2\frac{1}{2}$-in. split ring or $2\frac{5}{8}$-in. shear plate	0	$6\frac{3}{4}$	$3\frac{1}{2}$	$3\frac{1}{2}$
	15	6	$3\frac{3}{4}$	$3\frac{1}{2}$
	30	$5\frac{1}{8}$	$3\frac{7}{8}$	$3\frac{1}{2}$
	45	$4\frac{1}{4}$	$4\frac{1}{8}$	$3\frac{1}{2}$
	60–90	$3\frac{1}{2}$	$4\frac{1}{4}$	$3\frac{1}{2}$
4-in split ring or 4-in. shear plate.	0	9	5	5
	15	8	$5\frac{1}{4}$	5
	30	7	$5\frac{1}{2}$	5
	45	6	$5\frac{3}{4}$	5
	60–90	5	6	5

[a]Reproduced from the Uniform Building Code Standards, 1982 ed., Copyright 1982, with permission of the International Conference of Building Officials, Whittier, CA.

[b]Use straight-line interpolation for determining A and B for intermediate angles of load to grain.

Example. The following illustrates the use of Figure 6.14 to determine the required spacing for full load when the angle of axis of connectors is 40° and the angle of load to grain is 45°.

For full design value spacing, enter chart with an angle of axis of 40°. From where this line crosses the angle of load to grain curve at 45°, project downward and read parallel-to-grain spacing of approximately $3\frac{1}{4}$ in. Project horizontally to read perpendicular-to-grain spacing of approximately $2\frac{3}{4}$ in. The center-to-center spacing at 100% of design load is $4\frac{1}{4}$ in. measured from the intersection to the origin. This value can also be determined by calculating the square root of the sum of the squares of the parallel- and perpendicular-to-grain spacings. For minimum spacing at 50% of design load, determine where the 40° angle of axis line crosses the 50% of full design value line and project downward to approximately $2\frac{3}{4}$ in. for parallel-to-grain spacing and horizontally to approximately $2\frac{1}{4}$ in. for perpendicular-to-grain spacing. This results in a minimum spacing center to center of approximately $3\frac{1}{2}$ in. For intermediate design values, use straight-line interpolation between spacing at 50 and 100%.

Figures 6.14 and 6.15 each have five curves representing recommended spacing for full design value at the particular angle of load to grain noted on the curve. For intermediate angles of load to grain, straight-line interpolation may be used. The spacing for full design value is determined by locating the intersection of the proper angle of load-to-grain curve and the radial line representing angle of axis to grain. The distance from that point of intersection to the origin of the axes (lower left-hand corner of the chart) is the recommended spacing. The parallel-to-grain component of spacing may be read by projecting downward from the point of intersection to the bottom of the chart. The perpendicular-to-grain component of spacing may be read by projecting horizontally from the point of intersection to the left side of the chart.

The quarter circle on each spacing chart represents the spacing for 50% of design value for any angle of load to grain and is the minimum spacing permissable. For design values between 50% and full design value, interpolate radially on the straight line between the 50% curve and full design value curve representing the proper angle of load to grain.

When three or more connectors are used in one face of a member, such as is shown in Figure 6.3, the spacing between any two connectors should be checked. In the joint shown in Figure 6.3, the angle of load to grain is the same for all three connectors but the angle of axis of connectors to grain varies with each pair of connectors, therefore, the minimum spacing requirements are determined for each axis.

Edge Distance

Edge distances for connectors are given in Tables 6.12–6.14. They are also shown graphically in Figures 6.14 and 6.15.

The edge distance shown in Tables 6.12–6.14 are for the unloaded edge and the loaded edge. The unloaded edge distance is a minimum distance and is used both for parallel-to-grain loading and for the unloaded edge when the load acts at an angle to the grain. The edge distances for the loaded edge are shown as the minimum distance, and the distance required for full load. For intermediate spacings, use straight-line interpolation between the design values at minimum edge distance and full edge distance. For angles of load to grain of 45° to 90°, use the edge distance for 90°. For angles of load to grain between 0° and 45°, the required edge distance may be determined by straight-line interpolation. The capacity of a connector at minimum edge distance loaded at an angle of 45° to 90° is 83% of full load.

The edge distance charts in Figures 6.14 and 6.15 may be used to determine a modifying factor, C_e, to apply to the full design load. The minimum edge distance is given in the lower right corner of these charts. The dimension is the minimum edge distance for any direction of loading. Loaded edge distance varies with the angle of load to grain. Percentages of full design loads for various loaded edge distances and angles of load to grain are determined by projecting horizontally to either side from the point of intersection of the proper loaded edge distance line and angle of load-to-grain line. The right side of the edge distance charts is set at the minimum edge distance permitted. This distance allows 100% of full load when the angle of load to grain is 0° and varies to 83% for an angle of load to grain from 45° to 90°.

The edge distances for connectors were determined by test on lumber and design values for $2\frac{1}{2}$ in. and $2\frac{5}{8}$ in.connectors were based on edge distance that allow the use of a connector in nominal 4-in.-wide pieces (net $3\frac{1}{2}$ in.) and the use of 4-in. connectors in nominal 6-in.-wide pieces (net $5\frac{1}{2}$ in.). Glued laminated timbers are slightly smaller in width than sawn lumber of the same nominal width. A review

TABLE 6.17

Modification Factors C_e for Connectors with Reduced Edge Distances, Connectors Centered on Face

Connector Size		Parallel-to-Grain Loading[a]		Perpendicular-to-Grain Loading[b]	
$2\frac{1}{2}$-in. split ring or	Width	3 in.	$3\frac{1}{8}$ in.	3 in.	$3\frac{1}{8}$ in.
$2\frac{5}{8}$-in. shear plate		0.88	0.91	0.86	0.90
4-in. split ring or	Width	5 in.	$5\frac{1}{8}$ in.	5 in.	$5\frac{1}{8}$ in.
shear plate		0.93	0.95	0.93	0.95

[a]The modification factor C_e for parallel-to-grain loading should be applied to tabular design values in Tables 6.12 to 6.14.

[b]The modification factor C_e for perpendicular-to-grain loading should be applied to the tabular design values for minimum loaded edge distances in Tables 6.12 to 6.14.

of Technical Bulletin No. 865 *Timber Connector Joints; Their Strength and Design* (5),which contains the test data from which the design values were developed, indicates that a modification factor C_e as shown in Table 6.17 should be applied to members where the minimum tabular edge distances cannot be obtained. A $2\frac{1}{2}$-in. or $2\frac{5}{8}$-in. connector should not be used in a member less than 3 in. wide and a 4-in. wide connector should not be used in a member less than 5 in. wide. The reduction factors shown in Table 6.17 should be used when connectors are used on the narrow faces of nominal 4-in. and 6-in. wide glued laminated timbers.

End Distance

End distances and reduction factors for less than full end distance, C_n, are shown in Table 6.15 and Figures 6.14 and 6.15. The end distances in Table 6.15 are listed as minimum end distances in both tension and compression for parallel- and perpendicular-to-grain loading. When minimum end distance is used, the design value is 62.5% of the full design value. For end distances intermediate between the minimum and that required for full load, the percentage of full design value is determined by straight-line interpolation. End distances for members loaded at angles of grain other than 0° or 90° may be determined by straight-line interpolation.

End distance may also be determined from Figures 6.14 and 6.15. These charts are divided into two sections. End distance requirements depend on whether the member is in tension or compression. If the member is in tension, the allowable percentage of full design value is obtained by projecting horizontally from the intersection of the end distance line and the diagonal line. This process can be reversed by starting with the percentage of full design value and determining the required end distance. For members in compression, the angle of load to grain must also be considered. The use of the chart is the same, except that the diagonal line for the proper angle of load to grain should be used. Straight-line interpolation between lines should be used for intermediate angles.

The following example illustrates the design of a tension connection.

Example. For group B species, F_t = 1200 psi, used in dry location, 4 × 6 in. main member, 2 × 6 in. side members. Use $2\frac{1}{2}$ in. split rings as connectors.

Design load: Dead load = 5000 lb, snow load = 12,000 lb, C_D = 1.15.

$$F_t' = F_t C_D = (1200)(1.15) = 1380 \text{ psi}$$

Check net section of main and side members. From Table 6.9, projected area of one $2\frac{1}{2}$-in. split ring and bolt is 1.73 in.^{2} for side members.

$$\text{Net area} = (2)(8.25 - 1.73) = 13.04 \text{ in.}^2$$

For two $2\frac{1}{2}$-in. split rings and a bolt in the main member, projected area is 3.75 in.^{2}.

$$\text{Net area} = 19.25 - 3.75 = 15.50 \text{ in.}^2$$

Determine design load controlled by the strength of the wood.

$$\text{Design load} = (13.04)(1380) = 18{,}000 \text{ lb} > 17{,}000 \text{ lb} \quad \text{O.K.}$$

Determine connector spacings, end distance, and edge distance for full design value of each connector from Figure 6.14 or Tables 6.14 and 6.15.

End distance = $5\frac{1}{2}$ in.
Edge distance = $2\frac{3}{4}$ in.
Spacing = $6\frac{3}{4}$ in.

Determine number of connectors required. From Table 6.14, full design value for one connector,

$$P = 2730 \text{ lb}$$

$$P' = PC_D = (2730)(1.15) = 3140 \text{ lb}$$

$$\text{Number of split rings required} = 17{,}000/3140 = 5.4$$

Try 3 split rings in each side member and check for group action of connectors in a row. For this connection, there are two rows of three connectors in each row.

Area of side members A_2 = (2)(8.25) = 16.50 in.^{2}
Area of main member A_1 = 19.25 in.^{2}

Because $A_1/A_2 > 1$ use A_2/A_1 = 16.50/19.25 = 0.86. From Table 6.7, for three connectors in a row interpolate between A_1/A_2 = 0.5 and A_1/A_2 = 1.0.

Modification factor C_g = 0.97
Design value per connector $P' = PC_DC_g$ = (0.97)(1.15)(2730) = 3050 lb
Design value = (2)(3)(3050) = 18,300 lb > 17,000 lb O.K.

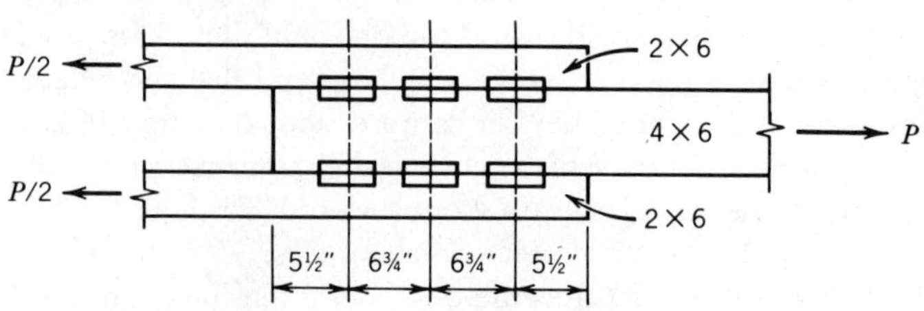

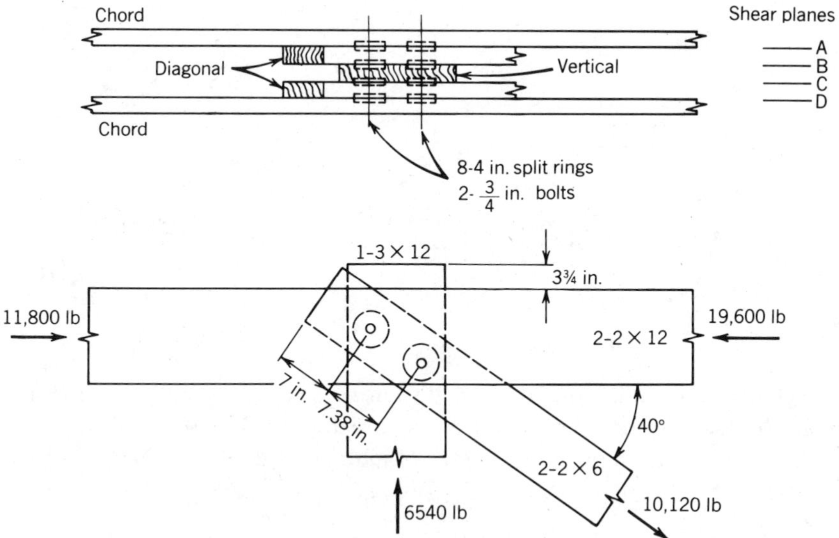

FIGURE 6.16 Top chord connection.

The final joint configuration is as shown in the sketch. Where clearance for the connection is critical, the end distance and spacing could be reduced by taking into account that the connectors are not fully loaded. The ratio of the actual load to design load in this example is 17,000/18,300 = 0.93. From Figure 6.14, the required end distance for 93% of full load is $4\frac{1}{4}$ in. and the required spacing is $6\frac{3}{8}$ in.

Example. The following example (Fig. 6.16) illustrates the transfer of loads, determination of angle of load to grain, angle of axis to grain, and the determination of required end and edge distances using multiple split rings in an existing truss joint.

Dry condition of use, snow load controls the design ($C_D = 1.15$). Select structural Douglas Fir-Larch (group B species).

Determine the loads transferred in each connector plane and the angle of load and angle of axis to grain of the members on each side of the plane being considered.

A 7800 lb (19,600 − 11,800) horizontal force is transferred by the connectors in planes *A* and *D* from the two-member top chord to the two-member diagonal. Assume the load is shared equally by the four connectors. The loads transferred on both planes *A* and *D* are as shown in Figure 6.17.

The angle of load to grain is 0° in the chord and 40° in the diagonal. The angle of axis to grain is 40° in the chord and 0° in the diagonal.

A 6540 lb force is transferred between the diagonal and the vertical member in planes *B* and *C*. The force is assumed to be divided equally between the four connectors as shown in Figure 6.18.

The angle of load to grain in the vertical is 0° and angle of axis to grain is 50°. The angle of load to grain in the diagonal is 50° and angle of axis to grain is 0°.

In some instances, the controlling member of the joint can be determined by

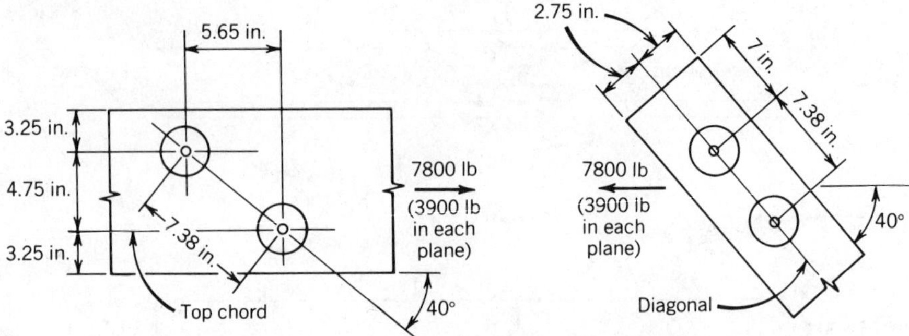

FIGURE 6.17 Load transfer in planes *A* and *D*.

inspection depending on end, edge distances, or angle of load to grain. In this example, for purposes of illustration, each member of the joint will be analyzed.

Determine spacing from geometry of joint spacing along axis

$$s = \sqrt{4.75^2 + 5.65^2} = 7.38 \text{ in.}$$

Top chord (planes *A* and *D*): Check load capacity as determined by connectors in chord. From Table 6.12, $P = 5160$ lb (connector in one face of a $1\frac{1}{2}$-in.-thick member). Actual edge distance = 3.25 in.

From Table 6.14, required edge distance for full load = $2\frac{3}{4}$ in., use $C_e = 1.00$. Actual end distance exceeds distance required for full load from Table 6.15 which is 7 in., use $C_n = 1.00$. Actual spacing = 7.38 in. at angle of 40° to grain.

From Figure 6.15, spacing required for full load at angle of axis of connectors of 40° = 5 in. parallel to grain and 4.2 in. perpendicular to grain. $s = \sqrt{5^2 + 4.2^2} = 6.53$ in. along connector axis. Because 7.38 in. > 6.5 in., $C_s =$ 1.00.

Since C_e, C_n, and C_s each = 1.00, no modification is necessary for edge distance, end distance, or spacing.

$$P' = PC_D = (5160)(1.15) = 5930 \text{ lb}$$

$$\text{Total load} = (4)(5930) = 23{,}700 \text{ lb} > 7800 \text{ lb} \quad \text{O.K.}$$

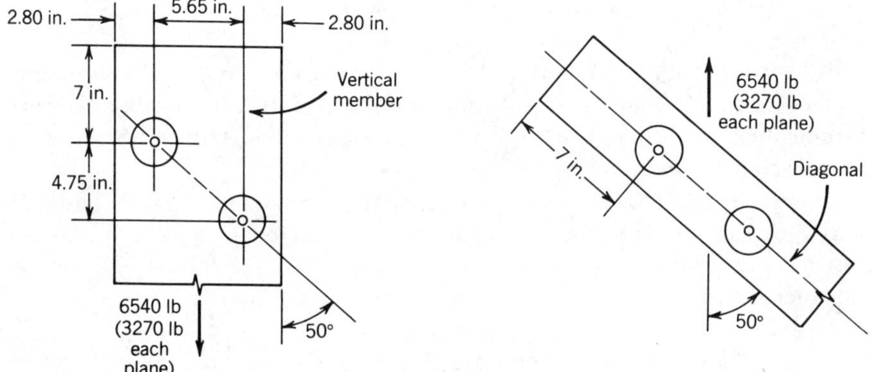

FIGURE 6.18 Vertical to diagonal connection.

Diagonal (planes A and D): Check load capacity of connectors in diagonal as determined by transfer of forces between chord and diagonal. From Table 6.14 $P = 3520$ lb and $Q = 2040$ lb. Actual loaded edge distance $= 2\frac{3}{4}$in. From Figure 6.15, minimum edge distance for 83% of full load at 45° angle of load to grain $= 2\frac{3}{4}$ in. When angle of load to grain is 40°, determine C_e by interpolation:

$$C_e = (5/45)\,(1.0 - 0.83) + 0.83 = 0.85$$

Actual end distance = 7 in.

From Table 6.15, end distance for required full load = 7 in.

$$C_n = 1.00$$

Actual spacing = 7.38 in. parallel to grain with angle of axis of connectors to grain of 0°.

From Figure 6.15, spacing for 100% full load at angle of load to grain of 40° is obtained by interpolation of values along the 0° angle of axis of connectors to grain at 30° and 45° angle of load to grain. This results in a required spacing of 6.3 in.

$$7.38 \text{ in.} > 6.3,\ C_s = 1.00$$

Determine split ring capacity N at an angle of load to grain of 40°.

$$N = \frac{PQ}{P \sin^2\theta + Q\cos^2\theta} = \frac{(3520)(2040)}{3520\sin^2 40° + 2040\cos^2 40°} = 2710 \text{ lb}$$

Edge distance controls, C_e is less than C_s and C_n

$$N' = NC_DC_e = (2710)(1.15)(0.85) = 2640 \text{ lb}$$
$$\text{Total load} = 4N' = (4)(2640) = 10{,}580 \text{ lb} > 7800 \text{ lb} \quad \text{O.K.}$$

Diagonal (planes B and C): Check load capacity of the connectors in the diagonals as determined by the transfer of the 6540 lb vertical load between the diagonals and the vertical member.

Check capacity of connectors in diagonal. From Table 6.14, (for connectors in two faces of a $1\frac{1}{2}$-in.-thick member)

$$P = 3520 \text{ lb} \qquad Q = 2040 \text{ lb}$$

Actual edge distance $= 2\frac{3}{4}$ in. From Figure 6.15, the capacity of a 4-in. split ring with minimum edge distance of $2\frac{3}{4}$ in. and loaded at an angle of load to grain of 45° to 90° is 83% of full design value. Diagonal is loaded at an angle of load to grain of 50°. $C_e = 0.83$.

Actual end distance = 7 in. From Table 6.15, end distance for full loading = 7 in. $C_n = 1.00$.

Actual spacing = 7.38 in. From Figure 6.15 for angle of load to grain of 50° and angle of axis of connectors to grain of 0°, required spacing is 5.67 in. along axis of connectors (by interpolation). 7.38 in. > 5.67 in. $C_s = 1.00$. $C_e = 0.83$ which is less than C_n and C_s.

$$N = \frac{(3520)(2040)}{3520 \sin^2 50° + 2040 \cos^2 50°} = 2470 \text{ lb}$$

$$N' = NC_DC_s = (2470)(1.15)(0.83) = 2360 \text{ lb}$$

$$\text{Total load} = 4N' = 9430 \text{ lb} > 6540 \text{ lb} \quad \text{O.K.}$$

Vertical (planes *B* and *C*): Check capacity of connectors in vertical member. From Table 6.14, $P = 5000$ lb (for connectors in two faces of a $2\frac{1}{2}$-in.-thick member). Actual edge distance = 2.80 in. From Table 6.14, the required edge distance for full load = $2\frac{3}{4}$ in. 2.80 in. > 2.75 in., $C_e = 1.00$.

Actual end distance = 7 in. From Table 6.15, the required end distance for full load = 7 in. $C_n = 1.00$.

Actual spacing = 7.38 in. From Figure 6.15, spacing for angle of load to grain of 0° and angle of axis of connectors to grain of 50° = 3.9 in. parallel to grain and 4.6 in. perpendicular to grain or 6.0 in. along axis of connectors. Since 7.38 > 6.0, $C_s = 1.00$. No modification required for edge distance, end distance, or spacing.

$$P' = PC_D = (5000)(1.15) = 5750 \text{ lb}$$

$$\text{Total load} = 4P' = (4)(5750) = 23{,}000 \text{ lb} > 6540 \text{ lb.} \qquad \text{O.K.}$$

In this example, the vertical and diagonals were shown with full end distances. This is recommended. If space above the top chord had been limited, the end distances could have been cut back. The connectors in the vertical are required to develop (6540/23,000)(100) = 28% of their capacity, therefore, the minimum end distance of $3\frac{1}{2}$ in., which gives 62.5% of full load and is adequate. The connectors in the diagonal are loaded to (6540/9424)(100) = 69% of their capacity.

$$\text{End distance for 69\% of capacity} = \frac{0.69 - 0.625}{1 - 0.625}(3.5) + 3.5 = 4.2 \text{ in.}$$

The 4.2-in. end distance is measured as shown in Figure 6.19. The top of the vertical must extend $\frac{1}{4}$ in. above the chord to obtain the $3\frac{1}{2}$-in. end distance. Cutting

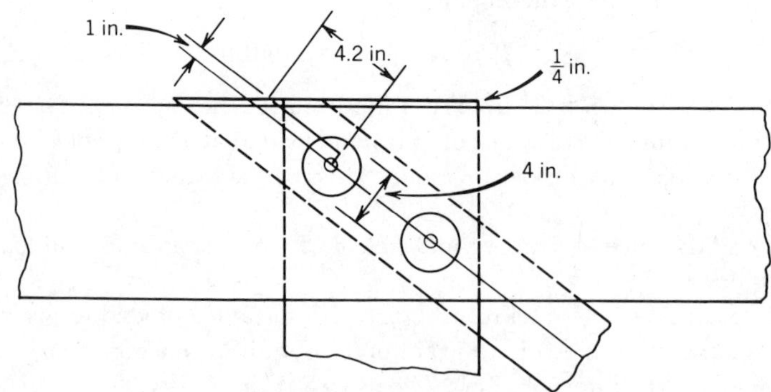

FIGURE 6.19 End distance on diagonal.

the diagonal horizontally and even with the top of the vertical as shown gives a distance slightly more than 4.2 in.

Check net sections:

$$F_t = 1200 \text{ psi}, \ F_t' = (1200)(1.15) = 1380 \text{ psi}$$

$$F_c = 1600 \text{ psi}, \ F_c' = 1840 \text{ psi}$$

Top chord $A = 16.88$ in.^{2}.

Because parallel-to-grain spacing is greater than the diameter of one connector, only the area of one split ring and bolt needs to be subtracted from the gross area. From Table 6.9, the projected area of the split rings and bolt within the member is 3.05 in.^{2}

$$A_{net} = 16.88 - 3.05 = 13.83 \text{ in.}^2$$

$$f_c = \frac{19{,}600}{(2)(13.83)} = 710 \text{ psi} < 1840 \text{ psi} \quad \text{O.K.}$$

Diagonal $A = 8.25$ in.^{2}. Area of connectors and bolt $= 4.88$ in.^{2}.

$$A_{net} = 8.25 - 4.88 = 3.37 \text{ in.}^2$$

$$f_t = \frac{10{,}180}{(2)(3.37)} = 1510 \text{ psi} > 1380 \text{ psi}$$

The diagonal member is 9.4% overstressed at the net section as determined by conventional calculations, however, in an existing structure the member can be examined and, if no knots occur at the critical section, the higher design value from Table 3.1 may be used. In this case $F_t = F_g = 2020$ psi and $F_t' = (2020)(1.15) = 2320$ psi.

Vertical $A = 28.12$ in.^{2}. Area of connectors and bolts $= 5.69$ in.^{2}.

$$A_{net} = 28.12 - 5.69 = 22.43 \text{ in.}^2$$

$$f_c = \frac{6540}{22.43} = 288 \text{ psi} < 1840 \text{ psi} \quad \text{O.K.}$$

Timber Connector Design Values in End Grain

When timber connectors are installed in a surface not parallel to the general direction of the grain of a member, such as the square cut or sloping surface at the end of a member, the design values in Tables 6.12–6.14 must be modified. Figure 6.20 illustrates these conditions of loading.

When the end of a member is square cut as shown in Figure 6.20*a*, the allowable design value, Q_{90}', is 60% of the perpendicular-to-grain design value regardless of the direction of the load, Q_{90}, in the plane of the cut.

$$Q_{90}' = 0.60Q' \tag{6-5}$$

where Q' = design value for a connector in a side grain surface loaded perpendicular to grain (in.), adjusted by applicable modifiers.

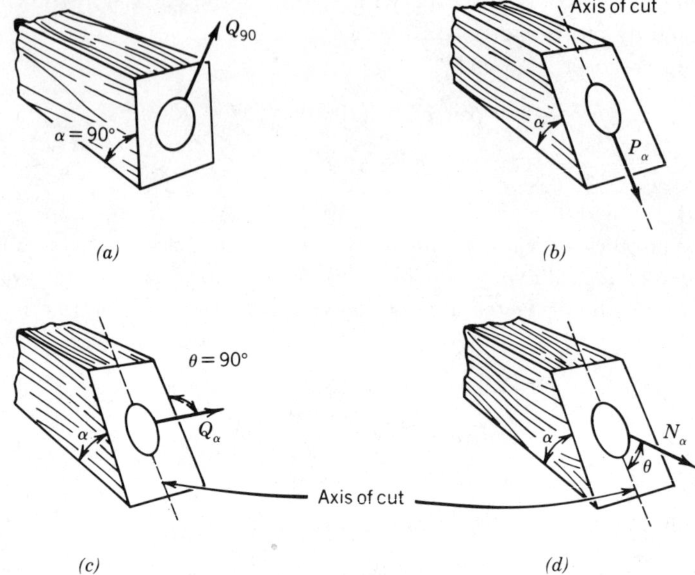

FIGURE 6.20 Timber connectors in end grain. (*a*) Square cut end, $\alpha = 90°$; (*b*) load parallel to axis of cut, $\theta = 0°$; (*c*) load perpendicular to axis of cut, $\theta = 90°$; (*d*) load at angle θ to axis of cut.

When the end of a member is a sloping surface forming an angle α with the general longitudinal grain direction of the member or laminations, the load P_α is applied parallel to the axis of cut and $\theta = 0°$, as shown in Figure 6.20*b*; the design value P'_α is determined by the Hankinson formula:

$$P'_\alpha = \frac{P'Q'_{90}}{P' \sin^2 \alpha + Q'_{90} \cos^2 \alpha} \tag{6-6}$$

where P'_α = design value for a connector in a sloping surface, cut at angle α and loaded parallel to the axis of cut (lb), adjusted by applicable modifiers,

P' = design value for a connector in a side grain surface, loaded parallel to grain (in.), adjusted by all applicable modifiers,

Q'_{90} = 60% of the design value for a connector in a side grain surface, Q', loaded perpendicular to grain (lb), adjusted by applicable modifiers, and

α = slope of cut surface (degrees), as shown in Figure 6.20*b*.

When the end of a member is a sloping surface forming an angle α with the general longitudinal grain direction of the member or laminations, the load Q_α is applied perpendicular to the axis of the cut and $\theta = 90°$, as shown in Figure 6.20*c*; the design value Q'_α is determined by the Hankinson formula:

$$Q'_\alpha = \frac{Q'Q'_{90}}{Q' \sin^2 \alpha + Q'_{90} \cos^2 \alpha} \tag{6-7}$$

where Q'_α = design load for a connector in a sloping surface, loaded perpendicular to the axis of cut (lb), adjusted by applicable modifiers,

Q' = design value for a connector in a side grain surface, loaded perpendicular to grain (lb), adjusted by applicable modifiers,

Q'_{90} = 60% of the design value for a connector in a side grain surface, loaded perpendicular to grain (lb), adjusted by applicable modifiers, and

α = slope of cut surface (degrees).

When the end of a member is a sloping surface forming an angle α with the general longitudinal grain direction of the member or laminations, and the load N_α is applied at an angle θ other than 0° or 90° to the axis of the cut, as shown in Figure 6.20*d*, the design value N'_α is determined by the Hankinson formula:

$$N'_\alpha = \frac{P'_\alpha Q'_\alpha}{P'_\alpha \sin^2 \theta + Q'_\alpha \cos^2 \theta} \tag{6-8}$$

where N'_α = design value for a connector in a sloping surface, loaded at an angle θ to the axis of the cut (lb),

and other terms are as defined previously.

The provisions for edge distance, end distance, and spacing shown for connectors in side grain shall be applied to the sloping cuts as follows:

1. For square cut ends the provisions for perpendicular-to-grain loading apply.
2. For sloping surfaces with angle α from 45° to 90° loaded in any direction, the provisions for perpendicular-to-grain loading apply.
3. For sloping surfaces with angle α less than 45° loaded parallel to the axis of the cut, the provisions for perpendicular-to-grain loading apply.
4. For sloping surfaces with angle α less than 45° loaded perpendicular to the axes of the cut, the provisions for perpendicular-to-grain loading apply.
5. For sloping surfaces with angle α less than 45° loaded at angle θ to axis of cut, the provisions for members loaded at angles of grain other than 0° or 90° apply.

The design values for connectors installed in end grain should also be checked for reduction in end shear caused by the notched beam effect as given on page 6-412. Only the component of load acting perpendicular to grain should be used in computing vertical reaction.

Example. The peak of an A-frame is fastened together with two 4-in. shear plates placed back to back and a $\frac{3}{4}$-in. through bolt with washers counterbored

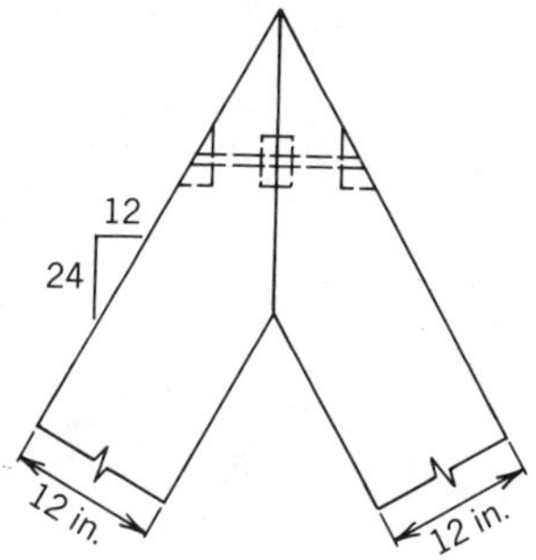

into the arch. Determine the vertical force P'_α which is primarily unbalanced snow load ($C_D = 1.15$) that can be transferred through the joint in either direction. The thickness of the member is assumed to be at the center of the connector = 6/2 = 3 in. From Table 6.12, the tabular design value for group B species for one 4-in. shear plate with wood side plates is

parallel-to-grain loading $P = 4360$ lb

perpendicular-to-grain loading with a loaded edge distance of 6 in. $Q = 3040$ lb. From Table 6.17, for $5\frac{1}{8}$-in. width (edge distance = 2.56 in.), $C_e = 0.95$ for both parallel- and perpendicular-to-grain loading.

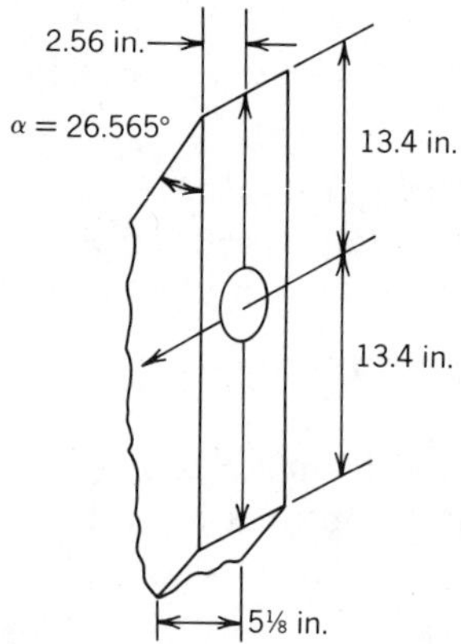

End distance in both directions equal 13.4 in., angle $\alpha < 45°$, therefore, use this distance for parallel-to-grain end distance. From Table 6.15 end distance equals 7 in. for full (100%) load (for both parallel- and perpendicular-to-grain loading).

$$C_n = 1.00 \ (13.4 \text{ in.} > 7 \text{ in.})$$

When modifiers are the same for both parallel- and perpendicular-to-grain loading, the modifiers can be applied directly to P and Q.

$$P' = PC_DC_e = (4360)(1.15)(0.95) = 4760 \text{ lb}$$

$$Q' = QC_DC_e = (3040)(1.15)(0.95) = 3320 \text{ lb}$$

$$Q'_{90} = 0.60\ Q' = (3320)(0.6) = 1990 \text{ lb}$$

$$P'_\alpha = \frac{P'Q'_{90}}{P' \sin^2 \alpha + Q'_{90} \cos^2 \alpha}$$

$$= \frac{(4760)(1990)}{4760 \sin^2 26.565° + 1990 \cos^2 26.565°} = 3720 \text{ lb}$$

BOLTS—GENERAL CONSIDERATIONS

The design values given herein are based on accurate drilling and placement of the bolt holes in both the main member and the side members.

A tight fit requiring forcible driving of bolts is not recommended. Bolt holes should be $\frac{1}{32}$ to $\frac{1}{16}$ in. larger than the bolt diameter, depending on the size of the bolt. Careful centering of holes in main members and splice plates is assumed. Washers of proper size or a metal plate or strap should be used between the wood and the bolt head and between the wood and the nut. Nuts should be tightened snugly, but not so tightly as to cause crushing of the wood under the washer or plate.

Lumber Species

The species groupings for design values for bolts in Tables 6.20–6.22 are based on species with similar strength characteristics. The species and growth rate for combinations of glued laminated timber for use in entering Tables 6.21 and 6.22 are obtained from Table 3, AITC 117—Design (7) included in Part III of this manual. When bolts are used in combinations containing lower-strength grades and/or species in the interior laminations, the design values are controlled by the lower-strength laminations, except where the bolts are located solely within a higher-strength zone of the combination.

Bolt Design Values

The design values given in Tables 6.20–6.22 are for one ASTM A307 bolt in double shear in a balanced three-member joint in sawn lumber or glued laminated timber. For glued laminated timber, the tabulated values apply to bolts installed either perpendicular or parallel to the wide face of the laminations. The design values for other types of joints are modified as shown in Table 6.19.

Design values for more than one bolt, either of the same or miscellaneous sizes, are the sum of the design values permitted for each bolt, provided that spacings, end distances, and edge distances are sufficient to develop the full strength of each bolt in such a joint; and provided that the design values are modified for group action in accordance with recommendations given on page 6-411. When end distance or spacing for parallel-to-grain loading is less than required for full load, the design value of the group is the total number of bolts in the group multiplied by the lowest bolt design value in the group.

Modification Factors for Bolts

The applicable modifiers shown in Table 6.3 must be applied to the tabular design values that have been adjusted for joint configuration.

Duration of load factor C_D is obtained from Figure 3.1 or Table 3.2.

Moisture content factor C_M are shown in Tables 6.4–6.6.

Fire retardant treatment factor C_R should be obtained from the manufacturer of the treatment for the species of lumber being fastened.

Temperature factor C_t is the same as for wood and is given on page 3-69.

End distance factor C_n is explained on page 6-402.

Spacing factor C_s is explained on page 6-402.

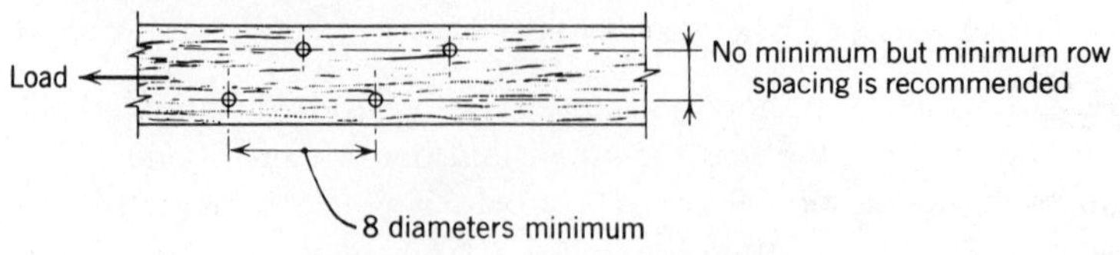

FIGURE 6.21 Staggered bolt spacing.

Group action factor C_g is given in Tables 6.7 and 6.8.

Steel side plate factor C_{st} for bolts has historically been 1.25. Recent research indicates that this value may be increased. See the current *National Design Specification for Wood Construction* (3) for increases in bolt values for sawn members with steel side plates. The tabulated values for bolts in glued laminated timber in Table 6.21 have been increased 25% over those for wood side plates and no further modification for the steel side plate factor is applicable.

Critical or Net Section

The net area at the critical section of a bolted joint is equal to the full cross-sectional area of the timber less the projected area of bolt holes at that section. For parallel-to-grain loading where bolts are staggered, adjacent bolts in a row with parallel-to-grain spacing less than eight times the bolt diameter, the nearest bolt in the next row is considered to occur at the same critical section (see Fig. 6.21).

The net cross-sectional area, in square inches, required at the critical section is determined by dividing the total load, in pounds, transferred through the critical section of the member by the appropriate unit tensile or compressive design value for the species. In other words, for tension members, divide by F_t', and for compression members, divide by F_c'.

Angle of Load to Grain

The Hankinson formula is used to determine the design value N of loads at an angle of load to grain, θ, between 0° and 90°.

Spacing of Bolts

A row of bolts is a number of bolts placed in a line parallel to the direction of load when the load is parallel or perpendicular to grain. Spacing of bolts in a row is measured in the direction of the load from center to center of bolts (dimension A, Figs. 6.22 and 6.23). Row spacing is measured between rows perpendicular to the direction of the load (dimension B, Figs. 6.22 and 6.23). Recommended bolt spacing values are given in Table 6.18.

End and Edge Distances

End distance is the distance from the end of a member to the center of the bolt nearest the end (dimension C, Fig. 6.22). Edge distance is the distance from the edge of the member to the center of the nearest bolt (dimension D, Fig. 6.22 and 6.23). For perpendicular-to-grain loading, the loaded edge distance is measured

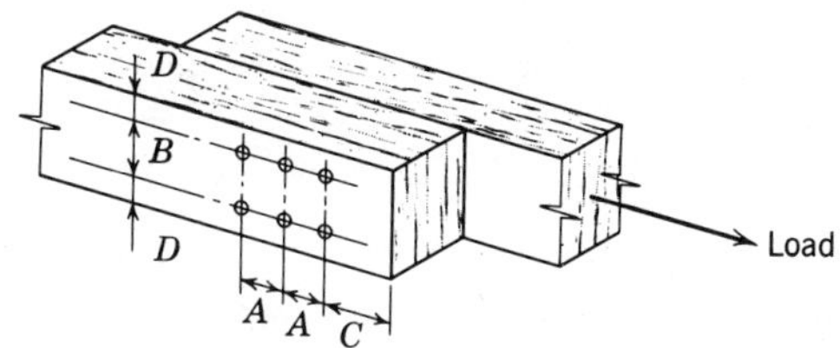

FIGURE 6.22 Load parallel to grain.

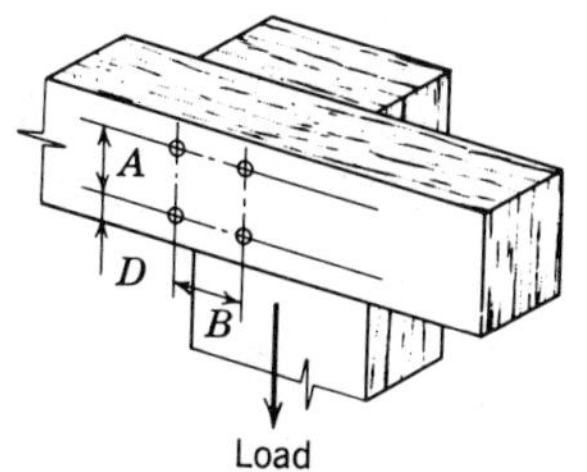

FIGURE 6.23 Load perpendicular to grain.

from the center of the bolt to the edge of the member in the direction of the load, and the unloaded edge distance is measured to the opposite edge of the member. End and edge distance values are given in Table 6.18.

For other than normal duration of loading, the design values from the tables should be multiplied by the appropriate modification factor C_D from Figure 3.1 or Table 3.2. The full design values may be used for lumber that is installed in seasoned (less than 19% moisture content) wood that will remain dry in service. They apply for bolted joints with wood side members having a single bolt and loaded parallel or perpendicular to grain, or having multiple rows of bolts loaded parallel to grain with separate splice plates for each row. The full design values may also be used for bolted joints with steel gusset plates on each side having a single row of bolts parallel to grain in each member and loaded parallel or perpendicular to grain (see Fig. 6.24). For joints in use under other service or seasoning conditions, the design values determined from Tables 6.20–6.22 are multiplied by the appropriate factor from Tables 6.4 and 6.5 for sawn lumber and Table 6.6 for glued laminated timber.

For other than three-member bolted joints, the appropriate length of bolt and load factors from Table 6.19 are applied.

Example. Design a tension connection using steel side plates to splice the ends of $6\frac{3}{4}$ in. × $7\frac{1}{2}$ in. combination 3 Western Species glued laminated tension members.

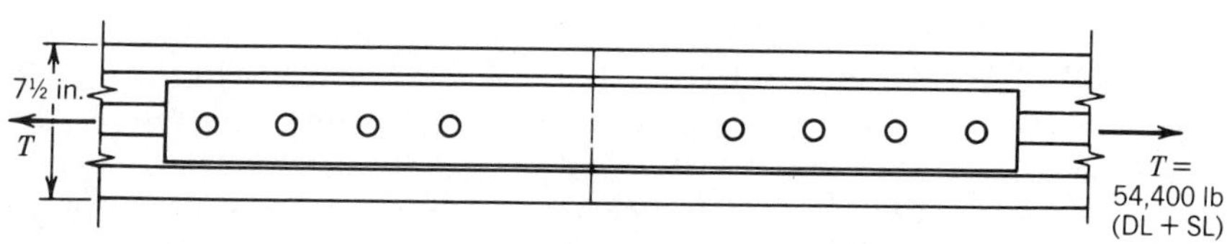

From Table 6.21, $P = 7460$ lb for a 1-in. diameter bolt.

$$P'' = PC_D = (7460)(1.15) = 8580 \text{ lb}$$

$$\text{Number of bolts required} = \frac{T}{P''} = \frac{54{,}500}{8580} = 6.3$$

TABLE 6.18

Recommended Spacing and Edge and End Distance Values for Bolts[a]

Dimension[b]	Parallel-to-Grain Loading	Perpendicular-to-Grain Loading
A, c–c spacing of bolts in a row	Four times bolt diameter for full design value. Minimum 3 times bolt diameter for 75% of full design value. Use straight-line interpolation for design values for intermediate spacing.	Design value limited by spacing requirements of attached member or members (whether of metal or of wood loaded parallel to grain).
Staggered bolts	Adjacent bolts are considered to be placed at critical section unless bolts in a row are spaced at a minimum of 8 times the bolt diameter (see Figure 6.21).	Staggering of bolts is desirable for members loaded perpendicular to grain.
B, row spacing	Minimum of $1\frac{1}{2}$ times bolt diameter.	$2\frac{1}{2}$ times bolt diameter for l/D[c] ratio of 2; 5 times bolt diameter for l/D ratios of 6 or more; use straight-line interpolation for l/D between 2 and 6.
	Spacing between rows paralleling a member may not exceed 5 in. unless separate splice plates are used for each row.	
C, end distance	In tension, 7 times bolt diameter for softwoods and 5 times bolt diameter for hardwoods for full design load. In compression, 4 times bolt diameter for full design load. Minimum end distance is $\frac{1}{2}$ that for full load for which the design value is 50% of that for full end distance. Interpolate for intermediate end distances.	4 times bolt diameter for full design load, 2 times bolt diameter for 50% of design load. Interpolate for intermediate loads. When members abut at a joint (not illustrated) the strength of the joint shall be evaluated also as a beam supported by fastenings (see page 6-410).
D, edge distance	$1\frac{1}{2}$ times the bolt diameter; except that for l/D ratios of more than 6, use $1\frac{1}{2}$ times bolt diameter or one-half the row spacing *B*, whichever is greater.	Minimum of 4 times bolt diameter at edge toward which load acts; minimum of $1\frac{1}{2}$ times bolt diameter at opposite edge.

[a]These are minimum values for bolt design values in Tables 6.20, 6.21, and 6.22.

[b]See Figures 6.22 and 6.23.

[c]Ratio of length of bolt in main member l to diameter of bolt D.

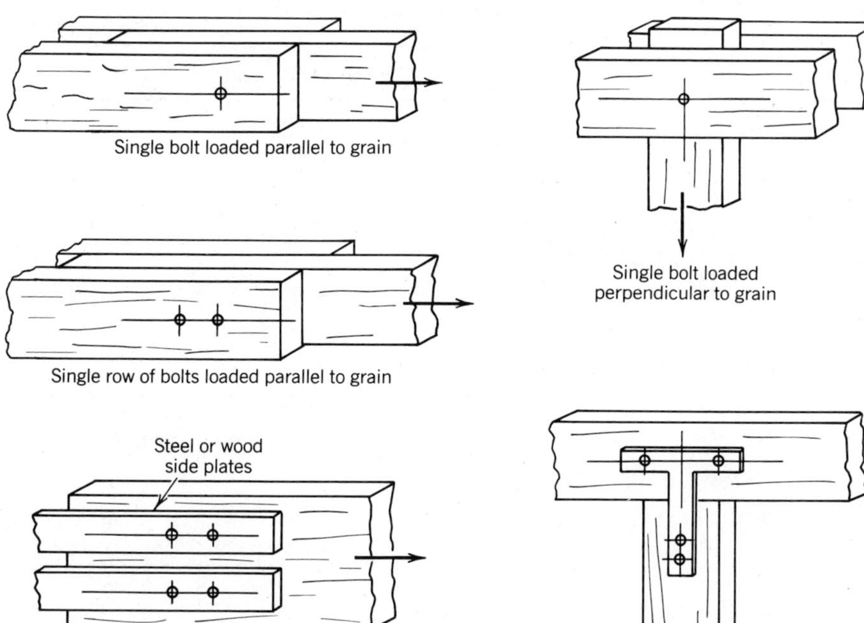

FIGURE 6.24 Examples of bolted joints for which full tabular design values may be used (further modification may be necessary due to conditions of use).

Try seven bolts in a row. Check group action factor C_g, assuming a $\frac{3}{8}$-in. × 4-in. steel side plate.

$$A_1 = (6.75)(7.5) = 50.6 \text{ in.}^2$$

$$A_2 = (0.375)(4)(2) = 3 \text{ in.}^2$$

$$\frac{A_1}{A_2} = \frac{50.6}{3} = 16.9$$

From Table 6.8, $C_g = 0.80$.

$$P' = PC_DC_g = (7460)(1.15)(0.80) = 6860 \text{ lb}$$

$$\text{Number of bolts required} = \frac{54{,}400}{6860} = 7.9$$

Seven bolts in a row are inadequate. Try two rows of four bolts each with $\frac{3}{8}$-in × 7-in. side plate.

$$A_2 = (0.375)(7)(2) = 5.25 \text{ in.}^3$$

$$\frac{A_1}{A_2} = \frac{50.3}{5.25} = 9.6$$

TABLE 6.19

Load Factors for Bolted Joints[a]

<table>
<tr><th>Type of Joint</th><th>Length l and Load Factors</th></tr>
<tr><td>Three-member joints
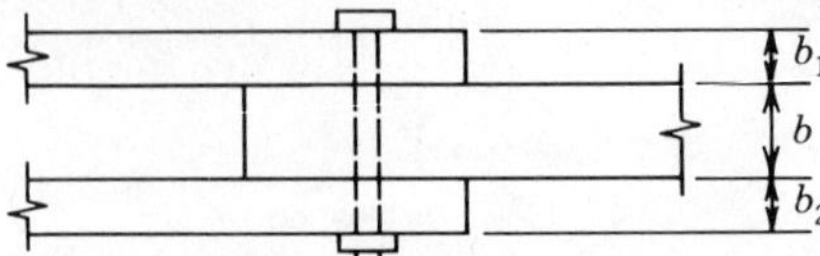
</td><td></td></tr>
<tr><td>Wood side members</td><td></td></tr>
<tr><td>$b_1 = b_2 = b/2$</td><td>Use tabulated design value.</td></tr>
<tr><td>b_1 and $b_2 > b/2$</td><td>Use tabulated design value.</td></tr>
<tr><td>$b_1 \leq b_2 < b/2$</td><td>Use tabulated design value for main member twice the thickness of b_1.</td></tr>
<tr><td>Steel side members</td><td>Use tabulated value for glued laminated timber, for sawn lumber modify tabulated value by C_{st}.</td></tr>
<tr><td>Two-member joints
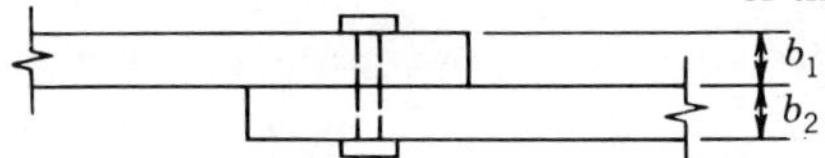

$b_1 = b_2$</td><td>Use $\frac{1}{2}$ tabulated design value for main member of thickness $= b_1$.</td></tr>
<tr><td>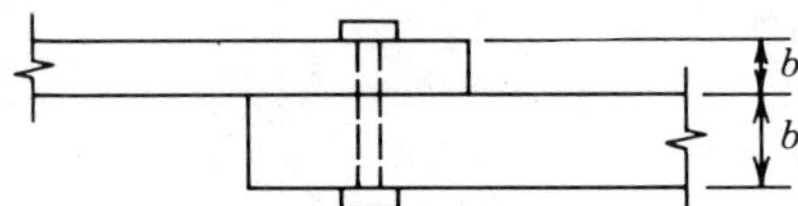

$b_1 < b_2$</td><td>Use lesser of $\frac{1}{2}$ tabulated design values for main member of thickness $= b_2$ or $2b_1$.</td></tr>
<tr><td>Steel side member</td><td>Use $\frac{1}{2}$ tabulated design value for a piece the thickness of the wood member (b_2) for glued laminated timber, for sawn lumber modify tabulated value by C_{st}.</td></tr>
<tr><td>Multiple-member joints (4 or more members)
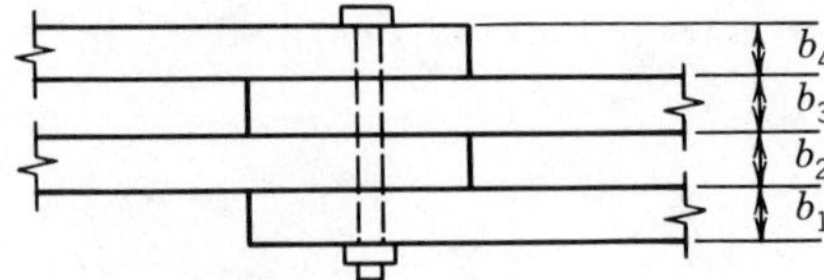

$b_1 = b_2 = b_3 = b_4$</td><td>Use tabulated value for main member of thickness $= b_1$ and multiply the load by $\frac{1}{2}$ the number of shear planes involved.</td></tr>
</table>

TABLE 6.19 (*Continued*)

Type of Joint	Length l and Load Factors
4 or more members of unequal thickness	Resolve the joint into the maximum number of contiguous three-member joints (in the illustration, two such three-member joints). For each three-member joint, determine the design value in the manner stated for a three-member joint and assign $\frac{1}{2}$ the value to each shear plane in the joint. For those shear planes to which two different values have been assigned, the design value is the lesser of the two values. For assemblies in which the load is shared equally among the members, or in which the distribution of load among members is indeterminate, the design value used should be the least design value for any one shear plane times the number of shear planes in the joint. For assemblies in which the load on each member is known, the bolt design value for any member in the joint will be the sum of the individual bolt design values for each of the two shear planes acting on that member determined as stated above.
Wood member attached to concrete masonry 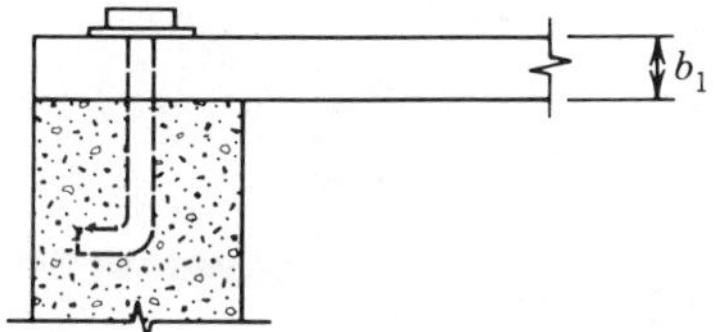	Use $\frac{1}{2}$ tabulated design values for a member twice the thickness of b_1 (the design load of the bolt in the concrete must be ample).
Two-member joints 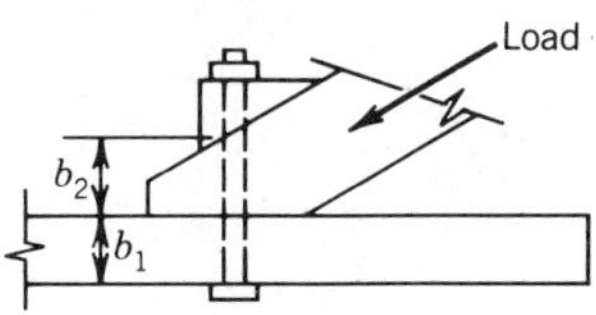 Load acting at angle with axis of bolt $b_1 < b_2$	Determine load component acting perpendicular to the axis of the bolt. Then use the applicable method for two-member joints with members of thicknesses b_1 and b_2. Sufficient bearing area must be provided under washers or plates to resist the load component acting parallel to the bolt's axis.
Steel plate used for main member 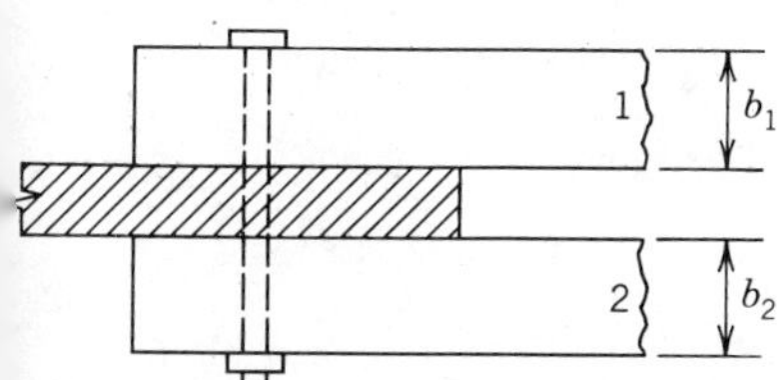	Determine design load for a member twice the thickness of the thinnest side member, $2b_1$. Steel plate shall be of ample strength.

[a]For use with Tables 6.20 to 6.22.

TABLE 6.20

Bolt Design Values, One Bolt Loaded at Both Ends, Double Shear, for Sawn Lumber[a,b]

Wood side plates

Group				1		2	
				Douglas Fir-Larch (Dense), Southern Pine (Dense)		Ash, Commercial White, Hickory	
Length of Bolt in Main Member l (in.)	Diameter of Bolt D (in.)	l/D	Projected Area of Bolts $A = l \times D$ (in.^{2})	Parallel to Grain P (lb)	Perpendicular to Grain Q (lb)	Parallel to Grain P (lb)	Perpendicular to Grain Q (lb)
$1\frac{1}{2}$	$\frac{1}{2}$	3.00	0.750	1100	500	1080	780
	$\frac{5}{8}$	2.40	0.938	1380	570	1360	880
	$\frac{3}{4}$	2.00	1.125	1660	630	1630	980
	$\frac{7}{8}$	1.71	1.313	1940	700	1910	1080
	1	1.50	1.500	2220	760	2180	1170
2	$\frac{1}{2}$	4.00	1.000	1370	670	1340	1030
	$\frac{5}{8}$	3.20	1.250	1810	760	1780	1170
	$\frac{3}{4}$	2.67	1.500	2200	840	2170	1300
	$\frac{7}{8}$	2.29	1.750	2580	930	2540	1430
	1	2.00	2.000	2960	1010	2910	1560
$2\frac{1}{2}$	$\frac{1}{2}$	5.00	1.250	1480	830	1450	1280
	$\frac{5}{8}$	4.00	1.563	2140	950	2100	1460
	$\frac{3}{4}$	3.33	1.875	2700	1050	2650	1630
	$\frac{7}{8}$	2.86	2.188	3210	1160	3150	1790
	1	2.50	2.500	3680	1270	3620	1960
3	$\frac{1}{2}$	6.00	1.500	1490	970	1460	1460
	$\frac{5}{8}$	4.80	1.875	2290	1130	2250	1750
	$\frac{3}{4}$	4.00	2.250	3080	1270	3020	1950
	$\frac{7}{8}$	3.43	2.625	3760	1390	3700	2150
	1	3.00	3.000	4390	1520	4310	2350
$3\frac{1}{2}$	$\frac{1}{2}$	7.00	1.750	1490	1120	1460	1460
	$\frac{5}{8}$	5.60	2.188	2320	1310	2280	2020
	$\frac{3}{4}$	4.67	2.625	3280	1470	3220	2270
	$\frac{7}{8}$	4.00	3.063	4190	1630	4120	2510
	1	3.50	3.500	5000	1770	4920	2740
4	$\frac{1}{2}$	8.00	2.000	1490	1010	1460	1460
	$\frac{5}{8}$	6.40	2.500	2330	1410	2290	2180
	$\frac{3}{4}$	5.33	3.000	3340	1690	3280	2600
	$\frac{7}{8}$	4.57	3.500	4440	1850	4360	2860
	1	4.00	4.000	5470	2030	5380	3130

TABLE 6.20 (*Continued*)

3		4		5		6	
California Redwood (Close grain), Douglas Fir-Larch, Southern Pine, Southern Cypress		Beech, Birch, Sweet & Yellow, Maple, Black & Sugar		Oak, Red & White		Douglas Fir South	
Parallel to Grain P (lb)	Perpendicular to Grain Q (lb)	Parallel to Grain P (lb)	Perpendicular to Grain Q (lb)	Parallel to Grain P (lb)	Perpendicular to Grain Q (lb)	Parallel to Grain P (lb)	Perpendicular to Grain Q (lb)
940	430	900	480	830	650	870	370
1180	490	1130	540	1050	730	1090	420
1420	540	1360	600	1260	820	1310	470
1660	600	1590	670	1470	900	1530	520
1890	650	1820	730	1690	980	1750	570
1170	570	1120	640	1040	870	1140	500
1550	650	1480	720	1380	980	1450	560
1880	720	1810	810	1680	1090	1750	630
2210	790	2120	890	1960	1200	2040	690
2520	870	2420	970	2250	1310	2330	750
1260	720	1210	800	1120	1070	1290	620
1820	810	1750	900	1620	1220	1780	710
2310	900	2210	1010	2050	1360	2180	790
2740	990	2630	1110	2440	1500	2550	860
3150	1080	3020	1210	2800	1640	2920	940
1270	860	1220	960	1130	1130	1330	750
1960	970	1880	1090	1740	1460	1980	850
2630	1080	2520	1210	2340	1630	2560	940
3220	1190	3080	1330	2860	1800	3040	1040
3750	1300	3600	1450	3340	1960	3500	1130
1270	980	1220	1090	1130	1130	1330	850
1980	1130	1900	1270	1770	1690	2060	990
2800	1260	2690	1410	2490	1900	2820	1100
3580	1390	3430	1550	3180	2100	3480	1210
4270	1520	4100	1690	3800	2290	4050	1320
1270	1010	1220	1130	1130	1130	1330	880
1990	1290	1910	1440	1770	1770	2080	1120
2850	1440	2740	1610	2540	2180	2950	1260
3790	1590	3640	1770	3370	2390	3800	1380
4670	1730	4480	1940	4160	2620	4540	1510

(*continued*)

TABLE 6.20 (*Continued*)

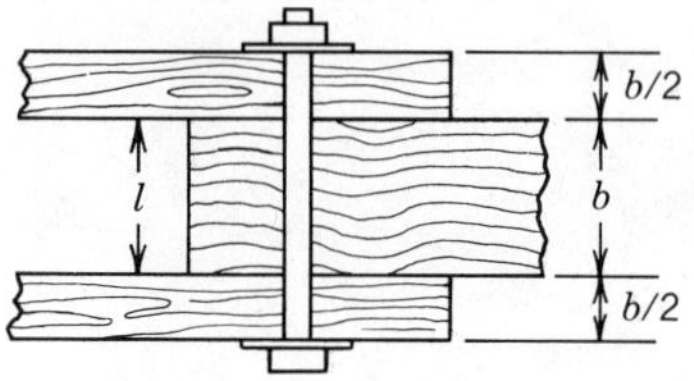

Wood side plates

b/2

l

b

b/2

Group				1		2	
				Douglas Fir-Larch (Dense), Southern Pine (Dense)		Ash, Commercial White, Hickory	
Length of Bolt in Main Member l (in.)	Diameter of Bolt D (in.)	l/D	Projected Area of Bolts $A = l \times D$ (in.^{2})	Parallel to Grain P (lb)	Perpendicular to Grain Q (lb)	Parallel to Grain P (lb)	Perpendicular to Grain Q (lb)
$4\frac{1}{2}$	$\frac{5}{8}$	7.20	2.813	2330	1440	2290	2230
	$\frac{3}{4}$	6.00	3.375	3350	1830	3290	2820
	$\frac{7}{8}$	5.14	3.938	4540	2110	4460	3260
	1	4.50	4.500	5770	2280	5670	3520
	$1\frac{1}{4}$	3.60	5.625	7970	2670	7830	4120
$5\frac{1}{2}$	$\frac{5}{8}$	8.80	3.438	2330	1390	2290	2150
	$\frac{3}{4}$	7.33	4.125	3350	1930	3300	2980
	$\frac{7}{8}$	6.29	4.813	4570	2400	4490	3710
	1	5.50	5.500	5930	2760	5830	4260
	$1\frac{1}{4}$	4.40	6.875	8930	3260	8780	5040
$7\frac{1}{2}$	$\frac{5}{8}$	12.00	4.688	2330	1210	2290	1870
	$\frac{3}{4}$	10.00	5.625	3350	1930	3290	2710
	$\frac{7}{8}$	8.57	6.563	4560	2410	4480	3720
	1	7.50	7.500	5950	3090	5850	4760
	$1\frac{1}{4}$	6.00	9.375	9310	4290	9150	6620
$9\frac{1}{2}$	$\frac{3}{4}$	12.67	7.125	3350	1570	3290	2420
	$\frac{7}{8}$	10.86	8.313	4560	2180	4480	3360
	1	9.50	9.500	5950	2890	5850	4460
	$1\frac{1}{4}$	7.60	11.875	9310	4510	9150	6960
	$1\frac{1}{2}$	6.33	14.250	13420	6070	13190	9370
$11\frac{1}{2}$	$\frac{7}{8}$	13.14	10.062	4560	1960	4490	3050
	1	11.50	11.500	5950	2650	5850	4080
	$1\frac{1}{4}$	9.20	14.375	9310	4280	9160	6610
	$1\frac{1}{2}$	7.67	17.250	13410	6210	13180	9590
$13\frac{1}{2}$	1	13.50	13.500	5960	2390	5850	3730
	$1\frac{1}{4}$	10.80	16.875	9300	3980	9150	6150
	$1\frac{1}{2}$	9.00	20.250	13400	5950	13180	9190

TABLE 6.20 (*Continued*)

3		4		5		6	
California Redwood (Close grain), Douglas Fir-Larch, Southern Pine, Southern Cypress		Beech, Birch, Sweet & Yellow, Maple, Black & Sugar		Oak, Red & White		Douglas Fir South	
Parallel to Grain P (lb)	Perpen-dicular to Grain Q (lb)	Parallel to Grain P (lb)	Perpen-dicular to Grain Q (lb)	Parallel to Grain P (lb)	Perpen-dicular to Grain Q (lb)	Parallel to Grain P (lb)	Perpen-dicular to Grain Q (lb)
1990	1400	1910	1560	1770	1770	2070	1220
2860	1620	2750	1810	2550	2360	2980	1410
3880	1790	3720	2000	3450	2720	3980	1560
4930	1950	4730	2180	4390	2940	4920	1700
6810	2280	6530	2550	6060	3450	6490	1990
1990	1410	1910	1570	1770	1770	2080	1220
2860	1880	2750	2100	2550	2490	2990	1630
3900	2180	3740	2430	3470	3100	4070	1900
5070	2380	4860	2660	4510	3560	5270	2080
7630	2790	7320	3120	6790	4210	7580	2430
1990	1260	1910	1410	1770	1560	2070	1100
2860	1820	2750	2030	2550	2260	2990	1580
3900	2420	3740	2700	3470	3110	4060	2110
5080	3030	4880	3390	4520	3980	5300	2640
7950	3800	7630	4250	7080	5530	8290	3310
2860	1640	2740	1840	2540	2030	2990	1430
3890	2270	3740	2540	3470	2810	4070	1980
5080	2960	4880	3310	4520	3730	5310	2580
7950	4450	7630	4970	7070	5820	8290	3870
11470	5520	11000	6170	10200	7830	11960	4810
3900	2060	3750	2300	3470	2550	4070	1790
5080	2770	4880	3090	4520	3410	5310	2410
7960	4360	7360	4870	7080	5530	8300	3800
11450	6140	10990	6860	10190	8010	11940	5350
5280	2530	4880	2830	4520	3120	5320	2190
7950	4160	7630	4650	7070	5140	8300	3620
11450	6040	10990	6750	10190	7680	11950	5260

(*continued*)

TABLE 6.20 (*Continued*)

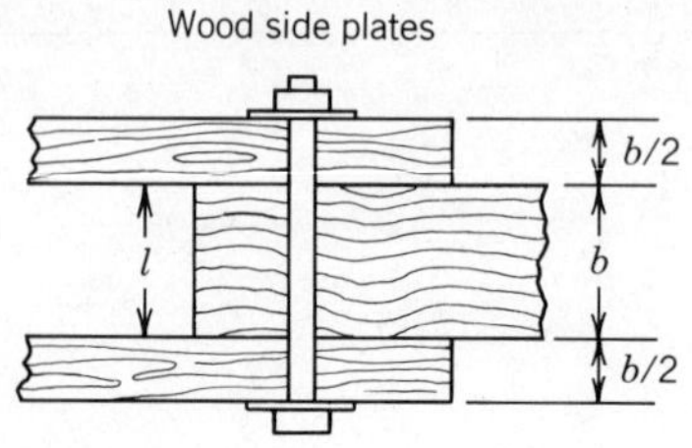

Group				7		8	
				Sweetgum & Tupelo		Eastern Hemlock-Tamarack, California Redwood (Open-grain), Hem-Fir, Western Hemlock	
Length of Bolt in Main Member l (in.)	Diameter of Bolt D (in.)	l/D	Projected Area of Bolts $A = l \times D$ (in.^{2})	Parallel to Grain P (lb)	Perpendicular to Grain Q (lb)	Parallel to Grain P (lb)	Perpendicular to Grain Q (lb)
$1\frac{1}{2}$	$\frac{1}{2}$	3.00	0.750	810	410	800	270
	$\frac{5}{8}$	2.40	0.938	1010	460	1000	310
	$\frac{3}{4}$	2.00	1.125	1220	510	1200	350
	$\frac{7}{8}$	1.71	1.313	1420	560	1400	380
	1	1.50	1.500	1620	620	1600	420
2	$\frac{1}{2}$	4.00	1.000	1050	540	1040	370
	$\frac{5}{8}$	3.20	1.250	1350	610	1330	410
	$\frac{3}{4}$	2.67	1.500	1620	680	1600	460
	$\frac{7}{8}$	2.29	1.750	1890	750	1870	510
	1	2.00	2.000	2160	820	2130	550
$2\frac{1}{2}$	$\frac{1}{2}$	5.00	1.250	1190	680	1180	460
	$\frac{5}{8}$	4.00	1.563	1640	770	1620	520
	$\frac{3}{4}$	3.33	1.875	2020	850	1990	580
	$\frac{7}{8}$	2.86	2.188	2360	940	2330	630
	1	2.50	2.500	2700	1030	2670	690
3	$\frac{1}{2}$	6.00	1.500	1230	810	1210	550
	$\frac{5}{8}$	4.80	1.875	1830	920	1810	620
	$\frac{3}{4}$	4.00	2.250	2370	1030	2340	690
	$\frac{7}{8}$	3.43	2.625	2820	1120	2780	760
	1	3.00	3.000	3240	1230	3200	830
$3\frac{1}{2}$	$\frac{1}{2}$	7.00	1.750	1230	920	1210	640
	$\frac{5}{8}$	5.60	2.188	1910	1070	1890	720
	$\frac{3}{4}$	4.67	2.625	2610	1200	2570	810
	$\frac{7}{8}$	4.00	3.063	3220	1320	3180	890
	1	3.50	3.500	3760	1440	3710	970
4	$\frac{1}{2}$	8.00	2.000	1230	960	1210	700
	$\frac{5}{8}$	6.40	2.500	1920	1220	1900	830
	$\frac{3}{4}$	5.33	3.000	2730	1370	2690	920
	$\frac{7}{8}$	4.57	3.500	3520	1500	3470	1010
	1	4.00	4.000	4210	1640	4150	1110

TABLE 6.20 (*Continued*)

9		10		11		12	
Mountain Hemlock, Western Cedars, Northern Pine		Spruce-Pine-Fir, Sitka Spruce, Yellow Poplar, Eastern Spruce, Lodgepole Pine		Red Pine, Western White Pine, Ponderosa Pine-Sugar Pine, Eastern White Pine, Balsam Fir, Idaho White Pine		Aspen, Eastern Cottonwood, Engelmann Spruce, Alpine Fir, Northern White Cedar	
Parallel to Grain *P* (lb)	Perpendicular to Grain *Q* (lb)	Parallel to Grain *P* (lb)	Perpendicular to Grain *Q* (lb)	Parallel to Grain *P* (lb)	Perpendicular to Grain *Q* (lb)	Parallel to Grain *P* (lb)	Perpendicular to Grain *Q* (lb)
750	300	680	280	630	190	530	210
930	340	850	320	790	210	660	230
1120	370	1020	350	950	240	800	260
1310	410	1190	390	1110	260	930	290
1490	450	1360	420	1260	290	1060	310
970	400	900	370	840	250	700	280
1240	450	1130	420	1050	280	880	310
1490	500	1360	470	1260	320	1060	350
1740	550	1580	520	1480	350	1240	380
1990	600	1810	560	1690	380	1420	420
1100	500	1080	470	1010	310	840	340
1510	560	1410	530	1310	360	1100	390
1860	620	1700	590	1580	400	1330	430
2180	690	1980	650	1840	440	1550	480
2490	750	2260	700	2110	480	1770	520
1130	590	1160	560	1080	380	910	410
1690	670	1640	630	1520	430	1280	470
2180	750	2030	700	1890	480	1590	520
2600	820	2380	780	2210	520	1860	570
2990	900	2710	850	2530	570	2120	620
1130	690	1160	650	1080	440	910	480
1760	780	1790	740	1660	500	1400	550
2400	870	2310	820	2150	560	1800	610
2970	960	2760	900	2570	610	2160	670
3460	1050	3170	990	2950	670	2480	730
1130	760	1160	720	1080	500	910	550
1770	900	1820	840	1690	570	1420	620
2510	1000	2520	940	2350	630	1970	690
3240	1100	3090	1030	2880	700	2420	760
3880	1200	3600	1130	3360	760	2820	830

(*continued*)

TABLE 6.20 (*Continued*)

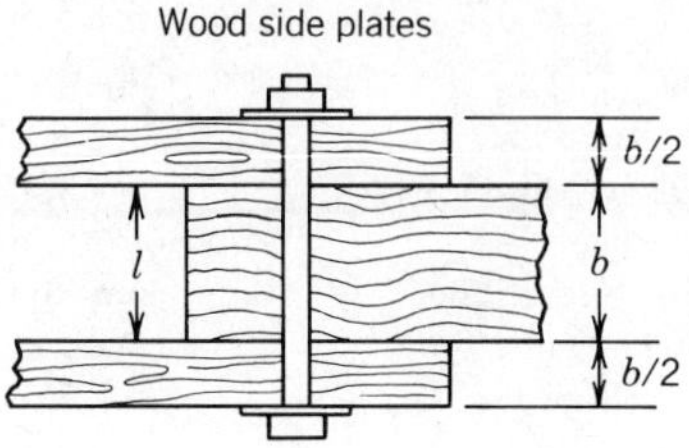

Wood side plates

Length of Bolt in Main Member l (in.)	Diameter of Bolt D (in.)	l/D	Projected Area of Bolts $A = l \times D$ (in.^{2})	Group 7: Sweetgum & Tupelo — Parallel to Grain P (lb)	Group 7: Sweetgum & Tupelo — Perpendicular to Grain Q (lb)	Group 8: Eastern Hemlock-Tamarack, California Redwood (Open-grain), Hem-Fir, Western Hemlock — Parallel to Grain P (lb)	Group 8: Eastern Hemlock-Tamarack, California Redwood (Open-grain), Hem-Fir, Western Hemlock — Perpendicular to Grain Q (lb)
$4\frac{1}{2}$	$\frac{5}{8}$	7.20	2.813	1920	1320	1900	930
	$\frac{3}{4}$	6.00	3.375	2760	1540	2730	1040
	$\frac{7}{8}$	5.14	3.938	3680	1690	3630	1140
	1	4.50	4.500	4560	1850	4500	1250
	$1\frac{1}{4}$	3.60	5.625	6010	2160	5930	1460
$5\frac{1}{2}$	$\frac{5}{8}$	8.80	3.438	1920	1330	1900	1010
	$\frac{3}{4}$	7.33	4.125	2770	1780	2730	1260
	$\frac{7}{8}$	6.29	4.813	3770	2060	3720	1400
	1	5.50	5.500	4890	2260	4820	1520
	$1\frac{1}{4}$	4.40	6.875	7020	2640	6930	1780
$7\frac{1}{2}$	$\frac{5}{8}$	12.00	4.688	1920	1200	1890	950
	$\frac{3}{4}$	10.00	5.625	2770	1720	2730	1320
	$\frac{7}{8}$	8.57	6.563	3770	2290	3720	1730
	1	7.50	7.500	4910	2870	4850	2060
	$1\frac{1}{4}$	6.00	9.375	7680	3600	7580	2430
$9\frac{1}{2}$	$\frac{3}{4}$	12.67	7.125	2770	1560	2730	1250
	$\frac{7}{8}$	10.86	8.313	3770	2150	3720	1660
	1	9.50	9.500	4920	2800	4850	2130
	$1\frac{1}{4}$	7.60	11.875	7680	4210	7580	3030
	$1\frac{1}{2}$	6.33	14.250	11080	5250	10930	3540
$11\frac{1}{2}$	$\frac{7}{8}$	13.14	10.062	3770	1950	3700	1590
	1	11.50	11.500	4920	2620	4860	2040
	$1\frac{1}{4}$	9.20	14.375	7690	4130	7590	3140
	$1\frac{1}{2}$	7.67	17.250	11070	5820	10920	4210
$13\frac{1}{2}$	1	13.50	13.500	4920	2400	4850	1970
	$1\frac{1}{4}$	10.80	16.875	7690	3940	7590	3030
	$1\frac{1}{2}$	9.00	20.250	11070	5720	10930	4340

[a]Source: *National Design Specification for Wood Construction* (3).

TABLE 6.20 (*Continued*)

9		10		11		12	
Mountain Hemlock, Western Cedars, Northern Pine		Spruce-Pine-Fir, Sitka Spruce, Yellow Poplar, Eastern Spruce, Lodgepole Pine		Red Pine, Western White Pine, Ponderosa Pine-Sugar Pine, Eastern White Pine, Balsam Fir, Idaho White Pine		Aspen, Eastern Cottonwood, Engelmann Spruce, Alpine Fir, Northern White Cedar	
Parallel to Grain P (lb)	Perpendicular to Grain Q (lb)	Parallel to Grain P (lb)	Perpendicular to Grain Q (lb)	Parallel to Grain P (lb)	Perpendicular to Grain Q (lb)	Parallel to Grain P (lb)	Perpendicular to Grain Q (lb)
1770	1010	1820	950	1690	640	1420	700
2550	1120	2610	1060	2440	710	2050	780
3390	1240	3360	1160	3130	790	2630	860
4200	1350	3990	1270	3710	860	3120	940
5540	1580	5080	1490	4740	1000	3980	1100
1770	1090	1820	1030	1690	750	1420	820
2550	1360	2620	1280	2440	870	2050	950
3470	1510	3560	1420	3320	960	2790	1050
4500	1650	4550	1550	4240	1050	3560	1150
6470	1930	6110	1820	5690	1230	4780	1340
1770	1030	1820	960	1690	730	1420	800
2550	1430	2620	1340	2440	1010	2050	1110
3470	1870	3560	1760	3320	1280	2780	1400
4520	2230	4650	2100	4330	1430	3640	1560
7070	2630	7260	2480	6770	1670	5680	1830
2550	1350	2620	1270	2440	970	2050	1060
3470	1790	3560	1690	3320	1280	2790	1400
4530	2300	4650	2170	4330	1630	3640	1780
7070	3280	7270	3090	6770	2120	5690	2320
10200	3830	10470	3610	9760	2440	8190	2660
3470	1730	3570	1630	3330	1230	2790	1350
4530	2210	4650	2080	4330	1580	3640	1730
7080	3400	7270	3200	6780	2380	5690	2600
10190	4550	10470	4280	9750	2950	8190	3230
4540	2140	4670	2020	4350	1530	3650	1680
7080	3280	7260	3080	6770	2340	5680	2560
10200	4700	10460	4420	9750	3280	8190	3580

TABLE 6.21

Bolt Design Values for Softwood Glued Laminated Timbers[a]

Group[b]				1		3		6		8		12	
Steel side plates (l, b)				Douglas Fir-Larch (dense) Southern Pine (dense)		Douglas Fir-Larch (medium grain) Southern Pine (medium grain) Cal. Redwood (close grain)		Douglas Fir South		Hem-Fir Douglas Fir-Larch (coarse grain) Southern Pine (coarse grain)		Western Woods[c]	
Length of Bolt in Main Member l (in.)	Diameter of Bolt D	l/D	Projected Area of Bolt, $A = l \times D$ (in.^{2})	Parallel to Grain P (lb)	Perp. to Grain Q (lb)	Parallel to Grain P (lb)	Perp. to Grain Q (lb)	Parallel to Grain P (lb)	Perp. to Grain Q (lb)	Parallel to Grain P (lb)	Perp. to Grain Q (lb)	Parallel to Grain P (lb)	Perp. to Grain Q (lb)
$2\frac{1}{8}$	$\frac{1}{2}$	4.25	1.063	1760	710	1500	610	1480	530	1350	390	930	290
	$\frac{5}{8}$	3.40	1.328	2380	810	2040	690	1930	600	1760	440	1180	330
	$\frac{3}{4}$	2.83	1.594	2920	900	2500	770	2320	670	2120	490	1410	370
3	$\frac{1}{2}$	6.00	1.500	1860	970	1590	860	1660	720	1520	550	1140	410
	$\frac{5}{8}$	4.80	1.875	2860	1140	2450	970	2470	850	2260	620	1600	470
	$\frac{3}{4}$	4.00	2.250	3840	1270	3280	1080	3200	940	2920	690	1980	520
	$\frac{7}{8}$	3.43	2.625	4700	1390	4020	1190	3800	1040	3480	760	2320	570
$3\frac{1}{8}$	$\frac{1}{2}$	6.25	1.563	1860	990	1590	890	1660	740	1520	570	1140	430
	$\frac{5}{8}$	5.00	1.953	2890	1180	2470	1010	2510	880	2300	650	1650	490
	$\frac{3}{4}$	4.17	2.344	3920	1320	3350	1130	3290	980	3010	720	2050	540
	$\frac{7}{8}$	3.57	2.734	4850	1450	4150	1240	3950	1080	3610	790	2420	600

$3\frac{1}{2}$	$\frac{1}{2}$	7.00	1.750	1860	1020	1590	980	1660	850	1520	640	1140	480
	$\frac{5}{8}$	5.60	2.188	2900	1310	2480	1130	2580	990	2360	720	1750	550
	$\frac{3}{4}$	4.67	2.625	4100	1480	3500	1260	3520	1100	3220	810	2250	610
	$\frac{7}{8}$	4.00	3.063	5230	1630	4470	1390	4350	1210	3980	890	2700	670
4	$\frac{1}{2}$	8.00	2.000	1860	1010	1590	1010	1660	880	1520	700	1140	550
	$\frac{5}{8}$	6.40	2.500	2910	1410	2490	1290	2590	1120	2370	830	1780	620
	$\frac{3}{4}$	5.35	3.000	4170	1680	3570	1440	3680	1260	3370	920	2460	690
	$\frac{7}{8}$	4.57	3.500	5550	1860	4740	1590	4750	1380	4340	1010	3020	760
$4\frac{1}{2}$	$\frac{1}{2}$	9.00	2.250	1860	990	1590	990	1660	860	1520	710	1140	590
	$\frac{5}{8}$	7.20	2.813	2910	1440	2490	1400	2590	1220	2370	930	1780	700
	$\frac{3}{4}$	6.00	3.375	4190	1830	3580	1620	3730	1410	3410	1040	2560	780
	$\frac{7}{8}$	5.14	3.938	5680	2080	4850	1790	4980	1560	4550	1140	3290	860
5	$\frac{5}{8}$	8.00	3.125	2910	1420	2490	1420	2590	1240	2370	1000	1780	780
	$\frac{3}{4}$	6.67	3.750	4190	1910	3580	1780	3740	1550	3420	1150	2560	870
	$\frac{7}{8}$	5.71	4.375	5700	2270	4870	1980	5070	1730	4630	1270	3440	950
	1	5.00	5.000	7390	2530	6310	2170	6440	1890	5890	1380	4220	1040
$5\frac{1}{8}$	$\frac{5}{8}$	8.20	3.203	2910	1420	2490	1420	2590	1240	2370	1000	1780	790
	$\frac{3}{4}$	6.83	3.844	4190	1920	3580	1810	3740	1580	3420	1180	2530	890
	$\frac{7}{8}$	5.86	4.484	5700	2310	4870	2030	5080	1770	4640	1300	3460	980
	1	5.13	5.125	7410	2590	6330	2220	6480	1930	5930	1420	4280	1070
6	$\frac{5}{8}$	9.60	3.750	2910	1380	2480	1380	2590	1200	2370	990	1780	830
	$\frac{3}{4}$	8.00	4.500	4190	1910	3580	1910	3730	1660	3410	1330	2560	1040
	$\frac{7}{8}$	6.86	5.250	5710	2460	4880	2330	5080	2030	4650	1520	3480	1150
	1	6.00	6.000	7450	2930	6360	2600	6630	2260	6060	1660	4550	1250
$6\frac{3}{4}$	$\frac{5}{8}$	10.80	4.219	2910	1330	2480	1330	2590	1160	2370	970	1780	820
	$\frac{3}{4}$	9.00	5.063	4190	1870	3580	1870	3730	1630	3410	1340	2560	1110
	$\frac{7}{8}$	7.71	5.906	5700	2460	4870	2440	5080	2130	4650	1680	3480	1290
	1	6.75	6.750	7460	3060	6370	2880	6640	2510	6070	1870	4550	1250

(continued)

TABLE 6.21 (*Continued*)

Steel side plates

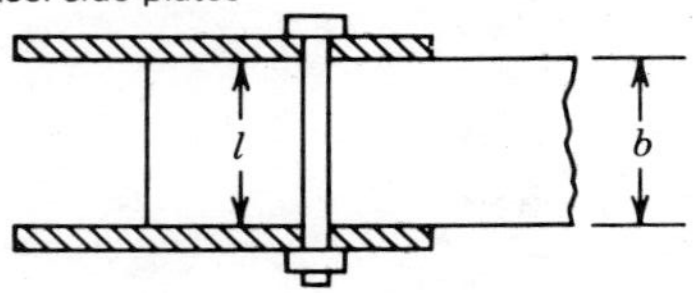

Group[b]				1		3		6		8		12	
				Douglas Fir-Larch (dense) Southern Pine (dense)		Douglas Fir-Larch (medium grain) Southern Pine (medium grain) Cal. Redwood (close grain)		Douglas Fir South		Hem-Fir Douglas Fir-Larch (coarse grain) Southern Pine (coarse grain)		Western Woods[c]	
Length of Bolt in Main Member l (in.)	Diameter of Bolt D	l/D	Projected Area of Bolt, $A = l \times D$ (in.^{2})	Parallel to Grain P (lb)	Perp. to Grain Q (lb)	Parallel to Grain P (lb)	Perp. to Grain Q (lb)	Parallel to Grain P (lb)	Perp. to Grain Q (lb)	Parallel to Grain P (lb)	Perp. to Grain Q (lb)	Parallel to Grain P (lb)	Perp. to Grain Q (lb)
$7\frac{1}{2}$	$\frac{5}{8}$	12.00	4.688	2910	1260	2490	1260	2590	1090	2370	940	1780	800
	$\frac{3}{4}$	10.00	5.625	4190	1820	3580	1820	3730	1580	3410	1320	2560	1110
	$\frac{7}{8}$	8.57	6.563	5700	2420	4870	2420	5080	2110	4640	1730	3480	1400
	1	7.50	7.500	7440	3090	6360	3030	6630	2640	6060	2060	4550	1560
$8\frac{3}{4}$	$\frac{3}{4}$	11.67	6.563	4190	1720	3580	1720	3730	1500	3410	1270	2560	1080
	$\frac{7}{8}$	10.00	7.656	5700	2330	4870	2330	5080	2030	4640	1690	3480	1420
	1	8.75	8.750	7450	3010	6370	3010	6640	2620	6070	2160	4550	1760

9	$\frac{3}{4}$	12.00	6.750	4190	1690	3580	1690	3730	1470	3410	1270	2560	1070
	$\frac{7}{8}$	10.29	7.875	5700	2310	4870	2310	5080	2010	4640	1680	3480	1410
	1	9.00	9.000	7450	2990	6360	2990	6640	2600	6070	2150	4550	1770
$10\frac{1}{2}$	$\frac{7}{8}$	12.00	9.188	5700	2170	4870	2170	5080	1890	4640	1630	3480	1370
	1	10.50	10.500	7450	2860	6360	2860	6630	2490	6060	2080	4540	1750
$10\frac{3}{4}$	$\frac{7}{8}$	12.29	9.406	5690	2140	4860	2140	5080	1870	4640	1620	3480	1370
	1	10.75	10.750	7450	2840	6360	2840	6640	2480	6070	2070	4550	1750
12	$\frac{7}{8}$	13.71	10.500	5740	2000	4900	2000	5050	1740	4620	1570	3470	1330
	1	12.00	12.000	7450	2700	6360	2700	6630	2350	6060	2030	4550	1710
$12\frac{1}{4}$	$\frac{7}{8}$	14.00	10.719	5740	1970	4910	1970	5000	1710	4570	1560	3460	1330
	1	12.25	12.250	7440	2680	6360	2680	6640	2330	6070	2020	4550	1710
$13\frac{1}{2}$	1	13.50	13.500	7480	2510	6390	2510	6630	2190	6060	1970	4540	1670
$14\frac{1}{4}$	1	14.25	14.250	7500	2420	6410	2420	6540	2110	5980	1940	4530	1640
15	1	15.00	15.000	7480	2320	6390	2310	6560	2010	6000	1910	4540	1620

[a]Design value in pounds for one bolt loaded at both ends (double shear, three-member joint) for the species listed.

[b]The numbers at the top of the species columns are the same as the group numbers for bolt values of sawn lumber. The perpendicular-to-grain bolt design values (Q) are in descending order as the group number increases. The parallel-to-grain bolt design values (P) are also in descending order except for Douglas Fir South, which ranks between group 1 and group 3.

[c]Species included in the Western Woods group for glued laminated timber are as defined in AITC 117—Manufacturing (9).

TABLE 6.22

Bolt Design Values for Softwood Glued Laminated Timbers[a]

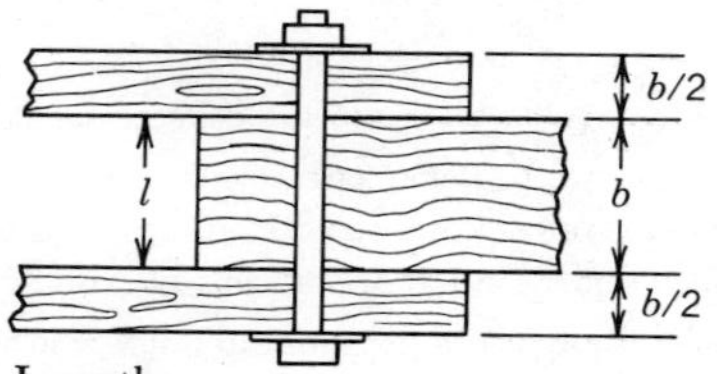
Wood side plates
b/2
b
b/2
l

Group[b]				1		3		6		8		12	
				Douglas Fir-Larch (dense) Southern Pine (dense)		Douglas Fir-Larch (medium grain) Southern Pine (medium grain) Cal. Redwood (close grain)		Douglas Fir South		Hem-Fir Douglas Fir-Larch (coarse grain) Southern Pine (coarse grain)		Western Woods[c]	
Length of Bolt in Main Member l (in.)	Diameter of Bolt D (in.)	l/D	Projected Area of Bolt $A = l \times D$ (in.^{2})	Parallel to Grain P (lb)	Perp. to Grain Q (lb)	Parallel to Grain P (lb)	Perp. to Grain Q (lb)	Parallel to Grain P (lb)	Perp. to Grain Q (lb)	Parallel to Grain P (lb)	Perp. to Grain Q (lb)	Parallel to Grain P (lb)	Perp. to Grain Q (lb)
$2\frac{1}{8}$	$\frac{1}{2}$	4.25	1.063	1410	710	1200	610	1180	530	1080	390	740	290
	$\frac{5}{8}$	3.40	1.328	1910	810	1630	690	1540	600	1410	440	940	330
	$\frac{3}{4}$	2.83	1.594	2340	900	2000	770	1860	670	1700	490	1130	370
3	$\frac{1}{2}$	6.00	1.500	1490	970	1270	860	1330	750	1210	550	910	410
	$\frac{5}{8}$	4.80	1.875	2290	1140	1960	970	1980	850	1810	620	1280	470
	$\frac{3}{4}$	4.00	2.250	3080	1270	2630	1080	2560	940	2340	690	1590	520
	$\frac{7}{8}$	3.43	2.625	3760	1390	3220	1190	3040	1040	2780	760	1860	570
$3\frac{1}{8}$	$\frac{1}{2}$	6.25	1.563	1490	990	1270	890	1330	780	1210	570	910	430
	$\frac{5}{8}$	5.00	1.953	2310	1180	1970	1010	2010	880	1840	650	1320	490
	$\frac{3}{4}$	4.17	2.344	3140	1320	2680	1130	2630	980	2400	720	1640	540
	$\frac{7}{8}$	3.57	2.734	3880	1450	3320	1240	3160	1080	2890	790	1930	600

$3\frac{1}{2}$	$\frac{1}{2}$	7.00	1.750	1490	1020	1270	980	1330	850	1210	640	910	480
	$\frac{5}{8}$	5.60	2.188	2320	1310	1980	1130	2060	990	1890	720	1400	550
	$\frac{3}{4}$	4.67	2.625	3280	1480	2800	1260	2820	1100	2570	810	1800	610
	$\frac{7}{8}$	4.00	3.063	4190	1630	3580	1390	3480	1210	3180	890	2160	670
4	$\frac{1}{2}$	8.00	2.000	1490	1010	1270	1010	1330	880	1210	700	910	550
	$\frac{5}{8}$	6.40	2.500	2330	1410	1990	1290	2080	1120	1900	830	1420	620
	$\frac{3}{4}$	5.33	3.000	3340	1680	2850	1440	2950	1260	2700	920	1970	690
	$\frac{7}{8}$	4.57	3.500	4440	1860	3790	1590	3800	1380	3470	1010	2420	760
$4\frac{1}{2}$	$\frac{1}{2}$	9.00	2.250	1490	990	1270	990	1330	860	1210	710	910	590
	$\frac{5}{8}$	7.20	2.813	2330	1440	1990	1400	2070	1220	1900	930	1420	700
	$\frac{3}{4}$	6.00	3.375	3350	1830	2860	1620	2980	1410	2730	1040	2050	780
	$\frac{7}{8}$	5.14	3.938	4540	2080	3880	1790	3980	1560	3640	1140	2630	860
5	$\frac{5}{8}$	8.00	3.125	2330	1430	1990	1430	2070	1240	1900	1000	1420	780
	$\frac{3}{4}$	6.67	3.750	3350	1910	2870	1780	2990	1550	2730	1150	2050	870
	$\frac{7}{8}$	5.71	4.375	4560	2270	3900	1980	4060	1730	3710	1270	2760	950
	1	5.00	5.000	5910	2530	5050	2170	5150	1890	4710	1380	3380	1040
$5\frac{1}{8}$	$\frac{5}{8}$	8.20	3.203	2330	1420	1990	1420	2080	1240	1900	1000	1420	790
	$\frac{3}{4}$	6.83	3.844	3350	1920	2870	1810	2990	1580	2730	1180	2020	890
	$\frac{7}{8}$	5.86	4.484	4560	2310	3900	2030	4060	1770	3710	1300	2770	980
	1	5.13	5.125	5930	2590	5060	2220	5190	1930	4740	1420	3430	1070
6	$\frac{5}{8}$	9.60	3.750	2330	1380	1990	1380	2070	1200	1900	990	1420	830
	$\frac{3}{4}$	8.00	4.500	3350	1910	2860	1910	2990	1660	2730	1330	2050	1040
	$\frac{7}{8}$	6.86	5.250	4570	2460	3900	2330	4070	2030	3720	1520	2790	1150
	1	6.00	6.000	5960	2930	5090	2600	5300	2260	4850	1660	3640	1250
$6\frac{3}{4}$	$\frac{5}{8}$	10.80	4.219	2330	1330	1990	1330	2070	1160	1900	970	1420	820
	$\frac{3}{4}$	9.00	5.063	3350	1870	2860	1870	2990	1630	2730	1340	2050	1110
	$\frac{7}{8}$	7.71	5.906	4560	2460	3900	2440	4070	2130	3720	1680	2790	1290
	1	6.75	6.750	5960	3060	5100	2880	5310	2510	4860	1870	3640	1410

TABLE 6.22 (Continued)

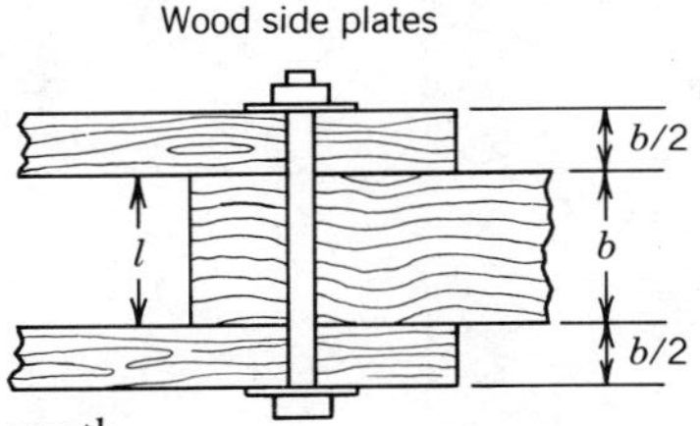
Wood side plates
b/2
b
b/2
l

Group[b]				1		3		6		8		12	
				Douglas Fir-Larch (dense) Southern Pine (dense)		Douglas Fir-Larch (medium grain) Southern Pine (medium grain) Cal. Redwood (close grain)		Douglas Fir South		Hem-Fir Douglas Fir-Larch (coarse grain) Southern Pine (coarse grain)		Western Woods[c]	
Length of Bolt in Main Member l (in.)	Diameter of Bolt D (in.)	l/D	Projected Area of Bolt $A = l \times D$ (in.^{2})	Parallel to Grain P (lb)	Perp. to Grain Q (lb)	Parallel to Grain P (lb)	Perp. to Grain Q (lb)	Parallel to Grain P (lb)	Perp. to Grain Q (lb)	Parallel to Grain P (lb)	Perp. to Grain Q (lb)	Parallel to Grain P (lb)	Perp. to Grain Q (lb)
$7\frac{1}{2}$	$\frac{5}{8}$	12.00	4.688	2330	1260	1990	1260	2070	1090	1890	940	1420	800
	$\frac{3}{4}$	10.00	5.625	3350	1820	2860	1820	2990	1580	2730	1320	2050	1110
	$\frac{7}{8}$	8.57	6.563	4560	2420	3900	2420	4060	2110	3720	1730	2780	1400
	1	7.50	7.500	5950	3090	5080	3030	5300	2640	4850	2060	3640	1560
$8\frac{3}{4}$	$\frac{3}{4}$	11.67	6.563	3350	1720	2860	1720	2990	1500	2730	1270	2050	1080
	$\frac{7}{8}$	10.00	7.656	4560	2330	3900	2330	4060	2030	3710	1690	2790	1420
	1	8.75	8.750	5960	3010	5090	3010	5310	2620	4860	2160	3640	1760
9	$\frac{3}{4}$	12.00	6.750	3350	1690	2860	1690	2980	1470	2730	1270	2050	1070
	$\frac{7}{8}$	10.29	7.875	4560	2310	3900	2310	4060	2010	3720	1680	2780	1410
	1	9.00	9.000	5960	2990	5090	2990	5310	2600	4860	2150	3640	1770

$10\frac{1}{2}$	$\frac{7}{8}$	12.00	9.188	4560	2170	3900	2170	4060	1890	3710	1630	2780	1370
	1	10.50	10.500	5960	2860	5090	2860	5300	2490	4850	2080	3640	1750
$10\frac{3}{4}$	$\frac{7}{8}$	12.29	9.406	4560	2140	3890	2140	4060	1870	3710	1620	2790	1370
	1	10.75	10.750	5960	2840	5090	2840	5310	2480	4850	2070	3640	1750
12	$\frac{7}{8}$	13.71	10.500	4590	2000	3920	2000	4040	1740	3690	1570	2780	1330
	1	12.00	12.000	5960	2700	5090	2700	5300	2350	4850	2030	3640	1710
$12\frac{1}{4}$	$\frac{7}{8}$	14.00	10.719	4590	1970	3920	1970	4000	1710	3660	1560	2770	1330
	1	12.25	12.250	5960	2680	5090	2680	5310	2330	4850	2020	3640	1710
$13\frac{1}{2}$	1	13.50	13.500	5980	2510	5110	2510	5310	2190	4850	1970	3630	1670
$14\frac{1}{4}$	1	14.25	14.250	6000	2420	5130	2420	5240	2110	4790	1940	3620	1640
15	1	15.00	15.000	5980	2320	5110	2310	5250	2010	4800	1910	3630	1620

[a]Design value in pounds for one bolt loaded at both ends (double shear, three-member joint) for the species listed.

[b]The numbers at the top of the species columns are the same as the group numbers for bolt values of sawn lumber. The perpendicular-to-grain bolt design values (Q) are in descending order as the group number increases. The parallel-to-grain bolt design values (P) are also in descending order except for Douglas Fir South, which ranks between group 1 and group 3.

[c]Species included in the Western Woods group for glued laminated timber are as defined in AITC 117—Manufacturing (9).

From Table 6.8, $C_g = 0.92$.

$$P' = (7460)(1.15)(0.92) = 7890 \text{ lb}$$

$$\text{Number of bolts required} = \frac{54{,}400}{7890} = 6.9 < 8 \quad \text{O.K.}$$

Check stress at net section of wood member.

$$A_{net} = 50.6 - (1.0625)(6.75)(2) = 36.3 \text{ in.}^2$$

From Table 2, AITC 117—Design, $F_t = 1450$ psi:

$$F_t' = F_t C_D = (1450)(1.15) = 1670 \text{ psi}$$

$$f_t = \frac{T}{A_{net}} = \frac{54{,}400}{36.3} = 1500 \text{ psi} < 1670 \text{ psi} \quad \text{O.K.}$$

Check stresses in steel side plate.

$$A_{gross} = (0.375)(7) = 2.63 \text{ in.}^2$$

$$A_{net} = 2.63 - (1.0625)(0.375)(2) = 1.83 \text{ in.}^2$$

$$f_t \text{ (on gross area)} = \frac{T}{A_{gross}} = \frac{54{,}400}{2.63} = 20{,}700 \text{ psi} < 22{,}000 \text{ psi} \quad \text{O.K.}$$

$$f_t \text{ (on net area)} = \frac{54{,}400}{1.83} = 29{,}700 \text{ psi} \approx 29{,}000 \text{ psi} \quad \text{O.K.}$$

From Table 6.18, minimum spacing parallel to grain for full load is 4 times bolt diameter, or 4 in. The minimum end distance for tension loading is 7 in. The minimum edge distance in the wood is $1\frac{1}{2}$ times bolt diameter or $1\frac{1}{2}$ in. The minimum row spacing for the wood is $1\frac{1}{2}$ times bolt diameter but this is less than the recommended 3-in. row spacing for the steel used. This results in an edge distance of 2 in. for the steel side plate, which is more than the $1\frac{3}{4}$-in. minimum requirement. The edge distance in the wood is $2\frac{1}{4}$ in., which is more than the minimum of $1\frac{1}{2}$ in.

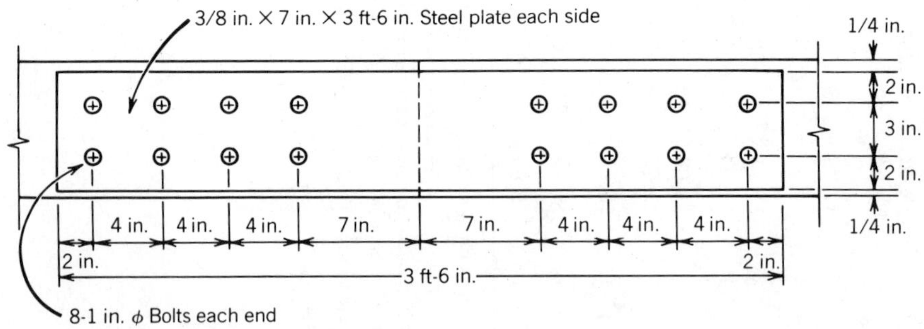

LAG BOLTS

The design values given herein are for lag bolts (or lag screws) made of ASTM A307 steel. Lag bolts require prebored lead holes. Lead hole diameters for the shank and threaded portions of lag bolts are listed in Table 6.23. Lead holes for the shank should have a depth equal to the length of the unthreaded portion in the main member. When attaching one wood member to another, the lead hole

TABLE 6.23

Lead Hole Diameters for Lag Bolts

	Diameter of Lead Hole (in.)			
		Threaded Portion		
Nominal Diameter of Lag Bolt (in.)	Shank (Unthreaded) Portion (in.)	Group I Species[a]	Group II Species[a]	Groups III and IV Species[a]
1/4	1/4	3/16	5/32	3/32
5/16	5/16	13/64	3/16	9/64
3/8	3/8	1/4	15/64	11/64
7/16	7/16	19/64	9/32	13/64
1/2	1/2	11/32	5/16	15/64
9/16	9/16	13/32	23/64	9/32
5/8	5/8	29/64	13/32	5/16
3/4	3/4	9/16	1/2	13/32
7/8	7/8	43/64	39/64	33/64
1	1	51/64	23/32	5/8
$1\frac{1}{8}$	$1\frac{1}{8}$	59/64	53/64	3/4
$1\frac{1}{4}$	$1\frac{1}{4}$	$1\frac{1}{16}$	15/16	7/8

[a] See Table 6.1 for species in each group.

for the shank in the attached member should be drilled $\frac{1}{32}$ to $\frac{1}{16}$ in. oversize, the same as for bolts. Lag bolts should be inserted in the lead hole by turning with a wrench, not by driving with a hammer. Soap or other lubricant should be used to facilitate insertion. Washers of proper size or a metal plate or strap should be installed between the wood and the bolt head. When lag bolts are loaded primarily in withdrawal, those $\frac{3}{8}$ in. and smaller in diameter may be inserted in a member made of Group III or Group IV species without a lead hole, provided spacing, end, and edge distances used are such that unusual splitting does not occur.

Lumber Species

Species load groups for lag bolt design values are given in Table 6.1 for sawn lumber and Table 6.2 for glued laminated timber.

Critical or Net Section

The net section requirements for lag bolt joints are the same as for bolted joints with bolts of a diameter equal to the shank diameter of the lag bolt used. Typical dimensions of standard lag bolts are given in Table 6.24.

Angle of Load to Grain

The Hankinson formula is used to determine the design value N of loads at an angle of load to grain θ between 0° and 90°,

Spacing, End Distance, and Edge Distance

Spacing, end distance, and edge distance requirements for lag bolts are the same as those for bolts of a diameter equal to the shank diameter of the lag bolt used.

TABLE 6.24

Typical Dimensions of Standard Lag Bolts for Wood[a,b]

Nominal Length L (in.)[c]	Item	Dimensions of Lag Bolt with Nominal Diameter D of											
		$\frac{1}{4}$	$\frac{5}{16}$	$\frac{3}{8}$	$\frac{7}{16}$	$\frac{1}{2}$	$\frac{9}{16}$	$\frac{5}{8}$	$\frac{3}{4}$	$\frac{7}{8}$	1	$1\frac{1}{8}$	$1\frac{1}{4}$
All lengths	$D_s = D$	0.250	0.3125	0.375	0.4375	0.500	0.5625	0.625	0.750	0.875	1.000	1.125	1.250
	D_r	0.173	0.227	0.265	0.328	0.371	0.435	0.471	0.579	0.683	0.780	0.887	1.012
	E	$\frac{3}{16}$	$\frac{1}{4}$	$\frac{1}{4}$	$\frac{9}{32}$	$\frac{5}{16}$	$\frac{3}{8}$	$\frac{3}{8}$	$\frac{7}{16}$	$\frac{1}{2}$	$\frac{9}{16}$	$\frac{5}{8}$	$\frac{3}{4}$
	H	$\frac{11}{64}$	$\frac{13}{64}$	$\frac{1}{4}$	$\frac{19}{64}$	$\frac{21}{64}$	$\frac{3}{8}$	$\frac{27}{64}$	$\frac{1}{2}$	$\frac{19}{32}$	$\frac{21}{32}$	$\frac{3}{4}$	$\frac{27}{32}$
	W	$\frac{3}{8}$	$\frac{1}{2}$	$\frac{9}{16}$	$\frac{5}{8}$	$\frac{3}{4}$	$\frac{7}{8}$	$\frac{15}{16}$	$1\frac{1}{8}$	$1\frac{5}{16}$	$1\frac{1}{2}$	$1\frac{11}{16}$	$1\frac{7}{8}$
	N	10	9	7	7	6	6	5	$4\frac{1}{2}$	4	$3\frac{1}{2}$	$3\frac{1}{4}$	$3\frac{1}{4}$
$1\frac{1}{2}$	S	$\frac{3}{8}$	$\frac{3}{8}$	$\frac{3}{8}$	$\frac{3}{8}$	$\frac{3}{8}$							
	T	$1\frac{1}{8}$	$1\frac{1}{8}$	$1\frac{1}{8}$	$1\frac{1}{8}$	$1\frac{1}{8}$							
	$T - E$	$\frac{15}{16}$	$\frac{7}{8}$	$\frac{7}{8}$	$\frac{27}{32}$	$\frac{13}{16}$							
2	S	$\frac{1}{2}$	$\frac{1}{2}$	$\frac{1}{2}$	$\frac{1}{2}$	$\frac{1}{2}$	$\frac{1}{2}$	$\frac{1}{2}$					
	T	$1\frac{1}{2}$	$1\frac{1}{2}$	$1\frac{1}{2}$	$1\frac{1}{2}$	$1\frac{1}{2}$	$1\frac{1}{2}$	$1\frac{1}{2}$					
	$T - E$	$1\frac{5}{16}$	$1\frac{1}{4}$	$1\frac{1}{4}$	$1\frac{7}{32}$	$1\frac{3}{16}$	$1\frac{1}{8}$	$1\frac{1}{8}$					
$2\frac{1}{2}$	S	1	$\frac{7}{8}$	$\frac{7}{8}$	$\frac{3}{4}$	$\frac{3}{4}$	$\frac{3}{4}$	$\frac{3}{4}$					
	T	$1\frac{1}{2}$	$1\frac{5}{8}$	$1\frac{5}{8}$	$1\frac{3}{4}$	$1\frac{3}{4}$	$1\frac{3}{4}$	$1\frac{3}{4}$					
	$T - E$	$1\frac{5}{16}$	$1\frac{3}{8}$	$1\frac{3}{8}$	$1\frac{15}{32}$	$1\frac{7}{16}$	$1\frac{3}{8}$	$1\frac{3}{8}$					
3	S	1	1	1	1	1	1	1	1	1	1		
	T	2	2	2	2	2	2	2	2	2	2		
	$T - E$	$1\frac{13}{16}$	$1\frac{3}{4}$	$1\frac{3}{4}$	$1\frac{23}{32}$	$1\frac{11}{16}$	$1\frac{5}{8}$	$1\frac{5}{8}$	$1\frac{9}{16}$	$1\frac{1}{2}$	$1\frac{7}{16}$		
4	S	$1\frac{1}{2}$	$1\frac{1}{2}$	$1\frac{1}{2}$	$1\frac{1}{2}$	$1\frac{1}{2}$	$1\frac{1}{2}$	$1\frac{1}{2}$	$1\frac{1}{2}$	$1\frac{1}{2}$	$1\frac{1}{2}$	$1\frac{1}{2}$	$1\frac{1}{2}$
	T	$2\frac{1}{2}$	$2\frac{1}{2}$	$2\frac{1}{2}$	$2\frac{1}{2}$	$2\frac{1}{2}$	$2\frac{1}{2}$	$2\frac{1}{2}$	$2\frac{1}{2}$	$2\frac{1}{2}$	$2\frac{1}{2}$	$2\frac{1}{2}$	$2\frac{1}{2}$
	$T - E$	$2\frac{5}{16}$	$2\frac{1}{4}$	$2\frac{1}{4}$	$2\frac{7}{32}$	$2\frac{3}{16}$	$2\frac{1}{8}$	$2\frac{1}{8}$	$2\frac{1}{16}$	2	$1\frac{15}{16}$	$1\frac{7}{8}$	$1\frac{3}{4}$
5	S	2	2	2	2	2	2	2	2	2	2	2	2
	T	3	3	3	3	3	3	3	3	3	3	3	3
	$T - E$	$2\frac{13}{16}$	$2\frac{3}{4}$	$2\frac{3}{4}$	$2\frac{23}{32}$	$2\frac{11}{16}$	$2\frac{5}{8}$	$2\frac{5}{8}$	$2\frac{9}{16}$	$2\frac{1}{2}$	$2\frac{7}{16}$	$2\frac{3}{8}$	$2\frac{1}{4}$

6	S	$2\frac{1}{2}$	$2\frac{1}{2}$	$2\frac{1}{2}$	$2\frac{1}{2}$	$2\frac{1}{2}$	$2\frac{1}{2}$	$2\frac{1}{2}$	$2\frac{1}{2}$	$2\frac{1}{2}$	$2\frac{1}{2}$	$2\frac{1}{2}$	$2\frac{1}{2}$
	T	$3\frac{1}{2}$	$3\frac{1}{2}$	$3\frac{1}{2}$	$3\frac{1}{2}$	$3\frac{1}{2}$	$3\frac{1}{2}$	$3\frac{1}{2}$	$3\frac{1}{2}$	$3\frac{1}{2}$	$3\frac{1}{2}$	$3\frac{1}{2}$	$3\frac{1}{2}$
	$T - E$	$3\frac{5}{16}$	$3\frac{1}{4}$	$3\frac{1}{4}$	$3\frac{7}{32}$	$3\frac{3}{16}$	$3\frac{1}{8}$	$3\frac{1}{8}$	$3\frac{1}{16}$	3	$2\frac{15}{16}$	$2\frac{7}{8}$	$2\frac{3}{4}$
7	S	3	3	3	3	3	3	3	3	3	3	3	3
	T	4	4	4	4	4	4	4	4	4	4	4	4
	$T - E$	$3\frac{13}{16}$	$3\frac{3}{4}$	$3\frac{3}{4}$	$3\frac{23}{32}$	$3\frac{11}{16}$	$3\frac{5}{8}$	$3\frac{5}{8}$	$3\frac{9}{16}$	$3\frac{1}{2}$	$3\frac{7}{16}$	$3\frac{3}{8}$	$3\frac{1}{4}$
8	S	$3\frac{1}{2}$	$3\frac{1}{2}$	$3\frac{1}{2}$	$3\frac{1}{2}$	$3\frac{1}{2}$	$3\frac{1}{2}$	$3\frac{1}{2}$	$3\frac{1}{2}$	$3\frac{1}{2}$	$3\frac{1}{2}$	$3\frac{1}{2}$	$3\frac{1}{2}$
	T	$4\frac{1}{2}$	$4\frac{1}{2}$	$4\frac{1}{2}$	$4\frac{1}{2}$	$4\frac{1}{2}$	$4\frac{1}{2}$	$4\frac{1}{2}$	$4\frac{1}{2}$	$4\frac{1}{2}$	$4\frac{1}{2}$	$4\frac{1}{2}$	$4\frac{1}{2}$
	$T - E$	$4\frac{5}{16}$	$4\frac{1}{4}$	$4\frac{1}{4}$	$4\frac{7}{32}$	$4\frac{3}{16}$	$4\frac{1}{8}$	$4\frac{1}{8}$	$4\frac{1}{16}$	4	$3\frac{15}{16}$	$3\frac{7}{8}$	$3\frac{3}{4}$
9	S	4	4	4	4	4	4	4	4	4	4	4	4
	T	5	5	5	5	5	5	5	5	5	5	5	5
	$T - E$	$4\frac{13}{16}$	$4\frac{3}{4}$	$4\frac{3}{4}$	$4\frac{23}{32}$	$4\frac{11}{16}$	$4\frac{5}{8}$	$4\frac{5}{8}$	$4\frac{9}{16}$	$4\frac{1}{2}$	$4\frac{7}{16}$	$4\frac{3}{8}$	$4\frac{1}{4}$
10	S	$4\frac{3}{4}$	$4\frac{3}{4}$	$4\frac{3}{4}$	$4\frac{3}{4}$	$4\frac{3}{4}$	$4\frac{3}{4}$	$4\frac{3}{4}$	$4\frac{3}{4}$	$4\frac{3}{4}$	$4\frac{3}{4}$	$4\frac{3}{4}$	$4\frac{3}{4}$
	T	$5\frac{1}{4}$	$5\frac{1}{4}$	$5\frac{1}{4}$	$5\frac{1}{4}$	$5\frac{1}{4}$	$5\frac{1}{4}$	$5\frac{1}{4}$	$5\frac{1}{4}$	$5\frac{1}{4}$	$5\frac{1}{4}$	$5\frac{1}{4}$	$5\frac{1}{4}$
	$T - E$	$5\frac{1}{16}$	5	5	$4\frac{31}{32}$	$4\frac{15}{16}$	$4\frac{7}{8}$	$4\frac{7}{8}$	$4\frac{13}{16}$	$4\frac{3}{4}$	$4\frac{11}{16}$	$4\frac{5}{8}$	$4\frac{1}{2}$
11	S	$5\frac{1}{2}$	$5\frac{1}{2}$	$5\frac{1}{2}$	$5\frac{1}{2}$	$5\frac{1}{2}$	$5\frac{1}{2}$	$5\frac{1}{2}$	$5\frac{1}{2}$	$5\frac{1}{2}$	$5\frac{1}{2}$	$5\frac{1}{2}$	$5\frac{1}{2}$
	T	$5\frac{1}{2}$	$5\frac{1}{2}$	$5\frac{1}{2}$	$5\frac{1}{2}$	$5\frac{1}{2}$	$5\frac{1}{2}$	$5\frac{1}{2}$	$5\frac{1}{2}$	$5\frac{1}{2}$	$5\frac{1}{2}$	$5\frac{1}{2}$	$5\frac{1}{2}$
	$T - E$	$5\frac{9}{32}$	$5\frac{1}{4}$	$5\frac{1}{4}$	$5\frac{7}{32}$	$5\frac{3}{16}$	$5\frac{1}{8}$	$5\frac{1}{8}$	$5\frac{1}{16}$	5	$4\frac{15}{16}$	$4\frac{7}{8}$	$4\frac{3}{4}$
12	S	6	6	6	6	6	6	6	6	6	6	6	6
	T	6	6	6	6	6	6	6	6	6	6	6	6
	$T - E$	$5\frac{13}{16}$	$5\frac{3}{4}$	$\frac{3}{4}$	$5\frac{23}{32}$	$5\frac{11}{16}$	$5\frac{5}{8}$	$5\frac{5}{8}$	$\frac{9}{16}$	$5\frac{1}{2}$	$5\frac{7}{16}$	$5\frac{3}{8}$	$5\frac{1}{4}$

[a] Source: *National Design Specification for Wood Construction* (3).
[b] All dimensions in inches

D = nominal diameter,
$D_s = D$ = diameter of shank,
D_r = diameter at root of thread,
W = width of head across flats,
H = height of head,

L = nominal length,
S = length of shank,
T = length of thread,
E = length of tapered tip, and
N = number of threads per inch.

[c] Length of thread T on intervening bolt lengths is the same as that of the next shorter length listed. The length of thread T on standard lag bolt lengths L in excess of 12 in. is equal to 1/2 the lag bolt length.

TABLE 6.25

Lateral Load Design Values for Lag Bolt Joints with Wood Side Pieces[a,b]

Thickness of Side Member (in.)	Length of Lag Bolt (in.)	Diameter of Lag Bolt Shank (in.)	Total Lateral Load per Lag Bolt in Single Shear (lb)[c]							
			Group I		Group II		Group III		Group IV	
			Parallel to Grain P	Perpendicular to Grain Q	Parallel to Grain P	Perpendicular to Grain Q	Parallel to Grain P	Perpendicular to Grain Q	Parallel to Grain P	Perpendicular to Grain Q
$1\frac{1}{2}$	4	$\frac{1}{4}$	200	200	170	170	130	130	100	100
		$\frac{5}{16}$	290	240	220	180	150	130	120	110
		$\frac{3}{8}$	330	250	250	190	180	140	140	110
		$\frac{7}{16}$	370	260	280	190	200	140	160	110
		$\frac{1}{2}$	390	250	290	190	210	140	170	110
		$\frac{5}{8}$	470	280	360	210	260	160	200	120
	5	$\frac{1}{4}$	240	230	200	200	180	180	160	160
		$\frac{5}{16}$	340	290	290	250	240	200	190	160
		$\frac{3}{8}$	440	340	380	290	270	210	220	170
		$\frac{7}{16}$	550	380	420	290	300	210	240	170
		$\frac{1}{2}$	580	380	440	280	310	200	250	160
		$\frac{5}{8}$	710	420	530	320	380	230	310	180
	6	$\frac{1}{4}$	270	260	230	220	210	200	180	180
		$\frac{5}{16}$	380	320	330	280	290	250	260	220
		$\frac{3}{8}$	490	370	420	320	370	280	300	230
		$\frac{7}{16}$	600	420	520	360	410	280	330	230
		$\frac{1}{2}$	700	460	600	390	430	280	340	220
		$\frac{5}{8}$	850	510	710	430	510	310	410	250

	7	$\frac{1}{4}$	280	270	240	230	210	210	190	180
		$\frac{5}{16}$	400	340	350	300	310	270	280	230
		$\frac{3}{8}$	530	400	460	350	410	310	360	270
		$\frac{7}{16}$	650	450	560	390	500	350	420	300
		$\frac{1}{2}$	760	490	660	430	550	360	440	290
		$\frac{5}{8}$	910	540	780	470	640	380	510	310
$2\frac{1}{2}$	6	$\frac{3}{8}$	450	340	380	290	270	210	220	170
		$\frac{7}{16}$	590	410	440	310	320	220	250	180
		$\frac{1}{2}$	620	410	470	310	340	220	270	180
		$\frac{5}{8}$	730	440	550	330	390	240	320	190
		$\frac{3}{4}$	830	460	630	350	450	250	360	200
		$\frac{7}{8}$	950	490	720	370	510	270	410	210
		1	1060	530	800	400	570	290	460	230
	7	$\frac{3}{8}$	500	380	430	330	380	290	300	230
		$\frac{7}{16}$	670	470	580	410	430	300	350	240
		$\frac{1}{2}$	830	540	650	420	460	300	370	240
		$\frac{5}{8}$	1000	600	750	450	540	320	430	260
		$\frac{3}{4}$	1120	620	850	470	610	330	490	270
		$\frac{7}{8}$	1280	660	970	500	690	360	550	290
		1	1440	720	1090	540	780	390	620	310
	8	$\frac{3}{8}$	560	420	480	370	430	330	380	290
		$\frac{7}{16}$	730	510	630	440	560	390	450	320
		$\frac{1}{2}$	890	580	770	500	600	390	480	310
		$\frac{5}{8}$	1230	740	970	580	700	420	560	340
		$\frac{3}{4}$	1440	790	1090	600	780	430	630	340
		$\frac{7}{8}$	1610	840	1220	630	870	450	700	360
		1	1810	910	1370	690	980	490	790	390

(continued)

TABLE 6.25 (*Continued*)

Thickness of Side Member (in.)	Length of Lag Bolt (in.)	Diameter of Lag Bolt Shank (in.)	Total Lateral Load per Lag Bolt in Single Shear (lb)[c]							
			Group I		Group II		Group III		Group IV	
			Parallel to Grain P	Perpendicular to Grain Q	Parallel to Grain P	Perpendicular to Grain Q	Parallel to Grain P	Perpendicular to Grain Q	Parallel to Grain P	Perpendicular to Grain Q
$2\frac{1}{2}$ (cont.)	9	$\frac{3}{8}$	600	460	520	400	470	350	410	310
		$\frac{7}{16}$	790	550	680	480	610	430	540	380
		$\frac{1}{2}$	960	630	830	540	740	480	600	390
		$\frac{5}{8}$	1310	790	1130	680	860	520	690	420
		$\frac{3}{4}$	1680	920	1350	740	960	530	770	430
		$\frac{7}{8}$	1940	1010	1470	760	1050	550	840	440
		1	2190	2000	1660	830	1190	590	950	480
$3\frac{1}{2}$	8	$\frac{3}{8}$	450	340	390	300	350	260	280	210
		$\frac{7}{16}$	610	430	530	370	410	290	330	230
		$\frac{1}{2}$	740	480	640	420	460	300	370	240
		$\frac{5}{8}$	890	530	770	470	560	330	450	270
		$\frac{3}{4}$	1030	570	890	490	640	350	510	280
		$\frac{7}{8}$	1140	600	980	510	700	360	560	290
		1	1240	620	1070	530	770	380	610	310
		$1\frac{1}{8}$	1370	680	1180	590	850	420	660	330
		$1\frac{1}{4}$	1470	740	1270	640	910	450	680	340
	9	$\frac{3}{8}$	500	380	430	330	380	290	340	260
		$\frac{7}{16}$	660	460	570	410	520	370	420	300
		$\frac{1}{2}$	860	560	740	480	590	380	470	310

		5/8	1160	700	1000	600	720	430	570	340
		3/4	1320	730	1140	650	820	450	660	360
		7/8	1460	760	1260	650	910	470	720	380
		1	1590	800	1370	680	990	490	770	390
		1 1/8	1750	880	1510	750	1080	540	850	420
		1 1/4	1890	940	1630	810	1170	580	880	440
	10	1/2	960	620	830	550	750	480	600	390
		5/8	1440	860	1240	740	900	540	720	430
		3/4	1660	910	1430	790	1030	560	820	450
		7/8	1820	950	1570	820	1130	580	900	470
		1	1980	990	1710	850	1230	610	980	490
		1 1/8	2130	1060	1840	920	1320	660	1060	530
		1 1/4	2300	1150	1990	990	1430	710	1100	550
	11	1/2	1060	690	920	600	830	540	730	470
		5/8	1580	950	1360	820	1110	660	880	530
		3/4	2020	1110	1740	960	1250	680	1000	550
		7/8	2210	1150	1910	990	1370	710	1100	570
		1	2400	1200	2070	1030	1490	740	1190	590
		1 1/8	2570	1280	2220	1110	1600	800	1270	640
		1 1/4	2710	1360	2340	1170	1690	840	1330	660
5 1/2	11	1/2	800	520	690	450	620	400	460	300
		5/8	1140	680	980	590	700	420	560	340
		3/4	1330	730	1150	640	830	450	660	360
		7/8	1530	800	1320	680	950	490	760	400
		1	1720	860	1490	740	1070	530	850	430
		1 1/8	1880	940	1620	810	1170	580	930	470
		1 1/4	1990	1000	1720	860	1230	610	990	490
	12	1/2	860	560	740	480	670	430	590	380
		5/8	1320	790	1140	680	870	520	690	420

(*continued*)

TABLE 6.25 (*Continued*)

Thickness of Side Member (in.)	Length of Lag Bolt (in.)	Diameter of Lag Bolt Shank (in.)	Total Lateral Load per Lag Bolt in Single Shear (lb)[c]							
			Group I		Group II		Group III		Group IV	
			Parallel to Grain P	Perpendicular to Grain Q	Parallel to Grain P	Perpendicular to Grain Q	Parallel to Grain P	Perpendicular to Grain Q	Parallel to Grain P	Perpendicular to Grain Q
$5\frac{1}{2}$ (cont.)	12 (cont.)	$\frac{3}{4}$	1640	900	1420	780	1020	560	820	450
		$\frac{7}{8}$	1890	980	1630	840	1170	610	930	490
		1	2120	1060	1830	910	1320	660	1050	530
		$1\frac{1}{8}$	2320	1160	2000	1000	1440	720	1150	570
		$1\frac{1}{4}$	2460	1230	2120	1060	1530	760	1220	610
	13	$\frac{5}{8}$	1400	840	1210	730	1040	620	830	500
		$\frac{3}{4}$	1970	1080	1700	930	1220	670	980	540
		$\frac{7}{8}$	2260	1180	1950	1010	1400	730	1120	580
		1	2540	1270	2190	1090	1580	790	1260	630
		$1\frac{1}{8}$	2770	1380	2390	1190	1720	860	1370	690
		$1\frac{1}{4}$	2940	1470	2540	1270	1830	910	1460	730
	14	$\frac{5}{8}$	1490	890	1290	770	1160	690	970	580
		$\frac{3}{4}$	2100	1160	1810	990	1440	790	1150	630
		$\frac{7}{8}$	2650	1380	2290	1190	1640	850	1320	680
		1	2960	1480	2560	1280	1840	920	1470	740
		$1\frac{1}{8}$	3230	1620	2790	1400	2010	1000	1610	800
		$1\frac{1}{4}$	3440	1720	2970	1480	2140	1070	1710	850

[a] Source ($1\frac{1}{2}$ and $2\frac{1}{2}$-in.-thick side members): *National Design Specification for Wood Construction* (3).

[b] Normal load duration, dry service conditions.

[c] See Table 6.1.

TABLE 6.26

Lateral Load Design Values for Lag Bolt Joints with Metal Side Pieces Up to ½-in. Thick[a, b]

		Total Lateral Load per Lag Bolt in Single Shear (lb)[c]							
		Group I		Group II		Group III		Group IV	
Length of Lag Bolt (in.)	Diameter of Lag Bolt Shank (in.)	Parallel to Grain P	Perpendicular to Grain Q	Parallel to Grain P	Perpendicular to Grain Q	Parallel to Grain P	Perpendicular to Grain Q	Parallel to Grain P	Perpendicular to Grain Q
3	$\frac{1}{4}$	240	190	210	160	160	120	130	100
	$\frac{5}{16}$	350	240	270	180	190	130	150	100
	$\frac{3}{8}$	420	250	320	190	230	140	180	110
	$\frac{7}{16}$	480	270	360	200	260	140	210	120
	$\frac{1}{2}$	540	280	400	210	290	150	230	120
	$\frac{5}{8}$	650	310	490	230	350	170	280	130
4	$\frac{1}{4}$[d]	270	210	240	180	210	160	190	150
	$\frac{5}{16}$	410	280	350	240	290	200	230	160
	$\frac{3}{8}$	570	350	480	290	340	210	280	170
	$\frac{7}{16}$	730	410	550	310	390	220	310	180
	$\frac{1}{2}$	810	420	610	320	440	230	350	180
	$\frac{5}{8}$	980	470	740	360	530	250	430	200
5	$\frac{5}{16}$	440	300	380	260	340	230	300	200
	$\frac{3}{8}$	620	380	530	320	470	290	380	230
	$\frac{7}{16}$	810	460	700	390	530	300	430	240
	$\frac{1}{2}$	1040	540	840	440	600	310	480	250
	$\frac{5}{8}$	1330	640	1010	480	720	350	580	280
	$\frac{3}{4}$	1550	680	1170	520	840	370	670	300

(continued)

TABLE 6.26 (*Continued*)

Length of Lag Bolt (in.)	Diameter of Lag Bolt Shank (in.)	Total Lateral Load per Lag Bolt in Single Shear (lb)[c]							
		Group I		Group II		Group III		Group IV	
		Parallel to Grain P	Perpendicular to Grain Q	Parallel to Grain P	Perpendicular to Grain Q	Parallel to Grain P	Perpendicular to Grain Q	Parallel to Grain P	Perpendicular to Grain Q
6	$\frac{5}{16}$[d]	450	300	390	260	340	230	300	210
	$\frac{3}{8}$	630	390	550	330	490	300	430	260
	$\frac{7}{16}$	850	480	730	410	660	370	540	300
	$\frac{1}{2}$	1100	570	950	490	760	400	610	320
	$\frac{5}{8}$	1640	790	1290	620	920	440	740	350
	$\frac{3}{4}$	1990	870	1500	660	1070	470	860	380
7	$\frac{3}{8}$[d]	640	390	560	340	500	300	440	270
	$\frac{7}{16}$	870	490	750	420	670	380	590	330
	$\frac{1}{2}$	1120	580	970	500	870	450	740	380
	$\frac{5}{8}$	1710	820	1480	710	1130	540	900	430
	$\frac{3}{4}$	2380	1050	1840	810	1310	580	1050	460
8	$\frac{7}{16}$[d]	880	490	760	420	680	380	600	330
	$\frac{1}{2}$	1140	590	980	510	880	460	780	400
	$\frac{5}{8}$	1750	840	1510	720	1320	630	1060	510
	$\frac{3}{4}$	2470	1090	2130	940	1560	690	1250	550
	$\frac{7}{8}$	3260	1360	2480	1030	1770	740	1420	590
9	$\frac{1}{2}$[d]	1150	600	990	510	890	460	780	410
	$\frac{5}{8}$	1770	850	1530	730	1370	660	1210	580
	$\frac{3}{4}$	2510	1100	2160	950	1790	790	1440	630
	$\frac{7}{8}$	3360	1400	2880	1200	2060	860	1650	690

10	$\frac{5}{8}^{d}$	1790	860	1550	740	1380	660	1220	590
	$\frac{3}{4}$	2550	1120	2200	970	1970	870	1630	720
	$\frac{7}{8}$	3430	1420	2960	1230	2340	970	1880	780
	1	4410	1770	3680	1470	2640	1050	2110	850
11	$\frac{3}{4}^{d}$	2570	1130	2220	980	1990	880	1750	770
	$\frac{7}{8}$	3470	1440	3000	1250	2620	1090	2100	870
	1	4490	1800	3880	1550	2960	1180	2370	950
12	$\frac{7}{8}$	3490	1450	3020	1260	2700	1120	2320	960
	1	4520	1810	3900	1560	3260	1310	2620	1050
	$1\frac{1}{8}$	5670	2270	4890	1960	3630	1450	2910	1170
13	$\frac{7}{8}^{d}$	3510	1460	3030	1260	2710	1130	2390	1000
	1	4550	1820	3930	1570	3520	1410	2870	1150
	$1\frac{1}{8}$	5710	2280	4930	1970	3980	1590	3200	1280
14	1	4570	1830	3950	1580	3530	1410	3110	1250
	$1\frac{1}{8}$	5750	2300	4960	1980	4330	1730	3470	1390
	$1\frac{1}{4}$	7030	2810	6070	2430	4750	1900	3810	1520
15	1	4590	1830	3960	1580	3540	1420	3130	1250
	$1\frac{1}{8}$	5770	2310	4990	1990	4460	1780	3750	1500
	$1\frac{1}{4}$	7070	2830	6110	2440	5130	2050	4120	1650
16	1^{d}	4590	1830	3960	1580	3540	1420	3130	1250
	$1\frac{1}{8}^{d}$	5790	2320	5000	2000	4480	1790	3950	1580
	$1\frac{1}{4}^{d}$	7120	2850	6150	2460	5500	2200	4430	1770

[a]Source: *National Design Specification for Wood Construction* (3).

[b]Normal load duration, dry service conditions.

[c]See Table 6.1 for sawn lumber and Table 6.2 for glued laminated timber.

[d]Greater lengths do not provide higher loads.

TABLE 6.27

Lag Bolts Withdrawal Design Values[a,b]

Specific gravity[c] G	Lag Bolt Diameter D (in.)											
	$\frac{1}{4}$, 0.250	$\frac{5}{16}$, 0.3125	$\frac{3}{8}$, 0.375	$\frac{7}{16}$, 0.4375	$\frac{1}{2}$, 0.500	$\frac{9}{16}$, 0.5625	$\frac{5}{8}$, 0.625	$\frac{3}{4}$, 0.750	$\frac{7}{8}$, 0.875	1, 1.000	$1\frac{1}{8}$, 1.125	$1\frac{1}{4}$, 1.250
0.75	413	489	560	629	695	759	822	942	1058	1169	1277	1382
0.68	357	422	484	543	600	656	709	813	913	1009	1103	1193
0.67	349	413	473	531	587	641	694	796	893	987	1078	1167
0.66	341	403	463	519	574	627	678	778	873	965	1054	1141
0.62	311	367	421	473	523	571	618	708	795	879	960	1039
0.55	260	307	352	395	437	477	516	592	664	734	802	868
0.54	253	299	342	384	425	464	502	576	646	714	780	844
0.51	232	274	314	353	390	426	461	528	593	656	716	775
0.49	218	258	296	332	367	401	434	498	559	617	674	730
0.48	212	250	287	322	356	389	421	482	542	599	654	708
0.47	205	242	278	312	345	377	408	467	525	580	634	686
0.46	199	235	269	302	334	365	395	453	508	562	613	664
0.45	192	227	260	292	323	353	382	438	492	543	594	642
0.44	186	220	252	283	312	341	369	423	475	525	574	621
0.43	179	212	243	273	302	330	357	409	459	508	554	600
0.42	173	205	235	264	291	318	344	395	443	490	535	579
0.41	167	198	226	254	281	307	332	381	428	473	516	559
0.40	161	190	218	245	271	296	320	367	412	455	497	538
0.39	155	183	210	236	261	285	308	353	397	438	479	518
0.38	149	176	202	227	251	274	296	340	381	422	461	498

0.37	143	169	194	218	241	263	285	326	367	405	443	479
0.36	137	163	186	209	231	253	273	313	352	389	425	460
0.35	132	156	179	200	222	242	262	300	337	373	407	441
0.33	121	143	164	184	203	222	240	275	309	341	373	403
0.31	110	130	149	167	185	202	218	250	281	311	339	367

[a] Source: *National Design Specification for Wood Construction* (3).

[b] Normal load duration, dry service conditions. Design values for load in withdrawal in lb/in. of penetration of threaded part into side grain of member holding point

[c] See Table 6.1 for specific gravity of various species. Specific gravity of the wood based on weight and volume when oven dry.

TABLE 6.28

Withdrawal Design Values for Lag Bolts in Softwood Glued Laminated Timber[a]

Species[b]	Growth Rate[b]	Lag Bolt Diameter D^c											
		$\frac{1}{4}$, 0.250	$\frac{5}{16}$, 0.3125	$\frac{3}{8}$, 0.375	$\frac{7}{16}$, 0.4375	$\frac{1}{2}$, 0.500	$\frac{9}{16}$, 0.5625	$\frac{5}{8}$, 0.625	$\frac{3}{4}$, 0.750	$\frac{7}{8}$, 0.875	1, 1.000	$1\frac{1}{8}$, 1.125	$1\frac{1}{4}$, 1.250
Southern Pine	Dense	260	307	352	395	437	477	516	592	664	734	802	868
	Medium Grain	260	307	352	395	437	477	516	592	664	734	802	868
	Coarse Grain	167	198	226	254	281	307	332	381	428	473	516	559
Douglas Fir-Larch	Dense	232	274	314	353	390	426	461	528	593	656	716	775
	Medium Grain	232	274	314	353	390	426	461	528	593	656	716	775
	Coarse Grain	167	198	226	254	281	307	332	381	428	473	516	559
Hem-Fir	Medium Grain	173	205	235	264	291	318	344	395	443	490	535	579
Western Woods		137	163	186	209	231	253	273	313	352	389	425	460
California Redwood		173	205	235	264	291	318	344	395	443	490	535	579
Douglas Fir South		212	250	287	322	356	389	421	482	542	599	654	708

[a] Normal load duration, dry service conditions. Design values for load in withdrawal in lb/in. of penetration of threaded part into side grain of the laminations receiving the point.

[b] Refer to Table 3, AITC 117—Design (7) for species, rate of growth, and location of lumber used in various combinations. For other species refer to Table 6.27 for withdrawal design values based on specific gravity obtained from Table 6.1.

[c] D = shank diameter.

Lag Bolt Design Values

Tabular design values in Tables 6.25–6.28 apply for one lag bolt in a two-member joint under normal duration of loading and dry-use conditions. The total design value for more than one lag bolt is the sum of the values for each lag bolt, provided that spacings, end distances, and edge distances are sufficient to develop the full strength of each lag bolt and that the values are modified for group action in accordance with the recommendations given on page 6-411. For other than normal durations of loading, the design values should be multiplied by the appropriate modification factor for duration of load C_D from Figure 3.1 or Table 3.2. The full design values may be used for lag bolts installed in wood that will remain dry in service. For other conditions of service, the appropriate modification factor C_M from Tables 6.4–6.6 should be applied. See Table 6.3 for applicability of modification factors for lag bolts.

Withdrawal Loads

Table 6.27 may be used to determine the withdrawal design values for lag bolts installed in the side grain of a member consisting of either sawn lumber or glued laminated timber. This table is based on specific gravity. Table 6.28 contains withdrawal values for softwood glued laminated timbers based on species and density. If possible, lag bolts should not be loaded in withdrawal from end grain. When this condition is unavoidable, the design value as determined from either table is multiplied by 0.75.

The withdrawal design value may not exceed the tensile strength of the lag bolt at its net (root) section. A depth of engagement of the lag bolt thread in its lead hole of 7 diameters for group I species, 8 diameters for group II, 10 diameters for group III, and 11 diameters for group IV species (see Table 6.1 for species in each group) will develop approximately the ultimate tensile strength of the lag bolt in axial withdrawal. Note that the length of the gimlet point is not included in calculating depth of engagement of the lag bolt thread. The length used is the distance $T-E$ as shown in Table 6.24.

Lateral Loads

Table 6.25 may be used to determine the lateral design values for one lag bolt installed in the side grain of a two-member joint comprised of seasoned wood with load applied either parallel or perpendicular to grain. Member thickness and shank and thread penetration in the main member have been taken into consideration in preparing Table 6.25. To determine lateral design values for any other angle of load to grain, the Hankinson formula or nomographs given in Figure 6.2 are used. Design values for lateral resistance, when the load acts perpendicular to grain and the lag bolt is inserted parallel to the fibers (i.e., in the end grain of the member), may be taken as two-thirds of those for lateral resistance when the load acts perpendicular to grain (i.e., in the side grain of the member).

The design values for lateral resistance, when lag bolts are installed with metal side members, are given in Table 6.26. If metal side members greater than $\frac{1}{2}$-in. thick are used, the loads for lag bolts should be reduced in proportion to the lesser penetration of the lag bolt. The allowable stresses for the metal side member should not be exceeded. Thinner metal side plates may be used without changing the tabular design values, provided the metal side plates are designed to carry the applied load.

Example. Lag bolts in withdrawal: Determine the capacity of two $\frac{3}{4}$ in. × 8 in. long lag bolts in withdrawal from the top of a 5 in. × 15 in. deep glued laminated beam made of combination 24F-V4 Southern Pine. The member is preservatively treated and is used in a wet location. The load is a wind load.

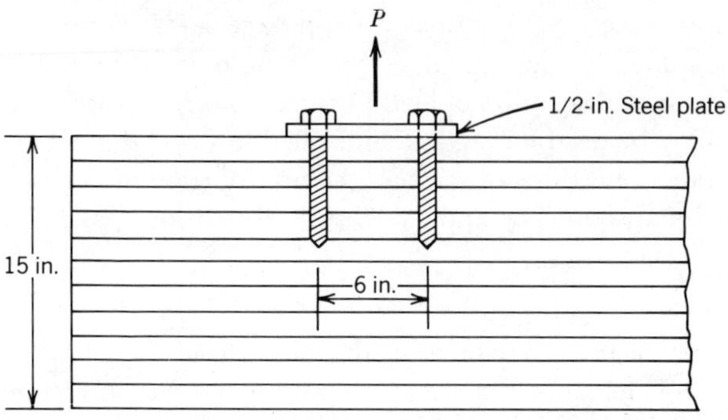

From Table 3.2, $C_D = 1.33$; from Table 6.5, $C_M = 0.67$; from Table 3, AITC 117—Design (7), the core laminations are coarse grain resulting in a group III species grouping for laminations receiving the point. From Table 6.28, the design value of the threaded portion, $P = 381$ lb/in. of penetration. From Table 6.24, the length of the threaded portion of the lag screw $T - E = 4\frac{1}{16}$ in.

$$P' = P(T - E)C_DC_M = (381)(4.0625)(1.33)(0.67) = 1380 \text{ lb/lag bolt}$$

Check tensile strength of lag bolt based on area of root diameter. From Table 6.24, $D_r = 0.579$ in.

$$A_n = \frac{\pi D_r^2}{4} = \frac{\pi(0.579)^2}{4} = 0.263 \text{ in.}^2$$

Capacity of lag bolt based on steel strength $= A_nF_t = (0.263)(20{,}000) = 5260 \text{ lb} > 1380$ lb, use 1380 lb.

$$2P' = 2760 \quad \text{lb.}$$

Example. Lag bolts under lateral wind load: Determine the lateral load that two $\frac{3}{4}$ in. × 6 in. long lag bolts can carry when used to attach a $\frac{3}{8}$ in. × 3 in. strap to a 6 × 12 in. Douglas Fir sawn timber as shown: No. 1 Beam and Stringer of Douglas Fir-Larch. The timber is unseasoned (wet) and will be partially seasoned when full load is applied. From Table 6.3, C_M (wet at time of fabrication and wet in service) $= 0.67$; for wet at time of installation and dry in service $C_M = 1.0$; for

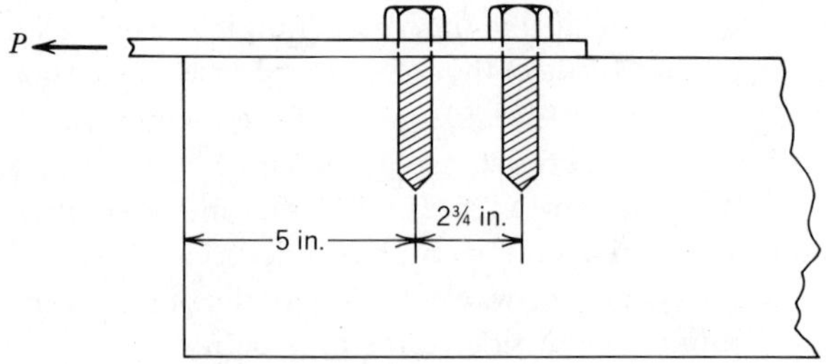

intermediate value use $(1.0 + 0.67)/2 = 0.835$. From Table 6.1, medium grain Douglas Fir-Larch is in group II species. From Table 6.26, $P = 1500$ lb. End distance for full load is

$$7D = (7)\left(\tfrac{3}{4}\right) = 5.25 \text{ in.}$$

Modification factor for end distance is

$$C_n = \frac{5.00}{5.25} = 0.952$$

Spacing for full load is

$$4D = (4)\left(\tfrac{3}{4}\right) = 3 \text{ in.}$$

Modification factor for spacing is

$$C_s = \frac{2.75}{3} = 0.917 < 0.952$$

C_s controls (C_n and C_s are not cumulative).

$$\text{Lateral load} = 2P' = 2PC_DC_MC_s = (2)(1500)(1.33)(0.835)(0.917) = 3050 \text{ lb}$$

Combined Lateral and Withdrawal Loads

When a lag bolt is subjected to combined lateral and withdrawal loads such as occurs when the lag bolt is inserted perpendicular to the wood fiber and the load acts at an angle between 0° and 90° with the surface, the design value should be determined by use of the Hankinson formula.

Example. Determine the load P_α that a $\frac{1}{2}$-in. × 5 in. long lag bolt in a $5\frac{1}{8}$-in. by 6-in. glued laminated timber can carry. Load is applied through a $\frac{3}{8}$-in. thick metal side plate, which is in the same plane that contains the longitudinal axis and is at an angle of 30° with the surface. The glued laminated timber is made with combination 47 (all No. 2 medium grain Southern Pine). The load is primarily a snow load and the fastening is used in a wet location.

From Table 3.2, $C_D = 1.15$. From Table 6.5, $C_M = 0.67$. End distance is greater than $7D$ in both directions, $C_n = 1$ for both parallel- and perpendicular-to-grain loading. Edge distance is

$$\frac{5.125}{2} = 2.56 \text{ in.} \qquad \frac{2.56}{0.5} = 5.12D > 4D$$

$C_e = 1$ for both parallel- and perpendicular-to-grain loading.

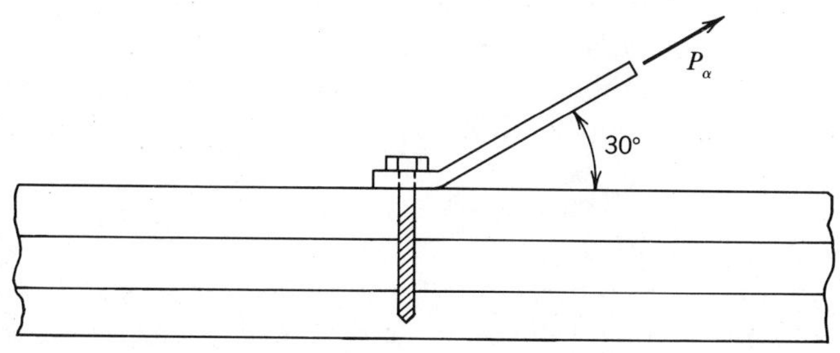

From Table 6.1, Southern Pine medium grain is group II species, 0.55 specific gravity. From Table 6.26, $P = 840$ lb; from Table 6.28, $P_w = 437$ lb/in.; from Table 6.24, $T-E = 2\text{-}\frac{11}{16}$ in.

$$P_w = (437)(2.6875) = 1175 \quad \text{lb}$$

$$P' = PC_DC_M = (840)(1.15)(0.67) = 650 \quad \text{lb}$$

$$P'_w = P_wC_DC_M = (1175)(1.15)(0.67) = 905 \quad \text{lb}$$

Use the Hankinson formula (6-6)

$$P'_\alpha = \frac{P'_w\, P'}{P'_w \sin^2\theta + P' \cos^2\theta} = \frac{(905)(650)}{905 \sin^2 30° + 650 \cos^2 30°} = 825 \quad \text{lb}$$

If the direction of the applied load in the above example had not been in the same plane as the longitudinal axis of the piece, the lateral resistance at the angle of load to grain, N', would need to be determined. For example, if the load P_α is in a plane that is 45° to the longitudinal axis of the piece, the following calculations apply.

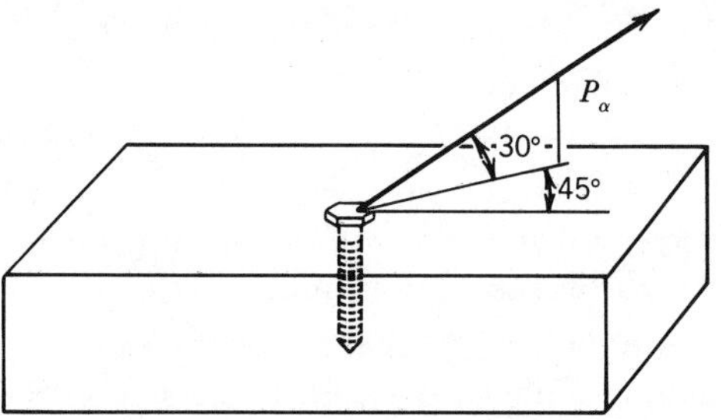

From Table 6.26, $Q = 440$ lb, $Q' = (440)(1.15)(0.67) = 340$ lb

$$N' = \frac{(650)(340)}{650 \sin^2 45° + 340 \cos^2 45°} = 445 \quad \text{lb}$$

$$P'_\alpha = \frac{(905)(445)}{905 \sin^2 30° + 445 \cos^2 30°} = 720 \quad \text{lb}$$

Modification Factors for Lag Bolts

The modification factors applicable to the tabular design values of lag bolts are shown in Table 6.3. For lateral loads, P represents the tabular design value for loads acting parallel to grain and Q represents the tabular design value for loads acting perpendicular to grain. Tabular withdrawal loads are shown as P_w. Reductions for use in end grain are applied in the same manner as for side grain.

WOOD SCREWS

The design values given herein are for any wood screw of sufficient strength to cause failure in the wood rather than in the metal. Wood screws require pre-bored lead holes to prevent splitting of the wood. Lead hole diameters for the shank

TABLE 6.29

Lead Hole Diameters for Wood Screws

		Diameter of Lead Hole (in.)					
		Withdrawal Loads		Lateral Loads			
				Group I Species[a]		Groups II, III, and IV Species[a]	
Gage of Screw	Shank Diameter of Screw (in.)	Group I Species[a]	Groups II, III, and IV Species[a, b]	Shank Portion	Threaded Portion	Shank Portion	Threaded Portion
6	0.138	5/64	1/16	9/64	3/32	1/8	5/64
7	0.151	3/32	5/64	5/32	7/64	1/8	3/32
8	0.164	7/64	5/64	5/32	7/64	9/64	3/32
9	0.177	7/64	5/64	11/64	1/8	5/32	7/64
10	0.190	7/64	3/32	3/16	1/8	11/64	7/64
12	0.216	9/64	7/64	7/32	9/64	3/16	1/8
14	0.242	5/32	7/64	1/4	5/32	7/32	9/64
16	0.268	11/64	1/8	17/64	3/16	15/64	5/32
18	0.294	3/16	9/64	19/64	13/64	1/4	11/64
20	0.320	13/64	5/32	5/16	7/32	9/32	3/16
24	0.372	15/64	3/16	3/8	1/4	21/64	15/64

[a] For species in each group see Table 6.1 for sawn lumber and Table 6.2 for glued laminated timber.
[b] For group III and IV species the screw may be inserted without a lead hole.

and threaded portions of wood screws are given in Table 6.29. Wood screws should be inserted by turning with a screwdriver and not by driving with a hammer. Soap or other lubricant may be used to aid insertion.

Lumber Species

Species load groups for wood screw design values are given in Table 6.1.

Spacing, End Distance, and Edge Distance

Spacing, end distance, and edge distance should be sufficient to prevent unusual splitting of the wood.

Wood Screw Design Values

The design values in Tables 6.30 and 6.31 apply for one wood screw in a two-member joint under normal duration of loading and dry-use conditions. The total design value for more than one wood screw is the sum of the values for each wood screw, provided that spacings, end distances, and edge distances are sufficient to develop the full strength of each wood screw. Tabular values apply to both sawn lumber and glued laminated timber. See Table 3, AITC 117—Design (7) for species groups for glued laminated timber.

For other than normal durations of loading, the design values should be mul-

TABLE 6.30

Wood Screws—Withdrawal Design Values[a,b,c]

Specific Gravity G^d	Gage of Screw										
	6	7	8	9	10	12	14	16	18	20	24
0.75	220	241	262	283	304	345	387	428	470	511	594
0.68	181	198	215	232	250	284	318	352	386	420	489
0.67	176	193	209	226	242	275	309	342	375	408	474
0.66	171	187	203	219	235	267	299	332	364	396	460
0.62	151	165	179	193	207	236	264	293	321	349	406
0.55	119	130	141	152	163	186	208	230	253	275	320
0.54	114	125	136	147	157	179	200	222	243	265	308
0.51	102	112	121	131	140	160	179	198	217	236	275
0.49	94	103	112	121	130	147	165	183	200	218	254
0.48	90	99	107	116	124	141	158	175	192	209	243
0.47	87	95	103	111	119	136	152	168	184	201	233
0.46	83	91	99	106	114	130	145	161	177	192	224
0.45	79	87	94	102	109	124	139	154	169	184	214
0.44	76	83	90	97	104	119	133	147	162	176	205
0.43	72	79	86	93	100	113	127	141	154	168	195
0.42	69	76	82	89	95	108	121	134	147	160	186
0.41	66	72	78	85	91	103	116	128	140	153	178
0.40	63	69	75	80	86	98	110	122	134	145	169
0.39	60	65	71	76	82	93	105	116	127	138	161
0.38	57	62	67	73	78	89	99	110	121	131	153

0.37	54	59	64	69	74	84	94	104	114	124	145
0.36	51	56	60	65	70	80	89	99	108	118	137
0.35	48	53	57	62	66	75	84	93	102	111	129
0.33	43	47	51	55	59	67	75	83	91	99	115
0.31	38	41	45	48	52	59	66	73	80	87	102

[a] Approximately two-thirds of the length of a standard wood screw is threaded. Normal load duration. Design values in withdrawal in lb/in. of penetration of threaded part into side grain of member holding point.

[b] Values based on the formula $P = 2850G^2D$ (6-10).

where G = specific gravity of the wood, based on weight and volume when oven-dry, and D = shank diameter, in.

[c] Source: *National Design Specification for Wood Construction* (3).

[d] See Table 6.1 for specific gravity values for sawn lumber and Table 6.2 for glued laminated timber

TABLE 6.31

Wood Screws—Lateral Load Design Values[a,b]

Gage of Screw		6	7	8	9	10	12	14	16	18	20	24
	D	0.138	0.151	0.164	0.177	0.190	0.216	0.242	0.268	0.294	0.320	0.372
	$7D$	0.966	1.057	1.148	1.239	1.330	1.512	1.694	1.876	2.058	2.240	2.604
	$4D$	0.552	0.604	0.656	0.708	0.760	0.864	0.968	1.072	1.176	1.280	1.488
Species Group[c]	I	91	109	129	150	173	224	281	345	415	492	664
	II	75	90	106	124	143	185	232	284	342	406	548
	III	62	74	87	101	117	151	190	233	280	332	448
	IV	48	58	68	79	91	118	148	181	218	258	349

[a] Source: *National Design Specification for Wood Construction* (3).

[b] Design values for lateral loads (shear), P, (lb) for screws embedded to approximately 7 times the shank diameter, D, into the member holding the point. For less penetration, reduce loads in proportion. Penetration should not be less than 4 times the shank diameter.

[c] See Tables 6.1 and 6.2 for species groups for sawn lumber and glued laminated timber, respectively.

tiplied by the appropriate modification factor C_D from Figure 3.1 or Table 3.2. For wood screw joints under other service or seasoning conditions, the design value determined is multiplied by the appropriate modification factor C_M from Tables 6.4–6.6. See Table 6.3 for applicability of modification factors for wood screws.

Withdrawal Loads

If possible, structural designs should be such that wood screws are not loaded in withdrawal. When such loading is unavoidable, the tensile strength of the wood screw at its net (root) section should not be exceeded. Loading in withdrawal from end grain is not permitted.

Table 6.30 may be used to determine withdrawal design values for one wood screw inserted in side grain of the member holding the screw point of a two-member joint. The effective depth of penetration used to determine the design value is the length of the threaded portion of the screw in the member receiving the point. Approximately two-thirds of the length of a wood screw is threaded.

Lateral Loads

Table 6.31 may be used to determine the lateral design values for one wood screw loaded in single shear at any angle of load to grain when the screw is inserted in the side grain and embedded approximately 7 diameters into the member receiving the point. For wood screws penetrating less than 7 diameters, the design value must be reduced in proportion to the actual depth of penetration. The penetration should not be less than 4 times the shank diameter. The modifying factor C_d to use for intermediate penetrations is

$$C_d = 0.143p \qquad (6\text{-}9)$$

where p = penetration in shank diameters.

The solution to this equation is shown graphically in Figure 6.25.

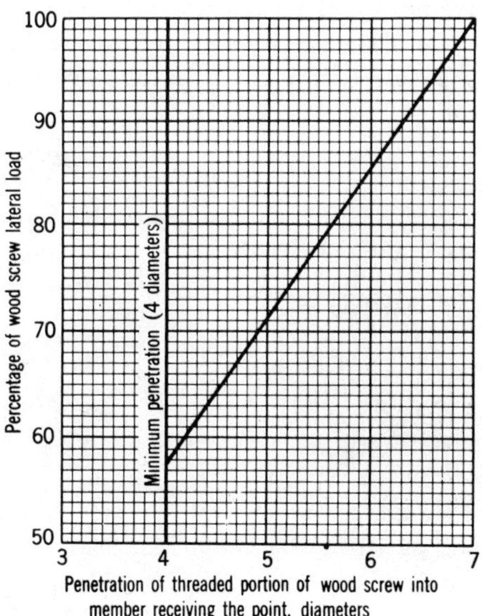

FIGURE 6.25 Modification of wood screw lateral load design values, C_d, for reduced penetrations.

When metal side plates are used rather than wood side pieces, the design values for wood screws in lateral resistance may be increased by 25% ($C_{st} = 1.25$).

When a wood screw is inserted in end grain and the load acts perpendicular to the grain, the design values for lateral resistance are two-thirds of those for lateral resistance given in Table 6.31.

Example. Determine the withdrawal value of a 3-in. No. 12 wood screw in Douglas Fir. From Table 6.1, G for Douglas Fir = 0.51.

$$P_w = (160)(3)(2/3) = 320 \text{ lb}$$

Example. Determine the design value, P'_w in withdrawal of a No. 16 wood screw when the threaded portion penetrates $1\frac{3}{4}$ in. in a medium grain Southern Pine glued laminated timber used under dry conditions. The load to be resisted is principally a snow load.

$$C_M = 1 \qquad C_D = 1.15$$

From Table 6.2, medium grain Southern Pine has a specific gravity of 0.55. From Table 6.31, the design value, P_w, is 230 lb/in. of penetration.

$$P'_w = 1.75 P_w C_D = (1.75)(230)(1.15) = 463 \text{ lb}$$

Example. Determine the lateral wind load that four No. 12 wood screws 2 in. long can carry in Hem-Fir with $\frac{1}{4}$-in. steel side plates. From Table 6.31, depth of penetration required for full load is

$7D = 1.512$ in.,
$2 - 0.25 = 1.75$ in. > 1.512 in. O.K.

Increase for steel side plates = 25%, $C_{st} = 1.25$. For wind load, $C_D = 1.33$. Hem-Fir group III species, $P = 151$ lb.

$$P' = 4PC_{st}C_D = (4)(151)(1.25)(1.33) = 1000 \text{ lb}$$

Example. Determine the lateral load for one No. 14 wood screw $1\frac{1}{2}$ in. long in seasoned Douglas Fir (dense), normal duration of load. The fastening is to be used in a wet location. From Table 6.31, depth of penetration required for full load ($7D$) is (7) (0.242) = 1.69 in. Actual penetration = $1\frac{1}{2} - \frac{1}{4} = 1.25$ in.

$$4D = (4)(0.242) = 0.97 \text{ in.} < 1.25 \text{ in.}$$

Depth of embedment modification factor C_d is 1.25/1.69 = 0.74. From Table 6.31, P for group I species = 281 lb. From Table 6.3, $C_M = 0.67$.

$$P' = P\,C_d C_M = (281)(0.74)(0.67) = 139 \text{ lb}$$

SPIRAL DOWELS

A spiral dowel is a twisted steel rod with spirally grooved ridges along its entire length, the lead of the spiral thread being sufficient to permit driving by any suitable means. The design values given in Table 6.33 are for any spiral dowel of sufficient strength to cause failure in the wood rather than in the metal. Ranges of diameters and lengths of stock sizes of spiral dowels are listed in Table 6.32. Spiral

TABLE 6.32

Dimensions and Lead Hole Diameters for Stock Sizes of Spiral Dowels in Group II Species[a]

Outside Diameter of Dowel (in.)	Minimum Length (in.)[b]	Maximum Length (in.)[b]	Maximum Driving Length (in.)[c,d]	Diameter of Lead Hole (in.)[c]
1/4	2-1/2	6		3/16
5/16	3	6-1/2		1/4
3/8	3-1/2	10		9/32
7/16	3-1/2	12	12	11/32
1/2	4	18	12	3/8
5/8	4	24	12	15/32
3/4	Lengths in these three sizes must be specially ordered		12	9/16
7/8			13	21/32
1			14	3/4

dowels require prebored lead holes of the diameters given in Table 6.32. The detail illustrated in Figure 6.26 is recommended for driving long dowels.

Lumber Species

The design values for spiral dowels are for medium grain Douglas Fir-Larch and medium grain Southern Pine (group II species) for either sawn lumber or glued laminated timber. The design values are also applicable to group I species. Design values for species groups III and IV have not been determined.

Spiral Dowel Design Values

The design values in Table 6.33 are for either sawn lumber or glued laminated timber and apply for one spiral dowel in a two-member joint under normal duration of loading and dry-use conditions. The total design value for more than one spiral dowel is the sum of the values for each spiral dowel, provided that spacings, end distances, and edge distances are sufficient to develop the full strength of each

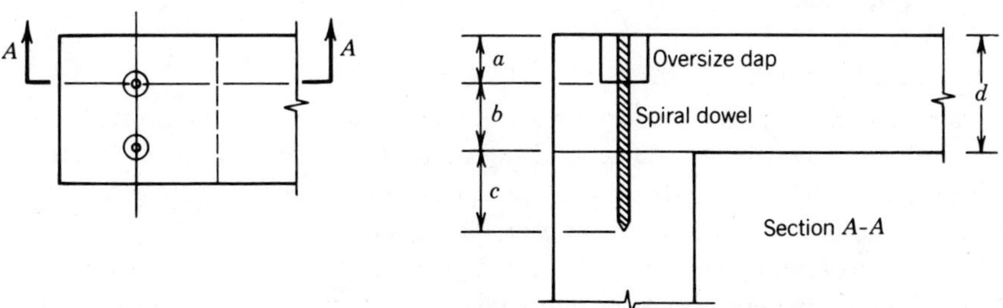

FIGURE 6.26 Driving of long spiral dowels. a = overall depth, d, of member, less $\frac{1}{2}$ maximum driving length: $b = c = \frac{1}{2}$ maximum driving length (see Table 6.32).

TABLE 6.33

Design Values for Spiral Dowels for Group II Species[a, b]

Outside Diameter of Dowel (in.)	Lateral Load Design Value (lb)	Withdrawal Load Design Value[c] (lb/in. of penetration)	
		Side Grain	End Grain
1/4	111	103	59
5/16	174	123	70
3/8	249	141	81
7/16	340	158	90
1/2	444	174	99
5/8	692	206	117
3/4	998	236	135
7/8	1,360	266	151
1	1,775	293	168

[a] All applicable provisions under the heading "Spiral Dowel Design Values" must be met to develop these values.

[b] For species included in group II species see Tables 6.1 and 6.2

[c] For spiral dowels installed or used in unseasoned material or exposed to the weather, use 25% of these design values ($C_M = 0.25$).

spiral dowel. The spacings, end distances, and edge distances required for bolts may be used as a guide.

For other than normal duration of loading, the design values should be multiplied by the appropriate modification factor C_D from Figure 3.1 or Table 3.2. For spiral dowel joints under other service or seasoning conditions, the value determined should be multiplied by the factor, $C_M = 0.25$. See Table 6.6 for modifiers applicable to spiral dowels.

Withdrawal Loads

Withdrawal load design values for one spiral dowel installed in side or end grain are given in Table 6.33.

Lateral Loads

Lateral load design values for one spiral dowel installed in side grain of a two-member joint, when the thickness of the side member is at least 5 times the diameter of the spiral dowel and the dowel penetrates at least 7 diameters into the member receiving the point, are given in Table 6.33.

Lateral strength of spiral dowels in end grain is 60% of the side grain lateral strength, and a penetration of 12 diameters into the end grain is required to develop that strength.

DRIFT BOLTS AND PINS

Drift bolts and pins are long unthreaded bolts or pins that are sometimes used to fasten large timbers. Design values for drift bolts and pins have not been studied as much as those for other fasteners and are not usually included in codes. However, an estimate of both the withdrawal design value and the lateral resistance design value can be obtained from information contained in the *Wood Handbook* (6).

Withdrawal Loads

The ultimate load formula in the *Wood Handbook* (6) can be divided by six and multiplied by 1.10 to obtain an estimated design value for loads of normal duration. This is the same procedure used for nails and spikes. The design value in withdrawal, P_w, is

$$P_w = 1200G^2D \tag{6-7}$$

where P_w = design value in withdrawal (lb/in. of penetration),
G = specific gravity based on oven dry weight and moisture content of 12%, and
D = diameter of drift bolt or pin (in.).

The design value is based on use of a predrilled hole $\frac{1}{8}$ in. less in diameter than the pin in seasoned wood.

Lateral Resistance

The lateral resistance of drift bolts or pins should be no greater than for bolts of the same diameter. Usually, a lesser value is used because of the lack of nuts and washers or heads in the case of a drift pin or bolt should be greater than that of a bolt to account for the lack of nuts and washers. See Table 6.3 for applicable modification factors.

Spiral dowels have a greater resistance to withdrawal than drift bolts or pins and their use is recommended where withdrawal is critical.

NAILS AND SPIKES

The design values given under this heading are for common wire nails and spikes, box nails, and threaded, hardened nails and spikes of the sizes listed in Table 6.34. Threaded, hardened nails and spikes as covered herein are made of high-carbon steel wire, headed, pointed, and annularly or helically threaded, and heat-treated and tempered to provide greater strength than is developed by common wire nails and spikes of corresponding sizes.

When it is necessary to avoid splitting of the wood, a prebored hole of a diameter not exceeding nine-tenths of the nail or spike diameter for group I species or three-fourths for group II, III, and IV species should be used. When these prebored holes are used the design values for the same size nail or spike taken from Tables 6.35 or 6.36 may be used.

TABLE 6.34

Nail and Spike Sizes[a]

		Wire Diameter (in.)			
Pennyweight	Length (in.)	Box Nails	Common Wire Nails	Threaded, Hardened Steel Nails	Common Wire Spikes
6*d*	2	0.099	0.113	0.120	
8*d*	$2\frac{1}{2}$	0.113	0.131	0.120	
10*d*	3	0.128	0.148	0.135	0.192
12*d*	$3\frac{1}{4}$	0.128	0.148	0.135	0.192
16*d*	$3\frac{1}{2}$	0.135	0.162	0.148	0.207
20*d*	4	0.148	0.192	0.177	0.225
30*d*	$4\frac{1}{2}$	0.148	0.207	0.177	0.244
40*d*	5	0.162	0.225	0.177	0.263
50*d*	$5\frac{1}{2}$		0.244	0.177	0.283
60*d*	6		0.263	0.177	0.283
70*d*	7			0.207	
80*d*	8			0.207	
90*d*	9			0.207	
$\frac{5}{16}$	7				0.312
$\frac{3}{8}$	$8\frac{1}{2}$				0.375

[a]Source: *National Design Specification for Wood Construction* (3).

Lumber Species

Species load groups for design values for nails and spikes are given in Tables 6.1 and 6.2.

Spacing, End Distance, and Edge Distance

Spacing, end distances, and edge distances should be sufficient to avoid unusual splitting of the wood.

Nail and Spike Design Values

Tables 6.35 and 6.36 are for one nail or spike in a two-member joint under normal duration of loading and dry-use conditions. Threaded, hardened steel nails may be used with the same design values given for common wire nails of the corresponding pennyweight, except as provided in Table 6.36. The total design value for more than one nail or spike is the sum of the values for each nail or spike.

For joints in use under other service or seasoning conditions, the values determined from the charts are multiplied by the appropriate modification factor from Table 6.4 and 6.6.

Withdrawal Loads

If possible, structural designs should be such that nails or spikes are not loaded in withdrawal. When such loading is unavoidable, the withdrawal load design values per inch of penetration for nails or spikes driven in side grain (perpendicular to grain) as determined from Table 6.36 apply. Loading of nails or spikes in withdrawal from end grain is not recommended.

The withdrawal load design values for toe-nailed joints, for all conditions of seasoning, are two-thirds those determined from Table 6.35. It is recommended that toe-nails be driven at an angle of approximately 30° to the piece and be started at approximately one-third the nail length from the end of the piece. Toe nails are usually not permitted to be used in withdrawal when seismic design codes apply.

The withdrawal resistance of clinched, common wire nails is considerably higher than that of unclinched fasteners. The ratio between values for clinched and unclinched nails varies with the moisture content of the wood, the difference in time between when the nail is driven and when the withdrawal load is applied, the species of wood, the size of nail, and the direction of clinch with respect to grain. In dry or green wood, a clinched nail provides from 45 to 170% more withdrawal resistance than does an unclinched nail withdrawn soon after driving. Nails clinched across the grain have approximately 20% more resistance to withdrawal than do nails clinched along the grain. The proper clinching of nails may be difficult to achieve under field conditions.

Lateral Loads

Table 6.36 may be used to determine the design values for loads in any lateral direction, for one nail or spike driven in the side grain of the main member to any depth of penetration of the point in the member receiving the point down to the minimum permissible penetration. The minimum permissible penetration is one-third the depth of penetration required for full design value. For intermediate depths of penetration, lateral load design values may be determined by straight-line interpolation. These values apply when side and main members have approximately the same density. When side and main members have different densities, the lighter-density member controls.

For nails or spikes not exceeding 12*d* in size which are driven through three members and extending for at least 3 diameters beyond the side members, the lateral load design values may be doubled when the nails are clinched, provided the side members are at least $\frac{3}{8}$-in. thick.

When properly designed metal side plates are used in place of wood side members, the lateral load design values are 1.25 times those from Table 6.36.

The lateral load design values for toe-nailed joints are five-sixths of those determined from Table 6.36. It is recommended that toe-nails be driven at an angle of approximately 30° to the piece and be started at approximately one-third the nail length from the end of the piece.

Design values for nails and spikes may be increased 30% for use in diaphragm construction. The diaphragm design values may be further increased by 33% for

TABLE 6.35

Withdrawal Load Design Values for Common Nails and Spikes[a, b]

Specific Gravity G^c	Pennyweight Diameter (in.)	Size of Common Nails 6d 0.113	8d 0.131	10d 0.148	12d 0.148	16d 0.162	20d 0.192	30d 0.207	40d 0.225	50d 0.244	60d 0.263
0.75		76	88	99	99	109	129	139	151	164	177
0.68		59	69	78	78	85	101	109	118	128	138
0.67		57	66	75	75	82	97	105	114	124	133
0.66		55	64	72	72	79	94	101	110	119	128
0.62		47	55	62	62	68	80	86	94	102	110
0.55		35	41	46	46	50	59	64	70	76	81
0.54		33	39	44	44	48	57	61	67	72	78
0.51		29	34	38	38	42	49	53	58	63	67
0.49		26	30	34	34	38	45	48	52	57	61
0.48		25	29	33	33	36	42	46	50	54	58
0.47		24	27	31	31	34	40	43	47	51	55
0.46		22	26	29	29	32	38	41	45	48	52
0.45		21	25	28	28	30	36	39	42	46	49
0.44		20	23	26	26	29	34	37	40	43	47
0.43		19	22	25	25	27	32	35	38	41	44
0.42		18	21	23	23	26	30	33	35	38	41
0.41		17	19	22	22	24	29	31	33	36	39
0.40		16	18	21	21	23	27	29	31	34	37
0.39		15	17	19	19	21	25	27	29	32	34
0.38		14	16	18	18	20	24	25	28	30	32
0.37		13	15	17	17	19	22	24	26	28	30
0.36		12	14	16	16	17	21	22	24	26	28
0.35		11	13	15	15	16	19	21	23	24	26
0.33		10	11	13	13	14	17	18	19	21	23
0.31		8	10	11	11	12	14	15	17	18	19

Specific Gravity G^c	Pennyweight Diameter (in.)	Size of Box Nail 6d 0.099	8d 0.113	10d 0.128	12d 0.128	16d 0.135	20d 0.148	30d 0.148	40d 0.162
0.75		67	76	86	86	91	99	99	109
0.68		52	59	67	67	71	78	78	85
0.67		50	57	65	65	68	75	75	82
0.66		48	55	63	63	66	72	72	79
0.62		41	47	53	53	56	62	62	68
0.55		31	35	40	40	42	46	46	50
0.54		29	33	38	38	40	44	44	48
0.51		25	29	33	33	35	38	38	42
0.49		23	26	30	30	31	34	34	38
0.48		22	25	28	28	30	33	33	36
0.47		21	24	27	27	28	31	31	34
0.46		20	22	25	25	27	29	29	32
0.45		19	21	24	24	25	28	28	30
0.44		18	20	23	23	24	26	26	29
0.43		17	19	21	21	23	25	25	27
0.42		16	18	20	20	21	23	23	26
0.41		15	17	19	19	20	22	22	24
0.40		14	16	18	18	19	21	21	23
0.39		13	15	17	17	18	19	19	21
0.38		12	14	16	16	17	18	18	20
0.37		11	13	15	15	16	17	17	19
0.36		11	12	14	14	14	16	16	17
0.35		10	11	13	13	14	15	15	16
0.33		9	10	11	11	12	13	13	14
0.31		7	8	9	9	10	11	11	12

(*continued*)

TABLE 6.35 (*Continued*)

Specific Gravity G^c	Pennyweight Diameter (in.)	Size of Threaded Nail[d] 30*d* 0.177	40*d* 0.177	50*d* 0.177	60*d* 0.177	70*d* 0.207	80*d* 0.207	90*d* 0.207
0.75		129	129	129	129	151	151	151
0.68		101	101	101	101	118	118	118
0.67		97	97	97	97	114	114	114
0.66		94	94	94	94	110	110	110
0.62		80	80	80	80	94	94	94
0.55		59	59	59	59	70	70	70
0.54		57	57	57	57	67	67	67
0.51		49	49	49	49	58	58	58
0.49		45	45	45	45	52	52	52
0.48		42	42	42	42	50	50	50
0.47		40	40	40	40	47	47	47
0.46		38	38	38	38	45	45	45
0.45		36	36	36	36	42	42	42
0.44		34	34	34	34	40	40	40
0.43		32	32	32	32	38	38	38
0.42		30	30	30	30	35	35	35
0.41		29	29	29	29	33	33	33
0.40		27	27	27	27	31	31	31
0.39		25	25	25	25	29	29	29
0.38		24	24	24	24	28	28	28
0.37		22	22	22	22	26	26	26
0.36		21	21	21	21	24	24	24
0.35		19	19	19	19	23	23	23
0.33		17	17	17	17	19	19	19
0.31		14	14	14	14	17	17	17

Specific Gravity G^c	Pennyweight Diameter (in.)	Size of Common Spike 10*d* 0.192	12*d* 0.192	16*d* 0.207	20*d* 0.225	30*d* 0.244	40*d* 0.263	50*d* 0.283	60*d* 0.283	$\frac{5}{16}$ in. 0.312	$\frac{3}{8}$ in. 0.375
0.75		129	129	139	151	164	177	190	190	210	252
0.68		101	101	109	118	128	138	149	149	164	197
0.67		97	97	105	114	124	133	144	144	158	190
0.66		94	94	101	110	119	128	138	138	152	183
0.62		80	80	86	94	102	110	118	118	130	157
0.55		59	59	64	70	76	81	88	88	97	116
0.54		57	57	61	67	72	78	84	84	92	111
0.51		49	49	53	58	63	67	73	73	80	96
0.49		45	45	48	52	57	61	66	66	72	87
0.48		42	42	46	50	54	58	62	62	69	83
0.47		40	40	43	47	51	55	59	59	65	78
0.46		38	38	41	45	48	52	56	56	62	74
0.45		36	36	39	42	46	49	53	53	58	70
0.44		34	34	37	40	43	47	50	50	55	66
0.43		32	32	35	38	41	44	47	47	52	63
0.42		30	30	33	35	38	41	45	45	49	59
0.41		29	29	31	33	36	39	42	42	46	56
0.40		27	27	29	31	34	37	40	40	44	52
0.39		25	25	27	29	32	34	37	37	41	49
0.38		24	24	25	28	30	32	35	35	38	46
0.37		22	22	24	26	28	30	33	33	36	43
0.36		21	21	22	24	26	28	30	30	33	40
0.35		19	19	21	23	24	26	28	28	31	38
0.33		17	17	18	19	21	23	24	24	27	32
0.31		14	14	15	17	18	19	21	21	23	28

[a] Source: *National Design Specification for Wood Construction* (3).

[b] For one nail or spike installed in side grain under normal duration of loading, in pounds per inch of penetration into the member receiving the point. Based on the the formula, $P = 7850\ G^{5/2}\ D$, where D = diameter of nail or spike (in.) and G = specific gravity of the wood, based on weight and volume when oven-dry.

[c] See Tables 6.1 and 6.2 for specific gravities of sawn lumber and glued laminated timber.

[d] Loads for threaded, hardened steel nails, in 6*d* to 20*d* sizes, are the same as for common nails.

TABLE 6.36

Lateral Load Design Values for Nails and Spikes[a,b]

Species Group[c]	Nail or Spike Size	Maximum Value (lb)	Required Penetration (in.)	Minimum Value (lb)	Minimum Penetration (in.)
Group I	6*d* box	64	0.99	21	0.33
	8*d* box		1.13		0.38
	6*d* common	77	1.13	26	0.38
	6*d* threaded		1.20		0.40
	10*d* and 12*d* box	93	1.28	31	0.43
	8*d* common	97	1.31	32	0.44
	8*d* threaded		1.20		0.40
	16*d* box	101	1.35	34	0.45
	20*d* and 30*d* box		1.48		0.49
	10*d* and 12*d* common	116	1.48	39	0.49
	10*d* and 12*d* threaded		1.35		0.45
	40*d* box		1.62		0.54
	16*d* common	133	1.62	44	0.54
	16*d* threaded		1.48		0.49
	20*d* common		1.92		0.64
	20*d*, 30*d*, 40*d*, 50*d*, and 60*d* threaded	172	1.77	57	0.59
	10*d* and 12*d* spike		1.92		0.64
	30*d* common	192	2.07	64	0.69
	16*d* spike				
	40*d* common		2.25		0.75
	70*d*, 80*d*, and 90*d* threaded	218	2.07	73	0.69
	20*d* spike		2.25		0.75

	50*d* common 30*d* spike	246	2.44	82	0.81
	60*d* common 40*d* spike	275	2.63	92	0.88
	50*d* and 60*d* spike	307	2.83	102	0.94
	5/16 in. spike	356	3.12	119	1.04
	3/8 in. spike	468	3.75	156	1.25
Group II	6*d* box	51	1.09	17	0.36
	8*d* box		1.24		0.41
	6*d* common	63	1.24	21	0.41
	6*d* threaded		1.32		0.44
	10*d* and 12*d* box	76	1.41	25	0.47
	8*d* common	78	1.44	26	0.48
	8*d* threaded		1.32		0.44
	16*d* box	82	1.49	27	0.50
	20*d* and 30*d* box		1.63		0.54
	10*d* and 12*d* common	94	1.63	31	0.54
	10*d* and 12*d* threaded		1.49		0.50
	40*d* box		1.78		0.59
	16*d* common	108	1.78	36	0.59
	16*d* threaded		1.63		0.54
	20*d* common		2.11		0.70
	20*d*, 30*d*, 40*d*, 50*d*, and 60*d* threaded	139	1.95	46	0.65
	10*d* and 12*d* spike		2.11		0.70
	30*d* common 16*d* spike	155	2.28	52	0.76
	40*d* common		2.48		0.83
	70*d*, 80*d*, and 90*d* threaded	176	2.28	59	0.76
	20*d* spike		2.48		0.83

(*continued*)

TABLE 6.36 (*Continued*)

Species Group[c]	Nail or Spike Size	Maximum Value (lb)	Required Penetration (in.)	Minimum Value (lb)	Minimum Penetration (in.)
Group II (cont.)	50*d* common	199	2.68	66	0.89
	30*d* spike				
	60*d* common	223	2.89	74	0.96
	40*d* spike				
	50*d* and 60*d* spike	248	3.11	83	1.04
	5/16 in. spike	288	3.43	96	1.14
	3/8 in. spike	379	4.13	126	1.38
Group III	6*d* box	42	1.29	14	0.43
	8*d* box		1.47		0.49
	6*d* common	51	1.47	17	0.49
	6*d* threaded		1.56		0.52
	10*d* and 12*d* box	62	1.66	21	0.55
	8*d* common	64	1.70	21	0.57
	8*d* threaded		1.56		0.52
	16*d* box	67	1.76	22	0.59
	20*d* and 30*d* box		1.92		0.64
	10*d* and 12*d* common	77	1.92	26	0.64
	10*d* and 12*d* threaded		1.76		0.59
	40*d* box		2.11		0.70
	16*d* common	88	2.11	29	0.70
	16*d* threaded		1.92		0.64
	20*d* common		2.50		0.83
	20*d*, 30*d*, 40*d*, 50*d*, and 60*d* threaded	114	2.30	38	0.77
	10*d* and 12*d* spike		2.50		0.83

	30*d* common	127	2.69	42	0.90
	16*d* spike				
	40*d* common		2.93		0.98
	70*d*, 80*d*, and 90*d* threaded	144	2.69	48	0.90
	20*d* spike		2.93		0.98
	50*d* common	163	3.17	54	1.06
	30*d* spike				
	60*d* common	182	3.42	61	1.14
	40*d* spike				
	50*d* and 60*d* spike	203	3.68	68	1.23
	5/16 in. spike	235	4.06	78	1.35
	3/8 in. spike	310	4.88	103	1.63
Group IV	6*d* box	34	1.39	11	0.46
	8*d* box		1.58		0.53
	6*d* common	41	1.58	14	0.53
	6*d* threaded		1.68		0.56
	10*d* and 12*d* box	49	1.79	16	0.60
	8*d* common	51	1.83	17	0.61
	8*d* threaded		1.68		0.56
	16*d* box	54	1.89	18	0.63
	20*d* and 30*d* box		2.07		0.69
	10*d* and 12*d* common	61	2.07	20	0.69
	10*d* and 12*d* threaded		1.89		0.63
	40*d* box		2.27		0.76
	16*d* common	70	2.27	23	0.76
	16*d* threaded		2.07		0.69
	20*d* common		2.69		0.90
	20*d*, 30*d*, 40*d*, 50*d*, and 60*d* threaded	91	2.48	30	0.83
	10*d* and 12*d* spike		2.69		0.90

(continued)

TABLE 6.36 (*Continued*)

Species Group[c]	Nail or Spike Size	Maximum Value (lb)	Required Penetration (in.)	Minimum Value (lb)	Minimum Penetration (in.)
Group IV (cont.)	30*d* common	102	2.90	34	0.97
	16*d* spike				
	40*d* common		3.15		1.05
	70*d*, 80*d*, and 90*d* threaded	115	2.90	38	0.97
	20*d* spike		3.15		1.05
	50*d* common	130	3.42	43	1.14
	30*d* spike				
	60*d* common	146	3.68	49	1.23
	40*d* spike				
	50*d* and 60*d* spike	163	3.96	54	1.32
	5/16 in. spike	188	4.37	63	1.46
	3/8 in. spike	248	5.25	83	1.75

[a] Based on *National Design Specification for Wood Construction* (3).

[b] For lateral load design values intermediate between the maximum required and the minimum penetration values, use straight-line interpolation.

[c] See Tables 6.1 and 6.2 for species in each group.

wind or earthquake loading. Diaphragm lateral load design values are also subject to the penetration requirements stated above and given in Table 6.36.

For nails or spikes driven in end grain and loaded laterally, design values are two-thirds of those determined from Table 6.36.

Special nails have been developed for use with strap-type joist and purlin hangers, tie straps on pipe-type hangers, and other metal ties and straps. Short nails of the same diameter given in Table 6.36 have been developed for use with plywood. Lateral load design values for such special nails used with a metal strap are given in Table 6.37.

TABLE 6.37

Lateral Load Design Values For Special Nails[a,b]

Size	Diameter	Length (in.)	Description	Lateral Load Design Value[c] (lb)
8*d*	11 gage	$1\frac{1}{4}$	Smooth shank	85
10*d*	9 gage	$1\frac{1}{2}$	Smooth shank	118
10*d*	9 gage	$2\frac{1}{8}$	Annular thread, stainless steel	118
16*d*	8 gage	$2\frac{1}{2}$	Smooth shank	135
16*d*	0.165 in.	$1\frac{3}{4}$	Annular thread, stainless steel	135
20*d*	0.192 in.	$1\frac{3}{4}$	Annular ring	174
20*d*	0.192 in.	$2\frac{1}{8}$	Annular ring	174
	0.250 in.	$2\frac{1}{2}$	Annular ring	257
	0.250 in.	3	Annular ring	257

[a]See paragraph above for definition of special nails.
[b]Used with a steel strap in group II species. See Tables 6.1 and 6.2 for species in this load group.
[c]Loads have been increased 25% for metal side plates.
[d]For more information on special nails and staples refer to (8).

STAPLES

Because staples and nails are similar in nature, the loads for staples may be determined in a manner similar to that for nails. The design value for one staple of a given diameter equals twice the value for a nail of equal diameter, provided that the staple leg spacing (or crown width) is adequate, and that the penetration of both legs of the staple into the member receiving the points is approximately two-thirds of their length. In general, nail penetration requirements and other provisions regarding seasoning of members, service conditions, and so on apply equally to staples.

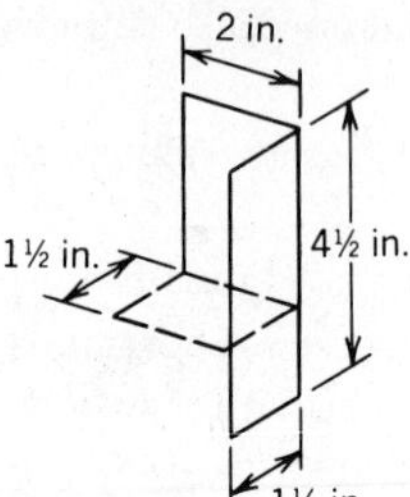

FIGURE 6.27 Typical framing anchor dimensions.

LIGHT METAL FRAMING DEVICES

Framing Anchors

Framing anchors are metal fittings used to provide a more positive connection between wood members than is obtained by toe-nailing. The several types of manufactured framing anchors are right-angle pieces formed from light gage (typically

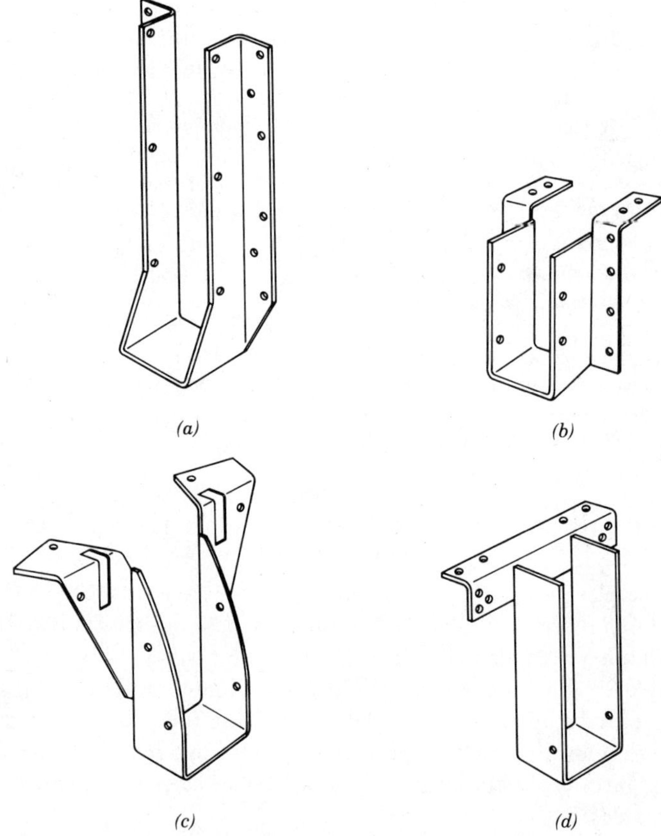

FIGURE 6.28 Typical joist and purlin hanger types. (*a*) Joist hanger; (*b*) joist hanger; (*c*) joist and purlin hanger; (*d*) joist and beam hanger.

18 gage) galvanized sheet steel (see Fig. 6.27). These may be bent to conform with use conditions. The outstanding legs may be of rectangular, triangular, or other shape, and they are predrilled or prepunched to receive special nails. Nails appropriate to the particular framing anchor should be used in all holes provided in the anchor to develop its full load-carrying capacity. Framing anchors are commonly used in pairs to avoid eccentricity.

Allowable load values for framing anchors are generally determined by test, but an estimate may be based on the lateral resistance of the nails used. Allowable loads for framing anchors are generally given for normal duration of load but may be adjusted for other durations of load. Before making adjustments for duration of load, the manufacturer's literature should be checked to ascertain the duration of load basis on which the recommended values were based and also to make sure that the stresses in the anchors or nails are not exceeded by increases for duration of load. The manufacturer's data for the particular type of framing anchor being used should be followed.

Joist and Purlin Hangers

Joist and purlin hangers are standard items fabricated by several manufacturers. The allowable loads shown in the manufacturer's literature are usually determined by tests. Several types of hangers are shown in Figure 6.28. The manufacturer's data for the particular type of hanger being used should be followed.

REFERENCES

1. American Wood-Preservers' Association, *Book of Standards*, Stevensville, MD, 1984.
2. National Lumber Grades Authority, *Standard Grading Rules for Canadian Lumber*, Ganges, British Columbia, 1980.
3. National Forest Products Association, *National Design Specification for Wood Construction*, Washington, DC, 1982.
4. American Institute of Timber Construction, *Typical Construction Details*, AITC 104, Englewood, CO, 1984.
5. United States Department of Agriculture, Forest Service, Forest Products Laboratory, *Timber-Connector Joints; Their Strength and Design*, Technical Bulletin No. 865, Madison, WI, 1944.
6. United States Department of Agriculture, Forest Service, Forest Products Laboratory, *Wood Handbook: Wood as an Engineering Material*, Agriculture Handbook No. 72, Madison, WI, 1974.
7. American Institute of Timber Construction, *Standard Specifications for Structural Glued Laminated Timber of Softwood Species*, AITC 117—Design, Englewood, CO, 1984.
8. National Evaluation Service, NER 272, *Pneumatic or Mechanically Driven Staples, Nails, P-Nails, and Allied Fasteners for Use in All Types of Building Construction*, issued to International Staple, Nail, and Tool Association, Chicago, IL, 1985
9. American Institute of Timber Construction, *Standard Specifications for Structural Glued Laminated Timber of Softwood Species*, AITC 117—Manufacturing, Englewood, CO, 1984.
10. Timber Engineering Company (TECO), *Design Manual for TECO Timber Connector Construction*, Washington, DC, 1973.
11. International Conference of Building Officials, *Uniform Building Code*, Whittier, CA, 1982.

Part III

REFERENCE

SECTION 7

REFERENCE INFORMATION

SECTION PROPERTIES

TABLE 7.1

Properties of Sawn Lumber and Timber

Nominal Size (in.) $b \times d$	Standard Dressed Size (in.) (S4S) $b \times d$	Area of Section A (in.^{2})	Moment of Inertia I (in.^{4})	Section Modulus S (in.^{3})	Weight[a] in Pounds per Linear Foot of Piece When Weight of Wood per Cubic Foot Equals: 25 lb	30 lb	35 lb	40 lb	45 lb	50 lb
1 × 3	$\frac{3}{4} \times 2\frac{1}{2}$	1.875	0.9766	0.7812	0.3	0.4	0.5	0.5	0.6	0.7
1 × 4	$\frac{3}{4} \times 3\frac{1}{2}$	2.625	2.680	1.531	0.5	0.5	0.6	0.7	0.8	0.9
1 × 6	$\frac{3}{4} \times 5\frac{1}{2}$	4.125	10.40	3.781	0.7	0.9	1.0	1.1	1.3	1.4
1 × 8	$\frac{3}{4} \times 7\frac{1}{4}$	5.438	23.82	6.570	0.9	1.1	1.3	1.5	1.7	1.9
1 × 10	$\frac{3}{4} \times 9\frac{1}{4}$	6.938	49.47	10.70	1.2	1.4	1.7	1.9	2.2	2.4
1 × 12	$\frac{3}{4} \times 11\frac{1}{4}$	8.438	88.99	15.82	1.5	1.8	2.1	2.3	2.6	2.9
2 × 3	$1\frac{1}{2} \times 2\frac{1}{2}$	3.750	1.953	1.563	0.7	0.8	0.9	1.0	1.2	1.3
2 × 4	$1\frac{1}{2} \times 3\frac{1}{2}$	5.250	5.359	3.063	0.9	1.1	1.3	1.5	1.6	1.8
2 × 5	$1\frac{1}{2} \times 4\frac{1}{2}$	6.750	11.39	5.063	1.2	1.4	1.6	1.9	2.1	2.3
2 × 6	$1\frac{1}{2} \times 5\frac{1}{2}$	8.250	20.80	7.563	1.4	1.7	2.0	2.3	2.6	2.9
2 × 8	$1\frac{1}{2} \times 7\frac{1}{4}$	10.88	47.64	13.14	1.9	2.3	2.6	3.0	3.4	3.8
2 × 10	$1\frac{1}{2} \times 9\frac{1}{4}$	13.88	98.93	21.39	2.4	2.9	3.4	3.9	4.3	4.8
2 × 12	$1\frac{1}{2} \times 11\frac{1}{4}$	16.88	178.0	31.64	2.9	3.5	4.1	4.7	5.3	5.9
2 × 14	$1\frac{1}{2} \times 12\frac{1}{4}$	19.88	290.8	43.89	3.5	4.1	4.8	5.5	6.2	6.9
3 × 1	$2\frac{1}{2} \times \frac{3}{4}$	1.875	0.08789	0.2344	0.3	0.4	0.5	0.5	0.6	0.7
3 × 2	$2\frac{1}{2} \times 1\frac{1}{2}$	3.750	0.7031	0.9375	0.7	0.8	0.9	1.0	1.2	1.3
3 × 4	$2\frac{1}{2} \times 3\frac{1}{2}$	8.750	8.932	5.104	1.5	1.8	2.1	2.4	2.7	3.0
3 × 5	$2\frac{1}{2} \times 4\frac{1}{2}$	11.25	18.98	8.438	2.0	2.3	2.7	3.1	3.5	3.9
3 × 6	$2\frac{1}{2} \times 5\frac{1}{2}$	13.75	34.66	12.60	2.4	2.9	3.3	3.8	4.3	4.8
3 × 8	$2\frac{1}{2} \times 7\frac{1}{4}$	18.12	79.39	21.90	3.1	3.8	4.4	5.0	5.7	6.3

(continued)

TABLE 7.1 (*Continued*)

Nominal Size (in.) b d	Standard Dressed Size (in.) (S4S) b d	Area of Section A (in.^{2})	Moment of Inertia I (in.^{4})	Section Modulus S (in.^{3})	Weight[a] in Pounds per Linear Foot of Piece When Weight of Wood per Cubic Foot Equals: 25 lb	30 lb	35 lb	40 lb	45 lb	50 lb
3 × 10	$2\frac{1}{2} \times 9\frac{1}{4}$	23.12	164.9	35.65	4.0	4.8	5.6	6.4	7.2	8.0
3 × 12	$2\frac{1}{2} \times 11\frac{1}{4}$	28.12	296.6	52.73	4.9	5.9	6.8	7.8	8.7	9.8
3 × 14	$2\frac{1}{2} \times 13\frac{1}{4}$	33.12	484.6	73.15	5.8	6.9	8.1	9.2	10.4	11.5
3 × 16	$2\frac{1}{2} \times 15\frac{1}{4}$	38.12	738.9	96.90	6.6	7.9	9.3	10.6	11.9	13.2
4 × 1	$3\frac{1}{2} \times \frac{3}{4}$	2.625	0.1230	0.3281	0.5	0.5	0.6	0.7	0.8	0.9
4 × 2	$3\frac{1}{2} \times 1\frac{1}{2}$	5.250	0.9844	1.313	0.9	1.1	1.3	1.5	1.6	1.8
4 × 3	$3\frac{1}{2} \times 2\frac{1}{2}$	8.750	4.557	3.646	1.5	1.8	2.1	2.4	2.7	3.0
4 × 4	$3\frac{1}{2} \times 3\frac{1}{2}$	12.25	12.50	7.146	2.1	2.6	3.0	3.4	3.8	4.3
4 × 5	$3\frac{1}{2} \times 4\frac{1}{2}$	15.75	26.53	11.81	2.7	3.3	3.8	4.4	4.9	5.5
4 × 6	$3\frac{1}{2} \times 5\frac{1}{2}$	19.25	48.53	17.65	3.3	4.0	4.7	5.3	6.0	6.7
4 × 8	$3\frac{1}{2} \times 7\frac{1}{4}$	25.38	111.1	30.66	4.4	5.3	6.2	7.0	7.9	8.8
4 × 10	$3\frac{1}{2} \times 9\frac{1}{4}$	32.38	230.8	49.91	5.6	6.7	7.9	8.9	10.1	12.2
4 × 12	$3\frac{1}{2} \times 11\frac{1}{4}$	39.38	415.3	73.83	6.8	8.2	9.6	10.9	12.3	13.7
4 × 14	$3\frac{1}{2} \times 13\frac{1}{4}$	46.38	678.5	102.4	8.0	9.7	11.3	12.9	14.5	16.1
4 × 16	$3\frac{1}{2} \times 15\frac{1}{4}$	53.38	1034	135.7	9.3	11.1	13.0	14.8	16.7	18.6
5 × 2	$4\frac{1}{2} \times 1\frac{1}{2}$	6.750	1.266	1.688	1.2	1.4	1.6	1.9	2.1	2.3
5 × 3	$4\frac{1}{4} \times 2\frac{1}{2}$	11.25	5.859	4.688	2.0	2.3	2.7	3.1	3.5	3.9
5 × 4	$4\frac{1}{2} \times 3\frac{1}{2}$	15.75	16.08	9.188	2.7	3.3	3.8	4.4	4.9	5.5
5 × 5	$4\frac{1}{2} \times 4\frac{1}{2}$	20.25	34.17	15.19	3.5	4.2	4.9	5.7	6.3	7.0
6 × 1	$5\frac{1}{2} \times \frac{3}{4}$	4.125	0.1933	0.5156	0.7	0.9	1.0	1.1	1.3	1.4
6 × 2	$5\frac{1}{2} \times 1\frac{1}{2}$	8.250	1.547	2.063	1.4	1.7	2.0	2.3	2.6	2.9

6 × 3	$5\frac{1}{2} \times 2\frac{1}{2}$	13.75	7.161	5.729	2.4	2.9	3.3	3.8	4.3	4.8
6 × 4	$5\frac{1}{2} \times 3\frac{1}{2}$	19.25	19.65	11.23	3.3	4.0	4.7	5.3	6.0	6.7
6 × 6	$5\frac{1}{2} \times 5\frac{1}{2}$	30.25	76.26	27.73	5.3	6.3	7.4	8.4	9.5	10.5
6 × 8	$5\frac{1}{2} \times 7\frac{1}{2}$	41.25	193.4	51.56	7.2	8.6	10.0	11.5	12.9	14.3
6 × 10	$5\frac{1}{2} \times 9\frac{1}{2}$	52.25	393.0	82.73	9.1	10.9	12.7	14.5	16.3	18.1
6 × 12	$5\frac{1}{2} \times 11\frac{1}{2}$	63.25	697.1	121.2	11.1	13.2	15.4	17.6	19.8	22.0
6 × 14	$5\frac{1}{2} \times 13\frac{1}{2}$	74.25	1128	167.1	12.9	15.5	18.0	20.6	23.2	25.8
6 × 16	$5\frac{1}{2} \times 15\frac{1}{2}$	85.25	1707	220.2	14.8	17.8	20.7	23.7	26.6	29.6
6 × 18	$5\frac{1}{2} \times 17\frac{1}{2}$	96.25	2456	280.7	16.7	20.1	23.4	26.7	30.1	33.4
6 × 20	$5\frac{1}{2} \times 19\frac{1}{2}$	107.2	3398	348.6	18.6	22.3	26.1	29.8	33.5	37.2
6 × 22	$5\frac{1}{2} \times 21\frac{1}{2}$	118.2	4555	423.7	20.5	24.6	28.7	32.8	37.0	41.1
6 × 24	$5\frac{1}{2} \times 23\frac{1}{2}$	129.2	5948	506.2	22.4	26.9	31.4	35.9	40.4	44.9
8 × 1	$7\frac{1}{4} \times \frac{3}{4}$	5.438	0.2549	0.6797	0.9	1.1	1.3	1.5	1.7	1.9
8 × 2	$7\frac{1}{4} \times 1\frac{1}{2}$	10.88	2.039	2.719	1.9	2.3	2.6	3.0	3.4	3.8
8 × 3	$7\frac{1}{4} \times 2\frac{1}{2}$	18.12	9.440	7.552	3.1	3.8	4.4	5.0	5.7	6.3
8 × 4	$7\frac{1}{4} \times 3\frac{1}{2}$	25.38	25.90	14.80	4.4	5.3	6.2	7.0	7.9	8.8
8 × 6	$7\frac{1}{2} \times 5\frac{1}{2}$	41.25	104.0	37.81	7.2	8.6	10.0	11.5	12.9	14.3
8 × 8	$7\frac{1}{2} \times 7\frac{1}{2}$	56.25	263.7	70.31	9.8	11.7	13.7	15.6	17.6	19.5
8 × 10	$7\frac{1}{2} \times 9\frac{1}{2}$	71.25	535.9	112.8	12.4	14.8	17.3	19.8	22.3	24.7
8 × 12	$7\frac{1}{2} \times 11\frac{1}{2}$	86.25	950.5	165.3	15.0	18.0	21.0	24.0	27.0	29.9
8 × 14	$7\frac{1}{2} \times 13\frac{1}{2}$	101.2	1538	227.8	17.6	21.1	24.6	28.1	31.6	35.2
8 × 16	$7\frac{1}{2} \times 15\frac{1}{2}$	116.2	2327	300.3	20.2	24.2	28.3	32.3	36.3	40.4
8 × 18	$7\frac{1}{2} \times 17\frac{1}{2}$	131.2	3350	382.8	22.8	27.3	31.9	36.5	41.0	45.6
8 × 20	$7\frac{1}{2} \times 19\frac{1}{2}$	146.2	4634	475.3	25.4	30.5	35.5	40.6	45.7	50.8
8 × 22	$7\frac{1}{2} \times 21\frac{1}{2}$	161.2	6211	577.8	28.0	33.6	39.2	44.8	50.4	56.0
8 × 24	$7\frac{1}{2} \times 23\frac{1}{2}$	176.2	8111	690.3	30.6	36.7	42.8	49.0	55.1	61.2
10 × 1	$9\frac{1}{4} \times \frac{3}{4}$	6.938	0.3252	0.8672	1.2	1.4	1.7	1.9	2.2	2.4
10 × 2	$9\frac{1}{4} \times 1\frac{1}{2}$	13.88	2.602	3.469	2.4	2.9	3.4	3.9	4.3	4.8
10 × 3	$9\frac{1}{4} \times 2\frac{1}{2}$	23.12	12.04	9.635	4.0	4.8	5.6	6.4	7.2	8.0
10 × 4	$9\frac{1}{4} \times 3\frac{1}{2}$	32.38	33.05	18.88	5.6	6.7	7.9	9.0	10.1	11.2

(continued)

TABLE 7.1 (*Continued*)

Nominal Size (in.) $b \times d$	Standard Dressed Size (in.) (S4S) $b \times d$	Area of Section A (in.^{2})	Moment of Inertia I (in.^{4})	Section Modulus S (in.^{3})	Weight[a] in Pounds per Linear Foot of Piece When Weight of Wood per Cubic Foot Equals: 25 lb	30 lb	35 lb	40 lb	45 lb	50 lb
10 × 6	$9\frac{1}{2} \times 5\frac{1}{2}$	52.25	131.7	47.90	9.1	10.9	12.7	14.5	16.3	18.1
10 × 8	$9\frac{1}{2} \times 7\frac{1}{2}$	71.25	334.0	89.06	12.4	14.8	17.3	19.8	22.3	24.7
10 × 10	$9\frac{1}{2} \times 9\frac{1}{2}$	90.25	678.8	142.9	15.7	18.8	21.9	25.1	28.2	31.3
10 × 12	$9\frac{1}{2} \times 11\frac{1}{2}$	109.2	1204	209.4	19.0	22.8	26.6	30.3	34.1	37.9
10 × 14	$9\frac{1}{2} \times 13\frac{1}{2}$	128.2	1948	288.6	22.3	26.7	31.2	35.6	40.1	44.5
10 × 16	$9\frac{1}{2} \times 15\frac{1}{2}$	147.2	2948	380.4	25.6	30.7	35.8	40.9	46.0	51.1
10 × 18	$9\frac{1}{2} \times 17\frac{1}{2}$	166.2	4243	484.9	28.9	34.6	40.4	46.2	52.0	57.7
10 × 20	$9\frac{1}{2} \times 19\frac{1}{2}$	185.2	5870	602.1	32.2	38.6	45.0	51.5	57.9	64.3
10 × 22	$9\frac{1}{2} \times 21\frac{1}{2}$	204.2	7868	731.9	35.5	42.6	49.6	56.7	63.8	70.9
10 × 24	$9\frac{1}{2} \times 23\frac{1}{2}$	223.2	10,270	874.4	38.8	46.5	54.3	62.0	69.8	77.5
12 × 1	$11\frac{1}{4} \times \frac{3}{4}$	8.438	0.3955	1.055	1.5	1.8	2.1	2.3	2.6	2.9
12 × 2	$11\frac{1}{4} \times 1\frac{1}{2}$	16.88	3.164	4.219	2.9	3.5	4.1	4.7	5.3	5.9
12 × 3	$11\frac{1}{4} \times 2\frac{1}{2}$	28.12	14.65	11.72	4.9	5.9	6.8	7.8	8.8	9.8
12 × 4	$11\frac{1}{4} \times 3\frac{1}{2}$	39.38	40.20	22.97	6.8	8.2	9.6	10.9	12.3	13.7
12 × 6	$11\frac{1}{2} \times 5\frac{1}{2}$	63.25	159.4	57.98	11.0	13.2	15.4	17.6	19.8	22.0
12 × 8	$11\frac{1}{2} \times 7\frac{1}{2}$	86.25	404.3	107.8	15.0	18.0	21.0	24.0	27.0	29.9
12 × 10	$11\frac{1}{2} \times 9\frac{1}{2}$	109.2	821.7	173.0	19.0	22.8	26.6	30.3	34.1	37.9
12 × 12	$11\frac{1}{2} \times 11\frac{1}{2}$	132.2	1458	253.5	23.0	27.6	32.1	36.7	41.3	45.9
12 × 14	$11\frac{1}{2} \times 13\frac{1}{2}$	155.2	2358	349.3	27.0	32.3	37.7	43.1	48.5	53.9
12 × 16	$11\frac{1}{2} \times 15\frac{1}{2}$	178.2	3569	460.5	30.9	37.1	43.3	49.5	55.7	61.9
12 × 18	$11\frac{1}{2} \times 17\frac{1}{2}$	201.2	5136	587.0	34.9	41.9	48.9	55.9	62.9	69.9

12 × 20	11½ × 19½	224.2	7106	728.8	38.9	46.7	54.5	62.3	70.1	77.9
12 × 22	11½ × 21½	247.2	9524	886.0	42.9	51.5	60.1	68.7	77.3	85.9
12 × 24	11½ × 23½	270.2	12,440	1058	46.9	56.3	65.7	75.1	84.5	93.8
14 × 2	13¼ × 1½	19.88	3.727	4.969	3.5	4.1	4.8	5.5	6.2	6.9
14 × 3	13¼ × 2½	33.12	17.25	13.80	5.8	6.9	8.1	9.2	10.4	11.5
14 × 4	13¼ × 3½	46.38	47.34	27.05	8.0	9.7	11.3	12.9	14.5	16.1
14 × 6	13½ × 5½	74.25	187.2	68.06	12.9	15.5	18.0	20.6	23.2	25.8
14 × 8	13½ × 7½	101.2	474.6	126.6	17.6	21.1	24.6	28.1	31.6	35.2
14 × 10	13½ × 9½	128.2	964.5	203.1	22.3	26.7	31.2	35.6	40.1	44.5
14 × 12	13½ × 11½	155.2	1711	297.6	27.0	32.3	37.7	43.1	48.5	53.9
14 × 14	13½ × 13½	182.2	2768	410.1	31.6	38.0	44.3	50.6	57.0	63.3
14 × 16	13½ × 15½	209.2	4189	540.6	36.3	43.6	50.9	58.1	65.4	72.7
14 × 18	13½ × 17½	236.2	6029	689.1	41.0	49.2	57.4	65.6	73.2	82.0
14 × 20	13½ × 19½	263.2	8342	855.6	45.7	54.8	64.0	73.1	82.2	91.4
14 × 22	13½ × 21½	290.2	11,180	1040	50.4	60.5	70.5	80.6	90.7	100.8
14 × 24	13½ × 23½	317.2	14,600	1243	55.1	66.1	77.1	88.1	99.1	110.2
16 × 3	15¼ × 2½	38.12	19.86	15.88	6.6	7.9	9.3	10.6	11.9	13.2
16 × 4	15¼ × 3½	53.38	54.49	31.14	9.3	11.1	13.0	14.8	16.7	18.5
16 × 6	15½ × 5½	85.25	214.9	78.15	14.8	17.8	20.7	23.7	26.6	29.6
16 × 8	15½ × 7½	116.2	544.9	145.3	20.2	24.2	28.3	32.3	36.3	40.4
16 × 10	15½ × 9½	147.2	1107	233.1	25.6	30.7	35.8	40.9	46.0	51.1
16 × 12	15½ × 11½	178.2	1964	341.6	30.9	37.1	43.3	49.5	55.7	61.9
16 × 14	15½ × 13½	209.2	3178	470.8	36.3	43.6	50.9	58.1	65.4	72.7
16 × 16	15½ × 15½	240.2	4810	620.6	41.7	50.1	58.4	66.7	75.1	83.4
16 × 18	15½ × 17½	271.2	6923	791.1	47.1	56.5	65.9	75.3	84.8	94.2
16 × 20	15½ × 19½	302.2	9578	982.3	52.5	63.0	73.4	84.0	94.5	104.9
16 × 22	15½ × 21½	333.2	12,840	1194	57.9	69.4	81.0	92.6	104.1	115.7
16 × 24	15½ × 23½	364.2	16,760	1427	63.2	75.9	88.5	101.2	113.8	126.5
18 × 6	17½ × 5½	96.25	242.6	88.23	16.7	20.1	23.4	26.7	30.1	33.4
18 × 8	17½ × 7½	131.2	615.2	164.1	22.8	27.3	31.9	36.5	41.0	45.6

(continued)

TABLE 7.1 (*Continued*)

Nominal Size (in.) $b \times d$	Standard Dressed Size (in.) (S4S) $b \times d$	Area of Section A (in.^{2})	Moment of Inertia I (in.^{4})	Section Modulus S (in.^{3})	Weight[a] in Pounds per Linear Foot of Piece When Weight of Wood per Cubic Foot Equals: 25 lb	30 lb	35 lb	40 lb	45 lb	50 lb
18 × 10	$17\frac{1}{2} \times 9\frac{1}{2}$	166.2	1250	263.2	28.9	34.6	40.4	46.2	52.0	57.7
18 × 12	$17\frac{1}{2} \times 11\frac{1}{2}$	201.2	2218	385.7	34.9	41.9	48.9	55.9	62.9	69.9
18 × 14	$17\frac{1}{2} \times 13\frac{1}{2}$	236.2	3588	531.6	41.0	49.2	57.4	65.6	73.8	82.0
18 × 16	$17\frac{1}{2} \times 15\frac{1}{2}$	271.2	5431	700.7	47.1	56.5	65.9	75.3	84.8	84.2
18 × 18	$17\frac{1}{2} \times 17\frac{1}{2}$	306.2	7816	893.2	53.2	63.8	74.4	85.1	95.7	106.3
18 × 20	$17\frac{1}{2} \times 19\frac{1}{2}$	341.2	10,813	1109	59.2	71.1	82.9	94.8	106.6	118.5
18 × 22	$17\frac{1}{2} \times 21\frac{1}{2}$	376.2	14,493	1348	65.3	78.4	91.4	104.5	117.6	130.6
18 × 24	$17\frac{1}{2} \times 23\frac{1}{2}$	411.2	18,926	1611	71.4	85.7	100.0	114.2	128.5	142.8
20 × 6	$19\frac{1}{2} \times 5\frac{1}{2}$	107.2	270.4	98.31	18.6	22.3	26.1	29.8	33.5	37.2
20 × 8	$19\frac{1}{2} \times 7\frac{1}{2}$	146.2	685.5	182.8	25.4	30.5	35.5	40.6	45.7	50.8
20 × 10	$19\frac{1}{2} \times 9\frac{1}{2}$	185.2	1393	293.3	32.2	38.6	45.0	51.5	57.9	64.3
20 × 12	$19\frac{1}{2} \times 11\frac{1}{2}$	224.2	2471	429.8	38.9	46.7	54.5	62.3	70.1	77.9
20 × 14	$19\frac{1}{2} \times 13\frac{1}{2}$	263.2	3998	592.3	45.7	54.8	64.0	73.1	82.3	91.4
20 × 16	$19\frac{1}{2} \times 15\frac{1}{2}$	302.2	6051	780.8	52.5	63.0	73.5	84.0	94.5	104.9
20 × 18	$19\frac{1}{2} \times 17\frac{1}{2}$	341.2	8709	995.3	59.2	71.1	82.9	94.8	106.6	118.5
20 × 20	$19\frac{1}{2} \times 19\frac{1}{2}$	380.2	12,050	1236	66.0	79.2	92.4	105.6	118.8	132.0
20 × 22	$19\frac{1}{2} \times 21\frac{1}{2}$	419.2	16,150	1502	72.8	87.3	101.9	116.5	131.0	145.6
20 × 24	$19\frac{1}{2} \times 23\frac{1}{2}$	458.2	21,090	1795	79.6	95.5	111.4	127.3	143.2	159.1
22 × 6	$21\frac{1}{2} \times 5\frac{1}{2}$	118.2	298.1	108.4	20.5	24.6	28.7	32.8	37.0	41.1
22 × 8	$21\frac{1}{2} \times 7\frac{1}{2}$	161.2	755.9	201.6	28.0	33.6	39.2	44.8	50.4	56.0
22 × 10	$21\frac{1}{2} \times 9\frac{1}{2}$	204.2	1536	323.4	35.5	42.6	49.6	56.7	63.8	70.9

22 × 12	21½ × 11½	247.2	2725	473.9	42.9	51.5	60.1	68.7	77.3	85.9
22 × 14	21½ × 13½	290.2	4408	653.1	50.4	60.5	70.5	80.6	90.7	100.8
22 × 16	21½ × 15½	333.2	6672	860.9	57.9	69.4	81.0	92.6	104.1	115.7
22 × 18	21½ × 17½	376.2	9602	1097	65.3	78.4	91.5	104.5	117.6	130.6
22 × 20	21½ × 19½	419.2	13,280	1363	72.8	87.3	101.9	116.5	131.0	145.6
22 × 22	21½ × 21½	462.2	17,810	1656	80.3	96.3	112.4	128.4	144.5	160.5
22 × 24	21½ × 23½	505.2	23,250	1979	87.7	105.3	122.8	140.3	157.9	175.4
24 × 6	23½ × 5½	129.2	325.8	118.5	22.4	26.9	31.4	35.9	40.4	44.9
24 × 8	23½ × 7½	176.2	826.2	220.3	30.6	36.7	42.8	49.0	55.1	61.2
24 × 10	23½ × 9½	223.2	1679	353.5	38.8	46.5	54.3	62.0	69.8	77.5
24 × 12	23½ × 11½	270.2	2978	518.0	46.9	56.3	65.7	75.1	84.5	93.8
24 × 14	23½ × 13½	317.2	4818	713.8	55.1	66.1	77.1	88.1	99.1	110.2
24 × 16	23½ × 15½	364.2	7293	941.0	63.2	75.9	88.5	101.2	113.8	126.5
24 × 18	23½ × 17½	411.2	10,500	1199	71.4	85.7	100.0	114.2	128.5	142.8
24 × 20	23½ × 19½	458.2	14,520	1489	79.6	95.5	111.4	127.3	143.2	159.1
24 × 22	23½ × 21½	505.2	19,460	1810	87.7	105.3	122.8	140.3	157.9	175.4
24 × 24	23½ × 23½	552.2	25,420	2163	95.9	115.1	134.2	153.4	172.6	191.8

[a]To obtain the weight of a species of wood at a given moisture content use the following formula:

$$W = \frac{G}{1 + G(0.009)MC}\left[1 + \frac{MC}{100}\right](62.4)$$

where W = weight of wood (lb/ft^3)

G = specific gravity of species obtained from Table 6.1

MC = moisture content of the lumber (%).

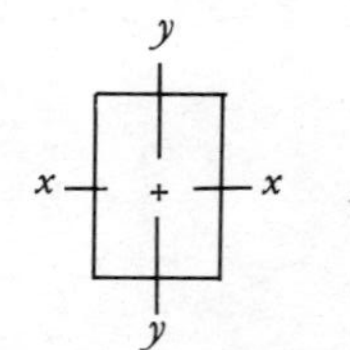

TABLE 7.2

Section Properties of Structural Glued Laminated Timber[a]

Number of $1\frac{1}{2}$ in. Laminations	Depth d (in.)	Size Factor C_F	Area A (in.^{2})	x–x Axis S_x (in.^{3})	x–x Axis S_n[b] (in.^{3})	x–x Axis I_x (in.^{4})	y–y Axis S_y (in.^{3})	y–y Axis I_y (in.^{4})	Volume per Ft (ft^3)	Weight (lb/ft) DFL[c] (35 pcf)	Weight (lb/ft) SP[c] (36 pcf)	Weight (lb/ft) HF/CR[c] (27 pcf)
$2\frac{1}{8}$ IN. WIDTH												
2	3	1.00	6.375	3.188	3.188	4.781	2.258	2.399	0.04	1.5	1.6	1.2
3	4.5	1.00	9.562	7.172	7.172	16.137	3.387	3.598	0.07	2.3	2.4	1.8
4	6	1.00	12.75	12.75	12.75	38.25	4.516	4.798	0.09	3.1	3.2	2.4
5	7.5	1.00	15.94	19.92	19.92	74.71	5.644	5.997	0.11	3.9	4.0	3.0
6	9	1.00	19.12	28.69	28.69	129.1	6.773	7.197	0.13	4.6	4.8	3.6
7	10.5	1.00	22.31	39.05	39.05	205.0	7.902	8.396	0.15	5.4	5.6	4.2
8	12	1.00	25.50	51.00	51.00	306.0	9.031	9.596	0.18	6.2	6.4	4.8
9	13.5	0.99	28.69	64.55	63.71	435.7	10.16	10.80	0.20	7.0	7.2	5.4
10	15	0.98	31.88	79.69	77.74	597.7	11.29	11.99	0.22	7.7	8.0	6.0
$3\frac{1}{8}$ IN. WIDTH[d]												
2	3	1.00	9.375	4.688	4.688	7.031	4.883	7.629	0.07	2.3	2.3	1.8
3	4.5	1.00	14.06	10.55	10.55	23.73	7.324	11.44	0.10	3.4	3.5	2.6
4	6	1.00	18.75	18.75	18.75	56.25	9.766	15.26	0.13	4.6	4.7	3.5
5	7.5	1.00	23.44	29.30	29.30	109.9	12.21	19.07	0.16	5.7	5.9	4.4
6	9	1.00	28.12	42.19	42.19	189.8	14.65	22.89	0.20	6.8	7.0	5.3

7	10.5	1.00	32.81	57.42	57.42	301.5	17.09	26.70	0.23	8.0	8.2	6.2
8	12	1.00	37.50	75.00	75.00	450.0	19.53	30.52	0.26	9.1	9.4	7.0
9	13.5	0.99	42.19	94.92	93.69	640.7	21.97	34.33	0.29	10.3	10.5	7.9
10	15	0.98	46.88	117.2	114.3	878.9	24.41	38.15	0.33	11.4	11.7	8.8
11	16.5	0.97	51.56	141.8	136.9	1170	26.86	41.96	0.36	12.5	12.9	9.7
12	18	0.96	56.25	168.8	161.3	1519	29.30	45.78	0.39	13.7	14.1	10.5
13	19.5	0.95	60.94	198.0	187.6	1931	31.74	49.59	0.42	14.8	15.2	11.4
14	21	0.94	65.62	229.7	215.8	2412	34.18	53.41	0.46	16.0	16.4	12.3
15	22.5	0.93	70.31	263.7	245.9	2966	36.62	57.22	0.49	17.1	17.6	13.2
16	24	0.93	75.00	300.0	277.8	3600	39.06	61.04	0.52	18.2	18.7	14.1
17	25.5	0.92	79.69	338.7	311.5	4318	41.50	64.85	0.55	19.4	19.9	14.9
18	27	0.91	84.38	379.7	347.0	5126	43.95	68.66	0.59	20.5	21.1	15.8
19	28.5	0.91	89.06	423.0	384.3	6028	46.39	72.48	0.62	21.6	22.3	16.7
20	30	0.90	93.75	468.8	423.4	7031	48.83	76.29	0.65	22.8	23.4	17.6
$5\frac{1}{8}$ IN. WIDTH[d]												
2	3	1.00	15.38	7.688	7.688	11.53	13.13	33.65	0.11	3.7	3.8	2.9
3	4.5	1.00	23.06	17.30	17.30	38.92	19.70	50.48	0.16	5.6	5.8	4.3
4	6	1.00	30.75	30.75	30.75	92.25	26.27	67.31	0.21	7.5	7.7	5.8
5	7.5	1.00	38.44	48.05	48.05	180.2	32.83	84.13	0.27	9.3	9.6	7.2
6	9	1.00	46.12	69.19	69.19	311.3	39.40	101.0	0.32	11.2	11.5	8.6
7	10.5	1.00	53.81	94.17	94.17	494.4	45.96	117.8	0.37	13.1	13.5	10.1
8	12	1.00	61.50	123.0	123.0	738	52.53	134.6	0.43	14.9	15.4	11.5
9	13.5	0.99	69.19	155.7	153.6	1051	59.10	151.4	0.48	16.8	17.3	13.0
10	15	0.98	76.88	192.2	187.5	1441	65.66	168.3	0.53	18.7	19.2	14.4
11	16.5	0.97	84.56	232.5	224.5	1919	72.23	185.1	0.59	20.6	21.1	15.9
12	18	0.96	92.25	276.8	264.6	2491	78.80	201.9	0.64	22.4	23.1	17.3
13	19.5	0.95	99.98	324.8	307.7	3167	85.36	218.7	0.69	24.3	25.0	18.7
14	21	0.94	107.6	376.7	354.0	3955	91.93	235.6	0.75	26.2	26.9	20.2
15	22.5	0.93	115.3	432.4	403.2	4865	98.50	252.4	0.80	28.0	28.8	21.6
16	24	0.93	123.0	492.0	455.5	5904	105.1	269.2	0.85	29.9	30.7	23.1

(continued)

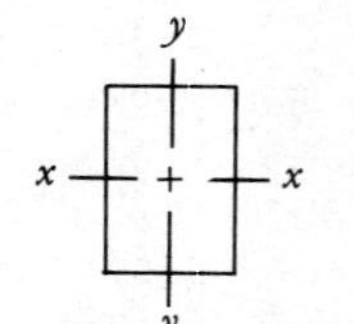

TABLE 7.2 (*Continued*)

Number of $1\frac{1}{2}$ in. Laminations	Depth d (in.)	Size Factor C_F	Area A (in.^{2})	x–x Axis: S_x (in.^{3})	x–x Axis: S_n[b] (in.^{3})	x–x Axis: I_x (in.^{4})	y–y Axis: S_y (in.^{3})	y–y Axis: I_y (in.^{4})	Volume per Ft (ft^3)	Weight (lb/ft): DFL[c] (35 pcf)	Weight (lb/ft): SP[c] (36 pcf)	Weight (lb/ft): HF/CR[c] (27 pcf)
$5\frac{1}{8}$ IN. WIDTH[d] (cont.)												
17	25.5	0.92	130.7	555.4	510.8	7082	111.6	286.0	0.91	31.8	32.7	24.5
18	27	0.91	138.4	622.7	569.0	8406	118.2	302.9	0.96	33.6	34.6	25.9
19	28.5	0.91	146.1	693.8	630.2	9887	124.8	319.7	1.01	35.5	36.5	27.4
20	30	0.90	153.8	768.8	694.3	11,530	131.3	336.5	1.07	37.4	38.4	28.8
21	31.5	0.90	161.4	847.5	761.4	13,350	137.9	353.4	1.12	39.2	40.4	30.3
22	33	0.89	169.1	930.2	831.3	15,350	144.5	370.2	1.17	41.1	42.3	31.7
23	34.5	0.89	176.8	1017	904.1	17,540	151.0	387.0	1.23	43.0	44.2	33.2
24	36	0.89	184.5	1107	979.8	19,930	157.6	403.8	1.28	44.8	46.1	34.6
$6\frac{3}{4}$ IN. WIDTH												
2	3	1.00	20.25	10.12	10.12	15.19	22.78	96.89	0.14	4.9	5.1	3.8
3	4.5	1.00	30.38	22.78	22.78	51.26	34.17	115.3	0.21	7.4	7.6	5.7
4	6	1.00	40.50	40.50	40.50	121.5	45.56	153.8	0.28	9.8	10.1	7.6
5	7.5	1.00	50.62	63.28	63.28	237.3	56.95	192.2	0.35	12.3	12.7	9.5
6	9	1.00	60.75	91.12	91.12	410.1	68.34	230.7	0.42	14.8	15.2	11.4
7	10.5	1.00	70.88	124.0	124.0	651.2	79.73	269.1	0.49	17.2	17.7	13.3
8	12	1.00	81.00	162.0	162.0	972.0	91.12	307.5	0.56	19.7	20.3	15.2
9	13.5	0.99	91.12	205.0	202.4	1384	102.5	346.0	0.63	22.1	22.8	17.1
10	15	0.98	101.3	253.1	246.9	1898	113.9	384.4	0.70	24.6	25.3	19.0

11	16.5	0.97	111.4	306.3	295.6	2527	125.3	422.9	0.77	27.1	27.8	20.9
12	18	0.96	121.5	364.5	348.4	3280	136.7	461.3	0.84	29.5	30.4	22.8
13	19.5	0.95	131.6	427.8	405.3	4171	148.1	499.8	0.91	32.0	32.9	24.7
14	21	0.94	141.8	496.1	466.2	5209	159.5	538.2	0.98	34.5	35.4	26.6
15	22.5	0.93	151.9	569.5	531.1	6407	170.9	576.7	1.05	36.9	38.0	28.5
16	24	0.93	162.0	648.0	600.0	7776	182.3	615.1	1.13	39.4	40.5	30.4
17	25.5	0.92	172.1	731.5	672.8	9327	193.6	653.5	1.20	41.8	43.0	32.3
18	27	0.91	182.3	820.1	749.5	11,070	205.0	692.0	1.27	44.3	45.6	34.2
19	28.5	0.91	192.4	913.8	830.0	13,020	216.4	730.4	1.34	46.8	48.1	36.1
20	30	0.90	202.5	1013	914.5	15,190	227.8	768.9	1.41	49.2	50.6	38.0
21	31.5	0.90	212.6	1117	1003	17,580	239.2	807.3	1.48	51.7	53.2	39.9
22	33	0.89	222.8	1225	1095	20,220	250.6	845.8	1.55	54.1	55.7	41.8
23	34.5	0.89	232.9	1339	1190	23,100	262.0	884.2	1.62	56.6	58.2	43.7
24	36	0.89	243.0	1458	1290	26,240	273.4	922.6	1.69	59.1	60.8	45.6
25	37.5	0.88	253.1	1582	1394	29,660	284.8	961.1	1.76	61.5	63.3	47.5
26	39	0.88	263.3	1711	1501	33,370	296.2	999.5	1.83	64.0	65.8	49.4
27	40.5	0.87	273.4	1845	1612	37,370	307.5	1038.0	1.90	66.4	68.3	51.3
28	42	0.87	283.5	1984	1727	41,670	318.9	1076.4	1.97	68.9	70.9	53.2
29	43.5	0.87	293.6	2129	1845	46,300	330.3	1114.9	2.04	71.4	73.4	55.1
30	45	0.86	303.8	2278	1967	51,260	341.7	1153.3	2.11	73.8	75.9	57.0
31	46.5	0.86	313.9	2432	2093	56,560	353.1	1191.7	2.18	76.3	78.5	58.9
32	48	0.86	324.0	2592	2222	62,210	364.5	1230.2	2.25	78.8	81.0	60.8
$8\frac{3}{4}$ in. Width												
2	3	1.00	26.25	13.12	13.12	19.69	38.28	167.5	0.18	6.4	6.6	4.9
3	4.5	1.00	39.38	29.53	29.53	66.45	57.42	251.2	0.27	9.6	9.8	7.4
4	6	1.00	52.50	52.50	52.50	157.5	76.56	335.0	0.36	12.8	13.1	9.8
5	7.5	1.00	65.62	82.03	82.03	307.6	95.70	418.7	0.46	16.0	16.4	12.3
6	9	1.00	78.75	118.1	118.1	531.6	114.8	502.4	0.55	19.1	19.7	14.8
7	10.5	1.00	91.88	160 8	160.8	844.1	134.0	586.2	0.64	22.3	23.0	17.2
8	12	1.00	105.0	210.0	210.0	1200	153.1	669.9	0.78	25.5	26.2	19.7

(continued)

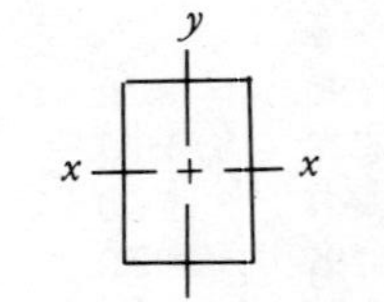

TABLE 7.2 (*Continued*)

Number of $1\frac{1}{2}$ in. Laminations	Depth d (in.)	Size Factor C_F	Area A (in.^{2})	x–x Axis: S_x (in.^{3})	x–x Axis: S_n[b] (in.^{3})	x–x Axis: I_x (in.^{4})	y–y Axis: S_y (in.^{3})	y–y Axis: I_y (in.^{4})	Volume per Ft (ft^3)	Weight (lb/ft): DFL[c] (35 pcf)	Weight (lb/ft): SP[c] (36 pcf)	Weight (lb/ft): HF/CR[c] (27 pcf)
$8\frac{3}{4}$ IN. WIDTH (cont.)												
9	13.5	0.99	118.1	265.8	262.3	1794	172.3	753.7	0.82	28.7	29.5	22.1
10	15	0.98	131.3	328.1	320.1	2461	191.4	837.4	0.91	31.9	32.8	24.6
11	16.5	0.97	144.4	397.0	383.2	3276	210.5	921.1	1.00	35.1	36.1	27.1
12	18	0.96	157.5	472.5	451.7	4252	229.7	1005	1.09	38.3	39.4	29.5
13	19.5	0.95	170.6	554.5	525.4	5407	248.8	1089	1.18	41.5	42.7	32.0
14	21	0.94	183.8	643.1	604.4	6753	268.0	1172	1.28	44.7	45.9	34.5
15	22.5	0.93	196.9	738.3	688.5	8306	287.1	1256	1.37	47.9	49.2	36.9
16	24	0.93	210.0	840.0	777.7	10,080	306.3	1340	1.46	51.0	52.5	39.4
17	25.5	0.92	223.1	948.3	872.1	12,090	325.4	1424	1.55	54.2	55.8	41.8
18	27	0.91	236.3	1063	971.5	14,350	344.5	1507	1.64	57.4	59.1	44.3
19	28.5	0.91	249.4	1184	1076	16,880	363.7	1591	1.73	60.6	62.3	46.8
20	30	0.90	262.5	1312	1186	19,690	382.8	1675	1.82	63.8	65.6	49.2
21	31.5	0.90	275.6	1447	1230	22,790	402.0	1758	1.91	67.0	68.9	51.7
22	33	0.89	288.8	1588	1419	26,200	421.1	1843	2.01	70.2	72.2	54.1
23	34.5	0.89	301.9	1736	1544	29,940	440.2	1926	2.10	73.4	75.5	56.6
24	36	0.89	315.0	1890	1673	34,020	459.4	2010	2.19	76.6	78.8	59.1
25	37.5	0.88	328.1	2051	1807	38,450	478.5	2094	2.28	79.8	82.0	61.5
26	39	0.88	341.3	2218	1946	43,250	497.7	2177	2.37	82.9	85.3	64.0
27	40.5	0.87	354.4	2392	2090	48,440	516.8	2261	2.46	86.1	88.6	66.4

28	42	0.87	367.5	2572	2238	54,020	535.9	2345	2.55	89.3	91.9	68.9
29	43.5	0.87	380.6	2760	2392	60,020	555.1	2428	2.64	92.5	95.2	71.4
30	45	0.86	393.8	2953	2550	66,440	574.2	2513	2.73	95.7	98.4	73.8
31	46.5	0.86	406.9	3154	2713	73,310	593.4	2596	2.83	98.9	101.7	76.3
32	48	0.86	420.0	3360	2880	80,640	612.5	2680	2.92	102.1	105.0	78.7
33	49.5	0.85	433.1	3573	3053	88,440	631.6	2763	3.01	105.3	108.3	81.2
34	51	0.85	446.3	3793	3230	96,720	650.8	2847	3.10	108.5	111.6	83.7
35	52.5	0.85	459.4	4020	3412	105,500	669.9	2931	3.19	111.7	114.8	86.1
36	54	0.85	472.5	4252	3598	114,800	689.1	3015	3.28	114.8	118.1	88.6
37	55.5	0.84	485.6	4492	3789	124,700	708.2	3098	3.37	118.0	121.4	91.1
38	57	0.84	498.8	4738	3985	135,000	727.3	3182	3.46	121.2	124.7	93.5
39	58.5	0.84	511.9	4991	4185	146,000	746.5	3266	3.55	124.4	128.0	96.0
40	60	0.84	525.0	5250	4390	157,500	765.6	3350	3.65	127.6	131.2	98.4
41	61.5	0.83	538.1	5516	4600	169,600	784.8	3433	3.74	130.8	134.5	100.9
42	63	0.83	551.3	5788	4814	182,300	803.9	3517	3.83	134.0	137.8	103.4
$10\frac{3}{4}$ in. Width												
2	3	1.00	32.25	16.12	16.12	24.19	57.78	310.6	0.22	7.8	8.1	6.0
3	4.5	1.00	48.38	36.28	36.28	81.63	86.67	465.9	0.34	11.8	12.1	9.1
4	6	1.00	64.50	64.50	64.50	193.5	115.6	621.1	0.45	15.7	16.1	12.1
5	7.5	1.00	80.62	100.8	100.8	377.9	144.5	776.4	0.56	19.6	20.2	15.1
6	9	1.00	96.75	145.1	145.1	653.1	173.3	931.7	0.67	23.5	24.2	18.1
7	10.5	1.00	112.9	197.5	197.5	1037	202.2	1087	0.78	27.4	28.2	21.2
8	12	1.00	129.0	258.0	258.0	1548	231.1	1242	0.90	31.4	32.2	24.2
9	13.5	0.99	145.1	326.5	322.3	2204	260.0	1398	1.01	35.3	36.3	27.2
10	15	0.98	161.3	403.1	393.3	3023	288.9	1553	1.12	39.2	40.3	30.2
11	16.5	0.97	177.4	487.8	470.8	4024	317.8	1708	1.23	43.1	44.3	33.3
12	18	0.96	193.5	580.5	554.9	4224	346.7	1863	1.34	47.0	48.4	36.3
13	19.5	0.95	209.6	681.3	645.5	6642	375.6	2019	1.46	51.0	52.4	39.3
14	21	0.94	225.8	790.1	742.5	8296	404.5	2174	1.57	54.9	56.4	42.3
15	22.5	0.93	241.9	907.0	845.8	10,200	433.4	2321	1.68	58.8	60.5	45.4
16	24	0.93	258.0	1032	955.5	12,380	462.3	2485	1.79	62.7	64.5	48.4
17	25.5	0.92	274.1	1165	1071	14,850	491.1	2640	1.90	66.6	68.5	51.4

(continued)

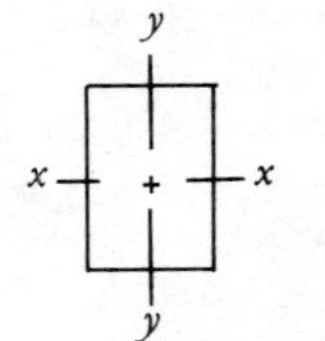

TABLE 7.2 (*Continued*)

				x–x Axis			y–y Axis			Weight (lb/ft)		
Number of $1\frac{1}{2}$ in. Laminations	Depth d (in.)	Size Factor C_F	Area A (in.^{2})	S_x (in.^{3})	S_n[b] (in.^{3})	I_x (in.^{4})	S_y (in.^{3})	I_y (in.^{4})	Volume per Ft (ft^3)	DFL[c] (35 pcf)	SP[c] (36 pcf)	HF/CR[c] (27 pcf)
$10\frac{3}{4}$ in. Width (cont.)												
18	27	0.91	290.3	1306	1194	17,630	520.0	2795	2.02	70.5	72.6	54.4
19	28.5	0.91	306.4	1455	1322	20,740	548.9	2950	2.13	74.5	76.6	57.4
20	30	0.90	322.5	1612	1456	24,190	577.8	3106	2.24	78.4	80.6	60.5
21	31.5	0.90	338.6	1778	1597	28,000	606.7	3261	2.35	82.3	84.7	63.5
22	33	0.89	354.8	1951	1744	32,190	635.6	3416	2.46	86.2	88.7	66.5
23	34.5	0.89	370.9	2132	1896	36,790	664.5	3572	2.58	90.1	92.7	69.5
24	36	0.89	387.0	2322	2055	41,800	693.4	3727	2.69	94.1	96.8	72.6
25	37.5	0.88	403.1	2520	2220	47,240	722.3	3882	2.80	98.0	100.8	75.6
26	39	0.88	419.3	2725	2391	53,140	751.2	4038	2.91	101.9	104.8	78.6
27	40.5	0.87	435.4	2939	2567	59,510	780.0	4193	3.02	105.8	108.8	81.6
28	42	0.87	451.5	3160	2750	66,370	808.9	4348	3.14	109.7	112.9	84.7
29	43.5	0.87	467.6	3390	2938	73,740	837.8	4503	3.25	113.7	116.9	87.7
30	45	0.86	483.8	3628	3133	81,630	866.7	4659	3.30	117.6	120.9	90.7
31	46.5	0.86	499.9	3874	3333	90,070	895.6	4814	3.47	121.5	125.0	93.7
32	48	0.86	516.0	4128	3539	99,070	924.5	4969	3.58	125.4	129.0	96.7
33	49.5	0.85	532.1	4390	3750	108,700	953.4	5124	3.70	129.3	133.0	99.8
34	51	0.85	548.3	4660	3968	118,900	982.3	5280	3.81	133.3	137.1	102.8
35	52.5	0.85	564.4	4938	4191	129,600	1011	5435	3.92	137.2	141.1	105.8

36	54	0.85	580.5	5224	4420	141,100	1041	5590	4.03	141.1	145.1	108.8
37	55.5	0.84	596.6	5519	4655	153,100	1069	5746	4.14	145.0	149.1	111.9
38	57	0.84	612.8	5821	4896	165,900	1098	5901	4.26	148.9	153.2	114.9
39	58.5	0.84	628.9	6132	5142	179,300	1127	6056	4.37	152.9	157.2	117.9
40	60	0.84	645.0	6450	5394	193,500	1156	6212	4.48	156.8	161.2	120.9
41	61.5	0.83	661.1	6776	5651	208,400	1184	6367	4.59	160.7	165.3	124.0
42	63	0.83	677.3	7111	5915	224,000	1213	6522	4.70	164.6	169.3	127.0
43	64.5	0.83	693.4	7454	6183	240,400	1242	6677	4.82	168.5	173.3	130.0
44	66	0.83	709.5	7804	6458	257,500	1271	6833	4.93	172.4	177.4	133.0
45	67.5	0.83	725.6	8163	6738	275,500	1300	6988	5.04	176.4	181.4	136.1
46	69	0.82	741.8	8530	7023	294,300	1329	7143	5.15	180.3	185.4	139.1
47	70.5	0.82	757.9	8905	7315	313,900	1358	7299	5.26	184.2	189.5	142.1
48	72	0.82	774.0	9288	7611	334,400	1387	7454	5.38	188.1	193.5	145.1
49	73.5	0.82	790.1	9679	7914	355,700	1416	7609	5.49	192.0	197.5	148.1
50	75	0.82	806.3	10,080	8222	377,900	1444	7764	5.60	196.0	201.6	151.2
$12\frac{1}{4}$ IN. WIDTH												
2	3	1.00	36.75	18.38	18.38	27.56	75.03	459.6	0.26	8.9	9.2	6.9
3	4.5	1.00	55.12	41.34	41.34	93.02	112.5	689.3	0.38	13.4	13.8	10.3
4	6	1.00	73.50	73.50	73.50	220.5	150.1	919.1	0.51	17.9	18.4	13.8
5	7.5	1.00	91.88	114.8	114.8	430.7	187.6	1149	0.64	22.3	23.0	17.2
6	9	1.00	110.3	165.4	165.4	744.2	225.1	1379	0.77	26.8	27.6	20.7
7	10.5	1.00	128.6	225.1	225.1	1182	262.6	1608	0.89	31.3	32.2	24.1
8	12	1.00	147.0	294.0	294.0	1764	300.1	1838	1.02	35.7	36.7	27.6
9	13.5	0.99	165.4	372.1	367.3	2512	337.6	2068	1.15	40.2	41.3	31.0
10	15	0.98	183.8	459.4	448.1	3445	375.2	2298	1.28	44.7	45.9	34.5
11	16.5	0.97	202.1	555.8	536.5	4586	412.7	2528	1.40	49.1	50.5	37.9
12	18	0.96	220.5	661.5	632.4	5953	450.2	2757	1.53	53.6	55.1	41.3
13	19.5	0.95	238.9	776.3	735.6	7569	487.7	2987	1.66	58.1	59.7	44.8
14	21	0.94	257.3	900.4	846.1	9454	525.2	3217	1.79	62.5	64.3	48.2
15	22.5	0.93	275.6	1034	963.9	11,630	562.7	3447	1.91	67.0	68.9	51.7
16	24	0.93	294.0	1176	1089	14,110	600.3	3676	2.04	71.5	73.5	55.1
17	25.5	0.92	312.4	1328	1221	16,930	637.8	3906	2.17	75.9	78.1	58.6

(continued)

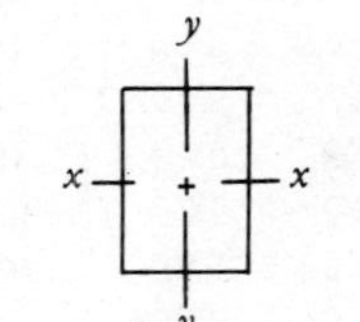

TABLE 7.2 (*Continued*)

Number of $1\frac{1}{2}$ in. Laminations	Depth d (in.)	Size Factor C_F	Area A (in.^{2})	x–x Axis S_x (in.^{3})	x–x Axis S_n[b] (in.^{3})	x–x Axis I_x (in.^{4})	y–y Axis S_y (in.^{3})	y–y Axis I_y (in.^{4})	Volume per Ft (ft^3)	Weight (lb/ft) DFL[c] (35 pcf)	Weight (lb/ft) SP[c] (36 pcf)	Weight (lb/ft) HF/CR[c] (27 pcf)
$12\frac{1}{4}$ in. Width (cont.)												
18	27	0.91	330.8	1488	1360	20,090	675.3	4136	2.30	80.4	82.7	62.0
19	28.5	0.91	349.1	1658	1506	23,630	712.8	4366	2.42	84.9	87.3	65.5
20	30	0.90	367.5	1838	1660	27,560	750.3	4596	2.55	89.3	91.9	68.9
21	31.5	0.90	385.9	2026	1820	31,910	787.8	4825	2.68	93.8	96.5	72.4
22	33	0.89	404.3	2223	1987	36,690	825.3	5055	2.81	98.3	101.1	75.8
23	34.5	0.89	422.6	2430	2161	41,920	862.9	5285	2.93	102.7	105.7	79.2
24	36	0.89	441.0	2646	2342	47,630	900.4	5515	3.06	107.2	110.3	82.7
25	37.5	0.88	459.4	2871	2530	53,830	937.9	5745	3.19	111.7	114.8	86.1
26	39	0.88	477.8	3105	2724	60,560	975.4	5974	3.32	116.1	119.4	89.6
27	40.5	0.87	496.1	3349	2926	67,810	1013.0	6204	3.45	120.6	124.0	93.0
28	42	0.87	514.5	3602	3134	75,630	1050.0	6434	3.57	125.1	128.6	96.5
29	43.5	0.87	532.9	3863	3348	84,030	1088	6664	3.70	129.5	133.2	99.9
30	45	0.86	551.3	4134	3570	93,020	1126	6894	3.83	134.0	137.8	103.4
31	46.5	0.86	569.6	4415	3798	102,600	1163	7123	3.96	138.5	142.4	106.8
32	48	0.86	588.0	4704	4032	112,900	1200	7353	4.08	142.9	147.0	110.2
33	49.5	0.85	606.4	5003	4274	123,800	1238	7583	4.21	147.4	151.6	113.7
34	51	0.85	624.8	5310	4522	135,400	1276	7813	4.34	151.8	156.2	117.1
35	52.5	0.85	643.1	5627	4776	147,700	1313	8042	4.47	156.3	160.8	120.6
36	54	0.85	661.5	5954	5037	160,700	1351	8272	4.59	160.8	165.4	124.0
37	55.5	0.84	679.9	6289	5305	174,500	1388	8502	4.72	165.2	170.0	127.5

38	57	0.84	698.3	6633	5579	189,100	1426	8732	4.85	169.7	174.6	130.9
39	58.5	0.84	716.6	6987	5859	204,400	1463	8962	4.98	174.2	179.2	134.4
40	60	0.84	735.0	7350	6146	220,500	1501	9191	5.10	178.6	183.7	137.8
41	61.5	0.83	753.4	7722	6440	237,500	1538	9421	5.23	183.1	188.3	141.3
42	63	0.83	771.8	8103	6740	255,300	1576	9651	5.36	187.6	192.9	144.7
43	64.5	0.83	790.1	8494	7046	274,000	1613	9881	5.49	192.0	197.5	148.1
44	66	0.83	808.5	8894	7359	293,500	1651	10,110	5.61	196.5	202.1	151.6
45	67.5	0.83	826.9	9302	7678	314,000	1688	10,340	5.74	201.0	206.7	155.0
46	69	0.82	845.3	9720	8003	335,400	1726	10,570	5.87	205.4	211.3	158.5
47	70.5	0.82	863.6	10,150	8335	357,700	1763	10,800	6.00	209.9	215.9	161.9
48	72	0.82	882.0	10,580	8673	381,000	1801	11,030	6.13	214.4	220.5	165.4
49	73.5	0.82	900.4	11,030	9018	405,300	1838	11,260	6.25	218.8	225.1	168.8
50	75	0.82	918.8	11,480	9369	430,700	1876	11,490	6.38	223.3	229.7	172.3
51	76.5	0.81	937.1	11,950	9726	457,000	1913	11,720	6.51	227.8	234.3	175.7
52	78	0.81	955.5	12,420	10,090	484,400	1951	11,950	6.64	232.2	238.9	179.2
53	79.5	0.81	973.9	12,900	10,460	512,900	1988	12,180	6.76	236.7	243.5	182.6
54	81	0.81	992.3	13,400	10,830	542,500	2026	12,410	6.89	241.2	248.1	186.0
55	82.5	0.81	1011	13,900	11,220	573,200	2063	12,640	7.02	245.6	252.7	189.5
56	84	0.81	1029	14,410	11,610	605,100	2101	12,870	7.15	250.1	257.2	192.9
$14\frac{1}{4}$ in. Width												
2	3	1.00	42.75	21.38	21.38	32.06	101.5	723.4	0.30	10.4	10.7	8.0
3	4.5	1.00	64.12	48.09	48.09	108.2	152.3	1085	0.45	15.6	16.0	12.0
4	6	1.00	85.50	85.50	85.50	256.5	203.1	1447	0.59	20.8	21.4	16.0
5	7.5	1.00	106.9	133.6	133.6	501.0	253.8	1808	0.74	26.0	26.7	20.0
6	9	1.00	128.3	192.4	192.4	865.7	304.6	2170	0.89	31.2	32.1	24.0
7	10.5	1.00	149.6	261.8	261.8	1375	355.4	2531	1.04	36.4	37.4	28.1
8	12	1.00	171.0	342.0	342.0	2052	406.1	2894	1.19	41.6	42.8	32.1
9	13.5	0.99	192.4	432.8	427.2	2922	456.9	3255	1.34	46.8	48.1	36.1
10	15	0.98	213.8	534.4	521.3	4008	507.7	3617	1.48	52.0	53.4	40.1
11	16.5	0.97	235.1	646.6	624.1	5334	558.4	3979	1.63	57.1	58.8	44.1
12	18	0.96	256.5	769.5	735.6	6925	609.2	4340	1.78	62.3	64.1	48.1
13	19.5	0.95	277.9	903.1	855.7	8805	660.0	4702	1.93	67.5	69.5	52.1

(continued)

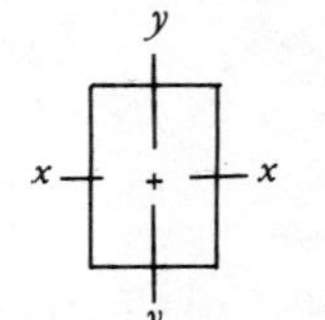

TABLE 7.2 (*Continued*)

Number of $1\frac{1}{2}$ in. Laminations	Depth d (in.)	Size Factor C_F	Area A (in.^{2})	x–x Axis S_x (in.^{3})	x–x Axis S_n[b] (in.^{3})	x–x Axis I_x (in.^{4})	y–y Axis S_y (in.^{3})	y–y Axis I_y (in.^{4})	Volume per Ft (ft^3)	Weight (lb/ft) DFL[c] (35 pcf)	Weight (lb/ft) SP[c] (36 pcf)	Weight (lb/ft) HF/CR[c] (27 pcf)
$14\frac{1}{4}$ IN. WIDTH (cont.)												
14	21	0.94	299.3	1047	984.2	11,000	710.7	5064	2.08	72.7	74.8	56.1
15	22.5	0.93	320.6	1202	1121	13,530	761.5	5426	2.23	77.9	80.2	60.1
16	24	0.93	342.0	1368	1267	16,420	812.3	5787	2.38	83.1	85.5	64.1
17	25.5	0.92	363.4	1544	1420	19,690	863.0	6149	2.52	88.3	90.8	68.1
18	27	0.91	384.8	1731	1582	23,370	913.5	6511	2.67	93.5	96.2	72.1
19	28.5	0.91	406.1	1929	1752	27,490	964.5	6872	2.82	98.7	101.5	76.1
20	30	0.90	427.5	2138	1930	32,060	1015	7234	2.97	103.9	106.9	80.2
21	31.5	0.90	448.9	2357	2117	37,120	1066	7596	3.12	109.1	112.2	84.2
22	33	0.89	470.3	2586	2311	42,680	1117	7958	3.27	114.3	117.6	88.2
23	34.5	0.89	491.6	2827	2514	48,760	1168	8319	3.41	119.5	122.9	92.2
24	36	0.89	513.0	3078	2724	55,400	1218	8681	3.56	124.7	128.3	96.2
25	37.5	0.88	534.4	3340	2943	62,620	1269	9043	3.71	129.9	133.6	100.2
26	39	0.88	555.8	3612	3169	70,440	1320	9404	3.86	135.1	138.9	104.2
27	40.5	0.87	577.1	3896	3403	78,890	1371	9766	4.01	140.3	144.3	108.2
28	42	0.87	598.5	4190	3645	87,980	1421	10,130	4.16	145.5	149.6	112.2
29	43.5	0.87	619.9	4494	3895	97,750	1472	10,490	4.30	150.7	155.0	116.2
30	45	0.86	641.3	4809	4153	108,200	1523	10,850	4.45	155.9	160.3	120.2
31	46.5	0.86	662.6	5135	4418	119,400	1574	11,210	4.60	161.1	165.7	124.2
32	48	0.86	684.0	5472	4690	131,300	1624	11,570	4.75	166.3	171.0	128.3
33	49.5	0.85	705.4	5819	4972	144,000	1675	11,940	4.90	171.4	176.3	132.3
34	51	0.85	726.8	6177	5260	157,500	1726	12,300	5.05	176.6	181.7	136.3

35	52.5	0.85	748.1	6546	5556	171,800	1777	12,660	5.20	181.8	187.0	140.3
36	54	0.85	769.5	6926	5860	187,000	1828	13,020	5.34	187.0	192.4	144.3
37	55.5	0.84	790.9	7316	6171	203,000	1878	13,380	5.49	192.2	197.7	148.3
38	57	0.84	812.3	7716	6490	219,900	1929	13,740	5.64	197.4	203.1	152.3
39	58.5	0.84	833.6	8128	6816	237,700	1980	14,110	5.79	202.6	208.4	156.3
40	60	0.84	855.0	8550	7150	256,500	2031	14,470	5.94	207.8	213.8	160.3
41	61.5	0.83	876.4	8983	7491	276,200	2081	14,830	6.09	213.0	219.1	164.3
42	63	0.83	897.8	9426	7840	296,900	2132	15,190	6.23	218.2	224.4	168.3
43	64.5	0.83	919.1	9881	8196	318,600	2183	15,550	6.38	223.4	229.8	172.3
44	66	0.83	940.5	10,350	8560	341,400	2234	15,920	6.53	228.6	235.1	176.3
45	67.5	0.83	961.9	10,820	8932	365,200	2284	16,280	6.68	233.8	240.5	180.4
46	69	0.82	983.3	11,310	9310	390,100	2335	16,640	6.83	239.0	245.8	184.4
47	70.5	0.82	1005	11,800	9696	416,100	2386	17,000	6.98	244.2	251.2	188.4
48	72	0.82	1026	12,310	10,090	443,200	2437	17,360	7.13	249.4	256.5	192.4
49	73.5	0.82	1047	12,830	10,490	471,500	2488	17,720	7.27	254.6	261.8	196.4
50	75	0.82	1069	13,360	10,900	501,000	2538	18,090	7.42	259.8	267.2	200.4
51	76.5	0.81	1090	13,900	11,310	531,600	2589	18,450	7.57	265.0	272.5	204.4
52	78	0.81	1112	14,450	11,740	563,500	2640	18,810	7.72	270.2	277.9	208.4
53	79.5	0.81	1133	15,010	12,170	596,700	2691	19,170	7.87	275.4	283.4	212.4
54	81	0.81	1154	15,580	12,600	631,100	2741	19,530	8.02	280.5	288.6	216.4
55	82.5	0.81	1176	16,160	13,050	666,800	2792	19,890	8.16	285.7	293.9	220.4
56	84	0.81	1197	16,760	13,500	703,800	2843	20,260	8.31	290.9	299.3	224.4
57	85.5	0.80	1218	17,360	13,960	742,200	2894	20,620	8.46	296.1	304.6	228.4
58	87	0.80	1240	17,980	14,420	782,000	2944	20,980	8.61	301.3	309.9	232.5
59	88.5	0.80	1261	18,600	14,900	823,100	2995	21,340	8.76	306.5	315.3	236.5
60	90	0.80	1282	19,240	15,380	865,700	3046	21,700	8.91	311.7	320.6	240.5
61	91.5	0.80	1304	19,880	15,870	909,700	3097	22,060	9.05	316.9	326.0	244.5
62	93	0.80	1325	20,540	16,360	955,200	3148	22,430	9.20	322.1	331.3	248.5
63	94.5	0.80	1347	31,210	16,860	1,002,000	3198	22,790	9.35	327.3	336.7	252.5
64	96	0.79	1368	21,890	17,370	1,051,000	3249	23,150	9.50	332.5	342.0	256.5

[a] For use with straight or slightly cambered members. Depths based on multiples of $1\frac{1}{2}$-in.-thick laminations.

[b] $S_n = S_x C_F$

[c] DFL = Douglas Fir-Larch; SP = Southern Pine; HF = Hem-Fir; CR = California Redwood.

[d] Normal widths are $3\frac{1}{8}$ and $5\frac{1}{8}$ in. for Western softwood glued laminated timbers and 3 and 5 in. for Southern Pine glued lamianted timbers. To determine alternate section properties, multiply A, S_x, and I_x values for $3\frac{1}{8}$ in. widths by 3/3.125 and for $5\frac{1}{8}$ in. widths by 5/5.125.

DESIGN VALUES

TABLE 7.3 Design Values for Visually Graded Structural Lumber (psi)*

*The design values listed in Tables 7.3 and 7.4 were reproduced from *Design Values for Wood Construction*, a supplement to the 1982 edition of the *National Design Specification for Wood Construction* by the National Forest Products Association (NFPA). This supplement is revised periodically, so the designer should check with NFPA for the latest information (Design values listed are for normal loading conditions. See other provisions in the footnotes and in Section 3 for adjustments of tabulated values.)

Species and Commercial Grade	Size Classification	Extreme Fiber in Bending, F_b: Single-Member Uses	Extreme Fiber in Bending, F_b: Repetitive-Member Uses	Tension Parallel to Grain, F_t	Horizontal Shear, F_v	Compression Perpendicular to Grain, $F_{c\perp}$	Compression Parallel to Grain, F_c	Modulus of Elasticity, E	Grading Rules Agency
ASPEN[a]									
Select structural	2–4 in. thick,	1300	1500	775	60	265	850	1,100,000	NELMA,
No. 1	2–4 in. wide	1100	1300	650	60	265	675	1,100,000	NHPMA,
No. 2		925	1050	525	60	265	550	1,000,000	WWPA[b-l]
No. 3		500	575	300	60	265	325	900,000	
Appearance		1100	1300	650	60	265	825	1,100,000	
Stud		500	575	300	60	265	325	900,000	
Construction	2–4 in. thick,	650	750	400	60	265	625	900,000	
Standard	4 in. wide	375	425	225	60	265	500	900,000	
Utility		175	200	100	60	265	325	900,000	
Select structural	2–4 in. thick,	1150	1300	750[d]	60	265	750	1,100,000	
No. 1	5 in. and	950	1100	650[d]	60	265	675	1,100,000	
No. 2	wider	775	900	425[d]	60	265	575	1,000,000	
No. 3		450	525	250[d]	60	265	375	900,000	
Appearance		950	1100	650[d]	60	265	825	1,100,000	
Stud		450	525	250[d]	60	265	375	900,000	

BALSAM FIR[a]									
Select structural	2–4 in. thick,	1750	2000	1000	70	305	1350	1,500,000	NELMA,
No. 1	2–4 in. wide	1450	1700	850	70	305	1050	1,500,000	NHPMA [b–m]
No. 2		1200	1400	700	70	305	850	1,300,000	
No. 3		675	775	400	70	305	525	1,200,000	
Appearance		1450	1700	850	70	305	1250	1,500,000	
Stud		675	775	400	70	305	525	1,200,000	
Construction	2–4 in. thick,	875	1000	525	70	305	950	1,200,000	
Standard	4 in. wide	500	575	275	70	305	775	1,200,000	
Utility		225	275	125	70	305	525	1,200,000	
Select structural	2–4 in. thick,	1500	1700	1000[d]	70	305	1200	1,500,000	
No. 1	5 in. and	1250	1450	850[d]	70	305	1050	1,500,000	
No. 2	wider	1050	1200	550[d]	70	305	900	1,300,000	
No. 3		600	700	325[d]	70	305	575	1,200,000	
Appearance		1250	1450	850[d]	70	305	1250	1,500,000	
Stud		600	700	325[d]	70	305	575	1,200,000	
Select structural	Beams and	1350	—	900	65	305	950	1,400,000	
No. 1	stringers	1100	—	750	65	305	800	1,400,000	
Select structural	Posts and	1250	—	825	65	305	1000	1,400,000	
No. 1	timbers	1000	—	675	65	305	875	1,400,000	
Select	Decking	—	1650	—	—	—	—	1,500,000	NELMA [b–m]
Commercial		—	1400	—	—	—	—	1,300,000	
BLACK COTTONWOOD[a]									
Select structural	2–3 in. thick,	1000	1200	600	50	180	725	1,200,000	NLGA
No. 1	2–4 in. wide	875	1000	500	50	180	575	1,200,000	(a Canadian
No. 2		725	825	425	50	180	450	1,100,000	agency) [b–m,p,q]

(continued)

TABLE 7.3 (*Continued*) Design Values for Visually Graded Structural Lumber

Species and Commercial Grade	Size Classification	Extreme Fiber in Bending, F_b: Single-Member Uses	Extreme Fiber in Bending, F_b: Repetitive-Member Uses	Tension Parallel to Grain, F_t	Horizontal Shear, F_v	Compression Perpendicular to Grain, $F_{c\perp}$	Compression Parallel to Grain, F_c	Modulus of Elasticity, E	Grading Rules Agency
BLACK COTTONWOOD[a] (cont.)									
No. 3	2–3 in. thick,	400	450	225	50	180	275	900,000	
Appearance	2–4 in. wide	875	1000	500	50	180	700	1,200,000	
Stud	(cont.)	400	450	225	50	180	275	900,000	
Construction	2–4 in. thick,	525	600	300	50	180	525	900,000	
Standard	4 in. wide	300	325	175	50	180	425	900,000	
Utility		150	150	75	50	180	275	900,000	
Select structural	2–4 in. thick,	875	1000	575[d]	50	180	650	1,200,000	
No. 1	5 in. and	750	875	500[d]	50	180	575	1,200,000	
No. 2	wider	625	700	325[d]	50	180	475	1,100,000	
No. 3		350	425	175[d]	50	180	300	900,000	
Appearance		750	875	500[d]	50	180	700	1,200,000	
Stud		350	425	175[d]	50	180	300	900,000	
CALIFORNIA REDWOOD[a]									
Clear heart structural	4 in. and less thick,	2300	2650	1550	145	650	2150	1,400,000	RIS[b-g,i-k,m]
Clear structural	any width	2300	2650	1550	145	650	2150	1,400,000	
Select structural	2–4 in. thick,	2050	2350	1200	80	650	1750	1,400,000	
Select structural, open grain	2–4 in. wide	1600	1850	950	80	425	1300	1,100,000	

No. 1		1700	1950	975	80	650	1400	1,400,000
No. 1, open grain		1350	1550	775	80	425	1050	1,100,000
No. 2		1400	1600	800	80	650	1100	1,250,000
No. 2, open grain		1100	1250	625	80	425	825	1,000,000
No. 3		800	900	475	80	650	675	1,100,000
No. 3, open grain		625	725	375	80	425	500	900,000
Stud		625	725	375	80	425	500	900,000
Construction	2–4 in. thick, 4 in. wide	825	950	475	80	425	925	900,000
Standard		450	525	250	80	425	775	900,000
Utility		225	250	125	80	425	500	900,000
Select structural	2–4 in. thick, 5 in. and wider	1750	2000	1150[d]	80	650	1550	1,400,000
Select structural, open grain		1400	1600	925[d]	80	425	1150	1,100,000
No. 1		1500	1700	975[d]	80	650	1400	1,400,000
No. 1, open grain		1150	1350	775[d]	80	425	1050	1,100,000
No. 2		1200	1400	650[d]	80	650	1200	1,250,000
No. 2, open grain		950	1100	500[d]	80	425	875	1,000,000
No. 3		700	800	375[d]	80	650	725	1,100,000
No. 3, open grain		550	650	350[d]	80	425	525	900,000
Stud		700	800	375[d]	80	650	725	1,100,000
Clear heart structural or clear structural	5 × 5 in. and larger	1850	—	1250	135	650	1650	1,300,000
Select structural		1400	—	950	95	650	1200	1,300,000
No. 1		1200	—	800	95	650	1050	1,300,000
No. 2		975	—	650	95	650	900	1,100,000
No. 3		550	—	375	95	650	550	1,000,000

(continued)

TABLE 7.3 (*Continued*) Design Values for Visually Graded Structural Lumber

Species and Commercial Grade	Size Classification	Extreme Fiber in Bending, F_b: Single-Member Uses	Extreme Fiber in Bending, F_b: Repetitive-Member Uses	Tension Parallel to Grain, F_t	Horizontal Shear, F_v	Compression Perpendicular to Grain, $F_{c\perp}$	Compression Parallel to Grain, F_c	Modulus of Elasticity, E	Grading Rules Agency
CALIFORNIA REDWOOD[a] (cont.)									
Select decking, close grain	Decking, 2 in. thick,	1850	2150	—	—	—	—	1,400,000	RIS[b, c, i, j]
Select decking	6 in. and	1450	1700	—	—	—	—	1,100,000	
Commercial decking	wider	1200	1350	—	—	—	—	1,000,000	
COAST SITKA SPRUCE[a]									
Select structural	2–3 in. thick,	1500	1700	875	65	455	1100	1,700,000	NLGA
No. 1	2–4 in. wide	1250	1450	750	65	455	875	1,700,000	(a Canadian
No. 2		1050	1200	625	65	455	700	1,500,000	agency)[b–m, p, q]
No. 3		575	675	350	65	455	425	1,300,000	
Appearance		1250	1450	725	65	455	1050	1,700,000	
Stud		575	675	350	65	455	425	1,300,000	
Construction	2–4 in. thick,	750	875	450	65	455	800	1,300,000	
Standard	4 in. wide	425	500	250	65	455	650	1,300,000	
Utility		200	225	125	65	455	425	1,300,000	
Select structural	2–4 in. thick,	1300	1500	850[d]	65	455	975	1,700,000	
No. 1	5 in. and	1100	1250	725[d]	65	455	875	1,700,000	
No. 2	wider	900	1050	475[d]	65	455	750	1,500,000	
No. 3		525	600	275[d]	65	455	475	1,300,000	
Appearance		1100	1250	725[d]	65	455	1050	1,700,000	
Stud		525	600	275[d]	65	455	475	1,300,000	

Select structural	Beams and stringers	1150	—	675	60	455	775	1,500,000	
No. 1		950	—	475	60	455	650	1,500,000	
Select structural	Posts and timbers	1100	—	725	60	455	825	1,500,000	
No. 1		875	—	575	60	455	725	1,500,000	
Select	Decking	1250	1450	—	—	455	—	1,700,000	
Commercial		1050	1200	—	—	455	—	1,500,000	
COAST SPECIES[a]									
Select structural	2–3 in. thick, 2–4 in. wide	1500	1700	875	65	370	1100	1,500,000	NLGA (a Canadian agency)[b–m, p, q]
No. 1		1250	1450	750	65	370	875	1,500,000	
No. 2		1050	1200	625	65	370	700	1,400,000	
No. 3		575	675	350	65	370	425	1,200,000	
Appearance		1250	1450	725	65	370	1050	1,500,000	
Stud		575	675	350	65	370	425	1,200,000	
Construction	2–4 in. thick, 4 in. wide	750	875	450	65	370	800	1,200,000	
Standard		425	500	250	65	370	650	1,200,000	
Utility		200	225	125	65	370	425	1,200,000	
Select structural	2–4 in. thick, 5 in. and wider	1300	1500	850[d]	65	370	975	1,500,000	
No. 1		1100	1250	725[d]	65	370	875	1,500,000	
No. 2		900	1050	475[d]	65	370	750	1,400,000	
No. 3		525	600	275[d]	65	370	475	1,200,000	
Appearance		1100	1250	725[d]	65	370	1050	1,500,000	
Stud		525	600	275[d]	65	370	475	1,200,000	
Select	Decking	1250	1450	—	—	370	—	1,500,000	
Commercial		1050	1200	—	—	370	—	1,400,000	

(continued)

TABLE 7.3 (*Continued*) Design Values for Visually Graded Structural Lumber

Species and Commercial Grade	Size Classification	Extreme Fiber in Bending, F_b: Single-Member Uses	Repetitive-Member Uses	Tension Parallel to Grain, F_t	Horizontal Shear, F_v	Compression Perpendicular to Grain, $F_{c\perp}$	Compression Parallel to Grain, F_c	Modulus of Elasticity, E	Grading Rules Agency
COTTONWOOD[a]									
Stud	2–3 in. thick, 2–4 in. wide	525	600	300	65	320	350	1,000,000	NHPMA[b–m]
Construction	2–4 in. thick, 4 in. wide	675	775	400	65	320	650	1,000,000	
Standard		375	425	225	65	320	525	1,000,000	
Utility		175	200	100	65	320	350	1,000,000	
DOUGLAS FIR-LARCH[a]									
Dense select structural	2–4 in. thick, 2–4 in. wide	2450	2800	1400	95	730	1850	1,900,000	WCLIB, WWPA[b–m]
Select structural		2100	2400	1200	95	625	1600	1,800,000	
Dense No. 1		2050	2400	1200	95	730	1450	1,900,000	
No. 1		1750	2050	1050	95	625	1250	1,800,000	
Dense No. 2		1700	1950	1000	95	730	1150	1,700,000	
No. 2		1450	1650	850	95	625	1000	1,700,000	
No. 3		800	925	475	95	625	600	1,500,000	
Appearance		1750	2050	1050	95	625	1500	1,800,000	
Stud		800	925	475	95	625	600	1,500,000	
Construction	2–4 in. thick, 4 in. wide	1050	1200	625	95	625	1150	1,500,000	
Standard		600	675	350	95	625	925	1,500,000	
Utility		275	325	175	95	625	600	1,500,000	

Dense select structural	2–4 in. thick, 5 in. and wider	2100	2400	1400[d]	95	730	1650	1,900,000	
Select structural		1800	2050	1200[d]	95	625	1400	1,800,000	
Dense No. 1		1800	2050	1200[d]	95	730	1450	1,900,000	
No. 1		1500	1750	1000[d]	95	625	1250	1,800,000	
Dense No. 2		1450	1700	775[d]	95	730	1250	1,700,000	
No. 2		1250	1450	650[d]	95	625	1050	1,700,000	
No. 3		725	850	375[d]	95	625	675	1,500,000	
Appearance		1500	1750	1000[d]	95	625	1500	1,800,000	
Stud		725	850	375[d]	95	625	675	1,500,000	
Dense select structural	Beams and stringers	1900	—	1100	85	730	1300	1,700,000	WCLIB [b–m]
Select structural		1600	—	950	85	625	1100	1,600,000	
Dense No. 1		1550	—	775	85	730	1100	1,700,000	
No. 1		1300	—	675	85	625	925	1,600,000	
Dense select structural	Posts and timbers	1750	—	1150	85	730	1350	1,700,000	
Select structural		1500	—	1000	85	625	1150	1,600,000	
Dense No. 1		1400	—	950	85	730	1200	1,700,000	
No. 1		1200	—	825	85	625	1000	1,600,000	
Select dex	Decking	1750	2000	—	—	625	—	1,800,000	
Commercial dex		1450	1650	—	—	625	—	1,700,000	
Dense select structural	Beams and stringers	1900	—	1250	85	730	1300	1,700,000	WWPA [b–n]
Select structural		1600	—	1050	85	625	1100	1,600,000	

(continued)

TABLE 7.3 (*Continued*) Design Values for Visually Graded Structural Lumber

Species and Commercial Grade	Size Classification	Extreme Fiber in Bending, F_b: Single-Member Uses	Extreme Fiber in Bending, F_b: Repetitive-Member Uses	Tension Parallel to Grain, F_t	Horizontal Shear, F_v	Compression Perpendicular to Grain, $F_{c\perp}$	Compression Parallel to Grain, F_c	Modulus of Elasticity, E	Grading Rules Agency
DOUGLAS FIR-LARCH[a] (cont.)									
Dense No. 1	Beams and	1550	—	1050	85	730	1100	1,700,000	
No. 1	stringers (cont.)	1350	—	900	85	625	925	1,600,000	
Dense select structural	Posts and timbers	1750	—	1150	85	730	1350	1,700,000	
Select structural		1500	—	1000	85	625	1150	1,600,000	
Dense No. 1		1400	—	950	85	730	1200	1,700,000	
No. 1		1200	—	825	85	625	1000	1,600,000	
Selected decking	Decking	—	2000	—	—	—	—	1,800,000	
Commercial decking		—	1650	—	—	—	—	1,700,000	
Selected decking	Decking	—	2150	Surfaced at 15% max. MC and			—	1,900,000	
Commercial decking		—	1800	used at 15% max. MC			—	1,700,000	
DOUGLAS FIR-LARCH (NORTH)[a]									
Select structural	2–3 in. thick,	2100	2400	1200	95	625	1550	1,800,000	NLGA
No. 1	2–4 in. wide	1750	2050	1050	95	625	1250	1,800,000	(a Canadian
No. 2		1450	1650	850	95	625	1000	1,700,000	agency[b–m, p, q]
No. 3		800	925	475	95	625	600	1,500,000	
Appearance		1750	2050	1050	95	625	1500	1,800,000	
Stud		800	925	475	95	625	600	1,500,000	

Construction	2–4 in. thick,	1050	1200	625	95	625	1150	1,500,000	
Standard	4 in. wide	600	675	350	95	625	925	1,500,000	
Utility		275	325	175	95	625	600	1,500,000	
Select structural	2–4 in. thick,	1800	2050	1200[d]	95	625	1400	1,800,000	
No. 1	5 in. and	1500	1750	1000[d]	95	625	1250	1,800,000	
No. 2	wider	1250	1450	650[d]	95	625	1050	1,700,000	
No. 3		725	850	375[d]	95	625	675	1,500,000	
Appearance		1500	1750	1000[d]	95	625	1500	1,800,000	
Stud		725	850	375[d]	95	625	675	1,500,000	
Select structural	Beams and	1600	—	950	85	625	1100	1,600,000	
No. 1	stringers	1300	—	675	85	625	925	1,600,000	
Select structural	Posts and	1500	—	1000	85	625	1150	1,600,000	
No. 1	timbers	1200	—	825	85	625	1000	1,600,000	
Select	Decking	1750	2000	—	—	625	—	1,800,000	
Commercial		1450	1650	—	—	625	—	1,700,000	
DOUGLAS FIR SOUTH[a]									
Select structural	2–4 in. thick,	2000	2300	1150	90	520	1400	1,400,000	WWPA[b–n]
No. 1	2–4 in. wide	1700	1950	975	90	520	1150	1,400,000	
No. 2		1400	1600	825	90	520	900	1,300,000	
No. 3		775	875	450	90	520	550	1,100,000	
Appearance		1700	1950	975	90	520	1350	1,400,000	
Stud		775	875	450	90	520	550	1,100,000	
Construction	2–4 in. thick,	1000	1150	600	90	520	1000	1,100,000	
Standard	4 in. wide	550	650	325	90	520	850	1,100,000	
Utility		275	300	150	90	520	550	1,100,000	

(continued)

TABLE 7.3 (*Continued*) Design Values for Visually Graded Structural Lumber

Species and Commercial Grade	Size Classification	Extreme Fiber in Bending, F_b: Single-Member Uses	Extreme Fiber in Bending, F_b: Repetitive-Member Uses	Tension Parallel to Grain, F_t	Horizontal Shear, F_v	Compression Perpendicular to Grain, $F_{c\perp}$	Compression Parallel to Grain, F_c	Modulus of Elasticity, E	Grading Rules Agency
DOUGLAS FIR SOUTH[a] (cont.)									
Select structural	2–4 in. thick,	1700	1950	1150[d]	90	520	1250	1,400,000	
No. 1	5 in. and	1450	1650	975[d]	90	520	1150	1,400,000	
No. 2	wider	1200	1350	625[d]	90	520	950	1,300,000	
No. 3		700	800	350[d]	90	520	600	1,100,000	
Appearance		1450	1650	975[d]	90	520	1350	1,400,000	
Stud		700	800	350[d]	90	520	600	1,100,000	
Select structural	Beams and	1550	—	1050	85	520	1000	1,200,000	
No. 1	stringers	1300	—	850	85	520	850	1,200,000	
Select structural	Posts and	1400	—	950	85	520	1050	1,200,000	
No. 1	timbers	1150	—	775	85	520	925	1,200,000	
Selected decking	Decking	—	1900	—	—	—	—	1,400,000	
Commercial decking		—	1600	—	—	—	—	1,300,000	
Selected decking	Decking	—	2050	Surfaced at 15% max. MC and used			—	1,500,000	
Commercial decking		—	1750	at 15% max, MC			—	1,300,000	

Eastern Hemlock[a]									
Select structural	2–4 in. thick,	1750	2050	1050	85	550	1350	1,200,000	NELMA,
No. 1	2–4 in. wide	1500	1750	875	85	550	1050	1,200,000	NHPMA[b–m]
No. 2		1250	1450	725	85	550	850	1,100,000	
No. 3		675	800	400	85	550	525	1,000,000	
Appearance		1500	1750	875	85	550	1250	1,200,000	
Stud		675	800	400	85	550	525	1,000,000	
Construction	2–4 in. thick,	900	1050	525	85	550	950	1,000,000	
Standard	4 in. wide	500	575	300	85	550	800	1,000,000	
Utility		250	275	150	85	550	525	1,000,000	
Select structural	2–4 in. thick,	1550	1750	1000[d]	85	550	1200	1,200,000	
No. 1	5 in. and	1300	1500	875[d]	85	550	1050	1,200,000	
No. 2	wider	1050	1250	550[d]	85	550	900	1,100,000	
No. 3		625	700	325[d]	85	550	575	1,000,000	
Appearance		1300	1500	875[d]	85	550	1250	1,200,000	
Stud		625	700	325[d]	85	550	575	1,000,000	
Select structural	Beams and	1350	—	925	80	550	950	1,200,000	
No. 1	stringers	1150	—	775	80	550	800	1,200,000	
Select structural	Posts and	1250	—	850	80	550	1000	1,200,000	
No. 1	timbers	1050	—	700	80	550	875	1,200,000	
Eastern Hemlock-Tamarack[a]									
Select structural	2–4 in. thick,	1800	2050	1050	85	555	1350	1,300,000	NELMA,
No. 1	2–4 in. wide	1500	1750	900	85	555	1050	1,300,000	NHPMA[b–m]
No. 2		1250	1450	725	85	555	850	1,100,000	
No. 3		700	800	400	85	555	525	1,000,000	

(continued)

TABLE 7.3 (*Continued*) Design Values for Visually Graded Structural Lumber

Species and Commercial Grade	Size Classification	Extreme Fiber in Bending, F_b: Single-Member Uses	Extreme Fiber in Bending, F_b: Repetitive-Member Uses	Tension Parallel to Grain, F_t	Horizontal Shear, F_v	Compression Perpendicular to Grain, $F_{c\perp}$	Compression Parallel to Grain, F_c	Modulus of Elasticity, E	Grading Rules Agency
EASTERN HEMLOCK-TAMARACK[a] (cont.)									
Apearance	2–4 in. thick,	1300	1500	900	85	555	1300	1,300,000	
Stud	2–4 in .wide (cont.)	700	800	400	85	555	525	1,000,000	
Construction	2–4 in. thick,	900	1050	525	85	555	950	1,000,000	
Standard	4 in. wide	500	575	300	85	555	800	1,000,000	
Utility		250	275	150	85	555	525	1,000,000	
Select structural	2–4 in. thick,	1550	1750	1050[d]	85	555	1200	1,300,000	
No. 1	5 in. and	1300	1500	875[d]	85	555	1050	1,300,000	
No. 2	wider	1050	1200	575[d]	85	555	900	1,100,000	
No. 3		625	725	325[d]	85	555	575	1,000,000	
Appearance		1300	1500	875[d]	85	555	1300	1,300,000	
Stud		625	725	325[d]	85	555	575	1,000,000	
Select structural	Beams and	1400	—	925	80	555	950	1,200,000	
No. 1	stringers	1150	—	775	80	555	800	1,200,000	
Select structural	Posts and	1300	—	875	80	555	1000	1,200,000	
No. 1	timbers	1050	—	700	80	555	875	1,200,000	
Select	Decking	1500	1700	—	—	—	—	1,300,000	NELMA[b-m]
Commercial		1250	1450	—	—	—	—	1,100,000	

EASTERN HEMLOCK-TAMARACK (NORTH)[a]									
Select structural	2–3 in. thick,	1800	2050	1050	85	555	1350	1,300,000	NLGA
No. 1	2–4 in. wide	1500	1750	900	85	555	1050	1,300,000	(a Canadian
No. 2		1250	1450	725	85	555	850	1,100,000	agency)[b–m, p, q]
No. 3		700	800	400	85	555	525	1,000,000	
Appearance		1500	1750	900	85	555	1300	1,300,000	
Stud		700	800	400	85	555	525	1,000,000	
Construction	2–4 in. thick,	900	1050	525	85	555	975	1,000,000	
Standard	4 in. wide	500	575	300	85	555	800	1,000,000	
Utility		250	275	150	85	555	525	1,000,000	
Select structural	2–4 in. thick,	1550	1750	1050[d]	85	555	1200	1,300,000	
No. 1	5 in. and	1300	1500	875[d]	85	555	1050	1,300,000	
No. 2	wider	1050	1200	575[d]	85	555	900	1,100,000	
No. 3		625	725	325[d]	85	555	575	1,000,000	
Appearance		1300	1500	875[d]	85	555	1300	1,300,000	
Stud		625	725	325[d]	85	555	575	1,000,000	
Select structural	Beams and	1450	—	850	85	555	950	1,300,000	
No. 1	stringers	1200	—	600	85	555	800	1,300,000	
Select structural	Posts and	1350	—	900	85	555	1000	1,300,000	
No. 1	timbers	1100	—	725	85	555	875	1,300,000	
Select	Decking	1500	1700	—	—	555	—	1,300,000	
Commercial		1250	1450	—	—	555	—	1,100,000	
EASTERN SOFTWOODS[a]									
Select structural	2–4 in. thick,	1350	1550	800	70	335	1050	1,200,000	NHPMA[b–m]
No. 1	2–4 in. wide	1150	1350	675	70	335	825	1,200,000	

(continued)

TABLE 7.3 (*Continued*) Design Values for Visually Graded Structural Lumber

Species and Commercial Grade	Size Classification	Extreme Fiber in Bending, F_b: Single-Member Uses	Extreme Fiber in Bending, F_b: Repetitive-Member Uses	Tension Parallel to Grain, F_t	Horizontal Shear, F_v	Compression Perpendicular to Grain, $F_{c\perp}$	Compression Parallel to Grain, F_c	Modulus of Elasticity, E	Grading Rules Agency
EASTERN SOFTWOODS[a] (cont.)									
No. 2	2-4 in. thick, 2-4 in. wide (cont.)	950	1100	550	70	335	650	1,100,000	
No. 3		525	600	300	70	335	400	1,000,000	
Stud	2–4 in. thick, 2–4 in. wide	525	600	300	70	335	400	1,000,000	NELMA, NHPMA[b–m]
Construction	2–4 in. thick, 4 in. wide	700	800	400	70	335	750	1,000,000	NHPMA[b–m]
Standard		375	450	225	70	335	625	1,000,000	
Utility		175	200	100	70	335	400	1,000,000	
Select structural	2–4 in. thick, 5 in. and wider	1150	1350	775[d]	70	335	925	1,200,000	
No. 1		1000	1150	675[d]	70	335	825	1,200,000	
No. 2		825	950	425[d]	70	335	700	1,100,000	
No. 3		475	550	250[d]	70	335	450	1,000,000	
Appearance		1000	1150	675[d]	70	335	1000	1,200,000	
Stud	2–4 in. thick, 5 in. and wider	475	550	250	70	335	450	1,000,000	NELMA, NHPMA[b–m]

EASTERN SPRUCE[a]									
Select structural	2–4 in. thick,	1400	1600	800	70	390	1050	1,500,000	NELMA,
No. 1	2–4 in. wide	1200	1350	700	70	390	825	1,500,000	NHPMA[b–m]
No. 2		975	1100	575	70	390	650	1,400,000	
No. 3		550	625	325	70	390	400	1,200,000	
Appearance		1200	1350	700	70	390	1000	1,500,000	
Stud		550	625	325	70	390	400	1,200,000	
Construction	2–4 in. thick,	700	800	400	70	390	750	1,200,000	
Standard	4 in. wide	400	450	225	70	390	625	1,200,000	
Utility		175	225	100	70	390	400	1,200,000	
Select structural	2–4 in. thick,	1200	1350	800[d]	70	390	925	1,500,000	
No. 1	5 in. and	1000	1150	675[d]	70	390	825	1,500,000	
No. 2	wider	825	950	425[d]	70	390	700	1,400,000	
No. 3		475	550	250[d]	70	390	450	1,200,000	
Appearance		1000	1150	675[d]	70	390	1000	1,500,000	
Stud		475	550	250[d]	70	390	450	1,200,000	
Select structural	Beams and	1050	—	725	65	390	750	1,400,000	
No. 1	stringers	900	—	600	65	390	625	1,400,000	
Select structural	Posts and	1000	—	675	65	390	775	1,400,000	
No. 1	timbers	800	—	550	65	390	675	1,400,000	
Select	Decking	—	1300	—	—	—	—	1,500,000	NELMA[b–m]
Commercial		—	1100	—	—	—	—	1,400,000	
EASTERN WHITE PINE[a]									
Select structural	2–4 in. thick,	1350	1550	800	70	350	1050	1,200,000	NELMA,
No. 1	2–4 in. wide	1150	1350	675	70	350	850	1,200,000	NHPMA[b–m]

(continued)

TABLE 7.3 (*Continued*) **Design Values for Visually Graded Structural Lumber**

Species and Commercial Grade	Size Classification	Extreme Fiber in Bending, F_b: Single-Member Uses	Extreme Fiber in Bending, F_b: Repetitive-Member Uses	Tension Parallel to Grain, F_t	Horizontal Shear, F_v	Compression Perpendicular to Grain, $F_{c\perp}$	Compression Parallel to Grain, F_c	Modulus of Elasticity, E	Grading Rules Agency
EASTERN WHITE PINE[a] (cont.)									
No. 2	2–4 in. thick,	950	1100	550	70	350	675	1,100,000	
No. 3	2–4 in. wide	525	600	300	70	350	400	1,000,000	
Appearance	(cont.)	1150	1350	675	70	350	1000	1,200,000	
Stud		525	600	300	70	350	400	1,000,000	
Construction	2–4 in. thick,	700	800	400	70	350	750	1,000,000	
Standard	4 in. wide	375	450	225	70	350	625	1,000,000	
Utility		175	200	100	70	350	400	1,000,000	
Select structural	2–4 in. thick,	1150	1350	775[d]	70	350	950	1,200,000	
No. 1	5 in. and	1000	1150	675[d]	70	350	850	1,200,000	
No. 2	wider	825	950	425[d]	70	350	700	1,100,000	
No. 3		475	550	250[d]	70	350	450	1,000,000	
Appearance		1000	1150	675[d]	70	350	1000	1,200,000	
Stud		475	550	250[d]	70	350	450	1,000,000	
Select structural	Beams and	1050	—	700	65	350	675	1,100,000	
No. 1	stringers	875	—	600	65	350	575	1,100,000	
Select structural	Posts and	975	—	650	65	350	725	1,100,000	
No. 1	timbers	800	—	525	65	350	625	1,100,000	

Select	Decking	1150	1300	—	—	—	—	1,200,000	NELMA [b–m]
Commercial		950	1100	—	—	—	—	1,100,000	
EASTERN WHITE PINE (NORTH)[a]									
Select structural	2–3 in. thick, 2–4 in. wide	1350	1550	800	65	350	1050	1,200,000	NLGA (a Canadian agency) [b–m, p, q]
No. 1		1150	1350	675	65	350	850	1,200,000	
No. 2		950	1100	550	65	350	675	1,100,000	
No. 3		525	600	300	65	350	400	1,000,000	
Appearance		1150	1350	675	65	350	1000	1,200,000	
Stud		525	600	300	65	350	400	1,000,000	
Construction	2–4 in. thick, 4 in. wide	700	800	400	65	350	750	1,000,000	
Standard		375	450	225	65	350	625	1,000,000	
Utility		175	200	100	65	350	400	1,000,000	
Select structural	2–4 in. thick, 5 in. and wider	1150	1350	775[d]	65	350	950	1,200,000	
No. 1		1000	1150	675[d]	65	350	850	1,200,000	
No. 2		825	950	425[d]	65	350	700	1,100,000	
No. 3		475	550	250[d]	65	350	450	1,000,000	
Appearance		1000	1150	675[d]	65	350	1000	1,200,000	
Stud		475	550	250[d]	65	350	450	1,000,000	
Select	Decking	900	1050	—	—	350	—	1,200,000	
Commercial		775	875	—	—	350	—	1,100,000	
EASTERN WOODS[a]									
Select structural	2–4 in. thick, 2–4 in. wide	1300	1500	775	60	270	850	1,100,000	NELMA, NHPMA [b–m]
No. 1		1100	1300	650	60	270	675	1,100,000	
No. 2		925	1050	525	60	270	550	1,000,000	
No. 3		500	575	300	60	270	325	900,000	
Appearance		1100	1300	650	60	270	825	1,100,000	
Stud		500	575	300	60	270	325	900,000	

(continued)

TABLE 7.3 (*Continued*) Design Values for Visually Graded Structural Lumber

Species and Commercial Grade	Size Classification	Extreme Fiber in Bending, F_b: Single-Member Uses	Repetitive-Member Uses	Tension Parallel to Grain, F_t	Horizontal Shear, F_v	Compression Perpendicular to Grain, $F_{c\perp}$	Compression Parallel to Grain, F_c	Modulus of Elasticity, E	Grading Rules Agency
EASTERN WOODS[a] (cont.)									
Construction	2–4 in. thick,	650	750	400	60	270	625	900,000	
Standard	4 in. wide	375	425	225	60	270	500	900,000	
Utility		175	200	100	60	270	325	900,000	
Select structural	2–4 in. thick,	1150	1300	750[d]	60	270	750	1,100,000	NHPMA [b–m]
No. 1	5 in. and	950	1100	650[d]	60	270	675	1,100,000	
No. 2	wider	775	900	425[d]	60	270	575	1,000,000	
No. 3		450	525	250[d]	60	270	375	900,000	
Appearance		950	1100	650[d]	60	270	825	1,100,000	
Stud		450	525	250[d]	60	270	375	900,000	
ENGELMANN SPRUCE–ALPINE FIR (ENGELMANN SPRUCE–LODGEPOLE PINE)[a]									
Select structural	2–4 in. thick,	1350	1550	800	70	320	950	1,300,000	WWPA [b–n]
No. 1	2–4 in. wide	1150	1350	675	70	320	750	1,300,000	
No. 2		950	1100	550	70	320	600	1,100,000	
No. 3		525	600	300	70	320	375	1,000,000	
Appearance		1150	1350	675	70	320	900	1,300,000	
Stud		525	600	300	70	320	375	1,000,000	
Construction	2–4 in. thick,	700	800	400	70	320	675	1,000,000	
Standard	4 in. wide	375	450	225	70	320	550	1,000,000	
Utility		175	200	100	70	320	375	1,000,000	

Select structural	2–4 in. thick,	1200	1350	775[d]	70	320	850	1,300,000	
No. 1	5 in. and	1000	1150	675[d]	70	320	750	1,300,000	
No. 2	wider	825	950	425[d]	70	320	625	1,100,000	
No. 3		475	550	250[d]	70	320	400	1,000,000	
Appearance		1000	1150	675[d]	70	320	900	1,300,000	
Stud		475	550	250[d]	70	320	400	1,000,000	
Select structural	Beams and	1050	—	700	65	320	675	1,100,000	
No. 1	stringers	875	—	600	65	320	550	1,100,000	
Select structural	Posts and	975	—	650	65	320	700	1,100,000	
No. 1	timbers	800	—	525	65	320	625	1,100,000	
Selected decking	Decking	—	1300	—	—	—	—	1,300,000	
Commercial decking		—	1100	—	—	—	—	1,100,000	
Selected decking	Decking	—	1400	Surfaced at 15% max. MC and used at 15% max. MC			—	1,300,000	
Commercial decking		—	1200				—	1,200,000	
HEM-FIR[a]									
Select structural	2–4 in. thick,	1650	1900	975	75	405	1300	1,500,000	WCLIB,
No. 1	2–4 wide	1400	1600	825	75	405	1050	1,500,000	WWPA[b–m]
No. 2		1150	1350	675	75	405	825	1,400,000	
No. 3		650	725	375	75	405	500	1,200,000	
Appearance		1400	1600	825	75	405	1250	1,500,000	
Stud		650	725	375	75	405	500	1,200,000	
Construction	2–4 in. thick,	825	975	500	75	405	925	1,200,000	
Standard	4 in. wide	475	550	275	75	405	775	1,200,000	
Utility		225	250	125	75	405	500	1,200,000	

(*continued*)

TABLE 7.3 (*Continued*) Design Values for Visually Graded Structural Lumber

Species and Commercial Grade	Size Classification	Extreme Fiber in Bending, F_b: Single-Member Uses	Extreme Fiber in Bending, F_b: Repetitive-Member Uses	Tension Parallel to Grain, F_t	Horizontal Shear, F_v	Compression Perpendicular to Grain, $F_{c\perp}$	Compression Parallel to Grain, F_c	Modulus of Elasticity, E	Grading Rules Agency
HEM-FIR[a] (cont.)									
Select structural	2–4 in. thick, 5 in. and wider	1400	1650	950[d]	75	405	1150	1,500,000	
No. 1		1200	1400	800[d]	75	405	1050	1,500,000	
No. 2		1000	1150	525[d]	75	405	875	1,400,000	
No. 3		575	675	300[d]	75	405	550	1,200,000	
Appearance		1200	1400	800[d]	75	405	1250	1,500,000	
Stud		575	675	300[d]	75	405	550	1,200,000	
Select structural	Beams and stringers	1300	—	750	70	405	925	1,300,000	WCLIB[b–m]
No. 1		1050	—	525	70	405	750	1,300,000	
Select structural	Posts and timbers	1200	—	800	70	405	975	1,300,000	
No 1		975	—	650	70	405	850	1,300,000	
Select dex	Decking	1400	1600	—	—	405	—	1,500,000	
Commercial dex		1150	1350	—	—	405	—	1,400,000	
Select structural	Beams and stringers	1250	—	850	70	405	925	1,300,000	WWPA[b–n]
No. 1		1050	—	725	70	405	775	1,300,000	
Select structural	Posts and timbers	1200	—	800	70	405	975	1,300,000	
No. 1		950	—	650	70	405	850	1,300,000	

Selected decking	Decking	—	1600	—	—	—	—	1,500,000	
Commercial decking		—	1350	—	—	—	—	1,400,000	
Selected decking	Decking	—	1700	Surfaced at 15% max. MC and used at 15% max. MC			—	1,600,000	
Commercial decking		—	1450				—	1,400,000	
HEM-FIR (NORTH) [a]									
Select structural	2–3 in. thick, 2–4 in. wide	1600	1800	925	75	370	1300	1,500,000	NLGA (a Canadian agency)[b–m, p, q]
No. 1		1350	1550	800	75	370	1050	1,500,000	
No. 2		1100	1300	650	75	370	800	1,400,000	
No. 3		625	700	350	75	370	500	1,200,000	
Appearance		1350	1550	800	75	370	1250	1,500,000	
Stud		625	700	350	75	370	500	1,200,000	
Construction	2–4 in. thick, 4 in. wide	800	925	475	75	370	925	1,200,000	
Standard		450	525	275	75	370	775	1,200,000	
Utility		225	250	125	75	370	500	1,200,000	
Select structural	2–4 in. thick, 5 in. and wider	1350	1550	900[d]	75	370	1150	1,500,000	
No. 1		1150	1350	775[d]	75	370	1050	1,500,000	
No. 2		950	1100	500[d]	75	370	850	1,400,000	
No. 3		550	650	300[d]	75	370	550	1,200,000	
Appearance		1150	1350	775[d]	75	370	1250	1,500,000	
Stud		550	650	300[d]	75	370	550	1,200,000	
Select structural	Beams and stringers	1250	—	725	70	370	900	1,300,000	
No. 1		1000	—	500	70	370	750	1,300,000	

(continued)

TABLE 7.3 (*Continued*) Design Values for Visually Graded Structural Lumber

Species and Commercial Grade	Size Classification	Extreme Fiber in Bending, F_b: Single-Member Uses	Extreme Fiber in Bending, F_b: Repetitive-Member Uses	Tension Parallel to Grain, F_t	Horizontal Shear, F_v	Compression Perpendicular to Grain, $F_{c\perp}$	Compression Parallel to Grain, F_c	Modulus of Elasticity, E	Grading Rules Agency
HEM-FIR[a] (NORTH) (cont.)									
Select structural	Posts and timbers	1150	—	775	70	370	950	1,300,000	
No. 1		925	—	625	70	370	850	1,300,000	
Select	Decking	1350	1500	—	—	370	—	1,500,000	
Commercial		1100	1300	—	—	370	—	1,400,000	
IDAHO WHITE PINE[a]									
Select structural	2–4 in. thick, 2–4 in. wide	1350	1550	775	70	315	1100	1,400,000	WWPA[b–n]
No. 1		1150	1300	650	70	315	875	1,400,000	
No. 2		925	1050	550	70	315	675	1,300,000	
No. 3		525	600	300	70	315	425	1,200,000	
Appearance		1150	1300	650	70	315	1050	1,400,000	
Stud		525	600	300	70	315	425	1,200,000	
Construction	2–4 in. thick, 4 in. wide	675	775	400	70	315	775	1,200,000	
Standard		375	425	225	70	315	650	1,200,000	
Utility		175	200	100	70	315	425	1,200,000	
Select structural	2–4 in. thick, 5 in. and wider	1150	1300	775[d]	70	315	950	1,400,000	
No. 1		975	1100	650[d]	70	315	875	1,400,000	
No. 2		800	925	425[d]	70	315	725	1,300,000	
No. 3		475	550	250[d]	70	315	450	1,200,000	
Appearance		975	1100	650[d]	70	315	1050	1,400,000	
Stud		475	550	250[d]	70	315	450	1,200,000	

Select structural	Beams and	1000	—	700	65	315	775	1,300,000	
No. 1	stringers	850	—	575	65	315	650	1,300,000	
Select structural	Posts and	950	—	650	65	315	800	1,300,000	
No. 1	timbers	775	—	525	65	315	700	1,300,000	
Selected decking	Decking	—	1300	—	—	—	—	1,400,000	
Commercial decking		—	1050	—	—	—	—	1,300,000	
Selected decking	Decking	—	1400	Surfaced at 15% max. MC and			—	1,500,000	
Commercial decking		—	1150	used at 15% max. MC			—	1,400,000	
LODGEPOLE PINE [a]									
Select structural	2–4 in. thick,	1500	1750	875	70	400	1150	1,300,000	WWPA [b–n]
No. 1	2–4 in. wide	1300	1500	750	70	400	900	1,300,000	
No. 2		1050	1200	625	70	400	700	1,200,000	
No. 3		600	675	350	70	400	425	1,000,000	
Appearance		1300	1500	750	70	400	1050	1,300,000	
Stud		600	675	350	70	400	425	1,000,000	
Construction	2–4 in. thick,	775	875	450	70	400	800	1,000,000	
Standard	4 in. wide	425	500	250	70	400	675	1,000,000	
Utility		200	225	125	70	400	425	1,000,000	
Select structural	2–4 in. thick,	1300	1500	875[d]	70	400	1000	1,300,000	
No. 1	5 in. and wider	1100	1300	750[d]	70	400	900	1,300,000	
No. 2		925	1050	475[d]	70	400	750	1,200,000	
No. 3		525	625	275[d]	70	400	475	1,000,000	
Appearance		1100	1300	750[d]	70	400	1050	1,300,000	
Stud		525	625	275[d]	70	400	475	1,000,000	

(continued)

TABLE 7.3 (*Continued*) Design Values for Visually Graded Structural Lumber

Species and Commercial Grade	Size Classification	Extreme Fiber in Bending, F_b: Single-Member Uses	Extreme Fiber in Bending, F_b: Repetitive-Member Uses	Tension Parallel to Grain, F_t	Horizontal Shear, F_v	Compression Perpendicular to Grain, $F_{c\perp}$	Compression Parallel to Grain, F_c	Modulus of Elasticity, E	Grading Rules Agency
LODGEPOLE PINE[a] (cont.)									
Select structural	Beams and stringers	1150	—	775	65	400	800	1,100,000	
No. 1		975	—	650	65	400	675	1,100,000	
Select structural	Posts and timbers	1100	—	725	65	400	850	1,100,000	
No. 1		875	—	600	65	400	725	1,100,000	
Selected decking	Decking	—	1450	—	—	—	—	1,300,000	
Commercial decking		—	1200	—	—	—	—	1,200,000	
Selected decking	Decking	—	1550	Surfaced at 15% max. MC and used at 15% max. MC			—	1,400,000	
Commercial decking		—	1300				—	1,200,000	
MOUNTAIN HEMLOCK[a]									
Select structural	2–4 in. thick, 2–4 in. wide	1750	2000	1000	95	570	1250	1,300,000	WCLIB, WWPA[b–m]
No. 1		1450	1700	850	95	570	1000	1,300,000	
No. 2		1200	1400	700	95	570	775	1,100,000	
No. 3		675	775	400	95	570	475	1,000,000	
Appearance		1450	1700	850	95	570	1200	1,300,000	
Stud		675	775	400	95	570	475	1,000,000	
Construction	2–4 in. thick, 4 in. wide	875	1000	525	95	570	900	1,000,000	
Standard		500	575	275	95	570	725	1,000,000	
Utility		225	275	125	95	570	475	1,000,000	

Select structural	2–4 in. thick,	1500	1700	1000[d]	95	570	1100	1,300,000	
No. 1	5 in. and wider	1250	1450	850[d]	95	570	1000	1,300,000	
No. 2		1050	1200	550[d]	95	570	825	1,100,000	
No. 3		625	700	325[d]	95	570	525	1,000,000	
Appearance		1250	1450	850[d]	95	570	1200	1,300,000	
Stud		625	700	325[d]	95	570	525	1,000,000	
Select structural	Beams and	1350	—	775	85	570	875	1,100,000	WCLIB [b–m]
No. 1	stringers	1100	—	550	85	570	725	1,100,000	
Select structural	Posts and	1250	—	825	85	570	925	1,100,000	
No. 1	timbers	1000	—	675	85	570	800	1,100,000	
Select dex	Decking	1450	1650	—	—	570	—	1,300,000	
Commercial dex		1200	1400	—	—	570	—	1,100,000	
Select structural	Beams and	1350	—	900	90	570	875	1,100,000	WWPA [b–n]
No. 1	stringers	1100	—	750	90	570	750	1,100,000	
Select structural	Posts and	1250	—	825	90	570	925	1,100,000	
No. 1	timbers	1000	—	675	90	570	800	1,100,000	
Selected decking	Decking	—	1650	—	—	—	—	1,300,000	
Commercial decking		—	1400	—	—	—	—	1,100,000	
Selected decking	Decking	—	1800	Surfaced at 15% max. MC and			—	1,300,000	
Commercial decking		—	1500	used at 15% max. MC			—	1,200,000	

(continued)

TABLE 7.3 (*Continued*) Design Values for Visually Graded Structural Lumber

Species and Commercial Grade	Size Classification	Extreme Fiber in Bending, F_b: Single-Member Uses	Extreme Fiber in Bending, F_b: Repetitive-Member Uses	Tension Parallel to Grain, F_t	Horizontal Shear, F_v	Compression Perpendicular to Grain, $F_{c\perp}$	Compression Parallel to Grain, F_c	Modulus of Elasticity, E	Grading Rules Agency
MOUNTAIN HEMLOCK–HEM-FIR [a]									
Select structural	2–4 in. thick,	1650	1900	975	75	405	1250	1,300,000	WWPA [b–n]
No. 1	2–4 in. wide	1400	1600	825	75	405	1000	1,300,000	
No. 2		1150	1350	675	75	405	775	1,100,000	
No. 3		650	725	375	75	405	475	1,000,000	
Appearance		1400	1600	825	75	405	1200	1,300,000	
Stud		650	725	375	75	405	475	1,000,000	
Construction	2–4 in. thick,	825	975	500	75	405	900	1,000,000	
Standard	4 in. wide	475	550	275	75	405	725	1,000,000	
Utility		225	250	125	75	405	475	1,000,000	
Select structural	2–4 in. thick,	1400	1650	950[d]	75	405	1100	1,300,000	
No. 1	5 in. and wider	1200	1400	800[d]	75	405	1000	1,300,000	
No. 2		1000	1150	525[d]	75	405	825	1,100,000	
No. 3		575	675	300[d]	75	405	525	1,000,000	
Appearance		1200	1400	800[d]	75	405	1200	1,300,000	
Stud		575	675	300[d]	75	405	525	1,000,000	
Select structural	Beams and	1250	—	850	70	405	875	1,100,000	
No. 1	stringers	1050	—	725	70	405	750	1,100,000	

Select structural	Posts and timbers	1200	—	800	70	405	925	1,100,000	
No. 1		950	—	650	70	405	800	1,100,000	
Selected decking	Decking	—	1600	—	—	—	—	1,300,000	
Commercial decking		—	1350	—	—	—	—	1,100,000	
Selected decking	Decking	—	1700	Surfaced at 15% max. MC and used at 15% max. MC			—	1,300,000	
Commercial decking		—	1450				—	1,200,000	
NORTHERN ASPEN [a]									
Select structural	2–3 in. thick, 2–4 in. wide	1300	1500	750	60	320	850	1,400,000	NLGA (a Canadian agency)[b–m, p, q]
No. 1		1100	1250	650	60	320	675	1,400,000	
No. 2		900	1050	525	60	320	525	1,200,000	
No. 3		500	575	275	60	320	325	1,100,000	
Appearance		1100	1250	650	60	320	800	1,400,000	
Stud		500	575	275	60	320	325	1,100,000	
Construction	2–4 in. thick, 4 in. wide	650	750	375	60	320	600	1,100,000	
Standard		350	425	200	60	320	500	1,100,000	
Utility		175	200	100	60	320	325	1,100,000	
Select structural	2–4 in. thick, 5 in. and wider	1100	1250	725[d]	60	320	750	1,400,000	
No. 1		950	1100	625[d]	60	320	675	1,400,000	
No. 2		775	900	400[d]	60	320	575	1,200,000	
No. 3		450	525	250[d]	60	320	350	1,100,000	
Appearance		950	1100	625[d]	60	320	800	1,400,000	
Stud		450	525	250[d]	60	320	350	1,100,000	
NORTHERN PINE [a]									
Select structural	2–4 in. thick, 2–4 in. wide	1650	1850	950	70	435	1200	1,400,000	NELMA, NHPMA [b–m]
No. 1		1400	1600	825	70	435	975	1,400,000	

(continued)

TABLE 7.3 (*Continued*) Design Values for Visually Graded Structural Lumber

Species and Commercial Grade	Size Classification	Extreme Fiber in Bending, F_b: Single-Member Uses	Extreme Fiber in Bending, F_b: Repetitive-Member Uses	Tension Parallel to Grain, F_t	Horizontal Shear, F_v	Compression Perpendicular to Grain, $F_{c\perp}$	Compression Parallel to Grain, F_c	Modulus of Elasticity, E	Grading Rules Agency
NORTHERN PINE [a] (cont.)									
No. 2	2–4 in. thick,	1150	1300	675	70	435	775	1,300,000	
No. 3	2–4 in. wide	625	725	375	70	435	475	1,100,000	
Appearance	(cont.)	1200	1400	800	70	435	1150	1,400,000	
Stud		625	725	375	70	435	475	1,100,000	
Construction	2–4 in. thick,	825	950	475	70	435	875	1,100,000	
Standard	4 in. wide	450	525	275	70	435	725	1,100,000	
Utility		225	250	125	70	435	475	1,100,000	
Select structural	2–4 in. thick,	1400	1600	950[d]	70	435	1100	1,400,000	
No. 1	5 in. and wider	1200	1400	800[d]	70	435	975	1,400,000	
No. 2		950	1100	525[d]	70	435	825	1,300,000	
No. 3		575	650	300[d]	70	435	525	1,100,000	
Appearance		1200	1400	800[d]	70	435	1150	1,400,000	
Stud		575	650	300[d]	70	435	525	1,100,000	
Select structural	Beams and	1250	—	850	65	435	850	1,300,000	
No. 1	stringers	1050	—	700	65	435	725	1,300,000	
Select structural	Posts and	1150	—	800	65	435	900	1,300,000	
No. 1	timbers	950	—	650	65	435	800	1,300,000	
Select	Decking	1350	1550	—	—	—	—	1,400,000	NELMA [b–m]
Commercial		1150	1300	—	—	—	—	1,300,000	

NORTHERN SPECIES [a]									
Select structural	2–3 in. thick, 2–4 in. wide	1350	1550	775	65	350	1050	1,100,000	NLGA (a Canadian agency)[b–m,p q]
No. 1		1150	1300	675	65	350	825	1,100,000	
No. 2		925	1050	550	65	350	650	1,000,000	
No. 3		525	600	300	65	350	400	900,000	
Appearance		1150	1300	675	65	350	975	1,100,000	
Stud		525	600	300	65	350	400	900,000	
Construction	2–4 in. thick, 4 in. wide	675	775	400	65	350	750	900,000	
Standard		375	425	225	65	350	600	900,000	
Utility		175	200	100	65	350	400	900,000	
Select structural	2–4 in. thick, 5 in. and wider	1150	1300	750[d]	65	350	900	1,100,000	
No. 1		975	1150	650[d]	65	350	825	1,100,000	
No. 2		800	925	425[d]	65	350	675	1,000,000	
No. 3		475	550	250[d]	65	350	425	900,000	
Appearance		975	1150	650[d]	65	350	975	1,100,000	
Stud		475	550	250[d]	65	350	425	900,000	
Select	Decking	900	1050	—	—	350	—	1,100,000	
Commercial		775	875	—	—	350	—	1,000,000	
NORTHERN WHITE CEDAR [a]									
Select structural	2–4 in. thick, 2–4 in. wide	1150	1350	700	65	370	875	800,000	NELMA [b–m]
No. 1		1000	1150	600	65	370	675	800,000	
No. 2		825	950	500	65	370	550	700,000	
No. 3		450	525	275	65	370	325	600,000	
Appearance		850	1000	575	65	370	825	800,000	
Stud		450	525	275	65	370	325	600,000	
Construction	2–4 in. thick, 4 in. wide	600	675	350	65	370	625	600,000	
Standard		325	375	200	65	370	500	600,000	
Utility		150	175	100	65	370	325	600,000	

(*continued*)

TABLE 7.3 (*Continued*) Design Values for Visually Graded Structural Lumber

Species and Commercial Grade	Size Classification	Extreme Fiber in Bending, F_b: Single-Member Uses	Extreme Fiber in Bending, F_b: Repetitive-Member Uses	Tension Parallel to Grain, F_t	Horizontal Shear, F_v	Compression Perpendicular to Grain, $F_{c\perp}$	Compression Parallel to Grain, F_c	Modulus of Elasticity, E	Grading Rules Agency
NORTHERN WHITE CEDAR [a] (cont.)									
Select structural	2–4 in. thick,	1000	1150	675[d]	65	370	775	800,000	
No. 1	5 in. and wider	850	1000	575[d]	65	370	675	800,000	
No. 2		700	825	375[d]	65	370	575	700,000	
No. 3		425	475	225[d]	65	370	375	600,000	
Appearance		850	1000	575[d]	65	370	825	800,000	
Stud		425	475	225[d]	65	370	375	600,000	
Select structural	Beams and	900	—	600	60	370	600	700,000	
No. 1	stringers	750	—	500	60	370	500	700,000	
Select structural	Posts and	850	—	575	60	370	650	700,000	
No. 1	timbers	675	—	450	60	370	550	700,000	
Select	Decking	975	1100	—	—	—	—	800,000	
Commercial		825	950	—	—	—	—	700,000	
PONDEROSA PINE [a]									
Select structural	2–3 in. thick,	1400	1650	825	70	535	1050	1,200,000	NLGA
No. 1	2–4 in. wide	1200	1400	700	70	535	850	1,200,000	(a Canadian
No. 2		1000	1150	575	70	535	675	1,100,000	agency)[b–m, p, q]
No. 3		550	625	325	70	535	400	1,000,000	
Appearance		1200	1400	700	70	535	1000	1,200,000	
Stud		550	625	325	70	535	400	1,000,000	

Construction	2–4 in. thick,	725	825	425	70	535	775	1,000,000	
Standard	4 in. wide	400	450	225	70	535	625	1,000,000	
Utility		200	225	100	70	535	400	1,000,000	
Select structural	2–4 in. thick,	1200	1400	825[d]	70	535	950	1,200,000	
No. 1	5 in. and wider	1050	1200	700[d]	70	535	850	1,200,000	
No. 2		850	975	450[d]	70	535	700	1,100,000	
No. 3		500	575	250[d]	70	535	450	1,000,000	
Appearance		1050	1200	700[d]	70	535	1000	1,200,000	
Stud		500	575	250[d]	70	535	450	1,000,000	
Select structural	Beams and	1100	—	725	65	535	750	1,100,000	
No. 1	stringers	925	—	500	65	535	625	1,100,000	
Select structural	Posts and	1000	—	675	65	535	800	1,100,000	
No. 1	timbers	825	—	550	65	535	700	1,100,000	
Select	Decking	1200	1450	—	—	535	—	1,300,000	
Commercial		1000	1250	—	—	535	—	1,100,000	
Ponderosa Pine–Sugar Pine (Ponderosa Pine–Lodgepole Pine) [a]									
Select structural	2–4 in. thick,	1400	1650	825	70	375	1050	1,200,000	WWPA [b–n]
No. 1	2–4 in. wide	1200	1400	700	70	375	850	1,200,000	
No. 2		1000	1150	575	70	375	675	1,100,000	
No. 3		550	625	325	70	375	400	1,000,000	
Appearance		1200	1400	700	70	375	1000	1,200,000	
Stud		550	625	325	70	375	400	1,000,000	
Construction	2–4 in. thick,	725	825	425	70	375	775	1,000,000	
Standard	4 in. wide	400	450	225	70	375	625	1,000,000	
Utility		200	225	100	70	375	400	1,000,000	

(continued)

TABLE 7.3 (*Continued*) Design Values for Visually Graded Structural Lumber

Species and Commercial Grade	Size Classification	Extreme Fiber in Bending, F_b: Single-Member Uses	Extreme Fiber in Bending, F_b: Repetitive-Member Uses	Tension Parallel to Grain, F_t	Horizontal Shear, F_v	Compression Perpendicular to Grain, $F_{c\perp}$	Compression Parallel to Grain, F_c	Modulus of Elasticity, E	Grading Rules Agency
PONDEROSA PINE–SUGAR PINE (PONDEROSA PINE–LODGEPOLE PINE) [a] (cont.)									
Select structural	2–4 in. thick,	1200	1400	825[d]	70	375	950	1,200,000	
No. 1	5 in. and wider	1050	1200	700[d]	70	375	850	1,200,000	
No. 2		850	975	450[d]	70	375	700	1,100,000	
No. 3		500	575	250[d]	70	375	450	1,000,000	
Appearance		1050	1200	700[d]	70	375	1000	1,200,000	
Stud		500	575	250[d]	70	375	450	1,000,000	
Select structural	Beams and	1100	—	725	65	375	750	1,100,000	
No. 1	stringers	925	—	625	65	375	625	1,100,000	
Select structural	Posts and	1000	—	675	65	375	800	1,100,000	
No. 1	timbers	825	—	550	65	375	700	1,100,000	
Selected decking	Decking	—	1350	—	—	—	—	1,200,000	
Commercial decking		—	1150	—	—	—	—	1,100,000	
Selected decking	Decking	—	1450	Surfaced at 15% max. MC and			—	1,300,000	
Commercial decking		—	1250	used at 15% max. MC			—	1,100,000	

Red Pine [a]									
Select structural	2–3 in. thick,	1400	1600	800	70	440	1050	1,300,000	NLGA
No. 1	2–4 in. wide	1200	1350	700	70	440	825	1,300,000	(a Canadian
No. 2		975	1100	575	70	440	650	1,200,000	agency)[b–m,p,q]
No. 3		525	625	325	70	440	400	1,000,000	
Appearance		1200	1350	700	70	440	975	1,300,000	
Stud		525	625	325	70	440	400	1,000,000	
Construction	2–4 in. thick,	700	800	400	70	440	750	1,000,000	
Standard	4 in. wide	400	450	225	70	440	600	1,000,000	
Utility		175	225	100	70	440	400	1,000,000	
Select structural	2–4 in. thick,	1200	1350	775[d]	70	440	900	1,300,000	
No. 1	5 in. and wider	1000	1150	675[d]	70	440	825	1,300,000	
No. 2		825	950	425[d]	70	440	675	1,200,000	
No. 3		500	550	250[d]	70	440	425	1,000,000	
Appearance		1000	1150	675[d]	70	440	975	1,300,000	
Stud		500	550	250[d]	70	440	425	1,000,000	
Select structural	Beams and	1050	—	625	65	440	725	1,100,000	
No. 1	stringers	875	—	450	65	440	600	1,100,000	
Select structural	Posts and	1000	—	675	65	440	775	1,100,000	
No. 1	timbers	800	—	550	65	440	675	1,100,000	
Select	Decking	1150	1350	—	—	440	—	1,300,000	
Commercial		975	1100	—	—	440	—	1,200,000	
Sitka Spruce [a]									
Select structural	2–4 in. thick,	1550	1800	925	75	435	1150	1,500,000	WCLIB [b–m]
No. 1	2–4 in. wide	1350	1550	775	75	435	925	1,500,000	
No. 2		1100	1250	650	75	435	725	1,300,000	

(continued)

TABLE 7.3 (*Continued*) Design Values for Visually Graded Structural Lumber

Species and Commercial Grade	Size Classification	Extreme Fiber in Bending, F_b: Single-Member Uses	Extreme Fiber in Bending, F_b: Repetitive-Member Uses	Tension Parallel to Grain, F_t	Horizontal Shear, F_v	Compression Perpendicular to Grain, $F_{c\perp}$	Compression Parallel to Grain, F_c	Modulus of Elasticity, E	Grading Rules Agency
SITKA SPRUCE [a] (cont.)									
No. 3		600	700	350	75	435	450	1,200,000	
Appearance		1350	1550	750	75	435	1100	1,500,000	
Stud		600	700	350	75	435	450	1,200,000	
Construction	2–4 in. thick,	800	925	475	75	435	825	1,200,000	
Standard	4 in. wide	450	500	250	75	435	675	1,200,000	
Utility		200	250	125	75	435	450	1,200,000	
Select structural	2–4 in. thick,	1350	1550	900[d]	75	435	1000	1,500,000	
No. 1	5 in. and wider	1150	1300	775[d]	75	435	925	1,500,000	
No. 2		925	1050	500[d]	75	435	775	1,300,000	
No. 3		525	600	275[d]	75	435	500	1,200,000	
Appearance		1150	1300	750[d]	75	435	1100	1,500,000	
Stud		525	600	275[d]	75	435	500	1,200,000	
Select structural	Beams and	1200	—	675	70	435	825	1,300,000	
No. 1	stringers	1000	—	500	70	435	675	1,300,000	
Select structural	Posts and	1150	—	750	70	435	875	1,300,000	
No. 1	timbers	925	—	600	70	435	750	1,300,000	
Select dex	Decking	1300	1500	—	—	435	—	1,500,000	
Commercial dex		1100	1250	—	—	435	—	1,300,000	

SOUTHERN PINE [u]									
Select structural	2–4 in. thick, 2–4 in. wide	2150	2500	1250	105	565	1800	1,800,000	SPIB [b, d–g, m, r–t]
Dense select structural		2500	2900	1500	105	660	2100	1,900,000	
No. 1		1850	2100	1050	105	565	1450	1,800,000	
No. 1 dense		2150	2450	1250	105	660	1700	1,900,000	
No. 2		1550	1750	900	95	565	1150	1,600,000	
No. 2 dense		1800	2050	1050	95	660	1350	1,700,000	
No. 3		850	975	500	95	565	675	1,500,000	
No. 3 dense		1000	1150	575	95	660	800	1,500,000	
Stud		850	975	500	95	565	675	1,500,000	
Construction	2–4 in. thick, 4 in. wide	1100	1250	650	105	565	1300	1,500,000	
Standard		625	725	375	95	565	1050	1,500,000	
Utility		275	300	175	95	565	675	1,500,000	
Select structural	2–4 in. thick, 5 in. and wider	1850	2150	1200[d]	95	565	1600	1,800,000	
Dense select structural		2200	2500	1450[d]	95	660	1850	1,900,000	
No. 1		1600	1850	1050[d]	95	565	1450	1,800,000	
No. 1 dense		1850	2150	1250[d]	95	660	1700	1,900,000	
No. 2		1300	1500	675[d]	95	565	1200	1,600,000	
No. 2 dense		1550	1750	800[d]	95	660	1400	1,700,000	
No. 3		750	875	400[d]	95	565	725	1,500,000	
No. 3 dense		875	1000	450[d]	95	660	850	1,500,000	
Stud		800	900	400[d]	95	565	725	1,500,000	
Dense standard decking	2–4 in. thick, 2 in. and wider decking	2150	2450	—	—	660	—	1,900,000	
Select decking		1550	1750	—	—	565	—	1,600,000	
Dense select decking		1800	2050	—	—	660	—	1,700,000	

(continued)

TABLE 7.3 (*Continued*) Design Values for Visually Graded Structural Lumber

Species and Commercial Grade	Size Classification	Extreme Fiber in Bending, F_b		Tension Parallel to Grain, F_t	Horizontal Shear, F_v	Compression Perpendicular to Grain, $F_{c\perp}$	Compression Parallel to Grain, F_c	Modulus of Elasticity, E	Grading Rules Agency
		Single-Member Uses	Repetitive-Member Uses						
SOUTHERN PINE [a] (cont.)									
Commercial decking	2–4 in. thick,	1550	1750	—	—	565	—	1,600,000	
Dense commercial	2 in. and wider								
decking	decking	1800	2050	—	—	660	—	1,700,000	
Dense structural 86	2–4 in. thick	2800	3250	1900	165	660	2300	1,900,000	
Dense structural 72		2400	2750	1600	135	660	1950	1,900,000	
Dense structural 65		2150	2450	1450	125	660	1750	1,900,000	
SOUTHERN PINE [v]									
Select structural	2–4 in. thick,	2000	2300	1150	100	565	1550	1,700,000	SPIB [b, d–g, m, r–t]
Dense select	2–4 in. wide								
structural		2350	2700	1350	100	660	1800	1,800,000	
No. 1		1700	1950	1000	100	565	1250	1,700,000	
No. 1 dense		2000	2300	1150	100	660	1450	1,800,000	
No. 2		1400	1650	825	90	565	975	1,600,000	
No. 2 dense		1650	1900	975	90	660	1150	1,600,000	
No. 3		775	900	450	90	565	575	1,400,000	
No. 3 dense		925	1050	525	90	660	675	1,500,000	
Stud		775	900	450	90	565	575	1,400,000	
Construction	2–4 in. thick,	1000	1150	600	100	565	1100	1,400,000	
Standard	4 in. wide	575	675	350	90	565	900	1,400,000	
Utility		275	300	150	90	565	575	1,400,000	

Select structural	2–4 in. thick, 5 in. and wider	1750	2000	1150[d]	90	565	1350	1,700,000	
Dense select structural		2050	2350	1300[d]	90	660	1600	1,800,000	
No. 1		1450	1700	975[d]	90	565	1250	1,700,000	
No. 1 dense		1700	2000	1150[d]	90	660	1450	1,800,000	
No. 2		1200	1400	625[d]	90	565	1000	1,600,000	
No. 2 dense		1400	1650	725[d]	90	660	1200	1,600,000	
No. 3		700	800	350[d]	90	565	625	1,400,000	
No. 3 dense		825	925	425[d]	90	660	725	1,500,000	
Stud		725	850	350[d]	90	565	625	1,400,000	
Dense standard decking	2–4 in. thick, 2 in. and wider decking	2000	2300	—	—	660	—	1,800,000	
Select decking		1400	1650	—	—	565	—	1,600,000	
Dense select decking		1650	1900	—	—	660	—	1,600,000	
Commercial decking		1400	1650	—	—	565	—	1,600,000	
Dense commercial decking		1650	1900	—	—	660	—	1,600,000	
Dense structural 86	2–4 in. thick	2600	3000	1750	155	660	2000	1,800,000	
Dense structural 72		2200	2550	1450	130	660	1650	1,800,000	
Dense structural 65		2000	2300	1300	115	660	1500	1,800,000	
SOUTHERN PINE [w]									
Select structural	$2\frac{1}{2}$–4 in. thick, $2\frac{1}{2}$–4 in. wide	1600	1850	925	95	375	1050	1,500,000	SPIB [b, d–g, m, r–t]
Dense select structural		1850	2150	1100	95	440	1200	1,600,000	
No. 1		1350	1550	800	95	375	825	1,500,000	
No. 1 dense		1600	1800	925	95	440	950	1,600,000	
No. 2		1150	1300	675	85	375	650	1,400,000	
No. 2 dense		1350	1500	775	85	440	750	1,400,000	
No. 3		625	725	375	85	375	400	1,200,000	
No. 3 dense		725	850	425	85	440	450	1,300,000	
Stud		625	725	375	85	375	400	1,200,000	

(continued)

TABLE 7.3 (*Continued*) Design Values for Visually Graded Structural Lumber

Species and Commercial Grade	Size Classification	Extreme Fiber in Bending, F_b: Single-Member Uses	Extreme Fiber in Bending, F_b: Repetitive-Member Uses	Tension Parallel to Grain, F_t	Horizontal Shear, F_v	Compression Perpendicular to Grain, $F_{c\perp}$	Compression Parallel to Grain, F_c	Modulus of Elasticity, E	Grading Rules Agency
SOUTHERN PINE [a] (cont.)									
Construction	$2\frac{1}{2}$–4 in. thick,	825	925	475	95	375	725	1,200,000	
Standard	4 in. wide	475	525	275	85	375	600	1,200,000	
Utility		200	250	125	85	375	400	1,200,000	
Select structural	$2\frac{1}{2}$–4 in. thick,	1400	1600	900[d]	85	375	900	1,500,000	
Dense select structural	5 in. and wider	1600	1850	1050[d]	85	440	1050	1,600,000	
No. 1		1200	1350	775[d]	85	375	825	1,500,000	
No. 1 dense		1400	1600	925[d]	85	440	950	1,600,000	
No. 2		975	1100	500[d]	85	375	675	1,400,000	
No. 2 dense		1150	1300	600[d]	85	440	800	1,400,000	
No. 3		550	650	300[d]	85	375	425	1,200,000	
No. 3 dense		650	750	350[d]	85	440	475	1,300,000	
Stud		575	675	300[d]	85	375	425	1,200,000	
Dense standard decking	$2\frac{1}{2}$–4 in. thick, 2 in. and wider	1600	1800	—	—	440	—	1,600,000	
Select decking	decking	1150	1300	—	—	375	—	1,400,000	
Dense select decking		1350	1500	—	—	440	—	1,400,000	
Commercial decking		1150	1300	—	—	375	—	1,400,000	
Dense commercial decking		1350	1500	—	—	440	—	1,400,000	

No. 1 SR	5 in. and thicker	1350	—	875	110	375	775	1,500,000	
No. 1 dense SR		1550	—	1050	110	440	925	1,600,000	
No. 2 SR		1100	—	725	95	375	625	1,400,000	
No. 2 dense SR		1250	—	850	95	440	725	1,400,000	
Dense structural 86	$2\frac{1}{2}$ in. and thicker	2100	2400	1400	145	440	1300	1,600,000	
Dense structural 72		1750	2050	1200	120	440	1100	1,600,000	
Dense structural 65		1600	1800	1050	110	440	1000	1,600,000	
SPRUCE-PINE-FIR [a]									
Select structural	2–3 in. thick, 2–4 in. wide	1450	1650	850	70	425	1100	1,500,000	NLGA (a Canadian agency)[b–m, p, q]
No. 1		1200	1400	725	70	425	875	1,500,000	
No. 2		1000	1150	600	70	425	675	1,300,000	
No. 3		550	650	325	70	425	425	1,200,000	
Appearance		1200	1400	725	70	425	1050	1,500,000	
Stud		550	650	325	70	425	425	1,200,000	
Construction	2–4 in. thick, 4 in. wide	725	850	425	70	425	775	1,200,000	
Standard		400	475	225	70	425	650	1,200,000	
Utility		175	225	100	70	425	425	1,200,000	
Select structural	2–4 in. thick, 5 in. and wider	1250	1450	825[d]	70	425	975	1,500,000	
No. 1		1050	1200	700[d]	70	425	875	1,500,000	
No. 2		875	1000	450[d]	70	425	725	1,300,000	
No. 3		500	575	275[d]	70	425	450	1,200,000	
Appearance		1050	1200	700[d]	70	425	1050	1,500,000	
Stud		500	575	275[d]	70	425	450	1,200,000	
Select structural	Beams and stringers	1100	—	650	65	425	775	1,300,000	
No. 1		900	—	450	65	425	625	1,300,000	

(*continued*)

TABLE 7.3 (*Continued*) Design Values for Visually Graded Structural Lumber

Species and Commercial Grade	Size Classification	Extreme Fiber in Bending, F_b: Single-Member Uses	Extreme Fiber in Bending, F_b: Repetitive-Member Uses	Tension Parallel to Grain, F_t	Horizontal Shear, F_v	Compression Perpendicular to Grain, $F_{c\perp}$	Compression Parallel to Grain, F_c	Modulus of Elasticity, E	Grading Rules Agency
SPRUCE-PINE-FIR [a] (cont.)									
Select structural	Posts and	1050	—	700	65	425	800	1,300,000	
No. 1	timbers	850	—	550	65	425	700	1,300,000	
Select	Decking	1200	1400	—	—	425	—	1,500,000	
Commercial		1000	1150	—	—	425	—	1,300,000	
VIRGINIA PINE-POND PINE [u]									
Select structural	2-4 in. thick,	2150	2500	1250	105	565	1750	1,600,000	SPIB [b, d-g, m,o, s, t]
No. 1	2-4 in. wide	1850	2150	1050	105	565	1400	1,600,000	
No. 2		1550	1800	900	95	565	1100	1,400,000	
No. 3		850	975	500	95	565	650	1,200,000	
Stud		850	975	500	95	565	650	1,200,000	
Construction	2-4 in. thick,	1100	1300	650	105	565	1250	1,200,000	
Standard	4 in. wide	625	725	375	95	565	1000	1,200,000	
Utility		275	325	175	95	565	650	1,200,000	
Select structural	2-4 in. thick,	1850	2150	1200[d]	95	565	1500	1,600,000	
No. 1	5 in. and wider	1600	1850	1050[d]	95	565	1400	1,600,000	
No. 2		1300	1500	675[d]	95	565	1150	1,400,000	
No. 3		750	875	400[d]	95	565	700	1,200,000	
Stud		800	925	400[d]	95	565	700	1,200,000	

Virginia Pine–Pond Pine[v]									
Select structural	2–4 in. thick,	2000	2300	1150	100	565	1500	1,500,000	SPIB[b, d–g, m,o, s, t]
No. 1	2–4 in. wide	1700	1950	1000	100	565	1200	1,500,000	
No.2		1400	1650	825	90	565	950	1,300,000	
No. 3		775	900	450	90	565	575	1,200,000	
Stud		775	900	450	90	565	575	1,200,000	
Construction	2–4 in. thick,	1000	1200	600	100	565	1050	1,200,000	
Standard	4 in. wide	575	675	350	90	565	875	1,200,000	
Utility		275	300	150	90	565	575	1,200,000	
Select structural	2–4 in. thick,	1750	2000	1150[d]	90	565	1300	1,500,000	
No. 1	5 in. and wider	1450	1700	975[d]	90	565	1200	1,500,000	
No. 2		1200	1400	625[d]	90	565	975	1,300,000	
No. 3		700	800	350[d]	90	565	600	1,200,000	
Stud		725	850	350[d]	90	565	600	1,200,000	
Virginia Pine–Pond Pine[w]									
Select structural	2–4 in. thick,	1600	1850	925	95	375	1000	1,300,000	SPIB[b, d–g, m,o, s, t]
No. 1	2–4 in. wide	1350	1550	800	95	375	800	1,300,000	
No. 2		1150	1300	675	85	375	625	1,200,000	
No. 3		625	725	375	85	375	375	1,000,000	
Stud		625	725	375	85	375	375	1,000,000	
Construction	2–4 in. thick,	825	950	475	95	375	700	1,000,000	
Standard	4 in. wide	475	550	275	85	375	575	1,000,000	
Utility		200	250	125	85	375	375	1,000,000	
Select structural	2–4 in. thick,	1400	1600	900[d]	85	375	875	1,300,000	
No. 1	5 in. and wider	1200	1350	775[d]	85	375	800	1,300,000	
No. 2		975	1100	500[d]	85	375	650	1,200,000	
No. 3		550	650	300[d]	85	375	400	1,000,000	
Stud		575	675	300[d]	85	375	400	1,000,000	

(continued)

TABLE 7.3 (*Continued*) Design Values for Visually Graded Structural Lumber

Species and Commercial Grade	Size Classification	Extreme Fiber in Bending, F_b: Single-Member Uses	Extreme Fiber in Bending, F_b: Repetitive-Member Uses	Tension Parallel to Grain, F_t	Horizontal Shear, F_v	Compression Perpendicular to Grain, $F_{c\perp}$	Compression Parallel to Grain, F_c	Modulus of Elasticity, E	Grading Rules Agency
WESTERN CEDARS[a]									
Select structural	2–4 in. thick,	1500	1750	875	75	425	1200	1,100,000	WCLIB,
No. 1	2–4 in. wide	1300	1500	750	75	425	950	1,100,000	WWPA[b–m]
No. 2		1050	1200	625	75	425	750	1,000,000	
No. 3		600	675	350	75	425	450	900,000	
Appearance		1300	1500	750	75	425	1100	1,100,000	
Stud		600	675	350	75	425	450	900,000	
Construction	2–4 in. thick,	775	875	450	75	425	850	900,000	
Standard	4 in. wide	425	500	250	75	425	700	900,000	
Utility		200	225	125	75	425	450	900,000	
Select structural	2–4 in. thick	1300	1500	875[d]	75	425	1050	1,100,000	
No. 1	5 in. and wider	1100	1300	750[d]	75	425	950	1,100,000	
No. 2		925	1050	475[d]	75	425	800	1,000,000	
No. 3		525	625	275[d]	75	425	500	900,000	
Appearance		1100	1300	750[d]	75	425	1100	1,100,000	
Stud		525	625	275[d]	75	425	500	900,000	
Select structural	Beams and	1150	—	675	70	425	875	1,000,000	WCLIB[b–m]
No. 1	stringers	975	—	475	70	425	725	1,000,000	
Select structural	Posts and	1100	—	725	70	425	925	1,000,000	
No. 1	timbers	875	—	600	70	425	800	1,000,000	

Select dex	Decking	1250	1450	—	—	425	—	1,100,000	
Commercial dex		1050	1200	—	—	425	—	1,000,000	
Select structural	Beams and stringers	1150	—	775	70	425	875	1,000,000	WWPA [b–n]
No. 1		975	—	650	70	425	725	1,000,000	
Select structural	Posts and timbers	1100	—	725	70	425	925	1,000,000	
No. 1		875	—	600	70	425	800	1,000,000	
Selected decking	Decking	—	1450	—	—	—	—	1,100,000	
Commercial decking		—	1200	—	—	—	—	1,000,000	
Selected decking	Decking	—	1550	Surfaced at 15% max. MC and used at 15% max. MC			—	1,100,000	
Commercial decking		—	1300				—	1,000,000	
WESTERN CEDARS (NORTH) [a]									
Select structural	2–3 in. thick, 2–4 in. wide	1450	1700	850	70	425	1200	1,100,000	NLGA (a Canadian agency) [b–m, p, q]
No. 2		1250	1450	725	70	425	950	1,100,000	
No. 2		1000	1200	600	70	425	750	1,000,000	
No. 3		575	650	325	70	425	450	900,000	
Appearance		1250	1450	725	70	425	1100	1,100,000	
Stud		575	650	325	70	425	450	900,000	
Construction	2–4 in. thick, 4 in. wide	750	850	425	70	425	850	900,000	
Standard		425	475	250	70	425	700	900,000	
Utility		200	225	125	70	425	450	900,000	
Select structural	2–4 in. thick, 5 in. and wider	1250	1450	825[d]	70	425	1050	1,100,000	
No. 1		1050	1250	725[d]	70	425	950	1,100,000	
No. 2		875	1000	475[d]	70	425	800	1,000,000	
No. 3		525	600	275[d]	70	425	500	900,000	
Appearance		1050	1250	725[d]	70	425	1100	1,100,000	
Stud		525	600	275[d]	70	425	500	900,000	

(continued)

TABLE 7.3 (*Continued*) Design Values for Visually Graded Structural Lumber

Species and Commercial Grade	Size Classification	Extreme Fiber in Bending, F_b: Single-Member Uses	Extreme Fiber in Bending, F_b: Repetitive-Member Uses	Tension Parallel to Grain, F_t	Horizontal Shear, F_v	Compression Perpendicular to Grain, $F_{c\perp}$	Compression Parallel to Grain, F_c	Modulus of Elasticity, E	Grading Rules Agency
WESTERN CEDARS (NORTH)[a] (cont.)									
Select structural	Beams and	1150	—	675	65	425	850	1,000,000	
No. 1	stringers	925	—	475	65	425	700	1,000,000	
Select structural	Posts and	1050	—	700	65	425	900	1,000,000	
No. 1	timbers	875	—	575	65	425	800	1,000,000	
Select	Decking	1200	1400	—	—	425	—	1,100,000	
Commercial		1050	1200	—	—	425	—	1,000,000	
WESTERN HEMLOCK[a]									
Select structural	2–4 in. thick,	1800	2100	1050	90	410	1450	1,600,000	WCLIB,
No. 1	2–4 in. wide	1550	1800	900	90	410	1150	1,600,000	WWPA[b–m]
No. 2		1300	1450	750	90	410	900	1,400,000	
No. 3		700	800	425	90	410	550	1,300,000	
Appearance		1550	1800	900	90	410	1350	1,600,000	
Stud		700	800	425	90	410	550	1,300,000	
Construction	2–4 in. thick,	925	1050	550	90	410	1050	1,300,000	
Standard	4 in. wide	525	600	300	90	410	850	1,300,000	
Utility		250	275	150	90	410	550	1,300,000	

Select structural	2–4 in. thick,	1550	1800	1050[d]	90	410	1300	1,600,000	
No. 1	5 in. and wider	1350	1550	900[d]	90	410	1150	1,600,000	
No. 2		1100	1250	575[d]	90	410	975	1,400,000	
No. 3		650	750	325[d]	90	410	625	1,300,000	
Appearance		1350	1550	900[d]	90	410	1350	1,600,000	
Stud		650	750	325[d]	90	410	625	1,300,000	
Select structural	Beams and	1400	—	825	85	410	1000	1,400,000	WCLIB [b–m]
No. 1	stringers	1150	—	575	85	410	850	1,400,000	
Select structural	Posts and	1300	—	875	85	410	1100	1,400,000	
No. 1	timbers	1050	—	700	85	410	950	1,400,000	
Select dex	Decking	1500	1750	—	—	410	—	1,600,000	
Commercial dex		1300	1450	—	—	410	—	1,400,000	
Select structural	Beams and	1400	—	950	85	410	1000	1,400,000	WWPA [b–n]
No. 1	stringers	1150	—	775	85	410	850	1,400,000	
Select structural	Posts and	1300	—	875	85	410	1100	1,400,000	
No. 1	timbers	1050	—	700	85	410	950	1,400,000	
Selected decking	Decking	—	1750	—	—	—	—	1,600,000	
Commercial decking		—	1450	—	—	—	—	1,400,000	
Selected decking	Decking	—	1900	Surfaced at 15% max. MC and			—	1,700,000	
Commercial decking		—	1600	used at 15% max. MC			—	1,500,000	
WESTERN HEMLOCK (NORTH)[a]									
Select structural	2–3 in. thick,	1800	2100	1050	75	410	1450	1,600,000	NLGA
No. 1	2–4 in. wide	1550	1800	900	75	410	1150	1,600,000	(a Canadian
No. 2		1300	1450	750	75	410	900	1,400,000	agency) [b–m, p, q]

(continued)

TABLE 7.3 (*Continued*) Design Values for Visually Graded Structural Lumber

Species and Commercial Grade	Size Classification	Extreme Fiber in Bending, F_b: Single-Member Uses	Extreme Fiber in Bending, F_b: Repetitive-Member Uses	Tension Parallel to Grain, F_t	Horizontal Shear, F_v	Compression Perpendicular to Grain, $F_{c\perp}$	Compression Parallel to Grain, F_c	Modulus of Elasticity, E	Grading Rules Agency
WESTERN HEMLOCK (NORTH)[a] (cont.)									
No. 3	2–3 in. thick,	700	800	425	75	410	550	1,300,000	
Appearance	2–4 in. wide	1550	1800	900	75	410	1350	1,600,000	
Stud	(cont.)	700	800	425	75	410	550	1,300,000	
Construction	2–4 in. thick,	925	1050	550	75	410	1050	1,300,000	
Standard	4 in. wide	525	600	300	75	410	850	1,300,000	
Utility		250	275	150	75	410	550	1,300,000	
Select structural	2–4 in. thick,	1550	1800	1050[d]	75	410	1300	1,600,000	
No. 1	5 in. and wider	1350	1550	900[d]	75	410	1150	1,600,000	
No. 2		1100	1250	575[d]	75	410	975	1,400,000	
No. 3		650	750	325[d]	75	410	625	1,300,000	
Appearance		1350	1550	900[d]	75	410	1350	1,600,000	
Stud		650	750	325[d]	75	410	625	1,300,000	
Select structural	Beams and	1400	—	825	70	410	1000	1,400,000	
No. 1	stringers	1150	—	575	70	410	850	1,400,000	
Select structural	Posts and	1300	—	875	70	410	1100	1,400,000	
No. 1	timbers	1050	—	700	70	410	950	1,400,000	
Select	Decking	1500	1750	—	—	410	—	1,600,000	
Commercial		1300	1450	—	—	410	—	1,400,000	

WESTERN WHITE PINE[a]									
Select structural	2–3 in. thick,	1350	1550	775	65	375	1100	1,400,000	NLGA
No. 1	2–4 in. wide	1150	1300	675	65	375	875	1,400,000	(a Canadian
No. 2		925	1050	550	65	375	675	1,300,000	agency)[b–m, p, q]
No. 3		525	600	300	65	375	425	1,200,000	
Appearance		1150	1300	675	65	375	1050	1,400,000	
Stud		525	600	300	65	375	425	1,200,000	
Construction	2–4 in. thick,	675	775	400	65	375	775	1,200,000	
Standard	4 in. wide	375	425	225	65	375	650	1,200,000	
Utility		175	200	100	65	375	425	1,200,000	
Select structural	2–4 in. thick,	1150	1300	750[d]	65	375	975	1,400,000	
No. 1	5 in. and wider	975	1150	650[d]	65	375	875	1,400,000	
No. 2		800	925	425[d]	65	375	725	1,300,000	
No. 3		475	550	250[d]	65	375	450	1,200,000	
Appearance		975	1150	650[d]	65	375	1050	1,400,000	
Stud		475	550	250[d]	65	375	450	1,200,000	
Select structural	Beams and	1050	—	600	60	375	775	1,300,000	
No. 1	stringers	850	—	425	60	375	625	1,300,000	
Select structural	Posts and	975	—	650	60	375	800	1,300,000	
No. 1	timbers	775	—	525	60	375	700	1,300,000	
Select	Decking	1100	1300	—	—	375	—	1,400,000	
Commercial		925	1050	—	—	375	—	1,300,000	
WHITE WOODS (WESTERN WOODS)[a]									
Select structural	2–4 in. thick,	1350	1550	775	70	315	950	1,100,000	WWPA[b–n]
No. 1	2–4 in. wide	1150	1300	650	70	315	750	1,100,000	

(continued)

Table 7.3 (*Continued*)

Design Values for Visually Graded Structural Lumber

Species and Commercial Grade	Size Classification	Extreme Fiber in Bending, F_b: Single-Member Uses	Extreme Fiber in Bending, F_b: Repetitive-Member Uses	Tension Parallel to Grain, F_t	Horizontal Shear, F_v	Compression Perpendicular to Grain, $F_{c\perp}$	Compression Parallel to Grain, F_c	Modulus of Elasticity, E	Grading Rules Agency
WHITE WOODS (WESTERN WOODS)[a] (cont.)									
No. 2	2–4 in. thick,	925	1050	550	70	315	600	1,000,000	
No. 3	2–4 in. wide	525	600	300	70	315	375	900,000	
Appearance	(cont.)	1150	1300	650	70	315	900	1,100,000	
Stud		525	600	300	70	315	375	900,000	
Construction	2–4 in. thick,	675	775	400	70	315	675	900,000	
Standard	4 in. wide	375	425	225	70	315	550	900,000	
Utility		175	200	100	70	315	375	900,000	
Select structural	2–4 in. thick,	1150	1300	775[d]	70	315	850	1,100,000	
No. 1	5 in. and wider	975	1100	650[d]	70	315	750	1,100,000	
No. 2		800	925	425[d]	70	315	625	1,000,000	
No. 3		475	550	250[d]	70	315	400	900,000	
Appearance		975	1100	650[d]	70	315	900	1,100,000	
Stud		475	550	250[d]	70	315	400	900,000	
Select structural	Beams and	1000	—	700	65	315	675	1,000,000	
No. 1	stringers	850	—	575	65	315	550	1,000,000	
Select structural	Posts and	950	—	650	65	315	700	1,000,000	
No. 1	timbers	775	—	525	65	315	625	1,000,000	

Selected decking	Decking	—	1300	—	—	—	—	1,100,000	
Commercial decking		—	1050	—	—	—	—	1,000,000	
Selected decking	Decking	—	1400	Surfaced at15% max. MC and used at 15% max. MC			—	1,100,000	
Commercial decking		—	1150				—	1,000,000	
YELLOW POPLAR[a]									
Select structural	2–3 in. thick, 2–4 in wide	1500	1700	875	80	420	1050	1,500,000	NHPMA [b–m]
No. 1		1250	1450	750	80	420	825	1,500,000	
No. 2		1050	1200	625	75	420	650	1,300,000	
No. 3		575	675	350	75	420	400	1,200,000	
Stud		575	675	350	75	420	400	1,200,000	
Construction	2–4 in. thick, 4 in. wide	750	875	450	80	420	750	1,200,000	
Standard		425	500	250	75	420	625	1,200,000	
Utility		200	225	125	75	420	400	1,200,000	
Select structural	2–4 in. thick, 5 in. and wider	1300	1500	850[d]	75	420	925	1,500,000	
No. 1		1100	1250	725[d]	75	420	825	1,500,000	
No. 2		900	1050	475[d]	75	420	700	1,300,000	
No. 3		525	600	275[d]	75	420	425	1,200,000	
Appearance		1100	1250	725[d]	75	420	1000	1,500,000	
Stud		525	600	275[d]	75	420	425	1,200,000	

*The design values listed in Tables 7.3 and 7.4 were reproduced from *Design Values for Wood Construction*, a supplement to the 1982 edition of the *National Design Specification for Wood Construction* by the National Forest Products Association (NFPA). This supplement is revised periodically, so the designer should check with NFPA for the latest information (Design values listed are for normal loading conditions. See other provisions in the footnotes and in Section 3 for adjustments of tabulated values.)

[a] Surfaced dry or surfaced green; used at 19% MC maximum.

Footnotes to Table 7.3 (*Continued*)

[b] Following is a list of agencies certified by the American Lumber Standards Committee Board of Review (as of 1982) for inspection and grading of untreated lumber under the rules indicated. For the most up-to-date list of certified agencies, write to American Lumber Standards Committee, P.O. Box 210, Germantown, Maryland 20874.

Rules-writing agencies	Rules for which grading authorized
Northeastern Lumber Manufacturers Association (NELMA), 4 Fundy Road, Falmouth, ME 04105	NELMA, NLGA
Northern Hardwood and Pine Manufacturers Association (NHPMA), Northern Bldg., Green Bay, WI 54301	NHPMA, WCLIB, WWPA, NLGA
Redwood Inspection Service (RIS), One Lombard St., San Francisco, CA 94111	RIS, WCLIB, WWPA
Southern Pine Inspection Bureau (SPIB), 4709 Scenic Highway, Pensacola, FL 32504	SPIB, NELMA
West Coast Lumber Inspection Bureau (WCLIB), 6980 SW Varnes Rd., P.O. Box 23145, Portland, OR 97223	WCLIB, RIS, WWPA, NLGA
Western Wood Products Association (WWPA), 1500 Yeon Building, Portland, OR 97204	WWPA, WCLIB, NLGA, RIS
National Lumber Grades Authority (NLGA), P.O. Box 97, Ganges, B.C., Canada VDS 1EO	
Non-rules-writing agencies	
California Lumber Inspection Service	RIS, WCLIB, WWPA, NLGA
Pacific Lumber Inspection Bureau, Inc.	RIS, WCLIB, WWPA, NLGA
Timber Products Inspection	RIS, SPIB, WCLIB, WWPA, NHPMA, NELMA, NLGA
Alberta Forest Products Association	NLGA
Canadian Lumbermans Association	NLGA
Cariboo Lumber Manufacturers Association	NLGA
Central Forest Products Association	NLGA
Council of Forest Industries of British Columbia	NLGA
Interior Lumber Manufacturers Association	NLGA
MacDonald Inspection	NLGA
Maritime Lumber Bureau	NLGA
Ontario Lumber Manufacturers Association	NLGA
Pacific Lumber Inspection Bureau	NLGA
Quebec Lumber Manufacturers Association	NLGA

[c] The design values herein are applicable to lumber that will be used under dry conditions such as in most covered structures. For 2–4 in.-thick lumber, the Dry surfaced size shall be used. In calculating design values, the natural gain in strength and stiffness that occurs as lumber dries has been taken into consideration as well as the reduction in size that occurs when unseasoned lumber shrinks. The gain in load-carrying capacity due to increased strength and stiffness resulting from drying more than offsets the design effect of size reductions due to shrinkage. For 5 in. and thicker lumber, the surfaced sizes also may be used because design values have been adjusted to compensate for any loss in size by shrinkage which may occur.

[d] Tabulated tension parallel to grain values for all species for 5 in. and wider, 2–4 in. thick (and $2\frac{1}{2}$–4 in. thick) size classification apply to 5 and 6 in. widths only, for grades of Select Structural, No. 1, No. 2, No. 3, Appearance and Stud (including dense grades). For lumber wider than 6 in. in these grades, the tabulated F_t values shall be multiplied by the following factors:

Grade (2–4 in. thick, 5 in. and wider) ($2\frac{1}{2}$–4 in. thick, 5 in. and wider) (Includes dense grades)	Multiply Tabulated F_t Values by		
	5 in. and 6 in.	8 in.	10 in. and wider
Select Structural	1.00	0.90	0.80
No. 1, No. 2, No. 3, and Appearance	1.00	0.80	0.60
Stud	1.00	—	—

[e] Design values for all species of Stud grade in 5 in. and wider size classifications apply to 5 and 6 in. widths only.

[f] Values for F_b, F_t, and F_c for all species of the grades of Construction, Standard, and Utility apply only to 4 in. widths. Design values for 2 and 3 in. widths of these grades are available from the grading rules agencies (see footnote b).

[g] The values in Table 7.3 for dimension lumber 2–4 in. in thickness are based on edgewise use. When such lumber is used flatwise, the design values for extreme fiber in bending for all species may be multiplied by the following factors:

Dimension Lumber Used Flatwise			
	Thickness (in.)		
Width	2	3	4
2 in.–4 in.	1.10	1.04	1.00
5 in. and wider	1.22	1.16	1.11

(*continued*)

Footnotes to Table 7.3 (*Continued*)

[h]The design values in Table 7.3 for extreme fiber in bending for decking may be increased by 10% for 2 in.-thick decking and by 4% for 3 in.-thick decking. (Not applicable to California Redwood.)

[i] When 2-4-in.-thick lumber is manufactured at a maximum moisture content of 15% and used in a condition where the moisture content does not exceed 15%, the design values for surfaced dry or surfaced green lumber shown in Table 7.3 may be multiplied by the following factors. (for Southern Pine and Virginia Pine–Pond Pine, use tabulated design values without adjustment):

2–4 in. thick lumber manufactured and used at 15% maximum moisture content (MC15)

Extreme Fiber in Bending, F_b	Tension Parallel to Grain, F_t	Horizontal Shear, F_v	Compression Perpendicular to Grain, $F_{c\perp}$	Compression* Parallel to Grain, F_c	Modulus* of Elasticity, E
C_M = 1.08	1.08	1.05	1.00	1.17	1.05
C_M =		*For Redwood only		1.15	1.04

[j] When 2–4 in.-thick lumber is designed for use where the moisture content will exceed 19% for an extended period of time, the design values shown herein shall be multiplied by the following factors, except that for Southern Pine and Virginia Pine–Pond Pine, footnote *s* applies:

2–4 in. thick lumber used where moisture content will exceed 19%

Extreme Fiber in Bending, F_b	Tension Parallel to Grain, F_t	Horizontal Shear, F_v	Compression Perpendicular to Grain, $F_{c\perp}$	Compression Parallel to Grain, F_c	Modulus of Elasticity, E
C_M = 0.86	0.84	0.97	0.67	0.70	0.97

[k] When lumber 5 in. and thicker is designed for use where the moisture content will exceed 19% for an extended period of time, the design values shown in Table 7.3 (except those for Southern Pine and Virginia Pine–Pond Pine) shall be multiplied by the following factors:

5 in. and thicker lumber used where moisture content will exceed 19%.

Extreme Fiber in Bending, F_b	Tension Parallel to Grain, F_t	Horizontal Shear, F_v	Compression Perpendicular to Grain, $F_{c\perp}$	Compression Parallel to Grain, F_c	Modulus of Elasticity, E
$C_M = 1.00$	1.00	1.00	0.67	0.91	1.00

[l] Specific horizontal shear values may be established by use of the following table when length of split or size of check or shake is known and no increase in them is anticipated. For California Redwood, Southern Pine, Virginia Pine–Pond Pine, or Yellow-Poplar, the provisions in this footnote apply only to the following F_v values: 75 psi, California Redwood; 95 psi, Southern Pine (KD-15); 90 psi, Southern Pine (S–Dry); 85 psi, Southern Pine (S–Green); 95 psi, Virginia Pine–Pond Pine (KD-15); 90 psi, Virginia Pine–Pond Pine (S–Dry); 85 psi, Virginia Pine–Pond Pine (S–Green); and 75 psi, Yellow-Poplar.

Shear Stress Modification Factor

Length of Split on Wide Face of 2 in. Lumber (nominal):	Multiply Tabulated F_v Value by:	Length of Split on Wide Face of 3 in. and Thicker Lumber (nominal):	Multiply Tabulated F_v Value by:	Size of Shake* in 3 in. and Thicker Lumber (nominal):	Multiply Tabulated F_v Value by:
No split	2.00	no split	2.00	no shake	2.00
$\frac{1}{2}$ × wide face	1.67	$\frac{1}{2}$ × narrow face	1.67	$\frac{1}{6}$ × narrow face	1.67
$\frac{3}{4}$ × wide face	1.50	1 × narrow face	1.33	$\frac{1}{3}$ × narrow face	1.33
1 × wide face	1.33	$1\frac{1}{2}$ × narrow face or more	1.00	$\frac{1}{2}$ × narrow face or more	1.00
$1\frac{1}{2}$ × wide face or more	1.00				

*Shake is measured at the end between lines enclosing the shake and parallel to the wide face.

[m] Stress-rated boards of nominal 1, $1\frac{1}{4}$, and $1\frac{1}{2}$ in. thickness, 2 in. and wider, of most species, are permitted the design values shown for Select Structural, No. 1, No. 2, No. 3, Construction, Standard, Utility, Appearance, Clear Heart Structural, and Clear Structural grades as shown in the 2–4 in.-thick categories herein, when graded in accordance with the stress-rated board provisions in the applicable grading rules. Information on stress-rated board grades applicable to the various species is available from the respective grading rules agencies. Information on additional design values may also be available from the respective grading agencies.

(continued)

Footnotes to Table 7.3 (*Continued*)

[n] When Decking graded to WWPA rules is surfaced at 15% maximum moisture content and used where the moisture content will exceed 15% for an extended period of time, the tabulated design values for decking surfaced at 15% maximum moisture content shall be multiplied by the following factors: extreme fiber in bending, F_b, $C_M = 0.79$; Modulus of elasticity, E; $C_M = 0.92$.

[o] To obtain a recommended design value for Spruce Pine, multiply the appropriate design value for Virginia Pine–Pond Pine by the corresponding conversion factor shown below and round to the nearest 100,000 psi for modulus of elasticity; to the next lower multiple of 5 psi for horizontal shear and compression perpendicular to grain; to the next lower multiple of 50 psi for bending, tension parallel to grain, and compression parallel to grain if 1000 psi or greater, 25 psi otherwise.

Conversion Factors for Determining Design Values for Spruce Pine

Design Category	Extreme Fiber in Bending, F_b		Tension Parallel to Grain, F_t	Horizontal Shear, F_v	Compression Perpendicular to Grain, $F_{c\perp}$	Compression Parallel to Grain, F_c	Modulus of Elasticity, E
	Single-Member uses	Repetitive-Member uses					
Conversion Factor	0.784	0.784	0.784	0.766	0.965	0.682	0.807

[p] National Lumber Grades Authority is the Canadian rules-writing agency responsible for preparation, maintenance, and dissemination of a uniform softwood lumber grading rule for all Canadian species.

[q] For species graded to NLGA rules, values shown in Table 7.3 for Select Structural, No. 1, No. 2, No. 3, and Stud grades are not applicable to 3 × 4 in. and 4 × 4 in. sizes.

[r] Repetitive-member design values for extreme fiber in bending for Southern Pine grades of Dense Structural 86, 72, and 65 apply to 2–4 in. thicknesses only.

[s] When 2–4 in. thick Southern Pine or Virginia Pine–Pond Pine lumber is surfaced dry or at 15% maximum moisture content (KD-15) and is designed for use where the moisture content will exceed 19% for an extended period of time, the design values in Table 7.3 for the corresponding grades of $2\frac{1}{2}$–4 in.-thick surfaced green Southern Pine lumber shall be used. The net green size may be used in such designs.

[t] When 2–4 in.-thick Southern Pine or Virginia Pine–Pond Pine lumber is surfaced dry or at 15% maximum moisture content (KD-15) and is designed for use under dry conditions, such as in most covered structures, the net Dry size shall be used in design. For other sizes and conditions of use, the net green size may be used in design.

[u] Surfaced at 15% MC maximum; KD-15; used at 15% MC maximum.

[v] Surfaced dry; used at 19% MC maximum.

[w] Surfaced green; used at any condition.

TABLE 7.4

Design Values for Machine Stress-Rated Structural Lumber (psi)[a–d]

Grade Designation[e]	Grading Rules Agency[f]	Size Classification	Extreme Fiber in Bending, F_b[g] Single-Member Uses	Extreme Fiber in Bending, F_b[g] Repetitive-Member Uses	Tension Parallel to Grain, F_t	Compression Parallel to Grain, F_c	Modulus of Elasticity, E
900f-1.0E	3, 4	Machine-	900	1050	350	725	1,000,000
1200f-1.2E	1, 2, 3, 4	rated	1200	1400	600	950	1,200,000
1350f-1.3E	2, 3, 4	lumber	1350	1550	750	1075[h]	1,300,000
1450f-1.3E	1, 3, 4	2 in. thick	1450	1650	800	1150	1,300,000
1500f-1.3E	2	or less,	1500	1750	900	1200	1,300,000
1500f-1.4E	1, 2, 3, 4	all widths	1500	1750	900	1200	1,400,000
1650f-1.4E	2		1650	1900	1020	1320	1,400,000
1650f-1.5E	1,2, 3, 4		1650	1900	1020	1320	1,500,000
1800f-1.6E	1,2, 3, 4		1800	2050	1175	1450	1,600,000
1950f-1.5E	2		1950	2250	1375	1550	1,500,000
1950f-1.7E	1, 2, 4		1950	2250	1375	1550	1,700,000
2100f-1.8E	1, 2, 3, 4		2100	2400	1575	1700	1,800,000
2250f-1.6E	2		2250	2600	1750	1800	1,600,000
2250f-1.9E	1, 2, 4		2250	2600	1750	1800	1,900,000
2400f-1.7E	2		2400	2750	1925	1925	1,700,000
2400f-2.0E	1, 2, 3, 4		2400	2750	1925	1925	2,000,000
2550f-2.1E	1, 2, 4		2550	2950	2050	2050	2,100,000
2700f-2.2E	1, 2, 3, 4		2700	3100	2150	2150	2,200,000
2850f-2.3E	2		2850	3300	2300	2300	2,300,000
3000f-2.4E	1, 2		3000	3450	2400	2400	2,400,000
3150f-2.5E	2		3150	3600	2500	2500	2,500,000
3300f-2.6E	2		3300	3800	2650	2650	2,600,000

(continued)

TABLE 7.4 (*Continued*)

Design Values for Machine Stress-Rated Structural Lumber (psi)[a–d]

Grade Designation[e]	Grading Rules Agency[f]	Size Classification	Extreme Fiber in Bending, F_b^g: Single-Member Uses	Extreme Fiber in Bending, F_b^g: Repetitive-Member Uses	Tension Parallel to Grain, F_t	Compression Parallel to Grain, F_c	Modulus of Elasticity, E
900f-1.0E	1, 2, 3	See footnote[i]	900	1050	350	725	1,000,000
900f-1.2E	1, 2, 3		900	1050	350	725	1,200,000
1200f-1.5E	1, 2, 3		1200	1400	600	950	1,500,000
1350f-1.8E	1, 2		1350	1550	750	1075	1,800,000
1500f-1.8E	3		1500	1750	900	1200	1,800,000
1800f-2.1E	1, 2, 3		1800	2050	1175	1450	2,100,000

[a]Stresses apply for lumber used at 19% maximum moisture content.

[b]When lumber 2 in. thick or less is designed for use where the moisture content will exceed 19% for an extended period of time, the design values shown herein shall be multiplied by the following factors:

Extreme Fiber in Bending, F_b	Tension Parallel to Grain, F_t	Horizontal Shear, F_v	Compression Perpendicular to Grain, $F_{c\perp}$	Compression Parallel to Grain, F_c	Modulus of Elasticity, E
$C_M = 0.86$	0.84	0.97	0.67	0.70	0.97

[c]Footnotes *b*, *c*, *l*, and *t* to Table 7.3 apply also to machine stress-rated lumber.

[d]Design values for horizontal shear, F_v, and compression perpendicular to grain, $F_{c\perp}$, for lumber used under dry conditions are the same as the values listed in Table 7.3 for No. 2 visually graded lumber of the appropriate species. For Mixed Species graded under WCLIB grading rules, $F_v = 70$ psi and $F_{c\perp} = 190$ psi.

[e]For any given value of fiber stress in bending, F_b, the average modulus of elasticity, E, may vary depending on species, timber source, and other variables. The E values included in the f-E grade designations in Table 7.4 are those usually associated with each F_b level. Grade stamps may show higher or lower E values (in increments of 100,000 psi) if machine rating indicates the assignment is appropriate. When an E value associated with a designated f level is lower or higher than those listed in Table 7.4, the tabulated F_b, F_t, and F_c values associated with the designated f value are applicable. The E value for design shall be that associated with the E value on the grade stamp.

[f]Key: (1) NLGA grading rules, see footnote *b*, Table 7.3. (2) SPIB grading rules, see footnote *b*, Table 7.3. (3) WCLIB grading rules, see footnote *b*, Table 7.3. (4) WWPA grading rules, see footnote *b*, Table 7.3.

[g]Tabulated extreme fiber in bending values, F_b, are applicable to lumber loaded on edge. When loaded flatwise, these values may be increased by multiplying by the following factors:

Nominal width (in.)	3	4	5	6	8	10	12	14
Factor	1.06	1.10	1.12	1.15	1.19	1.22	1.25	1.28

[h]When graded under WWPA grading rules, value shall be 1100 psi.

[i]Size classificatons for these grades are: NLGA—machine-rated lumber; 2 in. thick or less, all widths. SPIB—machine-rated lumber; 2 in. thick or less, all widths. WCLIB machine-rated joists; 2 in. thick or less, 6 in. and wider.

SPAN-LOAD TABLES

TABLE 7.5

Structural Glued Laminated Timber: Simple-Span Beam Span-Load Table[a]

Span (ft)	Spacing (ft)	Roof Beams Total Load (psf) 20	25	30	35	40	45	50	55	Floor Beams Total Load, 50 psf
8	4			*3-1/8 × 4-1/2	*3-1/8 × 4-1/2	*3-1/8 × 4-1/2	*3-1/8 × 4-1/2	*3-1/8 × 4-1/2	*3-1/8 × 6	*3-1/8 × 6
	6			*3-1/8 × 4-1/2	*3-1/8 × 6	*3-1/8 × 6	*3-1/8 × 6	*3-1/8 × 6	*3-1/8 × 6	*3-1/8 × 6
	8			*3-1/8 × 6	*3-1/8 × 6	*3-1/8 × 6	*3-1/8 × 6	*3-1/8 × 6	*3-1/8 × 6	*3-1/8 × 7-1/2
10	4			*3-1/8 × 6	*3-1/8 × 6	*3-1/8 × 6	*3-1/8 × 6	*3-1/8 × 6	*3-1/8 × 6	*3-1/8 × 7-1/2
	6			*3-1/8 × 6	*3-1/8 × 6	*3-1/8 × 6	*3-1/8 × 7-1/2	*3-1/8 × 7-1/2	*3-1/8 × 7-1/2	*3-1/8 × 7-1/2
	8			*3-1/8 × 6	*3-1/8 × 7-1/2	*3-1/8 × 7-1/2	*3-1/8 × 7-1/2	*3-1/8 × 7-1/2	*3-1/8 × 7-1/2	*3-1/8 × 9
	10			*3-1/8 × 7-1/2	*3-1/8 × 7-1/2	*3-1/8 × 7-1/2	*3-1/8 × 9	*3-1/8 × 9	*3-1/8 × 9	*3-1/8 × 9
12	6			*3-1/8 × 7-1/2	*3-1/8 × 7-1/2	*3-1/8 × 7-1/2	*3-1/8 × 7-1/2	*3-1/8 × 9	*3-1/8 × 9	*3-1/8 × 9
	8			*3-1/8 × 7-1/2	*3-1/8 × 7-1/2	*3-1/8 × 9	*3-1/8 × 9	*3-1/8 × 9	*3-1/8 × 9	*3-1/8 × 10-1/2
	10			*3-1/8 × 9	*3-1/8 × 9	*3-1/8 × 9	*3-1/8 × 10-1/2	*3-1/8 × 10-1/2	*3-1/8 × 10-1/2	*3-1/8 × 12
	12			*3-1/8 × 9	*3-1/8 × 9	*3-1/8 × 10-1/2	*3-1/8 × 10-1/2	*3-1/8 × 10-1/2	*3-1/8 × 12	*3-1/8 × 12
14	8			*3-1/8 × 9	*3-1/8 × 9	*3-1/8 × 10-1/2	*3-1/8 × 10-1/2	*3-1/8 × 10-1/2	*3-1/8 × 10-1/2	*3-1/8 × 12
	10			*3-1/8 × 9	*3-1/8 × 10-1/2	*3-1/8 × 10-1/2	*3-1/8 × 12	*3-1/8 × 12	*3-1/8 × 12	*3-1/8 × 13-1/2
	12			*3-1/8 × 10-1/2	*3-1/8 × 10-1/2	*3-1/8 × 12	*3-1/8 × 12	*3-1/8 × 13-1/2	*3-1/8 × 13-1/2	*3-1/8 × 13-1/2
	14			*3-1/8 × 10-1/2	*3-1/8 × 12	*3-1/8 × 12	*3-1/8 × 13-1/2	*3-1/8 × 13-1/2	*3-1/8 × 15	*3-1/8 × 15
16	8			*3-1/8 × 10-1/2	*3-1/8 × 10-1/2	*3-1/8 × 10-1/2	*3-1/8 × 12	*3-1/8 × 12	*3-1/8 × 12	*3-1/8 × 13-1/2
	12			*3-1/8 × 12	*3-1/8 × 12	*3-1/8 × 13-1/2	*3-1/8 × 13-1/2	*3-1/8 × 15	*3-1/8 × 15	*3-1/8 × 16-1/2
	14		*3-1/8 × 12	*3-1/8 × 12	*3-1/8 × 13-1/2	*3-1/8 × 13-1/2	*3-1/8 × 15	*3-1/8 × 16-1/2	*3-1/8 × 16-1/2	*3-1/8 × 16-1/2
	16		*3-1/8 × 12	*3-1/8 × 13-1/2	*3-1/8 × 13-1/2	*3-1/8 × 15	*3-1/8 × 16-1/2	*3-1/8 × 16-1/2	3-1/8 × 16-1/2	*5-1/8 × 15
18	8			*3-1/8 × 12	*3-1/8 × 12	*3-1/8 × 12	*3-1/8 × 13-1/2	*3-1/8 × 13-1/2	*3-1/8 × 13-1/2	*3-1/8 × 15
	12		*3-1/8 × 12	*3-1/8 × 13-1/2	*3-1/8 × 13-1/2	*3-1/8 × 15	*3-1/8 × 15	*3-1/8 × 16-1/2	3-1/8 × 16-1/2	3-1/8 × 16-1/2
	16		*3-1/8 × 13-1/2	*3-1/8 × 15	*3-1/8 × 16-1/2	*3-1/8 × 16-1/2	3-1/8 × 16-1/2	*5-1/8 × 15	*5-1/8 × 16-1/2	*5-1/8 × 16-1/2
	20		*3-1/8 × 15	*3-1/8 × 16-1/2	3-1/8 × 16-1/2	3-1/8 × 18	3-1/8 × 18	*5-1/8 × 16-1/2	5-1/8 × 16-1/2	5-1/8 × 16-1/2

20	8			*3-1/8 × 12	*3-1/8 × 13-1/2	*3-1/8 × 13-1/2	3-1/8 × 15	*3-1/8 × 15	*3-1/8 × 15	3-1/8 × 16-1/2
	12		*3-1/8 × 13-1/2	*3-1/8 × 15	*3-1/8 × 15	*3-1/8 × 16-1/2	3-1/8 × 16-1/2	3-1/8 × 16-1/2	*5-1/8 × 15	*5-1/8 × 16-1/2
	16		*3-1/8 × 15	*3-1/8 × 16-1/2	3-1/8 × 16-1/2	3-1/8 × 18	3-1/8 × 18	*5-1/8 × 16-1/2	5-1/8 × 16-1/2	5-1/8 × 18
	20		*3-1/8 × 16-1/2	3-1/8 × 16-1/2	3-1/8 × 18	*5-1/8 × 16-1/2	5-1/8 × 16-1/2	5-1/8 × 16-1/2	5-1/8 × 18	5-1/8 × 19-1/2
24	8			*3-1/8 × 15	*3-1/8 × 15	*3-1/8 × 16-1/2	3-1/8 × 16-1/2	3-1/8 × 16-1/2	3-1/8 × 18	5-1/8 × 16-1/2
	12		*3-1/8 × 16-1/2	3-1/8 × 16-1/2	3-1/8 × 16-1/2	3-1/8 × 18	*5-1/8 × 16-1/2	5-1/8 × 16-1/2	5-1/8 × 16-1/2	5-1/8 × 19-1/2
	16		3-1/8 × 16-1/2	3-1/8 × 18	5-1/8 × 16-1/2	5-1/8 × 16-1/2	5-1/8 × 18	5-1/8 × 18	5-1/8 × 19-1/2	5-1/8 × 21
	20		3-1/8 × 18	5-1/8 × 16-1/2	5-1/8 × 18	5-1/8 × 18	5-1/8 × 19-1/2	5-1/8 × 19-1/2	5-1/8 × 21	5-1/8 × 22-1/2
	24		3-1/8 × 19-1/2	5-1/8 × 18	5-1/8 × 18	5-1/8 × 19-1/2	5-1/8 × 21	5-1/8 × 22-1/2	5-1/8 × 24	5-1/8 × 24
28	8		*3-1/8 × 16-1/2	3-1/8 × 16-1/2	3-1/8 × 18	3-1/8 × 18	5-1/8 × 16-1/2	5-1/8 × 16-1/2	5-1/8 × 16-1/2	5-1/8 × 19-1/2
	12		3-1/8 × 18	5-1/8 × 16-1/2	5-1/8 × 16-1/2	5-1/8 × 18	5-1/8 × 18	5-1/8 × 19-1/2	5-1/8 × 19-1/2	5-1/8 × 22-1/2
	16		5-1/8 × 16-1/2	5-1/8 × 18	5-1/8 × 18	5-1/8 × 19-1/2	5-1/8 × 19-1/2	5-1/8 × 21	5-1/8 × 22-1/2	5-1/8 × 24
	20		5-1/8 × 18	5-1/8 × 19-1/2	5-1/8 × 19-1/2	5-1/8 × 21	5-1/8 × 22-1/2	5-1/8 × 24	5-1/8 × 25-1/2	5-1/8 × 25-1/2
	24		5-1/8 × 19-1/2	5-1/8 × 19-1/2	5-1/8 × 21	5-1/8 × 24	5-1/8 × 25-1/2	5-1/8 × 25-1/2	5-1/8 × 27	5-1/8 × 28-1/2
	28		5-1/8 × 19-1/2	5-1/8 × 22-1/2	5-1/8 × 24	5-1/8 × 25-1/2	5-1/8 × 27	5-1/8 × 28-1/2	5-1/8 × 31-1/2	5-1/8 × 30
32	8		3-1/8 × 18	5-1/8 × 16-1/2	5-1/8 × 16-1/2	5-1/8 × 18	5-1/8 × 18	5-1/8 × 19-1/2	5-1/8 × 19-1/2	5-1/8 × 22-1/2
	12		5-1/8 × 18	5-1/8 × 18	5-1/8 × 19-1/2	5-1/8 × 19-1/2	5-1/8 × 21	5-1/8 × 21	5-1/8 × 22-1/2	5-1/8 × 25-1/2
	16		5-1/8 × 19-1/2	5-1/8 × 19-1/2	5-1/8 × 21	5-1/8 × 22-1/2	5-1/8 × 22-1/2	5-1/8 × 24	5-1/8 × 25-1/2	5-1/8 × 27
	20	5-1/8 × 19-1/2	5-1/8 × 21	5-1/8 × 21	5-1/8 × 22-1/2	5-1/8 × 24	5-1/8 × 25-1/2	5-1/8 × 27	5-1/8 × 28-1/2	6-3/4 × 27
	24	5-1/8 × 19-1/2	5-1/8 × 21	5-1/8 × 22-1/2	5-1/8 × 25-1/2	5-1/8 × 27	5-1/8 × 28-1/2	5-1/8 × 30	6-3/4 × 27	5-1/8 × 33
	28	5-1/8 × 21	5-1/8 × 22-1/2	5-1/8 × 25-1/2	5-1/8 × 27	5-1/8 × 28-1/2	6-3/4 × 27	6-3/4 × 28-1/2	6-3/4 × 31-1/2	6-3/4 × 30
	32	5-1/8 × 22-1/2	5-1/8 × 24	5-1/8 × 27	5-1/8 × 28-1/2	6-3/4 × 27	6-3/4 × 28-1/2	6-3/4 × 30	6-3/4 × 31-1/2	6-3/4 × 33
36	12		5-1/8 × 19-1/2	5-1/8 × 21	5-1/8 × 21	5-1/8 × 22-1/2	5-1/8 × 24	5-1/8 × 24	5-1/8 × 25-1/2	6-3/4 × 25-1/2
	16		5-1/8 × 21	5-1/8 × 22-1/2	5-1/8 × 24	5-1/8 × 25-1/2	5-1/8 × 25-1/2	5-1/8 × 27	5-1/8 × 28-1/2	6-3/4 × 28-1/2
	20	5-1/8 × 21	5-1/8 × 22-1/2	5-1/8 × 24	5-1/8 × 25-1/2	5-1/8 × 27	5-1/8 × 30	6-3/4 × 27	6-3/4 × 28-1/2	6-3/4 × 30
	24	5-1/8 × 22-1/2	5-1/8 × 24	5-1/8 × 25-1/2	5-1/8 × 28-1/2	5-1/8 × 30	6-3/4 × 27	6-3/4 × 28-1/2	6-3/4 × 30	6-3/4 × 33
	28	5-1/8 × 24	5-1/8 × 25-1/2	5-1/8 × 28-1/2	5-1/8 × 30	6-3/4 × 28-1/2	6-3/4 × 30	6-3/4 × 31-1/2	6-3/4 × 33	6-3/4 × 34-1/2
	32	5-1/8 × 25-1/2	5-1/8 × 27	5-1/8 × 30	6-3/4 × 28-1/2	6-3/4 × 30	6-3/4 × 33	6-3/4 × 34-1/2	6-3/4 × 36	6-3/4 × 36
	36	5-1/8 × 25-1/2	5-1/8 × 30	6-3/4 × 28-1/2	6-3/4 × 30	6-3/4 × 33	6-3/4 × 34-1/2	6-3/4 × 36	6-3/4 × 39	6-3/4 × 39
40	12	5-1/8 × 19-1/2	5-1/8 × 21	5-1/8 × 22-1/2	5-1/8 × 24	5-1/8 × 25-1/2	5-1/8 × 25-1/2	5-1/8 × 27	6-3/4 × 25-1/2	6-3/4 × 28-1/2
	16	5-1/8 × 22-1/2	5-1/8 × 24	5-1/8 × 25-1/2	5-1/8 × 27	5-1/8 × 27	5-1/8 × 28-1/2	6-3/4 × 27	6-3/4 × 28-1/2	6-3/4 × 31-1/2
	20	5-1/8 × 24	5-1/8 × 25-1/2	5-1/8 × 27	5-1/8 × 28-1/2	6-3/4 × 27	6-3/4 × 28-1/2	6-3/4 × 30	6-3/4 × 31-1/2	6-3/4 × 34-1/2
	24	5-1/8 × 25-1/2	5-1/8 × 27	5-1/8 × 28-1/2	6-3/4 × 27	6-3/4 × 28-1/2	6-3/4 × 31-1/2	6-3/4 × 33	6-3/4 × 34-1/2	6-3/4 × 36
	28	5-1/8 × 25-1/2	5-1/8 × 28-1/2	6-3/4 × 27	6-3/4 × 30	6-3/4 × 31-1/2	6-3/4 × 33	6-3/4 × 36	6-3/4 × 37-1/2	6-3/4 × 39
	32	5-1/8 × 27	5-1/8 × 30	6-3/4 × 30	6-3/4 × 31-1/2	6-3/4 × 34-1/2	6-3/4 × 36	6-3/4 × 37-1/2	6-3/4 × 40-1/2	6-3/4 × 40-1/2

(continued)

TABLE 7.5 (*Continued*)

Structural Glued Laminated Timber: Simple-Span Beam Span-Load Table[d]

Span (ft)	Spacing (ft)	Roof Beams Total Load (psf) 20	25	30	35	40	45	50	55	Floor Beams Total Load, 50 psf
40	36	5-1/8 × 28-1/2	6-3/4 × 28-1/2	6-3/4 × 31-1/2	6-3/4 × 33	6-3/4 × 36	6-3/4 × 39	6-3/4 × 40-1/2	8-3/4 × 37-1/2	6-3/4 × 43
	40	5-1/8 × 30	6-3/4 × 30	6-3/4 × 33	6-3/4 × 36	6-3/4 × 37-1/2	6-3/4 × 40-1/2	8-3/4 × 37-1/2	8-3/4 × 39	8-3/4 × 40-1/2
44	12	5-1/8 × 22-1/2	5-1/8 × 24	5-1/8 × 25-1/2	5-1/8 × 27	5-1/8 × 27	5-1/8 × 28-1/2	6-3/4 × 27	6-3/4 × 28-1/2	
	16	5-1/8 × 24	5-1/8 × 25-1/2	5-1/8 × 27	5-1/8 × 28-1/2	5-1/8 × 30	6-3/4 × 28-1/2	6-3/4 × 30	6-3/4 × 31-1/2	
	20	5-1/8 × 25-1/2	5-1/8 × 28-1/2	5-1/8 × 30	6-3/4 × 28-1/2	6-3/4 × 30	6-3/4 × 31-1/2	6-3/4 × 33	6-3/4 × 34-1/2	
	24	5-1/8 × 27	5-1/8 × 30	6-3/4 × 28-1/2	6-3/4 × 30	6-3/4 × 31-1/2	6-3/4 × 34-1/2	6-3/4 × 36	6-3/4 × 37-1/2	
	28	5-1/8 × 28-1/2	6-3/4 × 28-1/2	6-3/4 × 30	6-3/4 × 33	6-3/4 × 34-1/2	6-3/4 × 37-1/2	6-3/4 × 39	8-3/4 × 36	
	32	5-1/8 × 30	6-3/4 × 30	6-3/4 × 33	6-3/4 × 34-1/2	6-3/4 × 37-1/2	6-3/4 × 40-1/2	8-3/4 × 37-1/2	8-3/4 × 39	
	36	6-3/4 × 28-1/2	6-3/4 × 31-1/2	6-3/4 × 34-1/2	6-3/4 × 37-1/2	6-3/4 × 40-1/2	8-3/4 × 37-1/2	8-3/4 × 39	8-3/4 × 40-1/2	
	40	6-3/4 × 30	6-3/4 × 33	6-3/4 × 36	6-3/4 × 39	8-3/4 × 37-1/2	8-3/4 × 39	8-3/4 × 42	8-3/4 × 43-1/2	
48	12	5-1/8 × 24	5-1/8 × 25-1/2	5-1/8 × 27	5-1/8 × 28-1/2	5-1/8 × 30	6-3/4 × 28-1/2	6-3/4 × 30	6-3/4 × 30	
	16	5-1/8 × 25-1/2	5-1/8 × 28-1/2	5-1/8 × 30	6-3/4 × 28-1/2	6-3/4 × 30	6-3/4 × 31-1/2	6-3/4 × 31-1/2	6-3/4 × 34-1/2	
	20	5-1/8 × 28-1/2	5-1/8 × 30	6-3/4 × 30	6-3/4 × 31-1/2	6-3/4 × 31-1/2	6-3/4 × 34-1/2	6-3/4 × 36	6-3/4 × 37-1/2	
	24	5-1/8 × 30	6-3/4 × 30	6-3/4 × 31-1/2	6-3/4 × 33	6-3/4 × 34-1/2	6-3/4 × 37-1/2	6-3/4 × 39	8-3/4 × 36	
	28	6-3/4 × 28-1/2	6-3/4 × 31-1/2	6-3/4 × 33	6-3/4 × 36	6-3/4 × 39	6-3/4 × 40-1/2	8-3/4 × 37-1/2	8-3/4 × 39	
	32	6-3/4 × 30	6-3/4 × 31-1/2	6-3/4 × 36	6-3/4 × 39	6-3/4 × 40-1/2	6-3/4 × 37-1/2	8-3/4 × 40-1/2	8-3/4 × 42	
	36	6-3/4 × 31-1/2	6-3/4 × 34-1/2	6-3/4 × 37-1/2	6-3/4 × 40-1/2	8-3/4 × 37-1/2	8-3/4 × 40-1/2	8-3/4 × 43-1/2	8-3/4 × 45	
	40	6-3/4 × 31-1/2	6-3/4 × 36	6-3/4 × 40-1/2	8-3/4 × 37-1/2	8-3/4 × 40-1/2	8-3/4 × 43-1/2	8-3/4 × 45	8-3/4 × 48	
52	12	5-1/8 × 25-1/2	5-1/8 × 27	5-1/8 × 30	5-1/8 × 31-1/2	6-3/4 × 30	6-3/4 × 30	6-3/4 × 31-1/2	6-3/4 × 33	
	16	5-1/8 × 28-1/2	5-1/8 × 30	6-3/4 × 30	6-3/4 × 31-1/2	6-3/4 × 33	6-3/4 × 33	6-3/4 × 34-1/2	6-3/4 × 37-1/2	
	20	5-1/8 × 30	6-3/4 × 30	6-3/4 × 31-1/2	6-3/4 × 33	6-3/4 × 34-1/2	6-3/4 × 37-1/2	6-3/4 × 39	8-3/4 × 36	
	24	6-3/4 × 30	6-3/4 × 31-1/2	6-3/4 × 33	6-3/4 × 36	6-3/4 × 37-1/2	6-3/4 × 40-1/2	8-3/4 × 37-1/2	8-3/4 × 39	
	28	6-3/4 × 31-1/2	6-3/4 × 33	6-3/4 × 36	6-3/4 × 39	8-3/4 × 36	8-3/4 × 39	8-3/4 × 40-1/2	8-3/4 × 43-1/2	
	32	6-3/4 × 33	6-3/4 × 34-1/2	6-3/4 × 39	8-3/4 × 36	8-3/4 × 39	8-3/4 × 42	8-3/4 × 43-1/2	8-3/4 × 46-1/2	
	36	6-3/4 × 33	6-3/4 × 37-1/2	6-3/4 × 40-1/2	8-3/4 × 39	8-3/4 × 42	8-3/4 × 43-1/2	8-3/4 × 46-1/2	8-3/4 × 49-1/2	
	40	6-3/4 × 34-1/2	6-3/4 × 39	8-3/4 × 37-1/2	8-3/4 × 40-1/2	8-3/4 × 43-1/2	8-3/4 × 46-1/2	8-3/4 × 49-1/2	8-3/4 × 51	

56	12	5-1/8 × 27	5-1/8 × 30	6-3/4 × 28-1/2	6-3/4 × 30	6-3/4 × 31-1/2	6-3/4 × 33	6-3/4 × 34-1/2	6-3/4 × 34-1/2
	16	5-1/8 × 30	6-3/4 × 30	6-3/4 × 31-1/2	6-3/4 × 33	6-3/4 × 34-1/2	6-3/4 × 36	6-3/4 × 37-1/2	8-3/4 × 36
	20	6-3/4 × 30	6-3/4 × 31-1/2	6-3/4 × 34-1/2	6-3/4 × 36	6-3/4 × 37-1/2	8-3/4 × 36	8-3/4 × 37-1/2	8-3/4 × 39
	24	6-3/4 × 31-1/2	6-3/4 × 34-1/2	6-3/4 × 36	6-3/4 × 39	8-3/4 × 36	8-3/4 × 39	8-3/4 × 40-1/2	8-3/4 × 42
	28	6-3/4 × 33	6-3/4 × 36	6-3/4 × 39	8-3/4 × 36	8-3/4 × 39	8-3/4 × 42	8-3/4 × 43-1/2	8-3/4 × 46-1/2
	32	6-3/4 × 34-1/2	6-3/4 × 37-1/2	8-3/4 × 36	8-3/4 × 39	8-3/4 × 42	8-3/ × 45	8-3/4 × 48	8-3/4 × 49-1/2
	36	6-3/4 × 36	6-3/4 × 40-1/2	8-3/4 × 39	8-3/4 × 42	8-3/4 × 45	8-3/4 × 46-1/2	8-3/4 × 51	8-3/4 × 52-1/2
	40	6-3/4 × 37-1/2	8-3/4 × 37-1/2	8-3/4 × 40-1/2	8-3/4 × 43-1/2	8-3/4 × 48	8-3/4 × 51	8-3/4 × 52-1/2	10-3/4 × 49-1/2
60	12	5-1/8 × 30	6-3/4 × 28-1/2	6-3/4 × 31-1/2	6-3/4 × 33	6-3/4 × 34-1/2	6-3/4 × 36	6-3/4 × 36	6-3/4 × 37-1/2
	16	6-3/4 × 30	6-3/4 × 31-1/2	6-3/4 × 34-1/2	6-3/4 × 36	6-3/4 × 37-1/2	6-3/4 × 39	8-3/4 × 37-1/2	8-3/4 × 37-1/2
	20	6-3/4 × 31-1/2	6-3/4 × 34-1/2	6-3/4 × 36	6-3/4 × 39	8-3/4 × 37-1/2	8-3/4 × 39	8-3/4 × 40-1/2	8-3/4 × 42
	24	6-3/4 × 34-1/2	6-3/4 × 36	6-3/4 × 39	8-3/4 × 37-1/2	8-3/4 × 39	8-3/4 × 42	8-3/4 × 43-1/2	8-3/4 × 45
	28	6-3/4 × 36	6-3/4 × 39	8-3/4 × 37-1/2	8-3/4 × 39	8-3/4 × 42	8-3/4 × 45	8-3/4 × 48	8-3/4 × 49-1/2
	32	6-3/4 × 37-1/2	6-3/4 × 40-1/2	8-3/4 × 39	8-3/4 × 42	8-3/4 × 45	8-3/4 × 48	8-3/4 × 51	10-3/4 × 48
	36	6-3/4 × 39	8-3/4 × 39	8-3/4 × 42	8-3/4 × 45	8-3/4 × 48	8-3/4 × 51	10-3/4 × 48	10-3/4 × 51
	40	6-3/4 × 40-1/2	8-3/4 × 40-1/2	8-3/4 × 43-1/2	8-3/4 × 48	8-3/4 × 51	10-3/4 × 48	10-3/4 × 51	10-3/4 × 54
64	12	6-3/4 × 28-1/2	6-3/4 × 31-1/2	6-3/4 × 33	6-3/4 × 34-1/2	6-3/4 × 36	6-3/4 × 37-1/2	6-3/4 × 39	6-3/4 × 40-1/2
	16	6-3/4 × 31-1/2	6-3/4 × 34-1/2	6-3/4 × 36	6-3/4 × 37-1/2	6-3/4 × 39	8-3/4 × 37-1/2	8-3/4 × 39	8-3/4 × 40-1/2
	20	6-3/4 × 34-1/2	6-3/4 × 36	6-3/4 × 39	6-3/4 × 40-1/2	8-3/4 × 39	8-3/4 × 40-1/2	8-3/4 × 42	8-3/4 × 45
	24	6-3/4 × 36	6-3/4 × 39	8-3/4 × 37-1/2	8-3/4 × 40-1/2	8-3/4 × 42	8-3/4 × 43-1/2	8-3/4 × 46-1/2	8-3/4 × 49-1/2
	28	6-3/4 × 37-1/2	6-3/4 × 40-1/2	8-3/4 × 40-1/2	8-3/4 × 42	8-3/4 × 45	8-3/4 × 48	8-3/4 × 51	10-3/4 × 48
	32	6-3/4 × 39	8-3/4 × 39	8-3/4 × 42	8-3/4 × 45	8-3/4 × 48	8-3/4 × 51	10-3/4 × 49-1/2	10-3/4 × 51
	36	6-3/4 × 40-1/2	8-3/4 × 40-1/2	8-3/4 × 45	8-3/4 × 48	8-3/4 × 51	10-3/4 × 49-1/2	10-3/4 × 52-1/2	10-3/4 × 54
	40	8-3/4 × 39	8-3/4 × 42	8-3/4 × 46-1/2	8-3/4 × 51	10-3/4 × 49-1/2	10-3/4 × 52-1/2	10-3/4 × 55-1/2	10-3/4 × 58-1/2
68	12	6-3/4 × 30	6-3/4 × 33	6-3/4 × 34-1/2	6-3/4 × 36	6-3/4 × 39	6-3/4 × 40-1/2	8-3/4 × 37-1/2	8-3/4 × 39
	16	6-3/4 × 33	6-3/4 × 36	6-3/4 × 39	6-3/4 × 40-1/2	8-3/4 × 39	8-3/4 × 40-1/2	8-3/4 × 42	8-3/4 × 43-1/2
	20	6-3/4 × 36	6-3/4 × 39	8-3/4 × 37-1/2	8-3/4 × 40-1/2	8-3/4 × 42	8-3/4 × 43-1/2	8-3/4 × 45	8-3/4 × 48
	24	6-3/4 × 39	8-3/4 × 37-1/2	8-3/4 × 40-1/2	8-3/4 × 42	8-3/4 × 45	8-3/4 × 46-1/2	8-3/4 × 49-1/2	8-3/4 × 52-1/2
	28	6-3/4 × 40-1/2	8-3/4 × 40-1/2	8-3/4 × 42	8-3/4 × 45	8-3/4 × 48	8-3/4 × 51	10-3/4 × 48	10-3/4 × 51
	32	8-3/4 × 39	8-3/4 × 42	8-3/4 × 45	8-3/4 × 48	8-3/4 × 51	10-3/4 × 49-1/2	10-3/4 × 52-1/2	10-3/4 × 54
	36	8-3/4 × 40-1/2	8-3/4 × 43-1/2	8-3/4 × 48	8-3/4 × 51	10-3/4 × 49-1/2	10-3/4 × 52-1/2	10-3/4 × 55-1/2	10-3/4 × 58-1/2
	40	8-3/4 × 42	8-3/4 × 45	8-3/4 × 49-1/2	10-3/4 × 48	10-3/4 × 52-1/2	10-3/4 × 55-1/2	10-3/4 × 58-1/2	10-3/4 × 61-1/2
72	12	6-3/4 × 33	6-3/4 × 34-1/2	6-3/4 × 37-1/2	6-3/4 × 39	6-3/4 × 40-1/2	8-3/4 × 39	8-3/4 × 40-1/2	8-3/4 × 42
	16	6-3/4 × 36	6-3/4 × 37-1/2	6-3/4 × 40-1/2	8-3/4 × 39	8-3/4 × 40-1/2	8-3/4 × 42	8-3/4 × 43-1/2	8-3/4 × 45
	20	6-3/4 × 37-1/2	8-3/4 × 37-1/2	8-3/4 × 40-1/2	8-3/4 × 42	8-3/4 × 43-1/2	8-3/4 × 46-1/2	8-3/4 × 48	8-3/4 × 51
	24	6-3/4 × 40-1/2	8-3/4 × 40-1/2	8-3/4 × 42	8-3/4 × 45	8-3/4 × 46-1/2	8-3/4 × 49-1/2	8-3/4 × 52-1/2	10-3/4 × 49-1/2

(continued)

TABLE 7.5 (*Continued*)

Structural Glued Laminated Timber: Simple-Span Beam Span-Load Table[d]

Span (ft)	Spacing (ft)	Roof Beams Total Load (psf) 20	25	30	35	40	45	50	55	Floor Beams Total Load, 50 psf
72	28	8-3/4 × 39	8-3/4 × 42	8-3/4 × 45	8-3/4 × 48	8-3/4 × 51	10-3/4 × 48	10-3/4 × 51	10-3/4 × 54	
(cont.)	32	8-3/4 × 40-1/2	8-3/4 × 43-1/2	8-3/4 × 46-1/2	8-3/4 × 51	10-3/4 × 49-1/2	10-3/4 × 52-1/2	10-3/4 × 55-1/2	10-3/4 × 58-1/2	
	36	8-3/4 × 42	8-3/4 × 45	8-3/4 × 49-1/2	10-3/4 × 48	10-3/4 × 52-1/2	10-3/4 × 55-1/2	10-3/4 × 58-1/2	10-3/4 × 61-1/2	
	40	8-3/4 × 43-1/2	8-3/4 × 48	8-3/4 × 52-1/2	10-3/4 × 51	10-3/4 × 55-1/2	10-3/4 × 58-1/2	10-3/4 × 61-1/2		
76	12	6-3/4 × 34-1/2	6-3/4 × 36	6-3/4 × 39	6-3/4 × 40-1/2	8-3/4 × 39	8-3/4 × 40-1/2	8-3/4 × 42	8-3/4 × 43-1/2	
	16	6-3/4 × 37-1/2	6-3/4 × 40-1/2	8-3/4 × 39	8-3/4 × 42	8-3/4 × 43-1/2	8-3/4 × 45	8-3/4 × 46-1/2	8-3/4 × 48	
	20	6-3/4 × 40-1/2	8-3/4 × 40-1/2	8-3/4 × 42	8-3/4 × 45	8-3/4 × 46-1/2	8-3/4 × 48	8-3/4 × 51	10-3/4 × 48	
	24	8-3/4 × 39	8-3/4 × 42	8-3/4 × 45	8-3/4 × 46-1/2	8-3/4 × 49-1/2	8-3/4 × 52-1/2	10-3/4 × 49-1/2	10-3/4 × 52-1/2	
	28	8-3/4 × 42	8-3/4 × 45	8-3/4 × 46-1/2	8-3/4 × 51	10-3/4 × 48	10-3/4 × 51	10-3/4 × 54	10-3/4 × 57	
	32	8-3/4 × 43-1/2	8-3/4 × 46-1/2	8-3/4 × 49-1/2	10-3/4 × 48	10-3/4 × 52-1/2	10-3/4 × 55-1/2	10-3/4 × 58-1/2	10-3/4 × 61-1/2	
	36	8-3/4 × 45	8-3/4 × 48	10-3/4 × 48	10-3/4 × 51	10-3/4 × 55-1/2	10-3/4 × 58-1/2	10-3/4 × 61-1/2		
	40	8-3/4 × 46-1/2	8-3/4 × 51	10-3/4 × 51	10-3/4 × 54	10-3/4 × 58-1/2	10-3/4 × 61-1/2			
80	12	6-3/4 × 36	6-3/4 × 39	6-3/4 × 40-1/2	8-3/4 × 39	8-3/4 × 42	8-3/4 × 43-1/2	8-3/4 × 45	8-3/4 × 46-1/2	
	16	6-3/4 × 39	8-3/4 × 39	8-3/4 × 42	8-3/4 × 43-1/2	8-3/4 × 45	8-3/4 × 46-1/2	8-3/4 × 49-1/2	8-3/4 × 51	
	20	8-3/4 × 39	8-3/4 × 42	8-3/4 × 45	8-3/4 × 46-1/2	8-3/4 × 49-1/2	8-3/4 × 51	10-3/4 × 49-1/2	10-3/4 × 51	
	24	8-3/4 × 42	8-3/4 × 45	8-3/4 × 46-1/2	8-3/4 × 49-1/2	8-3/4 × 52-1/2	10-3/4 × 51	10-3/4 × 52-1/2	10-3/4 × 55-1/2	
	28	8-3/4 × 43-1/2	8-3/4 × 46-1/2	8-3/4 × 49-1/2	8-3/4 × 52-1/2	10-3/4 × 51	10-3/4 × 54	10-3/4 × 57	10-3/4 × 60	
	32	8-3/4 × 45	8-3/4 × 49-1/2	8-3/4 × 52-1/2	10-3/4 × 51	10-3/4 × 55-1/2	10-3/4 × 58-1/2	10-3/4 × 61-1/2	10-3/4 × 64-1/2	
	36	8-3/4 × 46-1/2	8-3/4 × 51	10-3/4 × 51	10-3/4 × 54	10-3/4 × 58-1/2				
	40	8-3/4 × 49-1/2	10-3/4 × 49-1/2	10-3/4 × 52-1/2	10-3/4 × 57	10-3/4 × 61-1/2				
84	12	6-3/4 × 37-1/2	6-3/4 × 40-1/2	8-3/4 × 39	8-3/4 × 42	8-3/4 × 43-1/2	8-3/4 × 45	8-3/4 × 46-1/2	8-3/4 × 48	
	16	8-3/4 × 37-1/2	8-3/4 × 40-1/2	8-3/4 × 43-1/2	8-3/4 × 45	8-3/4 × 48	8-3/4 × 49-1/2	8-3/4 × 51	8-3/4 × 52-1/2	
	20	8-3/4 × 40-1/2	8-3/4 × 43-1/2	8-3/4 × 46-1/2	8-3/4 × 49-1/2	8-3/4 × 51	10-3/4 × 49-1/2	10-3/4 × 51	10-3/4 × 54	
	24	8-3/4 × 43-1/2	8-3/4 × 46-1/2	8-3/4 × 49-1/2	8-3/4 × 52-1/2	10-3/4 × 51	10-3/4 × 52-1/2	10-3/4 × 55-1/2	10-3/4 × 58-1/2	
	28	8-3/4 × 45	8-3/4 × 49-1/2	8-3/4 × 52-1/2	10-3/4 × 51	10-3/4 × 54	10-3/4 × 57	10-3/4 × 60	10-3/4 × 63	
	32	8-3/4 × 48	8-3/4 × 51	10-3/4 × 51	10-3/4 × 54	10-3/4 × 57	10-3/4 × 61-1/2	10-3/4 × 64-1/2		
	36	8-3/4 × 49-1/2	10-3/4 × 49-1/2	10-3/4 × 52-1/2	10-3/4 × 57	10-3/4 × 61-1/2	10-3/4 × 64-1/2			
	40	8-3/4 × 51	10-3/4 × 51	10-3/4 × 55-1/2	10-3/4 × 60	10-3/4 × 64-1/2				

88	12	6-3/4 × 39	8-3/4 × 39	8-3/4 × 42	8-3/4 × 43-1/2	8-3/4 × 45	8-3/4 × 46-1/2	8-3/4 × 49-1/2	8-3/4 × 51
	16	8-3/4 × 40-1/2	8-3/4 × 43-1/2	8-3/4 × 45	8-3/4 × 48	8-3/4 × 49-1/2	8-3/4 × 52-1/2	10-3/4 × 49-1/2	10-3/4 × 52-1/2
	20	8-3/4 × 43-1/2	8-3/4 × 46-1/2	8-3/4 × 49-1/2	8-3/4 × 51	10-3/4 × 49-1/2	10-3/4 × 52-1/2	10-3/4 × 54	10-3/4 × 55-1/2
	24	8-3/4 × 45	8-3/4 × 49-1/2	8-3/4 × 52-1/2	10-3/4 × 51	10-3/4 × 54	10-3/4 × 55-1/2	10-3/4 × 58-1/2	10-3/4 × 61-1/2
	28	8-3/4 × 48	8-3/4 × 51	10-3/4 × 51	10-3/4 × 54	10-3/4 × 57	10-3/4 × 60	10-3/4 × 63	
	32	8-3/4 × 49-1/2	10-3/4 × 49-1/2	10-3/4 × 54	10-3/4 × 57	10-3/4 × 60	10-3/4 × 64-1/2		
	36	8-3/4 × 52-1/2	10-3/4 × 52-1/2	10-3/4 × 55-1/2	10-3/4 × 60	10-3/4 × 64-1/2			
	40	10-3/4 × 49-1/2	10-3/4 × 54	10-3/4 × 58-1/2	10-3/4 × 63				
92	12	6-3/4 × 40-1/2	8-3/4 × 40-1/2	8-3/4 × 43-1/2	8-3/4 × 45	8-3/4 × 48	8-3/4 × 49-1/2	8-3/4 × 51	8-3/4 × 52-1/2
	16	8-3/4 × 42	8-3/4 × 45	8-3/4 × 48	8-3/4 × 49-1/2	8-3/4 × 52-1/2	10-3/4 × 51	10-3/4 × 52-1/2	10-3/4 × 54
	20	8-3/4 × 45	8-3/4 × 48	8-3/4 × 51	10-3/4 × 49-1/2	10-3/4 × 52-1/2	10-3/4 × 54	10-3/4 × 57	10-3/4 × 58-1/2
	24	8-3/4 × 48	8-3/4 × 51	10-3/4 × 51	10-3/4 × 54	10-3/4 × 55-1/2	10-3/4 × 58-1/2	10-3/4 × 61-1/2	10-3/4 × 64-1/2
	28	8-3/4 × 49-1/2	10-3/4 × 49-1/2	10-3/4 × 54	10-3/4 × 55-1/2	10-3/4 × 60	10-3/4 × 63		
	32	8-3/4 × 52-1/2	10-3/4 × 52-1/2	10-3/4 × 55-1/2	10-3/4 × 60	10-3/4 × 63			
	36	10-3/4 × 51	10-3/4 × 54	10-3/4 × 58-1/2	10-3/4 × 63				
	40	10-3/4 × 52-1/2	10-3/4 × 57	10-3/4 × 61-1/2					
96	12	8-3/4 × 39	8-3/4 × 42	8-3/4 × 45	8-3/4 × 48	8-3/4 × 49-1/2	8-3/4 × 51	8-3/4 × 52-1/2	10-3/4 × 51
	16	8-3/4 × 43-1/2	8-3/4 × 46-1/2	8-3/4 × 49-1/2	8-3/4 × 52-1/2	10-3/4 × 51	10-3/4 × 52-1/2	10-3/4 × 54	10-3/4 × 57
	20	8-3/4 × 46-1/2	8-3/4 × 49-1/2	8-3/4 × 52-1/2	10-3/4 × 52-1/2	10-3/4 × 54	10-3/4 × 57	10-3/4 × 58-1/2	10-3/4 × 61-1/2
	24	8-3/4 × 49-1/2	8-3/4 × 52-1/2	10-3/4 × 52-1/2	10-3/4 × 55-1/2	10-3/4 × 58-1/2	10-3/4 × 60	10-3/4 × 64-1/2	
	28	8-3/4 × 52-1/2	10-3/4 × 52-1/2	10-3/4 × 55-1/2	10-3/4 × 58-1/2	10-3/4 × 61-1/2			
	32	10-3/4 × 51	10-3/4 × 54	10-3/4 × 58-1/2	10-3/4 × 61-1/2				
	36	10-3/4 × 52-1/2	10-3/4 × 57	10-3/4 × 60					
	40	10-3/4 × 54	10-3/4 × 58-1/2	10-3/4 × 64-1/2					
100	12	8-3/4 × 40-1/2	8-3/4 × 43-1/2	8-3/4 × 46-1/2	8-3/4 × 49-1/2	8-3/4 × 51	10-3/4 × 49-1/2	10-3/4 × 52-1/2	10-3/4 × 54
	16	8-3/4 × 45	8-3/4 × 48	8-3/4 × 51	10-3/4 × 51	10-3/4 × 52-1/2	10-3/4 × 55-1/2	10-3/4 × 57	10-3/4 × 58-1/2
	20	8-3/4 × 48	8-3/4 × 52-1/2	10-3/4 × 52-1/2	10-3/4 × 54	10-3/4 × 57	10-3/4 × 58-1/2	10-3/4 × 61-1/2	10-3/4 × 64-1/2
	24	8-3/4 × 51	10-3/4 × 52-1/2	10-3/4 × 55-1/2	10-3/4 × 58-1/2	10-3/4 × 60	10-3/4 × 63		
	28	10-3/4 × 51	10-3/4 × 54	10-3/4 × 58-1/2	10-3/4 × 61-1/2	10-3/4 × 64-1/2			
	32	10-3/4 × 52-1/2	10-3/4 × 57	10-3/4 × 60	10-3/4 × 64-1/2				
	36	10-3/4 × 55-1/2	10-3/4 × 58-1/2	10-3/4 × 63					
	40	10-3/4 × 57	10-3/4 × 61-1/2						

[a]For preliminary design purposes only. This beam design table applies for straight, simply supported, laminated timber beams. Other beam support systems may be employed to meet varying design conditions. Roofs should have a minimum slope of $\frac{1}{4}$ in./ft to eliminate water ponding. Total load-carrying capacity includes beam weight. Floor beams are designed for uniform loads of 40 psf live load and 10 psf dead load. Allowable stresses: Bending stress, F_b = 2400 psi (reduced by size factor, C_F) except those marked *, in which cases F_b = 2000 psi (reduced by size factor, C_F). Shear stress, F_v = 165 psi. Modulus of elasticity, E = 1,700,000 psi except those marked * in which cases E = 1,500,000 psi. For roof beams, F_b and F_v were increased 15% for short duration of loading. Deflection limits: Roof beams—1/180 span for total load. Floor beams—1/360 span for 40 psf residential live load only. See Table 5.8 for other deflection limits for floor beams for commercial and other uses where increased stiffness is desired.

TABLE 7.6

Structural Glued Laminated Timber Beams: Simple-Span Beam Table, Applied Load Capacity (plf)[a]

Beam Size, (in.) b	d	Weight (plf)	Span, (ft) 8	9	10	12	14	16	18	20	24	28	32	36	40
ROOF BEAMS															
$3\frac{1}{8}$	6	4.6	445	338	245	140									
	$7\frac{1}{2}$	5.7	696	549	444	277	172	114							
	9	6.8	1004	792	640	442	301	199	138	99					
	$10\frac{1}{2}$	8.0	1320	1079	872	603	441	319	222	160					
	12	9.1	1572	1346	1141	789	578	440	334	241	136				
	$13\frac{1}{2}$	10.3		1571	1367	987	723	551	433	346	196	119			
	15	11.4			1570	1206	883	673	530	427	271	167	108		
	$16\frac{1}{2}$	12.5				1398	1147	971	765	617	414	256	167	114	
	18	13.7				1568	1280	1081	902	728	502	335	220	150	106
$5\frac{1}{8}$	6	7.5	729	555	403	230	142								
	$7\frac{1}{2}$	9.3	1142	900	727	454	282	186	128						
	9	11.2		1299	1050	726	493	327	226	162					
	$10\frac{1}{2}$	13.1			1431	990	724	523	364	262	146				
	12	14.9				1295	947	722	547	395	222	134			
	$13\frac{1}{2}$	16.8					1185	903	710	567	321	196	126		
	15	18.7					1448	1104	869	700	445	273	177	119	
	$16\frac{1}{2}$	20.6						1593	1254	1012	678	420	274	187	130
	18	22.4							1480	1195	823	549	360	246	174
	$19\frac{1}{2}$	24.3								1391	959	698	462	318	225
	21	26.2								1602	1105	805	582	401	285

	22½	28.0								1767	1260	918	697	497	355
	24	29.9									1425	1039	789	608	435
6¾	9	14.8			1382	956	649	430	298	213	117				
	10½	17.2				1303	953	689	479	345	192	115			
	12	19.7					1248	951	721	520	293	177	112		
	13½	22.1					1561	1190	936	747	423	258	166	110	
	15	24.6						1454	1144	922	586	360	233	156	107
	16½	27.1								1333	894	553	361	246	172
	18	29.5								1573	1084	723	475	325	229
	19½	32.0									1263	919	609	418	296
	21	34.5									1455	1060	766	528	376
	22½	36.9										1210	917	655	467
	24	39.4										1369	1039	800	573
	25½	41.8										1537	1167	913	692
	27	44.3											1302	1020	818
	28½	46.8											1445	1132	908
	30	49.2											1594	1249	1002
FLOOR BEAMS															
3⅛	6	4.6	301	210	152										
	7½	5.7	590	413	299	171	106								
	9	6.8	872	688	521	298	185	122							
	10½	8.0	1147	937	758	477	297	196	136						
	12	9.1	1366	1169	991	685	446	296	205	147					
	13½	10.3		1365	1187	857	627	424	295	212	118				
	15	11.4		1575	1364	1047	766	584	407	294	165	100			
	16½	12.5			1552	1214	996	843	619	448	254	155	100		
	18	13.7				1361	1111	938	783	584	332	204	132		
5⅛	6	7.5	493	344	249	141									
	7½	9.3	968	677	491	280	173	113							

(continued)

TABLE 7.6 (*Continued*)

Structural Glued Laminated Timber Beams: Simple-Span Beam Table, Applied Load Capacity (plf)[a]

Beam Size, (in.)			Span, (ft)												
b	d	Weight (plf)	8	9	10	12	14	16	18	20	24	28	32	36	40
FLOOR BEAMS (cont.)															
$5\frac{1}{8}$ (cont.)	9	11.2	1430	1128	854	489	304	200	137						
	$10\frac{1}{2}$	13.1		1537	1243	782	487	322	222	159					
	12	14.9				1124	732	486	337	241	133				
	$13\frac{1}{2}$	16.8				1406	1028	696	484	348	194	116			
	15	18.7					1257	958	668	482	271	164	104		
	$16\frac{1}{2}$	20.6						1382	1015	734	416	255	164	109	
	18	22.4						1539	1284	958	545	335	217	146	100
	$19\frac{1}{2}$	24.3							1466	1207	697	430	280	189	131
	21	26.2								1390	875	541	354	241	168
	$22\frac{1}{2}$	28.0								1533	1080	670	439	300	211
	24	29.9									1235	817	537	368	261

[a]For preliminary design purposes only.

Table specifications: This table applies for straight, simply supported laminated timber beams under dry-use service conditions. Roofs should have a minimum slope of $\frac{1}{4}$ in./ft to avoid water ponding (See Section 4). Tabulated values are for applied loading. The weight of the beam has been deducted from the load-carrying capacity. An approximate beam weight of 35 pcf was used to determine beam weights in plf shown in the table. Design values: F_b = 2000 psi (reduced by size factor, C_F) and E = 1,500,000 psi for beams 15 in. and less in depth. F_b = 2400 psi (reduced by size factor, C_F) and E = 1,700,000 psi for beams $16\frac{1}{2}$ in. and greater in depth. F_v = 165 psi (loads within a distance, d, from the support were neglected in the shear calculations). For roof beams, F_b and F_v were increased 15% for short-term duration of load. Deflection limits: For roof beams, 1/180 of span for total load; for floor beams, 1/360 of span for live load. Live load is assumed to be 80% of total load as is common for residential floors. When the live load exceeds 80% of the total load the member should be checked for a deflection limit of 1/360 of the span for live load. When the live load is less than $\frac{2}{3}$ the total load, the member should be checked for a deflection limit of 1/240 of the span for total load. See Table 5.8 for other deflection limits for floor beams for commercial and office uses where increased stiffness is desired.

TABLE 7.7

Structural Glued Laminated Timber Cantilever Beam Span-Load Table[a]

Main Support Spacing (ft)	Dead Load (psf)	Live Load (psf)	Two-Span System[b]		Three-Span System[c]		Three-Span System[d]	
			Suspended Beams	Cantilevered Beams	Suspended Beams	Cantilevered Beams	Suspended Beams	Double Cantilevered Beams
32	10	12	*$5\frac{1}{8} \times 16\frac{1}{2}$	$5\frac{1}{8} \times 16\frac{1}{2}$	*$5\frac{1}{8} \times 10\frac{1}{2}$	$5\frac{1}{8} \times 16\frac{1}{2}$	*$5\frac{1}{8} \times 15$	*$5\frac{1}{8} \times 15$
		20	$5\frac{1}{8} \times 16\frac{1}{2}$	$5\frac{1}{8} \times 19\frac{1}{2}$	*$5\frac{1}{8} \times 12$	$5\frac{1}{8} \times 19\frac{1}{2}$	$5\frac{1}{8} \times 18$	$5\frac{1}{8} \times 19\frac{1}{2}$
		30	$5\frac{1}{8} \times 19\frac{1}{2}$	$5\frac{1}{8} \times 25\frac{1}{2}$	*$5\frac{1}{8} \times 13\frac{1}{2}$	$5\frac{1}{8} \times 24$	$5\frac{1}{8} \times 21$	$5\frac{1}{8} \times 24$
	12	12	*$5\frac{1}{8} \times 16\frac{1}{2}$	$5\frac{1}{8} \times 16\frac{1}{2}$	*$5\frac{1}{8} \times 10\frac{1}{2}$	$5\frac{1}{8} \times 16\frac{1}{2}$	*$5\frac{1}{8} \times 16\frac{1}{2}$	$5\frac{1}{8} \times 16\frac{1}{2}$
		20	$5\frac{1}{8} \times 18$	$5\frac{1}{8} \times 21$	*$5\frac{1}{8} \times 12$	$5\frac{1}{8} \times 21$	$5\frac{1}{8} \times 18$	$5\frac{1}{8} \times 19\frac{1}{2}$
		30	$5\frac{1}{8} \times 21$	$5\frac{1}{8} \times 27$	*$5\frac{1}{8} \times 13\frac{1}{2}$	$5\frac{1}{8} \times 24$	$5\frac{1}{8} \times 21$	$5\frac{1}{8} \times 25\frac{1}{2}$
	15	20	$5\frac{1}{8} \times 19\frac{1}{2}$	$5\frac{1}{8} \times 22\frac{1}{2}$	*$5\frac{1}{8} \times 12$	$5\frac{1}{8} \times 21$	$5\frac{1}{8} \times 19\frac{1}{2}$	$5\frac{1}{8} \times 21$
		30	$5\frac{1}{8} \times 21$	$5\frac{1}{8} \times 28\frac{1}{2}$	*$5\frac{1}{8} \times 13\frac{1}{2}$	$5\frac{1}{8} \times 25\frac{1}{2}$	$5\frac{1}{8} \times 22\frac{1}{2}$	$5\frac{1}{8} \times 27$
36	10	12	$5\frac{1}{8} \times 16\frac{1}{2}$	$5\frac{1}{8} \times 18$	*$5\frac{1}{8} \times 12$	$5\frac{1}{8} \times 18$	$5\frac{1}{8} \times 16\frac{1}{2}$	$5\frac{1}{8} \times 16\frac{1}{2}$
		20	$5\frac{1}{8} \times 19\frac{1}{2}$	$5\frac{1}{8} \times 22\frac{1}{2}$	*$5\frac{1}{8} \times 13\frac{1}{2}$	$5\frac{1}{8} \times 22\frac{1}{2}$	$5\frac{1}{8} \times 19\frac{1}{2}$	$5\frac{1}{8} \times 21$
		30	$5\frac{1}{8} \times 22\frac{1}{2}$	$5\frac{1}{8} \times 28\frac{1}{2}$	*$5\frac{1}{8} \times 15$	$5\frac{1}{8} \times 27$	$5\frac{1}{8} \times 22\frac{1}{2}$	$5\frac{1}{8} \times 27$
	12	12	$5\frac{1}{8} \times 16\frac{1}{2}$	$5\frac{1}{8} \times 19\frac{1}{2}$	*$5\frac{1}{8} \times 12$	$5\frac{1}{8} \times 19\frac{1}{2}$	$5\frac{1}{8} \times 18$	$5\frac{1}{8} \times 18$
		20	$5\frac{1}{8} \times 19\frac{1}{2}$	$5\frac{1}{8} \times 24$	*$5\frac{1}{8} \times 13\frac{1}{2}$	$5\frac{1}{8} \times 24$	$5\frac{1}{8} \times 21$	$5\frac{1}{8} \times 21$
		30	$5\frac{1}{8} \times 22\frac{1}{2}$	$5\frac{1}{8} \times 30$	*$5\frac{1}{8} \times 15$	$5\frac{1}{8} \times 27$	$5\frac{1}{8} \times 24$	$5\frac{1}{8} \times 28\frac{1}{2}$
	15	20	$5\frac{1}{8} \times 21$	$5\frac{1}{8} \times 25\frac{1}{2}$	*$5\frac{1}{8} \times 13\frac{1}{2}$	$5\frac{1}{8} \times 24$	$5\frac{1}{8} \times 22\frac{1}{2}$	$5\frac{1}{8} \times 24$
		30	$5\frac{1}{8} \times 24$	$5\frac{1}{8} \times 33$	*$5\frac{1}{8} \times 16\frac{1}{2}$	$5\frac{1}{8} \times 28\frac{1}{2}$	$5\frac{1}{8} \times 25\frac{1}{2}$	$5\frac{1}{8} \times 31\frac{1}{2}$
40	10	12	$5\frac{1}{8} \times 18$	$5\frac{1}{8} \times 21$	*$5\frac{1}{8} \times 12$	$5\frac{1}{8} \times 21$	$5\frac{1}{8} \times 18$	$5\frac{1}{8} \times 18$
		20	$5\frac{1}{8} \times 21$	$5\frac{1}{8} \times 25\frac{1}{2}$	*$5\frac{1}{8} \times 15$	$5\frac{1}{8} \times 25\frac{1}{2}$	$5\frac{1}{8} \times 22\frac{1}{2}$	$5\frac{1}{8} \times 24$
		30	$5\frac{1}{8} \times 25\frac{1}{2}$	$5\frac{1}{8} \times 31\frac{1}{2}$	*$5\frac{1}{8} \times 16\frac{1}{2}$	$5\frac{1}{8} \times 30$	$5\frac{1}{8} \times 25\frac{1}{2}$	$5\frac{1}{8} \times 30$

(continued)

TABLE 7.7 (*Continued*)

Structural Glued Laminated Timber Cantilever Beam Span-Load Table[a]

Main Support Spacing (ft)	Dead Load (psf)	Live Load (psf)	Two-Span System[b]		Three-Span System[c]		Three-Span System[d]	
			Suspended Beams	Cantilevered Beams	Suspended Beams	Cantilevered Beams	Suspended Beams	Double Cantilevered Beams
40 (cont.)	12	12	$5\frac{1}{8} \times 18$	$5\frac{1}{8} \times 21$	*$5\frac{1}{8} \times 12$	$5\frac{1}{8} \times 21$	$5\frac{1}{8} \times 19\frac{1}{2}$	$5\frac{1}{8} \times 19\frac{1}{2}$
		20	$5\frac{1}{8} \times 22\frac{1}{2}$	$5\frac{1}{8} \times 25\frac{1}{2}$	*$5\frac{1}{8} \times 15$	$5\frac{1}{8} \times 25\frac{1}{2}$	$5\frac{1}{8} \times 22\frac{1}{2}$	$5\frac{1}{8} \times 24$
		30	$5\frac{1}{8} \times 25\frac{1}{2}$	$5\frac{1}{8} \times 33$	*$5\frac{1}{8} \times 16\frac{1}{2}$	$5\frac{1}{8} \times 30$	$5\frac{1}{8} \times 27$	$5\frac{1}{8} \times 31\frac{1}{2}$
	15	20	$5\frac{1}{8} \times 24$	$5\frac{1}{8} \times 28\frac{1}{2}$	*$5\frac{1}{8} \times 15$	$5\frac{1}{8} \times 27$	$5\frac{1}{8} \times 24$	$5\frac{1}{8} \times 27$
		30	$5\frac{1}{8} \times 27$	$5\frac{1}{8} \times 36$	$5\frac{1}{8} \times 16\frac{1}{2}$	$5\frac{1}{8} \times 31\frac{1}{2}$	$5\frac{1}{8} \times 28\frac{1}{2}$	$5\frac{1}{8} \times 34\frac{1}{2}$
44	10	12	$5\frac{1}{8} \times 19\frac{1}{2}$	$5\frac{1}{8} \times 22\frac{1}{2}$	*$5\frac{1}{8} \times 13\frac{1}{2}$	$5\frac{1}{8} \times 22\frac{1}{2}$	$5\frac{1}{8} \times 21$	$5\frac{1}{8} \times 21$
		20	$5\frac{1}{8} \times 24$	$5\frac{1}{8} \times 28\frac{1}{2}$	*$5\frac{1}{8} \times 16\frac{1}{2}$	$5\frac{1}{8} \times 28\frac{1}{2}$	$5\frac{1}{8} \times 25\frac{1}{2}$	$5\frac{1}{8} \times 25\frac{1}{2}$
		30	$5\frac{1}{8} \times 27$	$5\frac{1}{8} \times 34\frac{1}{2}$	$5\frac{1}{8} \times 16\frac{1}{2}$	$5\frac{1}{8} \times 33$	$5\frac{1}{8} \times 28\frac{1}{2}$	$5\frac{1}{8} \times 33$
	12	12	$5\frac{1}{8} \times 21$	$5\frac{1}{8} \times 22\frac{1}{2}$	*$5\frac{1}{8} \times 13\frac{1}{2}$	$5\frac{1}{8} \times 24$	$5\frac{1}{8} \times 21$	$5\frac{1}{8} \times 21$
		20	$5\frac{1}{8} \times 24$	$5\frac{1}{8} \times 28\frac{1}{2}$	*$5\frac{1}{8} \times 16\frac{1}{2}$	$5\frac{1}{8} \times 28\frac{1}{2}$	$5\frac{1}{8} \times 25\frac{1}{2}$	$5\frac{1}{8} \times 27$
		30	$5\frac{1}{8} \times 28\frac{1}{2}$	$5\frac{1}{8} \times 36$	$5\frac{1}{8} \times 18$	$5\frac{1}{8} \times 33$	$5\frac{1}{8} \times 30$	$5\frac{1}{8} \times 34\frac{1}{2}$
	15	20	$5\frac{1}{8} \times 25\frac{1}{2}$	$5\frac{1}{8} \times 31\frac{1}{2}$	$5\frac{1}{8} \times 16\frac{1}{2}$	$5\frac{1}{8} \times 30$	$5\frac{1}{8} \times 27$	$5\frac{1}{8} \times 30$
		30	$5\frac{1}{8} \times 30$	$6\frac{3}{4} \times 30$	$5\frac{1}{8} \times 18$	$5\frac{1}{8} \times 34\frac{1}{2}$	$5\frac{1}{8} \times 30$	$6\frac{3}{4} \times 30$
48	10	12	$5\frac{1}{8} \times 21$	$5\frac{1}{8} \times 24$	*$5\frac{1}{8} \times 13\frac{1}{2}$	$5\frac{1}{8} \times 24$	$5\frac{1}{8} \times 22\frac{1}{2}$	$5\frac{1}{8} \times 22\frac{1}{2}$
		20	$5\frac{1}{8} \times 25\frac{1}{2}$	$5\frac{1}{8} \times 30$	$5\frac{1}{8} \times 16\frac{1}{2}$	$5\frac{1}{8} \times 30$	$5\frac{1}{8} \times 27$	$5\frac{1}{8} \times 28\frac{1}{2}$
		30	$5\frac{1}{8} \times 30$	$5\frac{1}{8} \times 37\frac{1}{2}$	$5\frac{1}{8} \times 18$	$5\frac{1}{8} \times 36$	$5\frac{1}{8} \times 31\frac{1}{2}$	$5\frac{1}{8} \times 36$
	12	12	$5\frac{1}{8} \times 22\frac{1}{2}$	$5\frac{1}{8} \times 25\frac{1}{2}$	*$5\frac{1}{8} \times 15$	$5\frac{1}{8} \times 25\frac{1}{2}$	$5\frac{1}{8} \times 24$	$5\frac{1}{8} \times 24$
		20	$5\frac{1}{8} \times 27$	$5\frac{1}{8} \times 31\frac{1}{2}$	$5\frac{1}{8} \times 16\frac{1}{2}$	$5\frac{1}{8} \times 31\frac{1}{2}$	$5\frac{1}{8} \times 28\frac{1}{2}$	$5\frac{1}{8} \times 30$
		30	$5\frac{1}{8} \times 31\frac{1}{2}$	$6\frac{3}{4} \times 31\frac{1}{2}$	$5\frac{1}{8} \times 19\frac{1}{2}$	$5\frac{1}{8} \times 36$	$5\frac{1}{8} \times 31\frac{1}{2}$	$6\frac{3}{4} \times 30$

	15	20	$5\frac{1}{8} \times 28\frac{1}{2}$	$5\frac{1}{8} \times 34\frac{1}{2}$	$5\frac{1}{8} \times 18$	$5\frac{1}{8} \times 33$	$5\frac{1}{8} \times 30$	$5\frac{1}{8} \times 33$
		30	$5\frac{1}{8} \times 33$	$6\frac{3}{4} \times 33$	$5\frac{1}{8} \times 19\frac{1}{2}$	$5\frac{1}{8} \times 37\frac{1}{2}$	$5\frac{1}{8} \times 33$	$6\frac{3}{4} \times 31\frac{1}{2}$
52	10	12	$5\frac{1}{8} \times 24$	$5\frac{1}{8} \times 27$	*$5\frac{1}{8} \times 15$	$5\frac{1}{8} \times 27$	$5\frac{1}{8} \times 24$	$5\frac{1}{8} \times 25\frac{1}{2}$
		20	$5\frac{1}{8} \times 28\frac{1}{2}$	$6\frac{3}{4} \times 28\frac{1}{2}$	$5\frac{1}{8} \times 18$	$6\frac{3}{4} \times 28\frac{1}{2}$	$5\frac{1}{8} \times 30$	$5\frac{1}{8} \times 31\frac{1}{2}$
		30	$5\frac{1}{8} \times 33$	$6\frac{3}{4} \times 33$	$5\frac{1}{8} \times 19\frac{1}{2}$	$6\frac{3}{4} \times 34\frac{1}{2}$	$5\frac{1}{8} \times 34\frac{1}{2}$	$6\frac{3}{4} \times 31\frac{1}{2}$
	12	12	$5\frac{1}{8} \times 24$	$5\frac{1}{8} \times 27$	*$5\frac{1}{8} \times 16\frac{1}{2}$	$5\frac{1}{8} \times 27$	$5\frac{1}{8} \times 25\frac{1}{2}$	$5\frac{1}{8} \times 25\frac{1}{2}$
		20	$5\frac{1}{8} \times 30$	$6\frac{3}{4} \times 30$	$5\frac{1}{8} \times 18$	$6\frac{3}{4} \times 30$	$5\frac{1}{8} \times 30$	$5\frac{1}{8} \times 33$
		30	$5\frac{1}{8} \times 33$	$6\frac{3}{4} \times 34\frac{1}{2}$	$5\frac{1}{8} \times 21$	$6\frac{3}{4} \times 34\frac{1}{2}$	$5\frac{1}{8} \times 34\frac{1}{2}$	$6\frac{3}{4} \times 33$
	15	20	$5\frac{1}{8} \times 31\frac{1}{2}$	$6\frac{3}{4} \times 30$	$5\frac{1}{8} \times 19\frac{1}{2}$	$6\frac{3}{4} \times 31\frac{1}{2}$	$5\frac{1}{8} \times 33$	$5\frac{1}{8} \times 36$
		30	$5\frac{1}{8} \times 34\frac{1}{2}$	$6\frac{3}{4} \times 36$	$5\frac{1}{8} \times 21$	$6\frac{3}{4} \times 36$	$5\frac{1}{8} \times 36$	$6\frac{3}{4} \times 34\frac{1}{2}$
56	10	12	$5\frac{1}{8} \times 25\frac{1}{2}$	$5\frac{1}{8} \times 28\frac{1}{2}$	*$5\frac{1}{8} \times 16\frac{1}{2}$	$5\frac{1}{8} \times 28\frac{1}{2}$	$5\frac{1}{8} \times 27$	$5\frac{1}{8} \times 25\frac{1}{2}$
		20	$5\frac{1}{8} \times 31\frac{1}{2}$	$5\frac{1}{8} \times 36$	$5\frac{1}{8} \times 19\frac{1}{2}$	$6\frac{3}{4} \times 31\frac{1}{2}$	$5\frac{1}{8} \times 31\frac{1}{2}$	$5\frac{1}{8} \times 33$
		30	$5\frac{1}{8} \times 36$	$6\frac{3}{4} \times 36$	$5\frac{1}{8} \times 22\frac{1}{2}$	$6\frac{3}{4} \times 36$	$5\frac{1}{8} \times 37\frac{1}{2}$	$6\frac{3}{4} \times 34\frac{1}{2}$
	12	12	$5\frac{1}{8} \times 27$	$5\frac{1}{8} \times 30$	$5\frac{1}{8} \times 16\frac{1}{2}$	$5\frac{1}{8} \times 30$	$5\frac{1}{8} \times 27$	$5\frac{1}{8} \times 27$
		20	$5\frac{1}{8} \times 31\frac{1}{2}$	$6\frac{3}{4} \times 31\frac{1}{2}$	$5\frac{1}{8} \times 19\frac{1}{2}$	$6\frac{3}{4} \times 31\frac{1}{2}$	$5\frac{1}{8} \times 33$	$5\frac{1}{8} \times 34\frac{1}{2}$
		30	$5\frac{1}{8} \times 36$	$6\frac{3}{4} \times 37\frac{1}{2}$	$5\frac{1}{8} \times 22\frac{1}{2}$	$6\frac{3}{4} \times 37\frac{1}{2}$	$5\frac{1}{8} \times 37\frac{1}{2}$	$6\frac{3}{4} \times 34\frac{1}{2}$
	15	20	$5\frac{1}{8} \times 33$	$6\frac{3}{4} \times 33$	$5\frac{1}{8} \times 21$	$6\frac{3}{4} \times 33$	$5\frac{1}{8} \times 34\frac{1}{2}$	$6\frac{3}{4} \times 30$
		30	$5\frac{1}{8} \times 37\frac{1}{2}$	$6\frac{3}{4} \times 39$	$5\frac{1}{8} \times 24$	$6\frac{3}{4} \times 39$	$6\frac{3}{4} \times 34\frac{1}{2}$	$6\frac{3}{4} \times 37\frac{1}{2}$
60	10	12	$5\frac{1}{8} \times 27$	$5\frac{1}{8} \times 31\frac{1}{2}$	$5\frac{1}{8} \times 16\frac{1}{2}$	$5\frac{1}{8} \times 31\frac{1}{2}$	$5\frac{1}{8} \times 28\frac{1}{2}$	$5\frac{1}{8} \times 28\frac{1}{2}$
		20	$5\frac{1}{8} \times 33$	$6\frac{3}{4} \times 33$	$5\frac{1}{8} \times 21$	$6\frac{3}{4} \times 33$	$5\frac{1}{8} \times 34\frac{1}{2}$	$5\frac{1}{8} \times 36$
		30	$5\frac{1}{8} \times 37\frac{1}{2}$	$6\frac{3}{4} \times 39$	$5\frac{1}{8} \times 24$	$6\frac{3}{4} \times 39$	$6\frac{3}{4} \times 34\frac{1}{2}$	$6\frac{3}{4} \times 37\frac{1}{2}$
	12	12	$5\frac{1}{8} \times 28\frac{1}{2}$	$5\frac{1}{8} \times 31\frac{1}{2}$	$5\frac{1}{8} \times 18$	$5\frac{1}{8} \times 31\frac{1}{2}$	$5\frac{1}{8} \times 30$	$5\frac{1}{8} \times 30$
		20	$5\frac{1}{8} \times 34\frac{1}{2}$	$6\frac{3}{4} \times 34\frac{1}{2}$	$5\frac{1}{8} \times 21$	$6\frac{3}{4} \times 34\frac{1}{2}$	$5\frac{1}{8} \times 36$	$5\frac{1}{8} \times 37\frac{1}{2}$
		30	$6\frac{3}{4} \times 34\frac{1}{2}$	$8\frac{3}{4} \times 34\frac{1}{2}$	$5\frac{1}{8} \times 24$	$8\frac{3}{4} \times 34\frac{1}{2}$	$6\frac{3}{4} \times 34\frac{1}{2}$	$6\frac{3}{4} \times 37\frac{1}{2}$

(continued)

TABLE 7.7 (*Continued*)

Structural Glued Laminated Timber Cantilever Beam Span-Load Table[a]

Main Support Spacing (ft)	Dead Load (psf)	Live Load (psf)	Two-Span System[a]		Three-Span System[c]		Three-Span System[d]	
			Suspended Beams	Cantilevered Beams	Suspended Beams	Cantilevered Beams	Suspended Beams	Double Cantilevered Beams
60 (cont.)	15	20	$5\frac{1}{8} \times 36$	$6\frac{3}{4} \times 36$	$5\frac{1}{8} \times 22\frac{1}{2}$	$6\frac{3}{4} \times 36$	$5\frac{1}{8} \times 37\frac{1}{2}$	$6\frac{3}{4} \times 33$
		30	$6\frac{3}{4} \times 36$	$8\frac{3}{4} \times 36$	$5\frac{1}{8} \times 25\frac{1}{2}$	$8\frac{3}{4} \times 36$	$6\frac{3}{4} \times 36$	$8\frac{3}{4} \times 33$
64	10	12	$5\frac{1}{8} \times 28\frac{1}{2}$	$5\frac{1}{8} \times 33$	$5\frac{1}{8} \times 18$	$5\frac{1}{8} \times 33$	$5\frac{1}{8} \times 30$	$5\frac{1}{8} \times 30$
		20	$5\frac{1}{8} \times 36$	$6\frac{3}{4} \times 36$	$5\frac{1}{8} \times 21$	$6\frac{3}{4} \times 36$	$5\frac{1}{8} \times 36$	$6\frac{3}{4} \times 33$
		30	$6\frac{3}{4} \times 36$	$8\frac{3}{4} \times 36$	$5\frac{1}{8} \times 25\frac{1}{2}$	$8\frac{3}{4} \times 36$	$6\frac{3}{4} \times 36$	$8\frac{3}{4} \times 34\frac{1}{2}$
	12	12	$5\frac{1}{8} \times 30$	$5\frac{1}{8} \times 34\frac{1}{2}$	$5\frac{1}{8} \times 19\frac{1}{2}$	$5\frac{1}{8} \times 34\frac{1}{2}$	$5\frac{1}{8} \times 31\frac{1}{2}$	$5\frac{1}{8} \times 31\frac{1}{2}$
		20	$5\frac{1}{8} \times 36$	$6\frac{3}{4} \times 36$	$5\frac{1}{8} \times 22\frac{1}{2}$	$6\frac{3}{4} \times 36$	$6\frac{3}{4} \times 33$	$6\frac{3}{4} \times 33$
		30	$6\frac{3}{4} \times 36$	$8\frac{3}{4} \times 37\frac{1}{2}$	$5\frac{1}{8} \times 25\frac{1}{2}$	$8\frac{3}{4} \times 37\frac{1}{2}$	$6\frac{3}{4} \times 37\frac{1}{2}$	$8\frac{3}{4} \times 34\frac{1}{2}$
	15	20	$6\frac{3}{4} \times 33$	$6\frac{3}{4} \times 37\frac{1}{2}$	$5\frac{1}{8} \times 24$	$6\frac{3}{4} \times 39$	$6\frac{3}{4} \times 34\frac{1}{2}$	$6\frac{3}{4} \times 34\frac{1}{2}$
		30	$6\frac{3}{4} \times 37\frac{1}{2}$	$8\frac{3}{4} \times 39$	$5\frac{1}{8} \times 27$	$8\frac{3}{4} \times 39$	$6\frac{3}{4} \times 39$	$8\frac{3}{4} \times 36$

[a]For preliminary design purposes only. This beam design table applies for straight, cantilevered, laminated timber beams. Member sizes are governed by either bending or shear. Where building code deflection requirements apply, the member sizes must be checked. A minimum roof slope of $\frac{1}{4}$ in./ft should be provided to minimize water ponding. Specifications and allowable stresses: Beam spacing: 20 ft.. Bending stress, F_b = 2400 psi (reduced by size factor, C_F) except those marked * in which cases F_b = 2000 psi (reduced by size factor, C_f). Shear stress, F_v = 165 psi. Compression perpendicular to grain stress, $F_{c\perp}$ = 560 psi. Duration of load factor: 1.25 for 12 psf live loads and 1.15 for 20 and 30 psf live loads.

Member sizes are checked for full unbalanced live loading. Main supports are columns or bearing walls. Table is based on equal spacing of main supports. Dead load does not include weight of glulam.

[b]Two-span cantilever system. Cantilevered beam extends over center support with the length of cantilever, l', equal to approximately 0.20 × main support spacing, l.

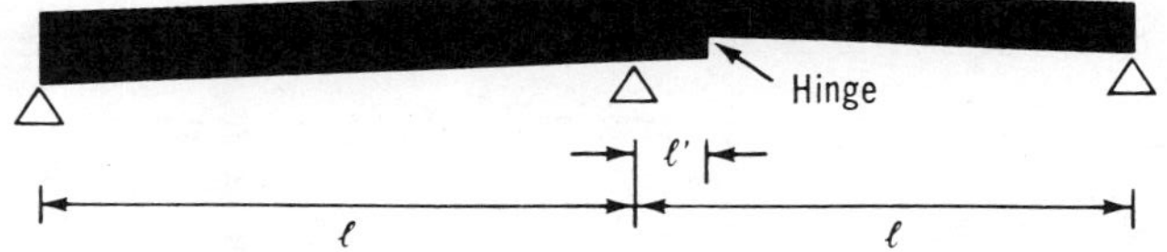

[c]Three-span cantilever system. End members cantilevered over intermediate column supports and carrying the suspended beam. Length of cantilevers, l', equal to approximately 0.25 × main support spacing, l.

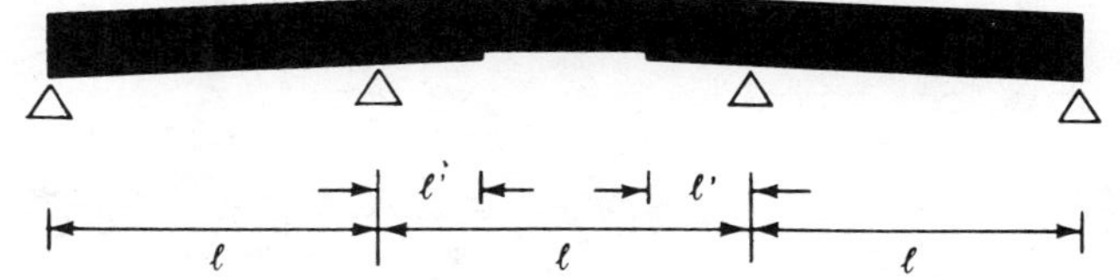

[d]Three-span cantilever system. Center member double cantilevered over intermediate column supports and carrying the suspended wall beams. Length of cantilevers, l', equal to approximately 0.17 × main support spacing, l.

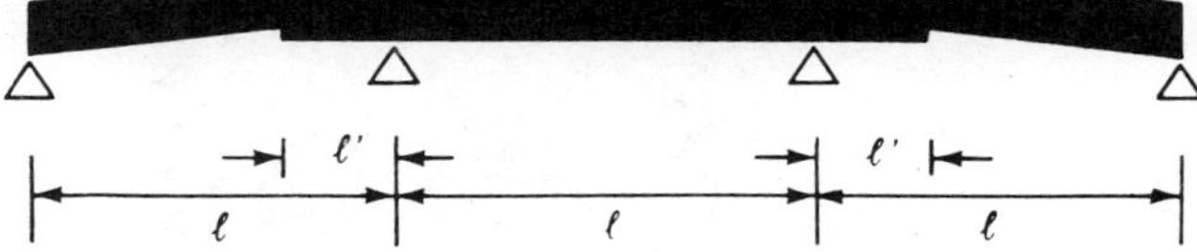

TABLE 7.8 Structural Glued Laminated

Loading	Roof Pitch	Wall Ht (ft)	30′ Span Width	30′ Span Base	30′ Span Lower Tang.	30′ Span Upper Tang.	30′ Span Crown	35′ Span Width	35′ Span Base	35′ Span Lower Tang.	35′ Span Upper Tang.	35′ Span Crown
Vertical dead + live load = 400 lb/ft	3/12	10	3 1/8	8 1/4	12	10 3/4	7 1/2	3 1/8	10 1/2	13 1/4	12	7 1/2
		12	5 1/8	7 1/2	11	10 3/4	7 1/2	5 1/8	7 1/2	12	12	7 1/2
		14	5 1/8	7 1/2	12	12	7 1/2	5 1/8	7 1/2	13 1/2	13 1/4	7 1/2
		16	5 1/8	7 1/2	13 1/4	13	7 1/2	5 1/8	7 1/2	14 3/4	14 1/2	7 1/2
		18	5 1/8	7 1/2	14 1/4	14 1/4	7 1/2	5 1/8	7 1/2	16	15 3/4	7 1/2
	4/12	10	3 1/8	7 1/2	11 3/4	12 3/4	7 1/2	3 1/8	9 3/4	13 1/2	12 3/4	7 1/2
		12	5 1/8	7 1/2	10 3/4	10 3/4	7 1/2	5 1/8	7 1/2	11 3/4	11 3/4	7 1/2
		14	5 1/8	7 1/2	12	12	7 1/2	5 1/8	7 1/2	13 1/4	13	7 1/2
		16	5 1/8	7 1/2	13 1/4	13	7 1/2	5 1/8	7 1/2	14 1/2	14 1/2	7 1/2
		18	5 1/8	7 1/2	14 1/4	14	7 1/2	5 1/8	7 1/2	15 3/4	15 1/2	7 1/2
	6/12	12	5 1/8	7 1/2	10 1/2	10 1/2	7 1/2	5 1/8	7 1/2	11 1/2	11 1/4	7 1/2
		14	5 1/8	7 1/2	11 3/4	11 3/4	7 1/2	5 1/8	7 1/2	12 3/4	12 3/4	7 1/2
		16	5 1/8	7 1/2	13	12 3/4	7 1/2	5 1/8	7 1/2	14	14	7 1/2
		18	5 1/8	7 1/2	14	14	7 1/2	5 1/8	7 1/2	15 1/4	15 1/4	7 1/2
	8/12	12	5 1/8	7 1/2	10 1/4	10	7 1/2	5 1/8	7 1/2	11	10 1/2	7 1/2
		14	5 1/8	7 1/2	11 1/2	11 1/4	7 1/2	5 1/8	7 1/2	12 1/4	12	7 1/2
		16	5 1/8	7 1/2	12 3/4	12 1/2	7 1/2	5 1/8	7 1/2	13 1/2	13 1/2	7 1/2
		18	5 1/8	7 1/2	13 3/4	13 1/2	7 1/2	5 1/8	7 1/2	14 3/4	14 3/4	7 1/2
Vertical dead + live load = 600 lb/ft	3/12	10	3 1/8	12	14 1/2	12 3/4	7 1/2	5 1/8	9 3/4	12	12 3/4	7 1/2
		12	5 1/8	7 1/2	12	12	7 1/2	5 1/8	8 1/2	13 1/2	13	7 1/2
		14	5 1/8	7 1/2	13 1/2	13 1/4	7 1/2	5 1/8	7 3/4	15 1/2	14 1/4	7 1/2
		16	5 1/8	7 1/2	14 3/4	14 3/4	7 1/2	5 1/8	7 1/2	17 1/4	15 1/4	7 1/2
		18	5 1/8	7 1/2	16	16	7 1/2	5 1/8	7 1/2	18 3/4	16 1/4	7 1/2
	4/12	10	3 1/8	11	15	15 1/2	7 1/2	5 1/8	9	10 1/4	15 1/2	7 1/2
		12	5 1/8	7 1/2	12	12	7 1/2	5 1/8	8	13 3/4	12 1/4	7 1/2
		14	5 1/8	7 1/2	13 1/2	13 1/4	7 1/2	5 1/8	7 1/2	15 1/2	13 3/4	7 1/2
		16	5 1/8	7 1/2	14 3/4	14 3/4	7 1/2	5 1/8	7 1/2	17	15	7 1/2
		18	5 1/8	7 1/2	16	16	7 1/2	5 1/8	7 1/2	18 1/2	16 1/4	7 1/2
	6/12	12	5 1/8	7 1/2	12	11 1/4	7 1/2	5 1/8	7 1/2	13 3/4	11 1/4	7 1/2
		14	5 1/8	7 1/2	13 1/2	12 3/4	7 1/2	5 1/8	7 1/2	15 1/2	12 1/2	7 1/2
		16	5 1/8	7 1/2	14 3/4	14	7 1/2	5 1/8	7 1/2	17	14	7 1/2
		18	5 1/8	7 1/2	16	15 1/4	7 1/2	5 1/8	7 1/2	18 1/4	15 1/2	7 1/2
	8/12	12	5 1/8	7 1/2	12	10 1/2	7 1/2	5 1/8	7 1/2	13 1/2	10 1/4	7 1/2
		14	5 1/8	7 1/2	13 1/4	12 1/4	7 1/2	5 1/8	7 1/2	15 1/4	11 1/2	7 1/2
		16	5 1/8	7 1/2	14 1/2	13 1/2	7 1/2	5 1/8	7 1/2	16 1/2	13 1/4	7 1/2
		18	5 1/8	7 1/2	15 3/4	14 3/4	7 1/2	5 1/8	7 1/2	18	14 1/4	7 1/2

Timber Three-Hinged Tudor Arch Span-Load Table[a]

40′ Span					50′ Span				
Width	Base	Lower Tang.	Upper Tang.	Crown	Width	Base	Lower Tang.	Upper Tang.	Crown
3⅛	13¼	14½	13¼	7½	5⅛	11¾	14	13¾	7½
5⅛	7½	13¼	13	7½	5⅛	10½	14¾	14½	7½
5⅛	7½	14¾	14½	7½	5⅛	9½	16¾	16¼	7½
5⅛	7½	16	16	7½	5⅛	8¾	18¾	17½	7½
5⅛	7½	17½	17¼	7½	5⅛	8	20½	19	7½
3⅛	12	15¼	12¾	7½	5⅛	10½	13	12¾	7½
5⅛	7½	12¾	12½	7½	5⅛	9½	14½	13½	7½
5⅛	7½	14¼	14¼	7½	5⅛	8¾	16¾	15	7½
5⅛	7½	15¾	15½	7½	5⅛	8	18½	16½	7½
5⅛	7½	17	17	7½	5⅛	7½	20¼	18	7½
5⅛	7½	12¼	11½	7½	5⅛	8	14¾	11¾	7½
5⅛	7½	13¾	13¼	7½	5⅛	7½	16½	12¾	7½
5⅛	7½	15	15	7½	5⅛	7½	18¼	14¼	7½
5⅛	7½	16½	16¼	7½	5⅛	7½	19¾	16	7½
5⅛	7½	12	10¼	7½	5⅛	7½	14¼	11½	7½
5⅛	7½	13½	11¾	7½	5⅛	7½	16	12¼	7½
5⅛	7½	14¾	13½	7½	5⅛	7½	17¾	13	7½
5⅛	7½	16	15	7½	5⅛	7½	19¼	14	7½
5⅛	12¼	14	13¾	12¼	5⅛	17½	17½	15½	12¼
5⅛	10¾	15	13½	7½	5⅛	15½	17½	16	7½
5⅛	9¾	17½	14½	7½	5⅛	14	21	16½	7½
5⅛	8¾	19¼	15¾	7½	5⅛	12¾	23½	17	7½
5⅛	8	21¼	16¾	7½	6¾	9	20½	20¼	7½
5⅛	11	11¼	15½	12	5⅛	15½	15½	15½	12
5⅛	9¾	15¼	12¾	7½	5⅛	14	18	15¼	7½
5⅛	9	17½	13¾	7½	5⅛	12¾	21	16	7½
5⅛	8¼	19¼	15¼	7½	5⅛	11¾	23½	16½	7½
5⅛	7½	21	16½	7½	5⅛	10¾	25½	16¾	7½
5⅛	8½	15¼	11	7½	5⅛	11¾	18¼	14½	7¾
5⅛	7¾	17¼	12½	7½	5⅛	10¾	20¾	15¼	7½
5⅛	7½	19	14	7½	5⅛	10	23	15¾	7½
5⅛	7½	20½	15¼	7½	5⅛	9½	25	16½	7½
5⅛	7½	15¼	11	7½	5⅛	10	18	14	9½
5⅛	7½	17	11¾	7½	5⅛	9¼	20¼	15	8
5⅛	7½	18½	12¾	7½	5⅛	8¾	22¼	15¾	7½
5⅛	7½	20¼	13¾	7½	5⅛	8¼	24¼	16¼	7½

(*continued*)

TABLE 7.8 (*Continued*) **Structural Glued Laminated**

Loading	Roof Pitch	Wall Ht (ft)	60′ Span Width	60′ Span Base	60′ Span Lower Tang.	60′ Span Upper Tang.	60′ Span Crown	70′ Span Width	70′ Span Base	70′ Span Lower Tang.	70′ Span Upper Tang.	70′ Span Crown
Vertical dead + live load = 400 lb/ft	3/12	12	$5\frac{1}{8}$	14	$16\frac{1}{4}$	$16\frac{3}{4}$	$7\frac{1}{2}$	$5\frac{1}{8}$	$17\frac{3}{4}$	$17\frac{3}{4}$	20	$7\frac{1}{2}$
		14	$5\frac{1}{8}$	$12\frac{3}{4}$	$19\frac{1}{2}$	$17\frac{1}{2}$	$7\frac{1}{2}$	$5\frac{1}{8}$	$16\frac{1}{4}$	22	$20\frac{3}{4}$	$7\frac{1}{2}$
		16	$5\frac{1}{8}$	$11\frac{3}{4}$	22	$17\frac{3}{4}$	$7\frac{1}{2}$	$5\frac{1}{8}$	15	25	$21\frac{1}{4}$	$7\frac{1}{2}$
		18	$5\frac{1}{8}$	$10\frac{3}{4}$	24	$19\frac{1}{4}$	$7\frac{1}{2}$	$6\frac{3}{4}$	$10\frac{3}{4}$	22	$21\frac{1}{4}$	$7\frac{1}{2}$
		20	$6\frac{3}{4}$	$7\frac{3}{4}$	22	22	$7\frac{1}{2}$	$6\frac{3}{4}$	10	$23\frac{1}{2}$	$23\frac{1}{4}$	$7\frac{1}{2}$
	4/12	12	$5\frac{1}{8}$	$12\frac{1}{2}$	$16\frac{3}{4}$	$15\frac{3}{4}$	$7\frac{1}{2}$	$5\frac{1}{8}$	$15\frac{3}{4}$	$18\frac{1}{2}$	$18\frac{3}{4}$	$7\frac{3}{4}$
		14	$5\frac{1}{8}$	$11\frac{1}{2}$	$19\frac{1}{2}$	$16\frac{1}{2}$	$7\frac{1}{2}$	$5\frac{1}{8}$	$14\frac{1}{2}$	22	$19\frac{1}{2}$	$7\frac{1}{2}$
		16	$5\frac{1}{8}$	$10\frac{1}{2}$	$21\frac{3}{4}$	17	$7\frac{1}{2}$	$5\frac{1}{8}$	$13\frac{1}{2}$	$24\frac{1}{2}$	$20\frac{1}{4}$	$7\frac{1}{2}$
		18	$5\frac{1}{8}$	10	$23\frac{3}{4}$	$17\frac{3}{4}$	$7\frac{1}{2}$	$6\frac{3}{4}$	$9\frac{3}{4}$	$21\frac{1}{2}$	$19\frac{1}{2}$	$7\frac{1}{2}$
		20	$5\frac{1}{8}$	$9\frac{1}{4}$	$25\frac{1}{2}$	$19\frac{1}{4}$	$7\frac{1}{2}$	$6\frac{3}{4}$	9	23	$21\frac{1}{2}$	$7\frac{1}{2}$
	6/12	12	$5\frac{1}{8}$	$10\frac{1}{4}$	$16\frac{3}{4}$	$14\frac{1}{4}$	$8\frac{1}{4}$	$5\frac{1}{8}$	$12\frac{3}{4}$	$18\frac{3}{4}$	$16\frac{3}{4}$	$10\frac{1}{2}$
		14	$5\frac{1}{8}$	$9\frac{1}{2}$	19	$15\frac{1}{4}$	$7\frac{1}{2}$	$5\frac{1}{8}$	12	$21\frac{1}{2}$	$17\frac{3}{4}$	9
		16	$5\frac{1}{8}$	9	21	16	$7\frac{1}{2}$	$5\frac{1}{8}$	$11\frac{1}{4}$	$23\frac{3}{4}$	$18\frac{3}{4}$	$7\frac{3}{4}$
		18	$5\frac{1}{8}$	$8\frac{1}{2}$	23	$16\frac{1}{2}$	$7\frac{1}{2}$	$6\frac{3}{4}$	8	$20\frac{3}{4}$	$17\frac{1}{4}$	$7\frac{1}{2}$
	8/12	12	$5\frac{1}{8}$	$8\frac{3}{4}$	$16\frac{1}{2}$	$13\frac{1}{2}$	10	$5\frac{1}{8}$	$10\frac{3}{4}$	$18\frac{1}{2}$	$15\frac{1}{2}$	$12\frac{3}{4}$
		14	$5\frac{1}{8}$	$8\frac{1}{4}$	$18\frac{1}{2}$	$14\frac{1}{2}$	$8\frac{3}{4}$	$5\frac{1}{8}$	$10\frac{1}{4}$	$20\frac{3}{4}$	$16\frac{3}{4}$	$11\frac{1}{4}$
		16	$5\frac{1}{8}$	$7\frac{3}{4}$	$20\frac{1}{2}$	$15\frac{1}{2}$	$7\frac{1}{2}$	$5\frac{1}{8}$	$9\frac{1}{2}$	23	$17\frac{3}{4}$	$9\frac{3}{4}$
		18	$5\frac{1}{8}$	$7\frac{1}{2}$	$22\frac{1}{4}$	$16\frac{1}{4}$	$7\frac{1}{2}$	$5\frac{1}{8}$	$9\frac{1}{4}$	$25\frac{1}{4}$	$18\frac{3}{4}$	$8\frac{3}{4}$
Vertical dead + live load = 600 lb/ft	3/12	12	$5\frac{1}{8}$	$20\frac{1}{2}$	$20\frac{1}{2}$	$20\frac{1}{4}$	$7\frac{1}{2}$	$5\frac{1}{8}$	$26\frac{1}{4}$	$26\frac{1}{4}$	24	$12\frac{1}{4}$
		14	$5\frac{1}{8}$	$18\frac{3}{4}$	24	21	$7\frac{1}{2}$	$6\frac{3}{4}$	$18\frac{1}{2}$	22	$22\frac{1}{4}$	$12\frac{1}{4}$
		16	$6\frac{3}{4}$	$13\frac{1}{4}$	22	19	$7\frac{1}{2}$	$6\frac{3}{4}$	$17\frac{1}{4}$	25	23	$12\frac{1}{4}$
		18	$6\frac{3}{4}$	$12\frac{1}{4}$	24	$20\frac{1}{4}$	$7\frac{1}{2}$	$6\frac{3}{4}$	16	$27\frac{1}{4}$	$23\frac{1}{2}$	$12\frac{1}{4}$
		20	$6\frac{3}{4}$	$11\frac{1}{2}$	$25\frac{3}{4}$	$21\frac{3}{4}$	$7\frac{1}{2}$	$6\frac{3}{4}$	$14\frac{3}{4}$	$29\frac{1}{2}$	24	$12\frac{1}{4}$
	4/12	12	$5\frac{1}{8}$	$18\frac{1}{4}$	$20\frac{1}{4}$	$19\frac{1}{4}$	$7\frac{1}{2}$	$5\frac{1}{8}$	$23\frac{1}{4}$	$23\frac{1}{4}$	$22\frac{3}{4}$	12
		14	$5\frac{1}{8}$	17	$24\frac{1}{4}$	20	$7\frac{1}{2}$	$6\frac{3}{4}$	$16\frac{1}{2}$	22	$21\frac{1}{4}$	12
		16	$6\frac{3}{4}$	12	$21\frac{3}{4}$	$18\frac{1}{4}$	$7\frac{1}{2}$	$6\frac{3}{4}$	$15\frac{1}{2}$	$24\frac{3}{4}$	$21\frac{3}{4}$	12
		18	$6\frac{3}{4}$	$11\frac{1}{4}$	$23\frac{1}{2}$	$18\frac{3}{4}$	$7\frac{1}{2}$	$6\frac{3}{4}$	$14\frac{1}{2}$	27	$22\frac{1}{2}$	12
		20	$6\frac{3}{4}$	$10\frac{1}{2}$	$25\frac{1}{4}$	$20\frac{1}{2}$	$7\frac{1}{2}$	$6\frac{3}{4}$	$13\frac{1}{2}$	$28\frac{3}{4}$	23	12
	6/12	12	$5\frac{1}{8}$	$15\frac{1}{4}$	21	$17\frac{1}{2}$	$10\frac{1}{4}$	$5\frac{1}{8}$	$18\frac{3}{4}$	$23\frac{1}{4}$	$20\frac{1}{2}$	$13\frac{1}{4}$
		14	$5\frac{1}{8}$	14	24	$18\frac{1}{2}$	$8\frac{3}{4}$	$6\frac{3}{4}$	$13\frac{1}{2}$	$21\frac{3}{4}$	$19\frac{1}{4}$	$11\frac{1}{4}$
		16	$6\frac{3}{4}$	$10\frac{1}{4}$	$21\frac{1}{4}$	$17\frac{1}{4}$	$7\frac{1}{2}$	$6\frac{3}{4}$	$12\frac{3}{4}$	24	20	$11\frac{1}{4}$
		18	$6\frac{3}{4}$	$9\frac{1}{2}$	$22\frac{3}{4}$	$17\frac{3}{4}$	$7\frac{1}{2}$	$6\frac{3}{4}$	12	26	21	$11\frac{1}{4}$
	8/12	12	$5\frac{1}{8}$	13	$20\frac{3}{4}$	$16\frac{1}{2}$	$12\frac{1}{2}$	$5\frac{1}{8}$	16	$23\frac{1}{4}$	19	16
		14	$5\frac{1}{8}$	$12\frac{1}{4}$	$23\frac{1}{4}$	$17\frac{3}{4}$	$10\frac{3}{4}$	$6\frac{3}{4}$	$11\frac{1}{2}$	21	18	12
		16	$6\frac{3}{4}$	$8\frac{3}{4}$	$20\frac{1}{2}$	$16\frac{1}{2}$	8	$6\frac{3}{4}$	11	23	$19\frac{1}{4}$	$10\frac{1}{2}$
		18	$6\frac{3}{4}$	$8\frac{1}{4}$	22	$17\frac{1}{2}$	$7\frac{1}{2}$	$6\frac{3}{4}$	$10\frac{1}{2}$	25	20	$10\frac{1}{2}$

Timber Three-Hinged Tudor Arch Span-Load Table[a]

80′ Span					90′ Span				
Width	Base	Lower Tang.	Upper Tang.	Crown	Width	Base	Lower Tang.	Upper Tang.	Crown
5⅛	21¾	21¾	23	7½	6¾	20	20	22¾	8
5⅛	20	24	24	7½	6¾	18½	21½	24	7½
6¾	14¼	22½	22	7½	6¾	17¼	24¾	24¾	7½
6¾	13¼	24½	22¾	7½	6¾	16	27	25¾	7½
6¾	12½	26½	23	7½	6¾	15	29¼	26	7½
5⅛	19¼	20	21½	9½	5⅛	22¾	22¾	24¼	12
5⅛	17¾	24¼	22¾	8¼	6¾	16¼	21½	22¼	8½
6¾	12¾	22	20¾	7½	6¾	15¼	24¼	23¼	7½
6¾	12	24	21¼	7½	6¾	14¼	26½	24¼	7½
6¾	11¼	25¾	22	7½	6¾	13½	28½	24¾	7½
5⅛	15¼	20½	19	13¼	5⅛	18	22¼	21¼	16
5⅛	14½	23¾	20¼	11½	6¾	13	21	19¾	12
6¾	10½	21¼	18¾	8½	6¾	12¼	23¼	21	10½
6¾	9¾	23	19½	7½	6¾	11¾	25¼	22	9¼
5⅛	12¾	20¼	17¼	16	5⅛	15	22	19¼	19¼
5⅛	12¼	23	18¾	14	5⅛	14¼	25¼	20¾	17
5⅛	11½	25½	20	12½	6¾	10¼	22¼	19½	12¾
6¾	8½	22	18½	9¼	6¾	10	24	20½	11½
6¾	25	25	24½	12¼	6¾	29¾	29¾	27½	12¼
6¾	23	23½	26	12¼	6¾	27½	27½	29¼	12¼
6¾	21¼	27½	26¾	12¼	6¾	25½	30	30¼	12¼
6¾	19¾	30½	27¼	12¼	6¾	23¾	33½	31	12¼
6¾	18½	33	28	12¼	8¾	17½	30	28¼	12¼
6¾	21¾	21¾	23	12	6¾	25¾	25¾	25½	12¾
6¾	20¼	24¼	24¼	12	6¾	24	26	27¼	12
6¾	18¾	27¼	25	12	6¾	22½	29¾	28¼	12
6¾	17¾	30	26	12	6¾	21¼	32¾	29¼	12
6¾	16¾	32¼	26½	12	8¾	15½	19¼	26¾	12
5⅛	22¾	25½	23½	16¼	6¾	20½	22¼	22¾	17
6¾	16¼	23¾	22	12	6¾	19¼	26	24¼	14¾
6¾	15½	26½	23	11¼	6¾	18¼	29	25½	13
6¾	14¾	28¾	23¾	11¼	6¾	17¼	31½	26¾	11¾
5⅛	19	25½	21¼	19¾	6¾	17	22¼	20½	20¼
6¾	13¾	23¼	20¼	14¾	6¾	16	25¼	22¼	18
6¾	13	25½	21½	13¼	6¾	15½	27¾	23¾	16
6¾	12½	27½	22½	11¾	6¾	14¾	30¼	25	14½

(continued)

TABLE 7.8 (*Continued*) Structural Glued Laminated

Loading	Roof Pitch	Wall Ht (ft)	30′ Span Width	30′ Span Base	30′ Span Lower Tang.	30′ Span Upper Tang.	30′ Span Crown	35′ Span Width	35′ Span Base	35′ Span Lower Tang.	35′ Span Upper Tang.	35′ Span Crown
Vertical dead + live load = 800 lb/ft	3/12	10	$5\frac{1}{8}$	10	$12\frac{1}{4}$	$12\frac{1}{4}$	$12\frac{1}{4}$	$5\frac{1}{8}$	13	$13\frac{3}{4}$	$13\frac{3}{4}$	$12\frac{1}{4}$
		12	$5\frac{1}{8}$	$8\frac{3}{4}$	$13\frac{3}{4}$	12	$7\frac{1}{2}$	$5\frac{1}{8}$	$11\frac{1}{4}$	$15\frac{3}{4}$	$12\frac{1}{2}$	$7\frac{1}{2}$
		14	$5\frac{1}{8}$	$7\frac{3}{4}$	$15\frac{3}{4}$	$12\frac{3}{4}$	$7\frac{1}{2}$	$5\frac{1}{8}$	10	$18\frac{1}{4}$	$13\frac{1}{4}$	$7\frac{1}{2}$
		16	$5\frac{1}{8}$	$7\frac{1}{2}$	$17\frac{1}{2}$	$13\frac{3}{4}$	$7\frac{1}{2}$	$5\frac{1}{8}$	$9\frac{1}{4}$	$20\frac{1}{4}$	14	$7\frac{1}{2}$
		18	$5\frac{1}{8}$	$7\frac{1}{2}$	$19\frac{1}{4}$	$14\frac{1}{4}$	$7\frac{1}{2}$	$5\frac{1}{8}$	$8\frac{1}{4}$	$22\frac{1}{4}$	$14\frac{3}{4}$	$7\frac{1}{2}$
	4/12	10	$5\frac{1}{8}$	$9\frac{1}{4}$	$9\frac{3}{4}$	$18\frac{1}{2}$	12	$5\frac{1}{8}$	$11\frac{3}{4}$	$11\frac{3}{4}$	$18\frac{1}{2}$	12
		12	$5\frac{1}{8}$	8	14	$11\frac{1}{2}$	$7\frac{1}{2}$	$5\frac{1}{8}$	$10\frac{1}{2}$	16	$11\frac{3}{4}$	$7\frac{1}{2}$
		14	$5\frac{1}{8}$	$7\frac{1}{2}$	16	$12\frac{1}{2}$	$7\frac{1}{2}$	$5\frac{1}{8}$	$9\frac{1}{2}$	$18\frac{1}{4}$	$12\frac{3}{4}$	$7\frac{1}{2}$
		16	$5\frac{1}{8}$	$7\frac{1}{2}$	$17\frac{1}{2}$	$13\frac{1}{2}$	$7\frac{1}{2}$	$5\frac{1}{8}$	$8\frac{1}{2}$	$20\frac{1}{4}$	$13\frac{3}{4}$	$7\frac{1}{2}$
		18	$5\frac{1}{8}$	$7\frac{1}{2}$	19	$14\frac{1}{2}$	$7\frac{1}{2}$	$5\frac{1}{8}$	$7\frac{3}{4}$	22	$14\frac{3}{4}$	$7\frac{1}{2}$
	6/12	12	$5\frac{1}{8}$	$7\frac{1}{2}$	$14\frac{1}{4}$	$10\frac{3}{4}$	$7\frac{1}{2}$	$5\frac{1}{8}$	9	$16\frac{1}{4}$	$10\frac{1}{2}$	$7\frac{1}{2}$
		14	$5\frac{1}{8}$	$7\frac{1}{2}$	16	$11\frac{3}{4}$	$7\frac{1}{2}$	$5\frac{1}{8}$	$8\frac{1}{4}$	$18\frac{1}{4}$	$11\frac{3}{4}$	$7\frac{1}{2}$
		16	$5\frac{1}{8}$	$7\frac{1}{2}$	$17\frac{1}{2}$	13	$7\frac{1}{2}$	$5\frac{1}{8}$	$7\frac{1}{2}$	20	13	$7\frac{1}{2}$
		18	$5\frac{1}{8}$	$7\frac{1}{2}$	19	14	$7\frac{1}{2}$	$5\frac{1}{8}$	$7\frac{1}{2}$	$21\frac{3}{4}$	14	$7\frac{1}{2}$
	8/12	12	$5\frac{1}{8}$	$7\frac{1}{2}$	$14\frac{1}{4}$	$9\frac{3}{4}$	$7\frac{1}{2}$	$5\frac{1}{8}$	8	16	$10\frac{3}{4}$	$7\frac{1}{2}$
		14	$5\frac{1}{8}$	$7\frac{1}{2}$	$15\frac{3}{4}$	$11\frac{1}{4}$	$7\frac{1}{2}$	$5\frac{1}{8}$	$7\frac{1}{2}$	18	$11\frac{1}{2}$	$7\frac{1}{2}$
		16	$5\frac{1}{8}$	$7\frac{1}{2}$	$17\frac{1}{4}$	$12\frac{1}{2}$	$7\frac{1}{2}$	$5\frac{1}{8}$	$7\frac{1}{2}$	$19\frac{3}{4}$	12	$7\frac{1}{2}$
		18	$5\frac{1}{8}$	$7\frac{1}{2}$	$18\frac{3}{4}$	$13\frac{1}{2}$	$7\frac{1}{2}$	$5\frac{1}{8}$	$7\frac{1}{2}$	$21\frac{1}{2}$	13	$7\frac{1}{2}$
Vertical dead + live load = 1000 lb/ft	3/12	10	$5\frac{1}{8}$	$12\frac{1}{2}$	$12\frac{1}{2}$	$18\frac{1}{2}$	$12\frac{1}{4}$	$5\frac{1}{8}$	16	16	$18\frac{1}{2}$	$12\frac{1}{4}$
		12	$5\frac{1}{8}$	$10\frac{3}{4}$	$15\frac{1}{2}$	$11\frac{1}{4}$	$7\frac{1}{2}$	$5\frac{1}{8}$	$13\frac{3}{4}$	$17\frac{1}{2}$	13	$7\frac{1}{2}$
		14	$5\frac{1}{8}$	$9\frac{1}{2}$	18	$11\frac{3}{4}$	$7\frac{1}{2}$	$5\frac{1}{8}$	$12\frac{1}{2}$	$20\frac{1}{2}$	$12\frac{3}{4}$	$7\frac{1}{2}$
		16	$5\frac{1}{8}$	$8\frac{3}{4}$	20	$12\frac{1}{4}$	$7\frac{1}{2}$	$5\frac{1}{8}$	$11\frac{1}{4}$	23	$12\frac{3}{4}$	$7\frac{1}{2}$
		18	$5\frac{1}{8}$	8	22	$12\frac{3}{4}$	$7\frac{1}{2}$	$5\frac{1}{8}$	$10\frac{1}{4}$	$25\frac{1}{4}$	$13\frac{1}{4}$	$7\frac{1}{2}$
	4/12	10	$5\frac{1}{8}$	$11\frac{1}{2}$	$11\frac{1}{2}$	23	12	$5\frac{1}{8}$	$14\frac{1}{2}$	$14\frac{1}{2}$	23	12
		12	$5\frac{1}{8}$	10	$15\frac{3}{4}$	11	$7\frac{1}{2}$	$5\frac{1}{8}$	$12\frac{3}{4}$	18	13	$7\frac{1}{2}$
		14	$5\frac{1}{8}$	9	18	$11\frac{3}{4}$	$7\frac{1}{2}$	$5\frac{1}{8}$	$11\frac{1}{2}$	$20\frac{3}{4}$	$12\frac{3}{4}$	$7\frac{1}{2}$
		16	$5\frac{1}{8}$	$8\frac{1}{4}$	20	$12\frac{1}{4}$	$7\frac{1}{2}$	$5\frac{1}{8}$	$10\frac{1}{2}$	23	$12\frac{1}{2}$	$7\frac{1}{2}$
		18	$5\frac{1}{8}$	$7\frac{1}{2}$	$21\frac{3}{4}$	$13\frac{1}{4}$	$7\frac{1}{2}$	$5\frac{1}{8}$	$9\frac{3}{4}$	$25\frac{1}{4}$	13	$7\frac{1}{2}$
	6/12	12	$5\frac{1}{8}$	$8\frac{3}{4}$	$16\frac{1}{4}$	10	$7\frac{1}{2}$	$5\frac{1}{8}$	11	$18\frac{1}{4}$	$11\frac{3}{4}$	$7\frac{1}{2}$
		14	$5\frac{1}{8}$	8	18	$11\frac{1}{4}$	$7\frac{1}{2}$	$5\frac{1}{8}$	$10\frac{1}{4}$	$20\frac{3}{4}$	$12\frac{1}{4}$	$7\frac{1}{2}$
		16	$5\frac{1}{8}$	$7\frac{1}{2}$	20	12	$7\frac{1}{2}$	$5\frac{1}{8}$	$9\frac{1}{2}$	$22\frac{3}{4}$	$12\frac{1}{2}$	$7\frac{1}{2}$
		18	$5\frac{1}{8}$	$7\frac{1}{2}$	$21\frac{3}{4}$	$12\frac{3}{4}$	$7\frac{1}{2}$	$5\frac{1}{8}$	$8\frac{3}{4}$	25	13	$7\frac{1}{2}$
	8/12	12	$5\frac{1}{8}$	$7\frac{3}{4}$	16	$9\frac{3}{4}$	$7\frac{1}{2}$	$5\frac{1}{8}$	$9\frac{3}{4}$	$18\frac{1}{4}$	12	$7\frac{1}{2}$
		14	$5\frac{1}{8}$	$7\frac{1}{2}$	18	$10\frac{1}{2}$	$7\frac{1}{2}$	$5\frac{1}{8}$	9	$20\frac{1}{4}$	$12\frac{3}{4}$	$7\frac{1}{2}$
		16	$5\frac{1}{8}$	$7\frac{1}{2}$	$19\frac{3}{4}$	$11\frac{1}{2}$	$7\frac{1}{2}$	$5\frac{1}{8}$	$8\frac{1}{2}$	$22\frac{1}{2}$	$13\frac{1}{4}$	$7\frac{1}{2}$
		18	$5\frac{1}{8}$	$7\frac{1}{2}$	$21\frac{1}{2}$	$12\frac{1}{4}$	$7\frac{1}{2}$	$5\frac{1}{8}$	8	$24\frac{1}{2}$	$13\frac{1}{2}$	$7\frac{1}{2}$

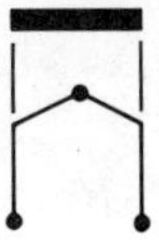

Timber Three-Hinged Tudor Arch Span-Load Table[a]

40′ Span					50′ Span				
Width	Base	Lower Tang.	Upper Tang.	Crown	Width	Base	Lower Tang.	Upper Tang.	Crown
$5\frac{1}{8}$	16	16	$14\frac{1}{2}$	$12\frac{1}{4}$	$5\frac{1}{8}$	$22\frac{3}{4}$	$22\frac{3}{4}$	$18\frac{1}{2}$	$12\frac{1}{4}$
$5\frac{1}{8}$	14	$17\frac{1}{4}$	$13\frac{1}{4}$	$7\frac{1}{2}$	$5\frac{1}{8}$	$20\frac{1}{2}$	$20\frac{1}{2}$	$18\frac{1}{2}$	$12\frac{1}{4}$
$5\frac{1}{8}$	$12\frac{3}{4}$	$20\frac{1}{2}$	$13\frac{3}{4}$	$7\frac{1}{2}$	$5\frac{1}{8}$	$18\frac{1}{2}$	$24\frac{1}{2}$	19	$12\frac{1}{4}$
$5\frac{1}{8}$	$11\frac{1}{2}$	$22\frac{3}{4}$	$14\frac{1}{2}$	$7\frac{1}{2}$	$6\frac{3}{4}$	13	22	$18\frac{3}{4}$	$12\frac{1}{4}$
$5\frac{1}{8}$	$10\frac{1}{2}$	25	$15\frac{1}{4}$	$7\frac{1}{2}$	$6\frac{3}{4}$	12	24	20	$12\frac{1}{4}$
$5\frac{1}{8}$	$14\frac{1}{2}$	$14\frac{1}{2}$	$18\frac{1}{2}$	12	$5\frac{1}{8}$	$20\frac{1}{2}$	$20\frac{1}{2}$	$18\frac{1}{2}$	12
$5\frac{1}{8}$	$12\frac{3}{4}$	$17\frac{3}{4}$	13	$7\frac{1}{2}$	$5\frac{1}{8}$	$18\frac{1}{2}$	$20\frac{3}{4}$	$17\frac{1}{2}$	12
$5\frac{1}{8}$	$11\frac{3}{4}$	$20\frac{1}{2}$	$13\frac{1}{2}$	$7\frac{1}{2}$	$5\frac{1}{8}$	$16\frac{3}{4}$	$24\frac{3}{4}$	$18\frac{1}{4}$	12
$5\frac{1}{8}$	$10\frac{3}{4}$	$22\frac{3}{4}$	$13\frac{3}{4}$	$7\frac{1}{2}$	$6\frac{3}{4}$	12	22	$17\frac{3}{4}$	12
$5\frac{1}{8}$	$9\frac{3}{4}$	25	$14\frac{1}{2}$	$7\frac{1}{2}$	$6\frac{3}{4}$	11	$23\frac{3}{4}$	$19\frac{1}{4}$	12
$5\frac{1}{8}$	11	18	$12\frac{1}{2}$	$7\frac{1}{2}$	$5\frac{1}{8}$	$15\frac{1}{4}$	$21\frac{1}{2}$	$16\frac{1}{2}$	$11\frac{1}{4}$
$5\frac{1}{8}$	10	$20\frac{1}{4}$	$13\frac{1}{4}$	$7\frac{1}{2}$	$5\frac{1}{8}$	$14\frac{1}{4}$	$24\frac{1}{2}$	$17\frac{1}{4}$	$11\frac{1}{4}$
$5\frac{1}{8}$	$9\frac{1}{4}$	$22\frac{1}{2}$	$13\frac{3}{4}$	$7\frac{1}{2}$	$6\frac{3}{4}$	$10\frac{1}{4}$	$21\frac{1}{2}$	16	$11\frac{1}{4}$
$5\frac{1}{8}$	$8\frac{3}{4}$	$24\frac{1}{2}$	$14\frac{1}{4}$	$7\frac{1}{2}$	$6\frac{3}{4}$	$9\frac{1}{2}$	23	$17\frac{1}{2}$	$11\frac{1}{4}$
$5\frac{1}{8}$	$9\frac{1}{2}$	$17\frac{3}{4}$	$12\frac{1}{2}$	8	$5\frac{1}{8}$	$13\frac{1}{4}$	$21\frac{1}{4}$	16	11
$5\frac{1}{8}$	9	20	$13\frac{1}{2}$	$7\frac{1}{2}$	$5\frac{1}{8}$	$12\frac{1}{2}$	24	17	$10\frac{1}{2}$
$5\frac{1}{8}$	$8\frac{1}{4}$	22	14	$7\frac{1}{2}$	$6\frac{3}{4}$	9	$20\frac{3}{4}$	16	$10\frac{1}{2}$
$5\frac{1}{8}$	$7\frac{3}{4}$	24	$14\frac{1}{2}$	$7\frac{1}{2}$	$6\frac{3}{4}$	$8\frac{1}{2}$	$22\frac{1}{2}$	$16\frac{1}{2}$	$10\frac{1}{2}$
$5\frac{1}{8}$	$19\frac{3}{4}$	$19\frac{3}{4}$	$18\frac{1}{2}$	$12\frac{1}{4}$	$5\frac{1}{8}$	$28\frac{1}{4}$	$28\frac{1}{4}$	23	17
$5\frac{1}{8}$	$17\frac{1}{2}$	19	16	$12\frac{1}{4}$	$5\frac{1}{8}$	25	25	$22\frac{1}{4}$	$12\frac{1}{4}$
$5\frac{1}{8}$	$15\frac{3}{4}$	23	$15\frac{3}{4}$	$12\frac{1}{4}$	$6\frac{3}{4}$	$17\frac{1}{2}$	22	$18\frac{3}{4}$	$12\frac{1}{4}$
$6\frac{3}{4}$	11	$20\frac{1}{2}$	$17\frac{1}{4}$	$12\frac{1}{4}$	$6\frac{3}{4}$	16	$24\frac{3}{4}$	19	$12\frac{1}{4}$
$6\frac{3}{4}$	10	$22\frac{1}{4}$	$18\frac{1}{2}$	$12\frac{1}{4}$	$6\frac{3}{4}$	$14\frac{3}{4}$	27	$19\frac{1}{2}$	$12\frac{1}{4}$
$5\frac{1}{8}$	$17\frac{3}{4}$	$17\frac{3}{4}$	23	12	$5\frac{1}{8}$	$25\frac{1}{4}$	$25\frac{1}{4}$	23	$16\frac{3}{4}$
$5\frac{1}{8}$	16	$19\frac{3}{4}$	$15\frac{3}{4}$	12	$5\frac{1}{8}$	$22\frac{1}{2}$	$22\frac{1}{2}$	$21\frac{3}{4}$	12
$5\frac{1}{8}$	$14\frac{1}{2}$	$23\frac{1}{4}$	$15\frac{1}{2}$	12	$6\frac{3}{4}$	16	$22\frac{1}{4}$	18	12
$6\frac{3}{4}$	$10\frac{1}{4}$	$20\frac{1}{2}$	$16\frac{3}{4}$	12	$6\frac{3}{4}$	$14\frac{3}{4}$	$24\frac{3}{4}$	$18\frac{1}{2}$	12
$6\frac{3}{4}$	$9\frac{1}{2}$	22	$18\frac{1}{4}$	12	$6\frac{3}{4}$	$13\frac{3}{4}$	$26\frac{3}{4}$	19	12
$5\frac{1}{8}$	$13\frac{3}{4}$	$20\frac{1}{2}$	$14\frac{1}{4}$	$11\frac{1}{4}$	$5\frac{1}{8}$	19	24	19	$11\frac{1}{4}$
$5\frac{1}{8}$	$12\frac{1}{2}$	$23\frac{1}{4}$	$14\frac{1}{2}$	$11\frac{1}{4}$	$6\frac{3}{4}$	$13\frac{1}{2}$	22	$17\frac{1}{4}$	$11\frac{1}{4}$
$5\frac{1}{8}$	$11\frac{3}{4}$	$25\frac{3}{4}$	15	$11\frac{1}{4}$	$6\frac{3}{4}$	$12\frac{3}{4}$	$24\frac{1}{4}$	$17\frac{3}{4}$	$11\frac{1}{4}$
$6\frac{3}{4}$	$8\frac{1}{4}$	$21\frac{3}{4}$	$17\frac{1}{4}$	$11\frac{1}{4}$	$6\frac{3}{4}$	$11\frac{3}{4}$	$26\frac{1}{4}$	$18\frac{1}{2}$	$11\frac{1}{4}$
$5\frac{1}{8}$	12	$20\frac{1}{4}$	14	$10\frac{1}{2}$	$5\frac{1}{8}$	$16\frac{1}{4}$	24	$17\frac{3}{4}$	$12\frac{1}{2}$
$5\frac{1}{8}$	11	$22\frac{3}{4}$	$14\frac{3}{4}$	$10\frac{1}{2}$	$6\frac{3}{4}$	$11\frac{3}{4}$	$21\frac{1}{2}$	$16\frac{3}{4}$	$10\frac{1}{2}$
$5\frac{1}{8}$	$10\frac{1}{4}$	$25\frac{1}{4}$	$15\frac{1}{2}$	$10\frac{1}{2}$	$6\frac{3}{4}$	11	$23\frac{1}{2}$	$17\frac{3}{4}$	$10\frac{1}{2}$
$6\frac{3}{4}$	$7\frac{1}{2}$	21	$16\frac{1}{2}$	$10\frac{1}{2}$	$6\frac{3}{4}$	$10\frac{1}{2}$	$25\frac{1}{4}$	$18\frac{1}{2}$	$10\frac{1}{2}$

(continued)

Loading	Roof Pitch	Wall Ht (ft)	60′ Span Width	60′ Span Base	60′ Span Lower Tang.	60′ Span Upper Tang.	60′ Span Crown	70′ Span Width	70′ Span Base	70′ Span Lower Tang.	70′ Span Upper Tang.	70′ Span Crown
Vertical dead + live load = 400 lb/ft	3/12	12	$5\frac{1}{8}$	$27\frac{1}{4}$	$27\frac{1}{4}$	23	$12\frac{1}{4}$	$6\frac{3}{4}$	$26\frac{3}{4}$	$26\frac{3}{4}$	$24\frac{1}{4}$	$12\frac{1}{4}$
		14	$6\frac{3}{4}$	$19\frac{1}{4}$	$22\frac{1}{2}$	$21\frac{1}{4}$	$12\frac{1}{4}$	$6\frac{3}{4}$	$24\frac{1}{2}$	$24\frac{1}{2}$	$25\frac{3}{4}$	$12\frac{1}{4}$
		16	$6\frac{3}{4}$	$17\frac{1}{2}$	$25\frac{3}{4}$	$21\frac{3}{4}$	$12\frac{1}{4}$	$6\frac{3}{4}$	$22\frac{1}{2}$	$28\frac{3}{4}$	$26\frac{1}{4}$	$12\frac{1}{4}$
		18	$6\frac{3}{4}$	$16\frac{1}{4}$	28	$22\frac{1}{4}$	$12\frac{1}{4}$	$6\frac{3}{4}$	21	$31\frac{3}{4}$	$26\frac{3}{4}$	$12\frac{1}{4}$
		20	$6\frac{3}{4}$	15	$30\frac{1}{4}$	$22\frac{1}{2}$	$12\frac{1}{4}$	$8\frac{3}{4}$	$15\frac{1}{4}$	$28\frac{1}{4}$	$24\frac{3}{4}$	$12\frac{1}{4}$
	4/12	12	$5\frac{1}{8}$	$24\frac{1}{4}$	$24\frac{1}{4}$	22	12	$6\frac{3}{4}$	$23\frac{1}{2}$	$23\frac{1}{2}$	23	12
		14	$6\frac{3}{4}$	$17\frac{1}{4}$	$22\frac{3}{4}$	$20\frac{1}{4}$	12	$6\frac{3}{4}$	$21\frac{3}{4}$	$25\frac{1}{4}$	$24\frac{1}{4}$	12
		16	$6\frac{3}{4}$	16	$25\frac{1}{2}$	21	12	$6\frac{3}{4}$	$20\frac{1}{4}$	$28\frac{3}{4}$	25	12
		18	$6\frac{3}{4}$	$14\frac{3}{4}$	$27\frac{3}{4}$	$21\frac{1}{2}$	12	$6\frac{3}{4}$	19	$31\frac{1}{2}$	$25\frac{3}{4}$	12
		20	$6\frac{3}{4}$	$13\frac{3}{4}$	$29\frac{3}{4}$	22	12	$6\frac{3}{4}$	$17\frac{3}{4}$	$33\frac{3}{4}$	$26\frac{1}{4}$	12
	6/12	12	$5\frac{1}{8}$	20	$24\frac{1}{4}$	$20\frac{1}{4}$	12	$6\frac{3}{4}$	19	22	21	13
		14	$6\frac{3}{4}$	$14\frac{1}{4}$	$22\frac{1}{2}$	$18\frac{3}{4}$	$11\frac{1}{4}$	$6\frac{3}{4}$	$17\frac{3}{4}$	$25\frac{1}{4}$	22	$11\frac{1}{4}$
		16	$6\frac{3}{4}$	$13\frac{1}{2}$	$24\frac{3}{4}$	$19\frac{3}{4}$	$11\frac{1}{4}$	$6\frac{3}{4}$	$16\frac{3}{4}$	28	23	$11\frac{1}{4}$
		18	$6\frac{3}{4}$	$12\frac{3}{4}$	$26\frac{3}{4}$	$20\frac{1}{2}$	$11\frac{1}{4}$	$6\frac{3}{4}$	16	$30\frac{1}{2}$	24	$11\frac{1}{4}$
	8/12	12	$5\frac{1}{8}$	17	$24\frac{1}{2}$	19	$14\frac{3}{4}$	$6\frac{3}{4}$	16	22	$19\frac{1}{4}$	16
		14	$6\frac{3}{4}$	$12\frac{1}{4}$	22	18	$10\frac{3}{4}$	$6\frac{3}{4}$	$15\frac{1}{4}$	$24\frac{1}{2}$	$20\frac{3}{4}$	14
		16	$6\frac{3}{4}$	$11\frac{1}{2}$	24	19	$10\frac{1}{2}$	$6\frac{3}{4}$	$14\frac{1}{2}$	27	22	$12\frac{1}{4}$
		18	$6\frac{3}{4}$	11	26	20	$10\frac{1}{2}$	$6\frac{3}{4}$	$13\frac{3}{4}$	$29\frac{1}{4}$	23	11
Vertical dead + live load = 600 lb/ft	3/12	12	$6\frac{3}{4}$	26	26	$22\frac{1}{2}$	$12\frac{1}{4}$	$6\frac{3}{4}$	$32\frac{3}{4}$	$32\frac{3}{4}$	$26\frac{1}{2}$	$12\frac{1}{4}$
		14	$6\frac{3}{4}$	$23\frac{1}{2}$	25	$23\frac{3}{4}$	$12\frac{1}{4}$	$6\frac{3}{4}$	30	30	$28\frac{1}{2}$	$12\frac{1}{4}$
		16	$6\frac{3}{4}$	$21\frac{3}{4}$	$28\frac{3}{4}$	$24\frac{1}{4}$	$12\frac{1}{4}$	$6\frac{3}{4}$	$27\frac{3}{4}$	$32\frac{1}{4}$	$29\frac{1}{4}$	$12\frac{1}{4}$
		18	$6\frac{3}{4}$	20	$31\frac{1}{2}$	$24\frac{3}{4}$	$12\frac{1}{4}$	$8\frac{3}{4}$	$20\frac{1}{4}$	$29\frac{3}{4}$	$26\frac{1}{2}$	$12\frac{1}{4}$
		20	$8\frac{3}{4}$	$14\frac{1}{2}$	28	$23\frac{1}{2}$	$12\frac{1}{4}$	$8\frac{3}{4}$	$18\frac{3}{4}$	$31\frac{3}{4}$	27	$12\frac{1}{4}$
	4/12	12	$6\frac{3}{4}$	23	23	$21\frac{1}{2}$	12	$6\frac{3}{4}$	29	29	$25\frac{1}{4}$	12
		14	$6\frac{3}{4}$	$21\frac{1}{4}$	$25\frac{1}{2}$	$22\frac{1}{2}$	12	$6\frac{3}{4}$	$26\frac{3}{4}$	28	27	12
		16	$6\frac{3}{4}$	$19\frac{3}{4}$	$28\frac{3}{4}$	$23\frac{1}{4}$	12	$6\frac{3}{4}$	25	$32\frac{1}{4}$	$27\frac{3}{4}$	12
		18	$6\frac{3}{4}$	$18\frac{1}{4}$	$31\frac{1}{4}$	24	12	$8\frac{3}{4}$	$18\frac{1}{4}$	$29\frac{1}{4}$	$25\frac{1}{2}$	12
		20	$6\frac{3}{4}$	$17\frac{1}{4}$	$33\frac{3}{4}$	$24\frac{1}{4}$	12	$8\frac{3}{4}$	$17\frac{1}{4}$	$31\frac{1}{4}$	26	12
	6/12	12	$6\frac{3}{4}$	19	22	$19\frac{3}{4}$	$11\frac{1}{2}$	$6\frac{3}{4}$	$23\frac{1}{2}$	$24\frac{1}{4}$	$23\frac{1}{4}$	$14\frac{3}{4}$
		14	$6\frac{3}{4}$	$17\frac{3}{4}$	$25\frac{1}{4}$	21	$11\frac{1}{4}$	$6\frac{3}{4}$	22	$28\frac{1}{4}$	$24\frac{1}{2}$	13
		16	$6\frac{3}{4}$	$16\frac{1}{2}$	28	22	$11\frac{1}{4}$	$6\frac{3}{4}$	$20\frac{3}{4}$	$31\frac{1}{2}$	$25\frac{3}{4}$	$11\frac{1}{4}$
		18	$6\frac{3}{4}$	$15\frac{3}{4}$	$30\frac{1}{2}$	$22\frac{3}{4}$	$11\frac{1}{4}$	$8\frac{3}{4}$	$15\frac{1}{4}$	$28\frac{1}{4}$	$23\frac{3}{4}$	$11\frac{1}{4}$
	8/12	12	$6\frac{3}{4}$	$16\frac{1}{4}$	22	$18\frac{3}{4}$	14	$6\frac{3}{4}$	20	$24\frac{1}{2}$	$21\frac{1}{2}$	18
		14	$6\frac{3}{4}$	$15\frac{1}{4}$	$24\frac{3}{4}$	$20\frac{1}{4}$	$12\frac{1}{4}$	$6\frac{3}{4}$	$18\frac{3}{4}$	$27\frac{3}{4}$	$23\frac{1}{4}$	$15\frac{3}{4}$
		16	$6\frac{3}{4}$	$14\frac{1}{4}$	$27\frac{1}{4}$	$21\frac{1}{4}$	$10\frac{3}{4}$	$6\frac{3}{4}$	18	$30\frac{1}{2}$	$24\frac{1}{2}$	14
		18	$6\frac{3}{4}$	$13\frac{3}{4}$	$29\frac{1}{2}$	$22\frac{1}{4}$	$10\frac{1}{2}$	$6\frac{3}{4}$	17	$33\frac{1}{4}$	$25\frac{3}{4}$	$12\frac{1}{2}$

Timber Three-Hinged Tudor Arch Span-Load Table[a]

80′ Span					90′ Span				
Width	Base	Lower Tang.	Upper Tang.	Crown	Width	Base	Lower Tang.	Upper Tang.	Crown
6¾	32½	32½	27¾	12¼	6¾	38¾	38¾	31	12½
6¾	30	30	29½	12¼	8¾	28¼	28¼	29¾	12¼
6¾	27¾	31¾	30¾	12¼	8¾	26¼	29	30¾	12¼
8¾	20¼	29½	27¾	12¼	8¾	24½	32¼	31¾	12¼
8¾	19	31¾	28½	12¼	8¾	23	34¾	32¼	12¼
6¾	28½	28½	26	12½	6¾	33¾	33¾	29¼	15¼
6¾	26½	27½	28	12	6¾	31½	31½	31¼	12¾
6¾	24¾	31¾	28¾	12	8¾	23	29	29	12
8¾	18¼	29	26½	12	8¾	21¾	31¾	29¾	12
8¾	17	31	27¼	12	8¾	20½	34	30¾	12
6¾	23	23¾	23¾	16¼	6¾	27	27	26¼	19¾
6¾	21½	27¾	25	14¼	6¾	25½	30	28	17½
6¾	20¼	31	26¼	12½	6¾	24	33¾	29½	15½
6¾	19¼	33¾	27½	11¼	8¾	17¾	30¼	27¼	11¾
6¾	19¼	24	21½	19¾	6¾	22½	25¾	23¾	23¾
6¾	18¼	27¼	23¼	17½	6¾	21¼	29½	25¾	21
6¾	17¼	30	24¾	15½	6¾	20¼	32¾	27¼	19
6¾	16½	32½	26	13¾	8¾	15¼	29	25½	14½
6¾	40½	40½	30¾	17	8¾	37¾	37¾	30¾	17
6¾	37¼	37¼	32¾	17	8¾	35	35	32¾	17
8¾	27	29¾	30½	17	8¾	32½	32½	34½	17
8¾	25¼	33	31	17	8¾	30¼	36	35½	17
8¾	23½	35½	31¾	17	8¾	28½	39	36	17
6¾	35¼	35¼	29¼	16¾	8¾	32¾	32¾	28¾	16¾
6¾	33	33	31	16¾	8¾	30½	30½	31	16¾
8¾	24	29¾	28½	16¾	8¾	28¾	32	32¼	16¾
8¾	22½	32½	29½	16¾	8¾	27	35½	33¼	16¾
8¾	21¼	35	30¼	16¾	8¾	25½	38¼	34¼	16¾
6¾	28½	28½	26¼	18½	6¾	33¼	33¼	29	22¾
6¾	26¾	31	28	16¼	8¾	24¾	28½	27¾	16¾
8¾	19¾	29	26	15¾	8¾	23½	31½	29¼	15¾
8¾	18¾	31¼	27	15¾	8¾	22¼	34¼	30¼	15¾
6¾	23¾	26¾	24¼	22¼	6¾	27¾	28¾	26¾	26¾
6¾	22½	30¾	26	19¾	6¾	26½	33¼	28¾	24
8¾	16¾	28	24¼	15	8¾	19¾	30½	27	18½
8¾	16	30	25¾	14¾	8¾	18¾	32¾	28½	16½

(*continued*)

TABLE 7.8 (*Continued*) Structural Glued Laminated

Loading	Roof Pitch	Wall Ht (ft)	30′ Span					35′ Span				
			Width	Base	Lower Tang.	Upper Tang.	Crown	Width	Base	Lower Tang.	Upper Tang.	Crown
Vertical dead = 240 lb/ft horizontal wind 320 lb/ft	10/12	8	$5\frac{1}{8}$	$7\frac{1}{2}$	$7\frac{1}{2}$	11	$7\frac{1}{2}$	$5\frac{1}{8}$	$7\frac{1}{2}$	$7\frac{1}{2}$	12	$10\frac{1}{4}$
		10	$5\frac{1}{8}$	$7\frac{1}{2}$	$9\frac{1}{4}$	$12\frac{1}{2}$	$12\frac{1}{2}$	$5\frac{1}{8}$	$7\frac{1}{2}$	10	$13\frac{1}{4}$	8
		12	$5\frac{1}{8}$	$7\frac{1}{2}$	$11\frac{1}{4}$	$13\frac{3}{4}$	$13\frac{3}{4}$	$5\frac{1}{8}$	$7\frac{1}{2}$	12	$14\frac{3}{4}$	$7\frac{1}{2}$
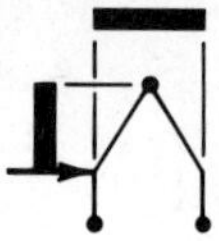	12/12	8	$5\frac{1}{8}$	$7\frac{1}{2}$	$7\frac{1}{2}$	$12\frac{1}{2}$	$8\frac{3}{4}$	$5\frac{1}{8}$	$7\frac{1}{2}$	$8\frac{1}{4}$	$13\frac{1}{2}$	12
		10	$5\frac{1}{8}$	$7\frac{1}{2}$	$9\frac{3}{4}$	$13\frac{3}{4}$	9	$5\frac{1}{8}$	$7\frac{1}{2}$	$10\frac{1}{2}$	15	$10\frac{1}{4}$
		12	$5\frac{1}{8}$	$7\frac{1}{2}$	$11\frac{3}{4}$	$15\frac{1}{4}$	$15\frac{1}{4}$	$5\frac{1}{8}$	$7\frac{1}{2}$	$12\frac{1}{2}$	$16\frac{1}{2}$	9
	14/12	8	$5\frac{1}{8}$	$7\frac{1}{2}$	$8\frac{3}{4}$	$13\frac{3}{4}$	$10\frac{1}{2}$	$5\frac{1}{8}$	$7\frac{1}{2}$	9	15	$13\frac{3}{4}$
		10	$5\frac{1}{8}$	$7\frac{1}{2}$	$11\frac{1}{4}$	$15\frac{1}{4}$	$8\frac{1}{4}$	$5\frac{1}{8}$	$7\frac{1}{2}$	$11\frac{1}{2}$	$16\frac{1}{2}$	$12\frac{1}{4}$
		12	$5\frac{1}{8}$	$7\frac{1}{2}$	$13\frac{1}{4}$	17	17	$5\frac{1}{8}$	$7\frac{1}{2}$	14	18	$8\frac{1}{4}$
	16/12	8	$5\frac{1}{8}$	$7\frac{1}{2}$	10	15	$12\frac{1}{2}$	$5\frac{1}{8}$	$7\frac{1}{2}$	$10\frac{1}{2}$	$16\frac{1}{4}$	$15\frac{3}{4}$
		10	$5\frac{1}{8}$	$7\frac{1}{2}$	$12\frac{1}{4}$	$16\frac{1}{2}$	$9\frac{1}{2}$	$5\frac{1}{8}$	$7\frac{1}{2}$	13	18	$14\frac{1}{2}$
		12	$5\frac{1}{8}$	$7\frac{1}{2}$	$14\frac{3}{4}$	$18\frac{1}{4}$	$7\frac{1}{2}$	$5\frac{1}{8}$	$7\frac{1}{2}$	$15\frac{1}{2}$	$19\frac{3}{4}$	$11\frac{1}{2}$
Vertical dead = 320 lb/ft horizontal wind 320 lb/ft	10/12	8	$5\frac{1}{8}$	$7\frac{1}{2}$	$7\frac{1}{2}$	$10\frac{3}{4}$	9	$5\frac{1}{8}$	$7\frac{1}{2}$	$8\frac{1}{4}$	$11\frac{3}{4}$	$11\frac{1}{2}$
		10	$5\frac{1}{8}$	$7\frac{1}{2}$	10	12	$7\frac{1}{2}$	$5\frac{1}{8}$	$7\frac{1}{2}$	$10\frac{3}{4}$	13	$10\frac{1}{2}$
		12	$5\frac{1}{8}$	$7\frac{1}{2}$	12	$13\frac{1}{4}$	$13\frac{1}{4}$	$5\frac{1}{8}$	$7\frac{1}{2}$	$12\frac{3}{4}$	$14\frac{1}{4}$	$8\frac{1}{2}$
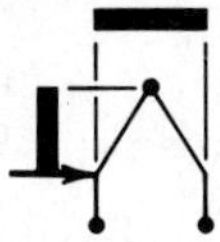	12/12	8	$5\frac{1}{8}$	$7\frac{1}{2}$	$8\frac{1}{4}$	$12\frac{1}{4}$	$10\frac{1}{4}$	$5\frac{1}{8}$	$7\frac{1}{2}$	9	$13\frac{1}{4}$	$13\frac{1}{4}$
		10	$5\frac{1}{8}$	$7\frac{1}{2}$	$10\frac{1}{2}$	$13\frac{1}{2}$	9	$5\frac{1}{8}$	$7\frac{1}{2}$	$11\frac{1}{4}$	$14\frac{3}{4}$	$12\frac{1}{4}$
		12	$5\frac{1}{8}$	$7\frac{1}{2}$	$12\frac{1}{2}$	15	9	$5\frac{1}{8}$	$7\frac{1}{2}$	$13\frac{1}{4}$	16	$10\frac{1}{4}$
	14/12	8	$5\frac{1}{8}$	$7\frac{1}{2}$	$8\frac{3}{4}$	$13\frac{1}{2}$	12	$5\frac{1}{8}$	$7\frac{1}{2}$	$9\frac{1}{4}$	$14\frac{3}{4}$	$14\frac{3}{4}$
		10	$5\frac{1}{8}$	$7\frac{1}{2}$	$10\frac{3}{4}$	15	$10\frac{1}{4}$	$5\frac{1}{8}$	$7\frac{1}{2}$	$11\frac{3}{4}$	$16\frac{1}{4}$	$14\frac{1}{4}$
		12	$5\frac{1}{8}$	$7\frac{1}{2}$	$12\frac{3}{4}$	$16\frac{1}{2}$	$8\frac{1}{4}$	$5\frac{1}{8}$	$7\frac{1}{2}$	$13\frac{3}{4}$	$17\frac{3}{4}$	$12\frac{1}{2}$
	16/12	8	$5\frac{1}{8}$	$7\frac{1}{2}$	$9\frac{1}{2}$	15	$13\frac{3}{4}$	$5\frac{1}{8}$	$7\frac{1}{2}$	$9\frac{3}{4}$	$16\frac{1}{4}$	$16\frac{1}{4}$
		10	$5\frac{1}{8}$	$7\frac{1}{2}$	$11\frac{3}{4}$	$16\frac{1}{2}$	$12\frac{1}{4}$	$5\frac{1}{8}$	$7\frac{1}{2}$	$12\frac{3}{4}$	18	$16\frac{1}{4}$
		12	$5\frac{1}{8}$	$7\frac{1}{2}$	$14\frac{1}{4}$	18	$7\frac{3}{4}$	$5\frac{1}{8}$	$7\frac{1}{2}$	$14\frac{3}{4}$	$19\frac{1}{2}$	$14\frac{1}{2}$
Vertical dead = 480 lb/ft horizontal wind = 320 lb/ft	10/12	8	$5\frac{1}{8}$	$7\frac{1}{2}$	9	$10\frac{1}{2}$	$10\frac{1}{2}$	$5\frac{1}{8}$	$7\frac{1}{2}$	$9\frac{3}{4}$	12	12
		10	$5\frac{1}{8}$	$7\frac{1}{2}$	$11\frac{1}{4}$	$11\frac{1}{2}$	$10\frac{1}{4}$	$5\frac{1}{8}$	$7\frac{1}{2}$	$12\frac{1}{4}$	$12\frac{1}{2}$	$12\frac{1}{2}$
		12	$5\frac{1}{8}$	$7\frac{1}{2}$	$13\frac{1}{2}$	$12\frac{3}{4}$	$8\frac{3}{4}$	$5\frac{1}{8}$	$7\frac{1}{2}$	$14\frac{3}{4}$	$13\frac{1}{2}$	$12\frac{1}{2}$
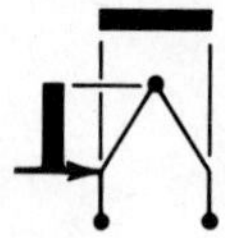	12/12	8	$5\frac{1}{8}$	$7\frac{1}{2}$	$9\frac{1}{4}$	12	12	$5\frac{1}{8}$	$7\frac{1}{2}$	$10\frac{1}{4}$	$13\frac{1}{2}$	$13\frac{1}{2}$
		10	$5\frac{1}{8}$	$7\frac{1}{2}$	$11\frac{3}{4}$	$13\frac{1}{4}$	$11\frac{1}{2}$	$5\frac{1}{8}$	$7\frac{1}{2}$	$12\frac{3}{4}$	$14\frac{1}{4}$	$14\frac{1}{4}$
		12	$5\frac{1}{8}$	$7\frac{1}{2}$	$13\frac{3}{4}$	$14\frac{1}{2}$	10	$5\frac{1}{8}$	$7\frac{1}{2}$	15	$15\frac{1}{2}$	14
	14/12	8	$5\frac{1}{8}$	$7\frac{1}{2}$	$9\frac{3}{4}$	$13\frac{1}{4}$	$13\frac{1}{4}$	$5\frac{1}{8}$	$7\frac{1}{2}$	$10\frac{1}{2}$	15	15
		10	$5\frac{1}{8}$	$7\frac{1}{2}$	12	$14\frac{3}{4}$	13	$5\frac{1}{8}$	$7\frac{1}{2}$	13	16	16
		12	$5\frac{1}{8}$	$7\frac{1}{2}$	$14\frac{1}{4}$	16	$11\frac{1}{2}$	$5\frac{1}{8}$	$7\frac{1}{2}$	$15\frac{1}{2}$	$17\frac{1}{2}$	$15\frac{1}{2}$
	16/12	8	$5\frac{1}{8}$	$7\frac{1}{2}$	10	$14\frac{3}{4}$	$14\frac{3}{4}$	$5\frac{1}{8}$	$7\frac{1}{2}$	$10\frac{3}{4}$	$16\frac{3}{4}$	$16\frac{3}{4}$
		10	$5\frac{1}{8}$	$7\frac{1}{2}$	$12\frac{1}{4}$	$16\frac{1}{4}$	$14\frac{1}{2}$	$5\frac{1}{8}$	$7\frac{1}{2}$	$13\frac{1}{4}$	$17\frac{3}{4}$	$17\frac{3}{4}$
		12	$5\frac{1}{8}$	$7\frac{1}{2}$	$14\frac{1}{2}$	$17\frac{3}{4}$	$13\frac{1}{4}$	$5\frac{1}{8}$	$7\frac{1}{2}$	$15\frac{3}{4}$	$19\frac{1}{4}$	$17\frac{1}{2}$

Timber Three-Hinged Tudor Arch Span-Load Table[a]

40′ Span					50′ Span				
Width	Base	Lower Tang.	Upper Tang.	Crown	Width	Base	Lower Tang.	Upper Tang.	Crown
$5\frac{1}{8}$	$7\frac{1}{2}$	$8\frac{1}{4}$	13	$12\frac{3}{4}$	$5\frac{1}{8}$	$7\frac{1}{2}$	$9\frac{1}{4}$	$15\frac{1}{2}$	$15\frac{1}{2}$
$5\frac{1}{8}$	$7\frac{1}{2}$	$10\frac{3}{4}$	$14\frac{1}{4}$	$11\frac{3}{4}$	$5\frac{1}{8}$	$7\frac{1}{2}$	12	$16\frac{1}{4}$	$16\frac{1}{4}$
$5\frac{1}{8}$	$7\frac{1}{2}$	$12\frac{3}{4}$	$15\frac{3}{4}$	$9\frac{3}{4}$	$5\frac{1}{8}$	$7\frac{1}{2}$	$14\frac{1}{4}$	$17\frac{1}{2}$	16
$5\frac{1}{8}$	$7\frac{1}{2}$	$8\frac{3}{4}$	$14\frac{3}{4}$	$14\frac{1}{2}$	$5\frac{1}{8}$	$7\frac{1}{2}$	$9\frac{3}{4}$	$17\frac{1}{2}$	$17\frac{1}{2}$
$5\frac{1}{8}$	$7\frac{1}{2}$	11	$16\frac{1}{4}$	$13\frac{3}{4}$	$5\frac{1}{8}$	$7\frac{1}{2}$	$12\frac{1}{2}$	$18\frac{1}{2}$	$18\frac{1}{2}$
$5\frac{1}{8}$	$7\frac{1}{2}$	$13\frac{1}{4}$	$17\frac{1}{2}$	$11\frac{3}{4}$	$5\frac{1}{8}$	$7\frac{1}{2}$	$14\frac{3}{4}$	20	$18\frac{1}{4}$
$5\frac{1}{8}$	$7\frac{1}{2}$	$9\frac{1}{2}$	$16\frac{1}{4}$	$16\frac{1}{4}$	$5\frac{1}{8}$	$7\frac{1}{2}$	$10\frac{1}{4}$	$19\frac{1}{2}$	$19\frac{1}{2}$
$5\frac{1}{8}$	$7\frac{1}{2}$	$12\frac{1}{2}$	18	$15\frac{3}{4}$	$5\frac{1}{8}$	$7\frac{3}{4}$	$13\frac{1}{4}$	$20\frac{1}{2}$	$20\frac{1}{2}$
$5\frac{1}{8}$	$7\frac{1}{2}$	$14\frac{1}{2}$	$19\frac{1}{2}$	$14\frac{1}{4}$	$5\frac{1}{8}$	$8\frac{1}{4}$	16	22	$20\frac{3}{4}$
$5\frac{1}{8}$	$7\frac{1}{2}$	11	18	18	$5\frac{1}{8}$	9	11	$21\frac{1}{2}$	$21\frac{1}{2}$
$5\frac{1}{8}$	$7\frac{3}{4}$	14	$19\frac{1}{2}$	18	$5\frac{1}{8}$	$9\frac{1}{4}$	$14\frac{3}{4}$	$22\frac{3}{4}$	$22\frac{3}{4}$
$5\frac{1}{8}$	$8\frac{1}{4}$	$16\frac{1}{4}$	$21\frac{1}{4}$	$16\frac{3}{4}$	$5\frac{1}{8}$	$9\frac{3}{4}$	$17\frac{1}{2}$	24	$23\frac{1}{2}$
$5\frac{1}{8}$	$7\frac{1}{2}$	9	13	13	$5\frac{1}{8}$	8	10	$15\frac{3}{4}$	$15\frac{3}{4}$
$5\frac{1}{8}$	$7\frac{1}{2}$	$11\frac{1}{2}$	14	$13\frac{1}{2}$	$5\frac{1}{8}$	$7\frac{3}{4}$	13	$16\frac{1}{2}$	$16\frac{1}{2}$
$5\frac{1}{8}$	$7\frac{1}{2}$	$13\frac{3}{4}$	$15\frac{1}{4}$	$12\frac{1}{2}$	$5\frac{1}{8}$	$7\frac{3}{4}$	$15\frac{3}{4}$	$17\frac{1}{4}$	$17\frac{1}{4}$
$5\frac{1}{8}$	$7\frac{1}{2}$	$9\frac{1}{2}$	$14\frac{3}{4}$	$14\frac{3}{4}$	$5\frac{1}{8}$	$7\frac{3}{4}$	$10\frac{1}{2}$	$17\frac{3}{4}$	$17\frac{3}{4}$
$5\frac{1}{8}$	$7\frac{1}{2}$	12	$15\frac{3}{4}$	$15\frac{1}{4}$	$5\frac{1}{8}$	$7\frac{3}{4}$	$13\frac{1}{2}$	$18\frac{3}{4}$	$18\frac{3}{4}$
$5\frac{1}{8}$	$7\frac{1}{2}$	$14\frac{1}{4}$	$17\frac{1}{4}$	$14\frac{1}{4}$	$5\frac{1}{8}$	$7\frac{3}{4}$	16	$19\frac{1}{2}$	$19\frac{1}{2}$
$5\frac{1}{8}$	$7\frac{1}{2}$	10	$16\frac{1}{2}$	$16\frac{1}{2}$	$5\frac{1}{8}$	$7\frac{3}{4}$	$10\frac{1}{2}$	$19\frac{3}{4}$	$19\frac{3}{4}$
$5\frac{1}{8}$	$7\frac{1}{2}$	$12\frac{1}{2}$	$17\frac{1}{2}$	$17\frac{1}{4}$	$5\frac{1}{8}$	$7\frac{3}{4}$	$13\frac{3}{4}$	$20\frac{3}{4}$	$20\frac{3}{4}$
$5\frac{1}{8}$	$7\frac{1}{2}$	$14\frac{3}{4}$	$19\frac{1}{4}$	$16\frac{1}{2}$	$5\frac{1}{8}$	8	$16\frac{1}{2}$	$21\frac{3}{4}$	$21\frac{3}{4}$
$5\frac{1}{8}$	$7\frac{1}{2}$	$10\frac{1}{4}$	$18\frac{1}{4}$	$18\frac{1}{4}$	$5\frac{1}{8}$	$8\frac{1}{4}$	$10\frac{3}{4}$	22	22
$5\frac{1}{8}$	$7\frac{1}{2}$	13	$19\frac{1}{4}$	$19\frac{1}{4}$	$5\frac{1}{8}$	$8\frac{1}{2}$	14	23	23
$5\frac{1}{8}$	$7\frac{1}{2}$	$15\frac{3}{4}$	21	$18\frac{3}{4}$	$5\frac{1}{8}$	9	$16\frac{3}{4}$	$24\frac{1}{4}$	$24\frac{1}{4}$
$5\frac{1}{8}$	8	$10\frac{1}{2}$	$13\frac{1}{2}$	$13\frac{1}{2}$	$5\frac{1}{8}$	$10\frac{1}{4}$	$11\frac{1}{4}$	$16\frac{1}{2}$	$16\frac{1}{2}$
$5\frac{1}{8}$	$7\frac{3}{4}$	$13\frac{1}{4}$	14	14	$5\frac{1}{8}$	10	$15\frac{1}{4}$	17	17
$5\frac{1}{8}$	$7\frac{3}{4}$	16	$14\frac{3}{4}$	$14\frac{3}{4}$	$5\frac{1}{8}$	$9\frac{3}{4}$	18	$17\frac{1}{2}$	$17\frac{1}{2}$
$5\frac{1}{8}$	$7\frac{3}{4}$	11	$15\frac{1}{4}$	$15\frac{1}{4}$	$5\frac{1}{8}$	$9\frac{3}{4}$	$11\frac{3}{4}$	$18\frac{1}{2}$	$18\frac{1}{2}$
$5\frac{1}{8}$	$7\frac{3}{4}$	$13\frac{3}{4}$	$15\frac{3}{4}$	$15\frac{3}{4}$	$5\frac{1}{8}$	$9\frac{3}{4}$	$15\frac{1}{2}$	19	19
$5\frac{1}{8}$	$7\frac{3}{4}$	$16\frac{1}{4}$	$16\frac{3}{4}$	$16\frac{3}{4}$	$5\frac{1}{8}$	$9\frac{1}{2}$	$18\frac{1}{4}$	$19\frac{3}{4}$	$19\frac{3}{4}$
$5\frac{1}{8}$	$7\frac{3}{4}$	$11\frac{1}{4}$	17	17	$5\frac{1}{8}$	$9\frac{1}{2}$	12	$20\frac{1}{2}$	$20\frac{1}{2}$
$5\frac{1}{8}$	$7\frac{3}{4}$	14	$17\frac{3}{4}$	$17\frac{3}{4}$	$5\frac{1}{8}$	$9\frac{1}{2}$	$15\frac{1}{2}$	$21\frac{1}{4}$	$21\frac{1}{4}$
$5\frac{1}{8}$	$7\frac{3}{4}$	$16\frac{1}{2}$	$18\frac{3}{4}$	$18\frac{3}{4}$	$5\frac{1}{8}$	$9\frac{1}{2}$	$18\frac{1}{2}$	$22\frac{1}{4}$	$22\frac{1}{4}$
$5\frac{1}{8}$	$7\frac{3}{4}$	$11\frac{1}{2}$	$18\frac{3}{4}$	$18\frac{3}{4}$	$5\frac{1}{8}$	$9\frac{1}{2}$	12	$22\frac{3}{4}$	$22\frac{3}{4}$
$5\frac{1}{8}$	$7\frac{3}{4}$	$14\frac{1}{4}$	$19\frac{1}{2}$	$19\frac{1}{2}$	$5\frac{1}{8}$	$9\frac{1}{2}$	$15\frac{1}{2}$	$23\frac{1}{2}$	$23\frac{1}{2}$
$5\frac{1}{8}$	$7\frac{3}{4}$	17	$20\frac{3}{4}$	$20\frac{3}{4}$	$5\frac{1}{8}$	$9\frac{1}{2}$	$18\frac{3}{4}$	$24\frac{1}{2}$	$24\frac{1}{2}$

(footnotes follow on p. 7-610)

Footnote to Table 7.8

[a]Values for preliminary design purposes only. Sizes are based on Douglas Fir laminated timber, developing an allowable shear stress of 165 psi and with a bending radius of 9 ft 4 in. For Southern Pine laminated timber, an allowable shear stress of 200 psi and a bending radius of 7 ft 0 in. may be used. For roof pitches less than $\frac{10}{12}$, the critical loading is generally the combined dead and live load on the horizontal projection of the full span. For roof pitches of $\frac{10}{12}$ or greater, the critical loading is generally a combination of dead load and horizontal wind load. Sizes shown are determined from a uniformly distributed wind load applied on the vertical projection of the roof arm with a concentrated wind load equal to one-half the total wind load on the wall height acting at the haunch. In the combined stress analysis, it was assumed that the bending portion of the loading exceeded the axial compression portion. The section sizes shown in this table are in inches and are based on the following design criteria: (1) Uniform loading. (2) Radius of curvature at the haunch = 9 ft 4 in. (3) Allowable stresses: Bending stress, F_b = 2400 psi (reduced by curvature factor when applicable); shear stress, F_v = 165 psi; compression parallel to grain stress, F_c = 1500 psi (adjusted for l/d ratio); modulus of elasticity, E = 1,600,000 psi. These stresses were increased 15% for short duration of loading and $33\frac{1}{3}$% for wind loading when applicable.

(4) Vertical arch legs are laterally supported. (5) Dead load equal to one-third of the total vertical load. Note: When arch deflection is a concern, arch sizes should be checked for deflection at the time of size selection.

TABLE 7.9

Controlled Random Layup Decking

ALLOWABLE UNIFORMLY DISTRIBUTED TOTAL ROOF LOAD LIMITED BY DEFLECTION

Species	Actual Size (in.)	Deflection Limit	8	9	10	11	12	13	14	15	16	17	18	19	20
GLUED LAMINATED TIMBER DECKING—CONTROLLED RANDOM LAYUP[a–f]															
Douglas Fir/ Larch and Southern Pine[e]	$3\frac{21}{32} \times 5\frac{3}{8}$	$l/180$							160	130	107	89	75	64	55
		$l/240$							120	97	80	67	56	48	41
	$2\frac{7}{8} \times 5\frac{3}{8}$	$l/180$					124	98	78	63	53	44			
		$l/240$					93	73	59	48	40	33			
	$2\frac{3}{16} \times 5\frac{3}{8}$	$l/180$	181	127	93	70	54	42							
		$l/240$	136	96	70	52	40	32							
Idaho White Pine and Inland White Fir	$3\frac{21}{32} \times 5\frac{3}{8}$	$l/180$								107	88	74	62	53	45
		$l/240$								80	66	55	46	40	34
	$2\frac{7}{8} \times 5\frac{3}{8}$	$l/180$					102	80	64	52	43	36			
		$l/240$					77	60	48	39	32	27			
	$2\frac{3}{16} \times 5\frac{3}{8}$	$l/180$	151	106	77	58	45	35							
		$l/240$	113	80	58	44	34	26							

TABLE 7.9 (*Continued*)

Species	Actual Size (in.)	Span (ft) 8	9	10	11	12	13	14	15	16	17	18	19	20
		Deflection Limit												
		l/180 l/240	l/180 l/240	l/180 l/240	l/180 l/240	l/180 l/240	l/180 l/240	l/180 l/240	l/180 l/240	l/180 l/240	l/180 l/240	l/180 l/240	l/180 l/240	l/180 l/240
Ponderosa Pine	$3\frac{3}{4} \times 7\frac{1}{8}$					192 144	151 113	120 90	98 74	81 61	68 51	57 43	48 36	41 31
	$3 \times 7\frac{1}{8}$					95 71	75 56	60 45	49 37	40 30	33	28		
	$2\frac{1}{4} \times 7\frac{1}{8}$	 101	95 71	69 52	52 39	40 30	31 23	25	20					
Inland Red Cedar	$3\frac{21}{32} \times 5\frac{3}{8}$								86 64	71 53	59 44	50 37	42 32	36 27
	$2\frac{7}{8} \times 5\frac{3}{8}$					82 61	64 48	51 39	42 31	34 26	29 22			
	$2\frac{3}{16} \times 5\frac{3}{8}$	121 91	85 64	62 46	47 35	36 27	28 21	22 17	18 14	15 11				
SOLID HEAVY TIMBER DECKING—CONTROLLED RANDOM LAYUP[g,d]														
Douglas Fir/ Larch— Select	$3\frac{1}{2} \times 5\frac{1}{4}$					200 150	157 118	126 94	102 77	84 63	70 53	59 44	50 38	43 32
	$2\frac{1}{2} \times 5\frac{1}{4}$		173 129	126 94	94 71	73 55	57 43	46 34	37 28	31 23	26 19	22 16	18 14	16 12
	$1\frac{1}{2} \times 5$	46 34	32 24	23 18	18 13	14 10								

Douglas Fir/ Larch— Commercial	$3\frac{1}{2} \times 5\frac{1}{4}$					189	148	119	97	80	66	56	48	41
						142	111	89	72	60	50	42	36	31
	$2\frac{1}{2} \times 5\frac{1}{4}$		163	119	89	69	54	43	35	29	24	20	17	15
			122	89	67	52	41	32	26	22	18	15	13	11
	$1\frac{1}{2} \times 5$	43	30	22	17	13								
		32	23	17	12	10								
Southern Pine— Select and Commercial	$3\frac{1}{2} \times 5\frac{1}{4}$					178	140	112	91	75	62	53	45	38
						133	105	84	68	56	47	39	34	29
	$2\frac{1}{2} \times 5\frac{1}{4}$		153	112	84	65	51	41	33	27	23	19	16	14
			115	84	63	49	38	31	25	20	17	14	12	10
	$1\frac{1}{2} \times 5$	41	28	21	16	12								
		30	21	16	12	9								
Hem-Fir— Select	$3\frac{1}{2} \times 5\frac{1}{4}$					166	131	105	85	70	59	49	42	36
						125	98	79	64	53	44	37	31	27
	$2\frac{1}{2} \times 5\frac{1}{4}$		144	105	79	61	48	38	31	26	21	18	15	13
			108	79	59	46	36	29	23	19	16	13	11	10
	$1\frac{1}{2} \times 5$	38	27	20	15	11								
		29	20	15	11	8								
Hem-Fir— Commercial	$3\frac{1}{2} \times 5\frac{1}{4}$					155	122	98	80	66	55	46	39	34
						117	92	73	60	49	41	35	29	25
	$2\frac{1}{2} \times 5\frac{1}{4}$	191	134	98	73	57	45	36	29	24	20	17	14	12
		143	101	73	55	42	33	27	22	18	15	13	11	9
	$1\frac{1}{2} \times 5$	36	25	18	14	10								
		27	19	14	10	8								

(continued)

TABLE 7.9 (*Continued*)

Species	Actual Size (in.)	Deflection Limit	8	9	10	11	12	13	14	15	16	17	18	19	20
Western Cedar—Select	$3\frac{1}{2} \times 5\frac{1}{4}$	$l/180$				158	122	96	77	63	52	43	36	31	26
		$l/240$			158	119	92	72	58	47	39	32	27	23	20
	$2\frac{1}{2} \times 5\frac{1}{4}$	$l/180$	150	105	77	58	44	35	28	23	19	16	13	11	10
		$l/240$	113	79	58	43	33	26	21	17	14	12	10	8	8
	$1\frac{1}{2} \times 5$	$l/180$	28	20	14	11									
		$l/240$	21	15	11	8									
Western Cedar—Commercial	$3\frac{1}{2} \times 5\frac{1}{4}$	$l/180$			192	144	111	87	70	57	47	39	33	28	24
		$l/240$		197	144	108	83	65	52	43	35	29	25	21	18
	$2\frac{1}{2} \times 5\frac{1}{4}$	$l/180$	136	96	70	52	40	32	25	21	17	14	12	10	
		$l/240$	102	72	52	39	30	24	19	16	13	11	9	8	
	$1\frac{1}{2} \times 5$	$l/180$	25	18	13	10									
		$l/240$	19	13	10										

Decking Species Data[g, h]

Laminated Timber Decking[i]

Species	Modulus of Elasticity, E (psi)	Actual Size (in.)	Weight (psf)	Coverage Factor[a, b] (bd ft/sq ft)
Douglas Fir/ Larch and Southern Pine[j]	1,800,000	$3\frac{21}{32} \times 5\frac{3}{8}$	10.5	5.58
		$2\frac{7}{8} \times 5\frac{3}{8}$	8.1	3.35
		$2\frac{3}{16} \times 5\frac{3}{8}$	6.5	3.35
Idaho White Pine and Inland White Fir	1,500,000	$3\frac{21}{32} \times 5\frac{3}{8}$	9.5	5.58
		$2\frac{7}{8} \times 5\frac{3}{8}$	7.3	3.35
		$2\frac{3}{16} \times 5\frac{3}{8}$	5.0	3.35
Ponderosa Pine	1,300,000	$3\frac{3}{4} \times 7\frac{1}{8}$	8.8	5.61
		$3 \times 7\frac{1}{8}$	7.0	4.49
		$2\frac{1}{4} \times 7\frac{1}{8}$	5.2	3.37
Inland Red Cedar	1,200,000	$3\frac{21}{32} \times 5\frac{3}{8}$	7.5	5.58
		$2\frac{7}{8} \times 5\frac{3}{8}$	5.8	3.35
		$2\frac{3}{16} \times 5\frac{3}{8}$	4.5	3.35

Solid Heavy Timber Decking

Species and Quality Grade	Modulus of Elasticity, E (psi)	Size (in.) Nominal	Size (in.) Actual	Weight (psf)	Coverage Factor (bd ft/sq ft)
Douglas Fir/ Larch—		4×6	$3\frac{1}{2} \times 5\frac{1}{4}$	10.1	4.57
Select	1,800,000	3×6	$2\frac{1}{2} \times 5\frac{1}{4}$	7.2	3.43
Commercial	1,700,000	2×6	$1\frac{1}{2} \times 5$	4.3	2.40
Hem-Fir—		4×6	$3\frac{1}{2} \times 5\frac{1}{4}$	8.0	4.57
Select	1,500,000	3×6	$2\frac{1}{2} \times 5\frac{1}{4}$	5.7	3.43
Commercial	1,400,000	2×6	$1\frac{1}{2} \times 5$	3.4	2.40
Southern Pine—		4×6	$3\frac{1}{2} \times 5\frac{1}{4}$	10.8	4.57
Select	1,600,000	3×6	$2\frac{1}{2} \times 5\frac{1}{4}$	7.7	3.43
Commercial	1,600,000	2×6	$1\frac{1}{2} \times 5$	4.6	2.40
Western Cedar—		4×6	$3\frac{1}{2} \times 5\frac{1}{4}$	7.3	4.57
Select	1,100,000	3×6	$2\frac{1}{2} \times 5\frac{1}{4}$	5.2	3.43
Commercial	1,000,000	2×6	$1\frac{1}{2} \times 5$	3.1	2.40

[a] Allowable uniformly distributed total roof load in pounds per square foot of roof surface for flat roofs, consisting of live load and dead load, including weight of deck (see Decking Species Data at end of Table 7.9). These roof loads do not exceed bending loads allowable under recognized bending formulas. Loads are for dry condition of use.

[b] All load values assume installation conforming to manufacturer's recommendation.

[c] Load values are based on manufacturer's recommendations. In all cases, data is subject to special requirements of local building codes.

[d] To determine allowable loads for $2\frac{5}{8}$ in. net thickness, multiply tabulated loads for $2\frac{1}{2}$ in. deck by 1.16. To determine allowable loads for $1\frac{5}{8}$ in. net thickness, multiply tabulated load for $1\frac{1}{2}$ in. deck by 1.27.

[e] Actual size for Southern pine is $2\frac{1}{4}$ in. × $5\frac{3}{8}$ in.: To calculate total loads for Southern pine for this size, multiply the listed valued by 1.075.

[f] Values may vary among manufacturers. See manufacturer's literature for more detailed information.

[g] To estimate board feet of decking required, multiply square feet of area to be covered by the coverage factor. Add for job site trimming and waste for irregular areas.

[h] For 8-in. nominal widths ($7\frac{1}{8}$ in. net), coverage factors are: $3\frac{3}{4}$ and $3\frac{21}{32}$ in.— 5.61; 3 in.—4.49; $2\frac{7}{8}$ and $2\frac{1}{4}$ in.—3.37.

[i] Laminated decking sizes may vary between manufacturers. The designer should check with the supplier to determine actual sizes and load-carrying capacities.

[j] Actual size for Southern Pine is $2\frac{1}{4} \times 5\frac{3}{8}$ in.

BEAM SYSTEM DIAGRAMS

TABLE 7.10

Beam Diagrams and Formulas[a]

1. SIMPLE BEAM—UNIFORMLY DISTRIBUTED LOAD

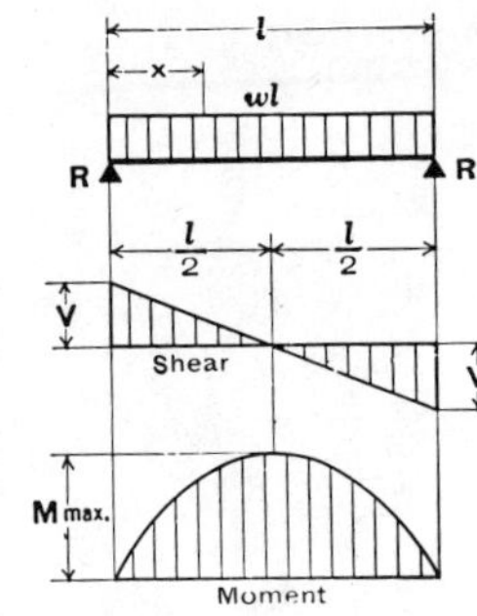

$$R = V \;.\;.\;.\;.\;.\;.\;.\;.\;.\;.\;. = \frac{wl}{2}$$

$$V_x \;.\;.\;.\;.\;.\;.\;.\;.\;.\;.\;. = w\left(\frac{l}{2} - x\right)$$

$$M\ max.\ \left(\text{at center}\right) \;.\;.\;.\;. = \frac{wl^2}{8}$$

$$M_x \;.\;.\;.\;.\;.\;.\;.\;.\;.\;.\;. = \frac{wx}{2}(l - x)$$

$$\Delta max.\ \left(\text{at center}\right) \;.\;.\;.\;. = \frac{5\,wl^4}{384\,EI}$$

$$\Delta x \;.\;.\;.\;.\;.\;.\;.\;.\;.\;.\;. = \frac{wx}{24EI}(l^3 - 2lx^2 + x^3)$$

2. SIMPLE BEAM—LOAD INCREASING UNIFORMLY TO ONE END

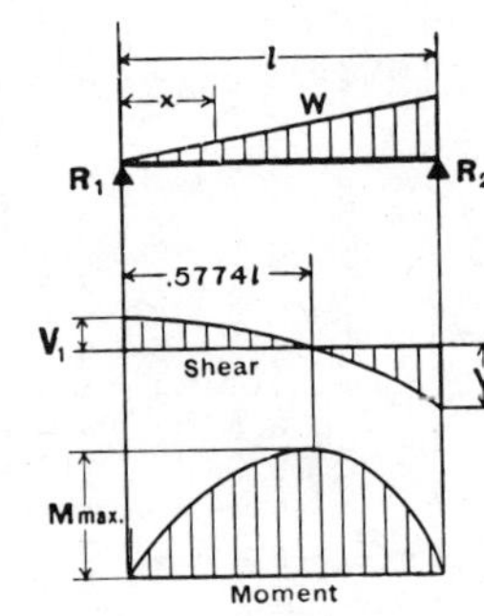

$$R_1 = V_1 \;.\;.\;.\;.\;.\;.\;.\;.\;.\;.\;. = \frac{W}{3}$$

$$R_2 = V_2\ max. \;.\;.\;.\;.\;.\;.\;.\;. = \frac{2W}{3}$$

$$V_x \;.\;.\;.\;.\;.\;.\;.\;.\;.\;.\;. = \frac{W}{3} - \frac{Wx^2}{l^2}$$

$$M\ max.\ \left(\text{at } x = \frac{l}{\sqrt{3}} = .5774l\right) \;.\;. = \frac{2Wl}{9\sqrt{3}} = .1283\,Wl$$

$$M_x \;.\;.\;.\;.\;.\;.\;.\;.\;.\;.\;. = \frac{Wx}{3l^2}(l^2 - x^2)$$

$$\Delta max.\ \left(\text{at } x = l\sqrt{1 - \sqrt{\frac{8}{15}}} = .5193l\right) = .01304\,\frac{Wl^3}{EI}$$

$$\Delta x \;.\;.\;.\;.\;.\;.\;.\;.\;.\;.\;. = \frac{Wx}{180EI\,l^2}(3x^4 - 10l^2x^2 + 7l^4)$$

3. SIMPLE BEAM—LOAD INCREASING UNIFORMLY TO CENTER

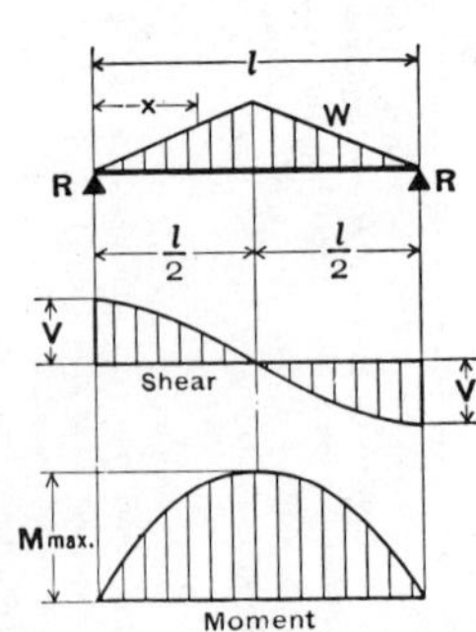

$$R = V \;.\;.\;.\;.\;.\;.\;.\;.\;.\;.\;. = \frac{W}{2}$$

$$V_x\ \left(\text{when } x < \frac{l}{2}\right) \;.\;.\;.\;. = \frac{W}{2l^2}(l^2 - 4x^2)$$

$$M\ max.\ \left(\text{at center}\right) \;.\;.\;.\;. = \frac{Wl}{6}$$

$$M_x\ \left(\text{when } x < \frac{l}{2}\right) \;.\;.\;.\;. = Wx\left(\frac{1}{2} - \frac{2x^2}{3l^2}\right)$$

$$\Delta max.\ \left(\text{at center}\right) \;.\;.\;.\;. = \frac{Wl^3}{60EI}$$

$$\Delta x \;.\;.\;.\;.\;.\;.\;.\;.\;.\;.\;. = \frac{Wx}{480\,EI\,l^2}(5l^2 - 4x^2)^2$$

TABLE 7.10 (*Continued*)

4. SIMPLE BEAM—UNIFORM LOAD PARTIALLY DISTRIBUTED

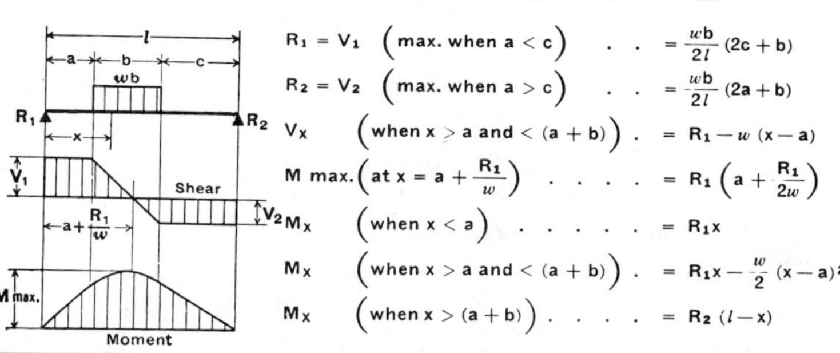

$$R_1 = V_1 \left(\text{max. when } a < c\right) \quad . \quad . \quad = \frac{wb}{2l}(2c + b)$$

$$R_2 = V_2 \left(\text{max. when } a > c\right) \quad . \quad . \quad = \frac{wb}{2l}(2a + b)$$

$$V_x \left(\text{when } x > a \text{ and } < (a + b)\right) \quad . \quad = R_1 - w\,(x - a)$$

$$M\ \text{max.}\left(\text{at } x = a + \frac{R_1}{w}\right) \quad . \quad . \quad . \quad . \quad = R_1\left(a + \frac{R_1}{2w}\right)$$

$$M_x \left(\text{when } x < a\right) \quad . \quad . \quad . \quad . \quad . \quad . \quad = R_1 x$$

$$M_x \left(\text{when } x > a \text{ and } < (a + b)\right) \quad . \quad = R_1 x - \frac{w}{2}(x - a)^2$$

$$M_x \left(\text{when } x > (a + b)\right) \quad . \quad . \quad . \quad . \quad = R_2\,(l - x)$$

5. SIMPLE BEAM—UNIFORM LOAD PARTIALLY DISTRIBUTED AT ONE END

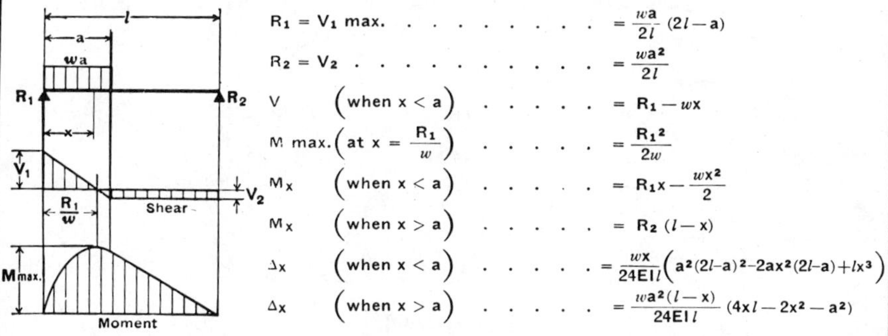

$$R_1 = V_1\ \text{max.} \quad . \quad . \quad . \quad . \quad . \quad . \quad . \quad . \quad . \quad = \frac{wa}{2l}(2l - a)$$

$$R_2 = V_2 \quad . \quad . \quad . \quad . \quad . \quad . \quad . \quad . \quad . \quad . \quad = \frac{wa^2}{2l}$$

$$V \left(\text{when } x < a\right) \quad . \quad . \quad . \quad . \quad . \quad . \quad = R_1 - wx$$

$$M\ \text{max.}\left(\text{at } x = \frac{R_1}{w}\right) \quad . \quad . \quad . \quad . \quad . \quad . \quad = \frac{R_1^2}{2w}$$

$$M_x \left(\text{when } x < a\right) \quad . \quad . \quad . \quad . \quad . \quad . \quad = R_1 x - \frac{wx^2}{2}$$

$$M_x \left(\text{when } x > a\right) \quad . \quad . \quad . \quad . \quad . \quad . \quad = R_2\,(l - x)$$

$$\Delta_x \left(\text{when } x < a\right) \quad . \quad . \quad . \quad . \quad . \quad . \quad = \frac{wx}{24EIl}\left(a^2(2l-a)^2 - 2ax^2(2l-a) + lx^3\right)$$

$$\Delta_x \left(\text{when } x > a\right) \quad . \quad . \quad . \quad . \quad . \quad . \quad = \frac{wa^2(l - x)}{24EIl}(4xl - 2x^2 - a^2)$$

6. SIMPLE BEAM—UNIFORM LOAD PARTIALLY DISTRIBUTED AT EACH END

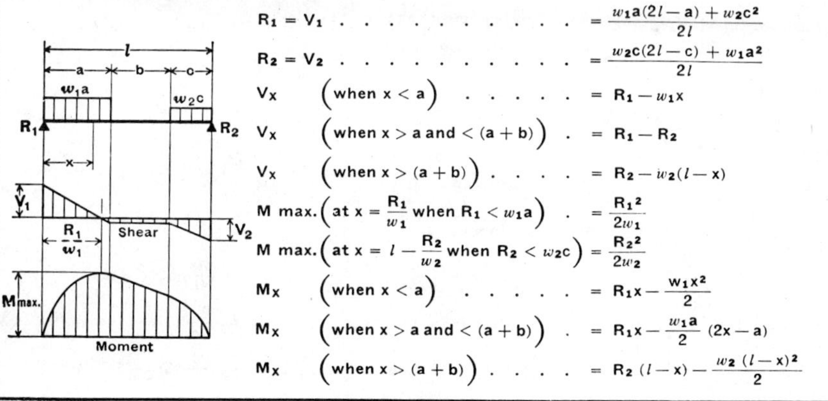

$$R_1 = V_1 \quad . \quad . \quad . \quad . \quad . \quad . \quad . \quad . \quad . \quad . \quad . \quad = \frac{w_1 a(2l - a) + w_2 c^2}{2l}$$

$$R_2 = V_2 \quad . \quad . \quad . \quad . \quad . \quad . \quad . \quad . \quad . \quad . \quad . \quad = \frac{w_2 c(2l - c) + w_1 a^2}{2l}$$

$$V_x \left(\text{when } x < a\right) \quad . \quad . \quad . \quad . \quad . \quad . \quad = R_1 - w_1 x$$

$$V_x \left(\text{when } x > a \text{ and } < (a + b)\right) \quad . \quad = R_1 - R_2$$

$$V_x \left(\text{when } x > (a + b)\right) \quad . \quad . \quad . \quad . \quad = R_2 - w_2(l - x)$$

$$M\ \text{max.}\left(\text{at } x = \frac{R_1}{w_1} \text{ when } R_1 < w_1 a\right) \quad . \quad = \frac{R_1^2}{2w_1}$$

$$M\ \text{max.}\left(\text{at } x = l - \frac{R_2}{w_2} \text{ when } R_2 < w_2 c\right) = \frac{R_2^2}{2w_2}$$

$$M_x \left(\text{when } x < a\right) \quad . \quad . \quad . \quad . \quad . \quad . \quad = R_1 x - \frac{w_1 x^2}{2}$$

$$M_x \left(\text{when } x > a \text{ and } < (a + b)\right) \quad . \quad = R_1 x - \frac{w_1 a}{2}(2x - a)$$

$$M_x \left(\text{when } x > (a + b)\right) \quad . \quad . \quad . \quad . \quad = R_2\,(l - x) - \frac{w_2\,(l - x)^2}{2}$$

TABLE 7.10 (*Continued*)

7. SIMPLE BEAM—CONCENTRATED LOAD AT CENTER

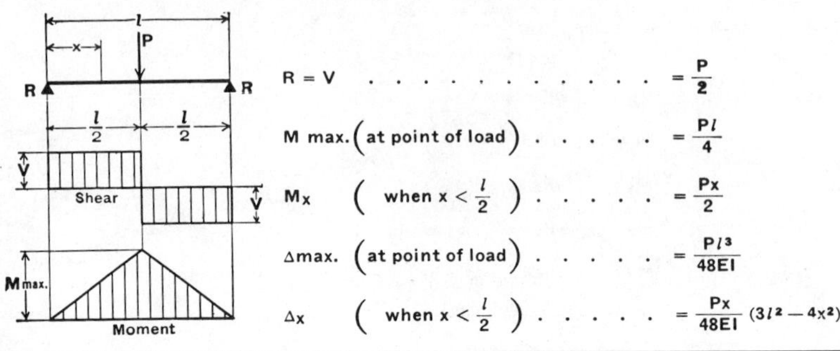

$$R = V \; \dots \; = \frac{P}{2}$$

$$M\ max.\left(\text{at point of load}\right) \; \dots \; = \frac{Pl}{4}$$

$$M_x \quad \left(\text{when } x < \frac{l}{2}\right) \; \dots \; = \frac{Px}{2}$$

$$\Delta max. \left(\text{at point of load}\right) \; \dots \; = \frac{Pl^3}{48EI}$$

$$\Delta x \quad \left(\text{when } x < \frac{l}{2}\right) \; \dots \; = \frac{Px}{48EI}(3l^2 - 4x^2)$$

8. SIMPLE BEAM—CONCENTRATED LOAD AT ANY POINT

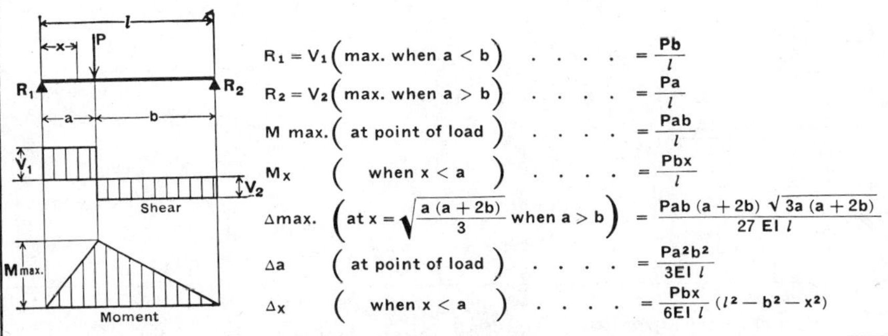

$$R_1 = V_1\left(\text{max. when } a < b\right) \; \dots \; = \frac{Pb}{l}$$

$$R_2 = V_2\left(\text{max. when } a > b\right) \; \dots \; = \frac{Pa}{l}$$

$$M\ max.\left(\text{at point of load}\right) \; \dots \; = \frac{Pab}{l}$$

$$M_x \quad \left(\text{when } x < a\right) \; \dots \; = \frac{Pbx}{l}$$

$$\Delta max. \left(\text{at } x = \sqrt{\frac{a\,(a + 2b)}{3}} \text{ when } a > b\right) = \frac{Pab\,(a + 2b)\,\sqrt{3a\,(a + 2b)}}{27\,EI\,l}$$

$$\Delta a \quad \left(\text{at point of load}\right) \; \dots \; = \frac{Pa^2b^2}{3EI\,l}$$

$$\Delta x \quad \left(\text{when } x < a\right) \; \dots \; = \frac{Pbx}{6EI\,l}(l^2 - b^2 - x^2)$$

9. SIMPLE BEAM—TWO EQUAL CONCENTRATED LOADS SYMMETRICALLY PLACED

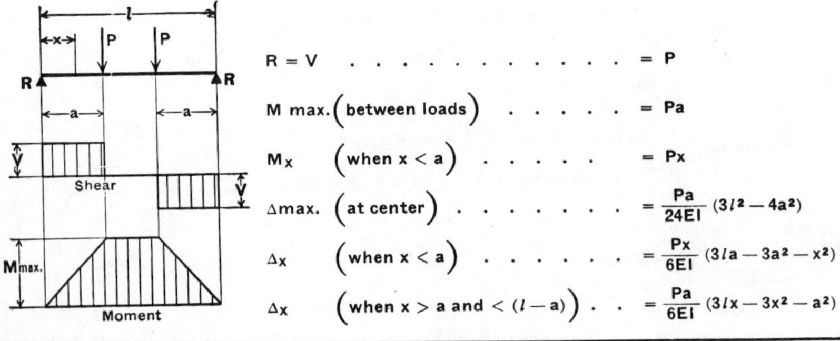

$$R = V \; \dots \; = P$$

$$M\ max.\left(\text{between loads}\right) \; \dots \; = Pa$$

$$M_x \quad \left(\text{when } x < a\right) \; \dots \; = Px$$

$$\Delta max. \left(\text{at center}\right) \; \dots \; = \frac{Pa}{24EI}(3l^2 - 4a^2)$$

$$\Delta x \quad \left(\text{when } x < a\right) \; \dots \; = \frac{Px}{6EI}(3la - 3a^2 - x^2)$$

$$\Delta x \quad \left(\text{when } x > a \text{ and } < (l - a)\right) \; \dots \; = \frac{Pa}{6EI}(3lx - 3x^2 - a^2)$$

10. SIMPLE BEAM—TWO EQUAL CONCENTRATED LOADS UNSYMMETRICALLY PLACED

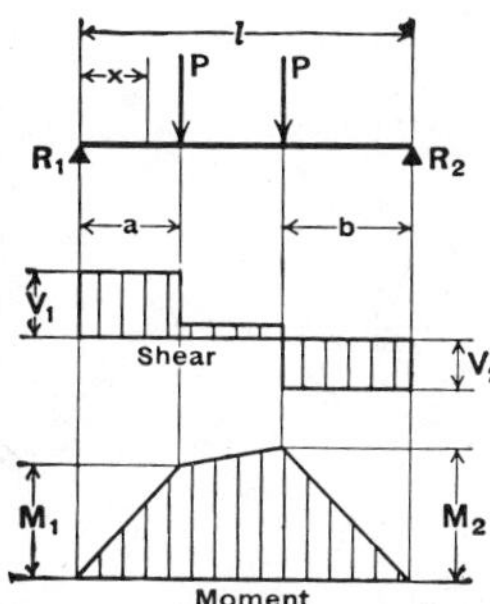

$R_1 = V_1$	(max. when $a < b$)	$= \frac{P}{l}(l - a + b)$
$R_2 = V_2$	(max. when $a > b$)	$= \frac{P}{l}(l - b + a)$
V_x	(when $x > a$ and $< (l - b)$)	$= \frac{P}{l}(b - a)$
M_1	(max. when $a > b$)	$= R_1 a$
M_2	(max. when $a < b$)	$= R_2 b$
M_x	(when $x < a$)	$= R_1 x$
M_x	(when $x > a$ and $< (l - b)$)	$= R_1 x - P(x - a)$

11. SIMPLE BEAM—TWO UNEQUAL CONCENTRATED LOADS UNSYMMETRICALLY PLACED

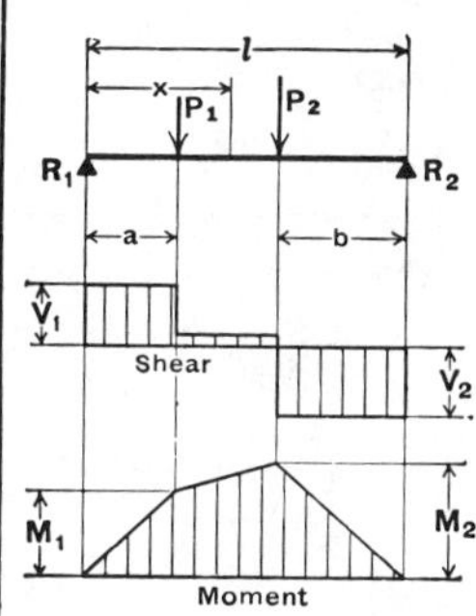

$R_1 = V_1$		$= \frac{P_1(l - a) + P_2 b}{l}$
$R_2 = V_2$		$= \frac{P_1 a + P_2(l - b)}{l}$
V_x	(when $x > a$ and $< (l - b)$)	$= R_1 - P_1$
M_1	(max. when $R_1 < P_1$)	$= R_1 a$
M_2	(max. when $R_2 < P_2$)	$= R_2 b$
M_x	(when $x < a$)	$= R_1 x$
M_x	(when $x > a$ and $< (l - b)$)	$= R_1 x - P_1(x - a)$

12. BEAM FIXED AT ONE END, SUPPORTED AT OTHER—UNIFORMLY DISTRIBUTED LOAD

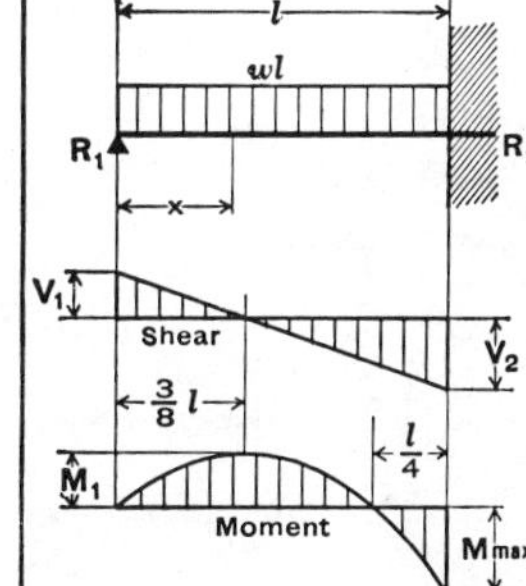

$R_1 = V_1$		$= \frac{3wl}{8}$
$R_2 = V_2$ max.		$= \frac{5wl}{8}$
V_x		$= R_1 - wx$
M max.		$= \frac{wl^2}{8}$
M_1	(at $x = \frac{3}{8}l$)	$= \frac{9}{128}wl^2$
M_x		$= R_1 x - \frac{wx^2}{2}$
Δmax.	(at $x = \frac{l}{16}(1 + \sqrt{33}) = .4215l$)	$= \frac{wl^4}{185EI}$
Δx		$= \frac{wx}{48EI}(l^3 - 3lx^2 + 2x^3)$

TABLE 7.10 (*Continued*)

13. BEAM FIXED AT ONE END, SUPPORTED AT OTHER— CONCENTRATED LOAD AT CENTER

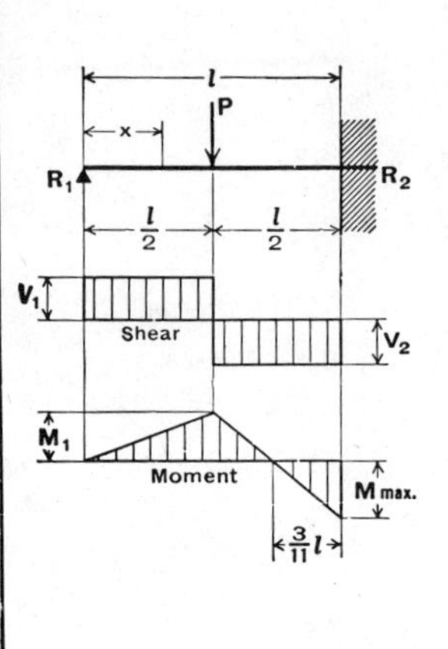

$R_1 = V_1$ . . . . $= \frac{5P}{16}$

$R_2 = V_2$ max. . . . . $= \frac{11P}{16}$

M max. $\left(\text{at fixed end}\right)$ . . . . . $= \frac{3Pl}{16}$

M_1 $\left(\text{at point of load}\right)$ . . . . $= \frac{5Pl}{32}$

M_x $\left(\text{when } x < \frac{l}{2}\right)$ . . . . . $= \frac{5Px}{16}$

M_x $\left(\text{when } x > \frac{l}{2}\right)$ . . . . . $= P\left(\frac{l}{2} - \frac{11x}{16}\right)$

Δmax. $\left(\text{at } x = l\sqrt{\frac{1}{5}} = .4472l\right)$ . . $= \frac{Pl^3}{48EI\sqrt{5}} = .009317\,\frac{Pl^3}{EI}$

Δx $\left(\text{at point of load}\right)$ . . . . $= \frac{7Pl^3}{768EI}$

Δx $\left(\text{when } x < \frac{l}{2}\right)$ . . . . . $= \frac{Px}{96EI}(3l^2 - 5x^2)$

Δx $\left(\text{when } x > \frac{l}{2}\right)$ . . . . . $= \frac{P}{96EI}(x - l)^2(11x - 2l)$

14. BEAM FIXED AT ONE END, SUPPORTED AT OTHER— CONCENTRATED LOAD AT ANY POINT

$R_1 = V_1$ . . . . $= \frac{Pb^2}{2l^3}(a + 2l)$

$R_2 = V_2$ . . . . $= \frac{Pa}{2l^3}(3l^2 - a^2)$

M_1 $\left(\text{at point of load}\right)$ . . . . $= R_1 a$

M_2 $\left(\text{at fixed end}\right)$ . . . . . $= \frac{Pab}{2l^2}(a + l)$

M_x $\left(\text{when } x < a\right)$ . . . . . $= R_1 x$

M_x $\left(\text{when } x > a\right)$ . . . . . $= R_1 x - P(x - a)$

Δmax. $\left(\text{when } a < .414l \text{ at } x = l\,\frac{l^2+a^2}{3l^2-a^2}\right) = \frac{Pa}{3EI}\,\frac{(l^2 - a^2)^3}{(3l^2 - a^2)^2}$

Δmax. $\left(\text{when } a > .414l \text{ at } x = l\sqrt{\frac{a}{2l+a}}\right) = \frac{Pab^2}{6EI}\sqrt{\frac{a}{2l + a}}$

Δa $\left(\text{at point of load}\right)$ . . . . $= \frac{Pa^2b^3}{12EIl^3}(3l + a)$

Δx $\left(\text{when } x < a\right)$ . . . . . $= \frac{Pb^2x}{12EIl^3}(3al^2 - 2lx^2 - ax^2)$

Δx $\left(\text{when } x > a\right)$ . . . . . $= \frac{Pa}{12EIl^3}(l-x)^2(3l^2x - a^2x - 2a^2l)$

15. BEAM FIXED AT BOTH ENDS—UNIFORMLY DISTRIBUTED LOADS

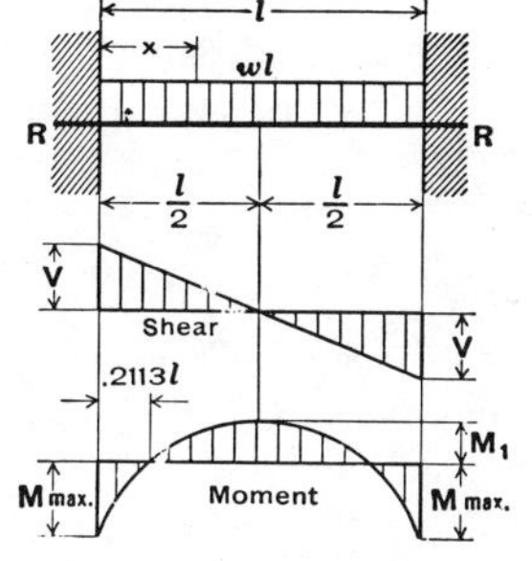

$R = V$	$= \frac{wl}{2}$
V_x	$= w\left(\frac{l}{2} - x\right)$
M max. (at ends)	$= \frac{wl^2}{12}$
M_1 (at center)	$= \frac{wl^2}{24}$
M_x	$= \frac{w}{12}(6lx - l^2 - 6x^2)$
Δmax. (at center)	$= \frac{wl^4}{384EI}$
Δx	$= \frac{wx^2}{24EI}(l - x)^2$

16 BEAM FIXED AT BOTH ENDS—CONCENTRATED LOAD AT CENTER

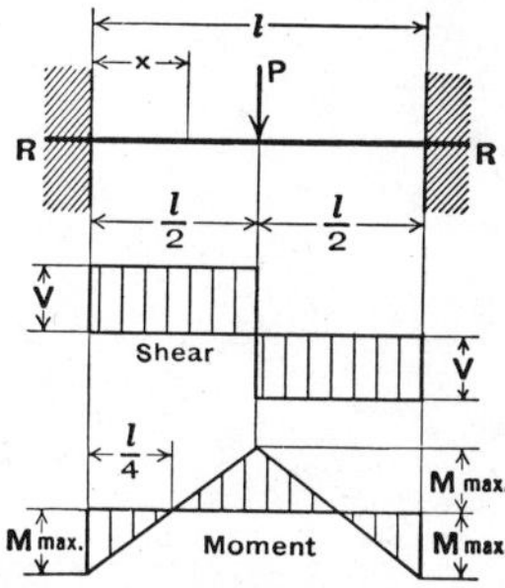

$R = V$	$= \frac{P}{2}$
M max. (at center and ends)	$= \frac{Pl}{8}$
M_x (when $x < \frac{l}{2}$)	$= \frac{P}{8}(4x - l)$
Δmax. (at center)	$= \frac{Pl^3}{192EI}$
Δx	$= \frac{Px^2}{48EI}(3l - 4x)$

17. BEAM FIXED AT BOTH ENDS—CONCENTRATED LOAD AT ANY POINT

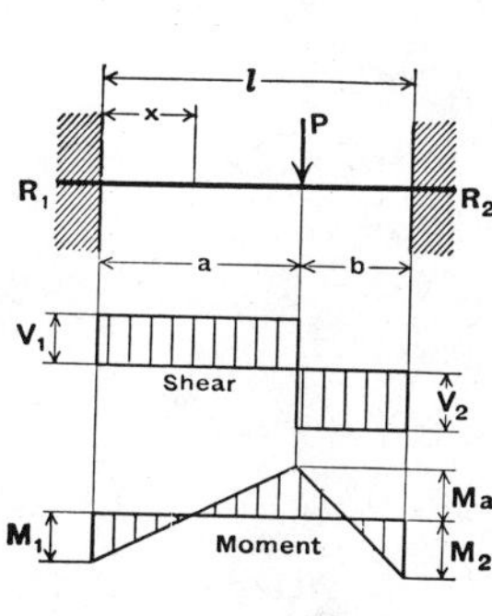

$R_1 = V_1$ (max. when $a < b$)	$= \frac{Pb^2}{l^3}(3a + b)$
$R_2 = V_2$ (max. when $a > b$)	$= \frac{Pa^2}{l^3}(a + 3b)$
M_1 (max. when $a < b$)	$= \frac{Pab^2}{l^2}$
M_2 (max. when $a > b$)	$= \frac{Pa^2b}{l^2}$
M_a (at point of load)	$= \frac{2Pa^2b^2}{l^3}$
M_x (when $x < a$)	$= R_1x - \frac{Pab^2}{l^2}$
Δmax. (when $a > b$ at $x = \frac{2al}{3a + b}$)	$= \frac{2Pa^3b^2}{3EI\,(3a + b)^2}$
Δa (at point of load)	$= \frac{Pa^3b^3}{3EIl^3}$
Δx (when $x < a$)	$= \frac{Pb^2x^2}{6EIl^3}(3al - 3ax - bx)$

TABLE 7.10 (*Continued*)

18. CANTILEVER BEAM—LOAD INCREASING UNIFORMLY TO FIXED END

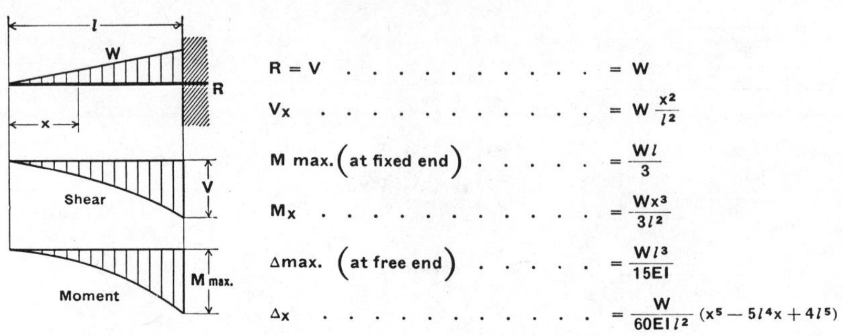

$R = V$ . . . . $= W$

V_x . . . . $= W \frac{x^2}{l^2}$

M max. (at fixed end) . . . . . . $= \frac{Wl}{3}$

M_x . . . . $= \frac{Wx^3}{3l^2}$

Δmax. (at free end) . . . . . . $= \frac{Wl^3}{15EI}$

Δx . . . . $= \frac{W}{60EIl^2}(x^5 - 5l^4x + 4l^5)$

19. CANTILEVER BEAM—UNIFORMLY DISTRIBUTED LOAD

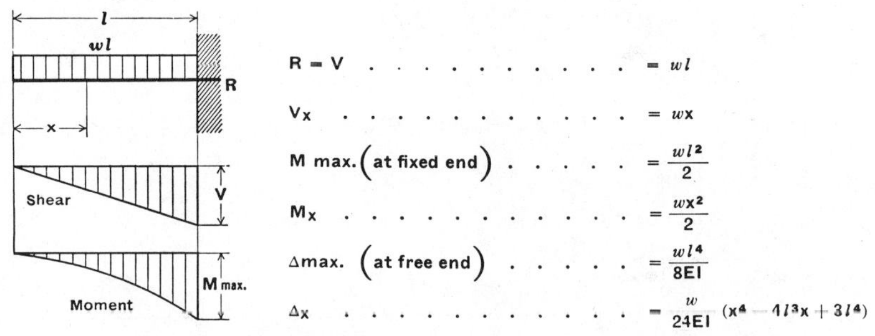

$R = V$ . . . . $= wl$

V_x . . . . $= wx$

M max. (at fixed end) . . . . . . $= \frac{wl^2}{2}$

M_x . . . . $= \frac{wx^2}{2}$

Δmax. (at free end) . . . . . . $= \frac{wl^4}{8EI}$

Δx . . . . $= \frac{w}{24EI}(x^4 - 4l^3x + 3l^4)$

20. BEAM FIXED AT ONE END, FREE TO DEFLECT VERTICALLY BUT NOT ROTATE AT OTHER—UNIFORMLY DISTRIBUTED LOAD

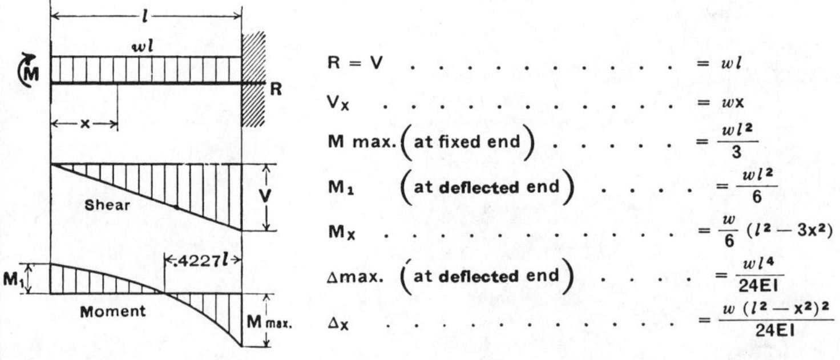

$R = V$ . . . . $= wl$

V_x . . . . $= wx$

M max. (at fixed end) . . . . . . $= \frac{wl^2}{3}$

M_1 (at deflected end) . . . . . $= \frac{wl^2}{6}$

M_x . . . . $= \frac{w}{6}(l^2 - 3x^2)$

Δmax. (at deflected end) . . . . . $= \frac{wl^4}{24EI}$

Δx . . . . $= \frac{w(l^2 - x^2)^2}{24EI}$

TABLE 7.10 (*Continued*)

21. CANTILEVER BEAM—CONCENTRATED LOAD AT ANY POINT

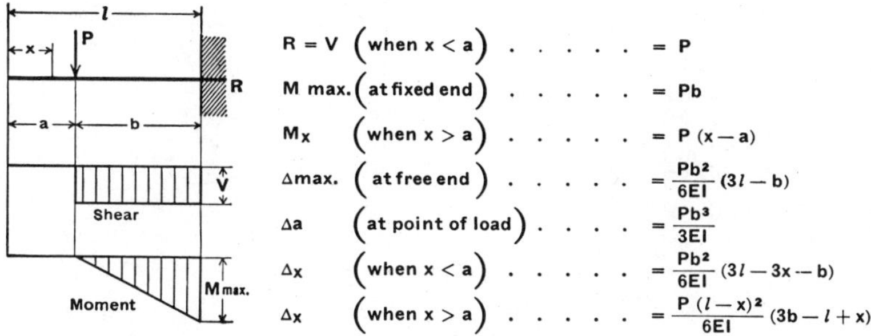

$R = V \left(\text{when } x < a\right) \ldots\ldots = P$

$M \text{ max.} \left(\text{at fixed end}\right) \ldots\ldots = Pb$

$M_x \left(\text{when } x > a\right) \ldots\ldots = P\,(x - a)$

$\Delta\text{max.} \left(\text{at free end}\right) \ldots\ldots = \frac{Pb^2}{6EI}(3l - b)$

$\Delta a \left(\text{at point of load}\right) \ldots\ldots = \frac{Pb^3}{3EI}$

$\Delta x \left(\text{when } x < a\right) \ldots\ldots = \frac{Pb^2}{6EI}(3l - 3x - b)$

$\Delta x \left(\text{when } x > a\right) \ldots\ldots = \frac{P\,(l - x)^2}{6EI}(3b - l + x)$

22. CANTILEVER BEAM—CONCENTRATED LOAD AT FREE END

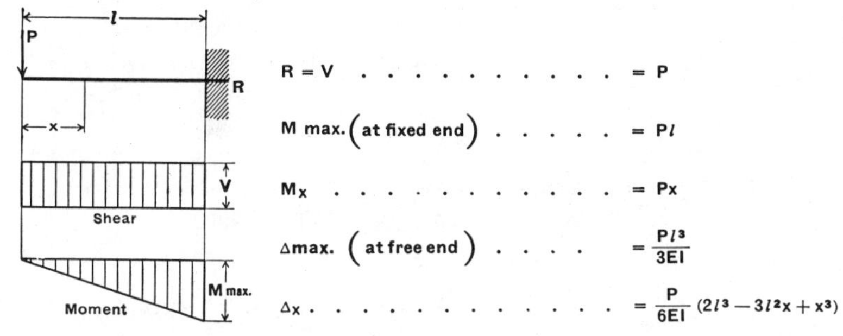

$R = V \ldots\ldots\ldots\ldots = P$

$M \text{ max.} \left(\text{at fixed end}\right) \ldots\ldots = Pl$

$M_x \ldots\ldots\ldots\ldots = Px$

$\Delta\text{max.} \left(\text{at free end}\right) \ldots\ldots = \frac{Pl^3}{3EI}$

$\Delta x \ldots\ldots\ldots\ldots = \frac{P}{6EI}(2l^3 - 3l^2x + x^3)$

23. BEAM FIXED AT ONE END, FREE TO DEFLECT VERTICALLY BUT NOT ROTATE AT OTHER—CONCENTRATED LOAD AT DEFLECTED END

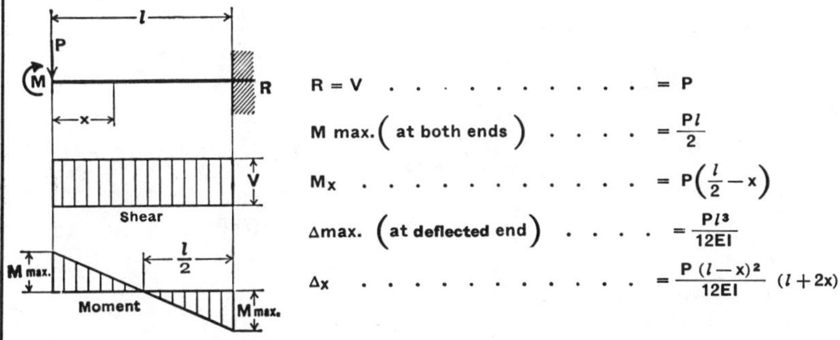

$R = V \ldots\ldots\ldots\ldots = P$

$M \text{ max.} \left(\text{at both ends}\right) \ldots\ldots = \frac{Pl}{2}$

$M_x \ldots\ldots\ldots\ldots = P\left(\frac{l}{2} - x\right)$

$\Delta\text{max.} \left(\text{at deflected end}\right) \ldots\ldots = \frac{Pl^3}{12EI}$

$\Delta x \ldots\ldots\ldots\ldots = \frac{P\,(l - x)^2}{12EI}(l + 2x)$

TABLE 7.10 (*Continued*)

24. BEAM OVERHANGING ONE SUPPORT—UNIFORMLY DISTRIBUTED LOAD

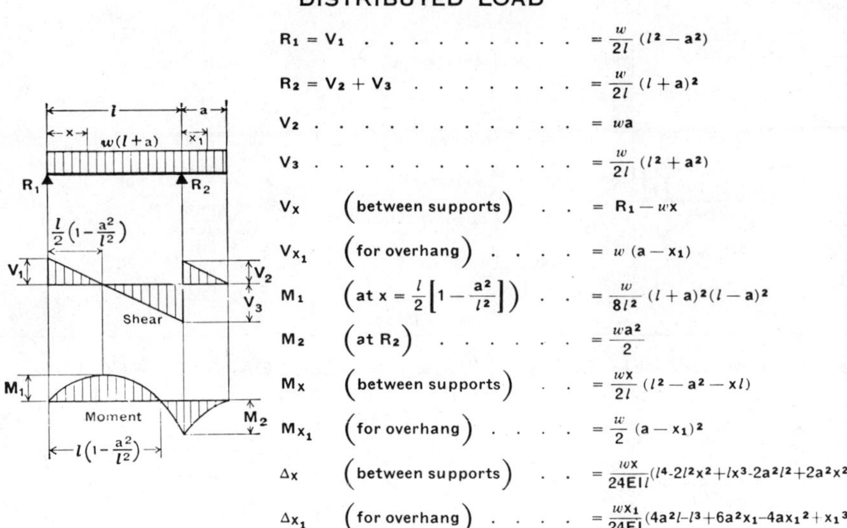

$R_1 = V_1$	. . . .	$= \frac{w}{2l}(l^2 - a^2)$
$R_2 = V_2 + V_3$	. . . .	$= \frac{w}{2l}(l + a)^2$
V_2	. . . .	$= wa$
V_3	. . . .	$= \frac{w}{2l}(l^2 + a^2)$
V_x	(between supports) . .	$= R_1 - wx$
V_{x_1}	(for overhang)	$= w(a - x_1)$
M_1	$\left(\text{at } x = \frac{l}{2}\left[1 - \frac{a^2}{l^2}\right]\right)$ . .	$= \frac{w}{8l^2}(l + a)^2(l - a)^2$
M_2	(at R_2)	$= \frac{wa^2}{2}$
M_x	(between supports) . .	$= \frac{wx}{2l}(l^2 - a^2 - xl)$
M_{x_1}	(for overhang)	$= \frac{w}{2}(a - x_1)^2$
Δx	(between supports) . .	$= \frac{wx}{24EIl}(l^4 - 2l^2x^2 + lx^3 - 2a^2l^2 + 2a^2x^2)$
Δx_1	(for overhang)	$= \frac{wx_1}{24EI}(4a^2l - l^3 + 6a^2x_1 - 4ax_1^2 + x_1^3)$

25. BEAM OVERHANGING ONE SUPPORT—UNIFORMLY DISTRIBUTED LOAD ON OVERHANG

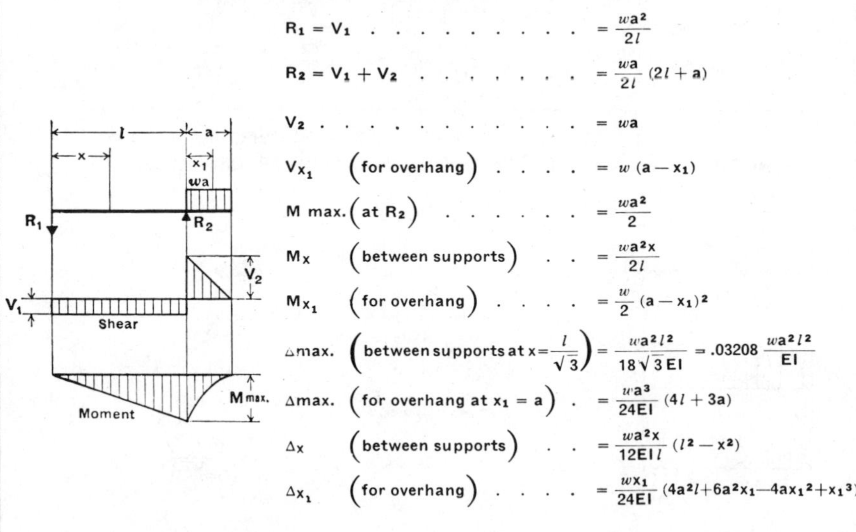

$R_1 = V_1$	. . . .	$= \frac{wa^2}{2l}$
$R_2 = V_1 + V_2$	. . . .	$= \frac{wa}{2l}(2l + a)$
V_2	. . . .	$= wa$
V_{x_1}	(for overhang)	$= w(a - x_1)$
M max.	(at R_2)	$= \frac{wa^2}{2}$
M_x	(between supports) . .	$= \frac{wa^2x}{2l}$
M_{x_1}	(for overhang)	$= \frac{w}{2}(a - x_1)^2$
Δmax.	$\left(\text{between supports at } x = \frac{l}{\sqrt{3}}\right)$	$= \frac{wa^2l^2}{18\sqrt{3}EI} = .03208\frac{wa^2l^2}{EI}$
Δmax.	(for overhang at $x_1 = a$) .	$= \frac{wa^3}{24EI}(4l + 3a)$
Δx	(between supports) . .	$= \frac{wa^2x}{12EIl}(l^2 - x^2)$
Δx_1	(for overhang)	$= \frac{wx_1}{24EI}(4a^2l + 6a^2x_1 - 4ax_1^2 + x_1^3)$

TABLE 7.10 (*Continued*)

26. BEAM OVERHANGING ONE SUPPORT—CONCENTRATED LOAD AT END OF OVERHANG

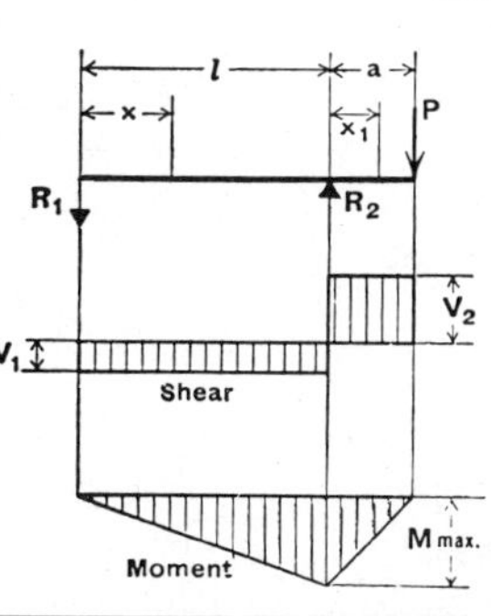

$R_1 = V_1$	$= \frac{Pa}{l}$
$R_2 = V_1 + V_2$	$= \frac{P}{l}(l + a)$
V_2	$= P$
M max. (at R_2)	$= Pa$
M_x (between supports) . .	$= \frac{Pax}{l}$
M_{x_1} (for overhang)	$= P(a - x_1)$
Δmax. (between supports at $x = \frac{l}{\sqrt{3}}$)	$= \frac{Pal^2}{9\sqrt{3}EI} = .06415 \frac{Pal^2}{EI}$
Δmax. (for overhang at $x_1 = a$) .	$= \frac{Pa^2}{3EI}(l + a)$
Δx (between supports) . .	$= \frac{Pax}{6EIl}(l^2 - x^2)$
Δx_1 (for overhang)	$= \frac{Px_1}{6EI}(2al + 3ax_1 - x_1^2)$

27. BEAM OVERHANGING ONE SUPPORT—UNIFORMLY DISTRIBUTED LOAD BETWEEN SUPPORTS

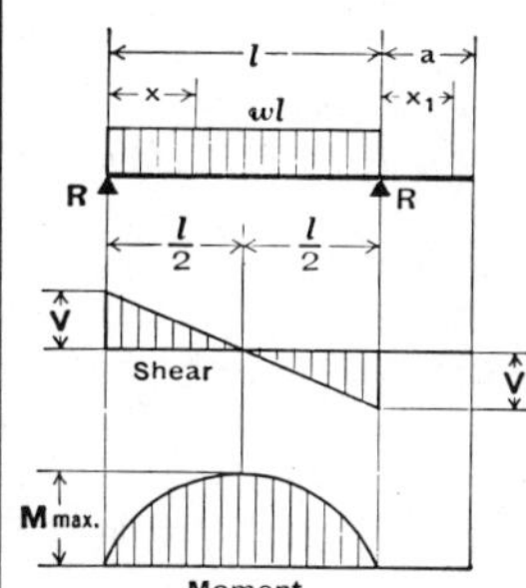

Equivalent Tabular Load . . .	$= wl$
$R = V$	$= \frac{wl}{2}$
V_x	$= w\left(\frac{l}{2} - x\right)$
M max. (at center)	$= \frac{wl^2}{8}$
M_x	$= \frac{wx}{2}(l - x)$
Δmax. (at center)	$= \frac{5wl^4}{384EI}$
Δx	$= \frac{wx}{24EI}(l^3 - 2lx^2 + x^3)$
Δx_1	$= \frac{wl^3x_1}{24EI}$

28. BEAM OVERHANGING ONE SUPPORT—CONCENTRATED LOAD AT ANY POINT BETWEEN SUPPORTS

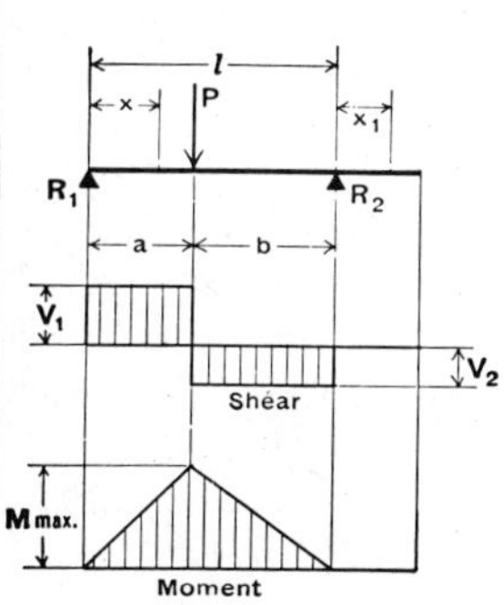

Equivalent Tabular Load . . .	$= \frac{8Pab}{l^2}$
$R_1 = V_1$ (max. when $a < b$) . . .	$= \frac{Pb}{l}$
$R_2 = V_2$ (max. when $a > b$) . . .	$= \frac{Pa}{l}$
M max. (at point of load) . . .	$= \frac{Pab}{l}$
M_x (when $x < a$)	$= \frac{Pbx}{l}$
Δmax. (at $x = \sqrt{\frac{a(a+2b)}{3}}$ when $a > b$)	$= \frac{Pab(a + 2b)\sqrt{3a(a + 2b)}}{27EIl}$
Δa (at point of load) . . .	$= \frac{Pa^2b^2}{3EIl}$
Δx (when $x < a$)	$= \frac{Pbx}{6EIl}(l^2 - b^2 - x^2)$
Δx (when $x > a$)	$= \frac{Pa(l - x)}{6EIl}(2lx - x^2 - a^2)$
Δx_1	$= \frac{Pabx_1}{6EIl}(l + a)$

TABLE 7.10 (*Continued*)

29. CONTINUOUS BEAM—TWO EQUAL SPANS—UNIFORM LOAD ON BOTH SPANS

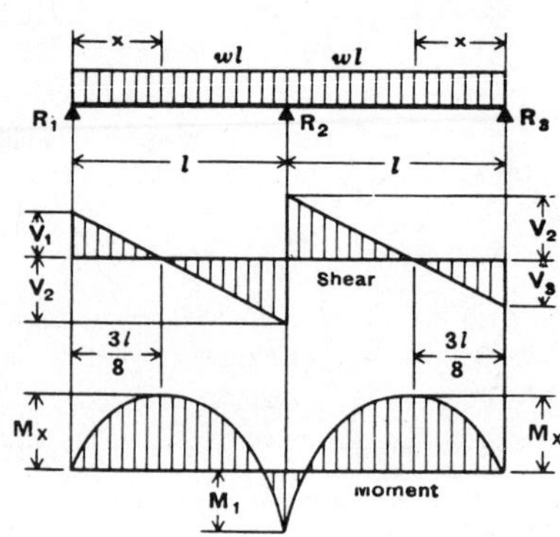

$R_1 = V_1 = R_3 = V_3 \quad . \quad . \quad = \frac{3}{8} wl$

$R_2 = 2V_2 \quad . \quad . \quad . \quad . \quad . \quad = \frac{10}{8} wl$

$V_2 \quad . \quad . \quad . \quad . \quad . \quad . \quad . \quad . \quad = \frac{5}{8} wl$

$M_x \quad . \quad . \quad . \quad . \quad . \quad . \quad . \quad = R_1 x - \frac{wx^2}{2}$

$M_x \left(\text{at } x = \frac{3l}{8}\right) \quad . \quad . \quad . \quad = \frac{9}{128} wl^2$

$M_1 \text{ (at support } R_2) \quad . \quad . \quad = -\frac{wl^2}{8}$

Δ Max. (0.4215l from R_1 or R_3) $= wl^4/185EI$

$\Delta_x = \frac{wx}{48EI} (l^3 - 3lx^2 + 2x^3)$

30. CONTINUOUS BEAM—TWO EQUAL SPANS—UNIFORM LOAD ON ONE SPAN

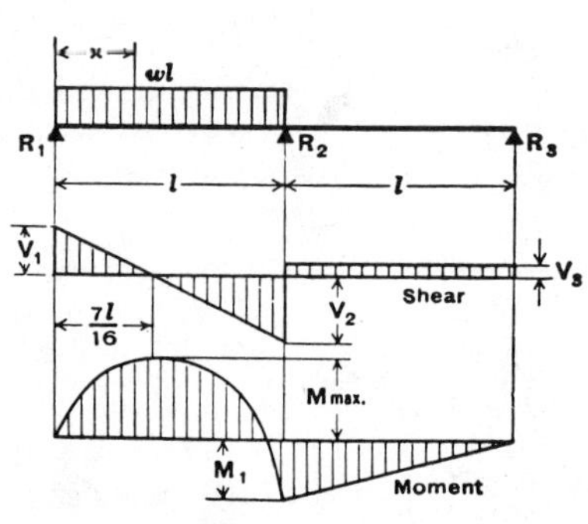

$R_1 = V_1 \quad . \quad . \quad . \quad . \quad . \quad . \quad = \frac{7}{16} wl$

$R_2 = V_2 + V_3 \quad . \quad . \quad . \quad . \quad = \frac{5}{8} wl$

$R_3 = V_3 \quad . \quad . \quad . \quad . \quad . \quad . \quad = -\frac{1}{16} wl$

$V_2 \quad . \quad . \quad . \quad . \quad . \quad . \quad . \quad . \quad = \frac{9}{16} wl$

$M \text{ Max.} \left(\text{at } x = \frac{7}{16} l\right) \quad . \quad = \frac{49}{512} wl^2$

$M_1 \text{ (at support } R_2) \quad . \quad . \quad = \frac{1}{16} wl^2$

$M_x \text{ (when } x < l) \quad . \quad . \quad = \frac{wx}{16} (7l - 8x)$

Δ Max. (0.472l from R_1) $= wl^4/109EI$

TABLE 7.10 (*Continued*)

31. CONTINUOUS BEAM—TWO EQUAL SPANS—CONCENTRATED LOAD AT CENTER OF ONE SPAN

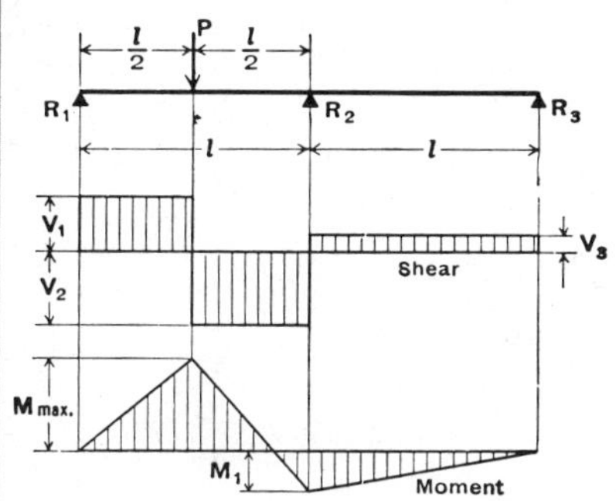

$R_1 = V_1$ . . . . $= \frac{13}{32} P$

$R_2 = V_2 + V_3$ . . . . . . $= \frac{11}{16} P$

$R_3 = V_3$ . . . . $= -\frac{3}{32} P$

V_2 . . . . $= \frac{19}{32} P$

M Max. (at point of load) . $= \frac{13}{64} Pl$

M_1 (at support R_2) . $= \frac{3}{32} Pl$

Δ Max. (0.480 l from R_1) = 0.015 Pl^3/EI

32. CONTINUOUS BEAM—TWO EQUAL SPANS—CONCENTRATED LOAD AT ANY POINT

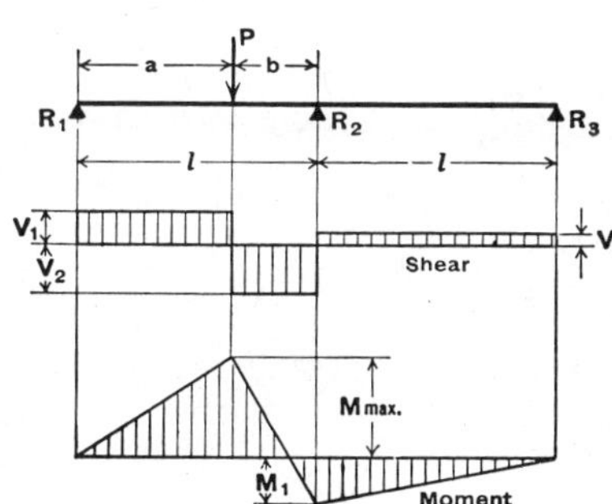

$R_1 = V_1$ . . . . $= \frac{Pb}{4l^3}\left(4l^2 - a(l+a)\right)$

$R_2 = V_2 + V_3$ . . . . . . $= \frac{Pa}{2l^3}\left(2l^2 + b(l+a)\right)$

$R_3 = V_3$ . . . . $= -\frac{Pab}{4l^3}(l + a)$

V_2 . . . . $= \frac{Pa}{4l^3}\left(4l^2 + b(l+a)\right)$

M max. (at point of load) . $= \frac{Pab}{4l^3}\left(4l^2 - a(l+a)\right)$

M_1 (at support R_2) . $= \frac{Pab}{4l^2}(l + a)$

33. BEAM—UNIFORMLY DISTRIBUTED LOAD AND VARIABLE END MOMENTS

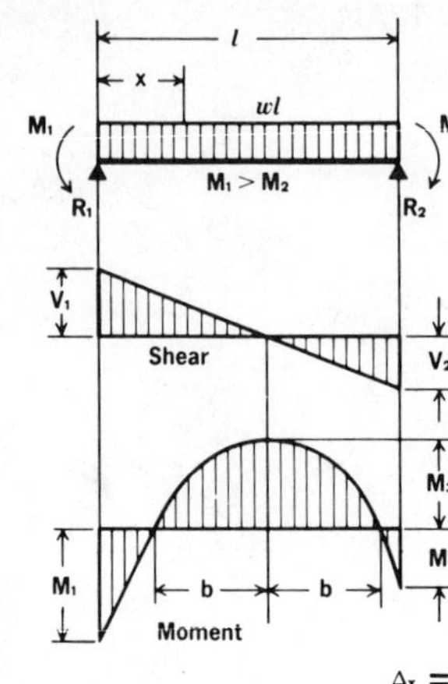

$$R_1 = V_1 = \frac{wl}{2} + \frac{M_1 - M_2}{l}$$

$$R_2 = V_2 = \frac{wl}{2} - \frac{M_1 - M_2}{l}$$

$$V_x = w\left(\frac{l}{2} - x\right) + \frac{M_1 - M_2}{l}$$

$$M_3\left(\text{at } x = \frac{l}{2} + \frac{M_1 - M_2}{wl}\right)$$

$$= \frac{wl^2}{8} - \frac{M_1 + M_2}{2} + \frac{(M_1 - M_2)^2}{2wl^2}$$

$$M_x = \frac{wx}{2}(l - x) + \left(\frac{M_1 - M_2}{l}\right)x - M_1$$

$$b\left(\begin{matrix}\text{To locate}\\ \text{inflection points}\end{matrix}\right) = \sqrt{\frac{l^2}{4} - \left(\frac{M_1 + M_2}{w}\right) + \left(\frac{M_1 - M_2}{wl}\right)^2}$$

$$\Delta_x = \frac{wx}{24EI}\left[x^3 - \left(2l + \frac{4M_1}{wl} - \frac{4M_2}{wl}\right)x^2 + \frac{12M_1}{w}x + l^3 - \frac{8M_1 l}{w} - \frac{4M_2 l}{w}\right]$$

34. BEAM—CONCENTRATED LOAD AT CENTER AND VARIABLE END MOMENTS

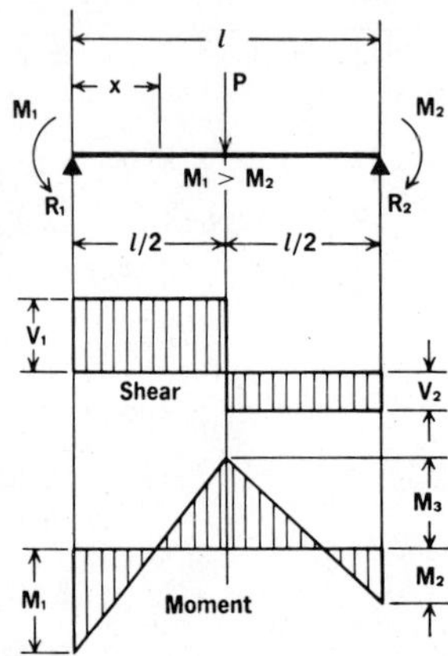

$$R_1 = V_1 = \frac{P}{2} + \frac{M_1 - M_2}{l}$$

$$R_2 = V_2 = \frac{P}{2} - \frac{M_1 - M_2}{l}$$

$$M_3\ (\text{At center}) = \frac{Pl}{4} - \frac{M_1 + M_2}{2}$$

$$M_x\left(\text{When } x < \frac{l}{2}\right) = \left(\frac{P}{2} + \frac{M_1 - M_2}{l}\right)x - M_1$$

$$M_x\left(\text{When } x > \frac{l}{2}\right) = \frac{P}{2}(l - x) + \frac{(M_1 - M_2)x}{l} - M_1$$

$$\Delta_x\left(\text{When } x < \frac{l}{2}\right) = \frac{Px}{48EI}\left(3l^2 - 4x^2 - \frac{8(l - x)}{Pl}[M_1(2l - x) + M_2(l + x)]\right)$$

TABLE 7.10 (*Continued*)

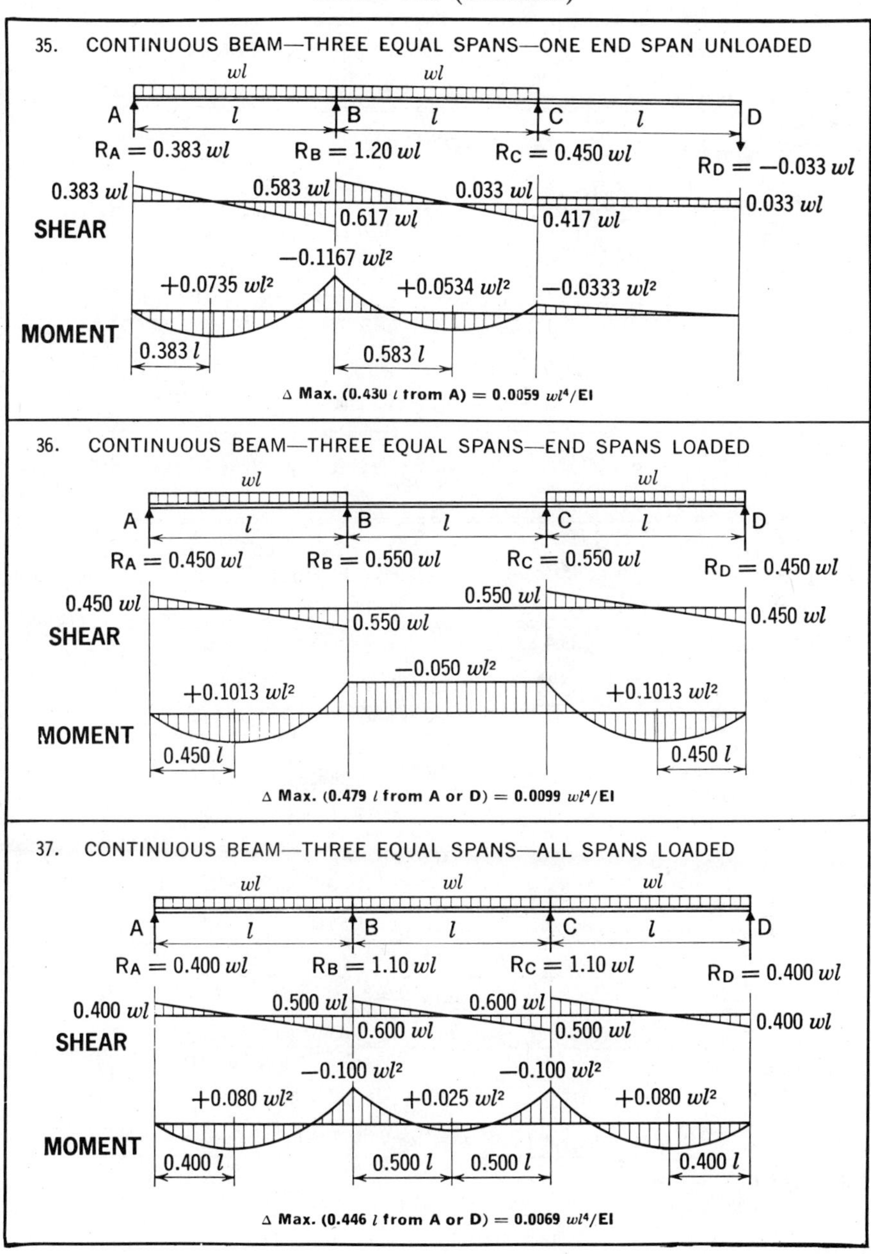

TABLE 7.10 (*Continued*)

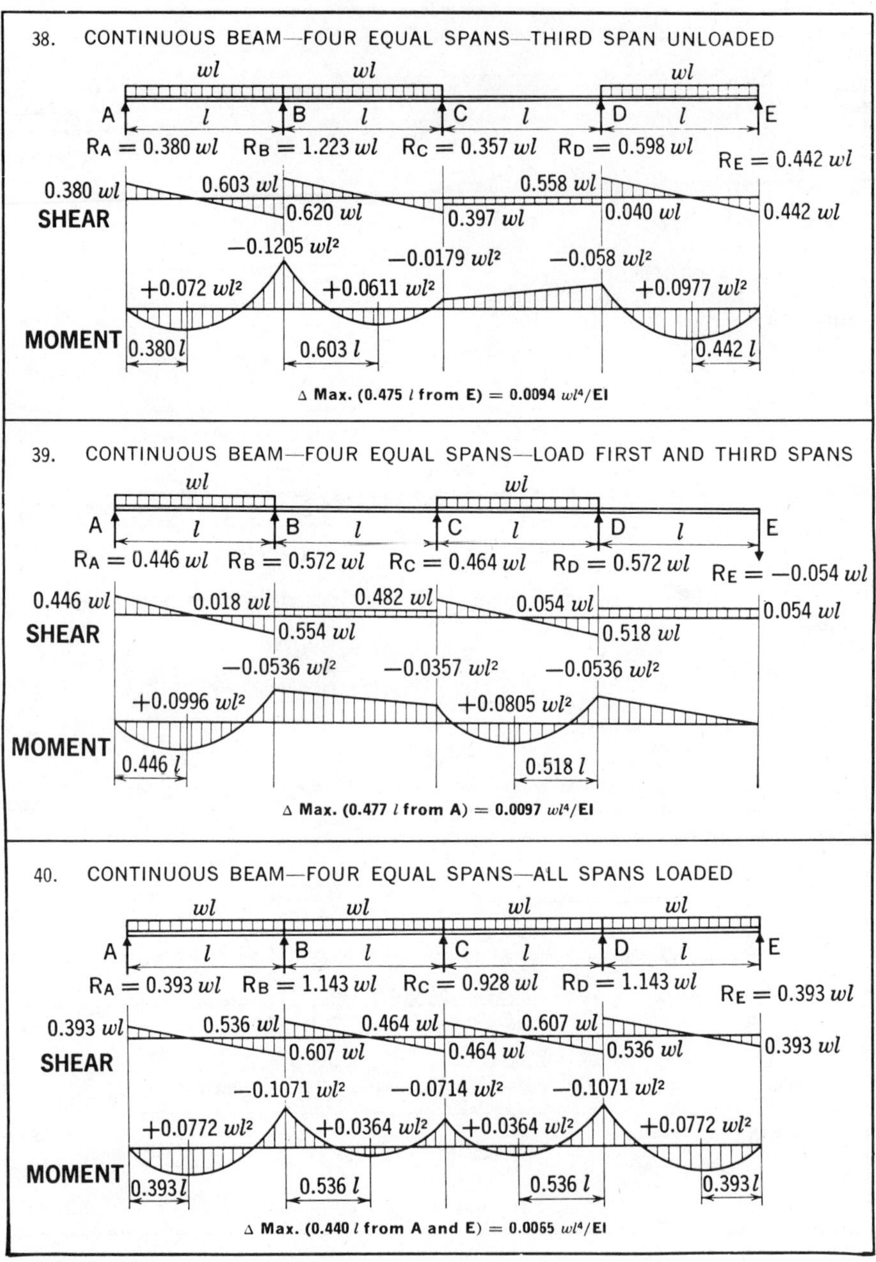

[a]For meaning of symbols, see General Nomenclature. (Source: American Institute of Steel Construction, Inc.)

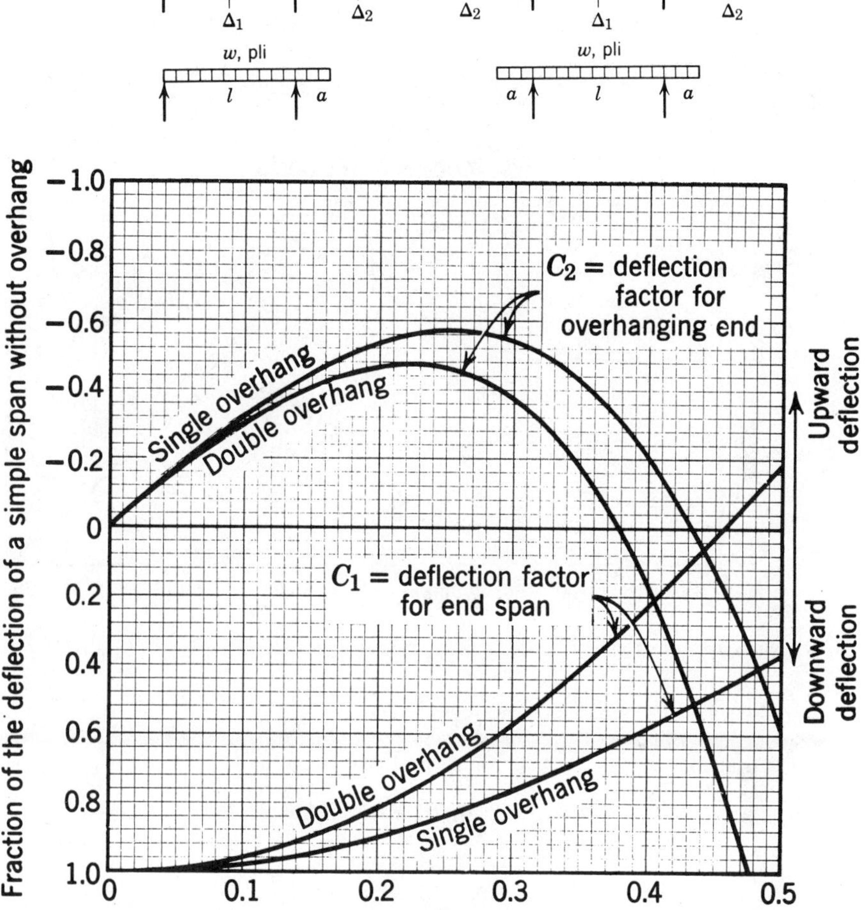

FIGURE 7.1 Effect of overhang on maximum deflection for a simple-span system.

Without overhang $\Delta = \dfrac{5wl^4}{384EI}$

Δ_1 = $C_1\Delta$, deflection between supports with overhang (in.)
Δ_2 = $C_2\Delta$, deflection of overhanging end (in.)
w = uniformly distributed load (pli)
l = span (in.)
a = overhang (in.)
Δ = deflection (in.)
E = modulus of elasticity (psi)
I = moment of inertia (in.^{4})

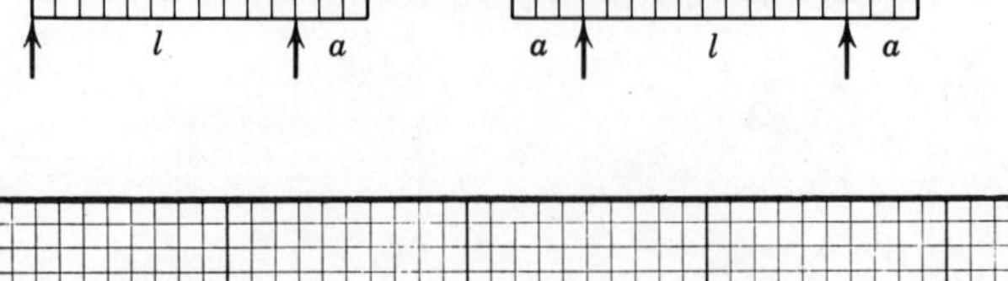

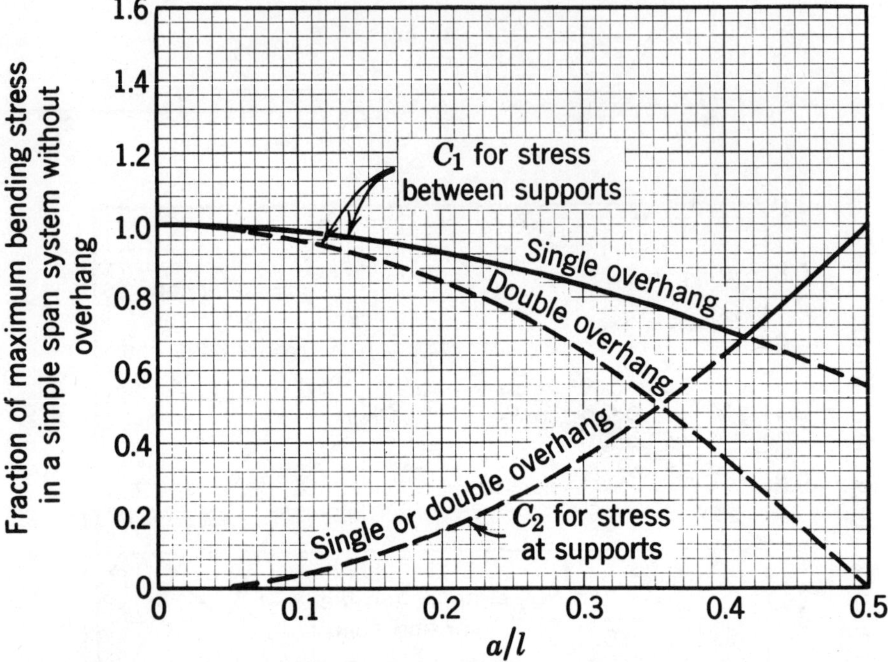

FIGURE 7.2. Effect of overhang on maximum bending stress for a simple-span system.

Maximum bending stress without overhangs: $f_b = \dfrac{wl^2c}{8I}$

Maximum bending stress with overhangs: $f_b \text{ max} = C_1 f_b$ for $0 < a/l < 0.408$

$f_b \text{ max} = C_2 f_b$ for $0.408 < a/l$

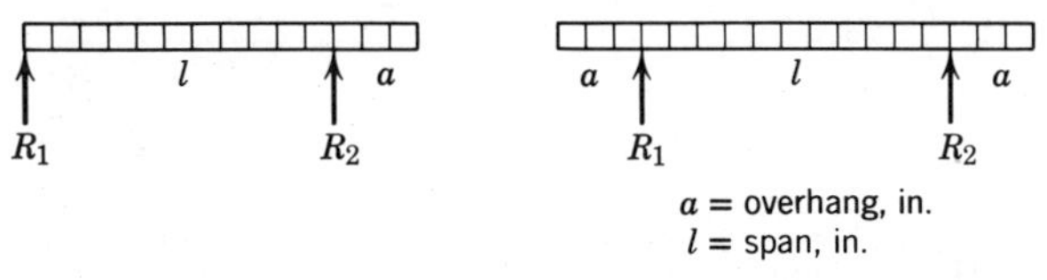

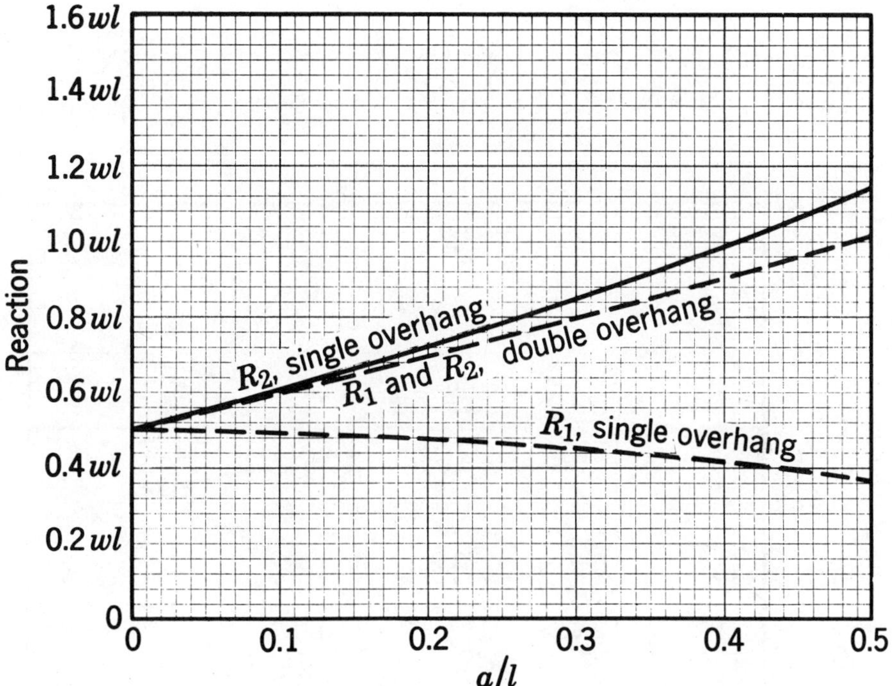

FIGURE 7.3. Effect of overhang on reactions for a simple-span system.

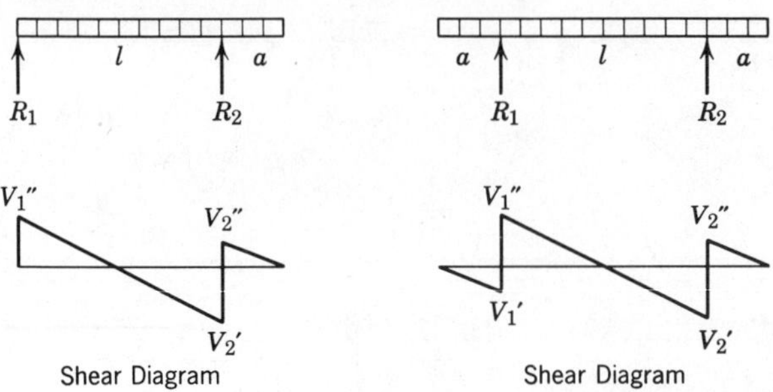

FIGURE 7.4. Effect of overhang on maximum vertical shear for a simple-span system.

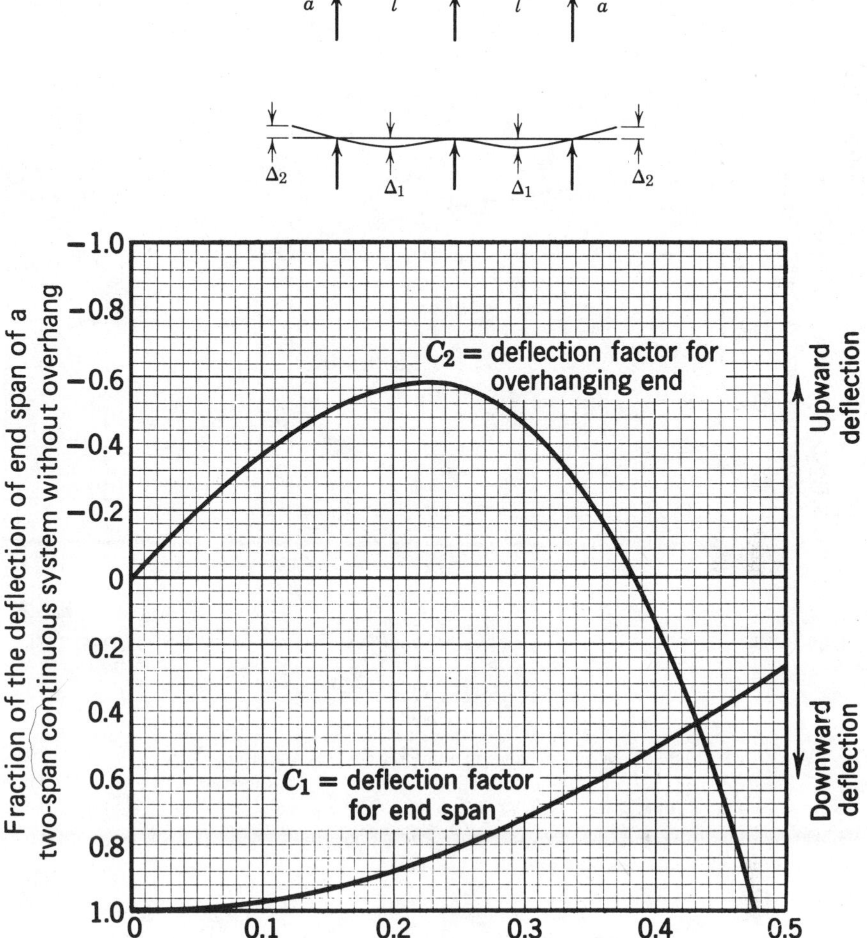

FIGURE 7.5. Effect of overhang on maximum deflection for a two-span continuous system.

Δ = deflection of end span of two-span continuous system without overhangs (in.)
Δ_1 = $C_1\Delta$ = deflection of end span with overhang (in.)
Δ_2 = $C_2\Delta$ = deflection of overhanging end (in.)

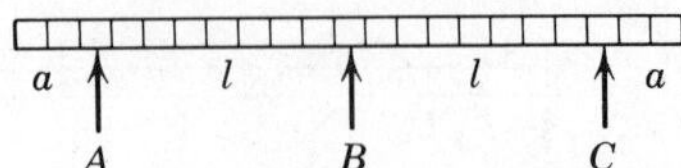

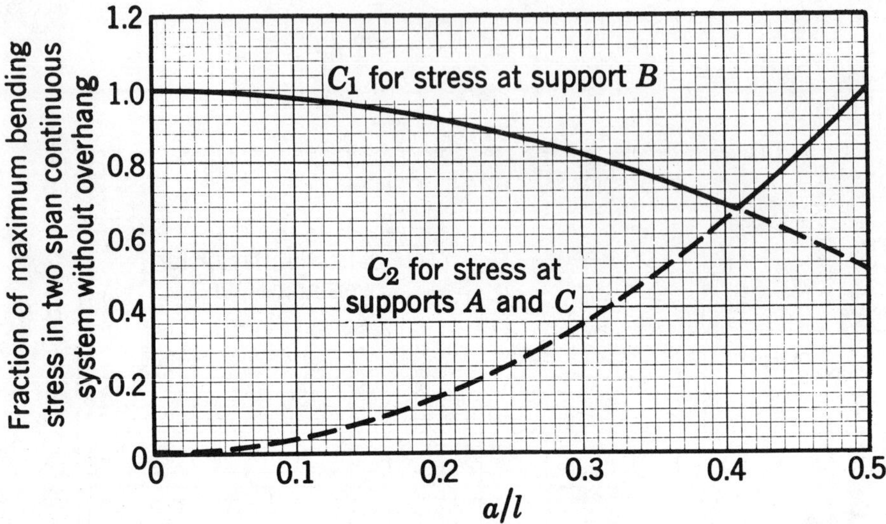

FIGURE 7.6. Effect of overhang on maximum bending stress for a two-span continuous system.

$$f_b = \frac{wl^2c}{8I}$$

Note: With no overhang, the maximum bending moment occurs at support B. With overhang, the bending moment at B decreases, and the bending moments at A and C increase. The bending moment at B is larger than at A and C for values of a/l less than 0.408. At a/l greater than 0.408, the bending moments at A and C exceed that at B. In Figure 7.6, the maximum imposed stress is expressed as a fraction of the bending stress over support B when $a = 0$.

$$f_{bmax} = C_1 f_b \quad \text{for } 0 < a/l < 0.408$$
$$f_{bmax} = C_2 f_b \quad \text{for } 0.408 < a/l$$

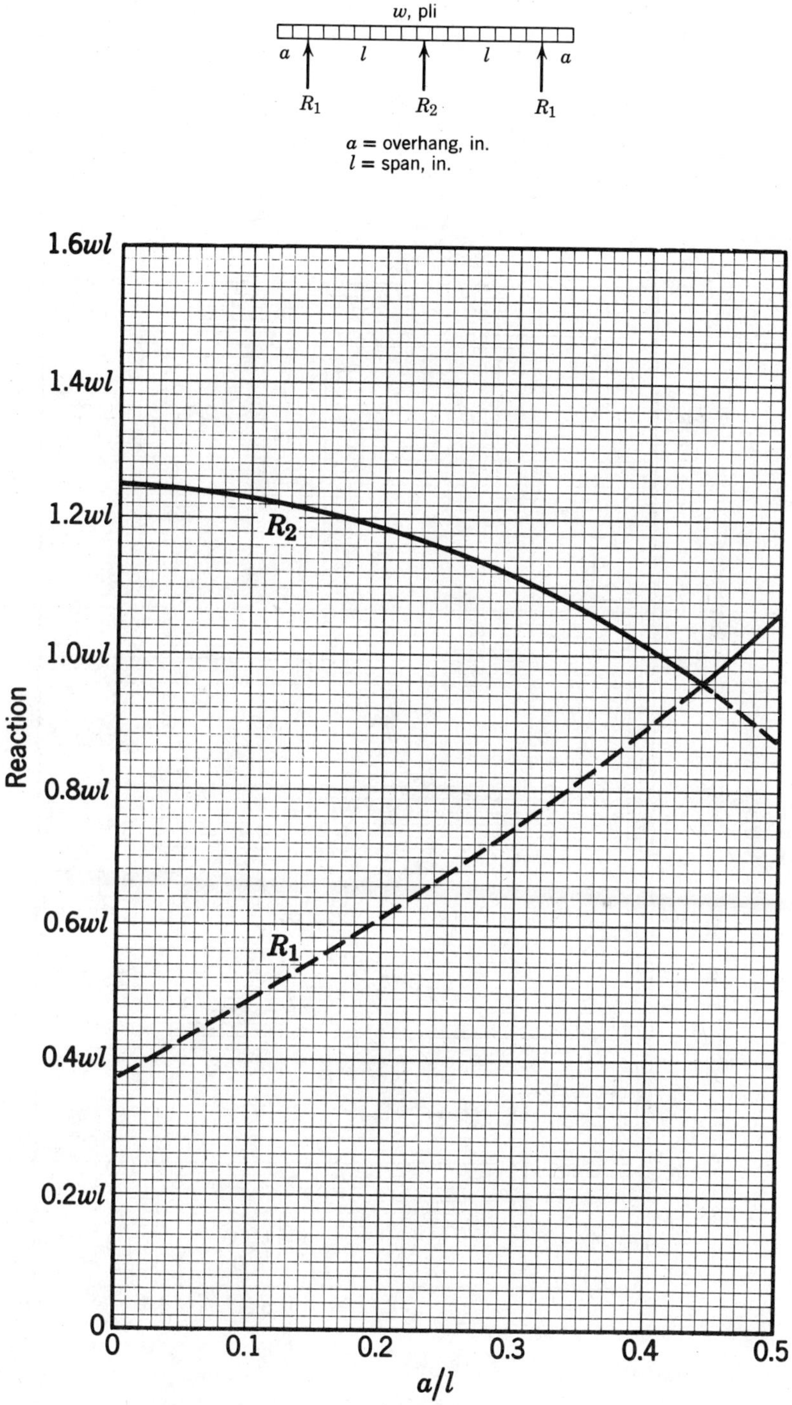

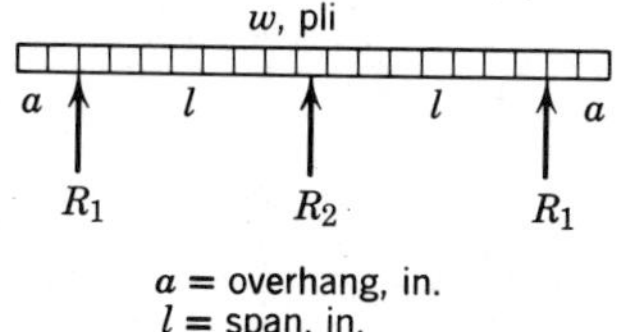

w, pli
a
l
l
a
R1
R2
R1
a = overhang, in.
l = span, in.
1.6wl
1.4wl
1.2wl
1.0wl
0.8wl
0.6wl
0.4wl
0.2wl
0
Reaction
R2
R1
0
0.1
0.2
0.3
0.4
0.5
a/l

FIGURE 7.7. Effect of overhang on reactions for a two-span continuous system.

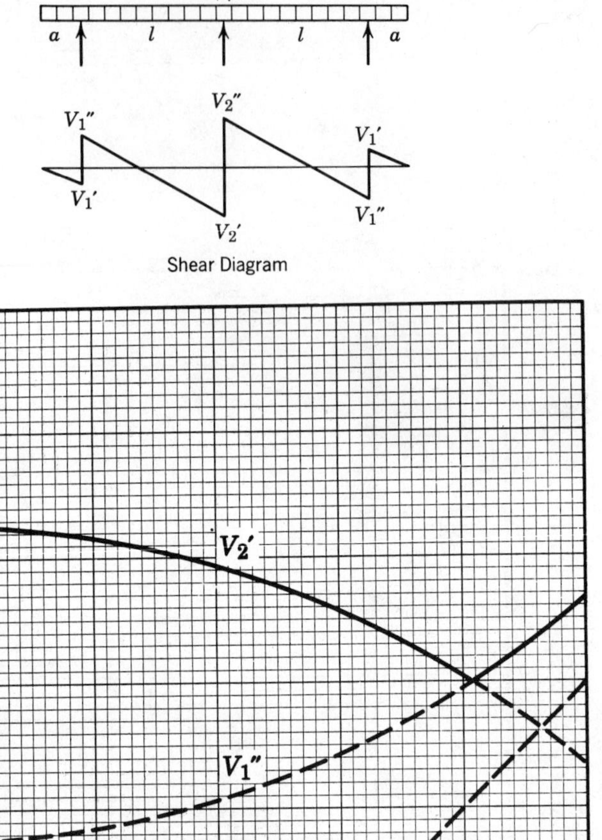

FIGURE 7.8. Effect of overhang on maximum vertical shear for a two-span continuous system. *Note:* Maximum shear occurs at support 2 for values of a/l less than 0.408. At higher values of a/l, maximum shear is at support 1.

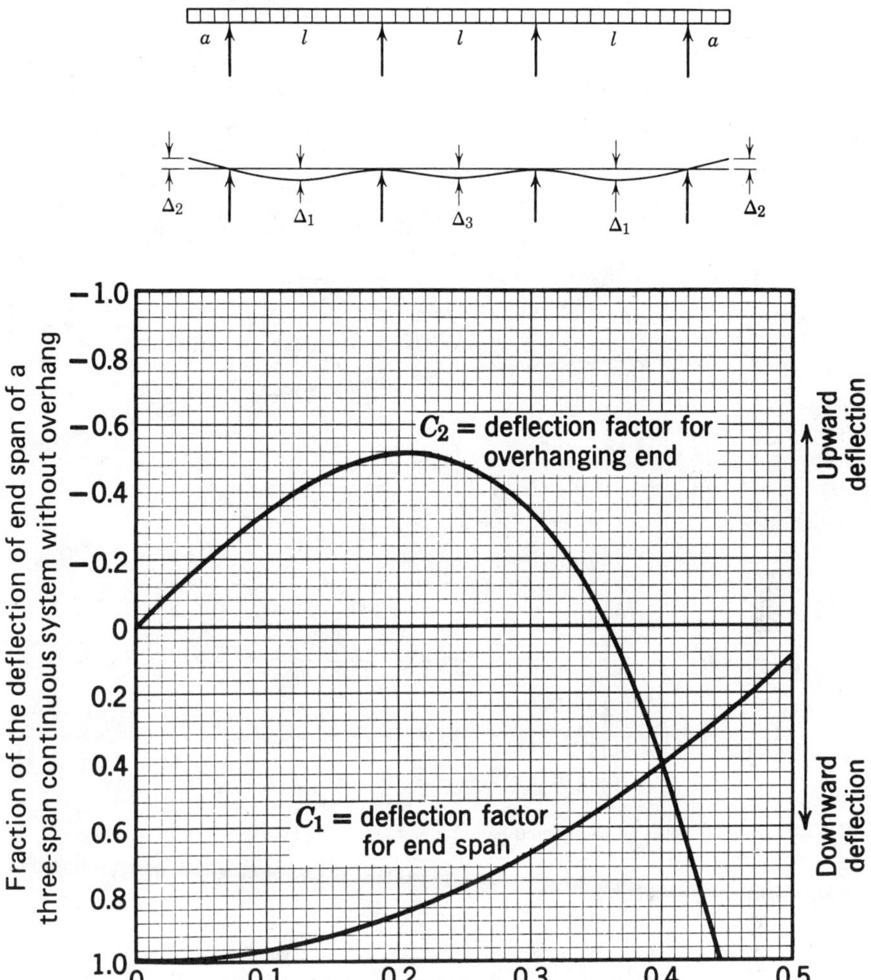

FIGURE 7.9. Effect of overhang on maximum deflection for a three-span continuous system.

Δ = deflection of end span of three-span continuous system, without overhangs (in.)

$\Delta_1 = C_1\Delta$ = deflection of end span with overhang (in.)

$\Delta_2 = C_2\Delta$ = deflection of overhanging end (in.)

$\Delta_1 > \Delta_3$

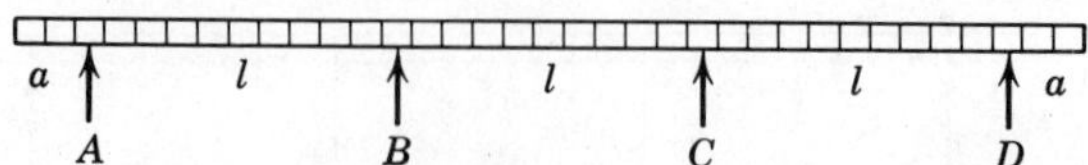

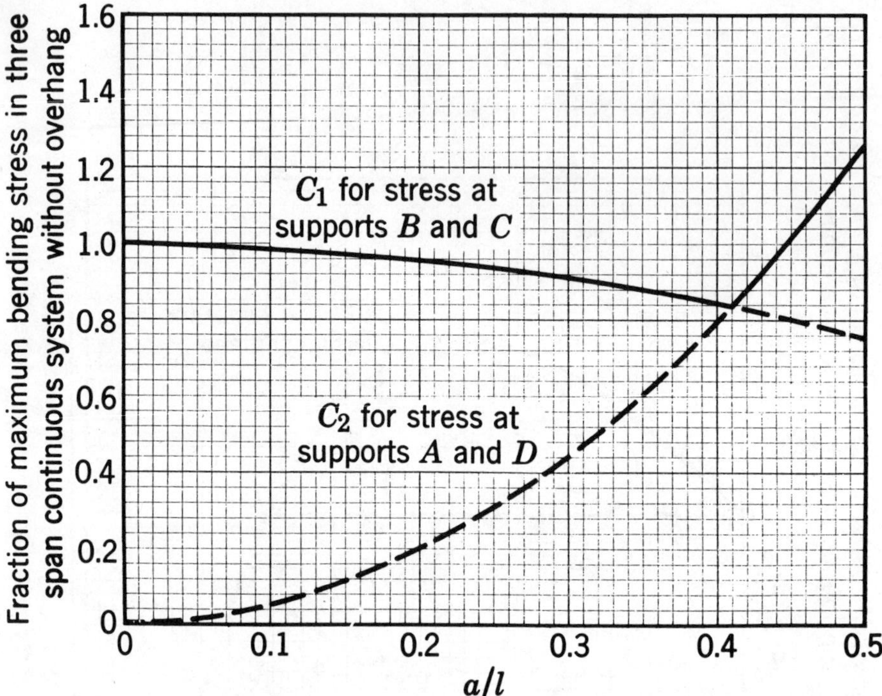

FIGURE 7.10. Effect of overhang on maximum bending stress for a three-span continuous system.

$$f_b = \frac{wl^2c}{10I}$$

Note: With no overhang, the maximum bending moment occurs at supports B and C. With overhang, the bending moments at B and C decrease, and the bending moments at A and D increase. The bending moments at B and C are larger than at A and D for values of a/l less than 0.408. At a/l greater than 0.408, the bending moments at A and D exceed those at B and C. In the Figure 7.10, the maximum imposed stress is expressed as a fraction of the bending stress over supports B and C when $a = 0$.

$$f_{b\,max} = C_1 f_b \quad \text{for } 0 < a/l < 0.408$$
$$f_{b\,max} = C_2 f_b \quad \text{for } 0.408 < a/l$$

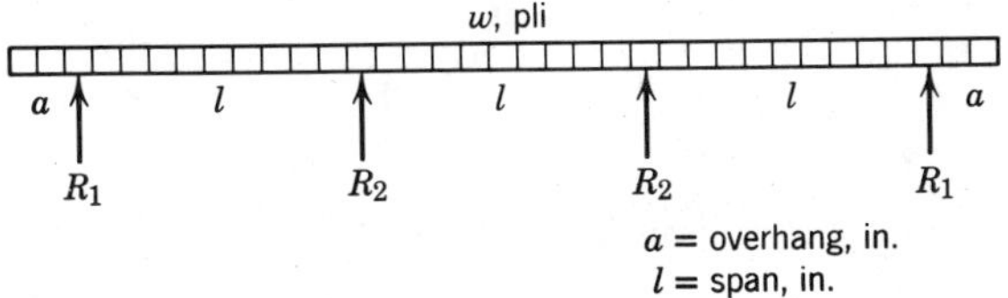

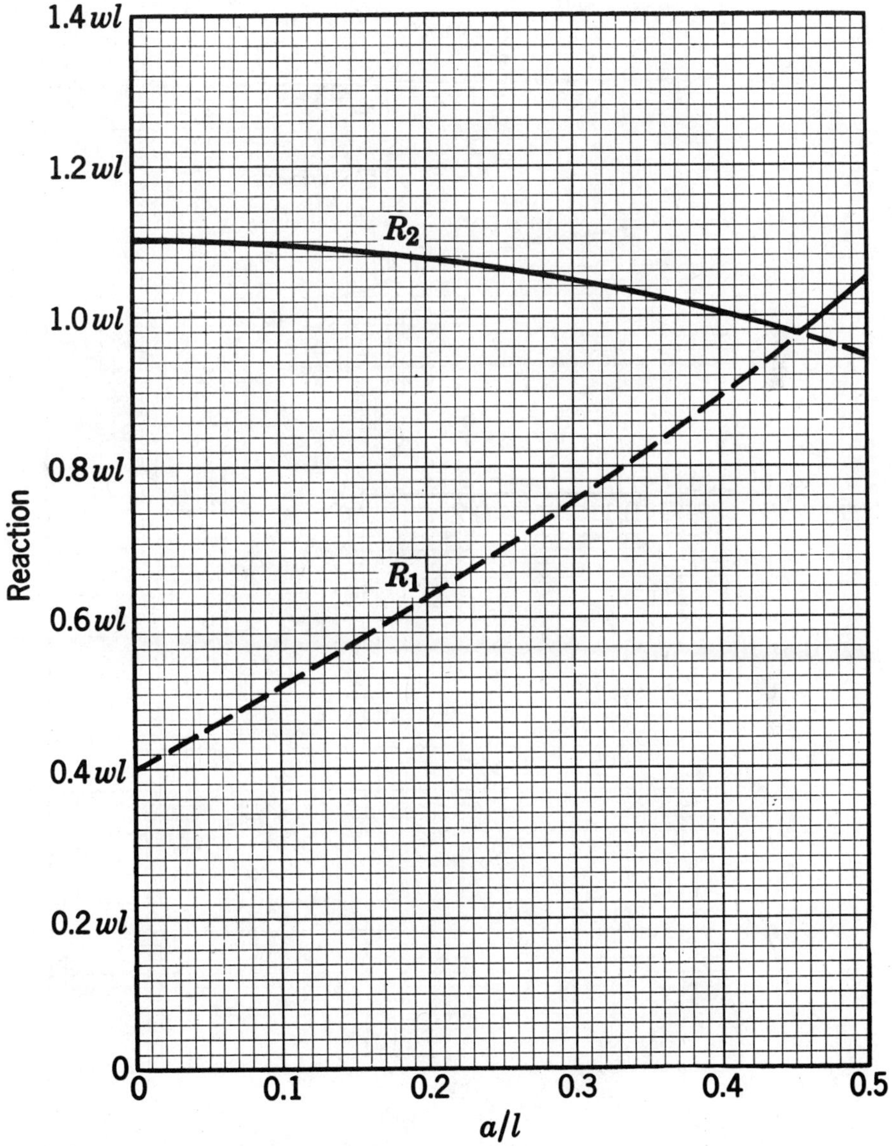

FIGURE 7.11. Effect of overhang on reactions for a three-span continuous system.

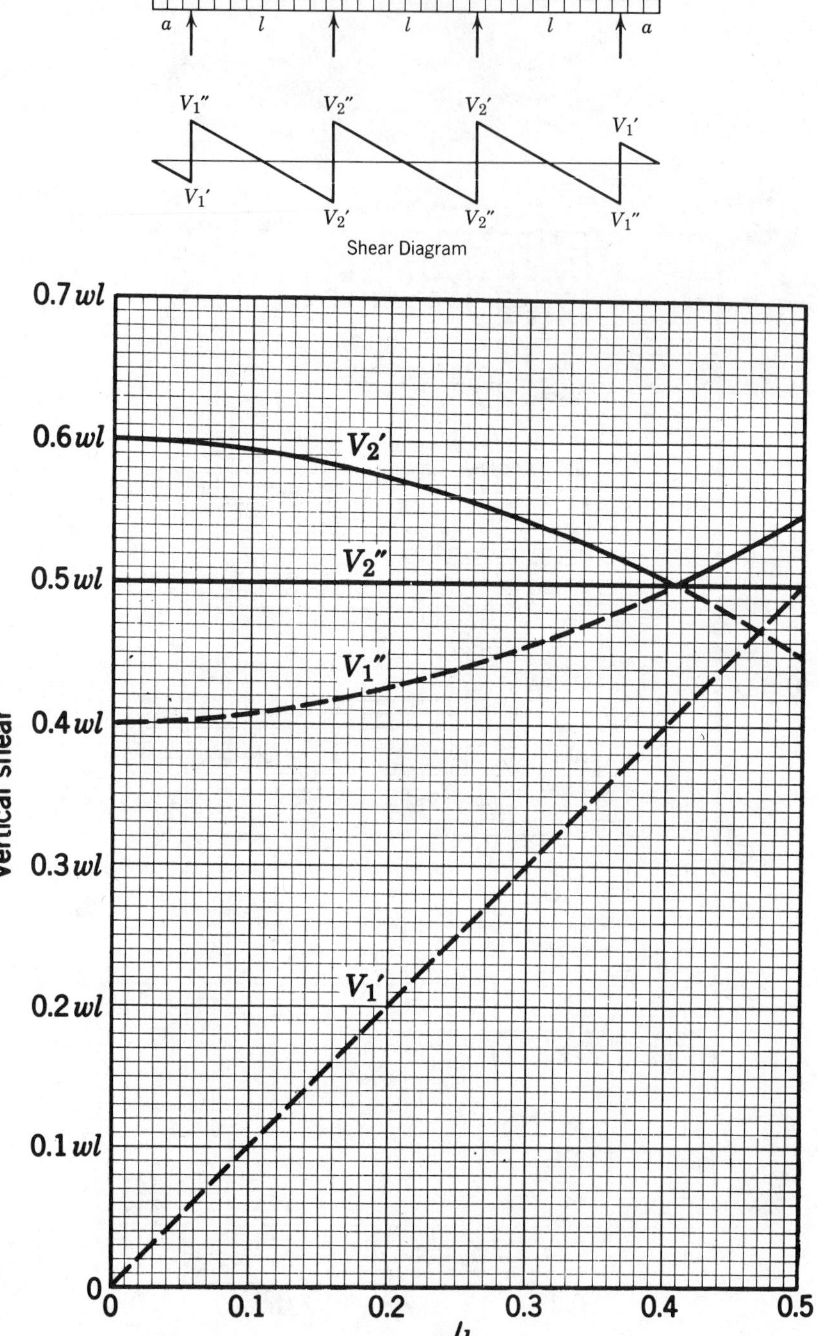

FIGURE 7.12. Effect of overhang on maximum vertical shear for a three-span continuous system. *Note:* Maximum shear occurs at support 2 for values of *a/l* less than 0.408. At higher values of *a/l* maximum shear is at support 1.

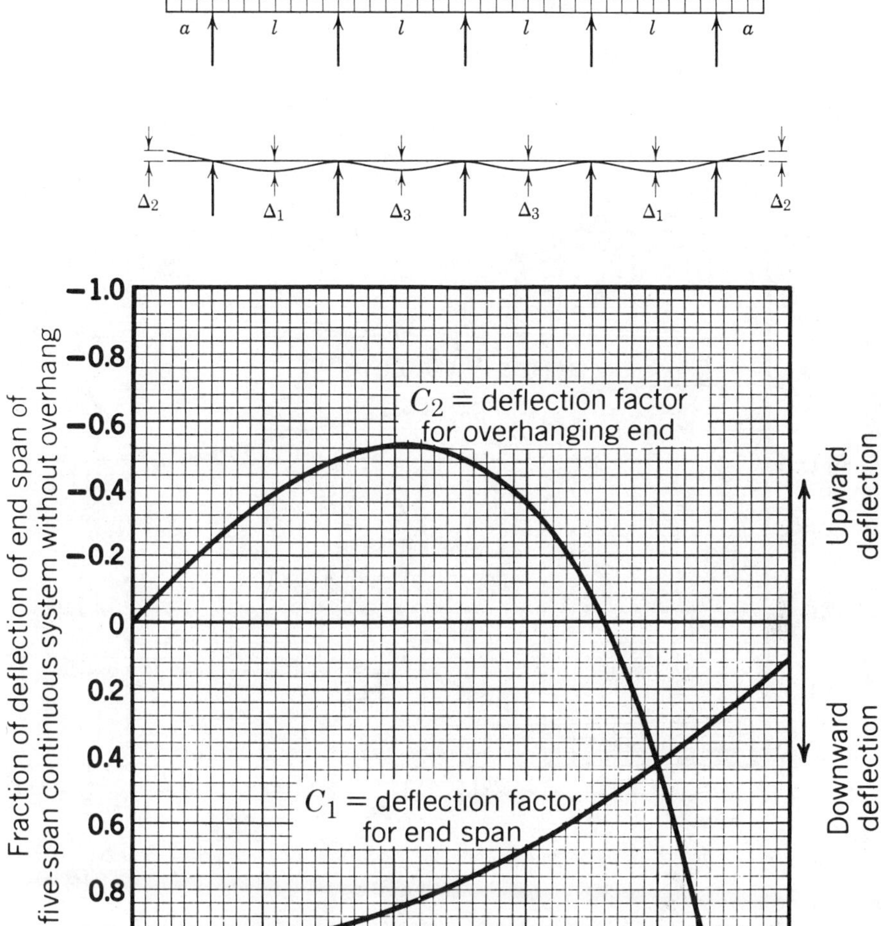

FIGURE 7.13. Effect of overhang on maximum deflection for a four-span continuous system.

Δ = deflection of end span of four-span continuous system, without overhangs (in.)

Δ_1 = $C_1\Delta$ = deflection of end span with overhang (in.)

Δ_2 = $C_2\Delta$ = deflection of overhanging end (in.)

$\Delta_1 > \Delta_3$

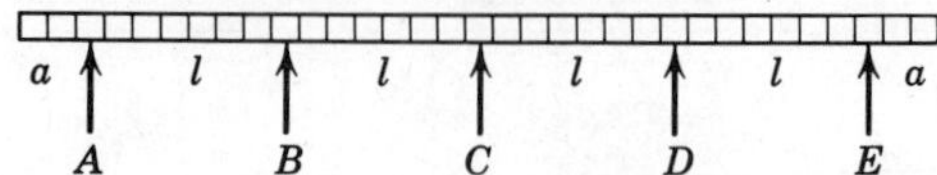

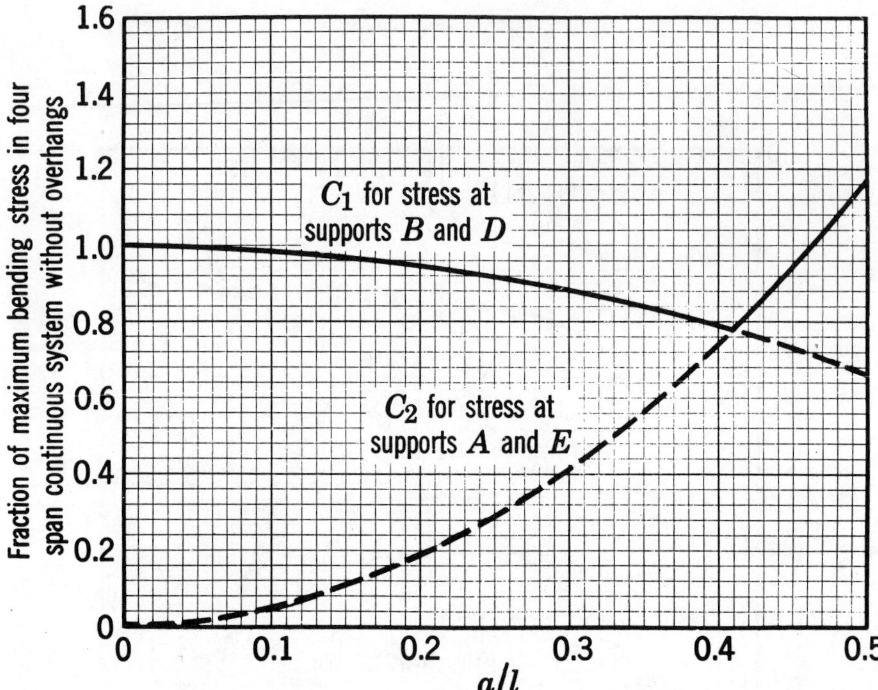

FIGURE 7.14. Effect of overhang on maximum bending stress for a four-span continuous system.

$$f_b = \frac{21wl^2c}{196I}$$

Note: With no overhang, the maximum bending moment occurs at supports B and D. With overhang, the bending moments at B and D decrease, and the bending moments at A and E increase. The bending moments at B and D are larger than at A and E for values of a/l less than 0.408. At a/l greater than 0.408, the bending moments at A and E exceed those at B and D. In Figure 7.14, the maximum imposed stress is expressed as a fraction of the bending stress over supports B and D when $a = 0$.

$$f_{b\max} = C_1 f_b \quad \text{for } 0 < a/l < 0.408$$

$$f_{b\max} = C_2 f_b \quad \text{for } 0.408 < a/l$$

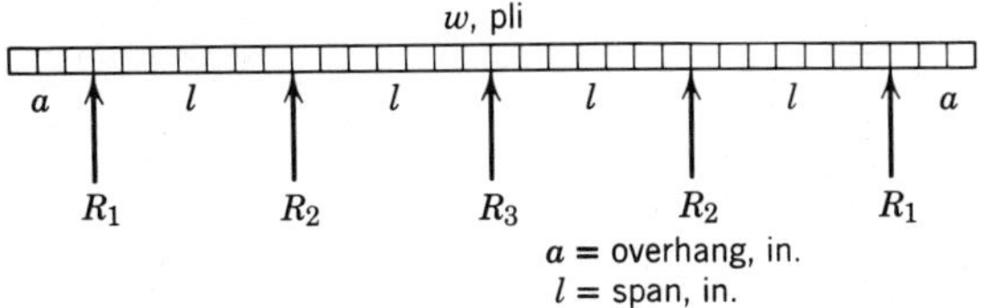

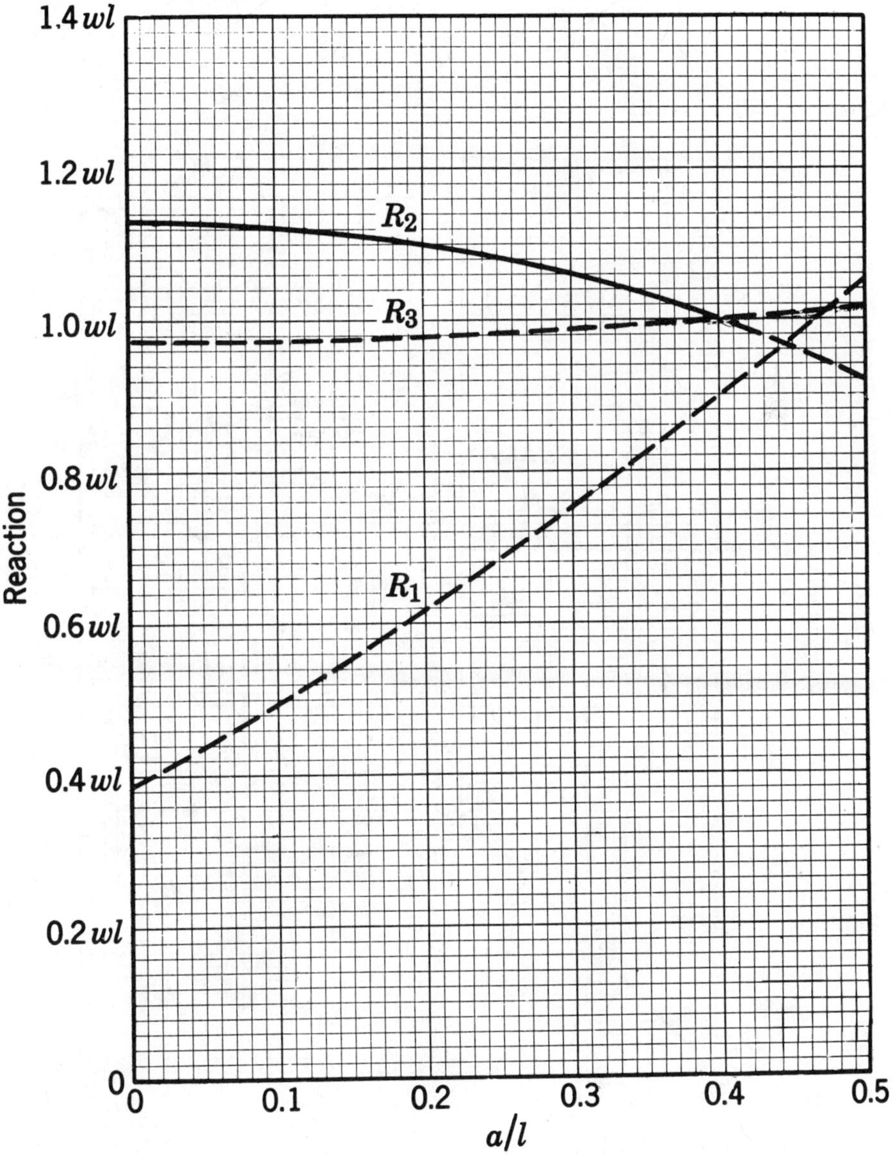

FIGURE 7.15. Effect of overhang on reactions for a four-span continuous system.

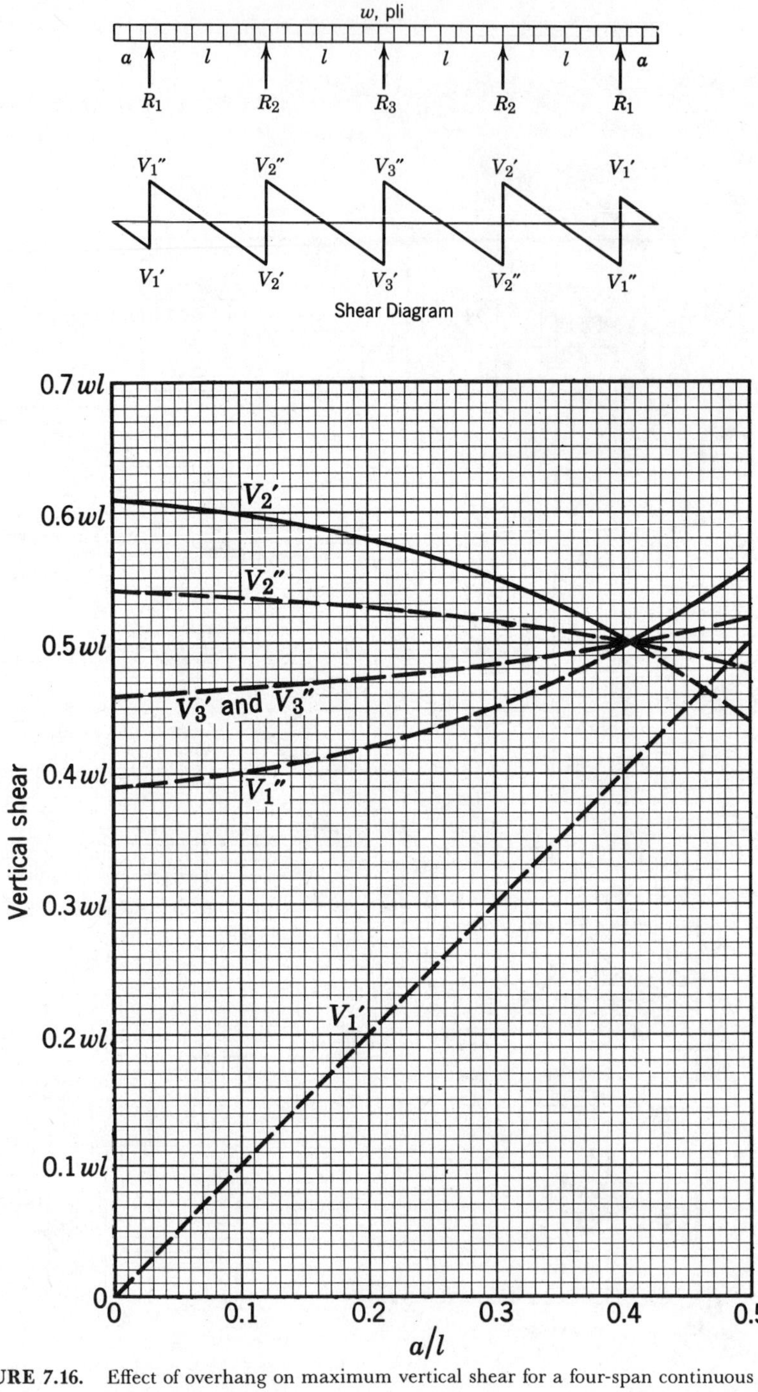

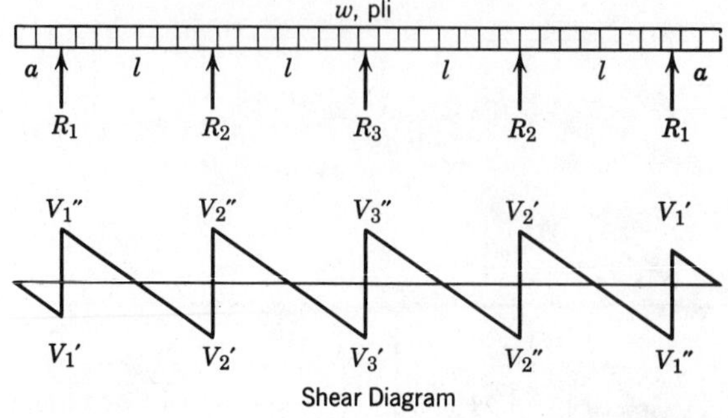

FIGURE 7.16. Effect of overhang on maximum vertical shear for a four-span continuous system. *Note:* Maximum shear occurs at support 2 for values of a/l less than 0.408. At higher values of a/l maximum shear is at support 1.

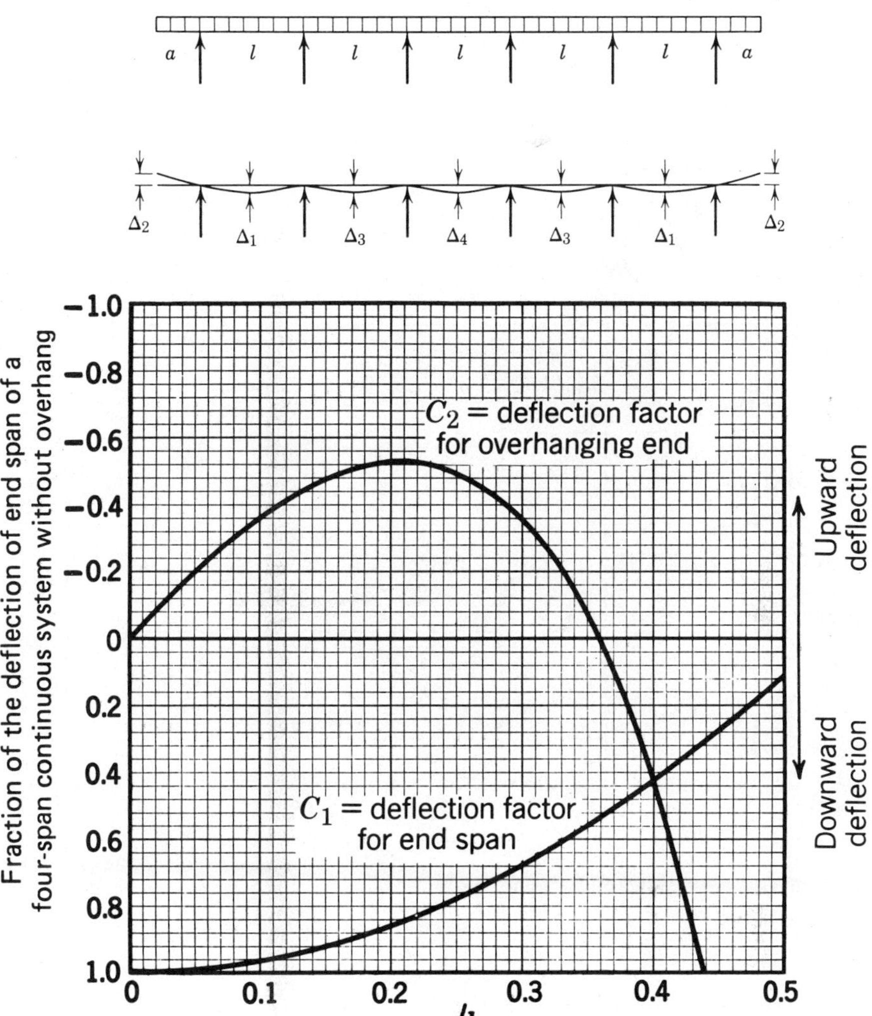

FIGURE 7.17. Effect of overhang on maximum deflection for a five-span continuous system.

Δ = deflection of end span of four-span continuous system, without overhangs (in.)

$\Delta_1 = C_1\Delta$ = deflection of end span with overhang (in.)

$\Delta_2 = C_2\Delta$ = deflection of overhanging end (in.)

$\Delta_1 > \Delta_3$

$\Delta_1 > \Delta_4$

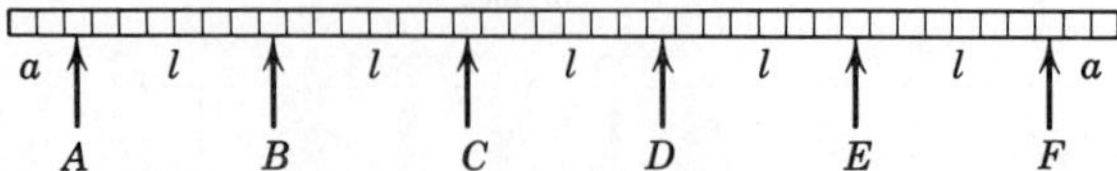

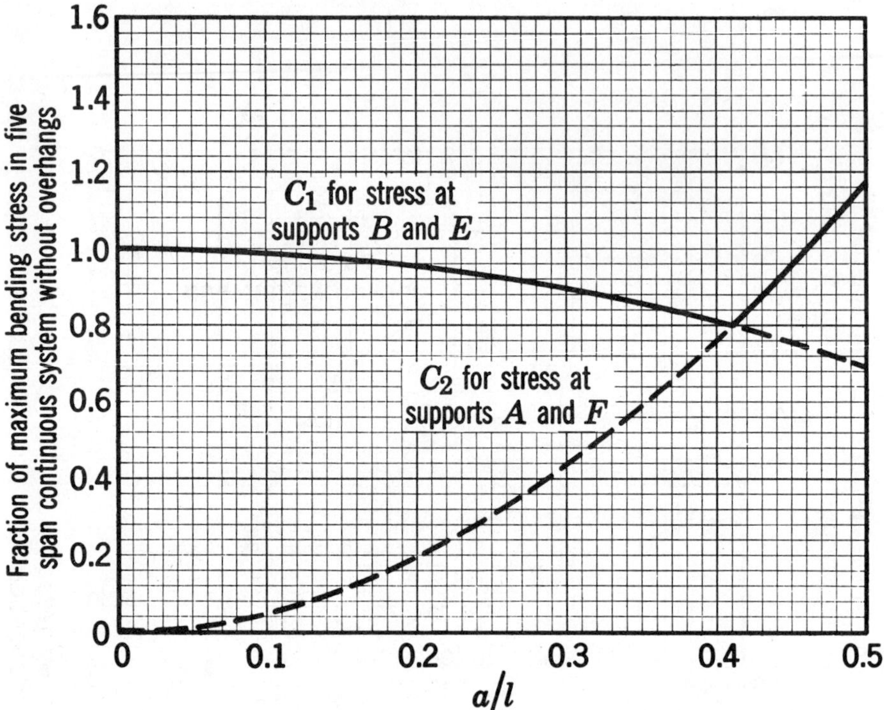

FIGURE 7.18. Effect of overhang on maximum bending stress for a five-span continuous system.

$$f_b = \frac{38wl^2c}{361I}$$

Note: With no overhang, the maximum bending moment occurs a supports B and E. With overhang, the bending moments at B and E decrease, and the bending moments at A and F increase. The bending moments at B and E are larger than at A and F for values of a/l less than 0.408. At a/l greater than 0.408, the bending moments at A and F exceed those at B and E. In Figure 7.18, the maximum imposed stress is expressed as a fraction of the bending stress over supports B and E when $a = 0$.

$$f_{b\,max} = C_1 f_b \quad \text{for } 0 < a/l < 0.408$$

$$f_{b\,max} = C_2 f_b \quad \text{for } 0.408 < a/l$$

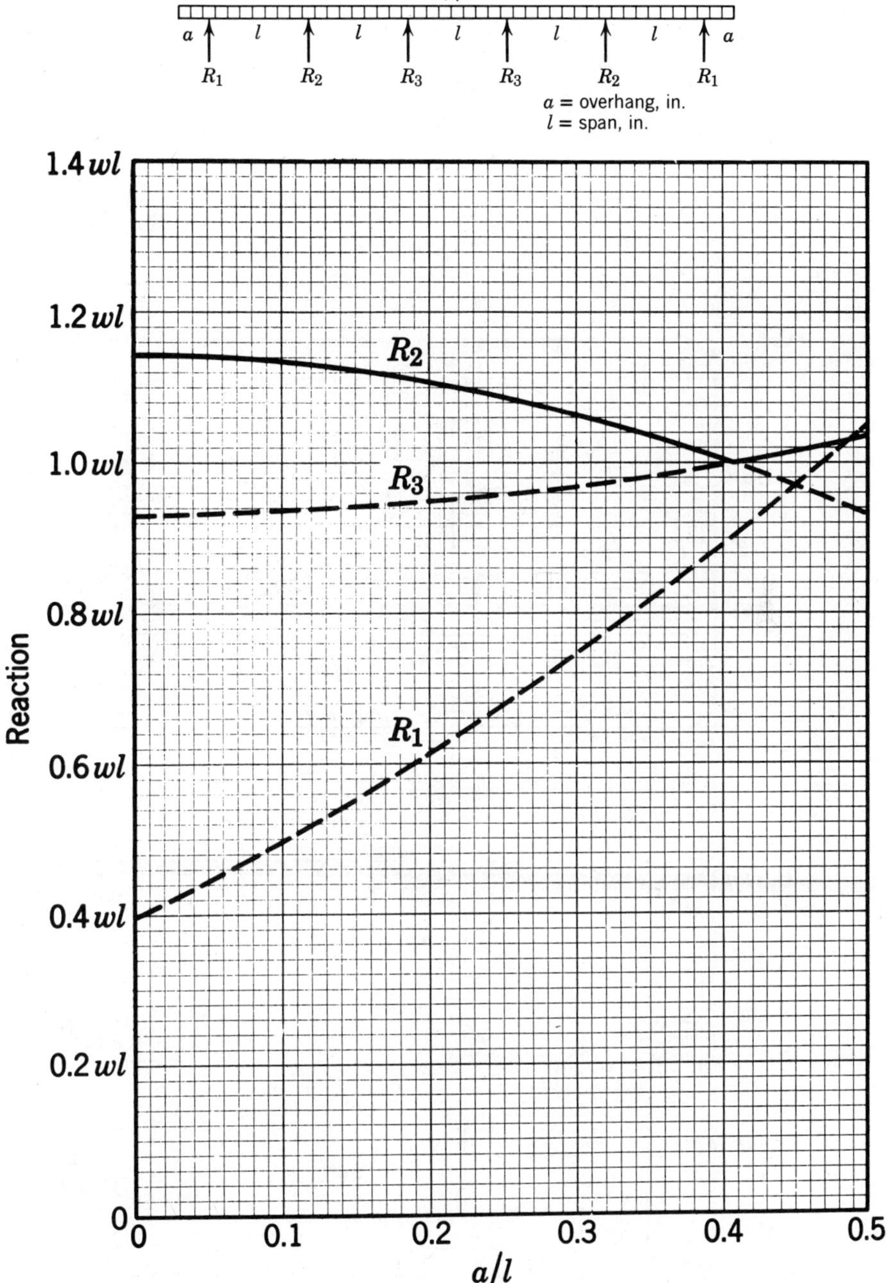

FIGURE 7.19. Effect of overhang on reactions for five-span continuous system.

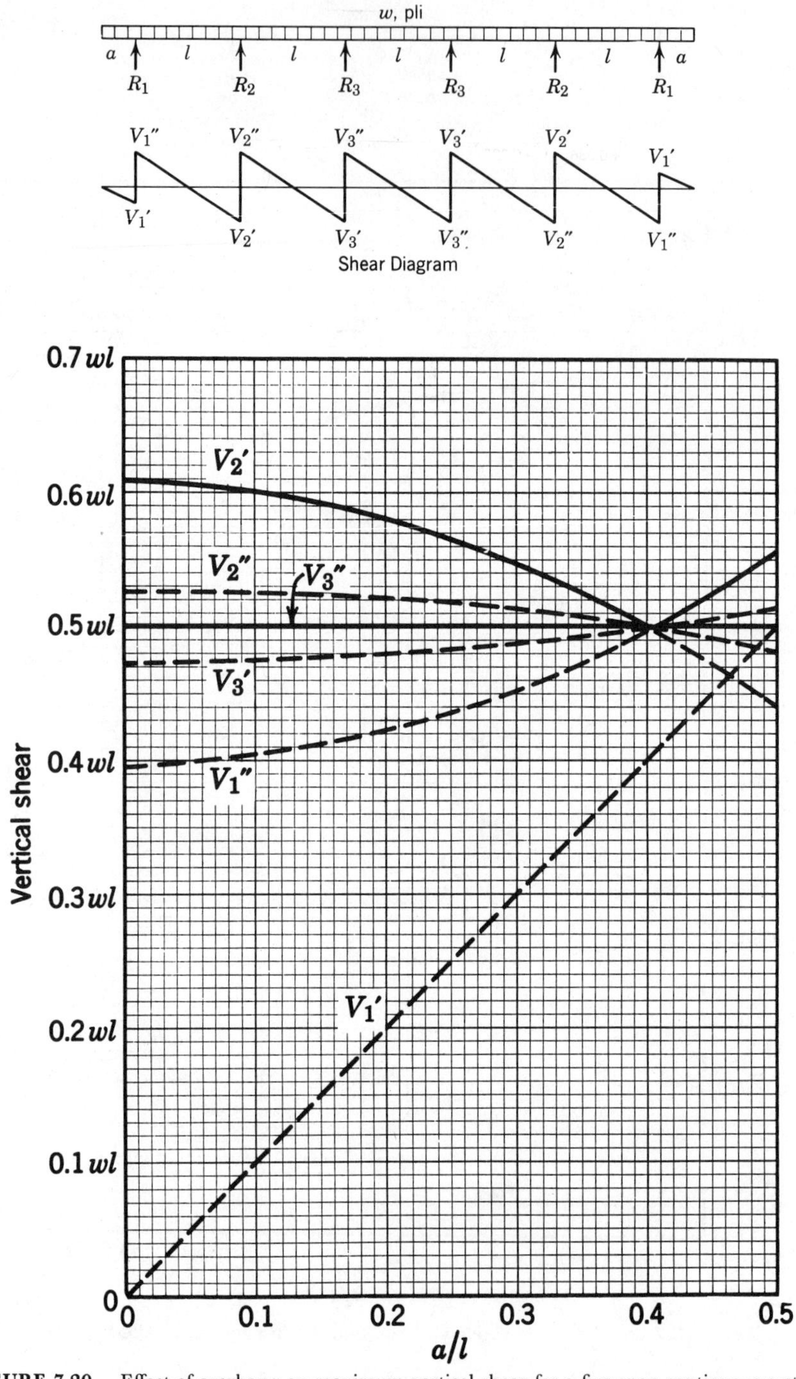

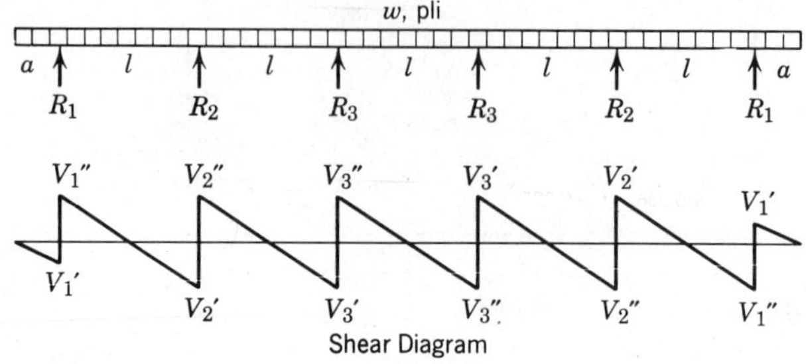

FIGURE 7.20. Effect of overhang on maximum vertical shear for a five-span continuous system. *Note:* Maximum shear occurs at support 2 for values of *a/l* less than 0.408. At higher values of *a/l*, maximum shear is at support 1.

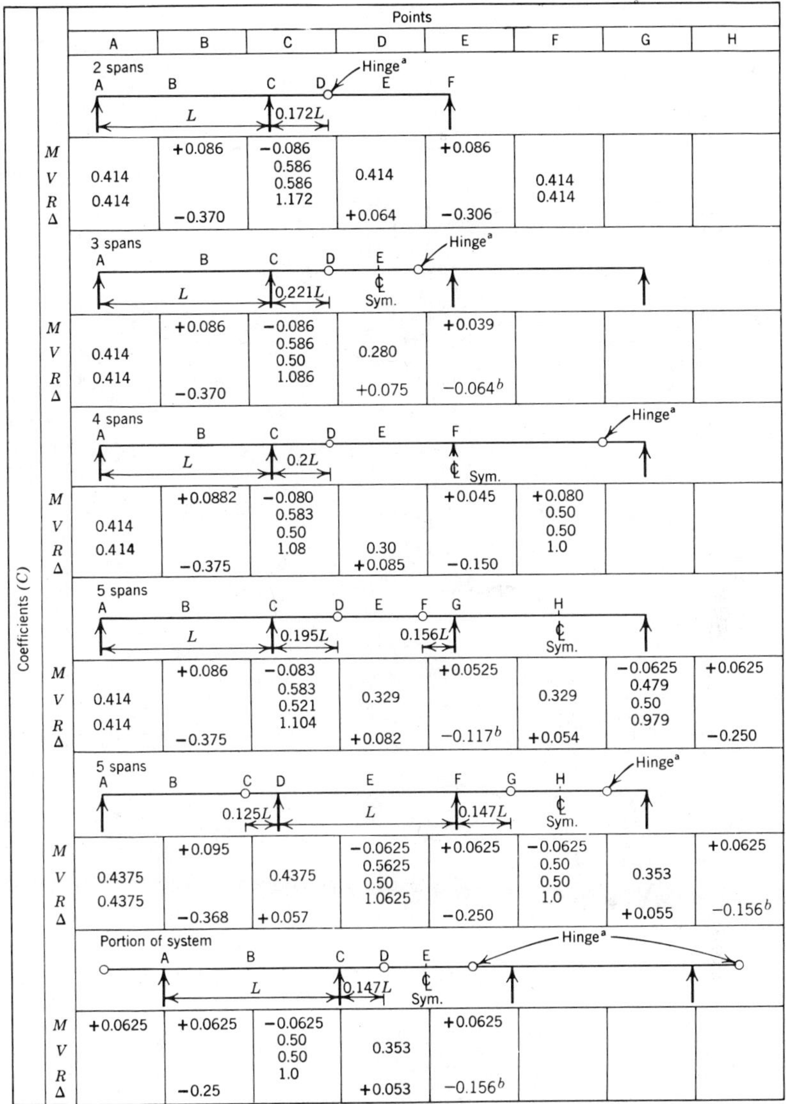

	A	B	C	D	E	F	G	H
2 spans								
M		+0.086	−0.086		+0.086			
V	0.414		0.586 / 0.586	0.414		0.414		
R	0.414		1.172			0.414		
Δ		−0.370		+0.064	−0.306			
3 spans								
M		+0.086	−0.086		+0.039			
V	0.414		0.586 / 0.50	0.280				
R	0.414		1.086					
Δ		−0.370		+0.075	−0.064[b]			
4 spans								
M		+0.0882	−0.080		+0.045	+0.080		
V	0.414		0.583 / 0.50			0.50 / 0.50		
R	0.414		1.08	0.30		1.0		
Δ		−0.375		+0.085	−0.150			
5 spans								
M		+0.086	−0.083		+0.0525		−0.0625	+0.0625
V	0.414		0.583 / 0.521	0.329		0.329	0.479 / 0.50	
R	0.414		1.104				0.979	
Δ		−0.375		+0.082	−0.117[b]	+0.054		−0.250
5 spans								
M		+0.095		−0.0625	+0.0625	−0.0625		+0.0625
V	0.4375		0.4375	0.5625 / 0.50		0.50 / 0.50	0.353	
R	0.4375			1.0625		1.0		
Δ		−0.368	+0.057		−0.250		+0.055	−0.156[b]
Portion of system								
M	+0.0625	+0.0625	−0.0625		+0.0625			
V			0.50 / 0.50	0.353				
R			1.0					
Δ		−0.25		+0.053	−0.156[b]			

FIGURE 7.21. Cantilever beam coefficients for balanced loading conditions. All spans equal, uniformly distributed load.

$$\text{Moment} = M = CwL^2 \qquad \text{Reaction} = R = CwL$$

$$\text{Shear} = V = CwL \qquad \text{Deflection} = \Delta = \frac{CwL^4}{48EI}$$

$$w = \text{load (plf)} \qquad L = \text{span (ft)}$$

[a]For typical hinge details see *Typical Construction Details,* AITC 104, in Part III of this manual.

[b]Deflecting of suspended beam span in from the ends of the simple-span beam between hinges, but the span L is used in making the calculation.

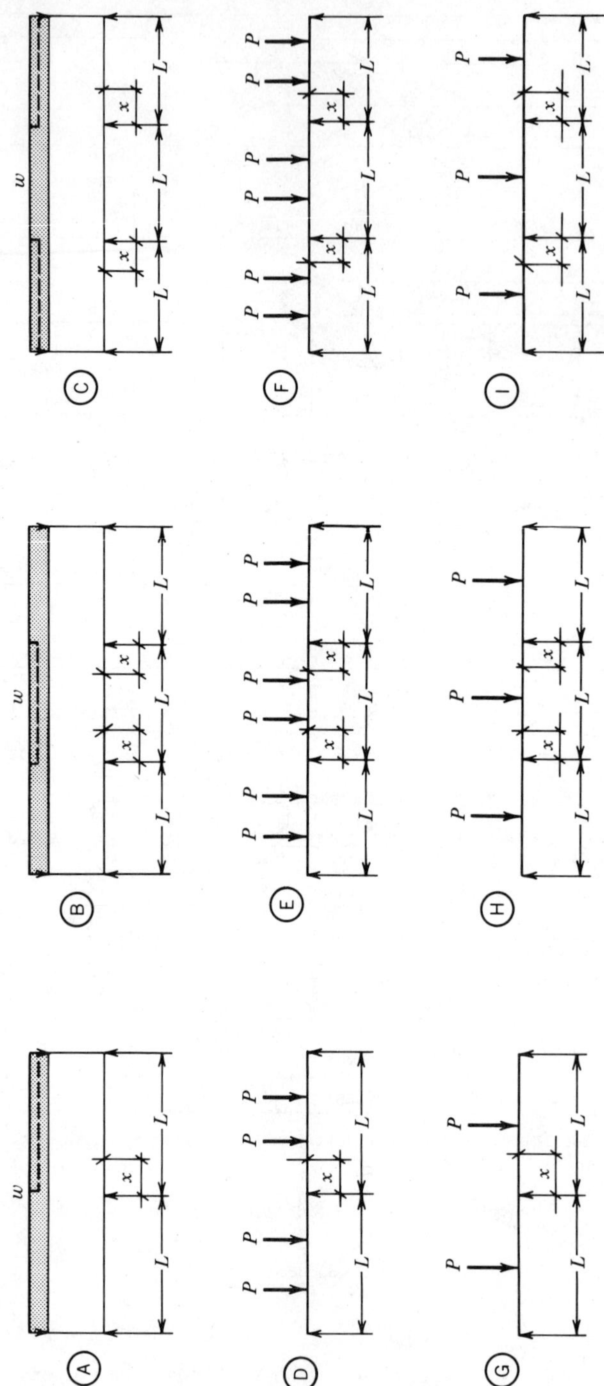
A
B
C
D
E
F
G
H
I
w
L
x
P

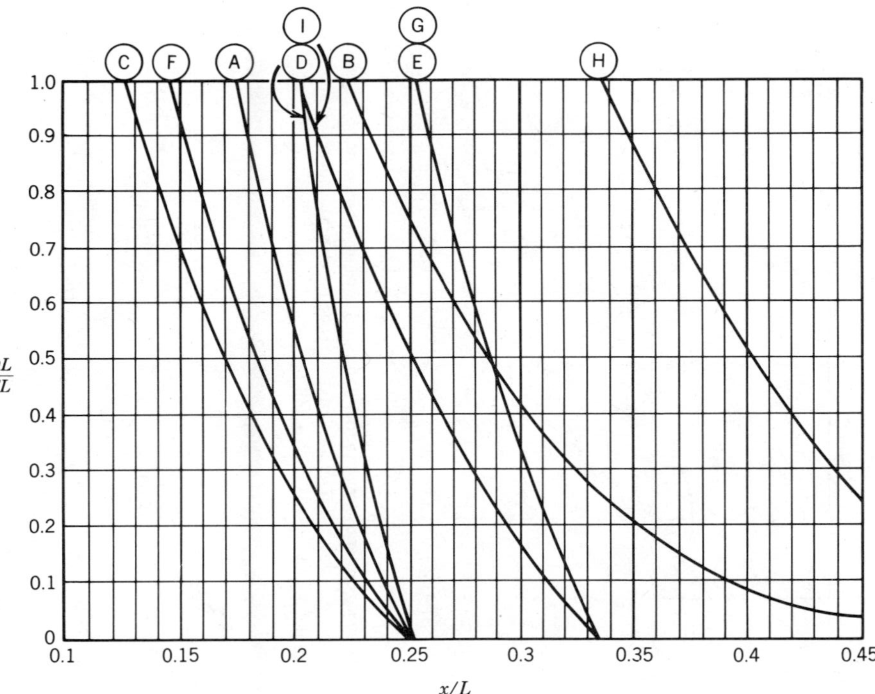

FIGURE 7.22. Cantilever beam coefficients (for unbalanced loading conditions). All spans equal, equal maximum positive, and negative moments.

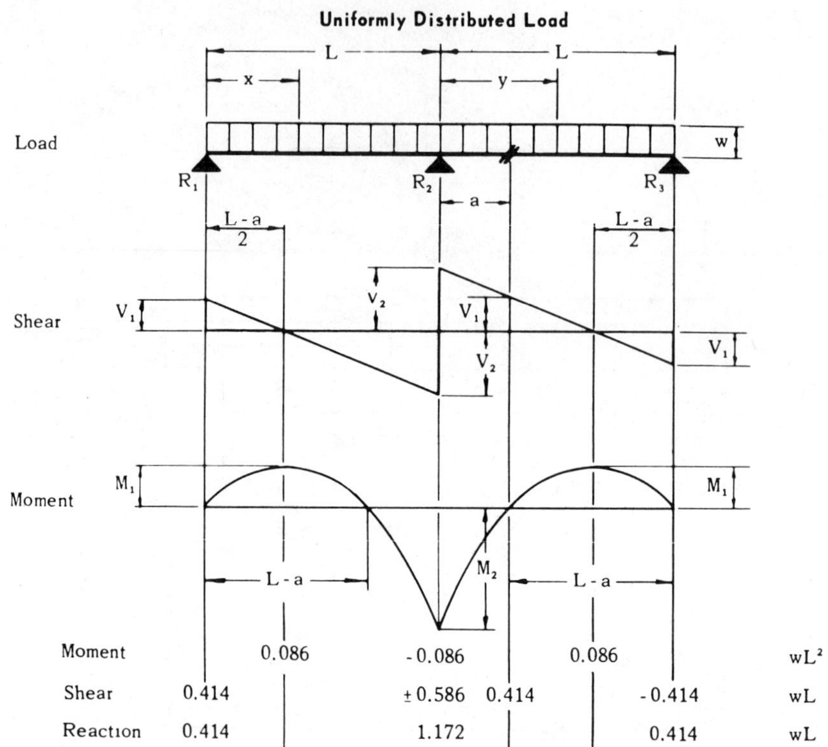

FIGURE 7.23. Cantilevered beam diagram. Two equal spans.

For maximum positive moment equal to maximum negative moment, $a = 0.172L$, and the above coefficients may be applied to wL^2 and wL to find the respective critical values of moment, shear, and reaction. Maximum deflection in either span will be

$$\Delta = 13.31 \frac{wL^4}{EI} \text{ in.}$$

General formulas are:

$$R_1 = R_3 = \frac{w}{2}(L - a) \qquad V_y = \frac{w}{2}(L + a - 2y)$$

$$R_2 = w(L + a) \qquad M_1 = \frac{w}{8}(L - a)^2$$

$$V_1 = \pm\frac{w}{2}(L - a) \qquad M_2 = -\frac{wLa}{2}$$

$$V_2 = \pm\frac{w}{2}(L + a) \qquad M_x = \frac{wx}{2}(L - a - x)$$

$$V_x = \frac{w}{2}(L - a - 2x) \qquad M_y = \frac{w}{2}(y - a)(L - y)$$

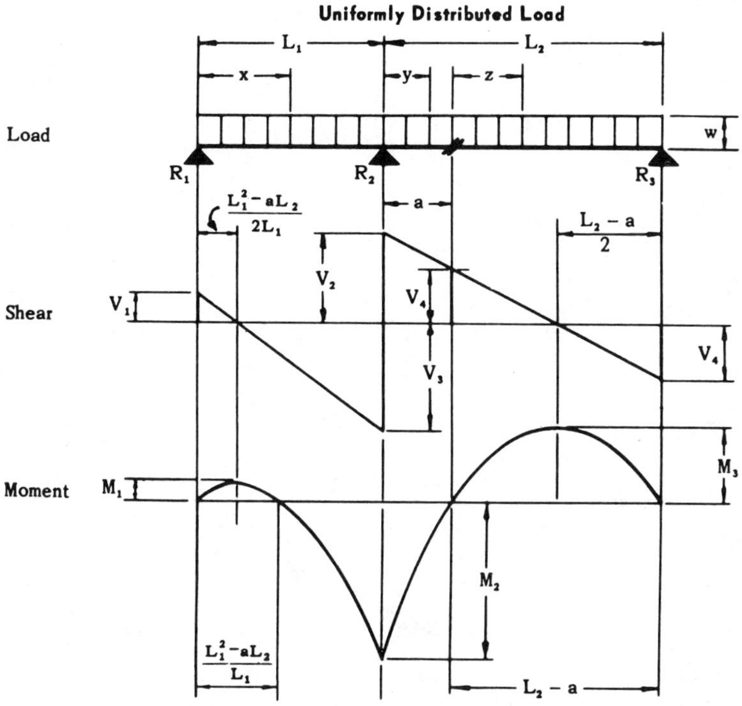

FIGURE 7.24. Cantilevered beam diagram: Two unequal spans.

For maximum positive and negative moments in the cantilevered portion to be equal, $a = 0.172L_1^2/L_2$. Under these conditions, $M_1 = M_2 = 0.086wL_1^2$, $R_1 = V_1 = 0.414wL_1$, and $V_3 = -0.586wL_1$. Other coefficients can be determined for the above or other values of a from the general formulas following:

$$R_1 = \frac{w}{2L_1}(L_1^2 - aL_2) \qquad V_3 = -\frac{w}{2L_1}(L_1^2 - aL_2) \qquad M_2 = -\frac{wL_2a}{2}$$

$$R_2 = \frac{w}{2L_1}(L_1 + a)(L_1 + L_2) \qquad V_4 = \pm\frac{w}{2}(L_2 - a) \qquad M_3 = \frac{w}{8}(L_2 - a)^2$$

$$R_3 = \frac{w}{2}(L_2 - a) \qquad V_x = \frac{w}{2L_1}(L_1^2 - aL_2) - wx \qquad M_x = \frac{wx}{2L_1}(L_1^2 - xL_1 - aL_2)$$

$$V_1 = \frac{w}{2L_1}(L_1^2 - aL_2) \qquad V_y = \frac{w}{2}(L_2 - a) - wy \qquad M_y = \frac{w}{2}(L_2 - y)(y - a)$$

$$V_2 = \frac{w}{2}(L_2 + a) \qquad V_z = \frac{w}{2}(L_2 - a) - wz \qquad M_z = \frac{w}{2}(L_2 - a - z)$$

$$M_1 = \frac{w}{8L_1^2}(L_1^2 - aL_2)^2$$

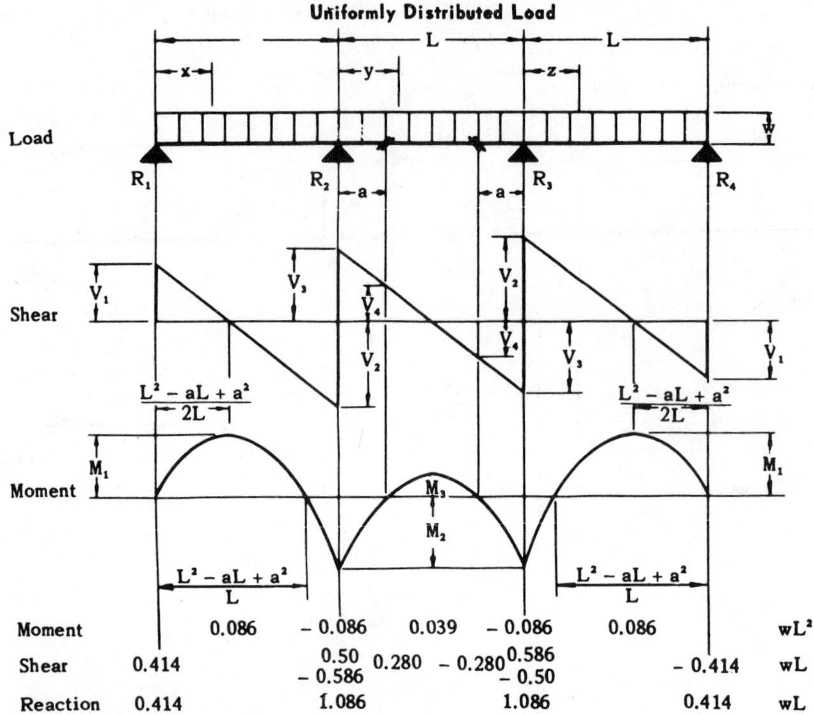

FIGURE 7.25. Cantilevered beam diagram: Three equal spans—single cantilever each end.

For the maximum positive and negative moments of the cantilevered portions of the beam to be equal, $a = 0.220L$, and the above coefficients may be applied to wL^2 and wL to find the respective critical values of moment, shear, and reaction. Maximum deflection in end spans will be

$$\Delta = 13.31 \frac{wL^4}{EI} \text{ in.}$$

General formulas are:

$$R_1 = R_4 = \frac{w}{2L}(L^2 - aL + a^2) \qquad M_1 = -\frac{w}{8L^2}(L^2 - aL + a^2)^2$$

$$R_2 = R_1 = \frac{w}{2L}(2L^2 + aL - a^2) \qquad M_2 = -\frac{w}{2}(aL - a^2)$$

$$V_1 = \pm\frac{w}{2L}(L^2 - aL + a^2) \qquad M_3 = \frac{w}{8}(L - 2a)^2$$

$$V_2 = \pm\frac{w}{2L}(L^2 + aL - a^2) \qquad V_x = \frac{w}{2L}(L^2 - aL + a^2) - wx$$

$$V_3 = \pm\frac{wL}{2} \qquad V_y = \frac{w}{2}(L - 2y)$$

$$V_4 = \pm\frac{w}{2}(L - 2a) \qquad V_z = \frac{w}{2L}(L^2 + aL - a^2) - wz$$

$$M_x = \frac{wx}{2L}(L^2 - aL + a^2) - \frac{wx^2}{2} \qquad M_y = \frac{w}{2}(y - a)(L - y - a) \qquad M_z = \frac{w}{2}(L - z)\left(\frac{a^2}{L} + z - a\right)$$

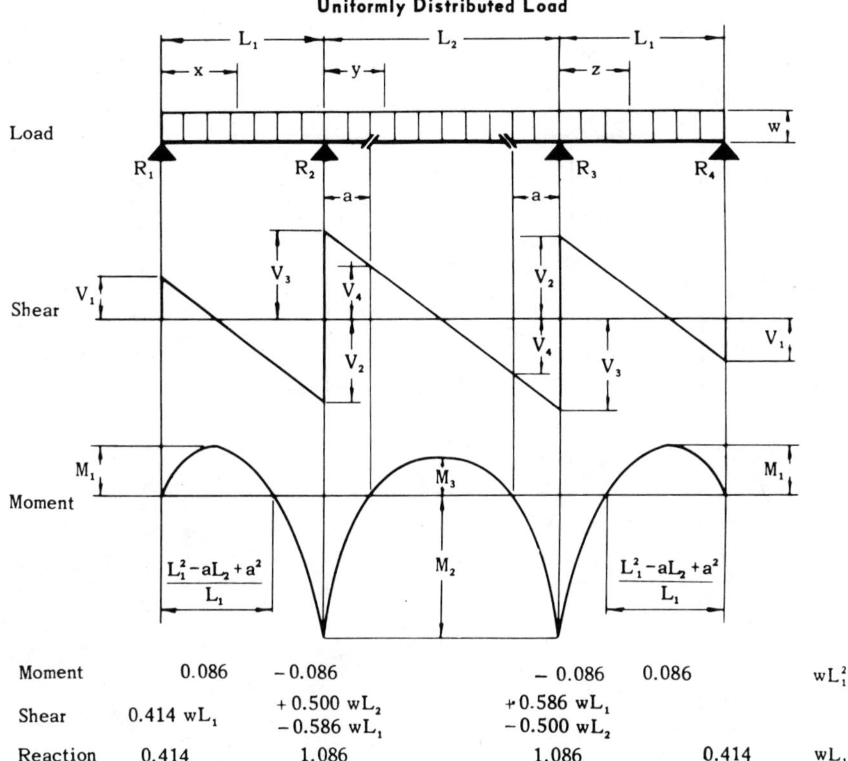

FIGURE 7.26. Cantilevered beam diagrams: Three spans—end spans equal—single cantilever each end

For the maximum positive and negative moments of the cantilevered portions of the beam to be equal, $a = \frac{1}{2}(L_2 - \sqrt{L_2^2 - 0.688L_1^2})$ and the above coefficients may be applied to find the respective critical values of moment, shear, and reaction. Coefficients are omitted when calculation using the general formula is simpler. General formulas are:

$$R_1 = R_4 = \frac{w}{2L_1}(L_1^2 - aL_2 + a^2)$$

$$R_2 = R_3 = \frac{w}{2L_1}(L_1 + a)(L_1 + L_2 - a)$$

$$V_1 = \pm \frac{w}{2L_1}(L_1^2 - aL_2 - a^2)$$

$$V_2 = \pm \frac{w}{2L_1}(L_1^2 - aL_2 + a^2)$$

$$V_3 = \pm \frac{wL_2}{2}$$

$$V_4 = \pm \frac{w}{2}(L_2 - 2a)$$

$$M_1 = \frac{w}{8L_1^2}(L_1^2 - aL_2 + a^2)^2$$

$$V_x = \frac{w}{2L_1}(L_1^2 - aL_2 + a^2) - wx$$

$$V_y = \frac{w}{2}(L_2 - 2y)$$

$$V_z = \frac{w}{2L_1}(L_1^2 - aL_2 + a^2) - wz$$

$$M_2 = -\frac{w}{2}(aL_2 - a^2)$$

$$M_3 = \frac{w}{8}(L_2 - 2a)^2$$

$$M_x = \frac{wx}{2L_1}(L_1^2 - aL_2 + a^2) - \frac{wx^2}{2}$$

$$M_y = \frac{w}{2}(y - a)(L_2 - y - a)$$

$$M_z = \frac{w}{2L_1}(L_1z - aL_2)(L_1 - z)$$

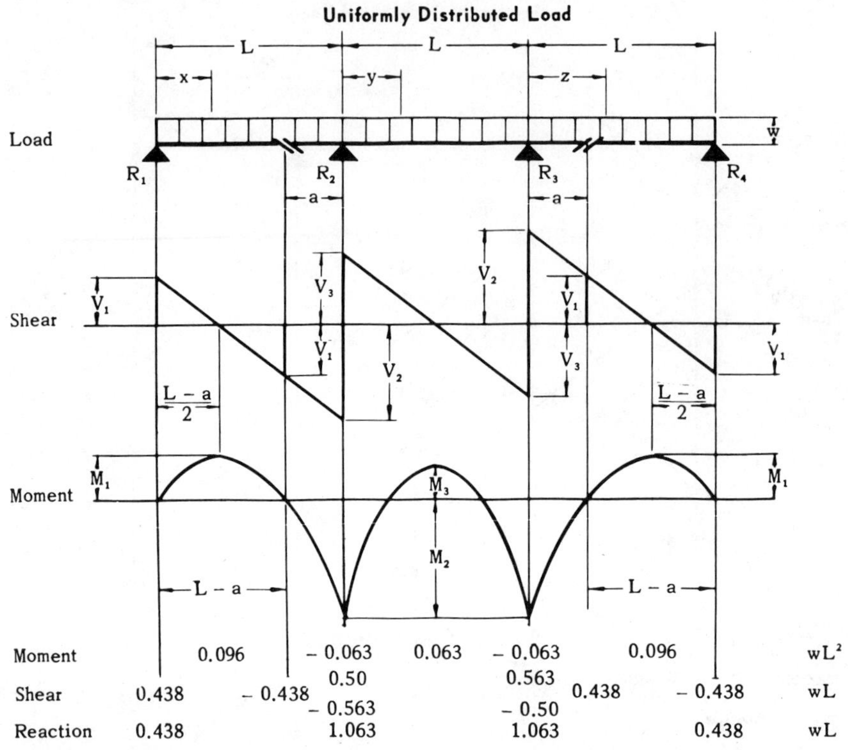

FIGURE 7.27. Cantilevered beam diagram: Three equal spans—double cantilever.

For the maximum positive and negative moments of the cantilevered portion of the beam to be equal, $a = L/8$, and the above coefficients may be applied to wL^2 and wL to find the respective critical values of moment, shear, and reaction. Maximum deflection in center span will be

$$\Delta = 8.99 \frac{wL^4}{EI}$$

General formulas are

$$R_1 = R_4 = \frac{w}{2}(L - a) \qquad V_x = \frac{w}{2}(L - a - 2x) \qquad M_3 = \frac{w}{8}(L^2 - 4aL)$$

$$R_2 = R_3 = \frac{w}{2}(2L + a) \qquad V_y = \frac{w}{2}(L - 2y) \qquad M_x = \frac{wx}{2}(L - a - x)$$

$$V_1 = \pm\frac{w}{2}(L - a) \qquad V_z = \frac{w}{2}(L + a - 2z) \qquad M_y = \frac{w}{2}(L_2 y - y^2 - La)$$

$$V_2 = \pm\frac{w}{2}(L + a) \qquad M_1 = \frac{w}{8}(L - a)^2 \qquad M_z = \frac{w}{2}(L - z)(z - a)$$

$$V_3 = \pm\frac{wL}{2} \qquad M_2 = -\frac{wLa}{2}$$

Points of zero moment in center span occur at

$$y = \frac{L_2 \pm \sqrt{L_2^2 - 4aL_1}}{2}$$

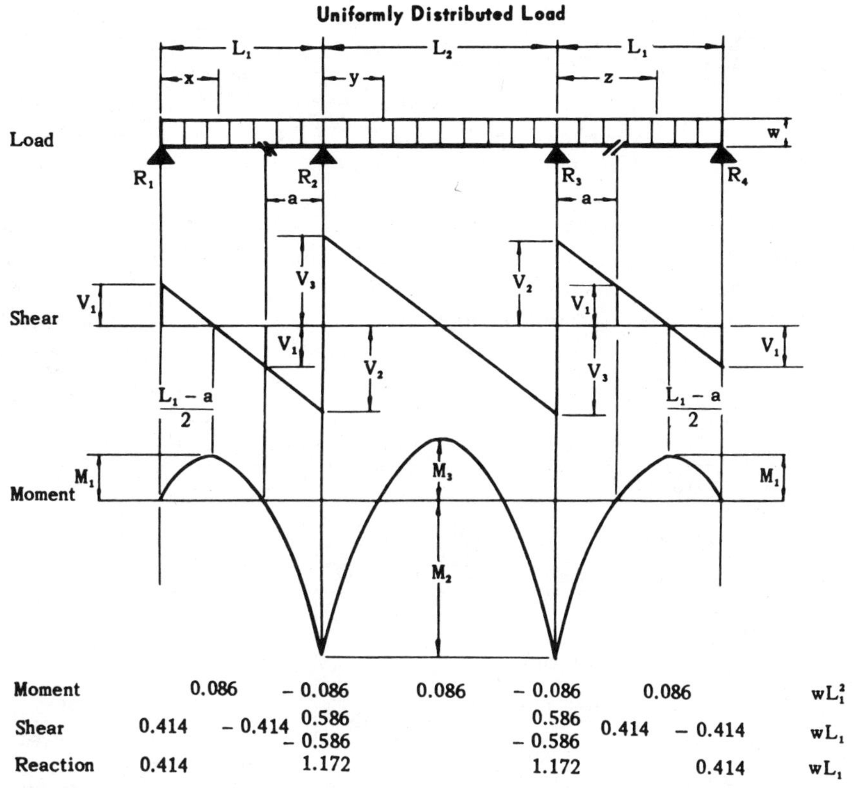

FIGURE 7.28. Cantilevered beam diagram: Three spans—end spans equal—double cantilever

For the maximum positive and negative moments of the cantilevered portion of the beam to be equal, $a = 0.125L_2^2/L_1$. For the special case where all maximum positive and negative moments are equal, that is, $M_1 = M_2 = M_3$, $a = 0.172L_1$ and $L_2 = 1.172L_1$, and the above coefficients may be applied to wL_1^2 and wL_1 to find the respective critical values of moment, shear, and rejection. General formulas are

$$R_1 = R_4 = \frac{w}{2}(L_1 - a) \qquad V_x = \frac{w}{2}(L_1 - a - 2x) \qquad M_3 = \frac{w}{8}(L_2^2 - 4aL_1)$$

$$R_2 = R_3 = \frac{w}{2}(L_1 + L_2 + a) \qquad V_y = \frac{w}{2}(L_2 - 2y) \qquad M_x = \frac{wx}{2}(L_1 - a - x)$$

$$V_1 = \pm (L_1 - a) \qquad V_z = \frac{w}{2}(L_1 + a - 2z) \qquad M_y = \frac{w}{2}(L_2 y - y^2 - aL_1)$$

$$V_2 = \pm \frac{w}{2}(L_1 + a) \qquad M_1 = \frac{w}{8}(L_1 - a)^2 \qquad M_z = \frac{w}{2}(z - L_1)(a - z)$$

$$V_3 = \pm \frac{wL_2}{2} \qquad M_2 = -\frac{wL_1 a}{2}$$

Points of zero moment in center span occur at

$$y = \frac{L_2 \pm \sqrt{L_2^2 - 4aL_1}}{2}$$

PANELIZED ROOF GRID SYSTEMS

The selection of an efficient column grid system is facilitated by the determination of an efficiency factor. The efficiency factor is the ratio of the board footage of glued laminated timber used in the primary framing system to the square feet of space enclosed under roof cover. In comparing glulam framing systems, the lower the ratio, the more efficient the system.

Figure 7.29 illustrates the calculation of the efficiency factors for a selected building size and framing system. The dead load used in this example is 10 psf (excluding the weight of the beam) with a live load of 20 psf.

Glulam and purlin sizes and corresponding efficiency factors are given in each

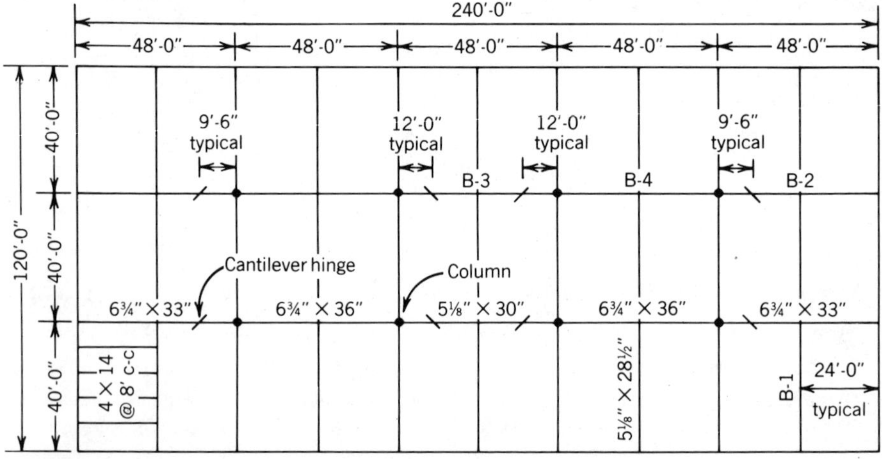

FIGURE 7.29. Calculation of efficiency factors:

B-1's fbm = 760 fbm/beam × 27 beams = 20,520 fbm
B-2's fbm = 1129 fbm/beam × 4 beams = 4,517 fbm
B-3's fbm = 480 fbm/beam × 2 beams = 960 fbm
B-4's fbm = 2224 fbm/beam × 4 beams = 34,893 fbm
Total = 34,893 fbm

$$\text{Area enclosed} = 28{,}800 \text{ ft}^2$$

$$\text{fbm glulam/ft}^2 = 34{,}893/28{,}800 = 1.21$$

$$\text{Solid sawn purlins; fbm} = 112 \text{ fbm/purlin} \times 120 \text{ purlins} = 13{,}440$$

$$\text{fbm solid sawn/ft}^2 = 13{,}440/28{,}800 = 0.47$$

Assumed design criteria and allowable stresses used in the calculation of the bending member sizes shown for the examples on pages 7-660 to 7-663 are:

Member sizes governed by either bending or shear
Full balanced or unbalanced live load, whichever controls
Dead load does not include weight of glulam
Bending stress, F_b = 2400 psi
Shear stress, F_v = 165 psi
Compression perpendicular to grain stress, $F_{c\perp}$ = 560 psi
Modulus of elasticity, E = 1,800,000 psi
Duration-of-load factor: 1.25 for 12-psf live loads; 1.15 for 20- and 30-psf live loads

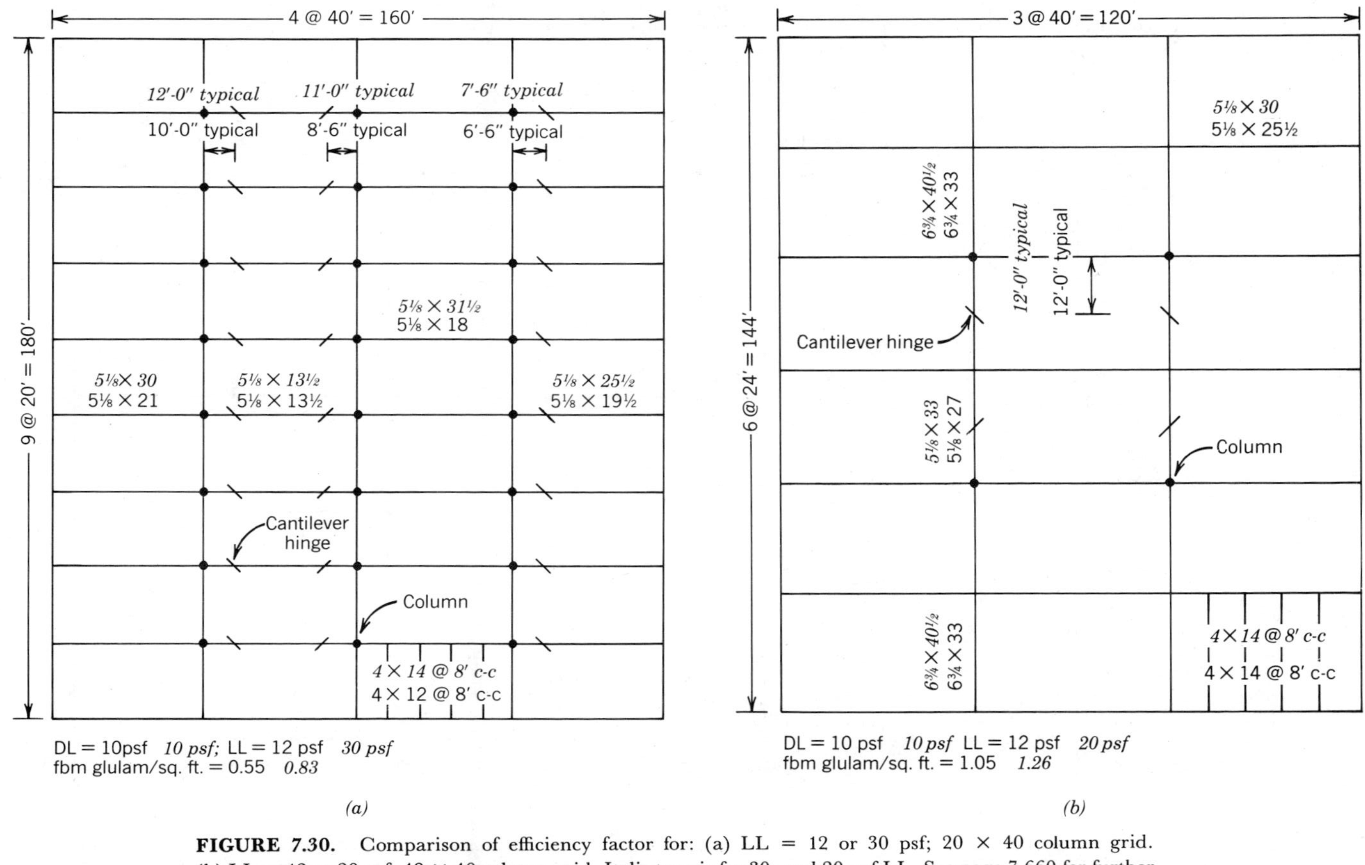

FIGURE 7.30. Comparison of efficiency factor for: (a) LL = 12 or 30 psf; 20 × 40 column grid. (b) LL = 12 or 20 psf; 48 × 40 column grid. Italic type is for 30- and 20-psf LL. See page 7-660 for further details (*Figure continued on next page.*)

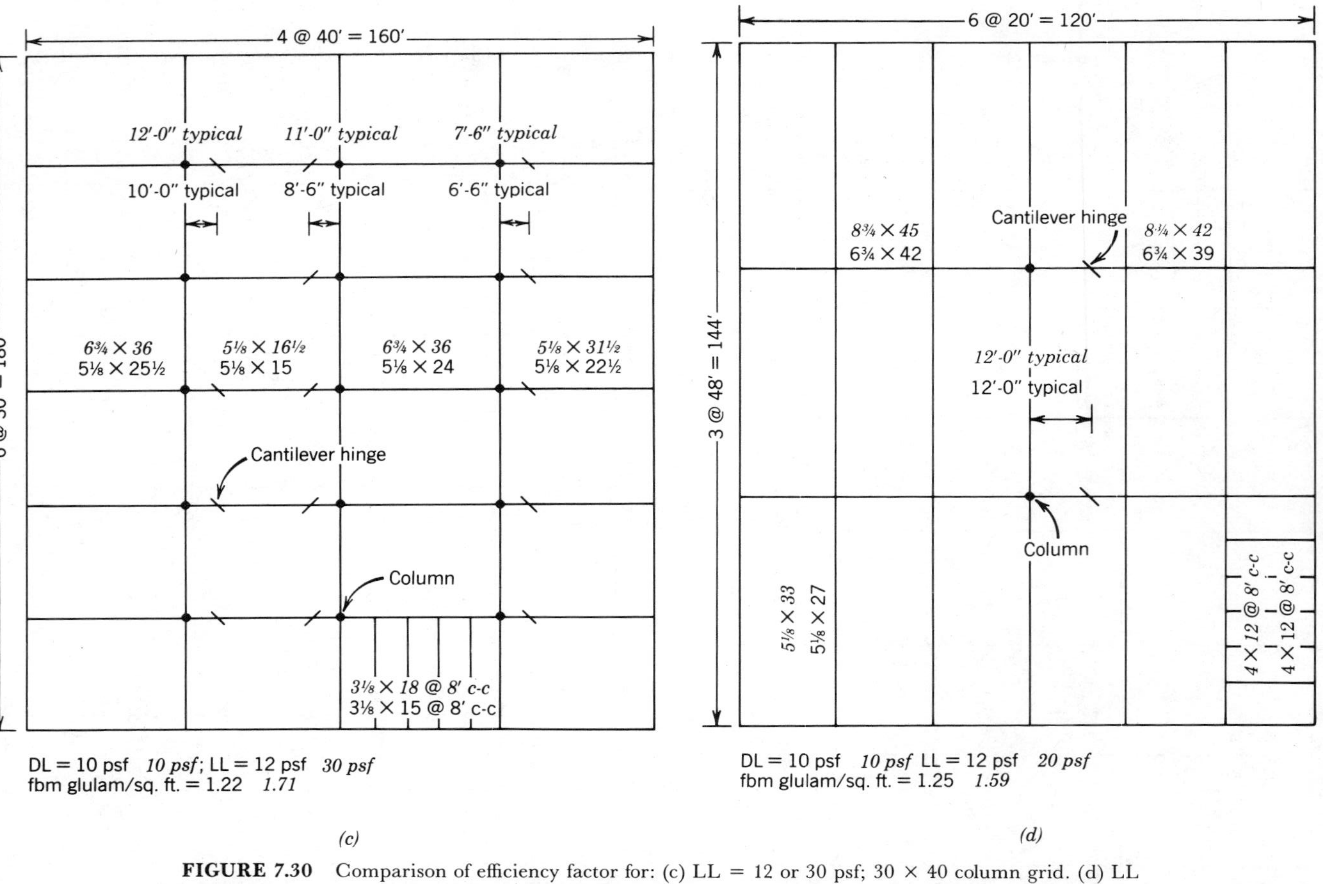

FIGURE 7.30 Comparison of efficiency factor for: (c) LL = 12 or 30 psf; 30 × 40 column grid. (d) LL = 12 or 20 psf; 48 × 60 column grid. Italic type is for 30- and 20-psf LL. See page 7-660 for further details.

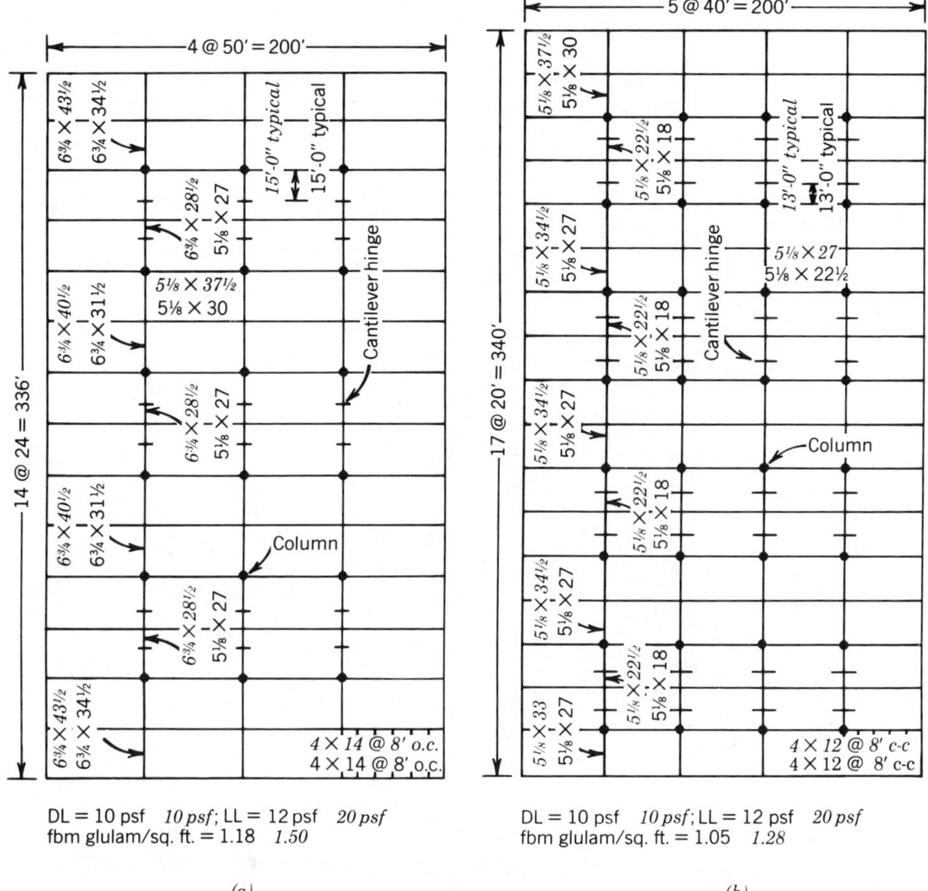

FIGURE 7.31. Comparison of efficiency of column grids for LL = 12 or 20 psf: (a) 48 × 50 column grid; (b) 40 × 40 column grid. Italic type is for 20-psf LL. See page 7-660 for further details.

layout for two different live loads. (*Note:* Regular type refers to one LL factor, with italic type for the other. See legend for complete explanation.) Member sizes shown are governed by either bending or shear. Sawn purlin sizes are based on the use of a No. 1 Douglas Fir grade. In all cases, the data is subject to specific design requirements of local codes or special conditions, including applicable deflection requirements. A minimum roof slope of $\frac{1}{4}$ in./ft should be provided in addition to camber to minimize water ponding. The examples shown are illustrative only and a complete design check should be made by competent engineering personnel.

TABLE 7.11

Minimum Design Dead Loads (psf)[a,b]

Component	Load
Ceilings	
Acoustical fiber tile	1
Gypsum board (per $\frac{1}{8}$ in. thick)	0.55
Mechanical duct allowance	4
Plaster on tile or concrete	5
Plaster on wood lath	8
Suspended steel channel system	2
Suspended metal lath and cement plaster	15
Suspended metal lath and gypsum plaster	10
Wood furring suspension system	2.5
Coverings, roof, and wall	
Asbestos-cement shingles	4
Asphalt shingles	2
Clay tile	16
Clay tile (for mortar add 10 lb):	
Book tile, 2 in.	12
Book tile, 3 in.	20
Ludowici	10
Roman	12
Spanish	19
Composition:	
Three-ply ready roofing	1
Four-ply felt and gravel	5.5
Five-ply felt and gravel	6
Copper or tin	1
Corrugated asbestos-cement roofing	4
Deck, metal, 20 gage	2.5
Deck, metal, 18 gage	3
Decking, 2 in. wood (Douglas fir)	5
Decking, 3 in. wood (Douglas fir)	8
Fiberboard, $\frac{1}{2}$ in.	0.75
Gypsum sheathing, $\frac{1}{2}$ in.	2
Insulation, roof boards (per in. thickness):	
Cellular glass	0.7
Fibrous glass	1.1
Fiberboard	1.5
Perlite	0.8
Polystyrene foam	0.2
Urethane foam with skin	0.5
Plywood (per $\frac{1}{8}$ in. thickness)	0.4
Coverings, roof, and wall (cont.)	
Rigid insulation, $\frac{1}{2}$ in.	0.75
Skylight, metal frame, $\frac{3}{8}$ in. wire glass	8
Slate, $\frac{3}{16}$ in.	7
Slate, $\frac{1}{4}$ in.	10
Waterproofing membranes:	
Bituminous, gravel covered	5.5
Bituminous, smooth surface	1.5
Liquid applied	1.0
Single-ply, sheet	0.7
Wood sheathing (per in. thickness)	3
Wood shingles	3
Floor fill	
Cinder concrete, per in.	9
Lightweight concrete, per in.	8
Sand, per in.	8
Stone concrete, per in.	12
Floors and floor finishes	
Asphalt block (2 in.), $\frac{1}{2}$ in. mortar	30
Cement finish (1 in.) on stone-concrete fill	32
Ceramic or quarry tile ($\frac{3}{4}$ in.) on $\frac{1}{2}$ in. mortar bed	16
Ceramic or quarry tile ($\frac{3}{4}$ in.) on 1 in. mortar bed	16
Concrete fill finish (per in. thick)	12
Hardwood flooring, $\frac{7}{8}$ in.	4
Linoleum or asphalt tile, $\frac{1}{4}$ in.	1
Marble and mortar on stone-concrete fill	33
Slate (per in. thickness)	15
Solid flat tile on 1 in. mortar base	23
Subflooring, $\frac{3}{4}$ in.	3
Terrazzo ($1\frac{1}{2}$ in.) directly on slab	19
Terrazzo (1 in.) on stone-concrete fill	32
Terrazzo (1 in.) on 2 in. stone-concrete	32
Wood block (3 in.) on mastic, no fill	10
Wood block (3 in.) on $\frac{1}{2}$ in. mortar base	16

TABLE 7.11 (*Continued*)

Floors, wood-joist (no plaster)-
Double wood floor

Joist sizes, in.	12 in. spacing (psf)	16 in. spacing (psf)	24 in. spacing (psf)
2 × 6	6	5	5
2 × 8	6	6	5
2 × 10	7	6	6
2 × 12	8	7	6

Frame partitions	
Movable steel partitions	4
Wood or steel studs, $\frac{1}{2}$ in. gypsum board each side	8
Wood studs, 2 × 4, unplastered	4
Wood studs, 2 × 4, plastered 1 side	12
Wood studs, 2 × 4, plastered 2 sides	20
Frame walls	
Exterior stud walls:	
2 × 4 at 16 in., $\frac{5}{8}$ in. gypsum, insulated, $\frac{3}{8}$ in. siding	11
2 × 6 at 16 in., $\frac{5}{8}$ in. gypsum, insulated, $\frac{3}{8}$ in. siding	12
Exterior stud walls with brick veneer	48
Windows, glass frame, and sash	8
Masonry partitions	
Clay tile:	
4 in.	18
6 in.	24
8 in.	
Concrete block, heavy aggregate:	
4 in.	30
6 in.	42
8 in.	55
12 in.	
Masonry partions (cont)	
Concrete block, light aggregate:	
4 in.	20
6 in.	28
8 in.	38
12 in.	55
Masonry walls	
Clay brick, medium absorption:	
4 in.	39
8 in.	79
$12\frac{1}{2}$ in.	115
17 in.	155
22 in.	
Concrete brick, heavy aggregate:	
4 in.	46
8 in.	89
$12\frac{1}{2}$ in.	130
17 in.	174
22 in.	
Concrete brick, light aggregate:	
4 in.	33
8 in.	68
$12\frac{1}{2}$ in.	98
17 in.	130
22 in.	160
Concrete block, heavy aggregate:	
8 in.	55
12 in.	85
Concrete block, light aggregate:	
8 in.	35
12 in.	55
Structural clay tile, load bearing:	
8 in.	42
12 in.	58
Brick, load-bearing structural clay tile backing:	
4 + 4 in.	60
4 + 8 in.	75
8 + 4 in.	102
Furring tile (2 in.) on one side of masonry wall: add to above figures	12

[a]Source: This material is reproduced with permission from American National Standard ANSI A58.1-1982, *Minimum Design Loads for Buildings and Other Structures*, copyright 1982 by the American National Standards Institute. Copies of this standard may be purchased from the American National Standards Institute at 1430 Broadway, New York, NY 10018.

[b]Weights of masonry include mortar but not plaster. For plaster, add 5 psf for each face plastered. Values given represent averages. In some cases, there is a considerable range of weight for the same construction.

TABLE 7.12

Minimum Design Loads for Materials[a] (pcf)

Material	pcf	Material	pcf
Bituminous products		Lead	710
Asphaltum	81	Lime, hydrated, loose	32
Graphite	135	Lime, hydrated, compacted	45
Paraffin	56	Masonry, ashlar:	
Petroleum, crude	55	Granite	165
Petroleum, refined	50	Limestone, crystalline	165
Petroleum, benzine	46	Limestone, oolitec	135
Petroleum, gasoline	42	Marble	173
Pitch	69	Sandstone	144
Tar	75	Masonry, brick:	
Brass	526	Hard (low absorption)	130
Bronze	552	Medium (medium absorption)	115
Cast-stone masonry (cement, stone,sand)	144	Soft (high absorption)	100
Cement, Portland, loose	90	Masonry, rubble mortar:	
Ceramic tile	150	Granite	153
Charcoal	12	Limestone, crystalline	147
Cinders, dry, in bulk	45	Limestone, oolitec	138
Cinder fill	57	Marble	156
Coal, anthracite, piled	52	Sandstone	137
Coal, bituminous, piled	47	Mortar, hardened	
Coal, lignite, piled	47	Cement	130
Coal, peat, dry, piled	23	Lime	110
Concrete, plain:		Particleboard	45
Cinder	108	Plywood	36
Expanded-slag aggregate	100	Riprap (not submerged):	
Haydite (burned-clay aggregate)	90	Limestone	83
Slag	132	Sandstone	90
Stone (including gravel)	144	Sand, clean and dry	90
Vermiculite and perlite aggregate,	25–50	Sand, river, dry	106
nonload-bearing		Slag, bank	70

Other light aggregate load-bearing	70–105	Slag, bank screenings	108
Concrete, reinforced:		Slag, machine	96
Cinder	111	Slag, sand	52
Slag	138	Slate	172
Stone (including gravel)	150	Steel, cold-drawn	489
Copper	556	Stone, quarried, piled:	
Cork, compressed	14.4	Basalt, granite, gneiss	96
Earth (not submerged):	63	Limestone, marble, quartz	95
Clay, dry	63	Sandstone	82
Clay, damp	110	Shale	92
Clay and gravel, dry	100	Greenstone, hornblende	107
Silt, moist, loose	78	Terra cotta, architectural:	
Silt, moist, packed	96	Voids filled	120
Silt, flowing	108	Voids unfilled	72
Sand and gravel, dry, loose	100	Tin	459
Sand and gravel, dry, packed	110	Water, fresh	62.4
Sand and gravel, wet	120	Water, sea	64
Earth (submerged):		Wood, seasoned[b]	
Clay	80	Ash, White	41
Soil	70	Cypress	34
River mud	90	Fir, Douglas	34
Sand and gravel	60	Hem-Fir	28
Sand or gravel, and clay	65	Oak, red and white	47
Gravel, dry	104	Pine, Southern	37
Gypsum, loose	70	Redwood	28
Gypsum wallboard	50	Spruce, red, white, and Sitka	29
Ice	57.2	Western Hemlock	32
Iron, cast	450	Zinc, rolled, sheet	449
Iron, wrought	480		

[a]Source: This material is reproduced with permission from American National Standard ANSI A58.1-1982, *Minimum Design Loads for Buildings and Other Structures*, copyright 1982 by the American National Standards Institute. Copies of this standard may be purchased from the American National Standards Institute at 1430 Broadway, New York, NY 10018.

[b]For additional information on weights of commercial lumber species, see Table 2.2

ENGLISH–METRIC CONVERSION FACTORS

TABLE 7.13

Multiply	By	To obtain
LENGTH		
Inches	25.4	Millimeters
Feet	0.3048	Meters
Yards	0.9144	Meters
Miles, statute	1.609347	Kilometers
Miles, nautical	1.852	Kilometers
Millimeters	0.0393701	Inches
Meters	3.280840	Feet
	1.093613	Yards
Kilometers	0.62137	Statute miles
	0.5399568	Nautical miles
AREA		
Square inches	645.16	Square millimeters
Square feet	0.092903	Square meters
Square yards	0.836127	Square meters
Square miles	2.58999	Square kilometers
Acres	0.0040469	Square kilometers
Square millimeters	0.0015500	Square inches
Square meters	10.76391	Square feet
	1.195990	Square yards
Square kilometers	0.3861022	Square miles
	247.1054	Acres
VOLUME AND CAPACITY		
Cubic inches	16,387.1	Cubic millimeters
Cubic feet	0.0283168	Cubic meters
Cubic yards	0.764555	Cubic meters
Ounces (U.S. fluid)	0.0295735	Liters
Quarts (U.S. liquid)	0.9463529	Liters
Gallons (U.S. liquid)	3.785412	Liters
Cubic centimeters	0.610237	Cubic inches
Cubic meters	35.31472	Cubic feet
	1.307951	Cubic yards
Liters	33.81406	Ounces (U.S. fluid)
	1.056688	Quarts (U.S. liquid)
	0.2641721	Gallons (U.S. liquid)
MASS		
Ounces (avdp.)	28.34952	Grams
Pounds	453.5924	Grams
Tons, short	907.1847	Kilograms
Grams	0.0352740	Ounces (avdp.)
	0.0022046	Pounds
Kilograms	0.0011023	Short tons

TABLE 7.13 (*Continued*)

Multiply	By	To obtain
FORCE (WEIGHT AT SEA LEVEL)		
Pounds (force)	4.44822	Newtons
Tons, short (force)	8.89644	Kilonewtons
Newtons	0.2248089	Pounds (force)
Kilonewtons	1.1240×10^{-4}	Tons, short (force)
PRESSURE		
Pounds (force) per square inch	6,894.757	Pascals (newtons per square meter)
Pounds (force) per square feet	47.88026	Pascals
Pascals	1.4504×10^{-4}	Pounds (force) per square foot
	0.0208854	Pounds (force) per square foot
TEMPERATURE		
Degrees, Fahrenheit (less 32°F)	0.5556	Degrees, Celsius
Degrees, Celsius	1.8	Degrees, Fahrenheit (less 32°F)
SPEED AND VELOCITY		
Feet per second	1.09728	Kilometers per hour
Miles per hour	1.609344	Kilometers per hour
Kilometers per hour	0.9113444	Feet per second
	0.6213712	Miles per hour

SELECTED AITC STANDARDS

On the following pages are reprints of selected AITC standards pertinent to this timber construction manual. The editions of the standards current at the time of this edition of the manual are included; however, the standards are frequently updated. Please check to determine the latest edition of the standards by writing or calling:

American Institute of Timber Construction
333 West Hampden Avenue
Englewood, Colorado 80110
Phone: (303) 761-3212

Additional AITC standards, not reprinted in this manual, are also available from the same address. They include:

Standard Specifications for Structural Glued Laminated Timber of Softwood Species, AITC 117—Manufacturing
Structural Glued Laminated Timber for Electric Utility Framing and Crossarms, AITC 114

AMERICAN INSTITUTE
333 West Hampden Avenue

TIMBER CONSTRUCTION
Englewood, Colorado 80110

AITC 104-84
TYPICAL CONSTRUCTION DETAILS

Adopted as Recommendations July 18, 1984

CONTENTS

1. INTRODUCTION

1.1. These typical construction details are intended as guides for architects and engineers. They have been developed and used by the engineered timber construction industry and, being based on judgment and experience, will help to assure a high quality of construction.

1.2 Warning. Because the details are to be used only as guides, dimensions have not been included and the drawings should not be scaled. Quantities and sizes of bolts, connectors and other fastening hardware are illustrative only. The actual quantities and sizes required will depend on the loads to be carried and the member sizes. End and edge distances, as well as spacing between fasteners, should be in accordance with the *National Design Specification for Wood Construction* by the National Forest Products Association. Sufficient clearance must be provided between sides of steel connection hardware and wood members to permit installation. This clearance should not exceed the member width plus $\frac{1}{4}$ in.

1.3 Designing for Strength. Connection details must effectively transfer loads, utilize durable materials and be as free from maintenance as possible. The strength of wood is different in the parallel and perpendicular to grain directions. Wood also has much less strength in tension perpendicular to grain than in compression perpendicular to grain. These facts influence design details.

Vertical loads should be transferred so as to take advantage of the high compression perpendicular to grain strength of wood. For example, a beam should bear on the top of a column or wall or be seated in a shoe or hanger. Such a detail is preferred to the support of a beam by bolts at its end, particularly where there are large numbers of bolts.

Beams should be anchored at the ends in order to carry induced horizontal and vertical loads. Vertical loads may be either gravity loads or net uplift loads. The connections typically shown in this standard are primarily for vertical gravity loads. Provisions should be made to resist uplift or lateral loads as required. The bolts or fasteners at the beam ends must be located near the bottom bearing of the beam to minimize the effect of shrinkage of the wood between the bottom of the beam and the fasteners.

In many cases, individual details do not include all structural elements such as lateral bracing ties to connect all the components of the building together.

Loads suspended from glued laminated timber beams or girders should preferably be suspended from the top of the member or at least above the neutral axis.

1.4 Consideration of Shrinkage and Swelling. In addition to designing connections to transfer loads, effort should be made to avoid splitting the member due to expansion and contraction of the wood. Consideration must be given to wood swelling and shrinking due to moisture content changes in service similar to the consideration given to details in metal construction which must accommodate the expanding and contracting metal due to changes in temperature.

Because wood swells and shrinks (primarily in the perpendicular to grain direction) due to moisture content changes, connections should not restrain this movement. Figure 1.1 illustrates typical shrinkage in a sawn member when drying from green to 8% moisture content and a glued laminated timber drying from 12% to 8% moisture content. Even in covered structures, large laminated timbers may shrink after installation due to moisture loss from low relative humidity conditions. Long rows of bolts perpendicular to grain fastened to a single cover plate should be avoided. Although relatively dry at time of manufacture, glued laminated timber can still shrink to reach equilibrium moisture content in service.

When possible, designers should avoid joint details that could loosen in service due to wood shrinking or that could cause problems when wood expands due to increased moisture content. Machine bolts should be used rather than lag bolts whenever possible. When lag bolts are used, correct lead hole sizes are important. Connections should be detailed to avoid loading lag bolts in withdrawal whenever possible.

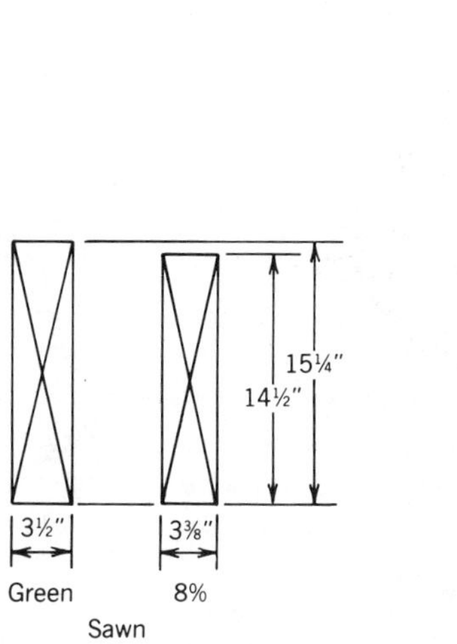

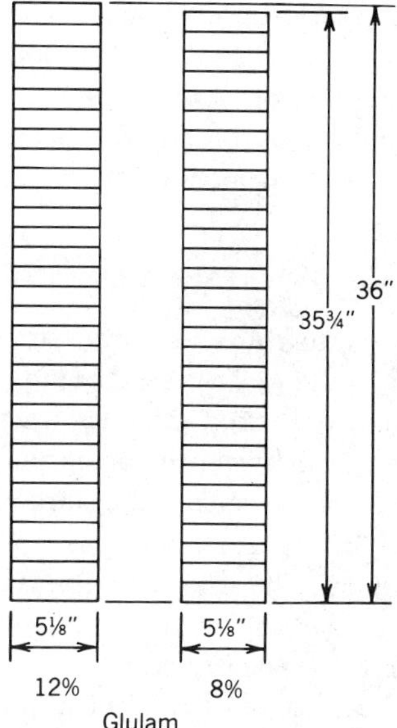

FIGURE 1.1 Shrinkage due to moisture loss.

1.5 Designing to Avoid Tension Perpendicular to Grain Stresses. Whenever possible, joints should be designed to avoid causing tension perpendicular to grain streses in wood members. Examples of connections that induce tension perpendicular to grain stresses are simple beams which have been notched at the ends on the tension side (see Detail A5).

Avoid long lines of fasteners spaced close together along the grain, particularly if the bolts are in tightly drilled holes. These types of connections may induce tension perpendicular to grain stresses due to prying actions from secondary moments.

1.6 Consideration of Decay. When proper construction details are used and other good design and construction practices are followed, wood is a permanent construction material. Moisture barriers, flashings and other protective features should be used to avoid moisture or free water being trapped (see Section 11, "Details to Protect Against Decay"). Preservative treatments are recommended when wood is fully exposed to the weather without roof cover. Provide adequate site drainage, protection during construction and protect metals from corrosion by use of corrosive-resistant metals or resistant coatings or platings. Do not embed wood columns or arch bases below finished concrete floor levels.

1.7. Other considerations are also necessary in the design of structures employing engineered timber construction. These are outlined in *Spec-Data Sheet for Structural Glued Laminated Timber*, and other AITC standards and recommendations.

1.8 End Rotation of Beams. Consideration should be given to end rotation of beams resulting from vertical load deflection. Location of fasteners that tend to create end fixity should be avoided. Splitting at fasteners can result unless such a connection is designed to develop a fixed end moment sufficient to resist end rotation due to deflection.

2. BEAM TO MASONRY ANCHORAGES

Figure 2.1 illustrates a common type of beam seat to resist both uplift and horizontal forces as well as vertical loads. In the case of uplift forces, the notched beam effect must be checked. The seat may be anchored in the concrete or masonry with one or more anchors. The beam may be fastened to the tabs on the seat with one or more bolts or, where forces are greater, with bolts and shear plates.

An important point illustrated in Figure 2.1, as well as other figures in this section, is that the wood member is separated from the masonry or concrete at the bearing surface by the bearing plate and along the vertical surfaces by a minimum of a $\frac{1}{2}$ in. clearance.

For beams sloped where a $\frac{1}{8}$ in. or greater gap might occur, the beam bearing should be detailed to obtain full contact between bearing surfaces (see Figures 2.6 and 2.7). Seat cuts at the top end of the slope should be checked for notched beam effect.

Figures 2.2 through 2.5 illustrate various beam-to-masonry anchorages.

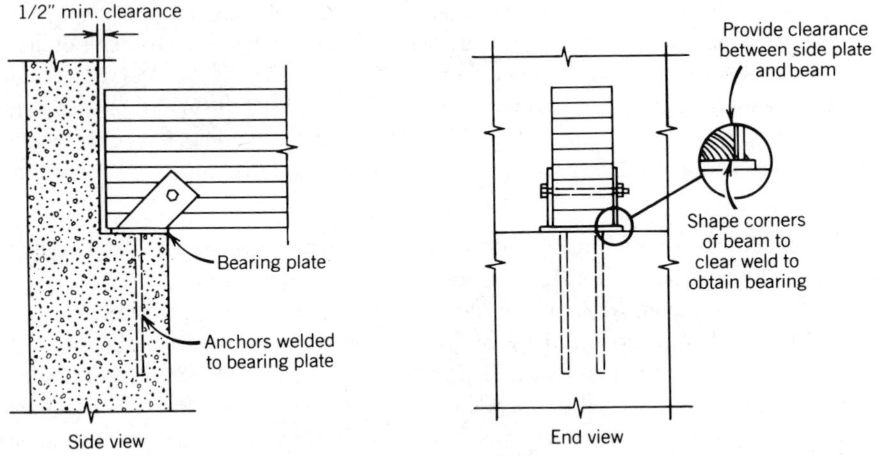

FIGURE 2.1

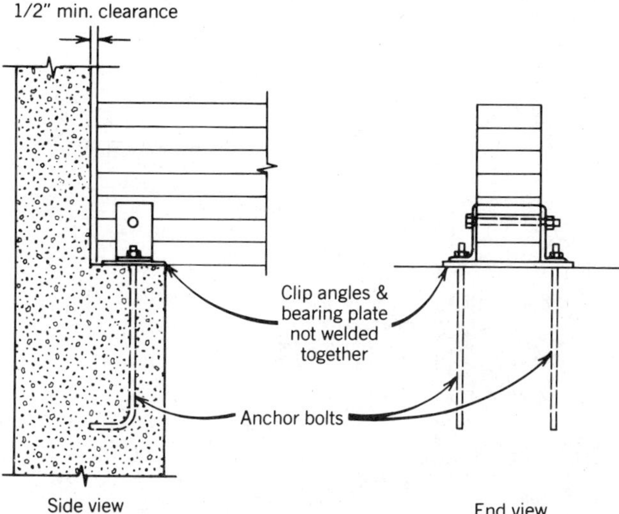

FIGURE 2.2 Shows steel angles with a separate bearing plate with only the anchor bolts being cast in place.

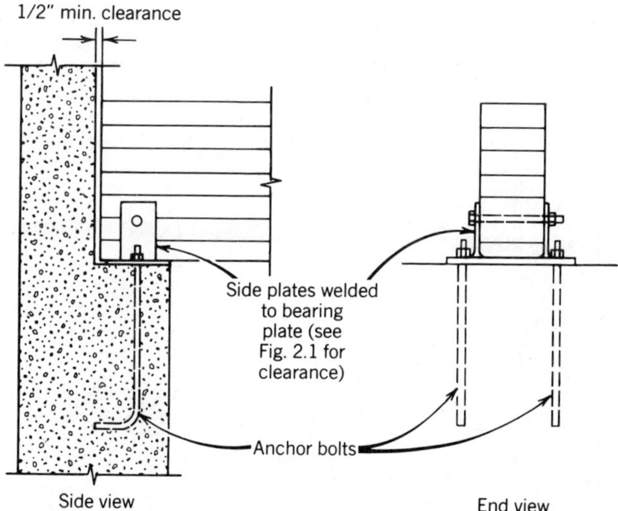

FIGURE 2.3 Similar to the detail in Figure 2.1 except the side plates are vertical which provides less end distance for the bolts and may lessen their resistance to horizontal forces.

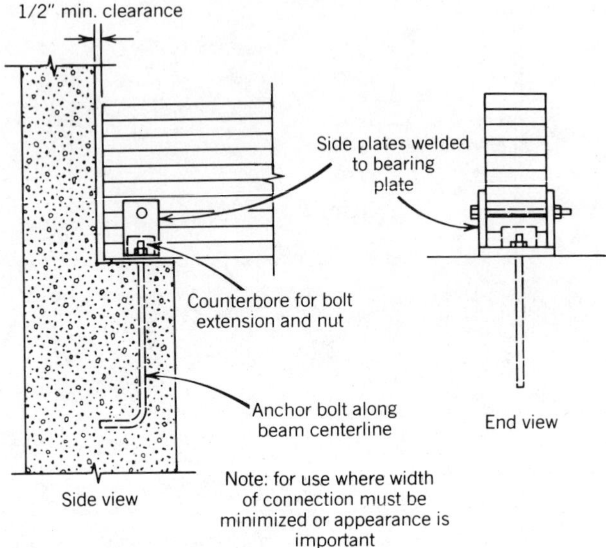

FIGURE 2.4 This detail may be used when the pilaster is not wide enough for outside anchor bolts. The anchor(s) may be welded to the underside of the bearing plate or the bolts and nuts may be located in holes counterbored into the bottom of the beam.

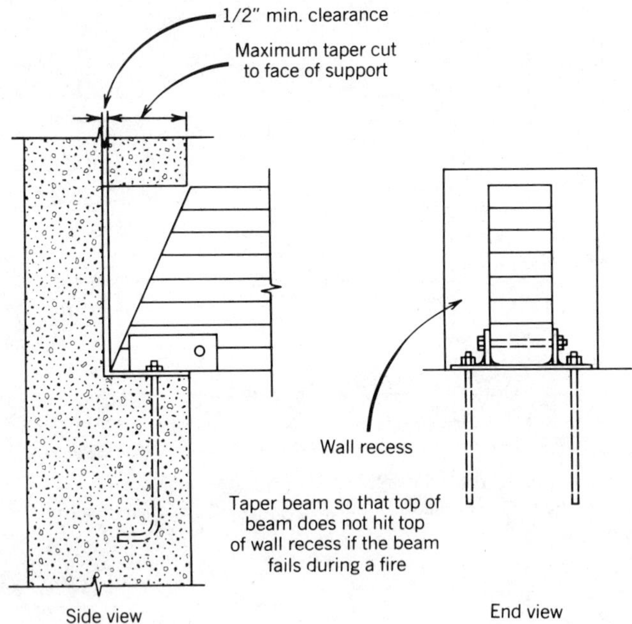

FIGURE 2.5 This detail illustrates a typical taper end cut sometimes referred to as a fire cut.

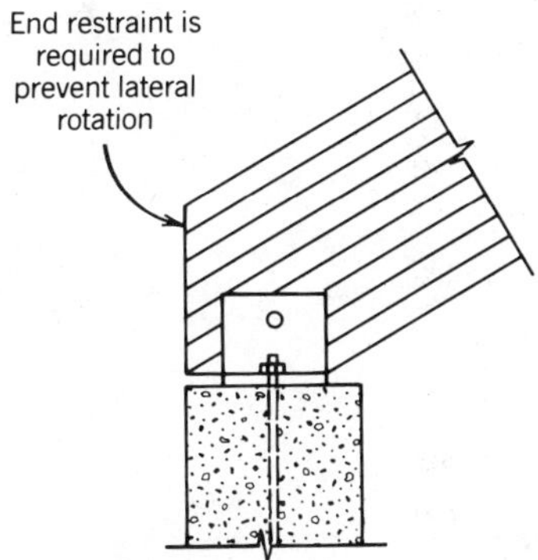

FIGURE 2.6 Sloped Beam—Lower End. The taper cut beam should be in bearing contact with the bearing plate. See Detail A6.

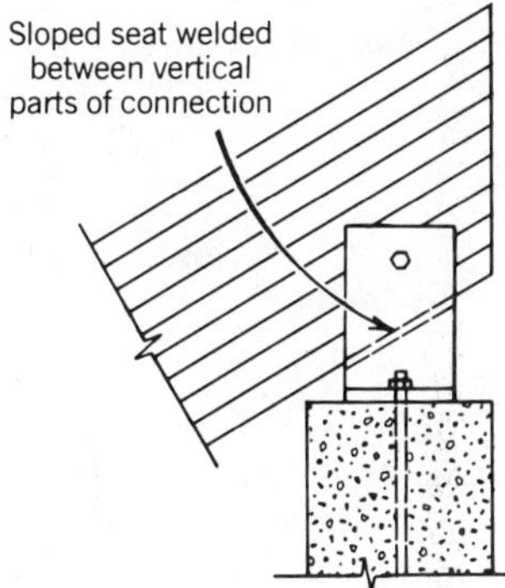

FIGURE 2.7 Sloped Beam—Upper End. The support at the top end of a sloped member should be designed with a sloping seat rather than a notched end. The bolt must be designed to resist the parallel-to-grain component of the vertical beam reaction. See Detail A7.

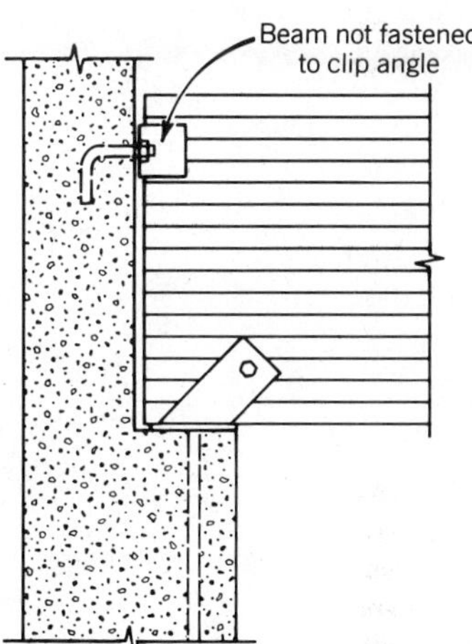

FIGURE 2.8 Lateral support of the ends of beams can be provided with the clip angles anchored to the wall but without a connection to the beam. This will not restain vertical movement due to shrinkage or horizontal movement due to end rotation. See Detail A8.

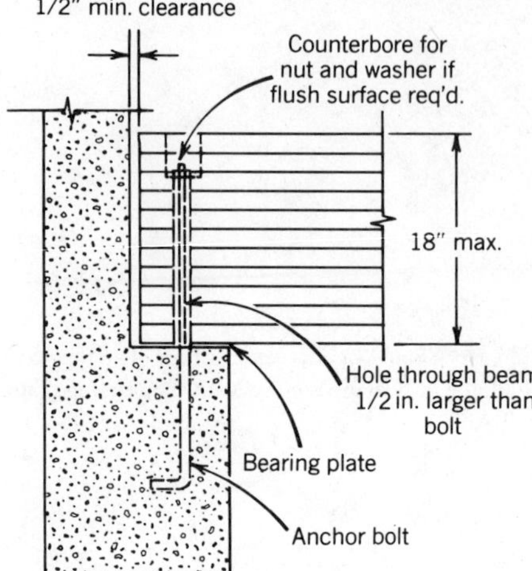

FIGURE 2.9 Simple Beam Anchorage. Resists uplift and small horizontal forces. Bearing plate or moisture barrier is recommended. Provide 1/2 in. minimum clearance from all wall contact surfaces, ends, sides and tops (if masonry exists above beam end).

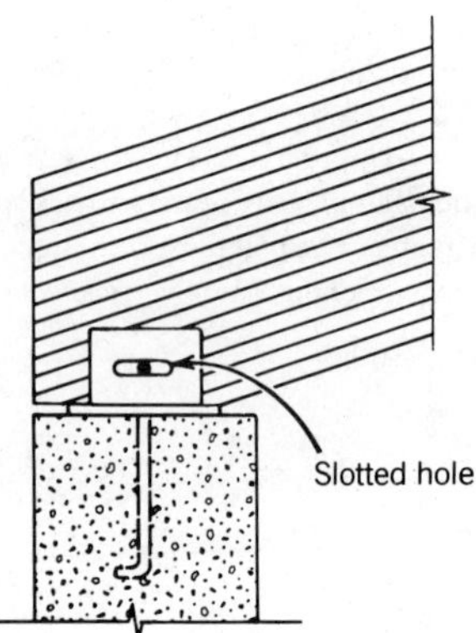

FIGURE 2.10 Curved or Pitched Beam Anchorage—Typical Slip Joint. Slotted or oversize holes at one or both ends of beam permit horizontal movement under lateral deflection or deformation. Length of slotted or oversized hole is based on calculated maximum horizontal deflection. Position bolt in slot to allow for anticipated movement. Bolt should be hand tightened only to permit movement. See Detail A6.

In certain seismic zones where beam provides lateral support for wall, this joint detail is not recommended.

3. CANTILEVER BEAM CONNECTIONS

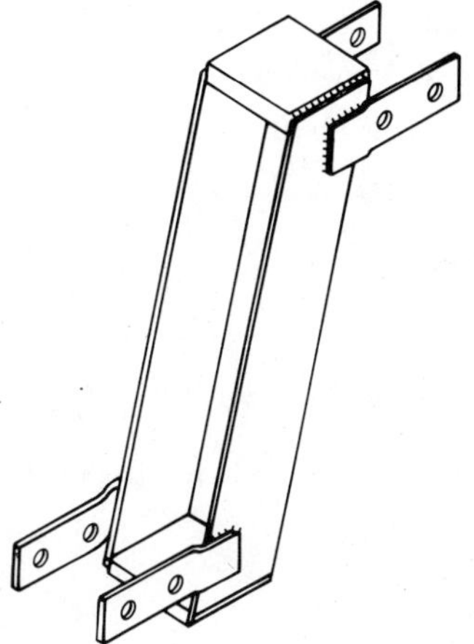

FIGURE 3.1 Cantilever Hinge Connection. Dee Detail A9. This is a common type cantilever beam connector. The details that follow are examples using this connector.

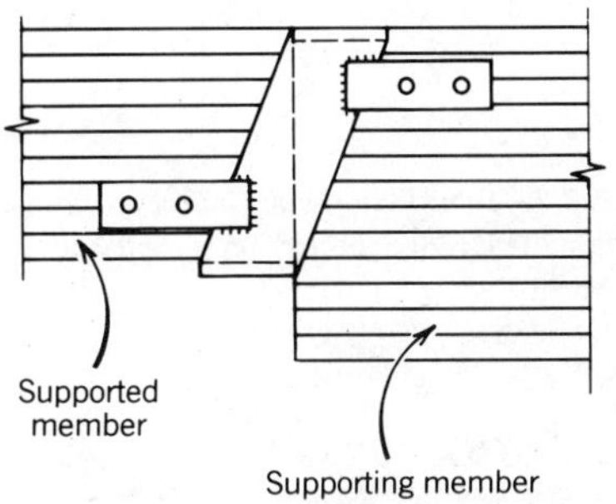

The vertical reaction of the supported member is carried by the side plates and transferred in bearing perpendicular to grain to the supporting member. The rotation due to the eccentric loading is resisted by the bolts through the tabs at the top and bottom. The connector may be installed with the top (and bottom) bearing plates dapped into the members to obtain a flush surface or may be installed without daps. Notching on the tension side should be minimized.

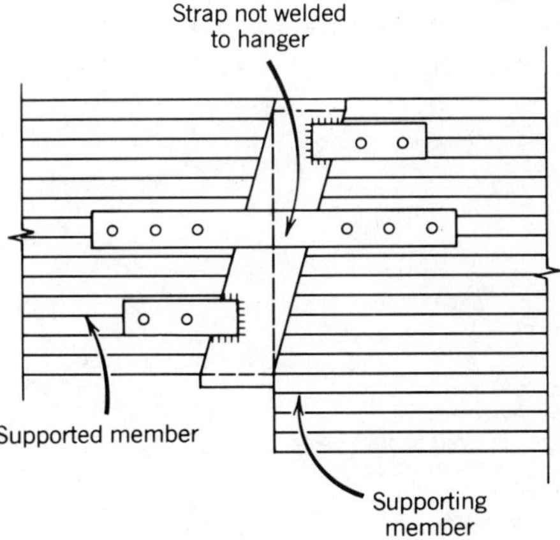

Where horizontal forces must be resisted by a hinge connection, loose tension ties may be installed on both sides of the beam. The tie shown is not fastened to the cantilever hanger.

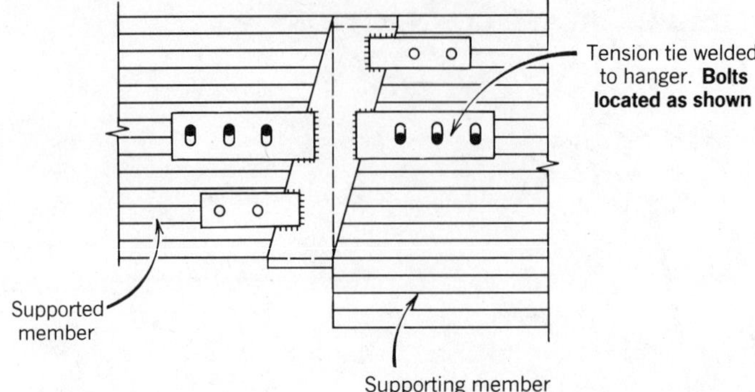

If tension ties are fastened to the cantilever hanger, vertically slotted holes are required in the tie and careful location of the bolts in the end of the slot farthest from the bearing seat is required to prevent splitting due to shrinkage and seating deformations. Bolts in slotted holes should be hand tightened only to permit movement.

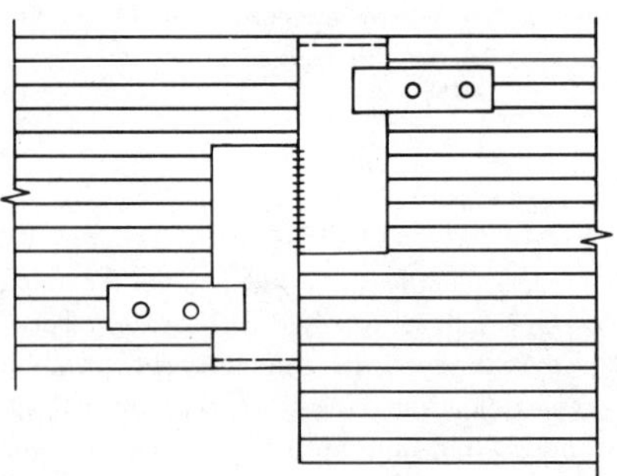

FIGURE 3.2 Bent Plate Type Cantilever Hinge Connection. This hinge connection is similar to the one shown in Figure 3.1 but is usually used with smaller members.

4. BEAM AND PURLIN HANGERS FOR ROOF SYSTEMS

In Figure 4.1 and similar details, locate fasteners as close as practical to the bearing surface to minimize splitting due to shrinkage (See Section 1.4). For floor systems, additional restrictions may be necessary to minimize the effects of differential shrinkage of connected members. See Detail A1.

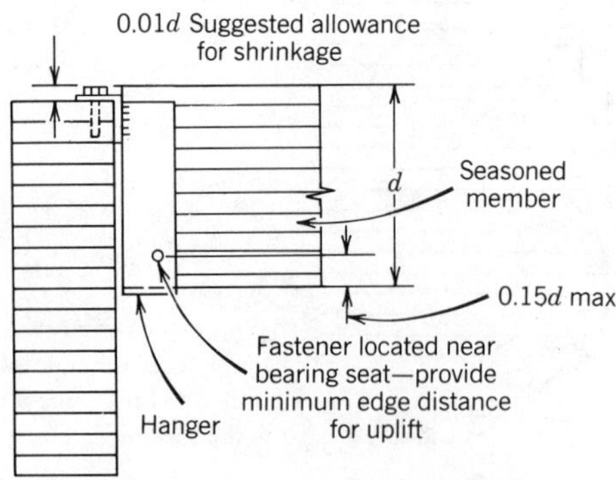

FIGURE 4.1 Seasoned Members. When supported members are of seasoned material, the top of the supported member may be set approximately flush with the top of the supporting member.

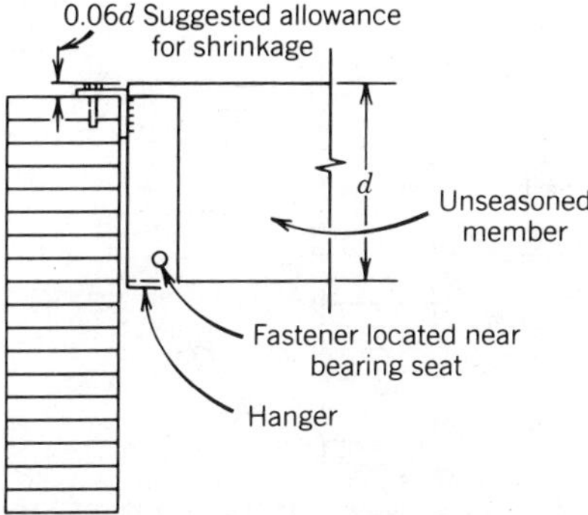

FIGURE 4.2 Unseasoned Members. When supported members are of unseasoned material, the hangers should be so dimensioned that the top edge of the supported member is raised above the top of the supporting member or the top of the hanger strap to allow for shrinkage as the members season in place. For supported members with moisture content at or above fiber saturation point when installed, the distance raised should be about 6% of the member's depth above its bearing point. Note: For main members loaded on one side as in Figures 4.1, 4.2 and 4.3, provide a tie between the beam and purlin to restrain potential rotation of the beam due to eccentricty of the hanger load.

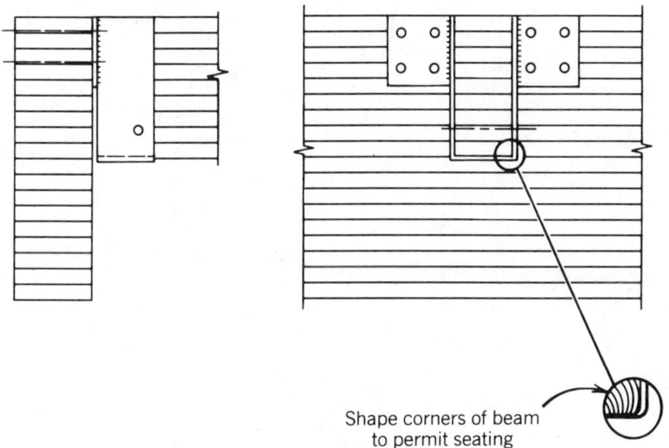

FIGURE 4.3 Welded Face Hanger. See Detail A2.

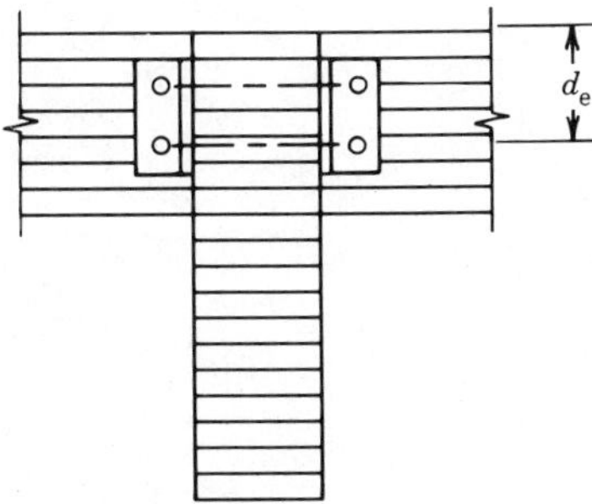

FIGURE 4.4 A clip angle connection without a bearing seat may be used for small beams and light loads. The connection should be designed for gravity loads as a notched beam using d_e as shown in the notched beam formula. The distance between bolts should be checked for possible effects of shrinkage. A bearing connection as shown in Figure 4.1 is preferred to the support of the beam by bolts. See Detail A3.

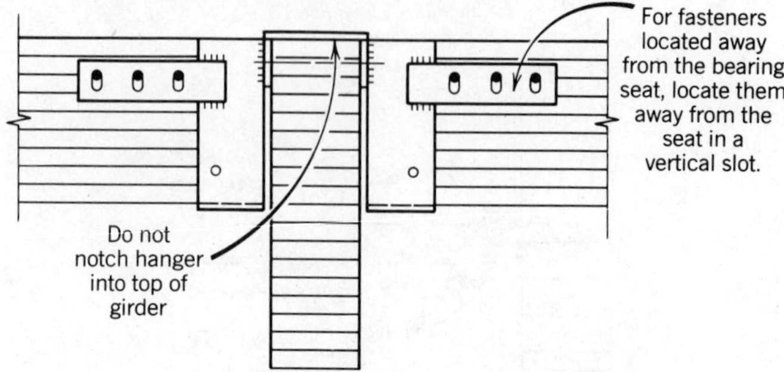

FIGURE 4.5 Welded and Bent Strap Hanger. A separate tension tie may be used across the top in lieu of the tabs to resist lateral forces.

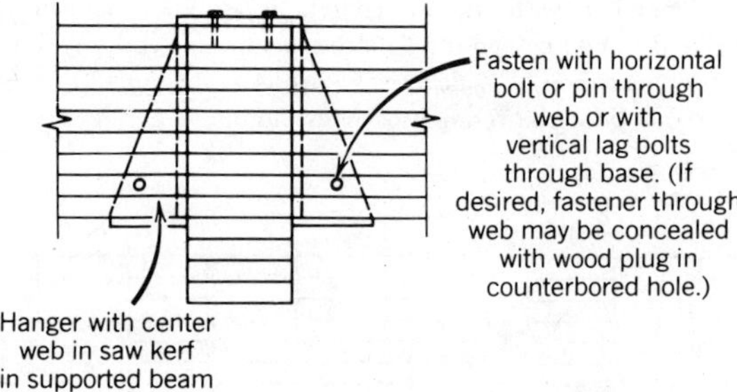

FIGURE 4.6 Partially Concealed Type. For moderate loads. Base may be let in flush with bottoms of purlins. See Detail A4.

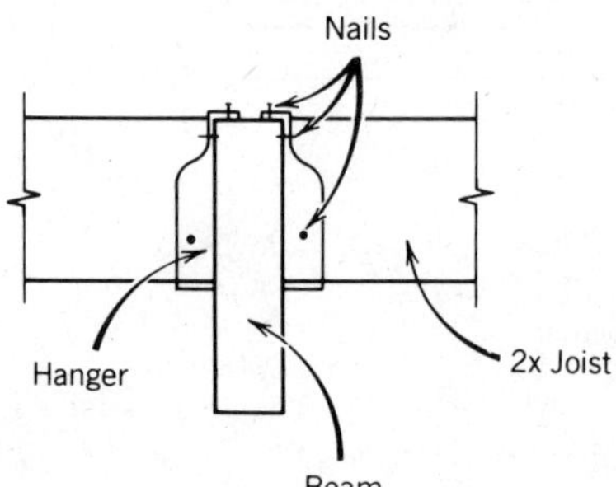

FIGURE 4.7 Stamped Joist Hanger. For light loads. Stamped from light-gage metal.

5. BEAM TO COLUMN CONNECTIONS

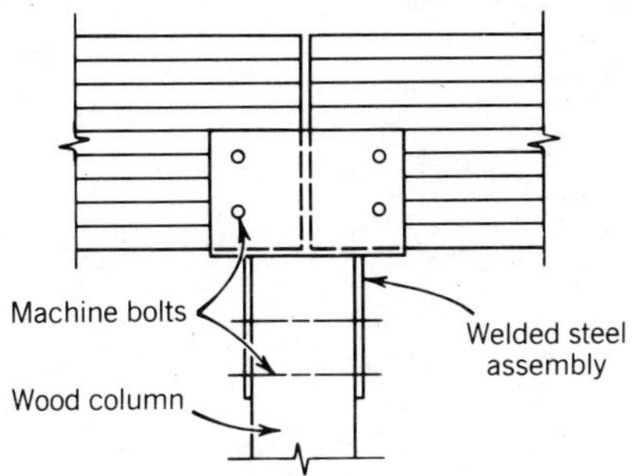

FIGURE 5.1 Beams to Wood Column—U-Plate. Welded steel assembly passes under abutting wood beams and is welded to steel side plate bolted to wood column.

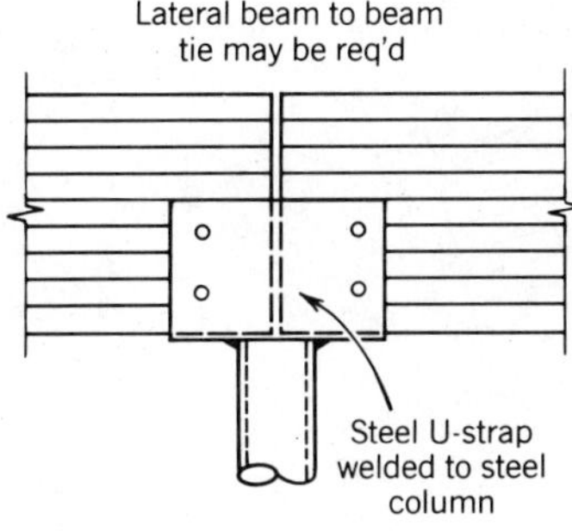

FIGURE 5.2 Beams to Steel Column. Similar to Figure 5.1.

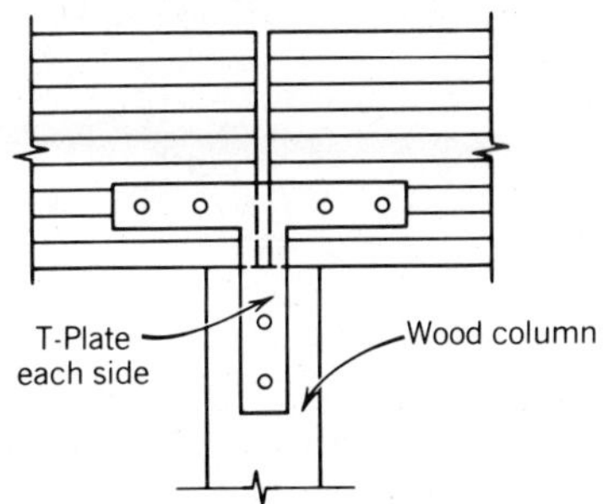

FIGURE 5.3 Beams to Wood Column—T-Plates. Steel T-plate is bolted to abutting wood beams and to wood column. Loose bearing plate may be used where column cross-sectional area is insufficient to provide bearing for beams in compression perpendicular to grain.

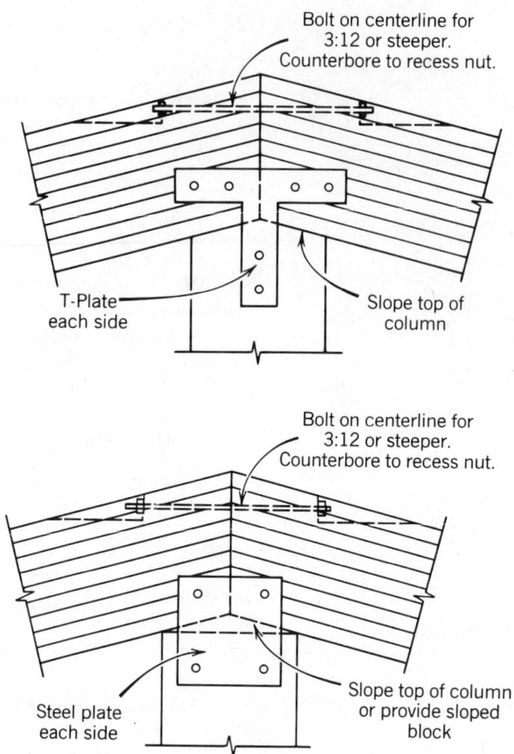

FIGURE 5.4 Shed Roof Type End Detail. For beam slopes of 1:12 maximum, beams may be notched at bottom to rest on column. Beams may be pitched away from both sides, or from only one side of column. See also Beam-to-Masonry Anchorages, Section 2.

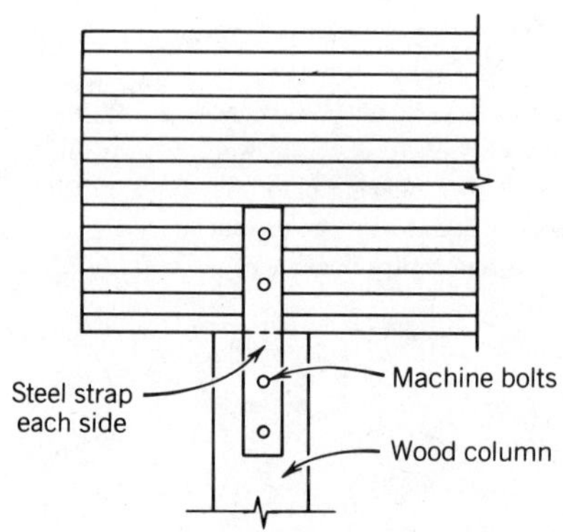

FIGURE 5.5 Beam to Wood Column. Connection provides for uplift. Metal bearing plate may be used where column cross-sectional area is insufficient to provide bearing for beam in compression perpendicular to grain.

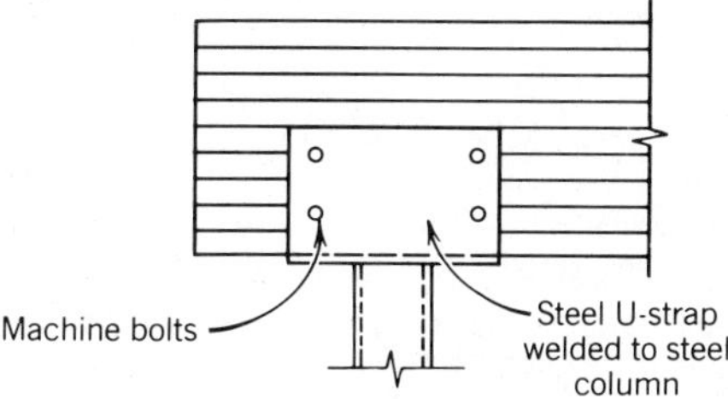

FIGURE 5.6 Beam to Steel Column. Steel U-strap passes under timber beam and is welded to top of steel column.

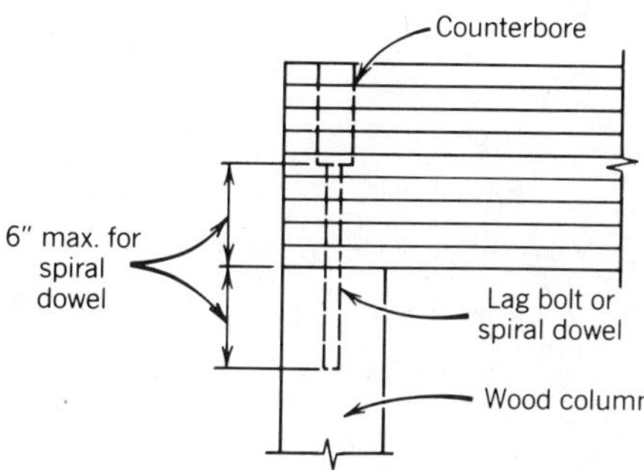

FIGURE 5.7 Concealed Type—Beam to Wood Column.

6. COLUMN ANCHORAGES

Column bearing elevation should be raised above finished floors which may be subjected to high moisture conditions. In locations where column base anchorages are subject to damange by moving vehicles, protection of the columns from such damage should be considered. Columns exposed to the weather should be treated in accordance with *Standard for Preservative Treatment of Structural Glued Laminated Timber*, AITC 109.

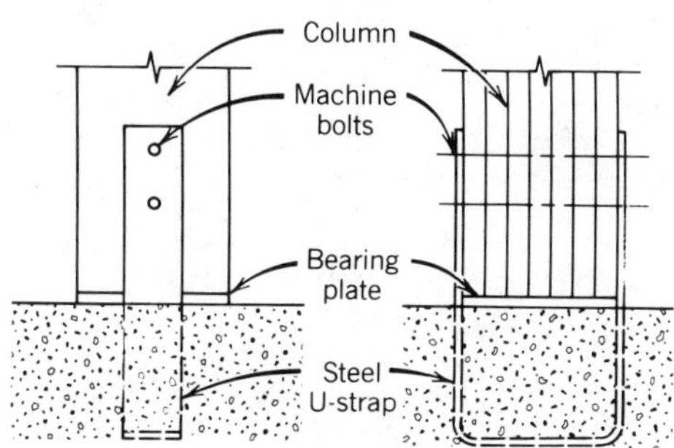

FIGURE 6.1 U-Strap Anchorage. Resists both horizontal forces and uplift. Bearing plate or moisture barrier is required. May be used with shear plates.

Do not place column below finished concrete floor level.

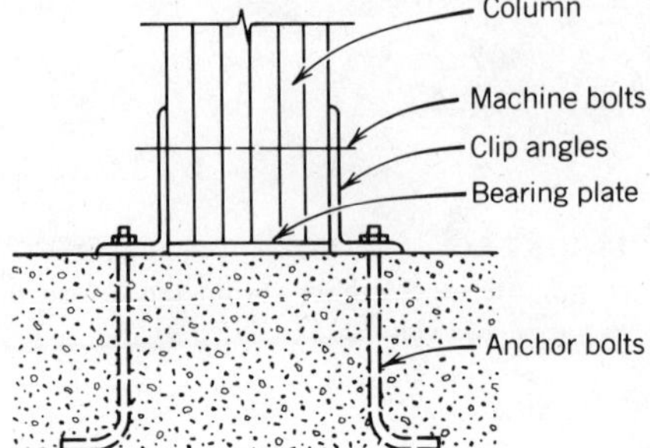

FIGURE 6.2 Clip Angle Anchorage to Concrete Base. Resists both horizontal forces and uplift. Bearing plate or moisture barrier is required.

Do not place column below finished concrete floor level.

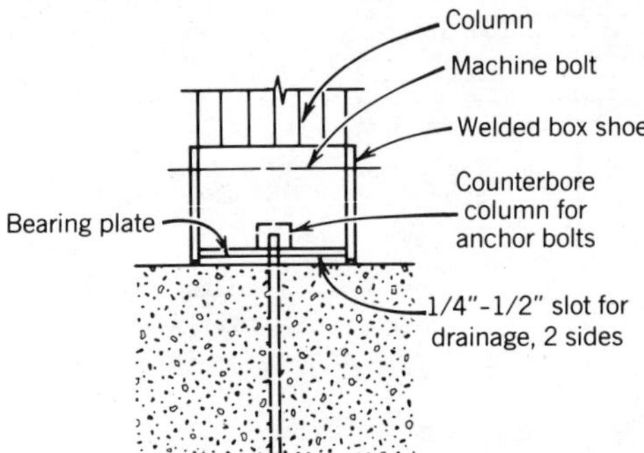

FIGURE 6.3 Box Shoe. For use when bottom of box shoe is flush with top of concrete floor.

Do not place column below finished concrete floor level.

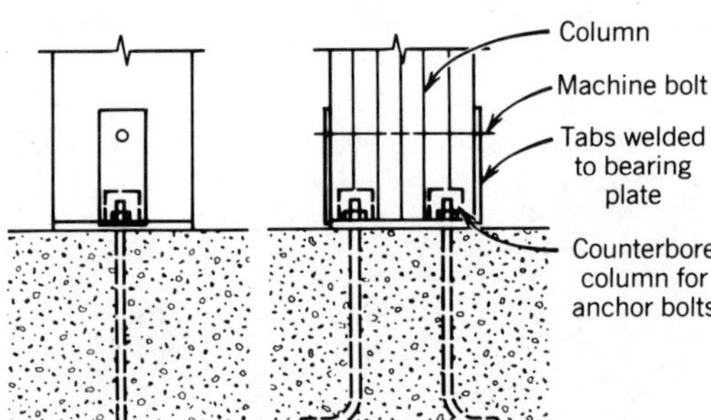

FIGURE 6.4 Semi-Concealed Column Anchorage. For use where concrete support area is limited in size. Resists both horizontal forces and uplift.

Do not place column below finished concrete floor level.

7. ARCH ANCHORAGES

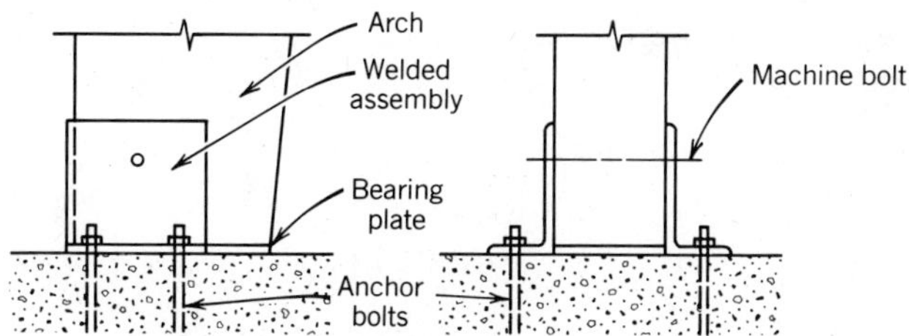

FIGURE 7.1 Arch Shoe With Exposed Anchor Bolts.

Do not place arch below finished concrete floor level.

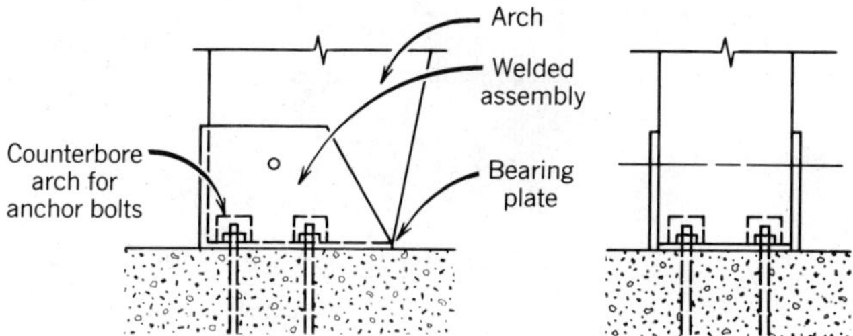

FIGURE 7.2 Arch Shoe With Concealed Anchor Bolts. Counterbores are provided in arch base for anchor bolt projections.

Do not place arch below finished concrete floor level.

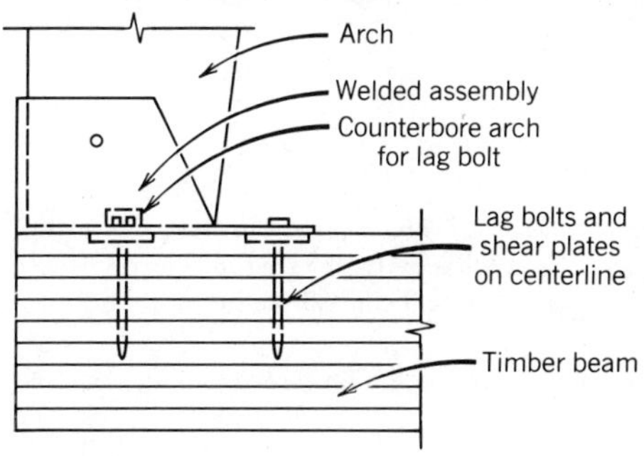

FIGURE 7.3 Arch Anchorage to Timber Beam. Vertical load is taken directly by bearing into timber beam. Vertical uplift and thrust are taken by the lag bolts and shear plates into the beam tie.

Do not place arch below finished concrete floor level.

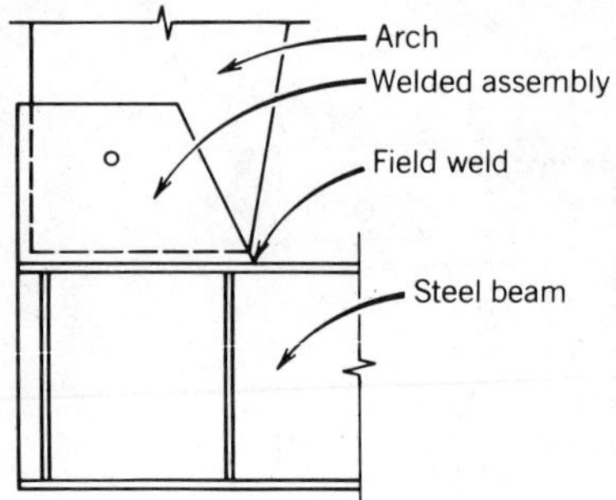

FIGURE 7.4 Arch Anchorage to Steel Girder.

Do not place arch below finished concrete floor level.

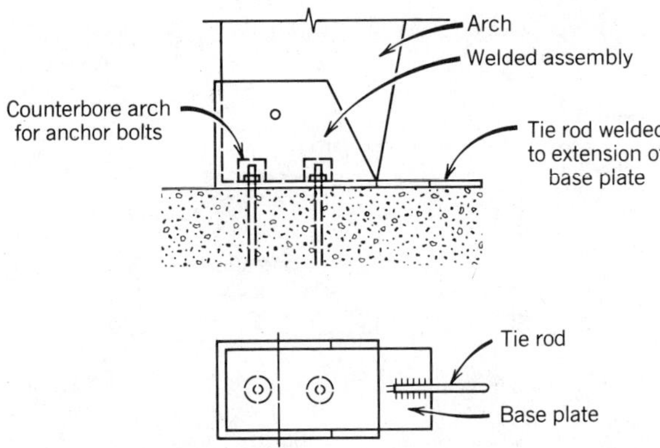

FIGURE 7.5 Tie Rod to Arch Shoe. Thrust due to vertical load is taken directly by the tie rod welded the arch shoe. This detail is intended for use with a raised joist floor where the tie rod can be concealed.

Do not place arch below finished concrete floor level.

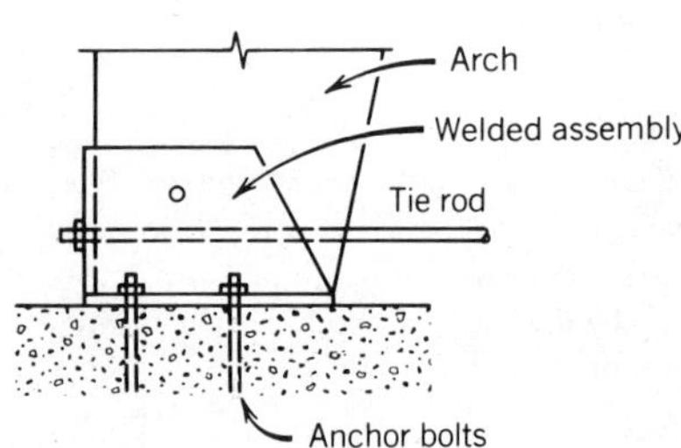

FIGURE 7.6 Tie Rod Arch. Thrust due to vertical load is taken directly by the tie rod. For use where raised joist floor will conceal the tie rod.

Do not place arch below finished concrete floor level.

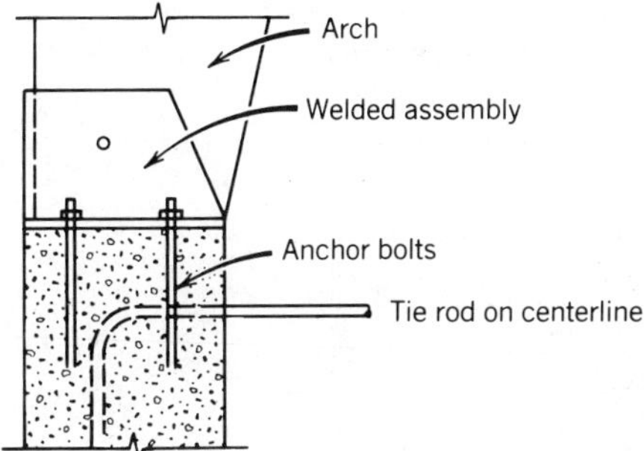

FIGURE 7.7 Tie Rod in Concrete. Thrust is taken by anchor bolts in shear into the concrete foundation and tie rod.

Do not place arch below finished concrete floor level.

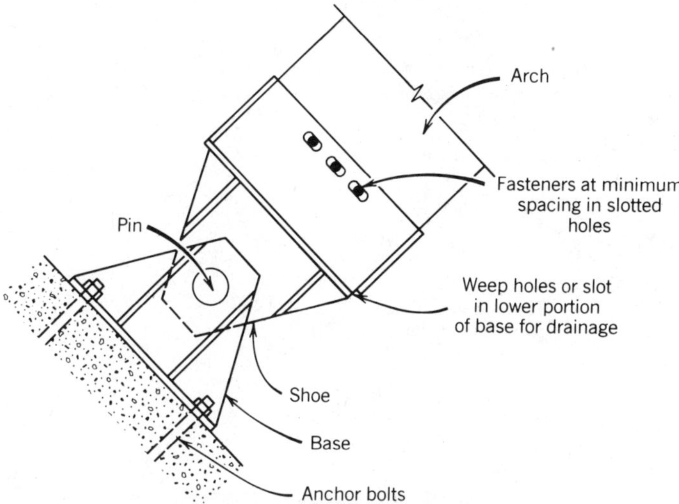

FIGURE 7.8 True Hinge Anchorage for Arches. Recommended for arches where true hinge action is desired (see Figure 11.6 for protection considerations).

Do not place arch below finished concrete floor level.

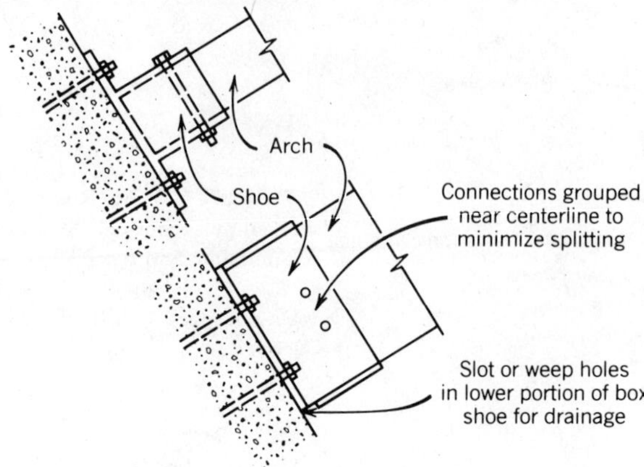

FIGURE 7.9 Arch Anchorage Where True Hinge Is Not Required. Recommened for arches where a true hinge is not required. Base shoe is anchored directly to buttress (see Figure 11.6 for protection considerations).

Do not embed arch in concrete floor.

8. ARCH CONNECTIONS

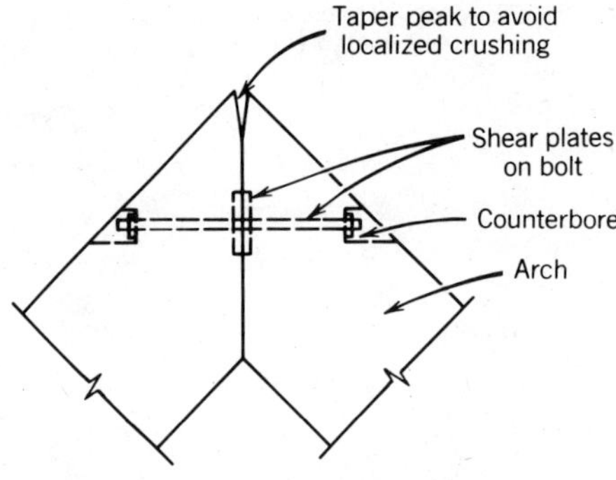

FIGURE 8.1 Arch Peak. For arches with slopes of 3:12 and greater. This connection will transfer both vertical forces (shear) and horizontal forces (tension and compression).

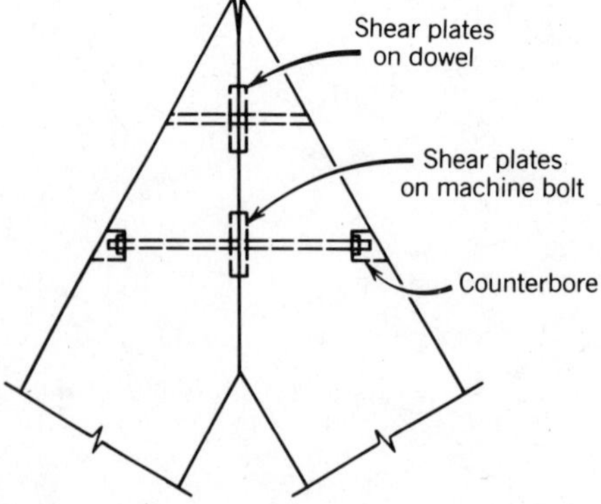

FIGURE 8.2 Arch Peak. When the vertical shear is too great for one pair of shear plates, or when deep sections would require extra shear plates for alignment, additional pairs of shear plates centered on dowels or machine bolts may be used.

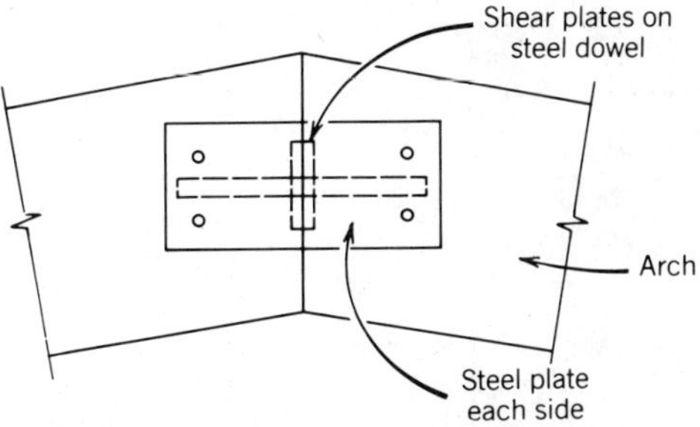

FIGURE 8.3 Low Pitched Arches. For arches with slopes that would require excessively long through bolts; shear plates back-to-back centered on a dowel are used in conjunction with a tie plate and through bolts.

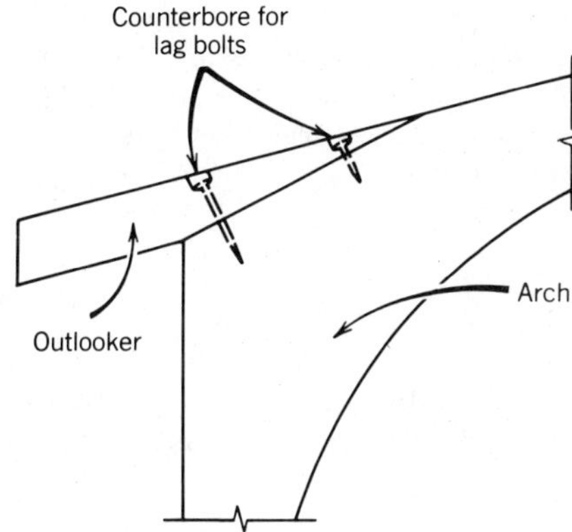

FIGURE 8.4 Outlooker Connection to Haunched Arch. Lag bolts used in this connection should be long enough so that the withdrawal resistance of the threads is in the main section of the arch. Connection must be designed to resist any cantilever action of the outlooker. Lag bolts may be counterbored when decking is applied.

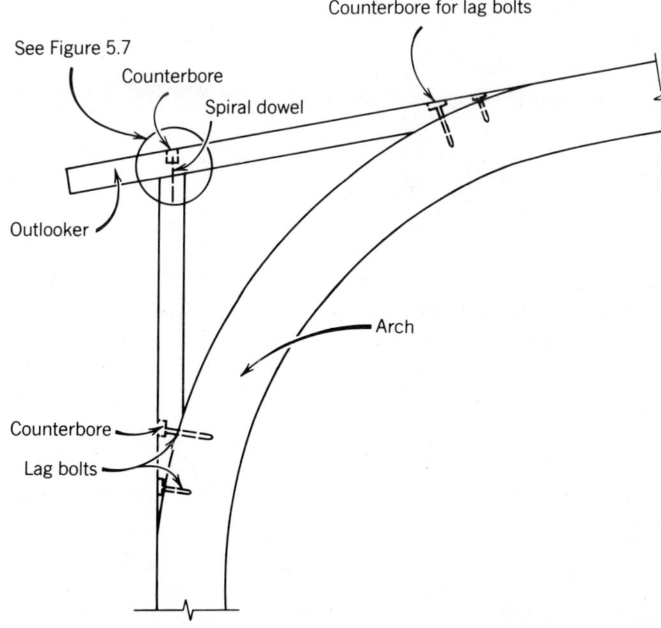

FIGURE 8.5 Outlooker Connection to Open Haunched Arch.

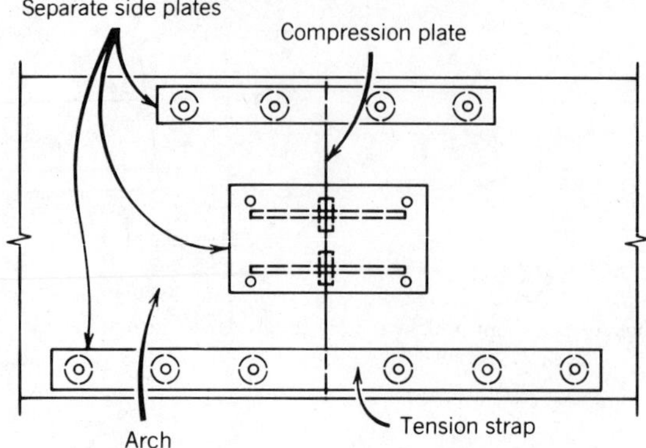

FIGURE 8.6 Arch Moment Splice. Drawing shows a typical moment splice. Compression stress is taken in bearing on the wood through a steel compression plate. Tension is taken across the splice by means of steel straps and shear plates. Side plates and straps are used to hold both sides and tops of members in position. Shear is taken by shear plates in end grain.

9. TRUSS CONNECTIONS

When unseasoned lumber is used in truss construction, periodic inspection is recommended along with retightening of hardware and connections as necessary.

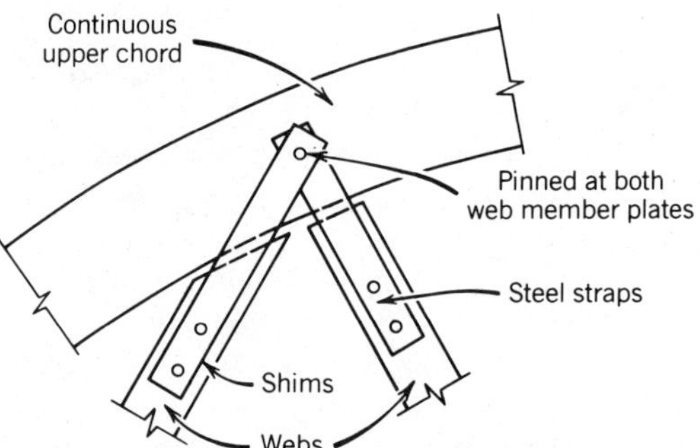

FIGURE 9.1 Monochord-Steel Straps. For trusses with continuous upper chord. Provide clearance between web ends and chord. Provide shims at web to prevent bending of plates.

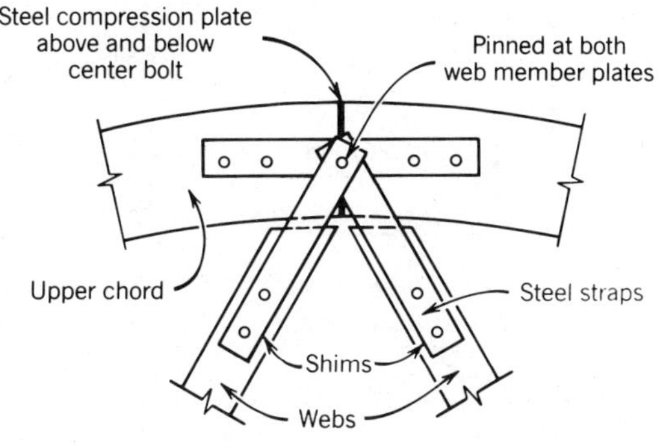

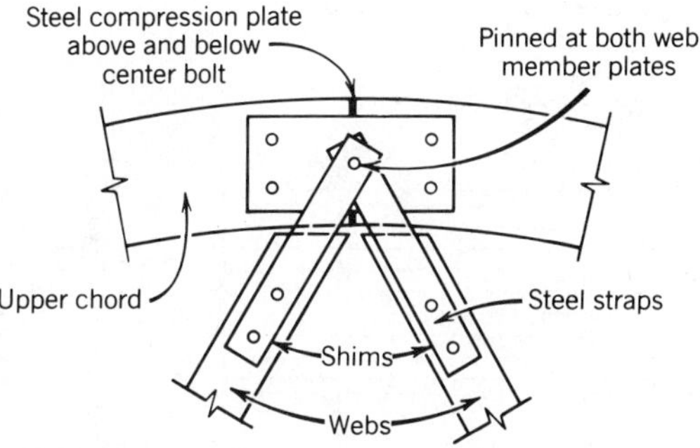

FIGURE 9.2 Monochord-Steel Strap Assembly. Similar to Figure 9.1. For use at ridge for upper chord splice. Provide shims at webs to prevent bending of plates.

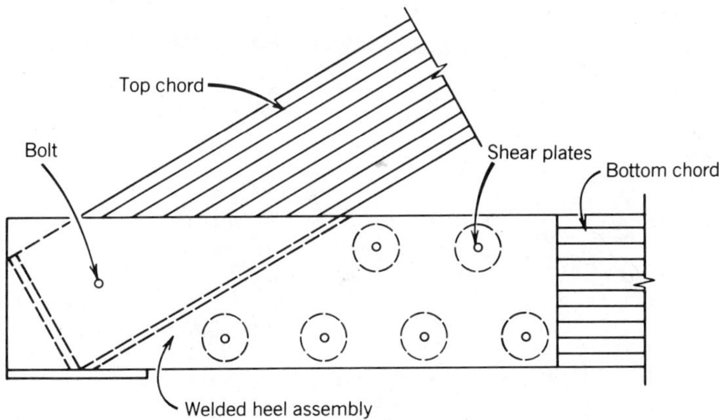

FIGURE 9.3 Truss Heel Connection. If substantial cross grain shrinkage is anticipated, double steel straps may be used in place of single plate along bottom chord.

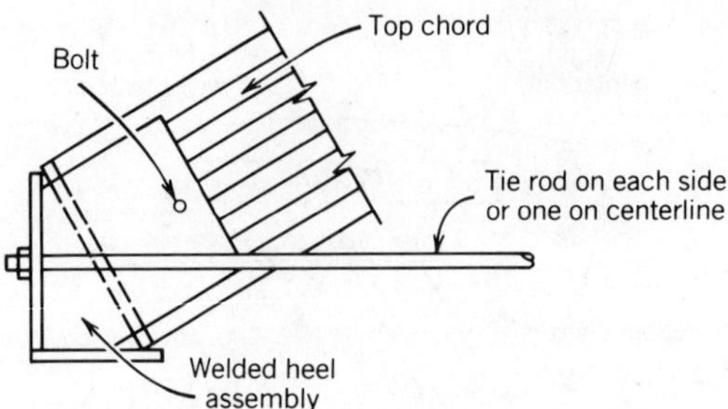

FIGURE 9.4 Rod-Tied Arch Heel Connection.

10. SUSPENDED LOADS

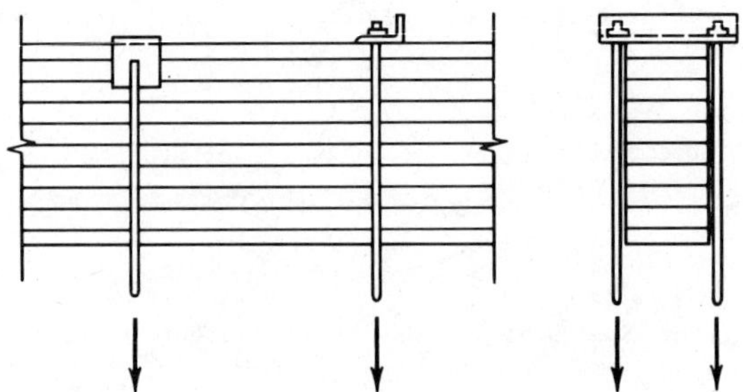

FIGURE 10.1 Loads suspended from glued laminated timber beams should be resisted from the top of the member or at least above the neutral axis.

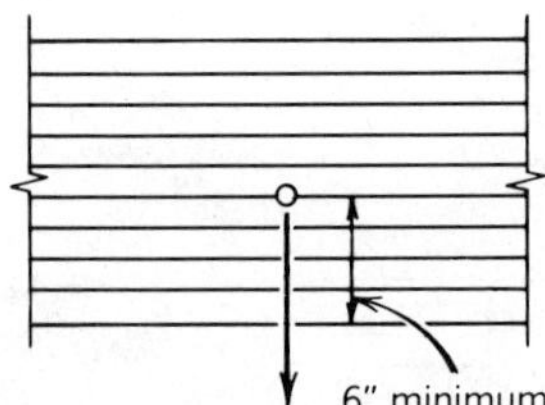

FIGURE 10.2 Light loads such as small conduit may be suspended with small fasteners near the bottom of glued laminated timber beams as shown.

11. DETAILS TO PROTECT AGAINST DECAY

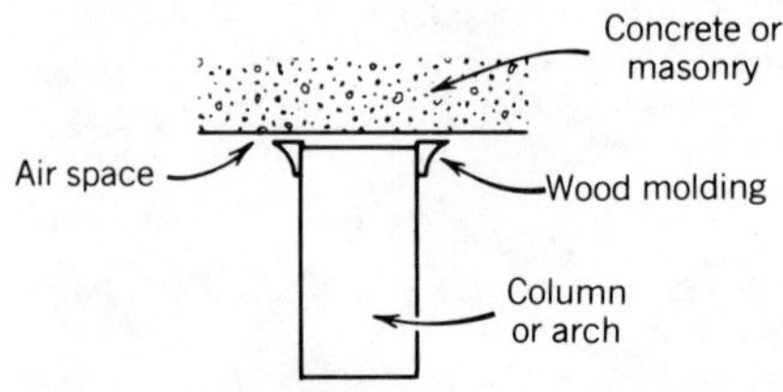

FIGURE 11.1 Wood Member Against Continuous Masonry Wall. Minimum of $\frac{1}{2}$ in. air space between member and wall or adequate moisture barrier must be provided. For arches, additional space may be required to permit outward deflection of the arch leg.

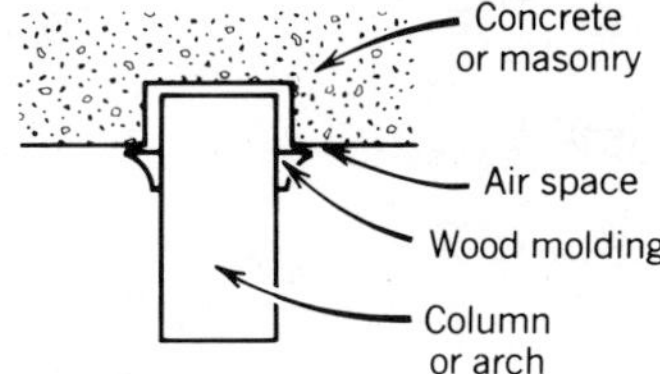

FIGURE 11.2 Wood Member Set in Masonry Wall Pocket. Minimum of $\frac{1}{2}$ in. air space between member and wall pocket or adequate moisture barrier must be provided. For arches, additional space may be required to permit outward deflection of the arch leg.

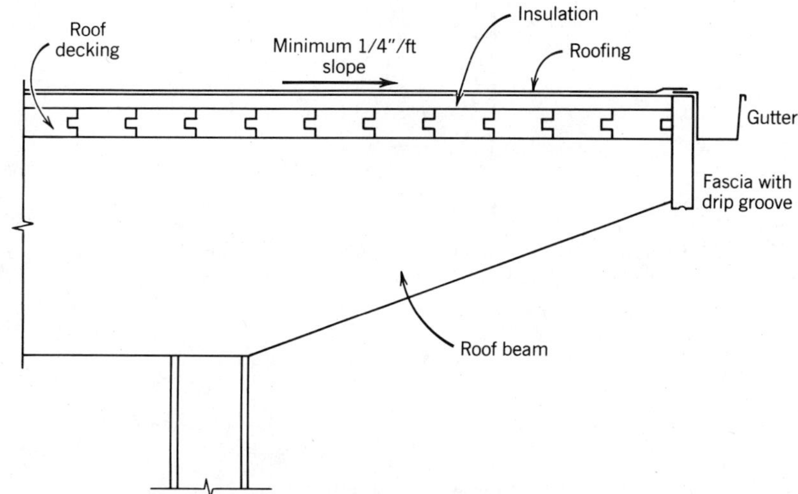

FIGURE 11.3 Protection Considerations for Building with Covered Overhang. Beam is protected from direct exposure to weather by fascia. Roof should be sloped for drainage or designed to prevent ponding of water. Fascia should be preservatively treated or made from decay-resistant species. Taper cut should be sealed.

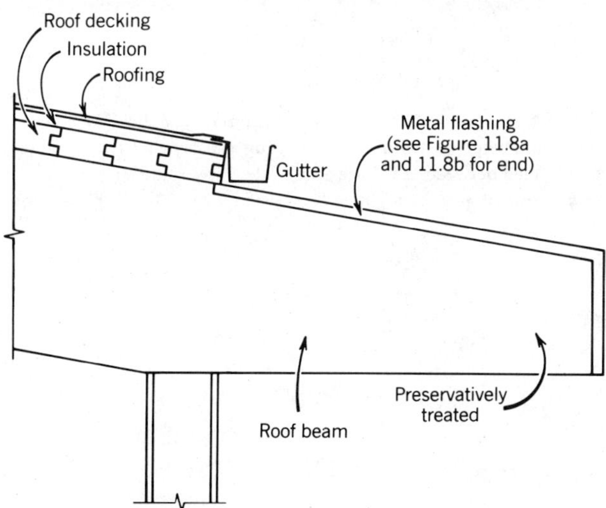

FIGURE 11.4 Protection Considerations for Building with Uncovered Overhang. Portion of beam extending outside of building should be protected by metal cap and preservative treatment. Periodic refinishing of the surfaces exposed to the weather may be required to maintain appearance.

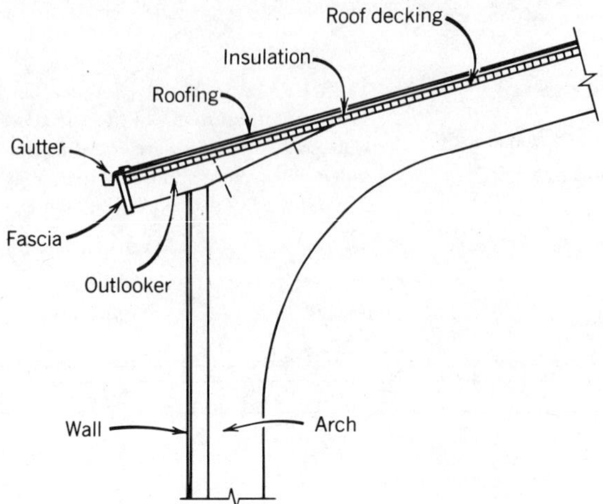

FIGURE 11.5 Protection Considerations for Arch Outlooker Overhang. Outlooker is protected from direct exposure to weather. Arch is protected by the wall from direct exposure to the weather.

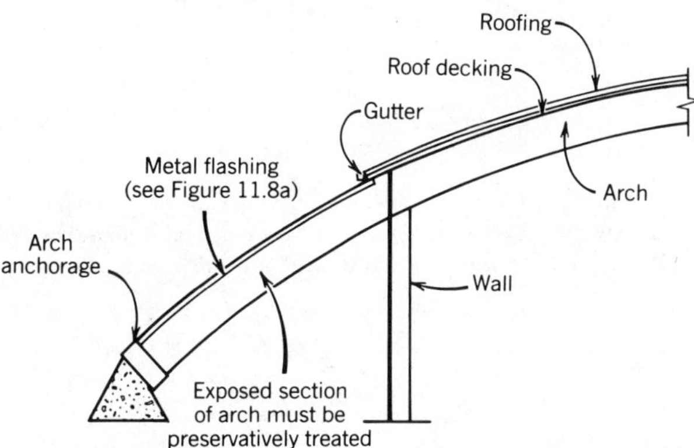

FIGURE 11.6 Protection Considerations for Partially Exposed Arches. Portion of arch leg extending outside of building should be protected by metal flashing and preservative treatment. At least 12 in. clearance must be provided between arch base and grade. Preservative treatment in accordance with *Standard for Preservative Treatment of Structural Glued Laminated Timber*, AITC 109, must be used for exposed portions of arch.

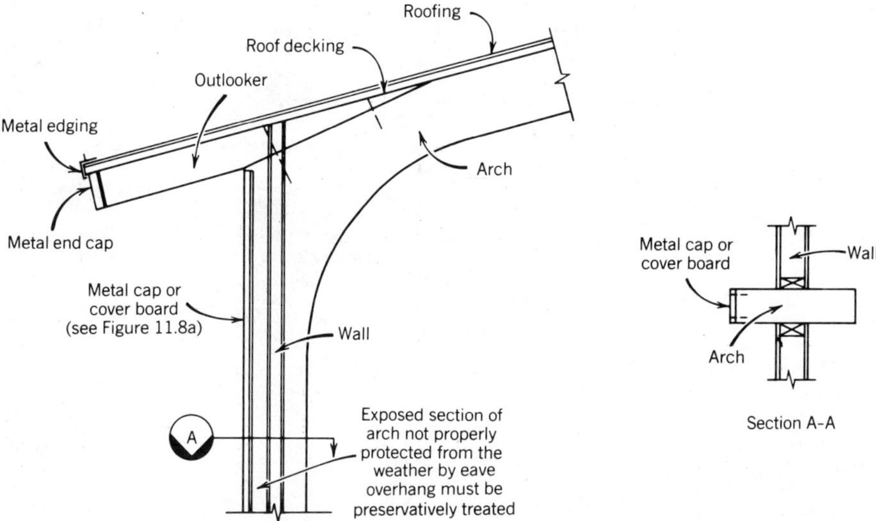

FIGURE 11.7 Arch Leg Protection. Metal end cap or treated cover board on edge of exterior portion of arch used in conjunction with preservative treatment of the arch leg. Metal cap is as illustrated in Figure 11.8. Cover board should be vertical grain material set in building sealant and attached with weatherproof nails or screws. All wood with exterior exposure must be adequately protected and maintained.

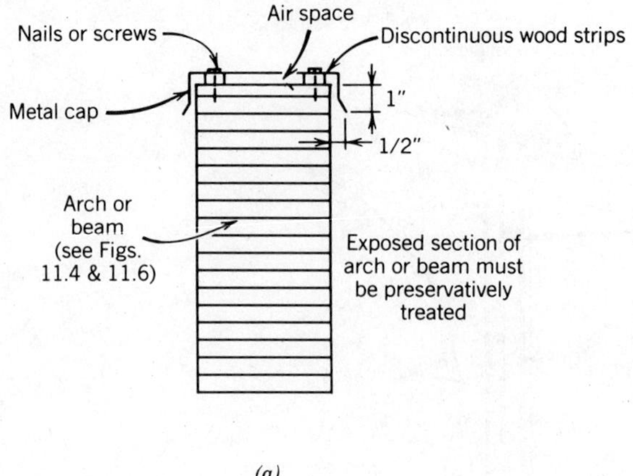

(a)

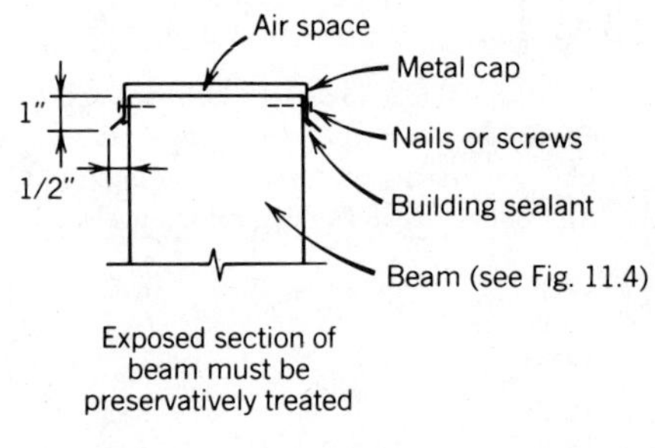

(b)

FIGURE 11.8 Protective Metal Cap or Flashing Details. Caps or flashings are made of 20-gage minimum thickness weatherproof metal. Nails or screws are weatherproofed, and heads are sealed with building sealant or neoprene washers. A minimum of $\frac{1}{2}$ in. air space must be provided between cap and the face of the wood section. For vertical use conditions, a continuous bead of building sealant is required. (a) Top cap for horizontal or sloped members. (b) End cap for exposed beams or vertical members.

APPENDIX TO AITC 104: CONNECTION DETAILS TO BE AVOIDED

The following are some examples of poor detailing practice and suggestions for improvement of the poor details.

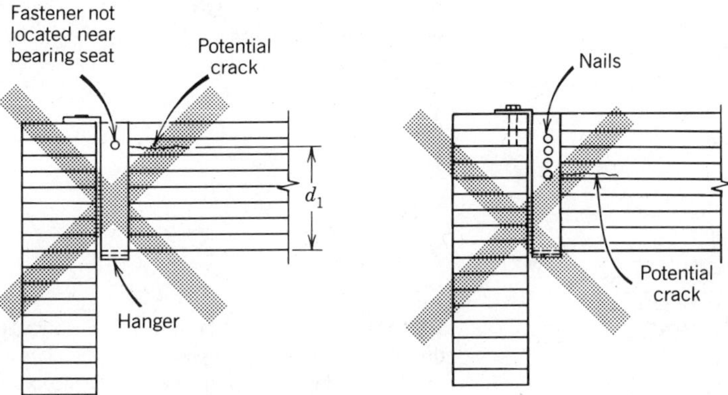

DETAIL A1 Glued laminated timbers, although relatively dry at the time of manufacture, may shrink as they reach equilibrium moisture content in service. When fasteners are not located near the bearing seat but in the upper portion of the beam, shrinkage in the beam over the depth, *d*, can cause the beam reaction to be carried by the fasteners rather than in bearing on the hanger. This induces notch shear and tension perpendicular to grain stresses that can cause splitting along the beam as shown.

SUGGESTED REVISION Detail the connection as shown in Figure 4.1 with the fasteners located near the bearing seat, or slot the hole in the steel hanger and place the fastener in the top of the slot.

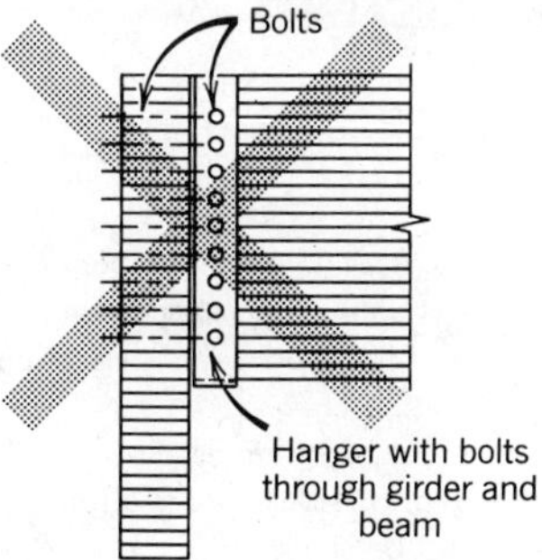

DETAIL A2 This detail is similar to Detail A1 in that shrinkage in the beam can result in the bearing being carried by the fasteners rather than in bearing on the hanger seat which, in turn, results in notch shear and tension perpendicular to grain stresses. Also, see Detail A3 concerning long rows of fasteners perpendicular to grain as it relates to the hanger-to-girder connection. Section 1.8 discusses the effect of end rotation on the fasteners in this type of connection.

SUGGESTED REVISION Detail the connection as illustrated in Figure 4.3 with the fasteners in the beam located near the bearing seat and the fasteners in the girder grouped near the top of the girder.

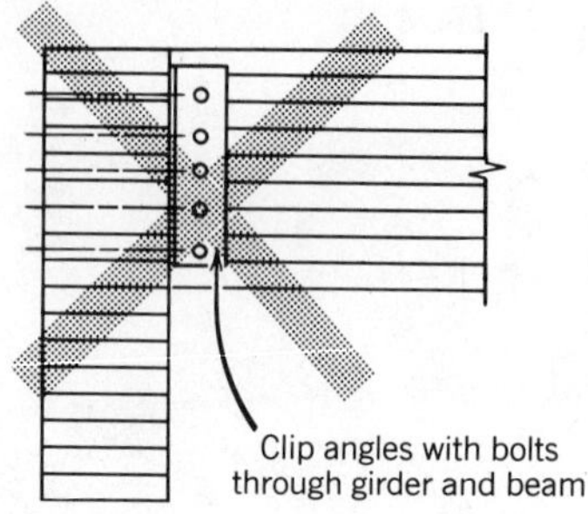

DETAIL A3 End connections which include long rows of fasteners perpendicular to grain through steel side members should be avoided. Shrinkage of the wood will be restrained by the steel and can result in notch shear and tension perpendicular to grain stresses at the end of the beam which may result in splitting of the member. See also Section 1.4
SUGGESTED REVISION Change to a bearing connection as shown in Figure 4.1 with the beam being supported in bearing and the fasteners located only near the bearing seat.

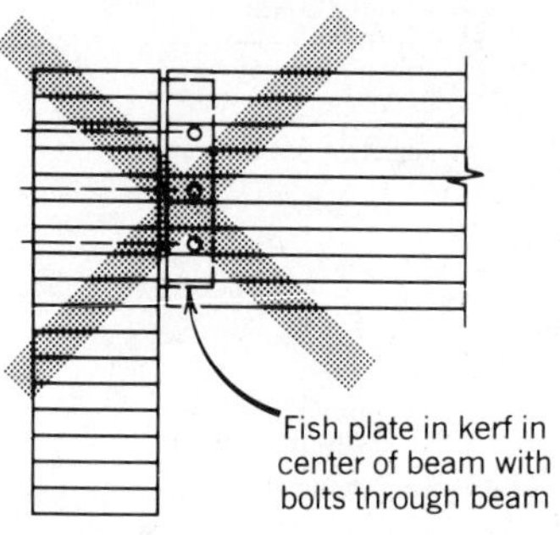

DETAIL A4 This detail is similar to Detail A3 except that the beam is supported by bolts through a plate located in a saw kerf in the center of the beam. See also Section 1.4.
SUGGESTED REVISION Detail the connection as shown in Figure 4.6 where a bearing seat has been added and the bolts away from the bearing seat have been omitted.

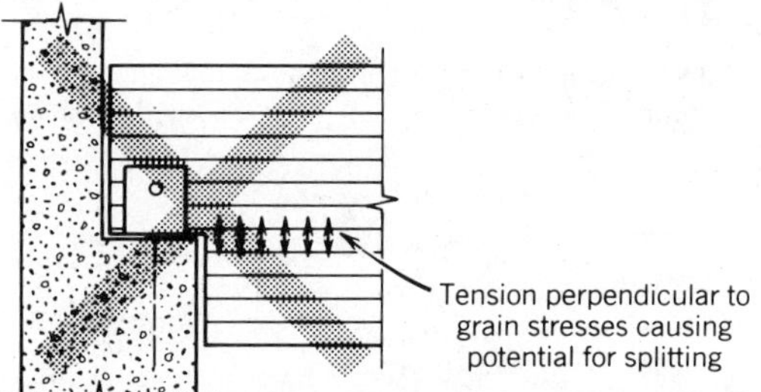

DETAIL A5 An abrupt notch in the end of a wood member creates two problems. One is that the effective shear strength of the member is reduced because of the end notch. The other is that the exposure of end grain in the notch will permit a more rapid migration of moisture in the upper portion of the member causing the indicated split.
SUGGESTED REVISION Detail the connection as shown in Figure 2.1 without the end notch. Notches are not recommended on the tension side of glued laminated timber, but if used, they should not exceed 10% of the depth and should be checked by the notched beam formula.

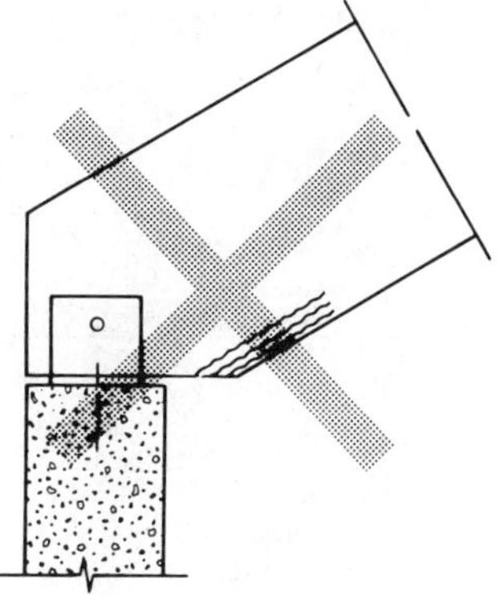

DETAIL A6 This condition is similar to that shown in Detail A5. The shear strength of the end of the member is reduced, tension perpendicular to grain stresses are induced and the exposed end grain may result in splitting because of rapid drying.

SUGGESTED REVISION Revise the taper cut to provide bearing as shown in Figure 2.6.

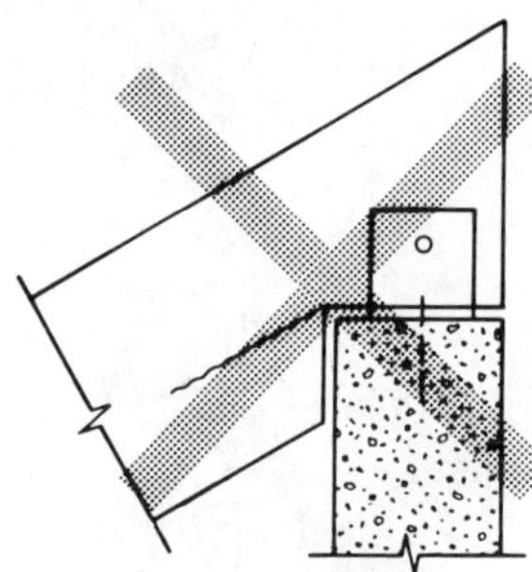

DETAIL A7 This detail at the upper end of a sloped beam is similar to the notched beam detail shown in Detail A5.

SUGGESTED REVISION Provide a sloping seat as shown in Figure 2.7.

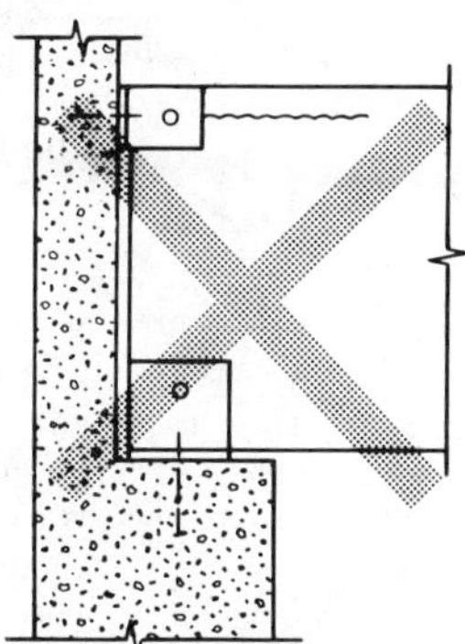

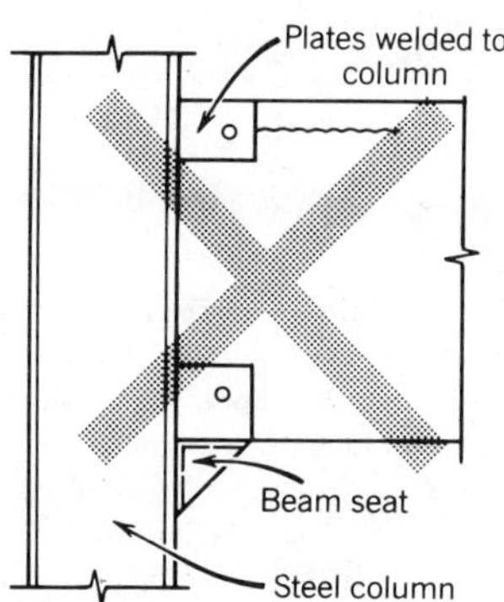

DETAIL A8 In this situation, the beam is bearing on the beam seat and the top is laterally supported by clip angles or similar hardware. In a deep beam, the shrinkage due to drying reduces the depth of the beam and will create a split at the upper connection.

SUGGESTED REVISION Provide restraint against lateral rotation without restraining the member against shrinkage as shown in Figure 2.8.

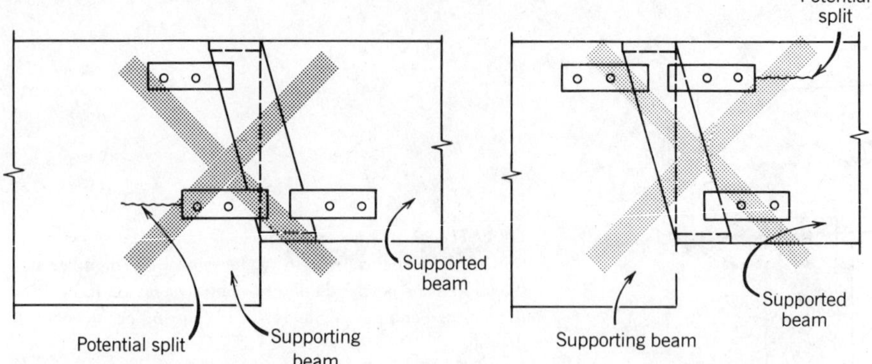

DETAIL A9 When a tension connection across a cantilever beam hanger is designed using integral tabs either at the top or bottom of the hanger, splitting may occur due to shrinkage between the bear ing point of the hanger and the bolts as shown.

SUGGESTED REVISION Detail as shown in Figure 3.1 for suggested cantilever hanger with a loose tension tie. If tabs are provided as shown, detail vertically slotted holes in the tabs and should be hand tightened only to permit movement.

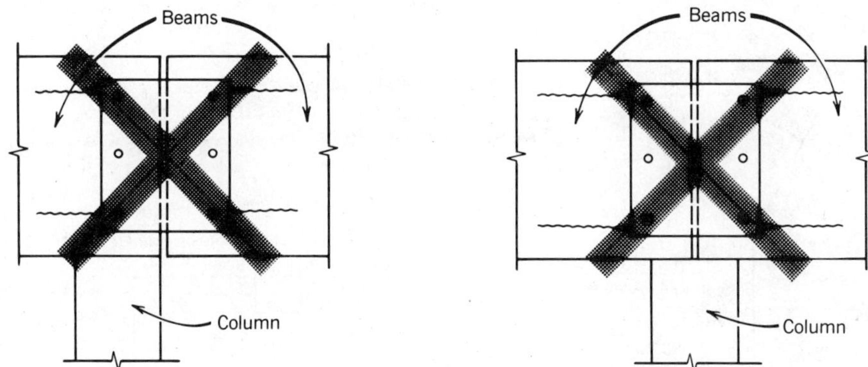

DETAIL A10 This situation is similar to Detail A3 where deep splice plates are applied to both faces of the beam. This may be a splice over a column or a situation where one beam is supporting the next one. As the wood shrinks, the steel side plates resist the shrinkage effect causing splits in the beams. This condition is particularly hazardous if one beam is supporting the next one as shown on the right or as a cantilever connection because the splits at the bolt holes will reduce effective strength of the beam.

SUGGESTED REVISION See details in Section 5 for recommended moment splice if continuity over column is desired. Moment splices are not recommended for beams.

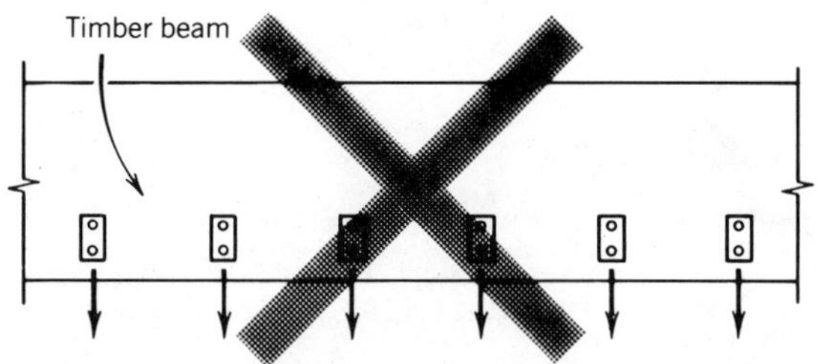

DETAIL A11 Loads suspended from beams as shown induce tension perpendicular to grain stresses.
SUGGESTED REVISION See Section 10 for recommendations for supporting loads from beams.

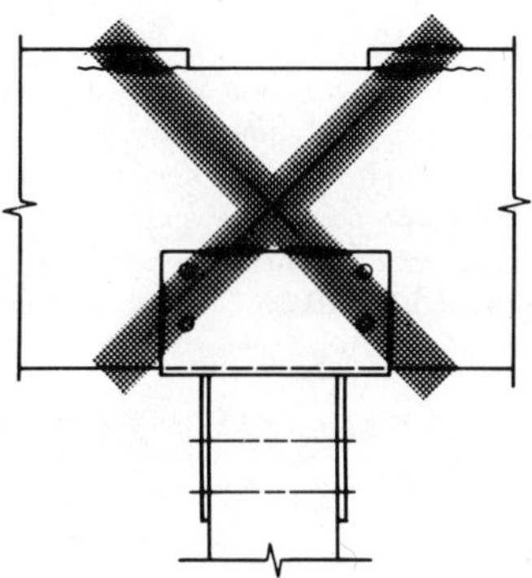

DETAIL A12 This shows a general condition where, particularly in continuous framing, the top tension fibers have been cut to provide for a recessed hardware connection or for the passage of conduit or other elements over the top of the beam. This is particularly serious in glulam construction since the tension laminations are critical to the proper performance of the structure.
SUGGESTED REVISION Detail the connection as shown in Figure 4.5 without the notch in the tension side of the cantilever member. If recessed hardware is required, provide mechanically fastened blocking.

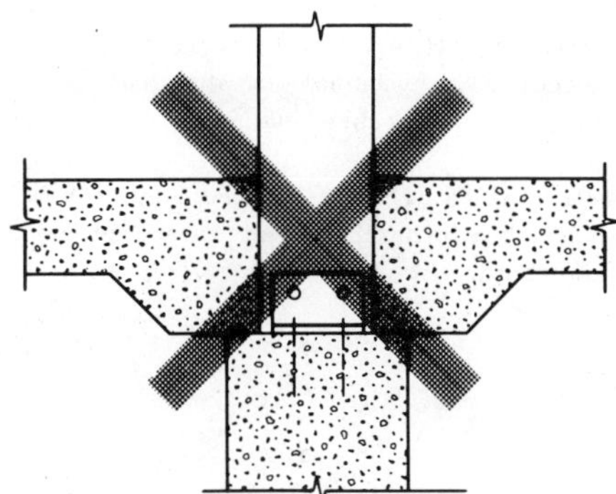

DETAIL A13 Some designers try to conceal the base of a column or an arch by placing concrete around the connection. Moisture may migrate into the lower portion of the wood and cause decay.
SUGGESTED REVISION Detail column bases as shown in Section 6 or arch anchorages as shown in Section 7.

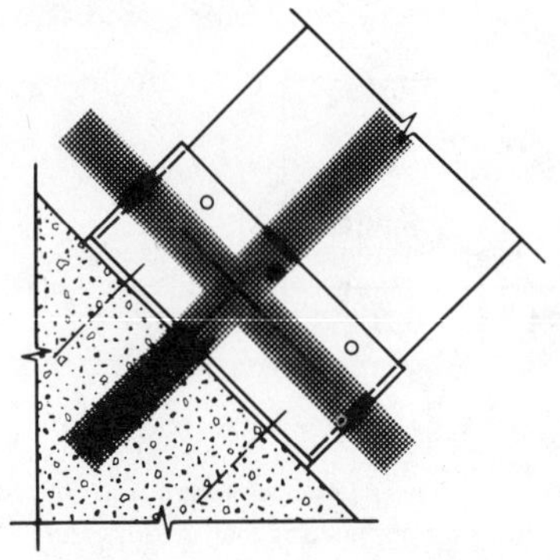

DETAIL A14 Similar to Detail A13 in that the base of an arch or a column is placed in a closed steel box where moisture may accumlate and cause decay.
SUGGESTED REVISION Detail arch base as shown in Figure 7.9 with connections grouped near the center of the arch and drainage provided to prevent collection of water in the shoe.

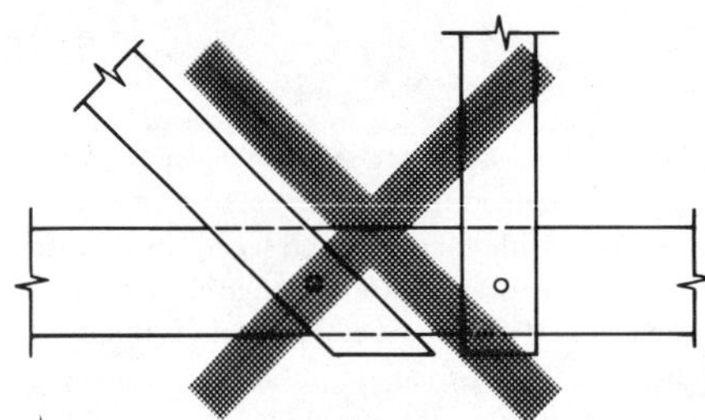

DETAIL A15 When the centerlines of members do not line up at a common point in a truss, considerable shear and moment stresses may result in the bottom chord. When these are combined with the presumably high tension stress in the member, failure may occur.
SUGGESTED REVISION Detail web to chord connections to be concentric as shown in Figure 9.1.

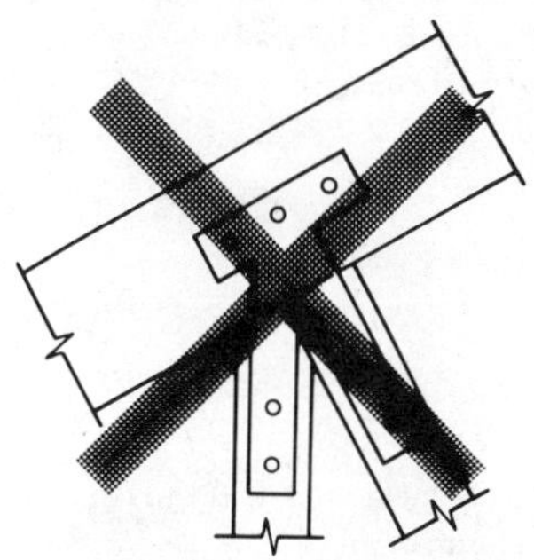

DETAIL A16 Truss chord to web connection made from plates welded rigidly together. Not recommended when truss deflections could produce rotation of members, which could cause splitting.
SUGGESTED REVISION Detail as shown in Figure 9.1 with pinned connection at the web and chord, and clearance provided between webs and chord.

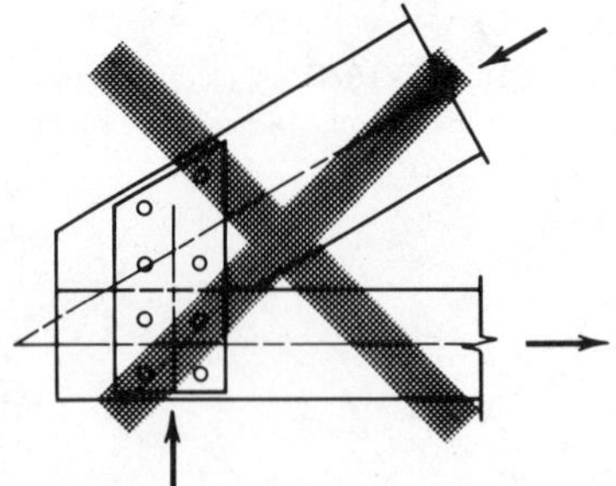

DETAIL A17 Truss heel connection with eccentric force lines that cause prying action which may result in splitting of the members.
SUGGESTED REVISION Detail connection similar to Figure 9.3.

AMERICAN INSTITUTE
333 West Hampden Avenue

TIMBER CONSTRUCTION
Englewood, Colorado 80110

AITC 108-80
STANDARD FOR HEAVY TIMBER CONSTRUCTION

Adopted as Recommendations June 27, 1979

CONTENTS

1. HEAVY TIMBER CONSTRUCTION

1.1 "Heavy timber construction" is that type in which fire resistance is attained by placing limitations on the minimum size, thickness, or composition of all load-carrying wood members as given in this section; by avoiding concealed spaces under floors or roofs; by using approved fastenings, construction details, and adhesives; and by providing the required degree of fire resistance in exterior and interior walls.

2. HEAVY TIMBER FRAMING

2.1 Columns

2.1.1 Wood columns may be sawn or glued laminated and shall be not less than 8 in., nominal, in any dimension when supporting floor loads and not less than 6 in., nominal, in width and not less than 8 in., nominal, in depth when supporting roof and ceiling loads only.

2.1.2 Columns shall be continuous or superimposed by means of reinforced concrete or metal caps with brackets, or shall be connected by properly designed steel or iron caps, with pintles and base plates, or by timber splice plates affixed to the columns by means of metal connectors housed within the contact faces, or by other approved methods.

2.2 Floor Framing

2.2.1 Beams and girders of wood may be sawn or glued laminated and shall be not less than 6 in., nominal, in width and not less than 10 in., nominal, in depth.

2.2.2 Framed or glued laminated arches which spring from grade or the floor line and support floor loads shall be not less than 8 in., nominal, in any dimension.

2.2.3 Framed timber trusses supporting floor loads shall have members of not less than 8 in., nominal, in any dimension.

2.3 Roof Framing

2.3.1 Framed or glued laminated arches for roof construction which spring from grade or the floor line and do not support floor loads shall have members not less than 6 in., nominal, in width and not less than 8 in., nominal, in depth for the lower half of the height and not less than 6 in., nominal, in depth for the upper half.

2.3.2 Framed or glued laminated arches for roof construction which spring from the top of walls or wall abutments, framed timber trusses, and other roof framing which do not support floor loads, shall have members not less than 4 in., nominal, in width and not less than 6 in., nominal, in depth. Spaced members may be composed of two or more pieces not less than 3 in., nominal, in thickness when blocked solidly throughout their intervening spaces or when such spaces are tightly closed by a continuous wood cover plate of not less than 2 in., nominal, in thickness, secured to the underside of the members. Splice plates shall be not less than 3 in., nominal, in thickness. When protected by approved automatic sprinklers under the roof deck, framing members shall be not less than 3 in., nominal, in width.

3. HEAVY TIMBER FLOORS

3.1 Floors shall be of sawn or glued laminated: (1) planks splined or tongue-and-groove, not less than 3 in., nominal, in thickness covered with 1 in., nominal, dimension tongue-and-groove flooring laid crosswise or diagonally to the plank or with other approved wearing surfaces, or (2) planks, not less than 4 in., nominal, in width set on edge close together and well spiked, and covered as for 3 in. thick plank. The planks shall be laid so that there is no continuous line of end joints except at points of support. Floors shall not extend closer than $\frac{1}{2}$ in. to walls to provide an expansion joint, but the joint shall be covered at top or bottom to avoid flue action.

4. HEAVY TIMBER ROOF DECKS

4.1 Roof decks shall be of sawn or glued laminated, splined or tongue-and-groove plank, not less than 2 in., nominal, in thickness, of tongue-and-groove $1\frac{1}{8}$ in., thick interior plywood (exterior glue), or of planks, not less than 3 in., nominal, in width, set on edge close together and laid as required for floors. Other wood and/ or wood-fiber based decking or other types of decking may be used if noncombustible.

5. WALLS

5.1 Bearing Walls. Bearing portions of exterior and interior walls shall be of approved noncombustible material and shall have a fire resistance rating of not less than 2 hours except that, where a horizontal separation of 3 ft or less is provided, bearing portions of exterior walls shall have a fire resistance rating of not less than 3 hours.

5.2 Nonbearing Walls. Nonbearing portions of exterior walls shall be of approved noncombustible materials except as otherwise noted, and:

5.2.1 Where a horizontal separation of 3 ft or less is provided, nonbearing exterior walls shall have a fire resistance rating of not less than 3 hours.

5.2.2 Where a horizontal separation of more than 3 ft but less than 20 ft is provided, nonbearing exterior walls shall have a fire resistance rating of not less than 2 hours.

5.2.3 Where a horizontal separation of 20 to 30 ft is provided, nonbearing exterior walls shall have a fire resistance rating of not less than 1 hour.

5.2.4 Where a horizontal separation of 30 ft or more is provided, no fire resistance rating is required.

5.2.5 Where a horizontal separation of 20 ft or more is provided, wood columns, arches, beams, and roof decks conforming with heavy timber sizes may be used externally.

6. CONSTRUCTION DETAILS

6.1 Wall plate boxes of self-releasing type or approved hangers shall be provided where beams and girders enter masonry. An air space of 1 in. shall be provided at the top, end, and sides of the member unless approved durable or treated wood is used.

6.2 Girders and beams shall be closely fitted around columns, and adjoining ends shall be cross-tied to each other, or intertied by caps or ties, to transfer horizontal loads across the joint. Wood bolsters may be placed on top of columns which support roof loads only.

6.3 Where intermediate beams are used to support a floor, they shall rest on the top of the girders, or shall be supported by ledgers or blocks securely fastened to the sides of the girders, or they may be supported by approved metal hangers into which the ends of the beams shall be closely fitted.

6.4 Wood beams and girders supported by walls required to have a fire resistance rating of 2 hours or more shall have not less than 4 in. of solid masonry between their ends and the outside face of the wall and between adjacent beams.

6.5 Columns, beams, girders, arches, trusses, and floor slabs of material other than wood shall have a fire resistance rating of not less than 1 hour.

6.6 Floors and roof decks shall be without concealed spaces, except that building service equipment may be enclosed provided the spaces between the equipment and enclosures are fire-stopped or protected by other acceptable means.

6.7 Adequate roof anchorage shall be provided.

7. STANDARD DIMENSIONS FOR HEAVY TIMBER

7.1 Excellent fire resistance is achieved with "heavy timber" construction. Minimum sawn lumber sizes have been long established. They are expressed in nominal dimensions and assume surfacing to *American Lumber Standard* net sizes.

7.2 For "heavy timber" construction, the net width of glued laminated structural members shall be the standard glued laminated net width for the nominal sawn width specified, and the net depth of glued laminated structural members shall be equal to or greater than the net finished depth specified by the following table.

Minimum Nominal Size			*Minimum Glued Laminated Net Size*		
Width, in.		Depth, in.	Width, in.		Depth, in.
8	×	8	$6\frac{3}{4}$	×	9
6	×	10	$5\frac{1}{8}$	×	$10\frac{1}{2}$
6	×	8	$5\frac{1}{8}$	×	9
6	×	6	$5\frac{1}{8}$	×	6
4	×	6	$3\frac{1}{8}$	×	$7\frac{1}{2}$

AMERICAN INSTITUTE TIMBER CONSTRUCTION
333 West Hampden Avenue Englewood, Colorado 80110

AITC 109-84
STANDARD FOR PRESERVATIVE TREATMENT OF STRUCTURAL GLUED LAMINATED TIMBER

Adopted as Recommendations July 18,1984

CONTENTS

1. PREFACE

1.1 This standard covers preservative treatment of structural glued laminated timber (glulam). It does not cover fire-retardant treatments.

1.2 Wood structures, properly designed and constructed, have performed in service with satisfaction for centuries. When the use of recognized design and construction principles do not eliminate the hazards due to decay, insect, or marine borer attack, then the wood must be preservatively treated.

1.3 The effectiveness of treatment is dependent upon an adequate penetration and retention of the chemical used. The specification of retentions for creosote or pentachlorophenol treatments in excess of those shown in Tables 2 and 3 may result in exudation of the treatment or its carrier. Such exudation may continue for a long period of time.

1.4 On July 13, 1984, the Environmental Protection Agency (EPA) announced restrictions on the uses of creosote, pentachlorophenol and inorganic arsenicals in Position Document (PD) No. 4. This document contains use, site, and handling precautions for the three types of treatments for all types of wood products. Restrictions on the use of preservatively treated glued laminated timber inside buildings are as follows:

1.4.1 Creosote. Should not be used in residential interiors nor in farm building interiors where there may be direct contact with humans or animals. Should only be used for industrial building components which are in ground contact and subject to decay or insect attack and where two coats of an appropriate sealer such as urethane, epoxy or shellac are applied. Do not use for structures used for storing food or where the treated wood may come into contact with drinking water.

1.4.2 Pentachlorophenol. May be used in residential, industrial, or commercial interiors where two coats of an appropriate sealer such as urethane, shellac, latex epoxy enamel, or varnish are applied. Do not use for building interiors where there may be direct contact with humans or animals. Do not use for structures used for storing food or where the treated wood may come into contact with drinking water.

1.4.3 Inorganic Arsenicals. May be used in residential interiors provided that all dust is vacuumed from the wood surface after installation. Do not use for structures used for storing food or where the treated wood may come into contact with public drinking water.

1.4.4 Complete information on the use and handling of preservatively treated wood products may be found in Consumer Information Sheets (CIS) accompanying shipments or in EPA PD No. 4.

2. GENERAL

2.1 Decay. Decay of wood is caused by low forms of plant life (fungi) that develop and grow from microscopic spores that are present wherever wood is used. The fungi convert wood substance into food required for development. If deprived of any one of the four essentials of life (food, air, moisture or favorable temperature), decay growth is prevented or stopped and the wood remains sound, retaining its existing strength with no further deterioration. Wood will not be attacked by decay fungi if it is submerged in water thereby excluding air, kept continuously below 20% moisture content or maintained at temperatures below freezing or much above 100°F. Growth can begin or resume whenever climatic conditions are favorable.

2.1.1 Examples of installations where wood structural members can be kept below 20% moisture content through proper design, construction details and maintenance or wet enough to exclude oxygen:

2.1.1.1 Enclosed buildings in which good roof coverage, proper roof maintenance, good joint details, adequate flashings to direct rain water, ventilation, properly located vapor or moisture barriers and a well-drained building site assure continuous moisture content of wood below 20%.

2.1.1.2 Any usage where wood is permanently and totally submerged in fresh water.

2.1.2 Examples of installations where wood structural members are protected from decay with difficulty and where particular care is required, especially in continuing proper maintenance (*the inability of the designer to effectively control maintenance should be carefully considered*):

2.1.2.1 Certain buildings, such as natatoriums and ice skating rinks in which wood structural members are subject to moisture condensation. Protection may be provided by means of proper moisture barriers, insulation and adequate ventilation. If the designer is sure that an adequate ventilation system will be installed and the system will remain operable during the life of the structure, untreated timbers and dry-use design values may be used, provided the moisture content of the wood is less than 16%.

2.1.2.2 Construction in which wood structural members are adjacent to masonry or concrete. Protection may be provided by the use of ventilated air spaces, vapor barriers, or other moisture barrier separations preventing direct contact with masonry or concrete.

2.1.3 Examples of installations where wood structural members require preservative treatment to be permanent:

2.1.3.1 Enclosed buildings housing "wet processes" or other high humidity environments where, despite ventilation, there is sufficient moisture remaining in the atmosphere to maintain equilibrium moisture content in the wood above 20%.

2.1.3.2 Outdoor exposures where there is no protective cover for wood structural members from the weather, such as in bridges, piers and wharves, towers and electrical utility structures.

2.1.3.3 Direct contact with the ground or with water, as for retaining walls, or for poles or piles where the wood extends out of the water and is not permanently and totally submerged.

2.1.3.4 Construction in which portions of wood structural members extend beyond the walls and roof coverage of the building. It is difficult to provide completely adequate flashing and ventilation details to insure less than 20% moisture content.

2.2 Insect Attack. Wood structural members that have been preservatively treated to definite retentions and penetrations are resistant to insect attack. The degree of resistance offered will depend upon the type of preservative as well as the retentions and penetrations obtained.

2.3 Marine Borers. Protection for glued laminated timbers against marine borers can be provided by the pressure treating process with recommended preservatives with retentions and penetrations as set forth in the current AWPA Standard C3, *Piles—Preservative Treatment by Pressure Processes.*

3. DESIGN

3.1 The allowable design values for preservatively treated wood, as included in the *National Design Specification for Wood Construction* (NDS) apply. NDS requires no reduction in design vaes for glued laminated timber or other wood products pressure-impregnated by processes and preservative approved by the American Wood-Preservers' Association (AWPA). When the moisture content of the wood in service is less than 16%, untreated timbers and dry-use design values may be used. When the moisture content of the wood in service is 16% or more, wet-use design values should be used and when the moisture content of the wood in service is 20% or more, glued laminated timbers should be preservatively treated and wet-use design values should be used.

3.2 Glued laminated timber may be manufactured with lumber treated prior to gluing or it may be treated after gluing. Arches and other sharply curved members must be manufactured with lumber treated prior to gluing. Straight or slightly curved members may be treated after gluing. The size of pressure-treating cylinders, which range up to 8 ft in diameter and 180 ft in length, must be considered when specifying treatment and when designing members to be treated after gluing. It is important to contact the supplier prior to initiating the design of large members, since special handling and treating problems may exist.

4. SPECIES

4.1 Lumber for glued laminated timber that is to be treated before or after gluing includes the following species: Coast region Douglas Fir; Hem-Fir; Western Hemlock; Southern Pine; Ponderosa Pine; Lodgepole Pine; Redwood; and Western Woods and Western Red Cedar treated to the equivalent of Hem-Fir treatments. Some species of lumber for glued laminated timber that is to be treated before or after gluing or lumber from some geographical areas may resist treatment.

5. FABRICATION

5.1 All fabrication should be performed prior to treatment for glued laminated members treated after gluing. When laminations are treated prior to gluing and there is fabrication after treatment and laminating, additional treatment should be applied in accordance with AWPA Standard M4, *Standard for the Care of Preservative-Treated Wood Products.*

6. TYPES OF PRESERVATIVE TREATMENTS

6.1 Various characteristics of preservative treatments affect their use with glued laminated timber. Some treatments are suitable for use in treating after gluing and some are suitable for use in treating before gluing. All preservative treatments to be used before gluing shall be in accordance with the current AWPA Standard C28, *Standard for Preservative Treatment of Structural Glued Laminated Members and Laminations Before Gluing of Southern Pine, Pacific Coast Douglas Fir, Hemfir and Western Hemlock by Pressure Processes.* Preservative treatments to be used after gluing may be as agreed upon between the buyer and seller, but the treatment process shall be in accordance with AWPA Standards. Preservative treatments are described in the following paragraphs. A summary of uses of these treatments is given in Table 1. *The specifier must select the preservative treatment suitable for a particular job.* Availability of each varies with geographic area and suppliers.

6.1.1 Creosote and Creosote/Coal Tar Solutions. Creosote is a coal tar product containing a multitude of chemical compounds toxic to decay fungi, insects, and most marine organisms. Creosote treated material is suitable for the most severe exposure conditions. It has a dark, oily surface appearance and cannot generally be stained or painted. It possesses an odor. It should not be used in contact with materials subject to staining such as plaster, wallboard, etc., nor in direct contact with roofing felt. As a practical consideration, the treatment is used only after gluing.

6.1.2 Oil-Borne Treatments. Oil-borne preservative treatments are toxic to decay fungi and insects, and utilize various hydrocarbon solvents as carriers. Pentachlorophenol (commonly referred to as "penta") is the most common oil-borne preservative. The solvents used have various effects on the finished products and are classified into the following four types:

AWPA P-9, Type A—Penta in oil. This treatment may become blotchy when exposed to the weather, although this condition diminishes with time. The surface is not readily paintable. This material should not be used in direct contact with roofing felt. As a practical consideration, the treatment is used only after gluing.

AWPA P-9, Type B—Penta in liquid petroleum gas. This treatment leaves a finished appearance close to that of untreated wood and can be stained or painted. When used after gluing, this treatment should be restricted to small members only.

AWPA P-9, Type C—Penta in light solvents. This treatment can leave a natural wood-appearing surface if the supplier is advised of appearance requirements. The surface can be stained or painted.

AWPA P-9, Type D—Penta in methylene chloride. The finished surface is slightly darker than untreated wood. The surface can be stained or painted. The solvent recovery process may result in raised grain and checking of the wood.

6.1.3 Waterborne Treatments. Waterborne preservative treatments utilize water-soluble chemicals which become fixed in the wood during the treating process. They are toxic to decay fungi, insects and most marine organisms. They will have a light green, grey-green or brown color, depending on the chemical used. When the surface is dry, the members can be stained or painted in accordance with the coating manufacturer's requirements. When used in high moisture areas, metal hardware in contact with the treated wood should be of corrosive-resistant metal in accordance with the chemical treatment manufacturer's recommendations. These treatments are generally used to treat the laminations prior to gluing. If glued laminated timber members are to be treated after gluing, di-

TABLE 1 Uses and Characteristics of Preservative

Treatment Type	Visual/Physical Appearance	When Architechtural Appearance is Important	Paintability	Treatment Before Gluing	Treatment After Gluing
Cresote and Cresote/Coal Tar Solutions	Dark, oily	NR	Cannot be painted or stained	NR	A
Oil-Borne Penta Treatments					
Type A Penta in Oil	Oily, may become blotchy when exposed to elements	NR	Not readily paintable	NR	A
Type B Penta in Liquid Petroleum Gas	Similar in appearance to untreated wood	A	Can be painted or stained	A[2]	L
Type C Penta in Light Solvent	Can be natural appearing if treater is advised of finished appearance requirements	A	Can be painted or stained after surface preparation as recommended by treater	A	A
Type D Penta in Methylene Chloride	Slightly darker than untreated wood	L	Can be painted or stained	A[2]	L
Waterborne Treatments	Light green, or gray-green or brown in color depending upon chemicals used	L	Can be painted or stained when surface is dry and prepared in accordance with coating manufacturer's recommendations	A[2]	L

A —Generally acceptable
L —Acceptable only within limitations. See comments under "Limitations" and in Section 6.
NR—Not recommended.

1. Some preservative treatments may cause irritation when in direct human contact. Contact American Wood Preserver's Institute (AWPI) for information.
2. Availability of treatments for use with specific species and in specific geographical regions should be verified before specifying.
3. See Table 3 for limitations for certain treatments.
4. For potable water, other restrictions may apply. Contact AWPI for information.

Treatments for Glued Laminated Timbers[1,2]

Ground Contact or Embedded in Concrete	Exposed to Wetting or in Contact with Masonry	Contact with Fresh Water[4]	Contact with Salt Water	Limitations
A[3]	A	A	A	Odor may be objectionable. Should not be used in direct contact with roofing felt. Should not be used in contact with materials subject to staining such as plaster, wallboard, etc.
A	A	A	NR	Should not be used in direct contact with roofing felt. Should not be used in contact with materials subject to staining, such as plaster, wallboard, etc.
A	A	A	NR	Should only be used to treat very small members when used after gluing. Softwood species used in laminating may exude resin after treatment although condition diminishes with time.
A	A	A	NR	Softwood species used in laminating may exude resin after treatment although condition diminishes with time.
A	A	A	NR	Solvent recovery process may result in raised grain and checking of wood surface. Softwood species used in laminating may exude resin after treatment although condition diminishes with time.
A[3]	A	A	A	Wetting and redrying process associated with treatment may result in dimensional changes, warping, checking or cracking of the members when treated after gluing. When used in high moisture conditions, all metallic connections should be of corrosive-resistant metal per the treater's recommendations.

mensional changes caused by wetting and subsequent redrying may result in warping, cracking or checking, and this should be considered prior to specifying.

6.1.3.1 When members are treated with waterborne preservatives and are used in chemical or industrial environments, care must be taken to insure compatibility of the treated material with the environment to prevent chemical degradation (not decay) of the member.

7. RETENTION, PENETRATION, CERTIFICATION AND MARKING REQUIREMENTS

7.1 Retention recommendations given in Tables 2 and 3, in pounds per cubic foot, and penetration requirements are those of AWPA Standard C28. Retention and penetration values from the latest edition of Standard C28 should be specified.

7.2 Glued laminated timber manufactured with lumber treated prior to gluing and represented as complying with this specification must have been manufactured from treated lumber certified and marked as follows:

A certificate must be furnished by the treater to the laminator stating the type of treatment and that the treatment has been performed in accordance with AWPA Standard C28. The laminator will maintain records of assays of retention and penetration which were made according to AWPA Standard C28. Each bundle or load shall be identified by the treater and this identification shall be by agreement between the laminator and the treater.

8. INCISING

8.1 Incising is not required for Southern Pine or Western Hemlock but is recommended for Coast Region Douglas Fir and Hem-Fir where the best treatment results are to be obtained in accordance with AWPA Standard C1, *All Timber Products—Preservative Treatment by Pressure Processes*. Incision is not recommended where appearance is a factor or where members are too large or irregular in shape to be machine incised. If incising is waived, the penetration and retention requirements for above ground and ground contact conditions of use still apply.

9. TREATING GLUED LAMINATED TIMBERS

9.1 Glued laminated timbers to be treated by pressure processes must be glued with wet-use adhesives conforming to American National Standard ANSI/AITC A190.1-1983, *Structural Glued Laminated Timber.* The pressure treatment of glued laminated timbers may be as agreed upon between the buyer and seller, but the treatment process shall be in accordance with AWPA standards.

10. TREATING INDIVIDUAL LAMINATIONS PRIOR TO GLUING

10.1 The net thickness of lumber to be treated should be not more than $\frac{1}{4}$ in. over the net thickness of the laminations when blanked lumber is used and $\frac{3}{8}$ in. when rough lumber is used. The width of either blanked or rough lumber should

TABLE 2

Structural Glued Laminated Timber Treated After Gluing

From AWPA Standard C28-84
RESULTS OF TREATMENT
Retention and Penetration

	Southern Pine		Pacific Coast Douglas Fir, Hemfir or Western Hemlock	
	Above Ground	Ground Contact	Above Ground	Ground Contact
Sampling for Assay[1]				
Zone Inches from Edge of Interior Laminations	0–3.00	0–3.00	0–0.60	0–0.60
Minimum Number of Borings Per Lot	20	20	20	20
Retention by Assay—pcf (min.)				
Creosote	6.0	12.0	6.0	12.0
Creosote–Coal Tar Solution	6.0	12.0	6.0	12.0
Creosote Petroleum	NR	NR	6.0	12.0
Pentachlorophenol	0.30	0.60	0.30	0.60
Determination of Penetration from Edge of Beams, Inches[2]	3.00 or 90%	3.00 or 90%	0.50	0.75

[1]More than one boring may be taken from the same piece, but not more than one from the same lamination unless there is an end joint separation. Using an increment borer core, 0.20 in. in diameter, twenty 0.60 in. long borings will provide an adequate sample for assay of pentachlorophenol. A minimum of forty-eight 0.60 in. long borings is required for creosote and creosote solutions. A 0.40 in. plug cutter may be used to increase the size of the sample taken for analysis or to reduce the number of borings required below forty-eight. The treated wood surface should be lightly scraped prior to taking a sample in order to remove surface deposits of preservative. The retention requirement for pentachlorophenol is based on an assay using the lime ignition method of analysis. When the copper pyridine method of analysis is used, multiply the result by 1.1 to convert to the lime ignition basis.

[2]Soil Contact Use. For members more than 75 sq in. in cross section at the groundline, every member shall be bored for penetration. For members 75 sq in. or less in cross section at the groundline, 20 members per charge shall be bored for penetration. Should the charge contain less than 20 members, each member shall be bored. When inspecting Southern Yellow Pine laminated timbers for penetration, borings shall be taken from two different laminations for each member. When boring Pacific Coast Douglas Fir or Western Hemlock laminated timbers for penetration, one boring shall be taken from each of the two face laminations and one boring shall be taken from each of two different interior laminations in each member. If any boring taken from any member fails to meet the penetration requirement, that member shall be rejected. If 90 percent or more of the members bored meet the above requirements for either size category, the charge shall be accepted. If less than 90 percent of the members bored meet the above requirements for either size category, the charge shall be rejected.

Above Ground Use. One boring from each of 20 members in a charge shall be taken for penetration. Should the charge contain less than 20 members, each member shall be bored. If any boring fails to meet the penetration requirement, that member shall be rejected. If 80 percent or more of the members bored meet the above requirement, the charge shall be accepted. If less than 80 percent of the members bored meet the above requirement, the charge shall be rejected.

TABLE 3

Structural Glued Laminated Timber Treated Before Gluing

From AWPA Standard C28-84
RESULTS OF TREATMENT
Retention and Penetration

	Southern Pine or Ponderosa Pine		Pacific Coast Douglas Fir, Hemfir, or Western Hemlock	
	Above Ground	Ground Contact	Above Ground	Ground Contact
Sampling for Assay[1,2]				
Zone Inches from Edge of Laminations	0.5–1.0	0.5–1.0	0.5–1.0	0.5–1.0
Minimum Number of Borings Per Lot	20	20	20	20
Retention by Assay—pcf (min.)				
Creosote	6.0	12.0	6.0	12.0
Creosote-Coal Tar	6.0	12.0	NR	NR
Creosote Petroleum	NR[4]	NR	6.0	12.0
Pentachlorophenol	0.30	0.60	0.30	0.60
Water-Borne Preservatives				
ACA	0.25	0.40	0.25	0.40
ACC	0.25	0.50	0.25	0.50
ACZA	0.25	0.40	0.25	0.40
CCA	0.25	0.40	0.25	0.40
CZC	0.45	NR	0.45	NR
FCAP	0.25	NR	0.25	NR
PAS	0.40	NR	0.40	NR
Determination of Penetration[3] Inches from Edge, 20 out of 20	3.00 or 90%	3.00 or 90%	1.00	1.25

[1]Using an increment borer core, 0.20 in. in diameter, twenty 0.50 in. long borings will provide an adequate sample for assay of pentachlorophenol and the water-borne preservatives. The method of analyses used for assaying retention of the water-borne preservatives should be based on the analytical methods of AWPA Standard A2. Forty-eight 0.50 in. long borings are required for creosote and creosote solutions. A 0.40 in. plug cutter may be used to increase the size of the sample taken for analysis or to reduce the number of borings required below forty-eight. The treated wood surface should be lightly scraped prior to taking a sample in order to remove surface deposits of preservative. The retention requirement for pentachlorophenol is based on an assay using the lime ignition method of analysis. When the copper pyridine method of analysis is used, multiply the result by 1.1 to convert to the lime ignition method.

[2]Laminated beams manufactured from material treated to meet the above requirements can be assayed by changing the zone samples from 0.50 to 1.00 in., as indicated above, to 0 to 0.50 in. Results of assay must meet 90% of the retention specified above.

[3]Laminated beams manufactured from material treated to meet the above requirements can be tested for penetration by taking samples from the edges of the laminated beams. The penetration required is 0.50 in. less than that required above on 18 out of 20 samples.

[4]NR—Not Recommended

be not more than $\frac{1}{2}$ in. over the net finished width of the laminated member. Lumber to be treated prior to gluing shall not exceed the maximum moisture content required for production of glued laminated timbers in accordance with American National Standard ANSI/AITC A190.1-1983, *Structural Glued Laminated Timber.*

10.2 Lumber treated with waterborne preservatives shall be dried after treatment to the moisture content required for gluing.

10.3 Pressure-treated lumber shall be resurfaced after treatment and prior to gluing. The surfacing of pressure-treated laminations shall remove as little wood as practical while making the surface clean planed and uniform in thickness for gluing. The time interval between surfacing and gluing of lumber treated with waterborne salts shall not exceed 24 hours. For lumber with other treatments, the maximum time between surfacing and gluing shall be established at the time of plant qualification.

10.4 In general, longer curing times or higher temperatures are required for treated than for untreated wood to obtain glue bonds of comparable quality. Treatment may influence the gluing surface requiring modifications in the adhesive spread and assembly times.

10.5 Certain combinations of lumber species, treatment and adhesive do not produce the same quality of glue bond as do other combinations even though used under the same procedures. It is, therefore, important that the use of any particular combination of species, treatment and adhesive be supported by adequate gluing data for the individual laminator's procedures. Each plant shall qualify the maximum retention it intends to use for each treatment glued.

10.5.1 The treating of individual laminations with creosote, creosote-coal tar, creosote-petroleum or pentachlorophenol in heavy petroleum oil treatments prior to gluing is not practical.

11. CARE AFTER TREATMENT

11.1 To assure best results, it is necessary to protect treated material from mechanical injury both in construction handling and under field service conditions. Cutting of treated material should be avoided whenever possible. When cuts are made, field treatments in accordance with AWPA Standard M4 should be used.

12. OTHER WOOD PRODUCTS. The principles of preservative treatment covered in this standard are generally applicable to other wood products. For specific requirements, see Section C of the AWPA *Book of Standards*.

13. EXUDATION OF RESIN

13.1 Most species of softwood commonly used in glued laminated timbers contain resins in some portions of the wood. It also may be present in resin pockets. Generally, this resin is "set" during drying and remains immobile during the service life of the product. Occasionally, some resin may exude from localized areas or from resin pockets which were not set during the drying process.

13.2 Wood which has been treated with preservatives dissolved in a solvent that also dissolves the resins, may tend to have more exudation of resins during the service life than untreated wood. This will vary with species, the type of treatment, and the amount of solvent remaining in the wood.

13.3 Exudation of resins usually decreases with time and the resin tends to harden as the solvents evaporate.

AMERICAN INSTITUTE
333 West Hampden Avenue

TIMBER CONSTRUCTION
Englewood, Colorado 80110

AITC 110-84
STANDARD APPEARANCE GRADES FOR STRUCTURAL GLUED LAMINATED TIMBER

Adopted as Recommendations July 18, 1984

CONTENTS

1. INTRODUCTION

1.1 Various grades of appearance are desired for different uses. These appearance grades apply to the surfaces of glued laminated members and include such items as growth characteristics, void filling and surfacing operations but not laminating procedures, stains, varnishes, or other finishes, nor wrappings, cratings, or other protective coverings. The appearance grades do not modify the design stresses, fabrication controls, grades of lumber used and other provisions of the standards for structural glued laminated timber.

1.1.1 These appearance grades are for the guidance of the designer so that a product consistent with the use of the structure may be specified and provide a suitable appearance at appropriate cost. The designer should specify the desired appearance grade to give a clear understanding between buyer and seller. Requirements given in appearance descriptions are intended to achieve a general and distinctive uniformity of appearance, and reasonable tolerance is permitted. Appearance grading should reflect good judgment. It should be kept in mind that

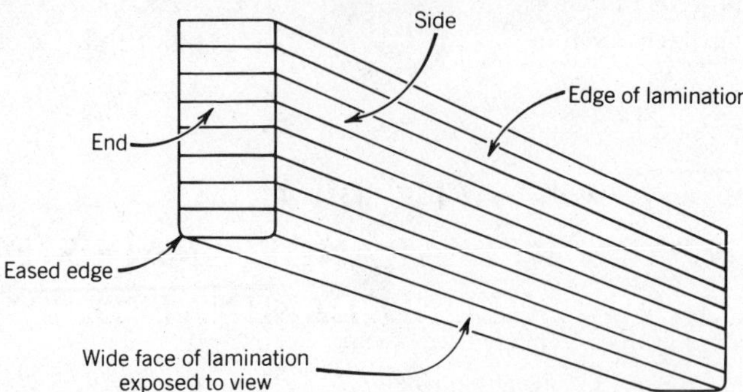

FIGURE 1 Illustration of terms related to appearance grades.

often the natural growth characteristics of the wood enhance the beauty of the member and avoid an artificial appearance.

1.1.2 Three appearance grades—industrial, architectural and premium—are applicable to all species or mixtures of species used in laminating. Some combinations in *Standard Specifications for Structural Glued Laminated Timber of Softwood Species*, AITC 117, permit the mixing of species within a given member. With this mixing of species, the potential for differences in color or grain of adjacent laminations must be recognized. For those appearance applications where such possible differences in color or grain might be important, the designer may specify a single species or species group with similar characteristics. In some cases, this may restrict availability.

1.1.3 When members containing wood filler are stained, the filler may not accept the stain in the same manner as the adjacent wood. In addition, filler exposed to the elements may weather differently than the adjacent wood. The buyer should check with the supplier concerning past experience in these matters before specifying.

1.1.4 Preservative treatment may affect the appearance of the finished laminated timber. See *Standard for Preservative Treatment of Structural Glued Laminated Timber*, AITC 109, for detailed information.

2. TEXTURED SURFACES

2.1 When specified by the designer, a textured surface may be used. Textured surfaces are produced by a variety of methods and the buyer should check with the supplier before specifying.

2.2 Textured surfaces may change the net finished sizes and tolerances given in *Standard for Dimensions of Glued Laminated Structural Members*, AITC 113. Depending upon the degree of texturing, it may be necessary for the designer to compensate for the resulting loss of cross section.

2.3 Texturing will change the appearance grade specified. Additional voids and splintering may result. Textured surfaces will readily absorb stain, but adjacent areas of wood filler may not accept stain in the same manner. Eased edges will not normally be furnished.

3. INDUSTRIAL APPEARANCE GRADE

3.1 Application. Industrial appearance grade is ordinarily suitable for construction in industrial plants, warehouses, garages, and for other uses where appearance is not of primary concern.

3.2 Description

3.2.1 Laminations may possess the natural growth characteristics of the lumber grade.

3.2.2 Voids appearing on the edge of laminations need not be filled.

3.2.3 Loose knots and open knot holes in the wide face of laminations exposed to view shall be filled. This restriction does not apply for glued laminated timber truss members. Edge joints appearing on the wide face of laminations exposed to view need not be filled.

3.2.4 Members are required to be surfaced on two sides only; appearance requirements apply to these sides. Occasional misses, low laminations or wane (limited to a maximum of $\frac{1}{4}$ in. measured across the width) are permitted on a cumulative basis. The cumulative depth of the misses, low laminations or wane shall not exceed 10% of the width of the member at any one glue line. The frequency of occurrence shall not exceed 1 in 10 pieces of lumber used. The maximum area of low laminations shall not exceed 5% of the surface area of a side; and no more than two low laminations shall be adjacent to one another.

3.2.5 In accordance with provisions in 3.2.4, wane (limited to $\frac{1}{4}$ in. measured across the width) is permitted in all combinations and is not limited in length. Wane permitted in specific laminating combinations up to one-sixth the lumber width on each side is not limited in length. Occasional wane approximately 1 ft in length and not exceeding the permissible depth of a low lamination shall be permitted in all combinations without regard to the cumulative effects indicated in 3.2.4.

4. ARCHITECTURAL APPEARANCE GRADE

4.1 Application. Architectural appearance grade is ordinarily suitable for construction where appearance is an important requirement. Any voids under $\frac{3}{4}$ in. in diameter shall be filled by others than the fabricator if the final decorative finish so requires.

4.2 Description

4.2.1 Laminations may possess the natural growth characteristics of the lumber grade.

4.2.2 In exposed surfaces, knot holes and other voids mesuring over $\frac{3}{4}$ in. shall be filled by the fabricator with a wood-tone colored filler which reasonably blends with the final product or with clear wood inserts. When inserts are used, they shall be selected with reasonable care for similarity of the grain and color of the wood insert to the adjacent wood.

4.2.2.1 For appearance grading purposes, measurement of knot holes and other voids is to be made on the basis of equivalent circular areas. Void measurement limitations apply only to the surfaces of the member exposed in the final structure. All characteristics must be considered, however, with respect to their effects on general appearance.

4.2.3 The wide face of laminations exposed to view shall be free of loose knots. Open knot holes shall be filled. Voids greater than $\frac{1}{16}$ in. wide in edge joints appearing on the wide face of laminations exposed to view shall be filled.

4.2.4 Exposed faces shall be surfaced smooth. Misses are not permitted.

4.2.5 The corners of the wide face of laminations exposed to view in the final structure shall be eased. Current industry practice for eased edges is for a radius between $\frac{1}{8}$ and $\frac{1}{2}$ in. Other radii for eased edges may be agreed upon between buyer and seller.

5. PREMIUM APPEARANCE GRADE

5.1 Application. Premium appearance grade is the highest standard appearance grade.

5.2 Description

5.2.1 Laminations may possess the natural growth characteristics of the lumber grade.

5.2.2 In exposed surfaces, knot holes and other voids shall be filled by the fabricator with a wood-tone colored filler which reasonably blends with the final product or with clear wood inserts. When inserts are used, they shall be selected with reasonable care for similarity of the grain and color of the wood insert to the adjacent wood.

5.2.3 The wide face of laminations exposed to view shall be selected for appearance and shall be free of loose knots. Voids shall be filled. Knot size shall be limited to 20% of the net face width of the lamination. Not over two maximum size knots or their equivalent shall occur in a 6 ft length. Voids greater than $\frac{1}{16}$ in. wide in edge joints appearing on the wide face of laminations exposed to view shall be filled.

5.2.4 Exposed faces shall be surfaced smooth. Misses are not permitted.

5.2.5 The corners of the wide face of laminations exposed to view in the final structure shall be eased. Current industry practice for eased edges is for a radius between $\frac{1}{8}$ and $\frac{1}{2}$ in. Other radii for eased edges may be agreed upon between buyer and seller.

6. SPECIAL APPEARANCE REQUIREMENTS

6.1 For special applications, the buyer may specify other requirements. Such requirements may limit availability.

6.1.1 Industrial Special appearance grade, indicated on the noncustom Quality Inspected stamp as IND-S, is applicable to nominal 4 in. wide glued laminated timbers surfaced "hit or miss" to $3\frac{1}{2}$ in. All requirements in Section 3.2 apply except the limitations on misses or low laminations, and the unsurfaced lamination may exhibit glue smears and discolorations associated with the production process.

7. SUMMARY OF APPEARANCE GRADE SPECIFICATIONS

Description Item	Industrial Appearance Grade	Architectural Appearance Grade	Premium Appearance Grade
Natural growth characteristics of lumber grade	Allowed	Allowed	Allowed
Paragraph reference	3.2.1	4.2.1	5.2.1
Filling of voids on edge of laminations	Not required	Required for voids over $\frac{3}{4}$ in. in diameter	Required for all voids
Paragraph reference	3.2.2	4.2.2, 4.2.2.1	5.2.2
Wide face of laminations exposed to view	Void filling required except for trusses	Free of loose knots, void filling required	Selected for appearance, free of loose knots, void filling required, knot sizes limited
Paragraph reference	3.2.3	4.2.3	5.2.3
Edge joints appearing on wide face of laminations exposed to view	Filling not required	Filling required for voids over $\frac{1}{16}$ in. wide	Filling required for voids over $\frac{1}{16}$ in. wide
Paragraph reference	3.2.3	4.2.3	5.2.3
Surfacing of sides	Required. Limited amounts of misses, low laminations and wane permitted	Required. Misses not permitted	Required. Misses not permitted
Paragraph reference	3.2.4, 3.2.5	4.2.4	5.2.4
Surfacing of wide face of laminations exposed to view	Not required	Required. Misses not permitted	Required. Misses not permitted
Paragraph reference	3.2.4	4.2.4	5.2.4
Eased edges	Not required	Required	Required
Paragraph reference	—	4.2.5	5.2.5

AMERICAN INSTITUTE  TIMBER CONSTRUCTION
333 West Hampden Avenue Englewood, Colorado 80110

AITC 111-79
RECOMMENDED PRACTICE FOR PROTECTION OF STRUCTURAL GLUED LAMINATED TIMBER DURING TRANSIT, STORAGE AND ERECTION

Adopted as Recommendations March 4, 1979

CONTENTS

1. INTRODUCTION

1.1 Protection of glued laminated timber structural members includes end sealers, surface sealers, primer coats and wrappings applied for the protection of members. End sealers, surface sealers, primer coats and wrappings offer a degree of protection, but they do not necessarily preclude damage resulting from negligence and other factors beyond the control of the laminator during shipment, handling, storage and placement.

1.2 The protection specified should be commensurate with the end use and final finish of the member. It may also vary with the method of shipment and with exposure to climatic and other conditions before construction is completed.

1.3 These recommended specifications are for the guidance of the designer so that the product may have protection consistent with the intended use of the member at appropriate cost. The designer should specify the desired protection in a way that establishes a clear understanding between the buyer and the seller. *This standard contains alternatives from which the specifier must make certain selections to suit the particular job.*

1.4 Experience has shown that the following protection methods have a sufficient range to fulfill normal requirements. The designer should select the methods of protection best suited to the particular job and include them in the job specification.

2. END SEALERS

2.1 End sealers retard moisture transmission and minimize end checking when use conditions are such that end checking is a major consideration.

2.2 Recommended Specifications.

2.2.1 A coat of sealer should be applied to the fresh-cut ends of all members after end trimming.

2.2.2 A colorless sealer shall be used on ends exposed to view in the completed structure.

3. SURFACE SEALERS

3.1 Surface sealers increase resistance to soiling, control grain raising, minimize checking and serve as a moisture retardant. Surface sealers fall into the two following classifications.

3.1.1 Translucent Penetrating Sealers. Translucent penetrating sealers have low solid content. They provide limited protection and are suitable for use when final finish requires staining.

3.1.1.1 Recommended Specifications. A penetrating sealer shall be applied to all surfaces before shipment.

3.1.2 Primer and Non-Penetrating Sealer Coats. Primer and non-penetrating sealer coats have higher solid content than penetrating sealers and provide maximum protection by sealing the surface of the wood. Primer and non-penetrating sealer coats should not be specified when final finish requires a natural or stained finish.

3.1.2.1 Recommended Specifications.

(a) A sealer (or primer) coat shall be applied to all surfaces before shipment.

(b) A non-penetrating sealer (or primer) shall have a minimum solid content of 25%.

4. WRAPPING

4.1 Wrapping the member with water-resistant covering or its equivalent for shipment provides additional protection from moisture, soiling and damage in handling. Wrapping is usually recommended when appearance is of prime importance and the additional protection is desired. Bundle or load wrapping may be specified in lieu of individual wrapping when further utilization of wrap after delivery is not desired. Time of removal of factory wrap is optional, but, it must be emphasized, factory-applied wrapping provides additional protection from damage in handling and in transit only. If further utilization of the wrap is desired for protection after shipment, the members should be inspected and provided with additional protection as necessary.

4.2 Water-resistant covering used for in-transit protection of individually wrapped members may be left in place until the members are enclosed within the building. If wrapping has to be removed at certain connection points during erection, it should be replaced after connection is made in order to prevent sun bleaching or water staining of parts of the member. If it is impractical to replace the wrapping, all of it should be removed. Individual wrapping should be slit or punctured on the lower side if there is evidence of moisture inside the wrapping.

4.3 Recommended Specifications

4.3.1 Individual Wrapping

4.3.1.1 Members shall be individually wrapped, covering all surfaces, with water-resistant paper, opaque polyethylene or their equivalent.

4.3.1.2 Wrapping shall be secured to the member by staples, tape or other suitable fastenings that do not damage exposed surfaces.

4.3.1.3 Seams of wrapping shall inhibit the passage of moisture.

4.3.2 Bundle Wrapping

4.3.2.1 Members shall be bundle wrapped, totally enclosing the bundle, with water-resistant paper, opaque polyethylene or their equivalent.

4.3.2.2 Convenience in handling shall determine the size of the bundle.

4.3.2.3 Wrapping shall be secured to the bundle by staples, tape or other suitable fastenings that do not damage exposed surfaces. Wherever possible, staples should be located in the bottom of the bundle.

4.3.2.4 Seams of wrapping shall inhibit the passage of moisture.

4.3.3 Load Wrapping

4.3.3.1 Members shall be load-wrapped, enclosing the top, sides and ends with water-resistant paper, opaque polyethylene or their equivalent.

4.3.3.2 Wrapping shall extend to the bottom of the members included in the load.

4.3.3.3 Wrapping shall be secured to the load by staples, tape or other suitable fastenings that do not damage the exposed surfaces.

4.3.3.4 Seams of wrapping shall inhibit the passage of moisture.

5. PROTECTION FOR PRESERVATIVELY TREATED MEMBERS

5.1 Preservative treatment of glued laminated timber can be divided into three general categories: (a) heavy treatments of members after gluing, such as creosote or pentachlorophenol (penta) in heavy oil; (b) light treatments of members after gluing such as penta in light solvent (penta in LPG and waterborne treatments are not recommended for treating glued laminated timber after gluing); and (c) treatments of laminations prior to gluing such as penta in LPG, penta in light solvent or waterborne salts.

5.1.1 Protection of glued laminated members to receive heavy treatments is generally limited to end sealers. Surface sealers may be recommended if the members are to be stored in an arid climate prior to treatment. If protection during transit and storage is desired, load wrapping should be satisfactory for members to receive heavy treatments. Where the time between fabrication and treatment is short, no special protection may be required.

5.1.2 Protection of glued laminated members to receive light treatments is usually more critical since they are generally intended for use where appearance is a factor. End sealers should be applied at the time the end cuts are made and following treatment to minimize end checking. Surface sealers may be recommended if the members are to be stored in an arid climate prior to treatment. Surface sealers may also be applied to the members after treatment for the same reasons they are applied to untreated members.

5.1.3 Protection of glued laminated members made from treated laminations is generally the same as for untreated members.

6. UNLOADING AND HANDLING

6.1 Recommended Specifications

6.1.1 Laminated members shall not be dragged or dropped. Care shall be taken in handling to prevent damage to finish surfaces. Cable slings or chokers should not be used to handle laminated materials unless adequate blocking is provided between the cable and the wood member. Web belting-type slings are recommended. Protection cleats or blocking shall be applied at pickup points to protect corners.

6.1.2 Spreader bars of suitable length should be used in lifting long members to reduce the probability of damage. The method of erection and handling should not overstress the member.

6.1.3 Whenever possible, members should be lifted on edge.

6.1.4 Extreme care should be taken to minimize impact forces during lifting.

7. JOB SITE STORAGE

7.1 Recommended Specifications

7.1.1 Laminated material stored at the job site shall be treated with care. A level area is required to avoid warpage. Members shall be supported with blocking so spaced as to provide uniform and adequate support. If covered storage is not available, the material shall be blocked well off the ground at a well drained location. Stored members shall be separated with stripping arranged vertically over the supports so that air circulates around all four sides of each member; the top and all sides shall be covered with moisture-resistant covering. If a paved surface is unavailable, the ground under the material shall be covered with polyethylene film. Clear polyethylene film shall not be used. Individual wrappings shall be slit full length or punctured on the lower side to permit drainage of water. Slats inserted between the member and wrapping minimize the marking of the wood. Long members shall be stored on edge.

8. ERECTION

8.1 Recommended Specifications

8.1.1 Padded or nonmarring slings shall be used, and corners shall be protected with wood blocking. (See Section 6 for additional information.)

NOTE: *Heat should not be fully turned on as soon as the structure is enclosed; otherwise, excessive checking may occur due to rapid lowering of the relative humidity in the building. A gradual seasoning period at moderate temperature should be provided.*

AMERICAN INSTITUTE
333 West Hampden Avenue

TIMBER CONSTRUCTION
Englewood, Colorado 80110

AITC 112*-81
STANDARD FOR TONGUE-AND-GROOVE HEAVY TIMBER ROOF DECKING

Adopted as Recommendations October 15, 1981
(Updated November 1983)

*Combines former AITC 112, *Standard for Heavy Timber Roof Decking*; and AITC 118, *Standard for Two Inch Nominal Thickness Lumber Roof Decking for Structural Applications*

CONTENTS

1. INTRODUCTION

1.1 This Standard applies to sawn tongue-and-groove decking only and does not apply to laminated, panelized or other special decking systems. It covers species, sizes, patterns, lengths, moisture content, application, specifications, weights, applicable unit stresses, allowable loads and slope conversion values for heavy timber roof decking in nominal 2, 3 and 4 in. thicknesses, using single or double tongues and grooves.

1.2 Heavy timber roof decking is a specialty lumber product, constituting an important part of modern timber construction, that can be used for many applications to provide an all wood appearance. Nominal three and four inch thick roof decking is especially well adapted for use with glued laminated arches and girders and is easily and quickly erected. To be suitable for purposes intended, heavy timber roof decking must be well manufactured to a low moisture content as described herein.

1.3 The lumber used in heavy timber roof decking shall be graded in accordance with the grading rules under which the species is customarily graded. The standard grading and dressing rules referenced in this Standard are:

(a) *Standard Grading Rules for Northeastern Lumber*, 1980, Northeastern Lumber Manufacturers Association, 4 Fundy Road, Falmouth, Maine 04105 (NeLMA)

(b) *Standard Specifications for Grades of California Redwood Lumber*, April 1982, Edition, Redwood Inspection Service, 591 Redwood Highway, #3100, Mill Valley, California 94941 (RIS)

(c) *Southern Pine Inspection Bureau Grading Rules*, 1977, Southern Pine Inspection Bureau, 4709 Scenic Highway, Pensacola, Florida 32504 (SPIB)

(d) *Standard Grading Rules for West Coast Lumber, No. 16*, September 1, 1970, Revised January 1, 1980, West Coast Lumber Inspection Bureau, P.O. Box 23145, Portland, Oregon 97223 (WCLIB)

(e) *Standard Grading Rules for Western Lumber*, Effective June 1, 1981, Western Wood Products Association, 1500 Yeon Building, Portland, Oregon 97204 (WWPA)

(f) *Standard Grading Rules for Canadian Lumber*, U.S. Edition, Effective December 1, 1980, National Lumber Grades Authority, P.O. Box 97, Ganges, B.C. V0S 1E0, Canada (a Canadian Agency)(NLGA)

Copies of these grading rules may be obtained from the respective grading rule agencies.

1.4 Moisture content requirements of the regional lumber grading rules may differ from this Standard. Unless this Standard is followed in all requirements, the product will not conform with this Standard.

2. SPECIES

2.1 The species usually available and currently used in this product, as well as the regional inspection agencies under which decking lumber is ordinarily graded, are given in Table 1.

3. SIZES AND PATTERNS

3.1 Two Inch Decking. The standard size is 2 × 6 in., nominal, dressed at the moisture content specified herein to the actual size and V-grooved pattern shown in Figure 1. Other thicknesses and widths are also available. See regional grading rules listed in paragraph 1.3 for dimensions for each species.

3.2 Three and Four Inch Decking. Standard sizes are 3 × 6 in. and 4 × 6 in., nominal, which are dressed to $2\frac{1}{2} \times 5\frac{9}{16}$ in. and $3\frac{1}{2} \times 5\frac{9}{16}$ in., respectively, at the moisture content specified herein. Finished face width overall for both thicknesses is $5\frac{1}{4}$ in. Figures 2 and 3 provide actual dimensions for 3 × 6 in. and 4 × 6 in. nominal decking, respectively, illustrating a V-joint pattern. Other thicknesses and widths may be available.

3.3 Other patterns are available, including grooved, striated and eased joint, and the regional grading rules agencies indicated in paragraph 1.3 should be contacted for further details concerning specific patterns and sizes.

3.4 Each piece shall be square end trimmed. When random lengths are furnished, each piece must be square end trimmed across the face so that at least 90% of the pieces will be within 3/64 in. of square. The vertical end cut may vary from square to the bevel cut shown in Figure 4.

3.5 End joints in 2 in. nominal thickness decking not occurring over supports when random length pieces are used should (a) be matched (T & G) or (b) have metal splines inserted at the ends so that loads may be distributed from end to end as well as across the planks.

4. LENGTHS

4.1 Decking pieces may be of specified length or may be random length. All layup arrangements except controlled random layup require that the specifier indicate the required lengths.

4.2 If pieces are for controlled random layup, odd or even lengths are permitted, and the minimum lengths based on fbm percentages shall be as follows:

4.2.1 ***Two Inch Decking***

Not less than 40% to be 14 ft and longer
Not over 10% to be less than 10 ft
Not over 1% to be 4 to 5 ft
Minimum length is limited to 75% of the span length (i.e., for 8 ft support spacing, 6 ft)

4.2.2 ***Three Inch Decking***

Not less than 40% to be 14 ft and longer with at least 20% equal to or greater in length than the maximum span.
Not over 10% to be less than 10 ft
Not over 1% to be 4 to 5 ft

4.2.3 ***Four Inch Decking***

Not less than 25% to be 16 ft and longer with at least 20% equal to or greater in length than the maximum span.
Not less than 50% to be 14 ft and longer
Not over 10% to be 5 to 10 ft
Not over 1% to be 4 to 5 ft

5. MOISTURE CONTENT

5.1 Two Inch Decking. The maximum moisture content shall be 15%.

5.2 Three and Four Inch Decking. The maximum moisture content shall be 19%.

5.3 Moisture content shall be determined by such methods as will assure these limitations.

TABLE 1

Heavy Timber Deck Species

Species	Grading Rules Under Which Graded	Paragraph Number of Grading Rules Under Which Graded[a]	
		Select Quality[b]	Commercial Quality[c]
Cedar, Northern White	NELMA	15.1	15.2
Cedars, Western	WWPA, WCLIB	55.11, 127-b	55.12, 127-c
Cedars, Western (North)	NLGA (Canadian)	127-b	127-c
Coast Species	NLGA (Canadian)	127-b	127-c
Douglas Fir-Larch	WWPA, WCLIB	55.11, 127-b	55.12, 127-c
Douglas Fir-Larch (North)	NLGA (Canadian)	127-b	127-c
Douglas Fir (South)	WWPA	55.11	55.12
Fir, Balsam	NELMA	15.1	15.2
Hem-Fir	WWPA, WCLIB	55.11, 127-b	55.12, 127-c
Hem-Fir (North)	NLGA (Canadian)	127-b	127-c
Hemlock, Eastern-Tamarack	NELMA	15.1	15.2
Hemlock, Eastern-Tamarack (North)	NLGA (Canadian)	127-b	127-c
Hemlock, Mountain	WWPA, WCLIB	55.11, 127-b	55.12, 127-c
Hemlock, Mountain-Hem-Fir	WWPA	55.11	55.12
Hemlock, Western	WWPA, WCLIB	55.11, 127-b	55.12, 127-c
Hemlock, Western (North)	NLGA (Canadian)	127-b	127-c
Northern Species	NLGA (Canadian)	127-b	127-c
Pine, Eastern White	NELMA	15.1	15.2
Pine, Eastern White (North)	NLGA (Canadian)	127-b	127-c
Pine, Idaho White	WWPA	55.11	55.12

Pine, Lodgepole	WWPA	55.11	55.12
Pine, Northern	NELMA	15.1	15.2
Pine, Ponderosa	NLGA (Canadian)	127-b	127-c
Pine, Ponderosa-Sugar	WWPA	55.11	55.12
Pine, Red	NLGA (Canadian)	127-b	127-c
Pine, Southern[d,e]	SPIB	432	433
Pine, Western White	NLGA (Canadian)	127-b	127-c
Redwood, California	RIS	315	316
Spruce, Coast Sitka	NLGA (Canadian)	127-b	127-c
Spruce, Eastern	NELMA	15.1	15.2
Spruce, Eastern-Balsam Fir	NELMA	15.1	15.2
Spruce, Engelmann-Alpine Fir	WWPA	55.11	55.12
Spruce-Pine-Fir	NLGA (Canadian)	127-b	127-c
Spruce, Sitka	WCLIB	127-b	127-c
White Woods (Western Woods)	WWPA	55.11	55.12

[a] When species may be graded under WCLIB and WWPA rules, the first paragraph number is for WWPA and the second for WCLIB rules.

[b] Select quality grades are as follows for the grading rules indicated:

WCLIB;	Select Dex	SPIB;	Select Decking
WWPA;	Selected Decking	RIS;	Select Decking
NELMA;	Selected Decking	NLGA;	Select Decking

[c] Commercial quality grades are as follows for the grading rules indicated:

WCLIB;	Commercial Dex	SPIB;	Commercial Decking
WWPA;	Commercial Decking	RIS;	Commercial Decking
NELMA;	Commercial Decking	NLGA;	Commercial Decking

[d] Southern Pine decking is also available in the following grades: Dense Standard Decking, para. 431; Dense Select Decking, para. 432.1; and Dense Commercial Decking, para. 433.1

[e] Southern Pine is limited to the botanical species of longleaf, slash, shortleaf and loblolly. Lumber cut from trees of these species is classified as "Southern Pine" in the SPIB Grading Rules.

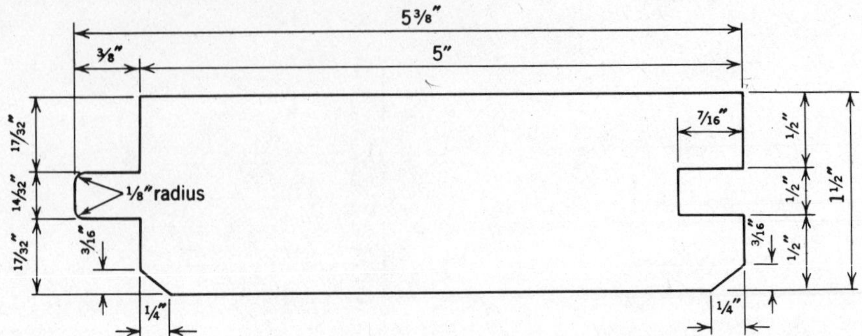

FIGURE 1 2 × 6 in., nominal, V-joint pattern. (See regional grading rules listed in paragraph 1.3 for dimensions for each species.)

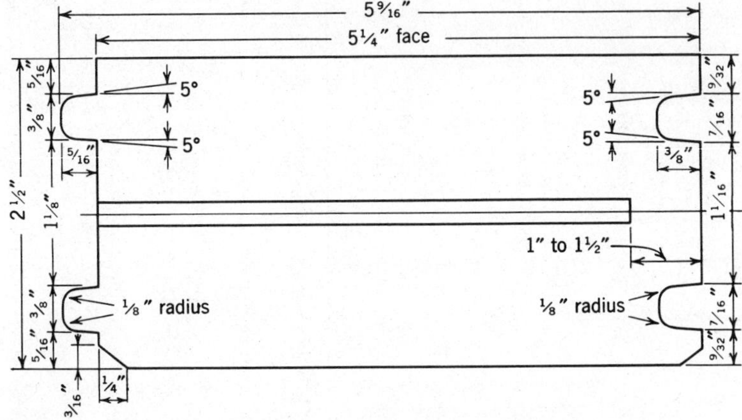

FIGURE 2 3 × 6 in., nominal, V-joint pattern. Note: Profile dimensions apply to all patterns. (See regional grading rules in paragraph 1.3 for dimensions for each species.)

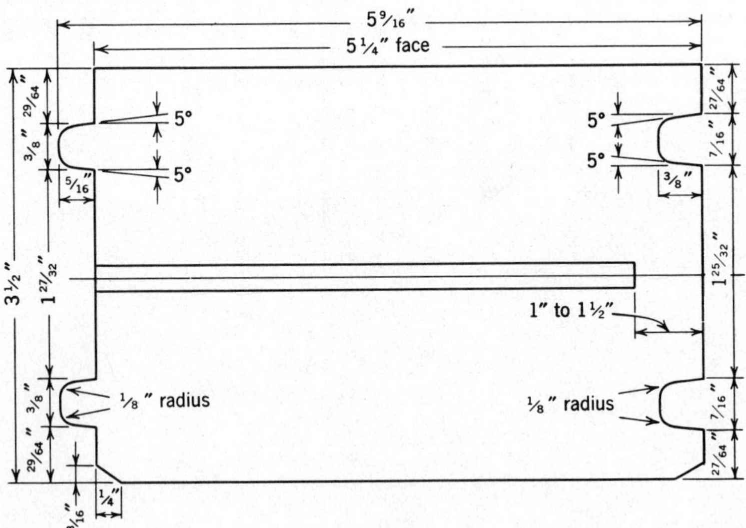

FIGURE 3 4 × 6 in., nominal, V-joint pattern. Note: Profile dimensions apply to all patterns. (See regional grading rules in paragraph 1.3 for dimensions for each species.)

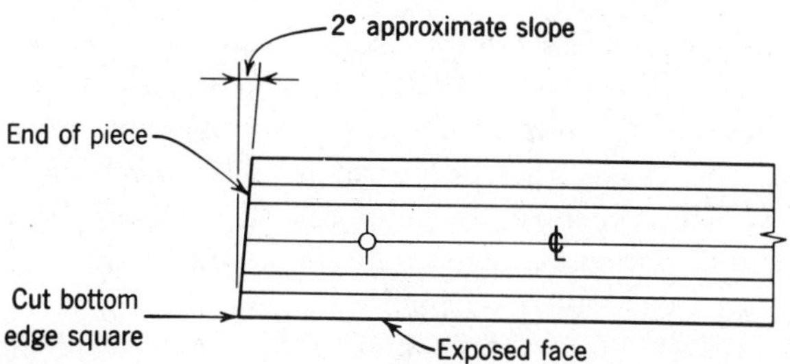

FIGURE 4 Beveled End Cut. (Beveled end cut is optional.)

6. APPLICATION

6.1 Tongue-and-groove wood decking is to be installed with tongues up on sloped or pitched roofs, and outward in direction of laying on flat roofs. It is to be laid with pattern faces down and exposed on the underside.

6.2 Nailing Schedules.

6.2.1 Two Inch Decking. Each piece should be toenailed through the tongue and face nailed with one nail per piece per support, using 16d common nails.

6.2.2 Three and Four Inch Decking. Each piece should be toenailed at each support with one 40d nail and face nailed with one 60d nail. Courses shall be spiked to each other with 8 in. spikes at intervals not to exceed 30 in. through predrilled edge holes and with one spike at a distance not exceeding 10 in. from each end of each piece. See Figure 5 for boring details.

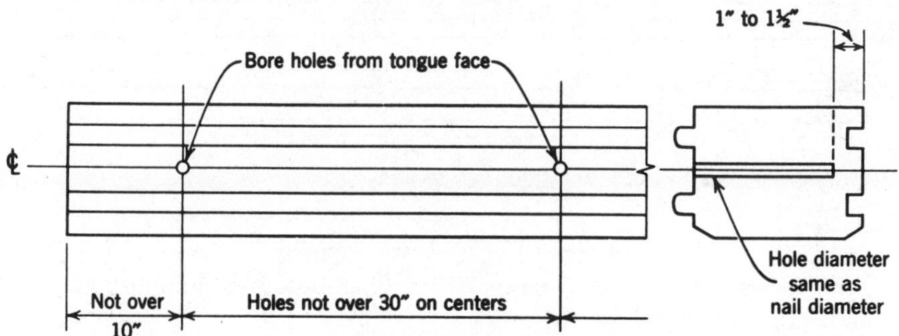

FIGURE 5 Boring Detail. Locate end holes not over 10 in. from end of piece.

6.3 Heavy timber decking may be laid in any of the following arrangements:

6.3.1 Simple Span. All pieces bear on two supports.

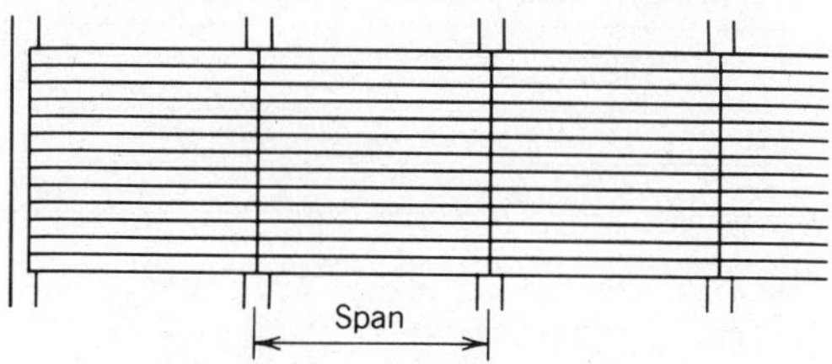

FIGURE 6 Simple Span Layup.

6.3.2 Controlled Random Layup. This arrangement is applicable to 4 or more supports (3 or more spans). (With less than 4 supports, a special pattern requiring specified lengths must be used.) Joints in the same general line (within 6 in. of being in line each way) shall be separated by at least two intervening courses. In the end bays each piece must rest on at least one support. When an end joint occurs in the end bay, the next piece in the same course must continue over the first inner support for at least 2 ft. For 3 and 4 in. decking in the interior bays, occasional pieces not resting over a support may occur provided the ends of the adjacent pieces in the same course are continued for at least 2 ft over the next support. This condition shall not occur more than once in every 6 courses in each interior bay.

6.3.2.1 Two Inch Decking. There shall be a minimum distance of 2 ft between end joints in adjacent courses. To provide a continuous tie for lateral restraint for the supporting member, the pieces in at least the first and second courses, and repeating at least after each group of seven intervening courses, must bear on at least two supports with end joints in these two courses occurring in alternate spans or on alternate supports, unless some other provision, such as plywood overlayment, is made to provide continuity.

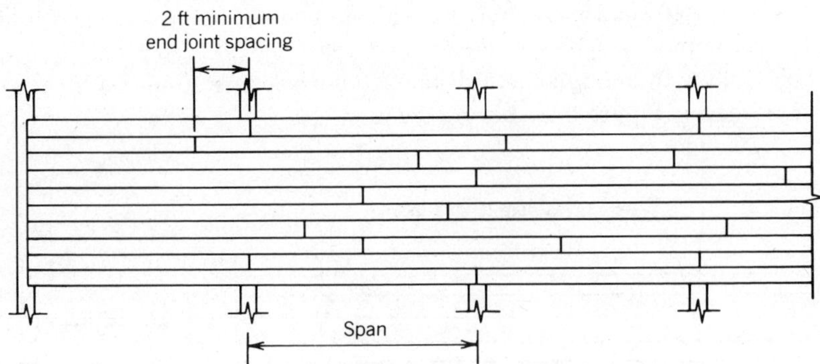

FIGURE 7 Controlled Random Layup. (Two inch decking.)

6.3.2.2 Three and Four Inch Decking. There shall be a minimum distance of 4 ft between end joints in adjacent courses.

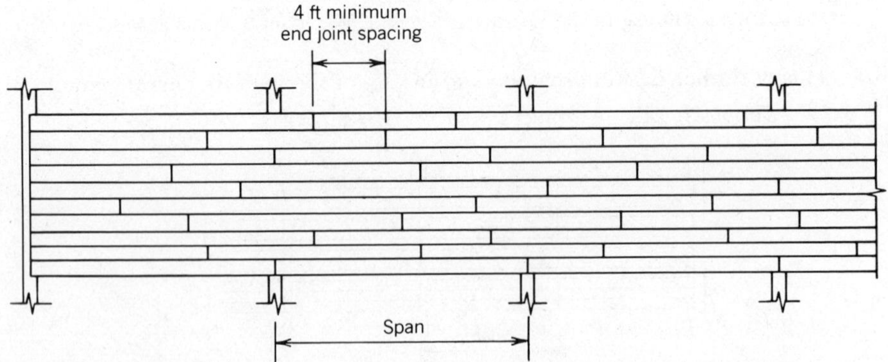

FIGURE 8 Controlled Random Layup. (Three and four inch decking.)

6.3.3 Cantilever Spans With Controlled Random Layup. When the overhang does not exceed $1\frac{1}{2}$ ft, 2 ft and 3 ft for nominal 2 in., 3 in. and 4 in. thick decking, no special considerations for layup are necessary. The maximum cantilever length for controlled random layup is limited to 0.3 times the length of the first adjacent interior span. For cantilever overhangs exceeding the normal overhang, but not exceeding the maximum, a structural fascia should be fastened to each decking piece to maintain a continuously straight roof line. Also, there shall be no end joints in the cantilevered portion or within $\frac{1}{2}$ the span ($L/2$) of the outer support.

6.3.4 Cantilevered Pieces Intermixed. This arrangement is applicable to 4 or more supports (3 or more spans). Pieces in the starter course and every third course are simple span. Pieces in other courses are cantilevered over the supports with end joints at alternate quarter or third points of the spans, and each piece rests on at least one support. A tie between supports is provided by the simple span courses of the arrangement.

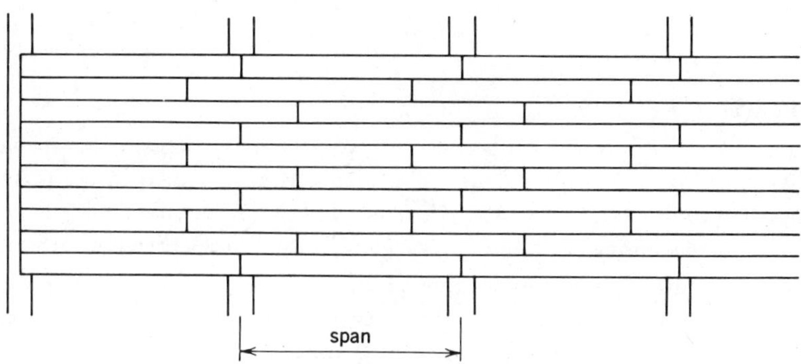

FIGURE 9 Cantilevered Pieces Intermixed Layup.

6.3.5 Combination Simple and Two-Span Continuous. Alternate pieces in end spans are simple spans; adjacent pieces are two-span continuous. End joints are staggered in adjacent courses and occur over supports.

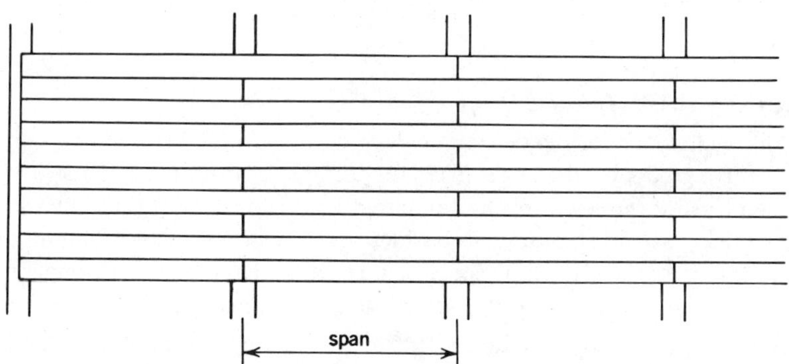

FIGURE 10 Combination Simple and Two-Span Continuous Layup.

6.3.6 Two-Span Continuous. All pieces bear on three supports. All end joints occur in line on every other support.

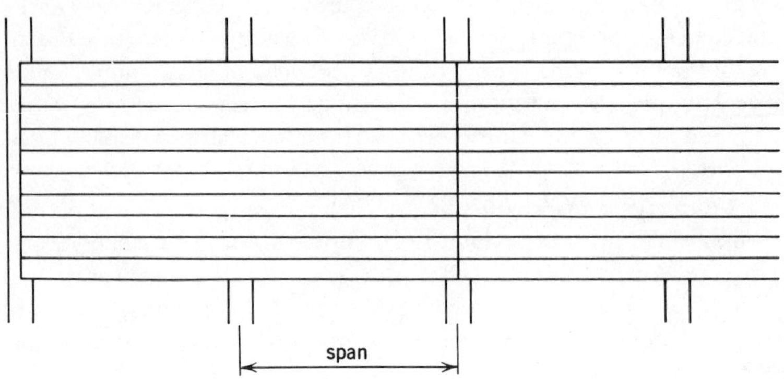

FIGURE 11 Two-Span Continuous Layup.

7. SPECIFICATIONS

7.1 The specifications for tongue-and-groove decking for the various species as well as inspection and shipping provisions shall be as specified in the standard grading rules under which the species is graded and shall be subject to such other provisions of the standard grading rules as may be applicable. (See paragraph 1.3.)

7.2 Select Quality. Decking of this quality is recommended for construction for which good strength and fine appearance are desired. Knots and other natural characteristics of specified limitations are permitted.

7.3 Commercial Quality. Decking of this quality is recommended and customarily used for the same purposes served by the higher quality when appearance requirements are less critical.

8. WEIGHTS OF INSTALLED DECKING (See Table 2, page 7-745.)

9. ALLOWABLE LOADS

9.1 Allowable loads for heavy timber decking may be determined by entering Tables 4 through 8 with the appropriate bending stress and modulus of elasticity values, and using the lower of the tabulated load values from the tables for the nominal thickness and span under consideration. Bending stress and modulus of elasticity values for wood decking species, as recommended by the regional lumber rules-writing agency by which the species is graded, are given in Table 3.

9.2 Allowable loads given in Tables 4 through 8 are for the simple span and controlled random layup arrangements illustrated under paragraphs 6.3.1 and 6.3.2.

TABLE 2

Weights of Installed Heavy Timber Decking in Pounds per Square Foot of Roof Surface[a]

Species	Thickness			
	$1\frac{1}{2}$ in. net[b] (2 in. nom.)	$2\frac{1}{2}$ in. net[c] (3 in. nom.)	$3\frac{1}{2}$ in. net (4 in. nom.)	Agency[d]
Cedar, Northern White	2.7	4.5	6.3	1
Cedars, Western[d]	3.0	4.9	6.9	3,4
Cedars, Western (North)[e]	2.9	4.8	6.7	2
Coast Species[e]	3.9	6.4	9.0	2
Douglas Fir-Larch[d]	4.3	7.2	10.1	3,4
Douglas Fir-Larch (North)[e]	4.4	7.3	10.3	2
Douglas Fir (South)	4.1	6.9	9.6	3
Fir, Balsam	3.2	5.4	7.5	1
Hem-Fir[e]	3.7	6.1	8.6	3,4
Hem-Fir (North)[e]	3.8	6.3	8.8	2
Hemlock, Eastern-Tamarack[e]	3.8	6.3	8.8	1
Hemlock, Eastern-Tamarack (North)[e]	4.0	6.7	9.4	2
Hemlock, Mountain	4.0	6.7	9.4	3,4
Hemlock, Mountain-Hem-Fir[f]	3.7–4.0	6.1–6.7	8.6–9.4	3
Hemlock, Western	4.0	6.7	9.4	3,4
Northern Species[f]	2.9–5.3	4.8–8.8	6.7–12.4	2
Pine, Eastern White	3.3	5.5	7.7	1
Pine, Eastern White (North)	3.4	5.7	8.0	2
Pine, Idaho White	3.5	5.8	8.2	3
Pine, Lodgepole	3.7	6.1	8.6	3
Pine, Northern[f]	3.8–4.5	6.3–7.5	8.8–10.5	1
Pine, Ponderosa	4.1	6.9	9.6	2
Pine, Ponderosa-Sugar[e]	3.6	6.0	8.4	3
Pine, Red	3.7	6.1	8.6	2
Pine, Southern[e]	4.6	7.6	10.7	5
Pine, Western White	3.4	5.7	8.0	2
Redwood, California	3.7	6.1	8.6	6
Spruce, Coast Sitka	3.3	5.5	7.7	2
Spruce, Eastern	3.6	6.0	8.4	1
Spruce, Eastern-Balsam Fir[e]	3.4	5.7	8.0	1
Spruce, Engelmann-Alpine Fir	3.0	5.1	7.1	3
Spruce-Pine-Fir[f]	3.0–4.0	5.1–6.7	7.1–9.4	2
Spruce, Sitka	3.6	6.0	8.4	4
White Woods (Western Woods)[f]	2.9–4.6	4.8–7.6	6.7–10.7	3

[a] All weights given in Table 2 are based on volume at 15% moisture content; rounded to nearest 0.1 lb. These weights may be reduced by 2% where 15% maximum moisture content is specified (which is an average of 12% M.C.).

[b] For a net thickness of $1\frac{5}{8}$ in., multiply tabulated weights by a factor of 1.08.

[c] For a net thickness of $2\frac{5}{8}$ in., multiply tabulated weights by a factor of 1.05.

[d] Species listed are as graded by the following grading rules agencies: NELMA (1), NLGA (Canadian) (2), WWPA (3), WCLIB (4), SPIB (5), and RIS (6).

[e] Weights given for this species grouping are based on the weighted average of the standing timber volume. Lumber from some areas or species within the group may vary slightly from the average.

[f] Weights given for this species grouping are the range of weights for species that could be included.

9.3 Controlled Random Layup Load Values

9.3.1 Two Inch Decking. The allowable loads for controlled random layup, limited by bending, for 2 in. nominal thickness decking as given in Table 4, are based on the standard engineering formula for a three-equal-span continuous, uniformly-loaded member; however, only $\frac{2}{3}$ of the moment of inertia for the cross section was used in calculating the loads. Loads limited by deflection as given in Table 5, are for the maximum deflections in the end spans.

9.3.2 Three and Four Inch Decking. The allowable loads for controlled random layup of 3 and 4 in. nominal thickness decking as given in Tables 6 through 8, are based on the standard engineering formula for a three-equal-span, continuous, uniformly-loaded member; however, only 80% of the moment of inertia for the cross section was used in calculating the loads. Loads limited by deflection, as given in Tables 7 and 8, are for the maximum deflections in the end spans.

9.3.3 The percentage adjustments in moment of inertia discussed in 9.3.1 and 9.3.2 take into account the differences between continuous decking without joints and the controlled random layup of decking as specified herein. The factors of $\frac{2}{3}$ for 2 inch and 80% for 3 and 4 in. decking were selected after careful evaluation of tests and previous experience.

9.3.4 When controlled random layup as specified herein is used for unequal spans, nonuniform loading, cantilever action, or conditions other than covered herein by the tabulated values, the same adjustment factors should be applied to the moment of inertia used in standard engineering formulas representing the actual conditions of load and span.

9.4 The allowable loads given in Tables 4 and 5 are based on a maximum moisture content of 15% for 2 in. decking. The allowable loads given in Table 6 through 8 are based on a maximum moisture content of 19% for 3 and 4 in. decking. If the maximum moisture content is limited to 15% for 3 and 4 in. decking, the allowable bending stress values given in Table 6 may be multiplied by 1.08 and the modulus of elasticity values given in Tables 7 and 8 may be multiplied by 1.05.

9.5 Allowable load values given in Tables 4 and 6 are based on normal duration of loading. If decking is used for purposes where other durations of load control, increase the tabulated values as follows:

15% for 2 months duration, as for snow;
25% for 7 days duration;
33% for wind or earthquake; or
100% for impact.

These increases are not cumulative.

9.6 The allowable load tables are for total uniformly distributed vertical loads, including dead and live, in pounds per square foot on a horizontal roof surface. When roofs have only a moderate slope (3 in 12 or less), dead and live load may be added together without adjustment for slope of roof.

TABLE 3

Bending Stress and Modulus of Elasticity Values for Heavy Timber Decking Species[a]

Species	Select Quality		Commercial Quality		Agency[d]
	Bending Stress[b], psi	Modulus of Elasticity[c], psi	Bending Stress[b], psi	Modulus of Elasticity[c], psi	
Cedar, Northern White	1100	800,000	950	700,000	1
Cedars, Western	1450	1,100,000	1200	1,000,000	3,4
Cedars, Western (North)	1400	1,100,000	1200	1,000,000	2
Coast Species	1450	1,500,000	1200	1,400,000	2
Douglas Fir-Larch	2000	1,800,000	1650	1,700,000	3,4
Douglas Fir-Larch (North)	2000	1,800,000	1650	1,700,000	2
Douglas Fir (South)	1900	1,400,000	1600	1,300,000	3
Fir, Balsam	1650	1,500,000	1400	1,300,000	1
Hem-Fir	1600	1,500,000	1350	1,400,000	3,4
Hem-Fir (North)	1500	1,500,000	1300	1,400,000	2
Hemlock, Eastern-Tamarack	1700	1,300,000	1450	1,100,000	1
Hemlock, Eastern-Tamarack (North)	1700	1,300,000	1450	1,100,000	2
Hemlock, Mountain	1650	1,300,000	1400	1,100,000	3,4
Hemlock, Mountain-Hem-Fir	1600	1,300,000	1350	1,100,000	3
Hemlock, Western	1750	1,600,000	1450	1,400,000	3,4
Hemlock, Western (North)	1750	1,600,000	1450	1,400,000	2
Northern Species	1050	1,100,000	875	1,000,000	2
Pine, Eastern White	1300	1,200,000	1100	1,100,000	1
Pine, Eastern White (North)	1050	1,200,000	875	1,100,000	2
Pine, Idaho White	1300	1,400,000	1050	1,300,000	3
Pine, Lodgepole	1450	1,300,000	1200	1,200,000	3
Pine, Northern	1550	1,400,000	1300	1,300,000	1

TABLE 3 (*Continued*)

Species	Select Quality		Commercial Quality		Agency[d]
	Bending Stress[b], psi	Modulus of Elasticity[c], psi	Bending Stress[b], psi	Modulus of Elasticity[c], psi	
Pine, Ponderosa	1450	1,300,000	1250	1,100,000	2
Pine, Ponderosa-Sugar	1350	1,200,000	1150	1,100,000	3
Pine, Red	1350	1,300,000	1100	1,200,000	2
Pine, Southern	1650	1,600,000	1650	1,600,000	5
Pine, Western White	1300	1,400,000	1050	1,300,000	2
Redwood, California[e]	1700	1,100,000	1350	1,000,000	6
Spruce, Coast Sitka	1450	1,700,000	1200	1,500,000	2
Spruce, Eastern	1300	1,500,000	1100	1,400,000	1
Spruce, Eastern-Balsam Fir	1300	1,500,000	1100	1,400,000	1
Spruce, Engelmann-Alpine Fir	1300	1,300,000	1100	1,100,000	3
Spruce-Pine-Fir	1400	1,500,000	1150	1,300,000	2
Spruce, Sitka	1500	1,500,000	1250	1,300,000	4
White Woods (Western Woods)	1300	1,100,000	1050	1,000,000	3

[a]Recommended by the regional lumber rules-writing ageny by which the species is graded, for decking used at 19% maximum moisture content (MC). When decking is manufactured and used at 15% maximum MC, bending stress values may be multiplied by a factor of 1.08, and modulus of elasticity values by a factor of 1.05 (1.04 for California Redwood). When decking is used where the moisture content will exceed 19% for an extended period of time, bending stress values should be multiplied by a factor of 0.86 and modulus of elasticity values by a factor of 0.97.

[b]Repetitive member use values.

[c]The tabulated values are the average for the species or species grouping. For information concerning coefficient of variation of modulus of elasticity, see the appropriate grading rules for the species.

[d]Stresses listed are as assigned by the following grading rules agencies: NELMA (1), NLGA (Canadian) (2), WWPA (3), WCLIB (4), SPIB (5) and RIS (6).

[e]If specified as "close grain," California Redwood select decking is assigned a bending stress value of 1850 psi and a modulus of elasticity value of 1,400,000 psi when used at 19% MC.

TABLE 4

Two Inch Nominal Thickness[a], Allowable Roof Load Limited by Bending

Bending Stress, psi	Allowable Uniformly Distributed Total Roof Load[b], psf													
	Simple Span, ft							Controlled Random Layup Span, ft						
	6	7	8	9	10	11	12	6	7	8	9	10	11	12
875	73	54	41	32	26	22	18	61	45	34	27	22	18	15
950	79	58	44	35	28	24	20	66	48	37	30	24	20	16
1000	83	61	47	37	30	25	21	69	51	39	31	25	21	17
1050	88	64	49	39	32	26	22	73	54	41	33	26	22	18
1100	92	67	52	41	33	27	23	76	56	43	34	28	23	19
1150	96	70	54	42	34	28	24	80	59	45	36	29	24	20
1200	100	73	56	44	36	30	25	83	61	47	38	30	25	21
1250	104	76	58	46	38	31	26	87	64	49	39	31	26	22
1300	108	80	61	48	39	32	27	90	66	51	41	32	27	22
1350	112	83	63	50	40	33	28	94	69	53	42	34	28	23
1400	117	86	66	52	42	35	29	97	71	55	44	35	29	24
1450	121	89	68	54	44	36	30	101	74	57	45	36	30	25
1500	125	92	70	56	45	37	31	104	76	58	47	38	31	26
1550	129	95	73	57	46	38	32	108	79	60	48	39	32	27
1600	133	98	75	59	48	40	33	111	82	62	50	40	33	28
1650	138	101	77	61	50	41	34	114	84	64	52	41	34	29
1700	142	104	80	63	51	42	35	118	87	66	53	42	35	30
1750	146	107	82	65	52	43	36	122	89	68	55	44	36	30
1900	158	116	89	70	57	47	40	132	97	74	60	48	39	33
2000	167	122	94	74	60	50	42	139	102	78	63	50	41	35

[a] $1\frac{1}{2}$ in. net thickness. To determine allowable loads for $1\frac{5}{8}$ in. net thickness, multiply tabulated values by 1.17.

[b] To determine allowable uniformly distributed total roof loads for other span conditions, use simple span load values for combination simple and two-span continuous, and two-span continuous layups; and use controlled random layup load values for cantilevered pieces intermixed layup.

TABLE 5

Two Inch Nominal Thickness[a], Allowable Roof Load Limited By Deflection

Modulus of Elasticity, psi	Deflection Limit[b]	Allowable Uniformly Distributed Total Roof Load[c], psf											
		Simple Span, ft.					Controlled Random Layup Span, ft.						
		6	7	8	9	10	6	7	8	9	10	11	12
700,000	*l*/180	32	20	14	10	7	42	27	18	12	9	7	5
	l/240	24	15	10	7	5	32	20	13	9	7	5	4
800,000	*l*/180	37	23	16	11	8	48	30	20	14	10	8	6
	l/240	28	17	12	8	6	36	23	15	11	8	6	4
900,000	*l*/180	42	26	18	12	9	54	34	23	16	12	9	7
	l/240	31	20	13	9	7	41	26	17	12	9	7	5
1,000,000	*l*/180	46	29	20	14	10	60	38	25	18	13	10	7
	l/240	35	22	15	10	8	45	28	19	13	10	7	6
1,100,000	*l*/180	51	32	21	15	11	66	42	28	20	14	11	8
	l/240	38	24	16	11	8	50	31	21	15	11	8	6
1,200,000	*l*/180	56	35	23	16	12	72	46	30	21	16	12	9
	l/240	42	26	18	12	9	54	34	23	16	12	9	7
1,300,000	*l*/180	60	38	25	18	13	78	49	33	23	17	13	10
	l/240	45	28	19	13	10	59	37	25	17	13	10	7
1,400,000	*l*/180	65	41	27	19	14	84	53	36	25	18	14	10
	l/240	49	31	20	14	10	63	40	27	19	13	10	8
1,500,000	*l*/180	69	44	29	20	15	90	57	38	27	20	15	11
	l/240	52	33	22	15	11	68	43	29	20	15	11	8
1,600,000	*l*/180	74	47	31	22	16	96	61	41	28	21	16	12
	l/240	56	35	23	16	12	72	46	30	21	16	12	9
1,700,000	*l*/180	79	50	33	23	17	102	64	43	30	22	17	13
	l/240	59	37	25	17	13	77	48	32	23	17	12	10
1,800,000	*l*/180	83	52	35	25	18	108	68	46	32	23	18	14
	l/240	62	39	26	18	14	81	51	34	24	18	13	10

[a] $1\frac{1}{2}$ in. net thickness. To determine allowable loads for $1\frac{5}{8}$ in. net thickness, multiply tabulated values by 1.27.

[b] For a deflection limit of *l*/360, use $\frac{1}{2}$ the tabulated value for a deflection limit of *l*/180.

[c] To determine allowable uniformly distributed total roof loads for other span conditions, multiply controlled random layup load values by the following factors:

Cantilevered pieces intermixed;	1.05
Combination simple span and two-span continuous;	1.31
Two-span continuous;	1.85

TABLE 6

Three and Four Inch Nominal Thickness, Allowable Roof Load Limited by Bending, Simple Span and Controlled Random Layups (3 or more spans)[a]

Bending Stress, psi	Uniformly Distributed Total Roof Load[a], psf																									
	3 in. Nominal Thickness[b], Span, ft[c]													4 in. Nominal Thickness[d], Span, ft[c]												
	8	9	10	11	12	13	14	15	16	17	18	19	20	8	9	10	11	12	13	14	15	16	17	18	19	20
875	114	90	73	60	51	43	37	32	28	25	22	20	18	223	176	143	118	99	84	73	64	56	49	44	40	36
950	124	98	79	65	55	47	40	35	31	27	24	22	20	242	192	155	128	108	92	79	69	61	54	48	43	39
1000	130	103	83	69	58	49	42	37	32	29	26	23	21	255	202	163	135	113	97	83	72	64	56	50	45	41
1050	137	108	88	72	61	52	45	39	34	30	27	24	22	262	212	172	142	119	101	88	76	67	59	53	48	43
1100	143	113	92	76	64	54	47	41	36	32	28	25	23	281	222	180	148	125	106	92	80	70	62	55	50	45
1150	150	118	96	79	66	57	49	42	37	33	30	26	24	293	232	188	155	130	111	96	83	73	65	58	52	47
1200	156	123	100	83	69	59	51	44	39	35	31	28	25	306	242	196	162	136	116	100	87	76	68	60	54	49
1250	163	129	104	86	72	62	53	46	41	36	32	29	26	319	252	204	169	142	121	104	91	80	71	63	56	51
1300	169	134	108	90	75	64	55	48	42	37	33	30	27	332	262	212	175	147	126	108	94	83	73	66	59	53
1350	176	139	112	93	78	66	57	50	44	39	35	31	28	344	272	220	182	153	130	112	98	86	76	68	61	55
1400	182	144	117	96	81	69	60	52	46	40	36	32	29	357	282	229	189	159	135	117	102	89	79	70	63	57
1450	189	149	121	100	84	71	62	54	47	42	37	33	30	370	292	237	196	164	140	121	105	92	82	73	66	59
1500	195	154	125	103	87	74	64	56	49	43	38	35	31	383	302	245	202	170	145	125	109	96	85	76	68	61
1550	202	159	129	107	90	76	66	57	50	45	40	36	32	396	312	253	209	176	150	129	112	99	88	78	70	63
1600	208	165	133	110	92	79	68	59	52	46	41	37	33	408	323	261	216	181	155	133	116	102	90	81	72	65
1650	215	170	138	114	95	81	70	61	54	48	42	38	34	421	333	270	223	187	159	138	120	105	93	83	75	67
1700	221	175	142	117	98	84	72	63	55	49	44	39	35	434	343	278	229	193	164	142	123	108	96	86	77	69
1750	228	180	146	120	101	86	74	65	57	50	45	40	36	447	353	286	236	198	169	146	127	112	99	88	79	71
1900	247	195	158	131	110	94	81	70	62	55	49	44	40	485	383	310	256	216	184	158	138	121	107	96	86	78
2000	260	206	167	138	116	99	85	74	65	58	51	46	42	510	403	327	270	227	193	167	145	128	113	101	90	82

[a] These load values may also be used for cantilevered pieces intermixed, combination simple span and two-span continuous, and two-span continuous layups.

[b] $2\frac{1}{2}$ in. net thickness. To determine allowable loads for $2\frac{5}{8}$ in. net thickness, multiply tabulated loads by 1.10.

[c] All spans to the right of the heavy line require special ordering of additional long lengths to assure that at least 20% of the decking is equal to the span length or longer.

[d] $3\frac{1}{2}$ in. net thickness.

TABLE 7 Three and Four Inch Nominal Thickness, Allowable Roof

Modulus of Elasticity, psi	Deflection Limit[a]	Allowable Uniformly Distributed Total Roof Load, psf								
		3 in. Nominal Thickness[b], Span, ft								
		8	9	10	11	12	13	14	15	16
700,000	*l*/180	63	44	32	24	19	15	12	10	8
	l/240	47	33	24	18	14	11	9	7	6
800,000	*l*/180	72	51	37	28	21	17	13	11	9
	l/240	54	38	28	21	16	13	10	8	7
900,000	*l*/180	81	57	42	31	24	19	15	12	10
	l/240	61	43	31	23	18	14	11	9	8
1,000,000	*l*/180	90	64	46	35	27	21	17	14	11
	l/240	68	48	35	26	20	16	13	10	8
1,100,000	*l*/180	99	70	51	38	29	23	19	15	12
	l/240	75	52	38	29	22	17	14	11	9
1,200,000	*l*/180	108	76	56	42	32	25	20	16	14
	l/240	81	57	42	31	24	19	15	12	10
1,300,000	*l*/180	117	83	60	45	35	27	22	18	15
	l/240	88	62	45	34	26	21	16	13	11
1,400,000	*l*/180	127	89	65	49	38	30	24	19	16
	l/240	95	67	49	37	28	22	18	14	12
1,500,000	*l*/180	136	95	69	52	40	32	25	21	17
	l/240	102	71	52	39	30	24	19	15	13
1,600,000	*l*/180	145	102	74	56	43	34	27	22	18
	l/240	109	76	56	42	32	25	20	16	14
1,700,000	*l*/180	154	108	79	59	46	36	29	23	19
	l/240	115	81	59	44	34	27	22	17	14
1,800,000	*l*/180	163	114	83	63	48	38	30	25	20
	l/240	122	86	62	47	36	28	23	19	15

[a] For a deflection limit of *l*/360, use $\frac{1}{2}$ the tabulated value for a deflection limit of *l*/180.

[b] $2\frac{1}{2}$ in. net thickness. To determine allowable loads for $2\frac{5}{8}$ in. net thickness, multiply tabulated loads by 1.16.

[c] $3\frac{1}{2}$ in. net thickness.

Load Limited by Deflection, Simple Span Layup

Allowable Uniformly Distributed Total Roof Load, psf												
4 in. Nominal Thickness[c], Span, ft												
8	9	10	11	12	13	14	15	16	17	18	19	20
174	122	89	67	51	40	32	26	22	18	15	13	11
130	91	67	50	38	30	24	20	16	14	11	10	8
198	139	102	76	59	46	37	30	25	21	17	15	13
149	104	76	57	44	35	28	22	19	16	13	11	10
223	157	114	86	66	52	42	34	28	23	20	17	14
167	118	86	64	50	39	31	25	21	17	15	13	11
248	174	127	95	74	58	46	38	31	26	22	19	16
186	131	95	72	55	43	35	28	23	19	16	14	12
273	192	140	105	81	64	51	41	34	28	24	20	17
205	144	105	79	61	48	38	31	26	21	18	15	13
298	209	153	114	88	69	56	45	37	31	26	22	19
223	157	114	86	66	52	42	34	28	23	20	17	14
322	227	165	124	96	75	60	49	40	34	28	24	21
242	170	124	93	72	56	45	37	30	25	21	18	15
347	244	178	134	103	81	65	53	43	36	30	26	22
261	183	133	100	77	61	49	40	33	27	23	19	17
372	261	191	143	110	87	69	56	47	39	33	28	24
279	196	143	107	83	65	52	42	35	29	25	21	18
397	279	203	153	118	93	74	60	50	41	35	30	25
298	209	152	115	88	69	56	45	37	31	26	22	19
422	296	216	162	125	98	79	64	53	44	37	31	27
316	222	162	122	94	74	59	48	40	33	28	24	20
446	314	229	172	132	104	83	68	56	47	39	33	29
335	235	172	129	99	78	62	51	42	35	29	25	21

TABLE 8 Three and Four Inch Nominal Thickness, Allowable Roof Load

Modulus of Elasticity, psi	Deflection Limit[b]	Allowable Uniformly Distributed Total Roof Load, psf												
		3 in. Nominal Thickness[c], Span, ft[d]												
		8	9	10	11	12	13	14	15	16	17	18	19	20
700,000	l/180	96	67	49	37	28	22	18	14	12	10	8	7	6
	l/240	72	50	37	28	21	17	13	11	9	7	6	5	4
800,000	l/180	109	77	56	42	32	25	20	16	14	11	10	8	7
	l/240	82	58	42	32	24	19	15	12	10	8	7	6	5
900,000	l/180	123	86	63	47	36	29	23	19	15	13	11	9	8
	l/240	92	65	47	35	27	21	17	14	12	10	8	7	6
1,000,000	l/180	136	96	70	52	40	32	25	21	17	14	12	10	9
	l/240	102	72	52	39	30	24	19	16	13	11	9	8	7
1,100,000	l/180	150	105	77	58	44	35	28	23	19	16	13	11	10
	l/240	113	79	58	43	33	26	21	17	14	12	10	8	7
1,200,000	l/180	164	115	84	63	49	38	31	25	20	17	14	12	10
	l/240	123	86	63	47	36	29	23	19	15	13	11	9	8
1,300,000	l/180	177	125	91	68	53	41	33	27	22	18	16	13	11
	l/240	133	93	68	51	39	31	25	20	17	14	12	10	9
1,400,000	l/180	191	134	98	73	57	45	36	29	24	20	17	14	12
	l/240	143	101	73	55	42	33	27	22	18	15	13	11	9
1,500,000	l/180	205	144	105	79	61	48	38	31	26	21	18	15	13
	l/240	154	108	79	59	46	36	29	23	19	16	13	11	10
1,600,000	l/180	218	153	112	84	65	51	41	33	27	23	19	16	14
	l/240	164	115	84	63	49	38	31	25	20	17	14	12	10
1,700,000	l/180	232	163	119	89	69	54	43	35	29	24	20	17	15
	l/240	174	122	89	67	52	41	32	26	22	18	15	13	11
1,800,000	l/180	246	173	126	94	73	57	46	37	31	26	22	18	16
	l/240	184	129	94	71	55	43	34	28	23	19	16	14	12

[a] To determine allowable uniformly distributed total roof loads for other span conditions, multiply controlled random layup load values by the following factors: Cantilevered pieces intermixed; 0.90; Combination simple span and two-span continuous; 1.13; Two-span continuous; 1.59.

[b] For a deflection limit of l/360, use $\frac{1}{2}$ the tabulated value for a deflection limit of l/180.

[c] $2\frac{1}{2}$ in. net thickness. To determine allowable loads for $2\frac{5}{8}$ in. net thickness, multiply tabulated loads by 1.16.

[d] All spans to the right of the heavy line require special ordering of additional long lengths to assure that at least 20% of the decking is equal to the span length or longer.

[e] $3\frac{1}{2}$ in. net thickness.

Limited by Deflection, Controlled Random Layup[a] (3 or more spans)

Allowable Uniformly Distributed Total Roof Load, psf												
4 in. Nominal Thickness[c], Span, ft[d]												
8	9	10	11	12	13	14	15	16	17	18	19	20
262	184	134	101	78	61	49	40	33	27	23	20	17
197	138	100	76	58	46	37	30	24	20	17	15	12
300	210	154	115	89	70	56	45	37	31	26	22	19
225	162	115	86	67	52	42	34	28	23	20	17	14
337	237	173	130	100	79	63	51	42	35	30	25	22
253	178	129	97	75	59	47	38	32	26	22	19	16
374	263	192	144	111	87	70	57	47	39	33	28	24
281	197	144	108	83	65	52	43	35	29	25	21	18
412	289	211	158	122	96	77	63	52	43	36	31	26
309	217	158	119	92	72	58	47	39	32	27	23	20
449	316	230	173	133	105	84	68	56	47	39	34	29
337	237	173	130	100	79	63	51	42	35	30	25	22
487	342	249	187	144	114	91	74	61	51	43	36	31
365	256	187	140	108	85	68	55	46	38	32	27	23
524	368	269	202	155	122	98	80	66	55	46	39	34
393	276	201	151	117	92	73	60	49	41	35	29	25
562	395	288	216	166	131	105	85	70	59	49	42	36
421	296	216	162	125	98	79	64	53	44	37	31	27
599	421	307	230	178	140	112	91	75	62	53	45	38
449	316	230	173	133	105	84	68	56	47	39	34	29
636	447	326	245	189	148	119	97	80	66	56	48	41
478	335	245	184	142	111	89	72	60	50	42	36	31
674	474	345	259	200	157	126	102	84	70	59	50	43
506	355	259	195	150	118	94	77	63	53	44	38	32

9.7 For steeper sloping roofs, it is customary to adjust the loads so as to express them in terms of square feet of roof surface. (See Figures 12 and 13.) For example, 10 lb dead load (6.7 lb for deck and 3.3 lb for roofing) is the vertical load of one square foot of sloping roof surface. Snow load is usually expressed in pounds per square foot of the horizontal projection of the sloping roof surface. Therefore, the vertical snow load must be converted to the vertical psf load of sloping roof surface. For example, a 60 psf snow load on the horizontal projection is equivalent to a vertical load of 46 psf on a 10 in 12 sloping roof surface. This combined with 10 psf dead load results in a total vertical load of 56 psf on the 10 in 12 sloping roof surface. The 56 psf total vertical load may then be converted to two components, one perpendicular or normal to the roof surface, and one parallel to the roof surface. In the example, the vertical load of 56 psf is equivalent to a component perpendicular to the roof of 43 psf and a component parallel to the roof of 37 psf.

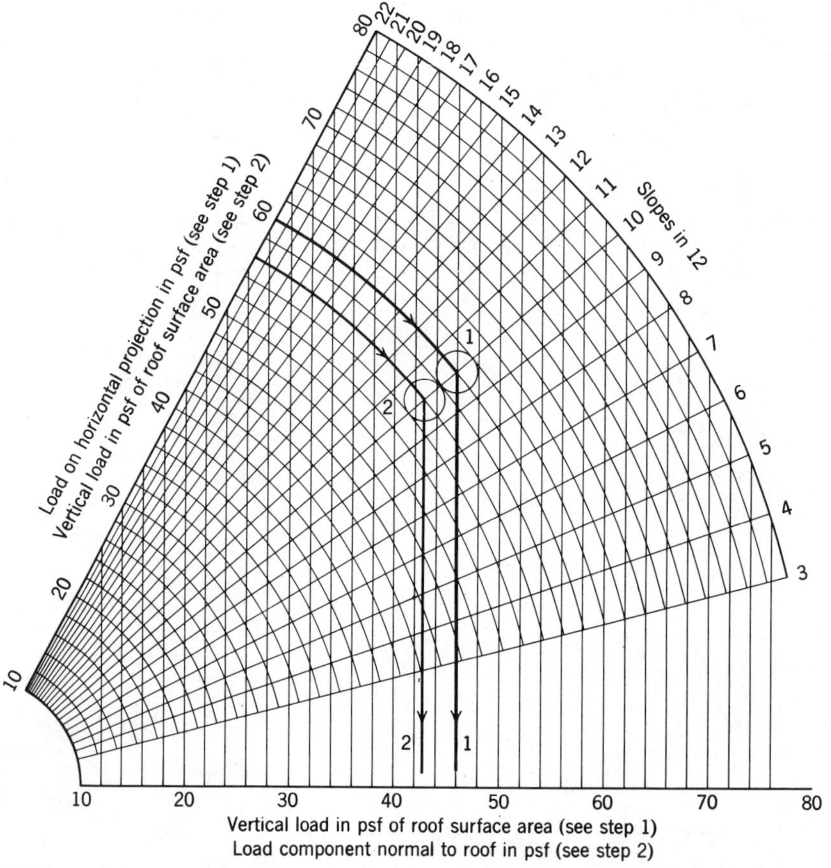

FIGURE 12 Load Conversion. Example: 60 psf live load and 10 psf dead load on 10 in 12 slope. Step 1: 60 psf live load on horizontal projection equals 46 psf of roof surface area vertical load on 10 in 12 roof slope. Step 2: 10 psf of roof surface area dead load plus 46 psf of roof surface area live load equals 56 psf of roof surface area combined load acting vertically; 56 psf of roof surface area vertical total load equals 43 psf normal to roof causing bending and deflection.

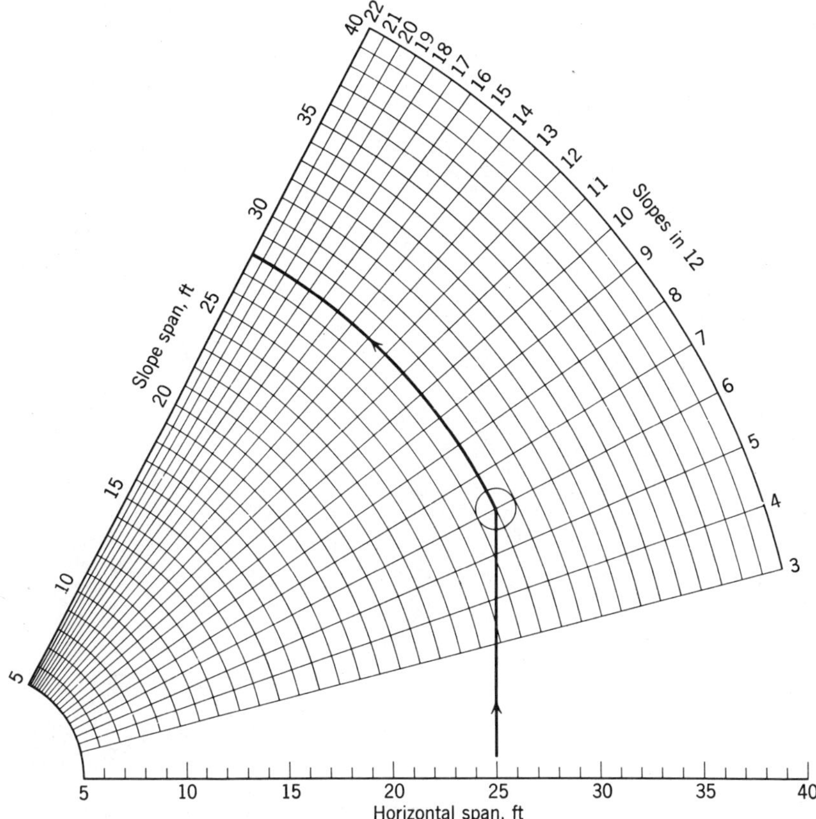

FIGURE 13 Span Conversion. Example: 25 ft horizontal span equals 28 ft slope span when slope is 6 in 12. Use 28 ft in determining footage.

9.8 Where decking is laid with the longitudinal axis parallel to the slope, the component perpendicular to the roof surface will produce bending and deflection; the parallel component will produce compression. The design value for compression parallel to grain may be taken as that of No. 2 structural joists and planks grade for the species. The decking must be designed for bending and axial stresses as well as deflection.

9.9 Where decking is laid with the longitudinal axis perpendicular to the slope, the component perpendicular to the roof surface produces bending and deflection; the parallel component, as may be induced by wind forces, is taken by diaphragm action.

AMERICAN INSTITUTE
333 West Hampden Avenue

TIMBER CONSTRUCTION
Englewood, Colorado 80110

AITC 113-83
STANDARD FOR DIMENSIONS OF STRUCTURAL GLUED LAMINATED TIMBER

Adopted as Recommendations June 22, 1978

CONTENTS

1. PREFACE

1.1 The most efficient and economical production of glued laminated structural members results when standard lumber sizes are used for the laminations. Industry recommended practice uses nominal 2 in. thick lumber of standard nominal width to produce straight members and curved members where the radius of curvature is within the bending radius limits for that thickness of the species. Nominal 1 in. thick boards are normally used when the bending radius is too sharp to permit use of nominal 2 in. thick laminations. These are standard practices subject to deviation to conform with specific job requirements and plant procedures. The use of nominal 1 in. and 2 in. thick laminations will generally be the most economical, and therefore, conformance with this standard is recommended for all normal uses. Exceptions should be made only when the shape of the structure requires nonstandard laminations. Textured surfaces for glued laminated timber are permitted in *Standard Appearance Grades for Structural Glued Laminated Timber*, AITC 110, in lieu of the surfaces specified in the AITC appearance grades. When textured surfaces are used, the net finished sizes and tolerances given herein and in AITC 117—DESIGN and ANSI/AITC 190.1-1983, *Structural Glued Laminated Timber*, may not be applicable. Depending upon the degree of texturing, it may be necessary for the designer to compensate for the resulting loss of cross section.

2. STANDARD DEPTHS OF MEMBERS

2.1 Proper gluing procedures require surfaces planed uniformly smooth to exact thickness. Normal standard practice is to surface nominal 2 in. laminations to a net $1\frac{1}{2}$ in. thickness, and nominal 1 in. laminations to a net $\frac{3}{4}$ in. thickness. Finished depths of members are thus increments of these net thicknesses.

No. of Laminations	Net Depth of Member, in. 1 in. Laminations	Net Depth of Member, in. 2 in. Laminations
4	3	6
5	$3\frac{3}{4}$	$7\frac{1}{2}$
6	$4\frac{1}{2}$	9
7	$5\frac{1}{4}$	$10\frac{1}{2}$
8	6	12
Etc.	Etc.	Etc.

2.2 The use of laminations of special thicknesses because of bending radius or the mixing of thicknesses for special purposes results in net finished depths other than those shown in the table.

3. STANDARD WIDTHS OF MEMBERS

3.1 It is necessary to surface the wide faces of members to remove the glue squeezeout and provide a uniformly smooth surface. Therefore, the net finished width of the glued laminated member is less than the net finished width of industry standard boards and dimension.

3.2 The normal standard net finished widths for glued laminated structural members are as follows. Other finished widths may be used to meet the size requirements of a design or to meet other special requirements.

Nominal width, in.	3	4	6	8	10	12	14	16
Net Finished width, in.	$2\frac{1}{8}$	$3\frac{1}{8}$	$5\frac{1}{8}$	$6\frac{3}{4}$	$8\frac{3}{4}$	$10\frac{3}{4}$	$12\frac{1}{4}$	$14\frac{1}{4}$

Note: $3\frac{1}{8}$ in. and $5\frac{1}{8}$ in. widths are normal for Western softwoods. 3 in. and 5 in. widths are normal for Southern Pine.

4. STANDARD DIMENSIONS FOR HEAVY TIMBER

4.1 Excellent fire resistance is achieved with "heavy timber" construction (see Standard for Heavy Timber Construction, AITC 108). Minimum sawn lumber sizes have been long established and are expressed in nominal dimensions and assume surfacing to *American Lumber Standard* net sizes.

4.2 For "heavy timber" construction, the net width of glued laminated structural members shall be the standard glued laminated net width for the nominal sawn width specified, and the net depth of glued laminated structural members shall be equal to or greater than the net finished depth specified in the following table.

Minimum Nominal Size			Minimum Glued Laminated Net Size		
Width, in.		Depth, in.	Width, in.		Depth, in.
8	×	8	$6\frac{3}{4}$	×	9
6	×	10	$5\frac{1}{8}$	×	$10\frac{1}{2}$
6	×	8	$5\frac{1}{8}$	×	9
6	×	6	$5\frac{1}{8}$	×	6
4	×	6	$3\frac{1}{8}$	×	$7\frac{1}{2}$

5. TOLERANCES

5.1 The following tolerances shall be permitted at the time of manufacturing.

5.1.1 Width. Plus or minus $\frac{1}{16}$ in. of the specified width.

5.1.2 Depth. Plus $\frac{1}{8}$ in. per foot of specified depth. Minus $\frac{1}{16}$ in. per foot of the specified depth or $\frac{1}{8}$ in. whichever is the larger.

5.1.3 Length. Plus or minus $\frac{1}{16}$ in. up to 20 ft and plus or minus $\frac{1}{16}$ in. per 20 ft of the specified length, except where length dimensions are not specified or critical.

5.1.4 Squareness. The cross section of all glued laminated structural members shall be square within plus or minus $\frac{1}{8}$ in. per foot of specified depth of member unless a specially shaped section is specified.

AMERICAN INSTITUTE  TIMBER CONSTRUCTION
333 West Hampden Avenue Englewood, Colorado 80110

AITC 117-84—DESIGN STANDARD SPECIFICATIONS FOR STRUCTURAL GLUED LAMINATED TIMBER OF SOFTWOOD SPECIES

Adopted as Recommendations February 24, 1984

Amended November 1, 1984

CONTENTS

Preface

These specifications consolidate, expand and update previously issued laminating specifications and supplements related to specific species or mixtures of species.

They represent the latest research available from the U.S. Forest Products Laboratory, various colleges and universities, and the American Institute of Timber Construction. With these specifications a designer can specify the required stress levels for a glued laminated timber member. It is the responsibility of a glued laminated timber manufacturer to produce a member with design values that meet or exceed those requirements. When the design stress level allows a choice, manufacturers will select laminating combinations to fit their varying raw material supplies, thus better utilizing available forest resources. A separate publication for the manufacture of glued laminated timber, AITC 117-84—*MANUFACTURING* (Reference 1), has been developed based on ASTM D 3737-83a, *Standard Method for Establishing Stresses for Structural Glued Laminated Timber* (*Glulam*) (Reference 2), as modified by subsequent research and by American National Standard ANSI/AITC A190.1-1983, *Structural Glued Laminated Timber* (Reference 3).

These specifications contain data relating to design values and modification of stresses for the design of glued laminated timber members. They are, however, neither a design manual nor an engineering textbook. For additional design information see the *Timber Construction Manual* prepared by AITC (Reference 4).

1. GENERAL

1.1 Structural Glued Laminated Timber

1.1.1 The term *structural glued laminated timber* as employed herein refers to an engineered, stress-rated product of a timber laminating plant, comprising assemblies of suitably selected and prepared wood laminations bonded together with adhesives. The grain of all laminations is approximately parallel longitudinally.

1.1.2 The individual laminations shall not exceed 2 in. in net thickness. They may be comprised of pieces end joined to form any length, of pieces placed or glued edge to edge to make any width, or of pieces bent to curved form during gluing.

1.1.3 These specifications are applicable to glued laminated timbers with the number of laminations indicated in Tables 1 and 2.

1.1.4 The production of structural glued laminated timber under these specifications shall be in accordance with AITC 117-84—*MANUFACTURING*, and American National Standard ANSI/AITC A190.1-1983, *Structural Glued Laminated Timber*.

1.1.5 End joints in laminated timber combinations listed herein may be plain scarf joints, finger joints or other types which qualify for the design values in accordance with the procedures in American National Standard ANSI/AITC A190.1-1983, *Structural Glued Laminated Timber*, and AITC 117-84—*MANUFACTURING*.

1.1.6 The design of glued laminated members and their fastenings should be in accordance with the provisions of these specifications and the *Timber Construction Manual*.

1.2 Design Values

1.2.1 The design values contained herein have been developed by AITC using procedures developed in cooperation with the U.S. Forest Products Laboratory, based on analytical studies confirmed by full-scale load tests.

1.3 Species

1.3.1 The softwood species most commonly used for laminating are included in these specifications. These are the Western species: Douglas Fir-Larch (DF), Douglas Fir South (DFS), Hem-Fir (HF), and Western woods and Canadian softwood species (WW), and the Southern Pine species (SP).

1.4 Specification of Design Values

1.4.1 ***Principal Stress—Bending.*** Table 1 is applicable to members consisting of 4 or more laminations stressed primarily in bending with the load applied perpendicular to the wide faces of the laminations. The 22*F* and 24*F* combinations 15 in. or less in depth require tension laminations and may not be readily available so it is recommended that the designer check on their availability prior to specifying. The 16*F* and 20*F* combinations are generally available in all depths. Table 1 contains 2 groups of species—Western species and Southern Pine species. The table includes combinations manufactured from visually graded lumber and combinations manufactured from *E*-rated lumber. There are four groupings of bending stress (*F*) levels with a number of options within groupings to give the same bending stress, but with some variations in the other design values shown for each option. Many designs can utilize more than one of the options listed within an F_b grouping. Where these other design values (F_t, F_c, $F_{c\perp}$, F_v and E) become critical in design, the designer should specify the stresses as required by design. Obviously, the specifying of values that are much higher than actually required will eliminate certain combinations and may result in a member that is not as readily available as would otherwise be the case. The arbitrary selection of the highest possible design values in all stress categories may result in a member impossible to manufacture under these specifications. It is also possible for the designer to specify a given combination that meets the design requirements, but this may limit availability. Note that increased values for horizontal shear and compression perpendicular to the grain for some combinations can be obtained by so specifying. See the footnotes following Table 1.

1.4.1.1 The design values in Table 1 are for loads applied perpendicular to the wide faces of the laminations which is the most common direction of loading for glued laminated timbers. For convenience, the design values for loads applied perpendicular to the wide faces of the laminations causing bending about the *X*–*X* axis are designated in the table by the subscript *X*. Two columns of design values are shown in Table 1 for bending with the load applied perpendicular to the wide faces of the laminations (F_{bx}). The first (column 3) is for the most common use of bending members where the tension portion of the bending stress occurs on the face of the member containing the tension zone laminations. The second (column 4) is for use where the face of the member containing the compression zone laminations is stressed in tension, such as a short overhang on a simple beam. For continuous beams or beams cantilevered over a support where high tensile stresses can exist on both the top and bottom of a member, see 1.4.1.3.

1.4.1.2 The design values for members stressed in bending about the *Y–Y* axis (loads applied parallel to the wide faces of the laminations) and members axially loaded are shown in Table 1. The design values for loads applied parallel to the wide faces of the laminations causing bending about the *Y–Y* axis are designed by the subscript *Y*. Neither the *X* nor *Y* subscripts are commonly used in wood references or textbooks.

1.4.1.3 The design values in bending with the load applied perpendicular to the wide faces of the laminations (F_{bx}) listed in column 3, Table 1, are for the most common installation of the member as a simple beam. This implies compressive stress occurring at the top of the member and tensile stress occurring at the bottom or soffit of the member (positive moment). For conditions where the beam support configuration and/or loading pattern produce negative moment which becomes significant and the resulting tensile stress on the top of the member exceeds the minimum design values listed in column 4 for the compression zone in tension (1,200 psi for 24*F* combinations, 1,100 psi for 22*F* combinations, 1,000 psi for 20*F* combinations and 800 psi for 16F combinations), *Tension Zone* grade requirements, including end joint spacing, must be applied to the top zones of the member so that the basic design values for bending listed in column 3, Table 1, may be utilized. A bending tensile stress in the negative moment area 200 psi higher than that tabulated in column 4 can be obtained by applying only tension zone end joint spacing restrictions to both top and bottom of the member. When specified with *Tension Zone* requirements both top and bottom, the design values in bending (F_{bx}) listed in column 3, Table 1, apply to either positive or negative moment loading conditions. Cantilever or continuous beams which are stressed higher in the negative moment area than the values listed in column 4, Table 1 (1,200 psi for 24*F* combinations, 1,100 psi for 22*F* combinations, 1,000 psi for 20*F* combinations and 800 psi for 16F combinations) should be identified by the designer. The manufacturer will then provide *Tension Zone* laminations in this area as required by the designer.

1.4.1.4 Balanced combinations for bending members which have equal or nearly equal positive and negative bending moments are included in Table 1.

1.4.1.5 The combinations in Table 2 are usually best suited for members with bending stresses caused by loads applied parallel to the wide faces of the laminations. Design values are also shown for members loaded perpendicular to the wide faces of the laminations. In addition, Table 2 contains combinations for members with 2 or 3 laminations. These combinations are applicable to members loaded either perpendicular or parallel to the wide faces of the laminations.

1.4.1.6 The design values in bending about the *x–x* axis (F_{bx}) in Column 3, Table 1, and Column 17, Table 2, are based on the use of special tension lamination(s) for most combinations in Table 1 and all combinations in Table 2 when these combinations are used for bending members. Special tension laminations are not required for arches. These special tension laminations may also be omitted for bending members provided the tabular design values in bending (F_{bx}) are reduced by multiplying by 0.75 for Table 1 combinations and for Table 2 combinations over 15 in. For Table 2 combinations 15 in. and less in depth use the design values in Column 16, Table 2.

1.4.2 ***Principal Stress—Axial.*** Table 2 contains combinations for members stressed primarily in axial tension or compression.

1.4.3 Members subjected to combined axial and bending stresses—When a combination of axial and bending stresses exists in a member, they should be checked by the interaction formula as shown in the *National Design Specification for Wood Construction*. The designer should specify the required tabular design values in bending, F_b, and compression parallel to grain, F_c, or tension parallel to grain, F_t; however, the stresses specified should be available in a single combination. When the predominant stress is axial, the combinations in Table 2 should be used in the design procedure. When the predominant stress is bending, the combinations in Table 1 may be more appropriate. The required tabular design values for axial and bending stresses should be specified regardless of whether the combination has been specified.

1.4.4 The design values for glued laminated timber made from visually graded Western Cedar are the same as those for Western woods (WW) in Table 2 except for modulus of elasticity, which is 100,000 psi less. Western Cedars are included in the Western woods (WW) species group for fastenings in Table 3.

2. LUMBER

2.1 Lumber shall be of any species or grade shown herein.

3. ADHESIVES

3.1 Adhesives used shall comply with the specifications contained in American National Standard ANSI/AITC A190.1-1983, *Structural Glued Laminated Timber*.

3.2 Wet-use adhesives may be specified for all moisture conditions but are required when the moisture content exceeds 16% for repeated or prolonged periods of service or when the wood is preservatively treated either before or after gluing.

4. DESIGN VALUES

4.1. General

4.1.1 For the design values given herein, or modifications thereof, lumber of the grades required shall be assembled in accordance with the zone requirements indicated in AITC 117-84—*MANUFACTURING*.

4.1.2 The design values given herein and the modifications required for other conditions of use and loading are also applicable to structural glued laminated timbers that have been pressure impregnated by an approved preservative process in accordance with *Standard for Preservative Treatment of Structural Glued Laminated Timber*, AITC 109 (Reference 5).

4.1.3 The design values for fire retardant treated glued laminated timber, treated before or after gluing, are dependent upon the species and treatment combinations involved. The effect on strength must be determined for each treatment; however, indications are that a 10 to 25% reduction in bending stress is applicable. The manufacturer of the treatment should be contacted for more specific information on stress adjustments for all design values.

4.1.4 The design values given herein are for normal durations of loading. Modifications for other durations of loading are given in 4.4.1.

4.1.5 The design values in bending, F_{bx}, given herein apply to a 12 in. deep member, uniformly loaded, with a span to depth ratio of 21. Modifications for other sizes and loading are given in 4.4.2 and in footnote **f** to Table 2.

4.1.6 The modulus of elasticity, E, values herein are the average values for the combination shown and reflect the effect of grade. The modulus of elasticity of wood of a given species is somewhat variable. The coefficient of variation of visually graded lumber of the same species is approximately 0.25 for species used in laminating. Tests and experience have shown that this variability is considerably reduced by the average effect of laminating. For glued laminated timber, the coefficient of variation (c.o.v.) decreases with an increasing number of laminations. In glued laminated timber made from 4 laminations of visually graded lumber, the c.o.v. is approximately 0.15, for 10 laminations 0.10 and for 16 or more laminations 0.08. The variation in modulus of elasticity is especially important in designs where stiffness is of prime importance such as in the design of long columns, lateral stability calculations or in calculations for ponding.

A standard deviation is the average value multiplied by the coefficient of variation. In a normal frequency distribution, about $\frac{2}{3}$ of the individual values will be within one standard deviation (above and below) the average value. Also about 95% of the individual values will be within two standard deviations of the average value. Thus, if a combination of glued laminated timber has an average E of 1,700,000 psi and the coefficient of variation is 0.10, $\frac{2}{3}$ of the members could be expected to have values between 1,530,000 and 1,870,000 psi and 95% could be expected to have values between 1,360,000 psi and 2,040,000 psi.

In a case where only the lower portion of the variation in E is of engineering importance, similar useful intepretations are possible. In a normal frequency distribution, $\frac{5}{6}$ of the individual values lie above a value located at one standard deviation below the mean (1,530,000 psi in the above example). In the same distribution, 95% of the individual values lie above a value located at 1.645 standard deviations below the mean (1,420,000 psi in the above example).

4.1.6.1 The tabulated E values shown for bending about the X–X axis of members in Table 1 are higher than those tabulated for bending about the Y–Y axis because the laminations in the outer zones have higher E values than those in the inner zones.

4.1.6.2 The modulus of elasticity values for bending members listed in Tables 1 and 2 are based on a span-to-depth ratio of approximately 21 and include an adjustment for shear deflection. These E values can be used for determining deflection for most designs without the necessity of calculating the shear deflection.

4.1.7 The tabulated compression perpendicular to grain design values in Tables 1 and 2 are based on a deformation limit of 0.04 in. obtained when testing in accordance with the standard method ASTM D 143 for compression perpendicular to grain. In special applications where deformation may be critical, use of a reduced compression perpendicular to grain design value may be appropriate. The following equation may be used for a deformation of 0.02 in.which is 50% of that associated with the values tabulated in Tables 1 and 2.

$$F_{c\perp(0.02)} = 0.73\ F_{c\perp}$$

where

$F_{c\perp(0.02)}$ = compression perpendicular to grain at 50% of deformation limit associated with tabulated $F_{c\perp}$ values (0.02 in.)

and

$F_{c\perp}$ = compression perpendicular to grain at 0.04 in. deformation limit.

4.2 Radial Tension or Compression

4.2.1 When a curved member is loaded in bending, radial stresses are induced.

4.2.2 When the bending moment (M) is in the direction tending to increase curvature (decrease the radius), the radial stress is compression across the grain (F_{rc}). The design value in radial compression (F_{rc}) is equal to the design value in compression perpendicular to grain ($F_{c\perp}$) of the grade and species being used. (For Douglas Fir-Larch, Douglas Fir South and Southern Pine, use F_{rc} = 560 psi; for Hem-Fir, use F_{rc} = 375; and for Western woods and Canadian softwood species use F_{rc} = 255 psi.)

4.2.3 When M is in the direction tending to decrease curvature (increase the radius), the radial stress is tension across the grain. The design value in radial tension perpendicular to grain (F_{rt}) shall be limited to $\frac{1}{3}$ the design value in horizontal shear for Southern Pine for all load conditions and for Douglas Fir-Larch, Douglas Fir South, Hem-Fir, Western woods and Canadian softwood species for wind or earthquake loads. The limit shall be 15 psi for Douglas Fir-Larch, Hem-Fir, Western woods and Canadian softwood species for other types of loading. These values are subject to modifications for duration of load and wet conditions of use. For wet conditions of use, the wet-use factor for radial tension is 0.875. If these values are exceeded,mechanical reinforcing shall be used and shall be sufficient to resist all radial tension stresses. For Douglas Fir-Larch, Douglas Fir South, Hem-Fir, Western woods and Canadian softwood species where mechanical reinforcement is provided to resist all radial tension stresses, the calculated radial tension stress shall not exceed $\frac{1}{3}$ the design value in horizontal shear. When mechanical reinforcing is used,the maximum moisture content of the laminations at the time of manufacture shall not exceed 12% for dry conditions of use.

4.3 Conditions of Use

4.3.1 Dry conditions of use design values shall be applicable when the moisture content in service is less than 16%, as in most covered structures.

4.3.2 Wet conditions of use design values shall be applicable when the moisture content in service is 16% or more, as may occur in members directly exposed to precipitation or in covered locations of high relative humidity.

4.4 Modification of Stresses

4.4.1 Duration of load

4.4.1.1 Normal load duration contemplates fully stressing a member to the design value by the application of the full design load for a duration of approximately 10 years (applied either continuously or cumulatively).

4.4.1.2 When a member is fully stressed by maximum design loads for long term loading conditions (greater than 10 years either continuously or cumulatively), the design values shall be 90% of the values tabulated.

4.4.1.3 When the duration of full design load (applied either continuously or cumulatively) does not exceed the period indicated, increase the tabulated design values, except for modulus of elasticity and compression perpendicular to grain, as follows:

15% for 2 months duration, as for snow;
25% for 7 days duration;
33% for wind or earthquake; and
100% for impact.
These increases are not cumulative.

4.4.2 Size Factor

4.4.2.1 When the depth of a rectangular beam exceeds 12 in., the tabulated design value for bending (F_b) shall be reduced by multiplying it by the size factor (C_F) as determined from the following relationship:

$$C_F = \left(\frac{12}{d}\right)^{1/9}$$

where C_F = size factor

d = depth of member (in.).

4.4.2.2 The size factor relationship as given in 4.4.2.1 is applicable to simply supported, uniformly loaded bending members with span-to-depth ratios (l/d) of 21. This factor can thus be applied with reasonable accuracy to most commonly encountered design situations. Where greater accuracy is desired for other sizes and conditions of loading, the percentage changes given in the following table may be applied directly to the size factor calculated for the basic conditions previously stated. Straight line interpolation may be used for other l/d ratios.

Span to Depth Ratio l/d	% Change
7	+6.3
14	+2.3
21	0
28	−1.6
35	−2.8

Loading Condition for Simply Supported Beams	% Change
Single Concentrated Load	+7.8
Uniform Load	0
Third Point Load	−3.2

4.4.2.3 For a more detailed analysis of the size factor and its application to the design of bending members, see the *Timber Construction Manual.*

4.4.3 *Lateral Stability*

4.4.3.1 The design values for bending contained in these specificiations are applicable to members which are adequately braced. When deep, slender members not adequately braced are used, a reduction to the tabulated design values in bending must be applied based on a computation of the slenderness factor of the member. In the check of lateral stability, the slenderness factor shall be applied in design as shown in the *Timber Construction Manual.*

4.4.3.2 A reduction in the design value in bending determined by applying the slenderness factor is not cumulative with a reduction in design value due to the application of size factor. In no case shall the design value in bending exceed the stress as determined by applying the size or slenderness factor, whichever governs.

4.4.4 *Curvature Factor*

4.4.4.1 For the curved portion of members, the design value in bending (F_b) shall be modified by multiplying it by the following curvature factor:

$$C_C = 1 - 2{,}000 \left(\frac{t}{R}\right)^2$$

where t = thickness of lamination (in.).

R = radius of curvature of lamination (in.).

No curvature factor need be applied to the design value in the straight portion of an assembly, regardless of curvature elsewhere.

5. CONNECTIONS AND FASTENERS

5.1 The design values for connections and fasteners for glued laminated timber are contained in the *Timber Construction Manual* and the *National Design Specification* (Reference 6).

5.2 The design values for fasteners used in glued laminated timber vary depending upon species, growth rate and the face of the member in which the fastener will be placed (i.e., tension, compression or side face). Table 3 contains this information for all combinations.

5.3 See *Typical Construction Details*, AITC 104 (Reference 7) for additional information on connections.

6. DIMENSIONS

6.1 Standard Sizes. American National Standard ANSI/AITC A190.1-1983 permits the use of any width or depth of glued laminated timber. The use of standard finished sizes, however, constitutes recommended practice to the extent that other considerations will permit. The laminator may use any thickness of lumber to develop the specified depth provided the volume of the higher grades of lumber equals or exceeds that specified in laminating combinations which are based on laminations of equal thickness. The depth and width of the glued laminated timber should be as agreed upon by buyer and seller.

6.2 Depth and Width

6.2.1 Straight and curved members shall be furnished in accordance with the width and depth dimensions required by the design.

6.2.2 The normal standard net finished widths are as follows:

Nominal Width, in.	Net Finished Width, in.	Nominal Width, in.	Net Finished Width, in.
3	$2\frac{1}{8}$	10	$8\frac{3}{4}$
4	$3\frac{1}{8}$*	12	$10\frac{3}{4}$
6	$5\frac{1}{8}$*	14	$12\frac{1}{4}$
8	$6\frac{3}{4}$	16	$14\frac{1}{4}$

*$3\frac{1}{8}$ in. and $5\frac{1}{8}$ in. are normal widths for Western softwood glued laminated timbers. 3 in. and 5 in. are normal widths for Southern Pine glued laminated timbers.

Other finished widths may be used to meet the size requirements of a design or to meet other special requirements.

6.3 Radius of Curvature

6.3.1 The ability to bend laminations is dependent upon many factors relating to both wood properties and manufacturing techniques, and it may be advisable to consult with the laminator prior to specifying. Two prime considerations are thickness of laminations t and bending radii, R. The t/R ratio should not exceed 1/100 for Southern Pine nor 1/125 for Douglas Fir-Larch and other softwoods.

6.3.2 Tudor arches utilize the majority of the most sharply curved laminations and usually include 5/8 in. to 1 in. actual thickness laminations. The normal standard radius of curvature used in the industry for these sharply curved members is 9 ft 4 in., but this may be reduced based on the lamination thickness or species. For less sharply curved members fabricated from $1\frac{1}{4}$ in. to $1\frac{5}{8}$ in. thick laminations, the normal radius of curvature is 27 ft 6 in.

7. APPEARANCE GRADES

7.1 Appearance grades shall be in accordance with the current *Standard Appearance Grades for Structural Glued Laminated Timber*, AITC 110 (Reference 8), unless otherwise specified on drawings or specifications.

7.2 For those combinations permitting the mixing of species, the potential for differences in color or grain of adjacent laminations must be recognized. For those architectural appearance applications where such possible differences in color or grain might be important, the designer may specify a combination symbol which will restrict the laminations to a single species or group of species with similar characteristics. In some cases, this may restrict availability.

8. INSPECTION AND QUALITY CONTROL

8.1 The assurance that quality materials and workmanship are used in structural glued laminated timber members shall be vested in the laminator's day-to-day quality control operations. Visual inspections and physical tests of samples of production are also required to assure conformance with AITC 117-84—*MANUFACTURING*, American National Standard ANSI/AITC A190.1-1983, *Structural Glued Laminated Timber*, and these specifications.

9. MARKING

9.1 The laminating combinations in Table 1 were developed primarily to resist bending loads. The grades of lumber in laminations on the compression side may not be the same as those on the tension side. Therefore, straight or slightly cambered glued laminated timber bending members shall be stamped **TOP** with letters approximately 2 in. high on the top at both ends of the member. Axially loaded members or bending members which are fabricated in such a manner that they cannot be installed upside down need not be marked.

10. PROTECTION DURING SHIPPING AND FIELD HANDLING

10.1 End sealers, surface sealers, primer coats and wrappings may be applied for the protection of the members. However, they do not necessarily preclude damage resulting from negligence and other factors beyond the control of the laminator during shipment, handling, storing and placing of the members. The protection specified should be commensurate with the end use and final finish of the member. It may also vary with the method of shipment and with exposure to climatic and other details. See *Recommended Practice for Protection of Structural Glued Laminated Timber During Transit, Storage and Erection*, AITC 111 (Reference 9).

Table 1 Design Values for Structural Glued Laminated Timber

For normal duration of load and dry conditions of use [a,b,c,d]

		Bending About X-X Axis						Bending About Y-Y Axis					Axially Loaded		
		Loaded Perpendicular to Wide Faces of Laminations						Loaded Parallel to Wide Faces of Laminations							
		Extreme Fiber in Bending, F_{bx}		Compression Perpendicular to Grain, $F_{c\perp x}$ [v]											
Combination Symbol [t]	Species-Outer Laminations/Core Laminations [e]	Tension Zone Stressed in Tension [f,w]	Compression Zone Stressed in Tension [g]	Tension Face	Compression Face	Horizontal Shear, F_{vx}	Modulus of Elasticity, E_x	Extreme Fiber in Bending, F_{by} [f,r]	Compression Perpendicular to Grain $F_{c\perp y}$ [v]	Horizontal Shear, F_{vy}	Horizontal Shear, Fvy, psi (For members with multiple piece laminations which are not edge glued) [u]	Modulus of Elasticity, E_y	Tension Parallel to Grain, F_t	Compression Parallel to Grain, F_c	Modulus of Elasticity, E
		psi	psi	psi	psi	psi	$\times 10^{-6}$psi	psi	psi	psi	psi	$\times 10^{-6}$psi	psi	psi	$\times 10^{-6}$psi
1	2	3	4	5	6	7	8	9	10	11	12	13	14	15	16
Visually Graded Western Species															
16F-V1	DF/WW	1600	800	560[h,l]	560[h,l]	140[s]	1.3	950	255	130[s]	65[s]	1.1	675	975	1.1
16F-V2	HF/HF			500[j]	375[j]	155	1.4	1250	375	135	70	1.3	875	1300	1.3
16F-V3	DF/DF			560[h,i]	560	165	1.5	1450	560	145	75	1.5	950	1550	1.5
16F-V8	DFS/DFS			650	500	165	1.2	1200	500	145	75	1.1	825	1350	1.1
The following two combinations are intended for straight or slightly cambered members for dry use and industrial appearance.[k]															
16F-V4	DF/N3WW	1600	800	650	560[h]	90[l,s]	1.5	900	255	130[s]	65[s]	1.3	650	600	1.3
16F-V5	DF/N3DF			650	560[h]	90[m]	1.6	1000	470	135	70	1.5	750	875	1.5
The following two combinations are balanced and are intended for members continuous or cantilevered over supports and provide equal capacity in both positive and negative bending.															
16F-V6	DF/DF	1600	1600	560[h,i]	560[h]	165	1.5	1450	560	145	75	1.4	950	1550	1.5
16F-V7	HF/HF			375[j]	375[j]	155	1.4	1200	375	135	70	1.3	850	1350	1.3

20F-V1	DF/WW	2000	1000	650	560[h]	140[s]	1.4	1000	255	130[s]	65[s]	1.2	750	1000	1.2
20F-V2	HF/HF			500[j]	375[j]	155	1.5	1200	375	135	70	1.4	950	1350	1.4
20F-V3	DF/DF			650	560[h]	165	1.6	1450	560	145	75	1.5	1000	1550	1.5
20F-V4	DF/DF			590[h,i]	560[h]	165	1.6	1450	560	145	75	1.6	1000	1550	1.6
20F-V10	DF/HF			650	560	155	1.5	1300	375	135	70	1.4	950	1500	1.4
20F-V11	DFS/DFS			650	500	165	1.3	1400	500	145	75	1.1	900	1400	1.1
The following two combinations are intended for straight or slightly cambered members for dry use and industrial appearance.[k]															
20F-V5	DF/N3WW	2000	1000	650	560[h]	90[l,s]	1.6	1000	255	135[s]	70[s]	1.3	750	725	1.3
20F-V6	DF/N3DF			650	560[h]	90[m]	1.6	1000	470	135	70	1.5	775	900	1.5
The following three combinations are balanced and are intended for members continuous or cantilevered over supports and provide equal capacity in both positive and negative bending															
20F-V7	DF/DF	2000	2000	650	650	165	1.6	1450	560	145	75	1.6	1000	1600	1.6
20F-V8	DF/DF			590[h,i]	590[h,i]	165	1.7	1450	560	145	75	1.6	1000	1600	1.6
20F-V9	HF/HF			500[j]	500[j]	155	1.5	1400	375	135	70	1.4	975	1400	1.4
22F-V1	DF/WW	2200	1100	650	560[h]	140[s]	1.6	1050	255	130[s]	65[s]	1.3	850	1100	1.3
22F-V2	HF/HF			500[j]	500[j]	155	1.5	1250	375	135	70	1.4	950	1350	1.4
22F-V3	DF/DF			650	560[j]	165	1.7	1450	560	145	75	1.6	1050	1500	1.6
22F-V4	DF/DF			590[h,i]	560[h]	165	1.7	1450	560	145	75	1.6	1000	1550	1.6
22F-V10	DF/DFS			650	560[h]	165	1.6	1600	500	145	75	1.3	1000	1400	1.3
The following two combinations are intended for straight or slightly cambered members for dry use and industrial appearance.[k]															
22F-V5	DF/N3WW	2200	1100	650	560[h]	90[l,s]	1.6	1100	255	135[s]	75[s]	1.4	800	725	1.4
22F-V6	DF/N3DF			650	560[h]	90[m]	1.7	1250	470	135	75	1.6	900	925	1.6
The following three combinations are balanced and are intended for members continuous or cantilevered over supports and provide equal capacity in both positive and negative bending															
22F-V7	DF/DF	2200	2200	650	650	165	1.8	1450	560	145	75	1.6	1100	1650	1.6
22F-V8	DF/DF			590[h,i]	590[h,i]	165	1.7	1450	560	145	75	1.6	1050	1650	1.6
22F-V9	HF/HF			500[j]	500[j]	155	1.5	1250	375	135	70	1.4	975	1400	1.4
Wet-use factors[b]		0.8	0.8	0.53	0.53	0.875	0.833	0.8	0.53	0.875	0.875	0.833	0.8	0.73	0.833

Table 1 Design Values for Structural Glued Laminated Timber (Cont.)

For normal duration of load and dry conditions of use [a,b,c,d]

		Bending About X-X Axis: Loaded Perpendicular to Wide Faces of Laminations						Bending About Y-Y Axis: Loaded Parallel to Wide Faces of Laminations					Axially Loaded		
		Extreme Fiber in Bending, F_{bx}		Compression Perpendicular to Grain, $F_{c\perp x}$ [v]											
Combination Symbol [t]	Species-Outer Laminations/Core Laminations [e]	Tension Zone Stressed in Tension [f,w] psi	Compression Zone Stressed in Tension [g] psi	Tension Face psi	Compression Face psi	Horizontal Shear, F_{vx} psi	Modulus of Elasticity, E_x $\times 10^{-6}$psi	Extreme Fiber in Bending, F_{by} [f,r] psi	Compression Perpendicular to Grain $F_{c\perp y}$ [v] psi	Horizontal Shear, F_{vy} psi	Horizontal Shear, Fvy, psi (For members with multiple piece laminations which are not edge glued) [u] psi	Modulus of Elasticity, E_y $\times 10^{-6}$psi	Tension Parallel to Grain, F_t psi	Compression Parallel to Grain, F_c psi	Modulus of Elasticity, E $\times 10^{-6}$psi
1	2	3	4	5	6	7	8	9	10	11	12	13	14	15	16
Visually Graded Western Species (continued)															
24F-V1	DF/WW	2400	1200	650	650	140[s]	1.7	1250	255	135[s]	70[s]	1.4	1000	1300	1.4
24F-V2	HF/HF			500[j]	500[j]	155	1.5	1250	375	135	70	1.4	950	1300	1.4
24F-V3	DF/DF			650	560[h]	165	1.7	1500	560	145	75	1.6	1100	1600	1.6
24F-V4	DF/DF			650	650	165	1.8	1500	560	145	75	1.6	1150	1650	1.6
24F-V5	DF/HF			650	650	155	1.7	1350	375	140	70	1.5	1100	1450	1.5
24F-V11	DF/DFS			650	560[h]	165	1.7	1600	500	145	75	1.4	1150	1700	1.4
The following two combinations are intended for straight or slightly cambered members for dry use and industrial appearance.[k]															
24F-V6	DF/N3WW	2400	1200	650	560[h]	90[l,s]	1.7	1200	255	140[s]	70[s]	1.5	950	800	1.5
24F-V7	DF/N3DF			650	560[h]	90[m]	1.7	1250	470	135	70	1.6	900	950	1.6
The following three combinations are balanced and are intended for members continuous or cantilevered over supports and provide equal capacity in both positive and negative bending.															
24F-V8	DF/DF	2400	2400	650	650	165	1.8	1450	560	145	75	1.6	1100	1650	1.6
24F-V9	HF/HF			500[j]	500[j]	155	1.5	1500	375	135	70	1.4	1000	1450	1.4
24F-V10	DF/HF			650	650	155	1.8	1400	375	140	70	1.6	1150	1600	1.6

E-Rated Western Species

16F-E1	WW/WW	1600	800	255[n]	255[n]	140[s]	1.3	1050	255	125[s]	65[s]	1.2	725	925	1.2
16F-E2[p]	HF/HF			500[o]	500[o]	155	1.4	1250	375	135	70	1.3	825	1200	1.3
16F-E3	DF/DF			650	650	165	1.6	1450	560	145	75	1.5	975	1600	1.5
The following two combinations are intended for straight or slightly cambered members for dry use and industrial appearance.[k]															
16F-E4	DF/N3WW	1600	800	650	650	90[l,s]	1.6	900	255	130[s]	65[s]	1.3	675	675	1.3
16F-E5	DF/N3DF			650	650	90[m]	1.6	1050	470	135	70	1.5	700	900	1.5
The following two combinations are balanced and are intended for members continuous or cantilevered over supports and provide equal capacity in both positive and negative bending.															
16F-E6	DF/DF	1600	1600	650	650	165	1.6	1500	560	145	75	1.5	1000	1600	1.5
16F-E7[p]	HF/HF			500[o]	500[o]	155	1.4	1250	375	135	70	1.3	850	1150	1.3
20F-E1	WW/WW	2000	1000	255[n]	255[n]	140[s]	1.6	1100	255	125[s]	65[s]	1.3	800	1050	1.3
20F-E2[p]	HF/HF			500[o]	500[o]	155	1.6	1400	375	135	70	1.4	925	1550	1.4
20F-E3	DF/DF			650	650	165	1.7	1550	560	145	75	1.6	1050	1650	1.6
The following two combinations are intended for straight or slightly cambered members for dry use and industrial appearance.[k]															
20F-E4	DF/N3WW	2000	1000	650	650	90[l,s]	1.6	1100	255	130[s]	65[s]	1.4	800	700	1.4
20F-E5	DF/N3DF			650	650	90[m]	1.7	1300	470	135	70	1.6	825	975	1.6
The following two combinations are balanced and are intended for members continuous or cantilevered over supports and provide equal capacity in both positive and negative bending.															
20F-E6	DF/DF	2000	2000	650	650	165	1.7	1600	560	145	75	1.6	1150	1650	1.6
20F-E7[p]	HF/HF			500[o]	500[o]	155	1.6	1500	375	135	70	1.4	1050	1550	1.4
Wet-use factors[b]		0.8	0.8	0.53	0.53	0.875	0.833	0.8	0.53	0.875	0.875	0.833	0.8	0.73	0.833

Table 1 Design Values for Structural Glued Laminated Timber (Cont.)

For normal duration of load and dry conditions of use [a,b,c,d]

Combination Symbol [t]	Species-Outer Laminations/Core Laminations [e]	Bending About X-X Axis; Loaded Perpendicular to Wide Faces of Laminations; Extreme Fiber in Bending, F_{bx}: Tension Zone Stressed in Tension [fw]	Extreme Fiber in Bending, F_{bx}: Compression Zone Stressed in Tension [g]	Compression Perpendicular to Grain, $F_{c\perp x}$ [v]: Tension Face	Compression Perpendicular to Grain, $F_{c\perp x}$ [v]: Compression Face	Horizontal Shear, F_{vx}	Modulus of Elasticity, E_x	Bending About Y-Y Axis; Loaded Parallel to Wide Faces of Laminations; Extreme Fiber in Bending, F_{by} [f,r]	Compression Perpendicular to Grain $F_{c\perp y}$ [v]	Horizontal Shear, F_{vy}	Horizontal Shear, Fvy, psi (For members with multiple piece laminations which are not edge glued) [u]	Modulus of Elasticity, E_y	Axially Loaded: Tension Parallel to Grain, F_t	Compression Parallel to Grain, F_c	Modulus of Elasticity, E
		psi	psi	psi	psi	psi	$\times 10^{-6}$psi	psi	psi	psi	psi	$\times 10^{-6}$psi	psi	psi	$\times 10^{-6}$psi
1	**2**	**3**	**4**	**5**	**6**	**7**	**8**	**9**	**10**	**11**	**12**	**13**	**14**	**15**	**16**
E-Rated Western Species (continued)															
22F-E1	DF/DF	2200	1100	650	650	165	1.7	1550	560	145	75	1.6	1050	1600	1.6
22F-E2 [p]	HF/HF			500[o]	500[o]	155	1.6	1400	375	135	70	1.4	950	1400	1.4
The following two combinations are intended for straight or slightly cambered members for dry use and industrial appearance.[k]															
22F-E3	DF/N3WW	2200	1100	650	650	90 [l,s]	1.7	1250	255	135[s]	70[s]	1.4	825	750	1.4
22F-E4	DF/N3DF			650	650	90 [m]	1.8	1350	470	135	70	1.6	950	950	1.6
The following two combinations are balanced and are intended for members continuous or cantilevered over supports and provide equal capacity in both positive and negative bending.															
22F-E5	DF/DF	2200	2200	650	650	165	1.7	1650	560	145	75	1.6	1100	1650	1.6
22F-E6 [p]	HF/HF			500[o]	500[o]	155	1.7	1550	375	135	70	1.5	1050	1500	1.5

24F-E1	DF/DF	2400	1200	650	650	165	1.8	1550	560	145	75	1.6	1100	1600	1.6
24F-E2 [p]	HF/HF			500 [o]	500 [o]	155	1.7	1450	375	135	70	1.5	1000	1400	1.5
24F-E3	DF/HF			650	500 [o]	155	1.8	1500	375	135	70	1.5	1050	1550	1.5
24F-E4	DF/DF			650	650	165	1.8	1650	560	145	75	1.7	1100	1700	1.7
24F-E5	DF/DF			650	650	165	1.8	1650	560	145	75	1.6	1100	1550	1.6
24F-E6 [p]	HF/WW			500 [o]	500 [o]	140 [s]	1.8	1250	255	130 [s]	65 [s]	1.4	925	1350	1.4
24F-E14	DF/DF			650	650	165	1.8	1600	560	145	75	1.6	1050	1600	1.6
24F-E15	HF/HF			500	500	155	1.8	1500	375	135	70	1.5	1000	1550	1.5
The following three combinations are intended for straight or slightly cambered members for dry use and industrial appearance.[k]															
24F-E7	DF/N3WW	2400	1200	650	650	90 [l,s]	1.9	1400	255	135 [s]	70 [s]	1.6	975	875	1.6
24F-E8	DF/N3DF			650	650	90 [m]	1.9	1400	470	135	70	1.7	1000	1050	1.7
24F-E9 [p]	HF/N3HF			500 [o]	500 [o]	90 [l]	1.8	1350	375	135	70	1.6	950	825	1.6
The following four combinations are balanced and are intended for members continuous or cantilevered over supports and provide equal capacity in both positive and negative bending.															
24F-E10	DF/DF	2400	2400	650	650	165	1.9	1850	560	145	75	1.7	1300	1750	1.7
24F-E11 [p]	HF/HF			500 [o]	500 [o]	155	1.8	1600	375	135	70	1.5	1150	1550	1.5
24F-E12	DF/HF			650	650	155	1.9	1750	375	135	70	1.6	1200	1600	1.6
24F-E13	DF/DF			650	650	165	1.8	1950	560	145	75	1.7	1250	1700	1.7
Wet-use factors [b]		0.8	0.8	0.53	0.53	0.875	0.833	0.8	0.53	0.875	0.875	0.833	0.8	0.73	0.833

Table 1 Design Values for Structural Glued Laminated Timber (Cont.)

For normal duration of load and dry conditions of use [a,b,c,d]

		Bending About X-X Axis						Bending About Y-Y Axis					Axially Loaded		
		Loaded Perpendicular to Wide Faces of Laminations						Loaded Parallel to Wide Faces of Laminations							
		Extreme Fiber in Bending, F_{bx}		Compression Perpendicular to Grain, $F_{c\perp x}$ [v]											
Combination Symbol [t]	Species-Outer Laminations/Core Laminations [e]	Tension Zone Stressed in Tension [f,w] psi	Compression Zone Stressed in Tension [g] psi	Tension Face psi	Compression Face psi	Horizontal Shear, F_{vx} psi	Modulus of Elasticity, E_x $\times 10^{-6}$psi	Extreme Fiber in Bending, F_{by} [f,r] psi	Compression Perpendicular to Grain $F_{c\perp y}$ [v] psi	Horizontal Shear, F_{vy} psi	Horizontal Shear, Fvy, psi (For members with multiple piece laminations which are not edge glued) [u] psi	Modulus of Elasticity, E_y $\times 10^{-6}$psi	Tension Parallel to Grain, F_t psi	Compression Parallel to Grain, F_c psi	Modulus of Elasticity, E $\times 10^{-6}$psi
1	2	3	4	5	6	7	8	9	10	11	12	13	14	15	16
Visually Graded Southern Pine															
16F-V1	SP/SP	1600	800	560 [h,i]	560 [h]	200	1.4	1450	560	175	90	1.3	950	1450	1.3
16F-V2	SP/SP			560 [h,i]	560 [h]	200	1.4	1600	560	175	90	1.4	1000	1550	1.4
16F-V3	SP/SP			650	650	200	1.4	1450	560	175	90	1.3	975	1450	1.3
The following combination is intended for straight or slightly cambered members for dry use and industrial appearance. [k]															
16F-V4	SP/SP	1600	800	560 [h,i]	560 [h]	90 [q]	1.3	975	470	150	75	1.2	650	950	1.2
The following combination is balanced and is intended for members continuous or cantilevered over supports and provides equal capacity in both positive and negative bending.															
16F-V5	SP/SP	1600	1600	560 [h,i]	560 [h,i]	200	1.4	1600	560	175	90	1.4	1000	1550	1.4

20F-V1	SP/SP	2000	1000	650	560[h]	200	1.5	1450	560	175	90	1.4	1000	1450	1.4
20F-V2	SP/SP			650	560[h]	200	1.6	1450	560	175	90	1.4	1050	1550	1.4
20F-V3	SP/SP			560[h,i]	560[h]	200	1.4	1600	560	175	90	1.4	1000	1500	1.4
The following combination is intended for straight or slightly cambered members for dry use and industrial appearance.[k]															
20F-V4	SP/SP	2000	1000	650	560[h]	90[q]	1.5	1100	470	150	75	1.3	725	950	1.3
The following combination is balanced and is intended for members continuous or cantilevered over supports and provides equal capacity in both positive and negative bending.															
20F-V5	SP/SP	2000	2000	650	650	200	1.6	1450	560	175	90	1.4	1050	1550	1.4
22F-V1	SP/SP	2200	1100	650	650	200	1.6	1600	560	175	90	1.5	1050	1650	1.5
22F-V2	SP/SP			560[h,i]	560[h]	200	1.4	1600	560	175	90	1.4	1000	1500	1.4
22F-V3	SP/SP			650	560[h]	200	1.6	1500	560	175	90	1.4	1050	1500	1.4
The following combination is intended for straight or slightly cambered members for dry use and industrial appearance.[k]															
22F-V4	SP/SP	2200	1100	650	560[h]	90[q]	1.6	1250	470	155	80	1.4	825	1000	1.4
The following combination is balanced and is intended for members continuous or cantilevered over supports and provides equal capacity in both positive and negative bending.															
22F-V5	SP/SP	2200	2200	650	650	200	1.6	1600	560	175	90	1.5	1050	1600	1.5
24F-V1	SP/SP	2400	1200	650	560[h]	200	1.7	1500	560	175	90	1.5	1100	1350	1.5
24F-V2	SP/SP			650	650	200	1.7	1600	560	175	90	1.5	1100	1600	1.5
24F-V3	SP/SP			650	650	200	1.8	1600	560	175	90	1.6	1150	1700	1.6
24F-V6	SP/SP			650	650	200	1.7	1500	560	175	90	1.5	1150	1750	1.5
The following combination is intended for straight or slightly cambered members for dry use and industrial appearance.[k]															
24F-V4	SP/SP	2400	1200	650	560[h]	90[q]	1.7	1250	470	155	80	1.4	850	1050	1.4
The following combination is balanced and is intended for members continuous or cantilevered over supports and provides equal capacity in both positive and negative bending.															
24F-V5	SP/SP	2400	2400	650	650	200	1.7	1600	560	175	90	1.5	1150	1700	1.5
Wet-use factors[b]		0.8	0.8	0.53	0.53	0.875	0.833	0.8	0.53	0.875	0.875	0.833	0.8	0.73	0.833

Table 1 Design Values for Structural Glued Laminated Timber (Cont.)

For normal duration of load and dry conditions of use [a,b,c,d]

Combination Symbol [t]	Species-Outer Laminations/Core Laminations [e]	Bending About X-X Axis: Loaded Perpendicular to Wide Faces of Laminations — Extreme Fiber in Bending, F_{bx}: Tension Zone Stressed in Tension [f,w] (psi)	Extreme Fiber in Bending, F_{bx}: Compression Zone Stressed in Tension [g] (psi)	Compression Perpendicular to Grain, $F_{c\perp x}$ [v]: Tension Face (psi)	Compression Perpendicular to Grain, $F_{c\perp x}$ [v]: Compression Face (psi)	Horizontal Shear, F_{vx} (psi)	Modulus of Elasticity, E_x ($\times 10^{-6}$ psi)	Bending About Y-Y Axis: Loaded Parallel to Wide Faces of Laminations — Extreme Fiber in Bending, F_{by} [f,r] (psi)	Compression Perpendicular to Grain $F_{c\perp y}$ [v] (psi)	Horizontal Shear, F_{vy} (psi)	Horizontal Shear, Fvy, psi (For members with multiple piece laminations which are not edge glued) [u] (psi)	Modulus of Elasticity, E_y ($\times 10^{-6}$ psi)	Axially Loaded: Tension Parallel to Grain, F_t (psi)	Compression Parallel to Grain, F_c (psi)	Modulus of Elasticity, E ($\times 10^{-6}$ psi)
1	2	3	4	5	6	7	8	9	10	11	12	13	14	15	16
E-Rated Southern Pine															
16F-E1	SP/SP	1600	800	650	650	200	1.6	1550	560	175	90	1.5	1050	1600	1.5
The following combination is intended for straight or slightly cambered members for dry use and industrial appearance.[k]															
16F-E2	SP/SP	1600	800	650	650	90 [q]	1.6	950	470	145	75	1.3	700	1050	1.3
The following combination is balanced and is intended for members continuous or cantilevered over supports and provides equal capacity in both positive and negative bending.															
16F-E3	SP/SP	1600	1600	650	650	200	1.6	1700	560	175	90	1.5	1100	1650	1.5
20F-E1	SP/SP	2000	1000	650	650	200	1.7	1600	560	175	90	1.5	1050	1600	1.5
The following combination is intended for straight or slightly cambered members for dry use and industrial appearance.[k]															
20F-E2	SP/SP	2000	1000	650	650	90 [q]	1.6	1100	470	150	75	1.4	750	1000	1.4
The following combination is balanced and is intended for members continuous or cantilevered over supports and provides equal capacity in both positive and negative bending.															
20F-E3	SP/SP	2000	2000	650	650	200	1.7	1800	560	175	90	1.5	1150	1700	1.5

22F-E1	SP/SP	2200	1100	650	650	200	1.7	1600	560	175	90	1.5	1050	1650	1.5
The following combination is intended for straight or slightly cambered members for dry use and industrial appearance.[k]															
22F-E2	SP/SP	2200	1100	650	650	90 [g]	1.6	1250	470	155	80	1.4	850	1050	1.4
The following combination is balanced and is intended for members continuous or cantilevered over supports and provides equal capacity in both positive and negative bending.															
22F-E3	SP/SP	2200	2200	650	650	200	1.7	1750	560	175	90	1.5	1150	1650	1.5
24F-E1	SP/SP	2400	1200	650	650	200	1.8	1600	560	175	90	1.6	1100	1750	1.6
24F-E2	SP/SP			650	650	200	1.9	1700	560	175	90	1.6	1150	1700	1.6
The following combination is intended for straight or slightly cambered members for dry use and industrial appearance.[k]															
24F-E3	SP/SP	2400	1200	650	650	90 [g]	1.8	1300	470	155	80	1.5	950	1100	1.5
The following combination is balanced and is intended for members continuous or cantilevered over supports and provides equal capacity in both positive and negative bending.															
24F-E4	SP/SP	2400	2400	650	650	200	1.8	2000	560	175	90	1.6	1250	1750	1.6
Wet-use factors [b]		0.8	0.8	0.53	0.53	0.875	0.833	0.8	0.53	0.875	0.875	0.833	0.8	0.73	0.833

[a]The combinations in this table are applicable to members consisting of 4 or more laminations and are intended primarily for members stressed in bending due to loads applied perpendicular to the wide faces of the laminations. Design values are tabulated, however, for loading both perpendicular and parallel to the wide faces of the laminations. For combinations and design values applicable to members loaded primarily axially or parallel to the wide faces of the laminations, see Table 2. For members of 2 or 3 laminations, see Table 2.

[b]The tabulated design values are for dry conditions of use. To obtain wet-use design values, multiply the tabulated values by the factors shown at the end of the table.

[c]The tabulated design values are for normal duration of loading. For other durations of loading, see 4.4.1.

[d]The 22F and 24F combinations for members 15 in. and less in depth may not be readily available and the designer should check on availability prior to specifying. The 16F and 20F combinations are generally available for members 15 in. and less in depth.

[e]The symbols used for species are DF = Douglas Fir-Larch, DFS = Douglas Fir South, HF = Hem-Fir, WW = Western woods or Canadian softwood species, and SP = Southern Pine. (N3 refers to No. 3 structural joists and planks or structural light framing grade.)

[f]The tabulated design values in bending are applicable to members 12 in. or less in depth. For members greater than 12 in. in depth, the requirements of 4.4.2 apply.

[g]Design values in this column are for extreme fiber stress in bending when the member is loaded such that the compression zone laminations are sub-

Footnotes for Table 1 continued

jected to tensile stresses. For more information, see 1.4.1.3. The values in this column may be increased 200 psi where end joint spacing restrictions are applied to the compression zone when stressed in tension.

[h]Where specified, this value may be increased to 650 psi by providing in the bearing area at least one dense 2 in. nominal thickness lamination of Douglas Fir-Larch for Western species combinations, or Southern Pine for Southern Pine combinations. These dense laminations must be backed by a medium grain lamination of the same species.

[i]For bending members greater than 15 in. in depth, the design value for compression stress perpendicular to grain is 650 psi on the tension face.

[j]Where specified, this value may be increased by providing at least two 2 in. nominal thickness Douglas Fir-Larch laminations in the bearing area. The compression perpendicular to grain design values for Douglas Fir-Larch are 560 psi for medium grain and 650 psi for dense.

[k]These combinations are for dry conditions of use only because they may contain wane. They are recommended for industrial appearance grade and for straight or slightly cambered members only. If wane is omitted these restrictions do not apply.

[l]Where specified, this value may be increased to 140 psi for Western woods and to 155 psi for Hem-Fir by prohibiting wane on both sides of the member; or to 115 psi for Western woods and to 130 psi for Hem-Fir by prohibiting wane on one side of the member.

[m]Where specified, this value may be increased to 110 psi by prohibiting coarse grain material; to 140 psi either by prohibiting wane on both sides of the member or by prohibiting both coarse grain material and wane on one side of the member; or to 165 psi by prohibiting both coarse grain material and wane on both sides of the member.

[n]The compression perpendicular to grain design value of 255 psi is based on the lowest strength species of the Western woods group. If at least one 2 in. nominal thickness lamination of E-rated Hem-Fir with the same E value, or E-rated Douglas Fir-Larch 200,000 psi higher in modulus of elasticity(E) than that specified is used in the bearing area of the face of the member subjected to the compression perpendicular to grain stress, $F_{c\perp}$, may be increased to 375 psi. If at least two 2 in. nominal thickness laminations of E-rated Hem-Fir with the same E value, or E-rated Douglas Fir-Larch 200,000 psi higher in modulus of elasticity than that specified are used in the bearing area on the face of the member subjected to the compression perpendicular to grain stress, $F_{c\perp}$, may be increased to 500 psi.

[o]Where specified this value may be increased to 650 psi by providing in the bearing area at least one 2 in. nominal thickness lamination of Douglas Fir-Larch for Western species combinations, or one 2 in. nominal thickness lamination of Southern Pine for Southern Pine combinations having a modulus of elasticity (E) value 200,000 psi higher than the E value specified.

[p]E-rated Douglas Fir-Larch 200,000 psi higher in modulus of elasticity may be substituted for the specified E-rated Hem-Fir.

[q]Where specified, this value may be increased to 140 psi either by prohibiting coarse grain material or by prohibiting wane on both sides of the member; to 165 psi by prohibiting both coarse grain material and wane on one side of the member; or to 200 psi by prohibiting both coarse grain material and wane on both sides of the member.

[r]Footnote f, to Table 2 also applies.

[s]Where Douglas Fir South is used in place of all of the western wood laminations required in western species combinations 16F-V1, 16F-V4, 20F-V1, 20F-V5, 22F-V1, 22F-V5, 24F-V1, 24F-V6, 16F-E1, 16F-E4,20F-E1, 20F-E4,22F-E3, 24F-E6 and 24F-E7 the design value for horizontal shear is the same as for combinations using all Douglas Fir-Larch. (F_{vx} = 165 psi and F_{vy} = 145 psi for L3 and F_{vx} = 90 psi and F_{vy} = 135 psi for N3.)

[t]The combination symbols relate to a specific combination of grades and species in AITC 117-84—MANUFACTURING that will provide the design values shown for the combination. The first two numbers in the combination symbol correspond to the design value in bending shown in Column 3. The letter in the combination symbol (either a "V" or an "E") indicates whether the combination is made from visually graded (V) or E-rated (*E*) lumber in the outer zones.

[u]These values for horizontal shear, F_{vy}, apply to members manufactured using multiple piece laminations with unbonded edge joints. For members manufactured using single piece laminations or using multiple piece laminations with bonded edge joints the horizontal shear values in column 11 apply.

[v]The compression perpendicular to grain design values in this Table are not subject to the duration of load modifications in 4.4.1.

[w]The design values in bending about the x-x axis (F_{bx}) in this column for bending members shall be multiplied by 0.75 when the member is manufactured without the required special tension lamination(s) (see 1.4.1.6).

Table 2 Design Values for Structural Glued Laminated Timber

For normal duration of load and dry conditions of use [a,b,c]

Combination Symbol	Species [d]	Grade [e]	Modulus of Elasticity, E	Compression Perpendicular to Grain, $F_{c\perp n}$	Axially Loaded: Tension Parallel to Grain, F_t	Axially Loaded: Compression Parallel to Grain, Fc		Loaded Parallel to Wide Faces of Laminations: Extreme Fiber in Bending [f] F_{by}		
					2 or More Lams.	4 or More Lams	2 or 3 Lams.	4 or More Lams	3 Lams	2 Lams
			$\times 10^{-6}$psi	psi	psi	psi	psi	psi	psi	psi
1	2	3	4	5	6	7	8	9	10	11
						Visually Graded Western Species				
1	DF	L3	1.5	560[k]	900	1550	1200	1450	1250	1000
2	DF	L2	1.7	560[k]	1250	1900	1600	1800	1600	1300
3	DF	L2D	1.8	650	1450	2300	1850	2100	1850	1550
4	DF	L1CL	1.9	590[k]	1400	2100	1900	2200	2000	1650
5	DF	L1	2.0	650	1600	2400	2100	2400	2100	1800
6	DF	N3C	1.4	470	350	875	550	550	550	550
7	DF	N3M	1.5	560	900	1550	700	1450	1250	1000
8	DF	N2	1.6	560[k]	1000	1550	1150	1600	1550	1300
9	DF	N2D	1.8	650	1150	1800	1350	1850	1800	1500
10	DF	N1	1.8	560[k]	1300	1950	1450	1950	1750	1500
11	DF	N1D	2.0	650	1500	2300	1700	2300	2100	1750
12	DF	SS	1.8	560[k]	1400	1950	1650	2100	1950	1650
13	DF	SSD	2.0	650	1600	2300	1950	2400	2300	1950
14	HF	L3	1.3	375[k]	800	1100	975	1200	1050	850
15	HF	L2	1.4	375[k]	1050	1350	1300	1500	1350	1100
16	HF	L1	1.6	375[k]	1200	1500	1450	1750	1550	1300
17	HF	L1D	1.7	500	1400	1750	1700	2000	1850	1550
18	HF	N3	1.3	375	425	900	575	700	700	700
19	HF	N2	1.4	375[k]	850	1300	975	1350	1300	1100
20	HF	N1	1.6	375[k]	975	1450	1250	1550	1500	1250
21	HF	SS	1.6	375[k]	1100	1450	1350	1750	1650	1400
22[o]	WW	L3	1.0	255	525	850	675	800	700	550
23[o]	WW	N3	1.0	255	275	625	450	450	450	450
24[o]	WW	N2	1.1	255	550	900	700	900	875	725
25[o]	WW	N1	1.2	255	650	1000	875	1050	1000	850
26[o]	WW	SS	1.2	255	750	1000	1000	1150	1100	925
59	DFS	L3	1.1	500	800	1400	1050	1200	1050	850
60	DFS	L2	1.3	500	1050	1750	1400	1750	1550	1150
61	DFS	L1	1.5	650	1350	2200	1850	2000	1800	1500
Wet-use factors [b]			0.833	0.53	0.8	0.73	0.73	0.8	0.8	0.8

Bending About Y-Y Axis				Bending About X-X Axis (Loaded Perpendicular to Wide Faces of Laminations)			
Horizontal Shear F_{vy}				Extreme Fiber in Bending[h], F_{bx}		Horizontal Shear,[g] F_{vx}	
4 or More Lams (For members with multiple piece lams)[m]	4 or More Lams.	3 Lams.	2 Lams.	2 Lams. to 15 in. Deep[l]	4 or More Lams.[l p]	2 or More Lams.	Combination Symbol
psi	psi	psi	psi	psi	psi	psi	
12	**13**	**14**	**15**	**16**	**17**	**18**	
75	145	135	125	1250	1500	165	1
75	145	135	125	1700	2000	165	2
75	145	135	125	2000	2300	165	3
75	145	135	125	1900	2200	165	4
75	145	135	125	2200	2400	165	5
60	120	115	105	450	—	140	6
75	145	135	125	1000	—	165	7
75	145	135	125	1350	1600	165	8
75	145	135	125	1600	1850	165	9
75	145	135	125	1750	2100	165	10
75	145	135	125	2100	2400	165	11
75	145	135	125	1900	2200	165	12
75	145	135	125	2200	2400	165	13
70	135	130	115	1100	1300	155	14
70	135	130	115	1450	1700	155	15
70	135	130	115	1600	1900	155	16
70	135	130	115	1900	2200	155	17
70	135	130	115	575		155	18
70	135	130	115	1150	1350	155	19
70	135	130	115	1350	1550	155	20
70	135	130	115	1500	1750	155	21
60	120	115	105	725	850	140	22
60	120	115	105	400	—	140	23
60	120	115	105	775	900	140	24
60	120	115	105	875	1050	140	25
60	120	115	105	1000	1150	140	26
75	145	135	125	1050	1250	165	59
75	145	135	125	1450	1700	165	60
75	145	135	125	1850	2200	165	61
0.875	0.875	0.875	0.875	0.8	0.8	0.875	

Table 2 Design Values for Structural Glued Laminated Timber (Cont.)

Combination Symbol	Species [d]	Grade [e]	Modulus of Elasticity, E	Compression Perpendicular to Grain, $F_{c\perp}$ n	Axially Loaded: Tension Parallel to Grain, F_t	Axially Loaded: Compression Parallel to Grain, Fc		Loaded Parallel to Wide Faces of Laminations: Extreme Fiber in Bending [f], F_{by}		
					2 or More Lams	4 or More Lams	2 or 3 Lams	4 or More Lams	3 Lams	2 Lams
			$\times 10^{-6}$ psi	psi	psi	psi	psi	psi	psi	psi
1	2	3	4	5	6	7	8	9	10	11
E-Rated Western Species										
27	DF	1/2-1.8E	1.8	650	900	1750	1200	1450	1250	1000
28	DF	1/2-2.0E	2.0	650	1100	2000	1400	1450	1250	1000
29	DF	1/2-2.2E	2.2	650	1250	2300	1550	1650	1400	1150
30	DF	1/6-1.8E	1.8	650	1550	2100	1700	2400	2400	2100
31	DF	1/6-2.0E	2.0	650	1800	2400	1900	2400	2400	2400
32	DF	1/6-2.2E	2.2	650	1800	2400	2100	2400	2400	2400
62	DF	1/2-2.1E	2.1	650	1150	2200	1500	1550	1350	1100
63	DF	1/6-2.1E	2.1	650	1800	2400	2000	2400	2400	2400
33	HF	1/2-1.5E	1.5	500	800	1050	950	1200	1050	850
34	HF	1/2-1.8E	1.8	500	900	1300	1200	1450	1250	1000
35	HF	1/2-2.0E	2.0	500	1100	1550	1400	1450	1250	1000
36	HF	1/4-1.5E	1.5	500	1200	1450	1300	2100	1900	1700
37	HF	1/6-1.8E	1.8	500	1550	1950	1700	2400	2400	2100
38	HF	1/6-2.0E	2.0	500	1800	2400	1900	2400	2400	2400
39	WW	1/2-1.5E	1.5	255	800	1200	950	1200	1050	850
40	WW	1/2-1.8E	1.8	255	900	1500	1200	1450	1250	1000
41	WW	1/2-2.0E	2.0	255	1100	1750	1400	1450	1250	1000
42	WW	1/4-1.5E	1.5	255	1200	1550	1300	2100	1900	1700
43	WW	1/6-1.8E	1.8	255	1550	1950	1700	2400	2400	2100
44	WW	1/6-2.0E	2.0	255	1800	2200	1900	2400	2400	2400
Visually Graded Southern Pine										
45	SP	N3C	1.1	470	325	850	550	550	550	550
46	SP	N3M	1.3	560	900	1500	675	1450	1250	1000
47	SP	N2M [l]	1.4	560 [k]	1200	1900	1150	1750	1550	1300
48	SP	N2D [l]	1.7	650	1400	2200	1350	2000	1800	1500
49	SP	N1M [l]	1.7	560 [k]	1350	2100	1450	1950	1750	1500
50	SP	N1D [l]	1.9	650	1550	2300	1700	2300	2100	1750
51	SP	SSM	1.7	560 [k]	1300	1900	1600	2100	1950	1650
52	SP	SSD	1.9	650	1500	2200	1850	2400	2300	1950
E-Rated Southern Pine										
53	SP	1/2-1.8E	1.8	650	900	1900	1200	1450	1250	1000
54	SP	1/2-2.0E	2.0	650	1100	2300	1400	1450	1250	1000
55	SP	1/2-2.2E	2.2	650	1250	2400	1550	1650	1400	1150
56	SP	1/6-1.8E	1.8	650	1550	1850	1700	2400	2400	2100
57	SP	1/6-2.0E	2.0	650	1800	2400	1900	2400	2400	2400
58	SP	1/6-2.2E	2.2	650	1800	2400	2100	2400	2400	2400
Wet-use factors [b]			0.833	0.53	0.8	0.73	0.73	0.8	0.8	0.8

For normal duration of load and dry conditions of use [a,b,c]

Bending About Y-Y Axis				Bending About X-X Axis			
				Loaded Perpendicular to Wide Faces of Laminations			
Horizontal Shear, F_{vy}				Extreme Fiber in Bending [h], F_{bx}		Horizontal Shear,[g] F_{vx}	
4 or More Lams (For members with multiple piece lams)[m]	4 or More Lams.	3 Lams.	2 Lams.	2 Lams. to 15 in. Deep [l]	4 or More Lams.[l p]	2 or More Lams.	Combination Symbol
psi	psi	psi	psi	psi	psi	psi	
12	**13**	**14**	**15**	**16**	**17**	**18**	
75	145	135	125	1250	1500	165	27
75	145	135	125	1500	1750	165	28
75	145	135	125	1700	2000	165	29
75	145	135	125	1800	2100	165	30
75	145	135	125	2100	2400	165	31
75	145	135	125	2300	2400	165	32
75	**145**	**135**	**125**	**1600**	**1900**	**165**	62
75	**145**	**135**	**125**	**2200**	**2400**	**165**	63
70	135	130	115	1100	1300	155	33
70	135	130	115	1250	1500	155	34
70	135	130	115	1500	1750	155	35
70	135	130	115	1400	1650	155	36
70	135	130	115	1800	2100	155	37
70	135	130	115	2100	2400	155	38
60	120	115	105	1100	1300	140	39
60	120	115	105	1250	1500	140	40
60	120	115	105	1500	1750	140	41
60	120	115	105	1400	1650	140	42
60	120	115	105	1800	2100	140	43
60	120	115	105	2100	2400	140	44
60	120	115	105	450	—	140	45
90	175	165	150	1000	—	200	46
90	175	165	150	1400	1600	200	47
90	175	165	150	1600	1900	200	48
90	175	165	150	1800	2100	200	49
90	175	165	150	2100	2400	200	50
90	175	165	150	1750	2100	200	51
90	175	165	150	2100	2400	200	52
90	175	165	150	1250	1500	200	53
90	175	165	150	1500	1750	200	54
90	175	165	150	1700	2000	200	55
90	175	165	150	1800	2100	200	56
90	175	165	150	2100	2400	200	57
90	175	165	150	2300	2400	200	58
0.875	0.875	0.875	0.875	0.8	0.8	0.875	

Footnotes for Table 2

[a]The combinations in this table are intended primarily for members loaded either axially or in bending with the loads acting parallel to the wide faces of the laminations. Design values for bending due to loading applied perpendicular to the wide faces of the laminations are also included, however, the combinations in Table 1 are usually better suited for this condition of loading. The design values for bendng about the X–X axis (F_{bx}) shown in Column 16 are for members from 2 laminations to 15 in. deep without tension laminations. Design values approximately 15% higher for members with 4 or more laminations are shown in Column 17. These higher design values, however, require special tension laminations which may not be readily available.

[b]The tabulated design values are for dry conditions of use. To obtain wet-use design values, multiply the tabulated values by the factors shown at the end of the table.

[c]The tabulated design values are for normal duration of loading. For other durations of loading see 4.4.1.

[d]The symbols used for species are DF = Douglas Fir-Larch, DFS = Douglas Fir South, HF = Hem-Fir, WW = Western woods and Canadian softwood species, and SP = Southern Pine.

[e]Grade designations are as follows:

Visually Graded Western Species

L1 is L1 laminating grade (dense for Douglas Fir-Larch and Douglas Fir South).
L1D is L1 dense laminating grade for Hem-Fir.
L1CL is L1 close grain laminating grade.
L2D is L2 laminating grade (dense).
L2 is L2 laminating grade (medium grain).
L3 is L3 laminating grade (medium grain for Douglas Fir-Larch, Douglas Fir South and Hem-Fir).
SSD is dense select structural, structural joists and planks, or structural light framing grade (dense).
SS is select structural, structural joists and planks, or structural light framing grade (medium grain for Douglas Fir-Larch).
N1D is dense No. 1 structural joists and planks, or structural light framing grade (dense).
N1 is No. 1 structural joists and planks, or structural light framing grade (medium grain for Douglas Fir-Larch).
N2D is dense No. 2 structural joists and planks, or structural light framing grade (dense).
N2 is No. 2 structural joists and planks or structural light framing grade (medium grain for Douglas Fir-Larch.
N3M is No. 3 structural joists and planks, or structural light framing grade (medium grain).
N3C is No. 3 structural joists and planks, or structural light framing grade (coarse grain).
N3 is No. 3 structural joists and planks, or structural light framing grade.

Visually Graded Southern Pine

SSD is dense select structural, structural joists and planks, or structural light framing grade (dense).
SSM is select structural, structural joists and planks, or structural light framing grade (medium grain).
N1D is No. 1 dense structural joists and planks, or structural light framing grade or No. 1 boards graded as dense.
N1M is No. 1 structural joists and planks, or structural light framing grade or No. 1 boards all with a medium grain rate of growth.
N2D is No. 2 dense structural joists and planks, or structural light framing grade or No. 2 boards graded as dense.
N2M is No. 2 structural joists and planks, or structural light framing grade or No. 2 boards all with a medium grain rate of growth.
N3M is No. 3 structural joists and planks, or structural light framing grade or No. 3 boards all with a medium grain rate of growth.
N3C is No. 3 structural joists and planks, or structural light framing grade or No. 3 boards all with a coarse grain rate of growth.

E-Rated Grades—All Species

$\frac{1}{6}$-2.2E has $\frac{1}{6}$ edge characteristic with 2.2E.
$\frac{1}{6}$-2.1E has $\frac{1}{6}$ edge characteristic with 2.1E.
$\frac{1}{6}$-2.0E has $\frac{1}{6}$ edge characteristic with 2.0E.
$\frac{1}{6}$-1.8E has $\frac{1}{6}$ edge characteristic with 1.8E.
$\frac{1}{4}$-1.5E has $\frac{1}{4}$ edge characteristic with 1.5E.
$\frac{1}{2}$-2.2E, $\frac{1}{2}$-2.1E, $\frac{1}{2}$-2.0E, $\frac{1}{2}$-1.5E are E-rated grades with edge characteristics occupying up to $\frac{1}{2}$ of cross section.

Footnotes for Table 2 continued

[f]The values of F_{by} were calculated based on members 12 in. in depth (bending about Y–Y axis). When the depth is less than 12 in., the values of F_{by} can be increased by multiplying by the following factors: 1.01 for 10.75 in., 1.04 for 8.75 in., 1.07 for 6.75 in., 1.10 for 5.125 in., and 1.16 for 3.125 in.

[g]The design values in horizontal shear contained in this table are based on members without wane.

[h]The tabulated design values in bending are applicable to members 12 in. or less in depth. For members greater than 12 in. in depth, the requirements of 4.4.2 apply.

[i]The design values in column 15 are for members of from 2 laminations to 15 in. in depth without tension laminations.

[j]The design values in column 16 are for members of 4 or more laminations in depth and require special tension laminations. When these values are used in design and the member is specified by combination symbol, the designer should also specify the required design value in bending.

[k]When tension laminations are used to obtain the design value for F_{bx} shown in column 16, the compression perpendicular to grain value, $F_{c\perp}$, for the tension face may be increased to 650 psi for Douglas Fir-Larch, Douglas Fir South, and Southern Pine, and to 500 psi for Hem-Fir because the tension laminations are required to be dense.

[l]Combinations 47, 48, 49 and 50 have more restrictive slope of grain requirements than the basic slope of grain of the grades of lumber used in order to obtain higher tension parallel to grain values and design values in bending when loaded perpendicular to the wide faces of the laminations. The slopes of grain used to calculate the design values in Table 2 were: Combination 47, 1 : 14; Combination 48, 1 : 14; Combination 49, 1 : 16; and Combination 50, 1 : 14. When design stresses are lower than the design values shown, or when a less restrictive slope of grain provides the same design value, a less restrictive slope of grain may be used. The following table gives the design values of these combinations for various slopes of grain: Values of F_{bx} in column 5 for members of 2 laminations to 15 in. depth without tension laminations and values in column 6 are for members of 4 or more laminations with tension laminations.

Slope of Grain.	Comb. No.	Tension Parallel to Grain, F_t 2 or More Lams. psi	Comp. Parallel to Grain, F_c 2 or 3 Lams. psi	Comp. Parallel to Grain, F_c 4 or More Lams. psi	Bending About the X–X Axis, F_{bx} 2 Lams to 15 in. psi	Bending About the X–X Axis, F_{bx} 4 or More Lams. psi	Bending About the Y–Y Axis, (F_{by}) 2 Lams. psi	Bending About the Y–Y Axis, (F_{by}) 3 Lams. psi	Bending About the Y–Y Axis, (F_{by}) 4 or More Lams. psi
1		2	3	4	5	6	7	8	9
1:12	47	1200	1150	1900	1400	1600	1300	1550	1750
	48	1400	1350	2200	1600	1900	1500	1800	2000
	49	1300	1450	1900	1750	2100	1500	1750	1950
	50	1550	1700	2200	2100	2400	1750	2100	2300
1:10	47	1150	1150	1700	1400	1600	1300	1550	1750
	48	1350	1350	2000	1600	1900	1500	1800	2000
	49	1150	1450	1700	1550	1850	1500	1750	1850
	50	1350	1700	2000	1800	2100	1750	2100	2100
1:8	47	1000	1150	1500	1350	1600	1300	1550	1600
	48	1150	1350	1750	1600	1850	1500	1800	1850
	49	—	—	—	—	—	—	—	—
	50	—	—	—	—	—	—	—	—

[m]These values for horizontal shear, F_{vy}, apply to members manufactured using multiple piece laminations with unbonded edge joints. For members using single piece laminations or using multiple piece laminations with bonded edge joints the horizontal shear values tabulated in columns 13, 14 and 15 apply.

[n]The compression perpendicular to grain values in this Table are not subject to the duration of load modifications in 4.4.1.

[o]The design values for Western Cedars are the same as shown for Western Woods except that modulus of elasticity is 100,000 psi lower.

[p]When special tension laminations are not used, the design values in bending about the x-x- axis (F_{bx}) shall be multiplied by 0.75 for bending members over 15 in. deep. For bending members 15 in. and less in depth, use the design values in column 16 (see 1.4.1.6).

Table 3 — Design — Species Grouping for Fasteners Used on the Faces of Glued Laminated Timbers

Table 1 Combinations [a, b, c]

Combination Symbol	Tension Face [d]				Side Face [e]				Compression Face			
	Species[b]	Growth Rate[c]	Timber Conn. Group	Lag Bolt & Driven Fastener Group	Species[b]	Growth Rate[c]	Timber Conn. Group	Lag Bolt & Driven Fastener Group	Species[b]	Growth Rate[c]	Timber Conn. Group	Lag Bolt & Driven Fastener Group
Visually Graded Western Species												
16F-V1	DF	M	B	II	WW	—	D	IV	DF	M	B	II
16F-V2	HF	—	C	III	HF	—	C	III	HF	—	C	III
16F-V3	DF	M	B	II	DF	M	B	II	DF	M	B	II
16F-V4	DF	D	A	II	WW	—	D	IV	DF	M	B	II
16F-V5	DF	D	A	II	DF	C	C	III	DF	M	B	II
16F-V6	DF	M	B	II	DF	M	B	II	DF	M	B	II
16F-V7	HF	—	C	III	HF	—	C	III	HF	—	C	III
16F-V8	**DFS**	**—**	**C**	**III**	**DFS**	**—**	**C**	**III**	**DFS**	**—**	**C**	**III**
20F-V1	DF	D	A	II	WW	—	D	IV	DF	M	B	II
20F-V2	HF	—	C	III	HF	—	C	III	HF	—	C	III
20F-V3	DF	D	A	II	DF	M	B	II	DF	M	B	II
20F-V4	DF	CL	B	II	DF	M	B	II	DF	M	B	II
20F-V5	DF	D	A	II	WW	—	D	IV	DF	M	B	II
20F-V6	DF	D	A	II	DF	C	C	III	DF	M	B	II
20F-V7	DF	D	A	II	DF	M	B	II	DF	D	A	II
20F-V8	DF	CL	B	II	DF	M	B	II	DF	CL	B	II
20F-V9	HF	—	C	III	HF	—	C	III	HF	—	C	III
20F-V10	**DF**	**D**	**A**	**II**	**HF**	**—**	**C**	**III**	**DF**	**M**	**B**	**II**
20F-V11	**DFS**	**—**	**C**	**III**	**DFS**	**—**	**C**	**III**	**DFS**	**—**	**C**	**III**
22F-V1	DF	D	A	II	WW	—	D	IV	DF	M	B	II
22F-V2	HF	—	C	III	HF	—	C	III	HF	—	C	III
22F-V3	DF	D	A	II	DF	M	B	II	DF	M	B	II
22F-V4	DF	CL	B	II	DF	M	B	II	DF	M	B	II
22F-V5	DF	D	A	II	WW	—	D	IV	DF	M	B	II
22F-V6	DF	D	A	II	DF	C	C	III	DF	M	B	II
22F-V7	DF	D	A	II	DF	M	B	II	DF	D	A	II

22F-V8	DF	CL	B	II	DF	M	B	II	DF	CL	B	II
22F-V9	HF	—	C	III	HF	—	C	III	HF	—	C	III
22F-V10	DF	D	A	II	DFS	—	C	III	DF	D	A	II
24F-V1	DF	D	A	II	WW	—	D	IV	DF	D	A	II
24F-V2	HF	—	C	III	HF	—	C	III	HF	—	C	III
24F-V3	DF	D	A	II	DF	M	B	II	DF	M	B	II
24F-V4	DF	D	A	II	DF	M	B	II	DF	D	A	II
24F-V5	DF	D	A	II	HF	—	C	III	DF	D	A	II
24F-V6	DF	D	A	II	WW	—	D	IV	DF	M	B	II
24F-V7	DF	D	A	II	DF	C	C	III	DF	M	B	II
24F-V8	DF	D	A	II	DF	M	B	II	DF	D	A	II
24F-V9	HF	—	C	III	HF	—	C	III	HF	—	C	III
24F-V10	DF	D	A	II	HF	—	C	III	DF	D	A	II
24F-V11	DF	D	A	II	DFS	—	C	III	DF	D	A	II
E-Rated Western Species												
16F-E1	WW	—	D	IV	WW	—	D	IV	WW	—	D	IV
16F-E2	HF	—	C	III	HF	—	C	III	HF	—	C	III
16F-E3	DF	—	B	II	DF	M	B	II	DF	—	B	II
16F-E4	DF	—	B	II	WW	—	D	IV	DF	—	B	II
16F-E5	DF	—	B	II	DF	C	C	III	DF	—	B	II
16F-E6	DF	—	B	II	DF	M	B	II	DF	—	B	II
16F-E7	HF	—	C	III	HF	—	C	III	HF	—	C	III
20F-E1	WW	—	D	IV	WW	—	D	IV	WW	—	D	IV
20F-E2	HF	—	C	III	HF	—	C	III	HF	—	C	III
20F-E3	DF	—	B	II	DF	M	B	II	DF	—	B	II
20F-E4	DF	—	A	II	WW	—	D	IV	DF	—	B	II
20F-E5	DF	—	B	II	DF	C	C	III	DF	—	B	II
20F-E6	DF	—	B	II	DF	M	B	II	DF	—	B	II
20F-E7	HF	—	C	III	HF	—	C	III	HF	—	C	III
22F-E1	DF	—	A	II	DF	M	B	II	DF	—	B	II
22F-E2	HF	—	C	III	HF	—	C	III	HF	—	C	III
22F-E3	DF	—	A	II	WW	—	D	IV	DF	—	B	II
22F-E4	DF	—	A	II	DF	C	C	III	DF	—	B	II
22F-E5	DF	—	A	II	DF	M	B	II	DF	—	A	II
22F-E6	HF	—	C	III	HF	—	C	III	HF	—	C	III

Table 3—Design—Species Grouping for Fasteners Used on the Faces of Glued Laminated Timber (Cont.)

Combination Symbol	Tension Face [d]				Side Face [e]				Compression Face			
	Species[b]	Growth Rate[c]	Timber Conn. Group	Lag Bolt & Driven Fastener Group	Species[b]	Growth Rate[c]	Timber Conn. Group	Lag Bolt & Driven Fastener Group	Species[b]	Growth Rate[c]	Timber Conn. Group	Lag Bolt & Driven Fastener Group
E-Rated Western Species (Cont.)												
24F-E1	DF	—	A	II	DF	M	B	II	DF	—	A	II
24F-E2	HF	—	C	III	HF	—	C	III	HF	—	C	III
24F-E3	DF	—	A	II	HF	—	C	III	HF	—	C	III
24F-E4	DF	—	A	II	DF	M	B	II	DF	—	A	II
24F-E5	DF	—	A	II	DF	M	B	II	DF	—	B	II
24F-E6	HF	—	C	III	WW	—	D	IV	HF	—	C	III
24F-E7	DF	—	A	II	WW	—	D	IV	DF	—	A	II
24F-E8	DF	—	A	II	DF	C	C	III	DF	—	A	II
24F-E9	HF	—	C	III	HF	—	C	III	HF	—	C	III
24F-E10	DF	—	A	II	DF	M	B	II	DF	—	A	III
24F-E11	HF	—	C	III	HF	—	C	III	HF	—	C	III
24F-E12	DF	—	A	II	HF	—	C	III	DF	—	A	II
24F-E13	DF	—	A	II	DF	M	B	II	DF	—	A	II
24F-E14	DF	—	A	II	DF	M	B	II	DF	—	B	II
24F-E15	HF	—	C	III	HF	—	C	III	HF	—	C	III
Visually Graded Southern Pine												
16F-V1	SP	M	B	II	SP	M	B	II	SP	M	B	II
16F-V2	SP	M	B	II	SP	M	B	II	SP	M	B	II
16F-V3	SP	D	A	II	SP	M	B	II	SP	D	A	II
16F-V4	SP	M	B	II	SP	C	C	III	SP	M	B	II
16F-V5	SP	M	B	II	SP	M	B	II	SP	M	B	II

20F-V1	SP	D	A	II	SP	M	B	II	SP	M	B	II
20F-V2	SP	D	A	II	SP	M	B	II	SP	M	B	II
20F-V3	SP	M	B	II	SP	M	B	II	SP	M	B	II
20F-V4	SP	D	A	II	SP	C	C	III	SP	M	B	II
20F-V5	SP	D	A	II	SP	M	B	II	SP	D	A	II
22F-V1	SP	D	A	II	SP	M	B	II	SP	D	A	II
22F-V2	SP	D	A	II	SP	M	B	II	SP	M	B	II
22F-V3	SP	D	A	II	SP	M	B	II	SP	M	B	II
22F-V4	SP	D	A	II	SP	C	C	III	SP	M	B	II
22F-V5	SP	D	A	II	SP	M	B	II	SP	D	A	II
24F-V1	SP	D	A	II	SP	M	B	II	SP	M	B	II
24F-V2	SP	D	A	II	SP	M	B	II	SP	D	A	II
24F-V3	SP	D	A	II	SP	M	B	II	SP	D	A	II
24F-V4	SP	D	A	II	SP	C	C	III	SP	M	B	II
24F-V5	SP	D	A	II	SP	M	B	II	SP	D	A	II
E-Rated Southern Pine												
16F-E1	SP	—	B	II	SP	M	B	II	SP	—	B	II
16F-E2	SP	—	B	II	SP	C	C	III	SP	—	B	II
16F-E3	SP	—	B	II	SP	M	B	II	SP	—	B	II
20F-E1	SP	—	B	II	SP	M	B	II	SP	—	B	II
20F-E2	SP	—	A	II	SP	C	C	III	SP	—	B	II
20F-E3	SP	—	B	II	SP	M	B	II	SP	—	B	II
22F-E1	SP	—	A	II	SP	M	B	II	SP	—	B	II
22F-E2	SP	—	A	II	SP	C	C	III	SP	—	B	II
22F-E3	SP	—	A	II	SP	M	B	II	SP	—	A	II
24F-E1	SP	—	A	II	SP	M	B	II	SP	—	B	II
24F-E2	SP	—	A	II	SP	M	B	II	SP	—	A	II
24F-E3	SP	—	A	II	SP	C	C	III	SP	—	A	II
24F-E4	SP	—	A	II	SP	M	B	II	SP	—	A	II

Table 3—Design—Species Grouping for Fasteners Used on the Faces of Glued Laminated Timber (Cont.)

Table 2 Combinations [a, c]

Combination Symbol[f]	Any Face: Growth Rate[c]	Any Face: Timber Connector Group	Any Face: Lag Bolt & Driven Fastener Group
Visually Graded Douglas Fir			
1	M	B	II
2	M	B	II
3	D	A	II
4	CL	B	II
5	D	A	II
6	C	C	III
7	M	B	II
8	M	B	II
9	D	A	II
10	M	B	II
11	D	A	II
12	M	B	II
13	D	A	II
Visually Graded Douglas Fir South			
59	—	C	III
60	—	C	III
61	D	B	II
Visually Graded Hem-Fir			
14	—	C	III
15	—	C	III
16	—	C	III
17	—	C	III
18	—	C	III
19	—	C	III
20	—	C	III
21	—	C	III

Combination Symbol[f]	Any Face: Growth Rate[c]	Any Face: Timber Connector Group	Any Face: Lag Bolt & Driven Fastener Group
E-Rated Western Woods			
39	—	D	IV
40	—	D	IV
41	—	D	IV
42	—	D	IV
43	—	D	IV
44	—	D	IV
Visually Graded Southern Pine			
45	C	C	III
46	M	B	II
47	M	B	II
48	D	A	II
49	M	B	II
50	D	A	II
51	M	B	II
52	D	A	II
E-Rated Southern Pine			
53	—	B	II
54	—	A	II
55	—	A	II
56	—	B	II
57	—	A	II
58	—	A	II

Visually Graded Western Woods			
22	—	D	IV
23	—	D	IV
24	—	D	IV
25	—	D	IV
26	—	D	IV
E-Rated Douglas Fir			
27	—	B	II
28	—	A	II
29	—	A	II
30	—	B	II
31	—	A	II
32	—	A	II
62	—	B	II
63	—	A	II
E-Rated Hem-Fir			
33	—	C	III
34	—	C	III
35	—	C	III
36	—	C	III
37	—	C	III
38	—	C	III

[a]For bolts, use the species and growth rate indicated in this table with tables in the *Timber Construction Manual* or the *National Design Specification*. For Western woods and Canadian softwood species (WW) use bolt design values for Engelmann Spruce—Alpine Fir as given in the *NDS* or *TCM*. Where coarse grain (C) rate of growth is indicated use bolt design values for Hem-Fir as given in the *NDS* or *TCM*.

[b]The symbols used for species are DF = Douglas Fir-Larch, DFS = Douglas Fir South, HF = Hem-Fir, WW = Western woods and Canadian softwood species and SP = Southern Pine.

[c]The symbols used for growth rate are D = dense, CL = close grain, M = medium grain, and C = coarse grain.

[d]The growth rate for the tension face is dense for visually graded Douglas Fir-Larch, Douglas Fir South, Hem-Fir and Southern Pine for members greater than 15 in. deep. Where the growth rate of dense for shallower members is known to the designer, this may be used.

[e]The species and growth rate for the side faces are for the core of the member. Where the species and growth rate in the outer zones are known to the designer and the fastenings are only within those zones, the appropriate fastener design values may be used.

[f]The numerical combination symbols shown apply to the combinations listed in Table 2. When dense tension laminations are used for combinations in Table 2, the appropriate fastener design values for dense wood may be used.

ANNEX A DESIGN VALUES FOR END GRAIN IN BEARING PARALLEL TO GRAIN

A-1 The following table lists the design values for end grain in bearing parallel to grain.

A-1.1 These design values apply to the net area in bearing and are subject to adjustments for duration of load.

A-1.2 When the stress in end-grain bearing exceeds 75% of the adjusted design value, bearing should be on a metal plate, strap or other durable, rigid, homogenous material of adequate strength.

A-1.3 These values are based on dry conditions of use. Multiply the tabulated values by 0.57 to obtain wet conditions of use design values.

A-1.4 The design values for end grain in bearing apply to end-to-end bearing of compression members provided there is adequate lateral support and the end cuts are accurately squared and parallel. When a rigid insert is required by A-1.2, it shall be of not less than 20 gage metal, or equivalent, inserted with a snug fit between abutting ends.

A-1.5 When the load in bearing is at an angle to grain, the maximum bearing value shall be determined by the Hankinson formula using the design value for end grain in bearing parallel to grain from this Annex, and the design value in compression perpendicular to grain as provided in Table 1 or 2.

A-1.6 The design values in the column *Bearing on Full Cross Section* are the average design values of the laminations in the combination. The design values in the column *Bearing on Partial Cross Section* are the design values for the core laminations which should be used for tapered members or members with end bearing on only part of the cross section.

Table A-1 Design Values for End Grain in Bearing Parallel to Grain
Design values are for normal load duration, dry conditions of use, and Table 1 Combinations [a,b]

Combination Symbol	Bearing On Full Cross Section, psi	Bearing On Partial Cross Section, psi
Visually Graded Western Species		
16F-V1	1690	1420
16F-V2	1960	1940
16F-V3	2350	2350
16F-V4	1830	1420
16F-V5	2140	1890
16F-V6	2350	2350
16F-V7	1940	1940
16F-V8	2130	2130
20F-V1	1720	1420
20F-V2	2010	1940
20F-V3	2370	2350
20F-V4	2360	2350
20F-V5	2020	1420
20F-V6	2210	1890
20F-V7	2390	2350
20F-V8	2370	2350
20F-V9	2010	1940
20F-V10	2040	1940
20F-V11	2170	2130

Combination Symbol	Bearing On Full Cross Section, psi	Bearing On Partial Cross Section, psi
22F-V1	1830	1420
22F-V2	2010	1940
22F-V3	2390	2350
22F-V4	2370	2350
22F-V5	1970	1420
22F-V6	2230	1890
22F-V7	2390	2350
22F-V8	2380	2350
22F-V9	2010	1940
22F-V10	2250	2130
24F-V1	2080	1420
24F-V2	2020	1940
24F-V3	2390	2350
24F-V4	2430	2350
24F-V5	2140	1940
24F-V6	2110	1420
24F-V7	2230	1890
24F-V8	2430	2350
24F-V9	2060	1940
24F-V10	2180	1940
24F-V11	2370	2130

Table A-1 (cont.) Design Values for End Grain in Bearing Parallel to Grain
Design values are for normal load duration, dry conditions of use, and Table 1 Combinations [a,b]

Combination Symbol	Bearing On Full Cross Section, psi	Bearing On Partial Cross Section, psi
E-Rated Western Species		
16F-E1	1620	1420
16F-E2	1920	1800
16F-E3	2360	2350
16F-E4	1810	1420
16F-E5	2130	1890
16F-E6	2360	2350
16F-E7	1920	1800
20F-E1	1770	1420
20F-E2	2030	1800
20F-E3	2370	2350
20F-E4	1880	1420
20F-E5	2190	1890
20F-E6	2370	2350
20F-E7	2020	2350
22F-E1	2400	2350
22F-E2	2070	1800
22F-E3	1950	1420
22F-E4	2230	1890
22F-E5	2400	2350
22F-E6	2100	1800

Combination Symbol	Bearing On Full Cross Section, psi	Bearing On Partial Cross Section, psi
20F-V1	2340	2300
20F-V2	2380	2300
20F-V3	2300	2300
20F-V4	2110	1840
20F-V5	2420	2300
22F-V1	2380	2300
22F-V2	2300	2300
22F-V3	2340	2300
22F-V4	2170	1840
22F-V5	2340	2300
24F-V1	2360	2300
24F-V2	2420	2300
24F-V3	2480	2300
24F-V4	2170	1840
24F-V5	2420	2300
24F-V6	2400	2300
E-Rated Southern Pine		
16F-E1	2310	2300
16F-E2	2060	1840
16F-E3	2320	2300

24F-E1	2460	2350
24F-E2	2160	1940
24F-E3	2230	1940
24F-E4	2450	2350
24F-E5	2440	2350
24F-E6	1990	1420
24F-E7	2140	1420
24F-E8	2260	1890
24F-E9	2270	1940
24F-E10	2510	2350
24F-E11	2210	1940
24F-E12	2260	1940
24F-E13	2450	2350
24F-E14	2450	2350
24F-E15	2070	1940
Visually Graded Southern Pine		
16F-V1	2300	2300
16F-V2	2300	2300
16F-V3	2340	2300
16F-V4	2070	1840
16F-V5	2300	2300
20F-E1	2330	2300
20F-E2	2130	1840
20F-E3	2330	2300
22F-E1	2360	2300
22F-E2	2200	1840
22F-E3	2360	2300
24F-E1	2380	2300
24F-E2	2480	2300
24F-E3	2240	1840
24F-E4	2420	2300
Visually Graded California Redwood		
B-16F-V1	1940	1940

[a]For loads of other duration, see 4.4.1.3.

[b]For wet conditions of use, multiply tabulated values by 0.57.

Table A-2 — Design Values for End Grain in Bearing Parallel to Grain
Design values are for normal load duration, dry conditions of use and Table 2 Combinations[a, b]

Combination Symbol	Bearing On Full Cross Section, psi
Visually Graded Douglas Fir	
1	2350
2	2350
3	2750
4	2510
5	2750
6	1890
7	2350
8	2350
9	2750
10	2350
11	2750
12	2350
13	2750
Visually Graded Douglas Fir South	
59	2130
60	2130
61	2490
Visually Graded Hem-Fir	
14	1940
15	1940
16	1940
17	2270
18	1940
19	1940
20	1940
21	1940
Visually Graded Western Woods	
22	1420
23	1420
24	1420
25	1420
26	1420
E-Rated Douglas Fir	
27	2350
28	2750
29	2750
30	2350
31	2750
32	2750
62	2750
63	2750
E-Rated Hem-Fir	
33	1940
34	2270
35	2270
36	1940
37	2270
38	2270
E-Rated Western Woods	
39	1420
40	1420
41	1420
42	1420
43	1420
44	1420
Visually Graded Southern Pine	
45	1840
46	2300
47	2300
48	2690
49	2300
50	2690
51	2300
52	2690
E-Rated Southern Pine	
53	2300
54	2690
55	2690
56	2300
57	2690
58	2690
Visually Graded California Redwood	
B-1	1940
B-2	1940
B-3	1940
B-4	1940
B-5	1940

[a]For loads of other duration, see 4.4.1.3.

[b]For wet conditions of use, multiply tabulated values by 0.57.

ANNEX B CALIFORNIA REDWOOD

B-1 General

B-1.1 Glued laminated timbers manufactured from California Redwood are used for special purposes where appearance or natural resistance to decay is a prime consideration. They may not be generally available and the designer should check on availability prior to specifying.

B-1.2 All sections of AITC 117-84—*DESIGN* are applicable to California Redwood except as modified or revised herein.

B-2 Lumber

B-2.1 Lumber shall be of the grades specified herein.

B-3 Adhesives

B-3.1 Wet-use adhesives conforming to American National Standard ANSI/AITC A190.1-1983, *Structural Glued Laminated Timber* are required for all moisture conditions of service in laminating California Redwood.

B-4 Design Stresses

B-4.1 The information on design stresses in Section 4 of the main body of these specifications is also applicable to California Redwood except as modified in B-4.2.

B-4.2 Radial Tension or Compression

B-4.2.1 The design value for radial compression, F_{rc}, is 315 psi.

B-4.2.2 The design value for radial tension, F_{rt}, is 42 psi ($\frac{1}{3}$ of horizontal shear design value).

Table B-1 — Design Values for Visually Graded California Redwood Glued Laminated Timber for Normal Loading Duration, Dry Conditions of Use [a,b,c]

		Bending About X-X Axis						Bending About Y-Y Axis				Axially Loaded		
		Loaded Parallel to the Wide Faces of the Laminations						Loaded Pependicular to Wide Faces of Lamination						
		Extreme Fiber in Bending		Compression Perp. To Grain										
Combination Symbol	Species Outer Laminations/Core Laminations [d]	Tension Zone Stressed In Tension [e] F_{bx} psi	Compression Zone Stressed In Tension [f] F_{bx} psi	Tension Face [g] $F_{c\perp x}$ psi	Compression Face [g] $F_{c\perp x}$ psi	Horizontal Shear F_{vx} psi	Modulus of Elasticity E_x $\times 10^{-6}$ psi	Extreme Fiber In Bending [e] F_{by} psi	Compression Perpendicular To Grain Side Faces [g] $F_{c\perp y}$ psi	Horizontal Shear F_{vy} psi	Modulus of Elasticity E_y $\times 10^{-6}$ psi	Tension Parallel To Grain F_t psi	Compression Parallel To Grain F_c psi	Modulus Of Elasticity E $\times 10^{-6}$ psi
1	**2**	**3**	**4**	**5**	**6**	**7**	**8**	**9**	**10**	**11**	**12**	**13**	**14**	**15**
B-16F V1	CR/CR	1600	800	315	315	125	1.1	1400	315	110	1.1	900	1400	1.1
Wet-Use factors [b]		0.8	0.8	0.53	0.53	0.875	0.833	0.8	0.53	0.875	0.833	0.8	0.73	0.833

[a]The combinations in this table are intended primarily for members stressed in bending due to loads applied perpendicular to the wide faces of the laminations for members with 4 or more laminations. Stresses are tabulated, however, for loading both perpendicular and parallel to the wide faces of the laminations. For combinations and stresses applicable to members loaded primarily axially or parallel to the wide faces of the laminations, see Table B-2. For members of 2 or 3 laminations, see Table B-2.

[b]The tabulated design values are for dry conditions of use. To obtain wet-use design values, multiply the tabulated values by the factors shown at the end of the table.

[c]The tabulated design values are for normal duration of loading. For other durations of loading, see 4.4.1 in the main body of these specifications.

[d]CR = California Redwood.

[e]The tabulated design values in bending are applicable to members 12 in. or less in depth. For members greater than 12 in. in depth, the requirements of 4.4.2 in the main body of these specifications apply.

[f]Design values in this column are for extreme fiber stress in bending when the member is loaded such that the compression zone laminations are subjected to tensile stresses. For more information, see 1.4.1.2 in the main body of these specifications. The values in this column may be increased to 1,200 psi when end joint spacing restrictions are applied to the compression zone when stressed in tension.

[g]The compression perpendicular to grain design values in this Table are not subject to the duration of load modifications in 4.4.1.

Table B-2 — Design Values for Visually Graded California Redwood Glued Laminated Timber for Normal Loading Duration, Dry Conditions of Use, psi [a, b, c]

Combination Symbol	Species[d]	Grade[e]	Modulus of Elasticity E × 10^{-6} psi	Comp. Perp. To Grain	Axially Loaded: Tension Parallel To Grain, 2 or More Lams.	Axially Loaded: Compression Parallel To Grain, 4 or More Lams.	Axially Loaded: Compression Parallel To Grain, 2 or 3 Lams.	Bending About Y-Y Axis, Loaded Parallel to Wide Faces of Laminations: Extreme Fiber in Bending, 4 or More Lams.	Extreme Fiber in Bending, 3 Lams.	Extreme Fiber in Bending, 2 lams.	Horizontal Shear, 4 or More Lams.	Horizontal Shear, 3 Lams.	Horizontal Shear, 2 Lams.	Bending About X-X Axis, Loaded Perpendicular In Wide Faces of Laminations: Extreme Fiber In Bending, 2 Lams. to 15" deep	Horizontal Shear, 2 or More Lams.
				$F_{c\perp}$	F_t	F_c	F_c	F_{by}	F_{by}	F_{by}	F_{vy}	F_{vy}	F_{vy}	F_{bx}	F_{vx}
1	**2**	**3**	**4**	**5**	**6**	**7**	**8**	**9**	**10**	**11**	**12**	**13**	**14**	**15**	**16**
B-1	CR	L5	1.0	315	875	1350	1350	1450	1300	1100	110	105	95	1200	125
B-2	CR	L4	1.0	315	875	1350	1350	1450	1300	1100	110	105	95	1200	125
B-3	CR	L3	1.2	315	1000	1550	1550	1450	1300	1100	110	105	95	1350	125
B-4	CR	L2	1.2	315	1000	1600	1600	1500	1350	1150	110	105	95	1350	125
B-5	CR	L1	1.2	315	1000	1600	1600	1600	1500	1250	110	105	95	1350	125
Wet-Use factors[b]			0.833	0.53	0.8	0.73	0.73	0.8	0.8	0.8	0.875	0.875	0.875	0.8	0.875

[a]The tabulated combinations in this table are intended primarily for members loaded either axially or in bending with the loads acting parallel to the wide faces of the laminations. Design values for bending due to loading applied perpendicular to the wide faces of the laminations are also included; however, the combination in Table B-1 is usually better suited for this condition of loading for members with 4 or more laminations.

[b]The tabulated design values are for dry conditions of use. To obtain wet-use design values, multiply the tabulated values by the factors shown at the end of the table.

[c]The tabulated values are for normal duration of loading. For other durations of loading, see 4.4.1 of the main body of these specifications.

[d]CR = California Redwood.

[e]Grade designations are as follows:
Visually Graded-California Redwood.
L1 is L1 laminating grade (close grain).
L2 is L2 laminating grade (close grain).
L3 is L3 laminating grade (close grain).
L4 is L4 laminating grade (close grain).
L5 is L5 laminating grade (close grain).

[f]The values of F_{by} were calculated based on members 12 in. in depth (bending about *Y–Y* axis). When the depth is less than 12 in., the values of F_{by} can be increased by multiplying by the following factors:

Depth, in.	*Multiplying Factor*
10.75	1.01
8.75	1.04
6.75	1.07
5.125	1.10
3.125	1.16

Footnotes for Table B-2 continued

[g]The tabulated design values for bending are applicable to members 12 in. or less in depth. For members greater than 12 in. in depth, the requirements of 4.4.2 in the main body of these specifications apply.

[h]The combinations in this table are not intended for deep bending members when loaded perpendicular to the wide faces of the laminations. However, if members over 15 in. in depth are necessary, AITC 302-24 tension laminations are required and the designer must specify that the member is for use in bending about the X–X axis; in which case, the design value F_{bx} is 1400 psi for combinations B-1 and B-2 and 1600 psi for B-3, B-4 and B-5.

[i]The compression perpendicular to grain design values in this Table are not subject to the duration of load modification in 4.4.1.

Table B-3 — Design Species Groupings for Fasteners Used on the Faces of Visually Graded California Redwood Glued Laminated Timber [a, b, c]

Combination Symbol	Species	Growth Rate	Timber Conn. Group	Lag Bolt & Driven Fastener Group
All Combinations	CR	CL	C	III

[a]CL = close grain growth rate.

[b]CR = California Redwood.

[c]For bolts, use the species and growth rate indicated in this table with tables in the *Timber Construction Manual* or the *National Design Specification*.

APPENDIX I GUIDE FOR SPECIFYING GLUED LAMINATED TIMBER

CONTENTS

I-1 General

I-1.1 AITC 117-84—*DESIGN* and its references include all of the information necessary to specify structural glued laminated timber. Design values for glued laminated timber are given in Tables 1 and 2.

Table 3 contains the information necessary to determine species groupings and rate of growth for use in the design of fastenings.

Annex A contains design values for end grain in bearing.

I-1.2 Table 1

I-1.2.1 Table 1 contains combinations of species and grades which are best suited for members resisting bending loads applied perpendicular to the wide faces of the laminations. The minimum depth of members made with all combinations in Table 1 is 4 laminations. The 22*F* and 24*F* combinations in depths of 15 in. or less require a special tension lamination which may decrease the availability of those shallow depth members. The designer should check on availability prior to specifying 22*F* and 24*F* combinations for members 15 in. and less in depth. It is strongly recommended that 20*F* combinations be used for members 15 in. and less in depth wherever possible because these combinations are generally available in these depths.

The combination symbol indicates the design value in bending about the X–X axis. For instance, the 24 F combinations have a design value of 2400 psi in bending (F_{bx}). Table 1 contains 15 columns. The first column (1) contains the combination designations, and the second column (2) indicates the species. In column 2, for most combinations, the first species designation indicates the species used in the outer zones, and the second indicates the species used in the core or inner zones. For example, DF/WW indicates Douglas Fir-Larch in the outer laminations and Western woods in the core.

The next group of columns (3–8) contains the design values for members loaded in bending about the X–X axis caused by loads applied perpendicular to the wide faces of the laminations—the most common loading condition. These are bending about the *X–X* axis (F_{bx}) compression perpendicular to grain ($F_{c\perp}$) on both the tension and compression faces, horizontal shear (F_{vx}) and modulus of elasticity (E_x).

When members made with Table 1 combinations are stressed in bending about the *Y–Y* axis (load applied parallel to the wide faces of the laminations), the corresponding design values are generally lower than those for the same members stressed in bending about the *X–X* axis.

The next group of columns (9–13) contains the design values when the members are loaded in bending about the *Y–Y* axis caused by loads applied parallel to the wide faces of the laminations. These values are listed for use primarily when a member is loaded in bending about both axes. When the principal load in bending is about the *Y–Y* axis, the combinations in Table 2 may be more suitable.

The last group of columns (14–16) lists axial stresses. These strength properties are listed primarily for members which are loaded in both bending and axial compression or tension. When the load is primarily compression or tension, the combinations in Table 2 may be more suitable. However, the combinations in Table 1 are generally used for arches.

I-1.3 Table 2

I-1.3.1 Table 2 contains combinations that are best suited for axially loaded members (either in tension or compression) and members stressed in bending about the *Y–Y* axis (loads applied parallel to the wide faces of the laminations). It is divided into two parts according to species—Western species and Southern Pine species. It is further subdivided into visually graded combinations and E-rated combinations.

This large number of grades is listed to provide greater flexibility. Each combination contains only one grade of lumber. The design values listed are for dry conditions of use. The line at the end of the table contains multiplying factors to apply to the tabulated values to obtain wet conditions of use values.

Table 2 contains 18 columns divided into 5 groups. The first group (1–3) indicates the combination symbol (1), the species (2) and the grade (3).

The second group of columns (4–5) contains the modulus of elasticity (4) and the compression perpendicular to grain design values (5) which are applicable to all directions of loading.

The third group of columns (6–8) contains strength properties for axially loaded members.

The fourth group of columns (9–15) contains design values for bending about the *Y–Y* axis when loads are applied parallel to the wide faces of the laminations. Note that strength properties may vary with the number of laminations.

The fifth group of columns (16–18) contains design values for members loaded in bending about the *X–X* axis. These design values are for members loaded in bending about both axes or for axially loaded members also loaded in bending about the *X–X* axis. For members loaded primarily in bending about the *X–X* axis, the combinations in Table 1 are more suitable than those shown in this table.

The design values shown for bending about the *X–X* axis (F_{bx}) column 16, Table 2—DESIGN are for members from 2 laminations to 15 in. in depth and should suffice for most purposes. Column 17 contains design values for members of 4 or more laminations loaded in bending about the X–X axis when special tension laminations are used.

These values are approximately 15% higher than the Table 2 values. Note that Table 2 combinations are generally not as well suited for bending about the X–X axis as Table 1 combinations and bending members requiring special tension laminations may not be readily available. Column 18 contains the design values in shear (F_{vx}) when the members are loaded in bending about the X–X axis.

I-2 Guide for Specifying Structural Glued Laminated Timber

I-2.1 Principal Stress Is Bending with Loads Applied Perpendicular to Wide Faces of Laminations

I-2.1.1 Glued laminated timber may be specified in several ways. It is recommended that the specifier list the lowest required stresses for each strength property. This will allow the manufacturer to select the most readily available grades and species to meet the design requirements. It is important that the designer check the design values very carefully and not arbitrarily specify the highest possible values in all categories since this will tend to reduce the manufacturing options and in some cases, result in a member impossible to manufacture.

I-2.1.2 A suggested form of specification when the principal stress is bending is as follows:

Glued laminated timber shall be manufactured from species and grades of lumber which will produce design values equal to or exceeding the following when loaded perpendicular to the wide faces of the laminations:
Bending (F_b) = ________*psi*
Horizontal shear (F_v) = ________*psi*
Modulus of elasticity (E) = ________*psi*
Compression perpendicular to grain (*tension face*) ($F_{c\perp}$) = ________*psi*
Compression perpendicular to grain (*compression face*) ($F_{c\perp}$) = ________*psi*
Compression parallel to grain (F_c) = ________*psi*
Tension parallel to grain (F_t) = ________*psi*

All of these values may not be necessary for design and when they are not, they need not be specified. For example, tension parallel to grain (F_t), and compression parallel to grain (F_c) may not be needed in the design of a beam. For other designs, additional design values may need to be specified such as for a member loaded in bending about the Y–Y axis as well as the X–X axis; in which case the design value in bending (F_{by}) must be specified.

Glued laminated timbers may also be specified by species or by species combinations. This method is not recommended since it tends to limit the availability.

Where California Redwood is desired for appearance or special use conditions, it should be specified by species. See Annex B for details.

I-2.1.3 A second alternative is for the designer to specify a species group. The species groupings are arranged in combinations using visually graded lumber or E-rated lumber. It is recommended that the manufacturer have the option of supplying combinations made either of visually graded lumber or E-rated lumber. Specifying by species groups can be accomplished by adding a notation such as *Western species* or *Southern Pine species* to the suggested specification in I-2.1.2.

I-2.1.4 Glued laminated timbers may also be obtained by specifying the combinations in Table 1, such as 24F-V1 DF/WW or 24F-V1 SP/SP. This procedure can limit availability and should be used only where there is a specific reason for so doing. Both visually graded and E-rated combinations are shown in Table 1. Visually graded combinations are generally available; however, the designer should check with suppliers on availability of E-rated combinations prior to specifying.

I-2.1.5 California Redwood can be specified by either method since only one combination is listed in Table B-1 (B-16F-V1 (CR/CR). If *all heart* redwood is desired, the designer should add *all heart* to the specification. For special appearance purposes, higher grade combinations listed in Table B-2 may also be used.

I-2.1.6 Design values for small members of 2 or 3 laminations with loads applied perpendicular to the wide faces of the laminations are listed in Table 2. It is recommended that the designer specify the design values needed and allow the manufacturer to pick the grade and species to meet the design requirements. If it is necessary to specify species, the manufacturer should be allowed the option of selecting the grade in order to increase availability.

I-2.2 Principal Stress Is Bending with Loads Applied Parallel to the Wide Faces of the Laminations

I-2.2.1 When members are loaded parallel to the wide faces of the laminations, the design values in bending are designated in Tables 1 and 2 as F_{by}. When the primary bending stresses of a member are in this direction, the combinations in Table 2 are often more suitable. It is recommended, however, that the designer specify the design values required and let the manufacturer select the combination of species and grades of lumber. If specification of a species is necessary, the manufacturer should be allowed the option of selecting the grade of lumber to increase availability.

I-2.3 Axially Loaded Members

I-2.3.1 The combinations in Table 2 are generally best suited for members where the primary load is axial—either tension or compression. It is recommended that the designer specify the required axial design value as well as the other required values including the bending design value in both the X–X and Y–Y directions if needed. The manufacturer will then select species and grades from available stock to manufacture a member which will meet or exceed the design requirements.

I-2.4 Preservatively Treated Glued Laminated Timber

I-2.4.1 The designer should refer to AITC 109 for information on treatments and treatability of species. The species specified should be those recommended for preservative treatments and the type of preservative and retention should also be specified.

I-3 Example of Design and Specification of Glued Laminated Timber

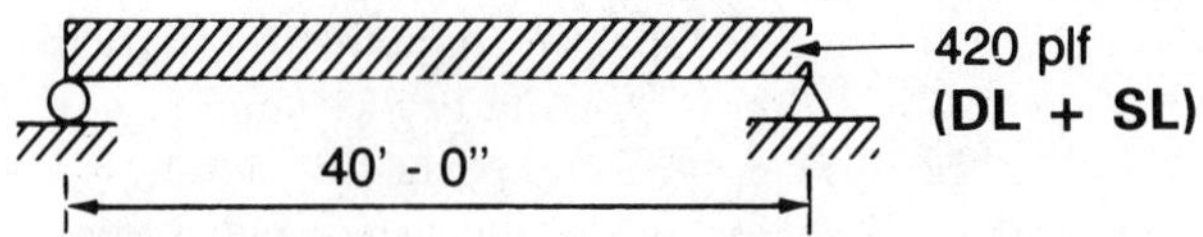

Example Design a roof beam to support the above loads. The beam is laterally supported continuously by decking applied directly to the top of the beam and the ends are restrained from rotation. Determine the proper size and specify the proper stresses. (Roof slope equals $\frac{1}{4}$ in. per foot.)

Assume:
$F_b' = (2{,}000)\ (1.15) = 2{,}300$ psi
Deflection limit $= l/180 = (40)\ (12)/180 = 2.67$ in.

Bending:
$M = wL^2/8 = (420)\ (40)^2\ (12)/8 = 1{,}008{,}000$ in.-lb
$S = M/F_b = 1{,}008{,}000/2{,}300 = 438$ in.^{3}
Try a $5\frac{1}{8}$ in. × 24 in. member, A = 123 in.^{2},
$S = 492$ in.^{3}, $I = 5{,}904$ in.^{4}
$C_F = 0.93$
$f_b = M/SC_F = 1{,}008{,}000/(492)\ (0.93) = 2{,}203$ psi
F_b (REQ'D) $= 2{,}203/1.15 =$ **1,916** psi

Shear:
Neglect all loads within a distance equal to depth of beam from the end.
$R_v = (420)\ (40\text{-}4)/2 = 7{,}560$ lb
$f_v = 3R_v/2bd = (3)\ (7{,}560)/(2)(5.125)\ (24) = 92$ psi
F_v (REQ'D) $= 92/1.15 =$ **80** psi

Bearing:
Compression perpendicular to grain on tension face
Assume 6 in. bearing at each end
$R_v = (420)(40)/2 = 8{,}400$ lb

$$f_{c\perp} = \frac{8{,}400}{(5.125)(6)} = 273 \text{ psi}$$

$F_{c\perp}$ (REQ'D) $=$ **273** psi

Note: No special specification is needed for the compression face because the roof decking is applied directly to the top of the beam and the lowest value listed for $F_{c\perp}$ is 255 psi. When purlin hangers or other connectors are used on the top of the beam, the compression perpendicular to grain on the top of the member should be checked and the designer should specify the required design value of $F_{c\perp}$ for this side of the member.

Deflection:

$$E\ (\text{REQ'D}) = \frac{5wl^4}{384\ (l/180)(I)}$$

$$= \frac{(5)(420)(40)^4(12)^3}{(384)(2.67)(5{,}904)}$$

$$= \mathbf{1{,}530{,}000}\ \text{psi}$$

The recommended method of specifying this member is:
Glued laminated timber shall be manufactured from species and grades of lumber which will

produce design values equal to or exceeding the following when loaded perpendicular to the wide faces of the laminations:
Bending $F_b = 2{,}000$ *psi*
Horizontal shear $F_v = 90$ *psi*
Compression perpendicular to grain on the tension face $F_{c\perp} = 375$ *psi*
Modulus of elasticity $E = 1{,}600{,}000$ *psi.*

Ordinarily the bending about the *Y–Y* axis, axial compression, or axial tension are not significant stresses in the design of a roof beam. However, if there is a special need for those design values, the stresses should be checked and the design values specified.

Note: A complete specification should also cover such items as camber, adhesives, appearance grades, protection, quality marks and certificates, and preservative treatment (if required).

REFERENCES

1. AITC 117-84—*MANUFACTURING, Standard Specifications for Structural Glued Laminated Timber of Softwood Species*, American Institute of Timber Construction.
2. ASTM D 3737-83a, *Standard Method for Establishing Stresses for Structural Glued Laminated Timber (Glulam)*, American Society for Testing and Materials.
3. American National Standard ANSI/AITC A190.1-1983, *Structural Glued Laminated Timber*, American National Standards Institue.
4. *Timber Construction Manual*, Third Edition, 1985, American Institute of Timber Construction, published by John Wiley & Sons, Inc.
5. AITC 109-84, *Standard for Preservative Treatment of Structural Glued Laminated Timber*, American Institute of Timber Construction.
6. *National Design Specification for Wood Construction*, 1982, National Forest Products Association.
7. AITC 104-84, *Typical Construction Details*, American Institute of Timber Construction.
8. AITC 110-84, *Standard Appearance Grades for Structural Glued Laminated Timber*, American Institute of Timber Construction.
9. AITC 111-79, *Recommended Practice for Protection of Structural Glued Laminated Timber During Transit, Storage, and Erection*, American Institute of Timber Construction.

AMERICAN INSTITUTE
333 West Hampden Avenue

TIMBER CONSTRUCTION
Englewood, Colorado 80110

AITC 119-85
STANDARD SPECIFICATIONS FOR HARDWOOD GLUED LAMINATED TIMBER

Adopted as Recommendations October 15, 1981

CONTENTS

1. GENERAL

1.1 Structural Glued Laminated Timber

1.1.1 The term "structural glued laminated timber" as employed herein refers to an engineered, stress-rated product of a timber laminating plant, comprising assemblies of suitably selected and prepared wood laminations bonded together with adhesives. The grain of all laminations is approximately parallel longitudinally.

1.1.2 The separate laminations shall not exceed 2 in. in net thickness. They may be comprised of pieces joined to form any length, of pieces placed or glued edge to edge to make wider ones, or of pieces bent to curved form during gluing.

1.1.3 This specification is applicable to laminated members of four or more laminations when the load is applied perpendicular to the wide faces of the laminations and to bending values for members of three or more laminations when the load is applied parallel to the wide faces of the laminations.

1.1.4 The production of structural glued laminated timber under these specifications shall be in accordance with the current American National Standard ANSI/AITC A190.1-1983, *Structural Glued Laminated Timber* (Reference 1).

1.2 Lumber

1.2.1 General

1.2.1.1 Slope of grain shall be limited in the full length of each lamination and shall be measured over a distance sufficiently great to determine the general slope, disregarding local deviations, except as noted for tension laminations.

1.2.1.2 For that portion of the cross section that is not a structural part of the member, the strength provisions of this specification need not apply.

1.2.1.3 When a top or bottom lamination is specially selected to meet appearance requirements, the basic structural requirements of the required grade still apply.

1.2.1.4 Appearance requirements shall be in accordance with the current *Standard Appearance Grades for Structural Glued Laminated Timber,* AITC 110.

1.2.2 Species. This specification is applicable to members laminated from any of the hardwood species listed in Table 1.

1.2.3 Grading

1.2.3.1 The grading of the lumber shall be based on the knot size and slope of grain requirements given in Table 2 for a specific allowable unit stress required and, in addition, shall meet such other requirements as may be necessary for a particular use.

1.2.3.2 The slope of grain requirements for a member stressed principally in bending apply only to the laminations in the outer 10% of the depth of the member, measured for each face at any cross section of the member as finally installed. The slope of grain for the balance of the laminations shall be no steeper than 1 in 8. For tension or compression members, the requirements given in Table 2 apply to all laminations.

1.2.3.3 When a member is designed such that the bending stress is produced by a load applied parallel to the wide faces of the laminations, the slope of grain requirements given in Table 2 apply to all laminations.

1.3 Adhesives

1.3.1 Dry-use adhesives are those which perform satisfactorily when the moisture content of wood does not exceed 16% for repeated or prolonged periods of service and are to be used only when these conditions exist.

1.3.2 Wet-use adhesives will perform satisfactorily for all moisture conditions, including exposure to weather, marine use and where approved pressure treatments are used either before or after gluing. They may be used for all moisture conditions of service, but are required when the moisture content exceeds 16% for repeated or prolonged periods of service.

TABLE 1

Stress Factors for use in Converting the Values of Table 2 to Design Values for Normal Duration of Load

Species	Extreme Fiber in Bending or Tension Parallel to Grain		Compression Parallel to Grain		Horizontal Shear		Compression Perpendicular to Grain		Modulus of Elasticity	
	Factor		Factor		Factor		Factor		Factor	
	Dry	Wet	Dry	Wet	Dry	Wet	Dry	Wet	Dry	Wet
Hickory, True and Pecan	3.85	3.10	3.05	2.20	0.26	0.23	1.00	0.53	1.80	1.50
Beech, American	3.05	2.45	2.45	1.80	0.23	0.21	0.92	0.49	1.70	1.40
Birch, Sweet and Yellow	3.05	2.45	2.45	1.80	0.23	0.21	0.86	0.46	1.90	1.60
Elm, Rock	3.05	2.45	2.45	1.80	0.23	0.21	0.92	0.49	1.40	1.30
Maple, Black and Sugar (Hard Maple)	3.05	2.45	2.45	1.80	0.23	0.21	0.77	0.41	1.70	1.40
Ash, Commercial White	2.80	2.25	2.20	1.60	0.23	0.21	0.83	0.44	1.70	1.40
Oak, Commercial Red and White	2.80	2.25	2.05	1.50	0.23	0.21	0.80	0.43	1.60	1.30
Elm, American and Slippery (White or Soft Elm)	2.20	1.75	1.60	1.15	0.19	0.17	0.59	0.31	1.40	1.20
Sweetgum (Red or Sap Gum)	2.20	1.75	1.60	1.15	0.19	0.17	0.59	0.31	1.40	1.20
Tupelo, Black (Blackgum)	2.20	1.75	1.60	1.15	0.19	0.17	0.62	0.33	1.20	1.00
Tupelo, Water	2.20	1.75	1.60	1.15	0.19	0.17	0.59	0.31	1.30	1.10
Ash, Black	2.00	1.60	1.30	0.95	0.17	0.14	0.56	0.30	1.30	1.00
Yellow-Poplar	2.00	1.60	1.45	1.05	0.15	0.13	0.41	0.22	1.50	1.20
Cottonwood, Eastern	1.55	1.20	1.20	0.90	0.11	0.10	0.32	0.17	1.20	1.00

TABLE 2

Values for Use in Computing Working Stresses with the Factors of Table 1 Together with Limitations Required to Permit the Use of Such Stresses

Combination Symbol[1]	Ratio of Size of Maximum Permitted Knot to Finished Width of Lamination	Number of Laminations[2]	Extreme Fiber in Bending		Tension Parallel to Grain		Modulus of Elasticity in Bending	Compression Parallel to Grain		Horizontal Shear	Compression Perpendicular to Grain	Tension Lamination Required for Members Greater than 15 in. in Depth
			Stress Module, psi	Steepest Grain Slope	Stress Module, psi	Steepest Grain Slope	Stress Module, psi	Stress Module, psi	Steepest Grain Slope	Stress Module, psi	Stress Module, psi	
A	0.1	4 to 14	800[3]	1:16	500[3]	1:16	1,000,000	970	1:15	1,000	1,000	302-24
	0.1	15 or more	800[3]	1:16	500[3]	1:16	1,000,000	970	1:15	1,000	1,000	302-24
B	0.2	4 to 14	770[3]	1:16	500[3]	1:16	1,000,000	920	1:15	1,000	1,000	302-24
	0.2	15 or more	800[3]	1:16	500[3]	1:16	1,000,000	930	1:15	1,000	1,000	302-24
C	0.3	4 to 14	600	1:12	450	1:15	900,000	860	1:14	1,000	1,000	302-24
	0.3	15 or more	660	1:12	450	1:16	900,000	870	1:14	1,000	1,000	302-24
D	0.4	4 to 14	450	1:8	350	1:10	800,000	780	1:12	1,000	1,000	302-20
	0.4	15 or more	520	1:8	350	1:12	800,000	810	1:12	1,000	1,000	302-20
E	0.5	4 to 14	300	1:8	300	1:8	800,000	690	1:10	1,000	1,000	302-20
	0.5	15 or more	380	1:8	300	1:8	800,000	730	1:10	1,000	1,000	302-20

[1]The tabulated combinations are applicable to arches, compression members and tension members. For bending members, the outermost tension lamination representing 5% of the total depth of the member shall meet the grading requirements given in the last column of Table 2 for depths over 15 in. For depths of 12 in. to 15 in., a 302-22 tension lamination may be used for combinations A, B, and C and for depths of two laminations to less than 12 in., 302-20 tension laminations are required. For combinations D and E, 302-20 tension laminations are required for all depths.

[2]When laminations of different thicknesses are used, divide the depth of the member by the thickness of the thickest lamination used and then assume the quotient to be the number of laminations in the member for use in determining the allowable stress.

[3]Stress modules for combinations A and B are identical in the cases of extreme fiber in bending and in tension parallel to grain because a slope of grain of 1:16 is a more restrictive limitation than knot size.

1.3.3 Adhesives used shall comply with the specifications contained in American National Standard ANSI/AITC A190.1-1983, *Structural Glued Laminated Timber*.

2. DESIGN VALUES

2.1 General

2.1.1 Stress factors for the principal commercial species of hardwoods are listed in Table 1. The use of only one species in the member is assumed. Species having substantially similar strength properties are grouped together and groups are listed in order of decreasing bending strength.

2.1.2 Design values are computed by multiplying the stress modules given in Table 2 by the appropriate stress factor shown in Table 1 for a specified combination of species and condition of use.

2.1.3 The design values determined from Tables 1 and 2 are for loads of normal duration. Modifications for other conditions of loading are given in Section 2.6.1.

2.1.4 The design values as determined in Tables 1 and 2, and the modifications required for other conditions of loading are applicable also to structural glued laminated members that have been pressure impregnated by an approved preservative process in accordance with *Standard for Preservative Treatment of Structural Glued Laminated Timber*, AITC 109 (Reference 2).

2.1.5 The design values in bending (F_b) as determined from Tables 1 and 2 apply to a 12 in. deep member, uniformly loaded with a span-to-depth ratio of 21 to 1. Modifications for other sizes and loading conditions are given in Section 2.6.2.

2.1.6 The modulus of elasticity (E) values herein are the average values for the combination shown and reflect the effect of grade. The modulus of elasticity of wood of a given species is somewhat variable. The coefficient of variation of visually graded lumber of the same species is approximately 0.25 for species used in laminating. Tests and experience have shown that this variability is considerably reduced by the averaging effect of laminating. For glued laminated timber, the coefficient of variation (c.o.v.) decreases with an increasing number of laminations. In glued laminated timber made from four laminations of visually graded lumber, the c.o.v. is approximately 0.15, for 10 laminations 0.10 and for 16 or more laminations 0.08. The variation in modulus of elasticity is especially important in designs where stiffness is of prime importance such as in the design of long columns, lateral stability calculations or in calculations for ponding.

A standard deviation is the average value multiplied by the coefficient of variation. In a normal frequency distribution, about $\frac{2}{3}$ of the individual values will be within one standard deviation (above and below) the average value. Also about 95% of the individual values will be within two standard deviations of the average value. Thus, if a combination of glued laminated timber has an average E of 1,700,000 psi and the coefficient of variation is 0.10, $\frac{2}{3}$ of the members could be expected to have values between 1,530,000 and 1,870,000 psi and 95% could be expected to have values between 1,360,000 psi and 2,040,000 psi.

In a case where only the lower portion of the variation in E is of engineering importance, similar useful interpretations are possible. In a normal frequency distribution, $\frac{5}{6}$ of the individual values lie above a value located at one standard deviation below the mean (1,530,000 psi in the above example). In the same distribution, 95% of the individual values lie above a value located at 1.645 standard deviations below the mean (1,420,000 psi in the above example).

2.1.6.1 The modulus of elasticity determined by the use of Tables 1 and 2 applies to members loaded in bending by a load applied either perpendicular or parallel to the wide faces of the laminations. It also applies to axially-loaded members, tension or compression.

2.1.7 End joints may be plain scarf joints, finger joints or other types which qualify for the design values in accordance with the procedure recommended in American National Standard ANSI/AITC A190.1-1983, *Structural Glued Laminated Timber*.

2.1.8 The design of glued laminated members and their fastenings shall be in accordance with the provisions of this specification and the *National Design Specification for Wood Construction* (Reference 3).

2.2 Members Stressed in Bending

2.2.1 Design values in bending determined from Tables 1 and 2 are based on an arrangement of laminations where the direction of loading is perpendicular to the wide faces of the laminations. This arrangement is best suited when the principal stress is in bending.

2.2.2 When a member is designed such that the bending stress is produced by a load applied parallel to the wide faces of the laminations, F_{by}, the design values as determined from Tables 1 and 2 are applicable provided knots at the edge of the laminations do not exceed $\frac{1}{2}$ the size of those permitted by Table 2, but may increase proportionally to a size at the center of the lamination width equivalent to that permitted by Table 2.

2.3 Members Stressed Principally in Axial Compression or Axial Tension

Design values determined from Tables 1 and 2 for members stressed principally in axial compression or tension are based on all laminations having a slope of grain no steeper than the values listed in Table 2 for a specified combination.

2.4 Condition of Use

2.4.1 Dry condition of use design values shall be applicable when the moisture content in service is less than 16%, as in most covered structures.

2.4.2 Wet condition of use design values shall be applicable when the moisture content in service is 16% or more, as may occur in members directly exposed to precipitation or in covered locations of high humidity.

2.5 Radial Tension or Compression

2.5.1 When a curved member is loaded in bending, radial stresses are induced.

2.5.2 When the bending moment (M) is in the direction tending to increase curvature (decrease the radius), the radial stress is compression across the grain (F_{rc}). The design value in radial compression (F_{rc}) is equal to the design value in compression perpendicular to grain ($F_{c\perp}$) of the grade and species being used.

2.5.3 When M is in the direction tending to decrease curvature (increase the radius), the radial stress is tension across the grain. The design value in radial tension perpendicular to grain (F_{rt}) shall be limited to $\frac{1}{3}$ the design value in horizontal shear (F_v). These values are subject to modifications for duration of load and wet conditions of use. For wet conditions of use, the wet-use factor for radial tension is 0.875.

2.6 Modification of Stresses

2.6.1 Duration of Load

2.6.1.1 Normal load duration contemplates fully stressing a member to the design value by the application of the full design load for a duration of approximately 10 years (applied either continuously or cumulatively).

2.6.1.2 When a member is fully stressed by maximum design loads for long-term loading conditions (greater than 10 years either continuously or cumulatively), the design values shall be 90% of the values tabulated.

2.6.1.3 When the duration of full design load (applied either continuously or cumulatively) does not exceed the period indicated, increase the tabulated design values, except for modulus of elasticity as follows:

15% for 2 months duration, as for snow;
25% for 7 days duration;
33% for wind or earthquake; and
100% for impact.

These increases are not cumulative.

2.6.1.4 The duration of load factors do not apply to compression perpendicular to grain, $F_{c\perp}$.

2.6.2 ***Size Factor***

2.6.2.1 When the depth of a rectangular beam exceeds 12 in., the tabulated design value for bending, F_b, shall be reduced by multiplying it by the size factor, C_F, as determined from the following relationship:

$$C_F = \left(\frac{12}{d}\right)^{1/9}$$

where C_F = size factor
d = depth of member (in.).

2.6.2.2 The size factor relationship as given in 2.6.2.1 is applicable to simply-supported, uniformly-loaded bending members with span-to-depth ratios (l/d) of 21. This factor can thus be applied with reasonable accuracy to most commonly encountered design situations. Where greater accuracy is desired for other sizes and conditions of loading, the percentage changes given in the following table may be applied directly to the size factor calculated for the basic conditions previously stated. Straight line interpolation may be used for other l/d ratios.

Span to Depth Ratio l/d	% Change
7	+6.3
14	+2.3
21	0
28	−1.6
35	−2.8

Loading Condition for Simply Supported Beams	% Change
Single Concentrated Load	+7.8
Uniform Load	0
Third Point Load	−3.2

2.6.3 *Lateral Stability*

2.6.3.1 The design values for bending contained in these specifications are applicable to members which are adequately braced. When deep, slender members not adequately braced are used, a reduction to the tabulated design values in bending must be applied based on a computation of the slenderness factor of the member. In the check of lateral stability, the slenderness factor shall be applied in design as shown in the *National Design Specification for Wood Construction*.

2.6.3.2 A reduction in the design value in bending determined by applying the slenderness factor is not cumulative with a reduction in design value due to the application of size factor. In no case shall the design value in bending exceed the stress as determined by applying the size factor or slenderness factor, whichever governs.

2.6.4 *Curvature Factor*

2.6.4.1 For the curved portion of members, the design value in bending F_b shall be modified by multiplying it by the following curvature factor:

$$C_C = 1 - 2000\ (t/R)^2$$

where t = thickness of lamination (in.).
R = radius of curvature of lamination (in.) and
t/R should not exceed 1/100.

No curvature factor need be applied to the design value in the straight portion of an assembly, regardless of curvature elsewhere.

3. CONNECTIONS AND FASTENERS

3.1 The design values for connections and fasteners for glued laminated timber are contained in the *National Design Specification for Wood Construction.*

3.2 See *Typical Construction Details*, AITC 104 (Reference 4), for additional information on connections.

4. DIMENSIONS

4.1 Standard Sizes. American National Standard ANSI/AITC A190.1-1983, permits the use of any width or depth of glued laminated timber. The use of standard finished sizes, however, constitutes recommended practice to the extent that other considerations will permit. The laminator may use any thickness of lumber to develop the specified depth provided the volume of the higher grades of lumber equals or exceeds that specified in laminating combinations which are based on laminations of equal thickness. The depth and width of the glued laminated timber should be as agreed upon by buyer and seller.

4.2 Depth and Width

4.2.1 Straight and curved members shall be furnished in accordance with the width and depth dimensions required by the design.

4.2.2 The nominal standard net finished widths are as follows:

Nominal Width, in.	Net Finished Width, in.
3	$2\frac{1}{8}$
4	$3\frac{1}{8}$
6	$5\frac{1}{8}$
8	$6\frac{3}{4}$
10	$8\frac{3}{4}$
12	$10\frac{3}{4}$
14	$12\frac{1}{4}$
16	$14\frac{1}{4}$

Other finished widths may be used to meet the size requirements of a design or to meet other special requirements.

4.3 Radius of Curvature

The recommended minimum radii of curvature for curved structural glued laminated hardwood timbers are 6 ft 3 in. for a lamination thickness of $\frac{3}{4}$ in.; and 12 ft 6 in. for a lamination thickness of $1\frac{1}{2}$ in. Other radii of curvature may be used with these thicknesses and other radius-thickness combinations may be used provided the t/R ratio does not exceed $\frac{1}{100}$.

5. APPEARANCE GRADES

5.1 Appearance grades shall be in accordance with the current *Standard Appearance Grades for Structural Glued Laminated Timber*, AITC 110 (Reference 5), unless otherwise specified on drawings or specifications.

5.2 For those combinations permitting the mixing of species, the potential for differences in color or grain of adjacent laminations must be recognized. For those architectural appearance applications where such possible differences in color or grain might be important, the designer may specify a combination symbol which will restrict the laminations to a single species or group of species with similar characteristics. In some cases, this may restrict availability.

6. INSPECTION AND QUALITY CONTROL

6.1 The assurance that quality materials and workmanship are used in structural glued laminated timber members shall be vested in the laminator's day-to-day quality control operations. Visual inspections and physical tests of samples of production are also required to assure conformance with this Standard and American National Standard, ANSI/AITC A190.1-1983, *Structural Glued Laminated Timber*.

7. MARKING

7.1 The laminating combinations in Table 1 were developed primarily to resist bending loads. The grades of lumber in laminations on the compression side may not be the same as those on the tension side. Therefore, straight or slightly cambered glued laminated timber bending members shall be stamped "TOP" with letters approximately 2 in. high on the top at both ends of the member. Axially-loaded members or bending members which are fabricated in such a manner that they cannot be installed upside down need not be marked.

8. PROTECTION DURING SHIPPING AND FIELD HANDLING

8.1 End sealers, surface sealers, primer coats and wrappings may be applied for the protection of the members. However, they do not necessarily preclude damage resulting from negligence and other factors beyond the control of the laminator during shipping, handling, storing and placing of the members. The protection specified should be commensurate with the end use and final finish of the member. It may also vary with the method of shipment and with exposure to climatic and other details. See *Recommended Practice for Protection of Structural Glued Laminated Timber During Transit, Storage and Erection,* AITC 111 (Reference 6).

REFERENCES

1. American National Standard ANSI/AITC A190.1-1983, *Structural Glued Laminated Timber*.
2. AITC 109-84, *Standard for Preservative Treatment of Structural Glued Laminated Timber*, American Institute of Timber Construction.
3. *National Design Specification for Wood Construction*, 1982, National Forest Products Association.
4. AITC 104-84, *Typical Construction Details*, American Institute of Timber Construction.
5. AITC 110-84, *Standard Appearance Grades for Structural Glued Laminated Timber*, American Institute of Timber Construction.
6. AITC 111-79, *Recommended Practice for Protection of Structural Glued Laminated Timber During Transit, Storage, and Erection*, American Institute of Timber Construction.

ANNEX A GRADING REQUIREMENTS EXCERPTED FROM AITC 302-20, 302-22, 302-24 TENSION LAMINATION RECOMMENDATIONS

CONTENTS

A1. AITC 302-20 TENSION LAMINATION—MEMBERS IN BENDING

A1.1 General Provisions. In addition to the basic requirements of the grades tabulated in these specifications, the following limitations shall apply:

A1.1.1 A one-foot length of a lamination shall be considered as a cross section.

A1.1.2 Knots shall not occur within two knot diameters of any finger joint.

A1.1.3 Knots shall not occupy more than $\frac{1}{4}$ of the cross section.

A1.1.4 The general slope of grain shall not exceed 1:12. Where more restrictive slope of grain requirements are required by the laminating combinations, these shall apply.

A1.1.5 Any cross section shall have at least 50% clear wood free of strength-reducing characteristics with a slope of grain no steeper than 1:12. (Knots plus associated localized cross grain, or knots plus associated localized cross grain plus localized cross grain not associated with a knot, or localized cross grain not associated with a knot may occupy up to $\frac{1}{2}$ of the cross section.)

A1.1.6 Pieces shall have near average or above average specific gravity for the species.

A1.1.7 Pieces containing wide-ringed or lightweight pith associated wood at the ends of the piece occupying over $\frac{1}{8}$ of the cross section shall be excluded. (The next inch of wood outside the area of the pith associated wood shall be of the same rate of growth as the remainder of the wood located away from the pith. The line along which measurement of this inch is made shall correspond to the line used in the standard grading rules for rate of growth and percentage of summerwood. If a distance of 1 in. is not available along this line, the measurement will be made over such lesser portion as exists.)

A2. AITC 302-22 TENSION LAMINATION—MEMBERS IN BENDING

A2.1 General Provisions. In addition to the basic requirements of the grades tabulated in these specifications, the following limitations shall apply:

A2.1.1 A one-foot length of a lamination shall be considered as a cross section.

A2.1.2 Knots shall not occur within two knot diameters of any finger joint.

A2.1.3 Knots shall not occupy more than $\frac{1}{4}$ of the cross section.

A2.1.4 Any cross section shall have at least 60% clear wood free of strength-reducing characteristics with a slope of grain no steeper than 1:16 (Knots plus associated localized cross grain, or knots plus associated localized cross grain plus localized cross grain not associated with a knot, or localized cross grain not associated with a knot may occupy up to 40% of the cross section.)

A2.1.5 The general slope of grain shall not exceed 1:16. Where more restrictive slope of grain requirements are required by the laminating combinations, these shall apply.

A2.1.6 Pieces shall have near average or above average specific gravity for the species.

A2.1.7 Pieces containing wide-ringed or lightweight pith associated wood at the ends of the piece occupying over $\frac{1}{8}$ of the cross section shall be excluded. (The next inch of wood outside the area of the pith associated wood shall be of the same rate of growth as the remainder of the wood located away from the pith. The line along which measurement of this inch is made shall correspond to the line used in the standard grading rules for rate of growth and percentage of summerwood. If a distance of 1 in. is not available along this line, the measurement will be made over such lesser portion as exists.)

A3. AITC 302-24 TENSION LAMINATION—MEMBERS IN BENDING

A3.1 General Provisions. In addition to the basic requirements of the grades tabulated in these specifications, the following limitations shall apply:

A3.1.1 A one-foot length of a lamination shall be considered as a cross section.

A3.1.2 Knots shall not occur within two knot diameters of any finger joint.

A3.1.3 Knots shall not occupy more than $\frac{1}{5}$ of the cross section.

A3.1.4 Any cross section shall have at least $\frac{2}{3}$ clear wood free of strength-reducing characteristics with a slope of grain no steeper than 1:16. (Knots plus associated localized cross grain, or knots plus associated localized cross grain plus localized cross grain not associated with a knot, or localized cross grain not associated with a knot may occupy up to $\frac{1}{3}$ of the cross section.)

A3.1.5 Maximum size single strength-reducing characteristics when not in the same horizontal projection must be at least 2 ft apart measured center to center.

A3.1.6 The general slope of grain shall not exceed 1:16. Where more restrictive slope of grain requirements are required by the laminating combinations, these shall apply.

A3.1.7 Pieces shall have near average or above average specific gravity for the species.

A3.1.8 Pieces containing wide-ringed or lightweight pith associated wood at the ends of the piece occupying over $\frac{1}{8}$ of the cross section shall be excluded. (The next inch of wood outside the area of the pith associated wood shall be of the same rate of growth as the remainder of the wood located away from the pith. The line along which measurement of this inch is made shall correspond to the line used in the standard grading rules for rate of growth and percentage of summerwood. If a distance of 1 in. is not available along this line, the measurement will be made over such lesser portion as exists.)

INDEX